Fundamentals of Robotics

In an era where robotics is reshaping industries and redefining possibilities, "Fundamentals of Robotics: Applied Case Studies with MATLAB® & Python" emerges as an essential guide for both aspiring engineers and seasoned professionals. This comprehensive book bridges the gap between theoretical knowledge and practical application, driving advancements in robotics technology that mimic the complexity and grace of biological creatures.

Explore the intricate world of serial robots, from their kinematic and dynamic foundations to advanced control systems. Discover how the precise movements of a magician's fingers or the poised posture of a king cobra inspire the mathematical principles that govern robotic motion. The book delves into the Denavit-Hartenberg method, screw theory, and the Jacobian matrix, providing a thorough understanding of robot design and analysis.

Unique to this text is the integration of MATLAB® and Python, offering readers practical experience through step-by-step solutions and ready-to-use code. Each chapter is enriched with real-world case studies, including the 6-DOF Stanford robot and the Fanuc S-900w, allowing readers to apply theoretical concepts to tangible problems. The inclusion of biological examples enhances the relevance and accessibility of complex topics, illustrating the natural elegance of robotics.

Ideal for senior undergraduate and graduate students, as well as industry professionals, this book covers a wide range of topics, including linear and nonlinear control methods, trajectory planning, and force control. The dynamic models and control strategies discussed are crucial for anyone involved in the design, operation, or study of industrial robots.

"Fundamentals of Robotics: Applied Case Studies with MATLAB® & Python" is more than a textbook; it is a vital resource that provides the knowledge and tools needed to succeed in the dynamic field of robotics. Join the journey towards mastering robotic technology and contribute to the future of intelligent machines.

Fundamentals of Robotics

Applied Case Studies with MATLAB® & Python

Hamid D. Taghirad

CRC Press
Taylor & Francis Group
Boca Raton London New York

CRC Press is an imprint of the
Taylor & Francis Group, an informa business

First edition published 2025
by CRC Press
2385 NW Executive Center Drive, Suite 320, Boca Raton FL 33431

and by CRC Press
4 Park Square, Milton Park, Abingdon, Oxon, OX14 4RN

CRC Press is an imprint of Taylor & Francis Group, LLC

ISBN: 978-1-032-79305-4 (hbk)
ISBN: 978-1-032-79300-9 (pbk)
ISBN: 978-1-003-49141-5 (ebk)

DOI: 10.1201/9781003491415

Typeset in Latin Modern font
by KnowledgeWorks Global Ltd.

Publisher's note: This book has been prepared from camera-ready copy provided by the authors.

Access the Instructor and Student Resources/Support Material: www.routledge.com/9781032793054

To my dear wife Azam
and
My dear daughter Matineh.

Contents

IV Control 353

Preface

In today's world, driven by advances in human knowledge, robotics technology plays a crucial role in automation industries and large-scale industrial production. The field of robotics focuses on creating intelligent machines that imitate human movements, providing the illusion of independent intent. These machines can perceive their environment and actively contribute to society. Human fascination with robotics is seen in the ongoing efforts to create machines that resemble themselves, driving global advancements in robotics technology. Turning this aspiration into reality requires the accumulation of knowledge, technological progress, and extensive research across multiple scientific disciplines. Robotics, encompassing a wide range of areas, is a multidisciplinary field focused on advancing this knowledge. As a result, capturing even a fraction of this vast knowledge within a single book is a challenging endeavor.

This book addresses the theoretical foundations and key aspects of the technological development essential for the analysis and design of serial robots. This first generation of industrial robots have played a pivotal role in the mass production of industrial products. The subsequent generation of industrial robots features a parallel structure, demanding familiarity with the knowledge and skills applicable to the analysis and design of serial robots. Hence, this book concentrates on the analysis and design of serial robots. For those interested in parallel robots, it is advisable to explore another book written by the author of this work, exclusively dedicated to the field of parallel robots [1].

The scope of this book encompasses the analysis and design of serial robots in the fields of kinematics, dynamics, and control. A comprehensive understanding of the topics presented here equips readers with the essential skills to analyze complex structures with multi-degrees-of-freedom kinematics. Given its primary focus on serial robots, the book initially addresses the mathematical representation of motion in three-dimensional space, providing precise definitions of position and orientation, along with various methods for their representation.

In the existing literature on the kinematic analysis of robots, two perspectives and strategies have emerged: one aligns with computer science, employing numerical and recursive methods [2], while the other aligns with engineering conceptualization, utilizing geometric and analytical methods [3]. This book intentionally focuses on the latter perspective, aiming to introduce mathematical methods grounded in relevant physical concepts. Following this, the kinematic analysis of serial robots is explored through two distinct methods: the Denavit-Hartenberg method and the application of the screw theory. The representation of the three-dimensional motion components of robots using screw theory holds unique characteristics in the kinematic analysis of robot motion, notably aligning with fundamental physical concepts of motion.

The introduction of the Jacobian matrix in robots and the analysis of differential kinematics are other pivotal topics extensively covered in this book. Many crucial aspects in the design of serial robots are intricately linked to the analysis of velocities and the Jacobian matrix. Biological examples, such as the skillful and intricate movements of a magician's fingers, the poised posture of a king cobra's head and neck just before launching an attack on its prey, and the sensation of fatigue experienced in arms while opening a curtain, provide tangible demonstrations of intuitive concepts. These examples draw multiple connections to the inherent features of the robot's Jacobian matrix, emphasizing the relevance and applicability of mathematical principles to real-world physical concepts. By establishing a connection between the mathematics governing the kinematic analysis of robots and their physical features, and drawing inspiration from optimal design in biological creations, this book unveils the beauty of contemplation in this knowledge more clearly than ever.

The analysis of robot dynamics is a meticulously addressed subject in this book, placing a special emphasis on achieving a necessary compromise between computational complexity and the depth of physical content. This distinctive feature sets it apart from similar books [2, 4], where such emphasis is less pronounced. The book predominantly focuses on the Euler-Lagrange method, detailing the procedure for succinctly expressing the robot's dynamic equations. This method which is based on energy analysis in stable dynamic systems, leverages computational efficiency and is not only comprehensible for mechanical engineers but also accessible to researchers with another background especially in electrical engineering and computer science. The application of the Euler-Lagrange method for determining motion dynamics in robots is versatile, offering a closed-form formulation for the dynamic equations applicable to any serial robot. Furthermore, the book thoroughly explores the characteristics of dynamic formulations and provides insights into their utilization in the design and analysis of model-based control methods.

The first and immediate application of dynamic motion formulations in robots lies in their utilization for computer simulations. In this context, it becomes crucial to address suitable methods for motion trajectory planning in robots. When considering industrial robots, the use of teaching pendants to specify the motion path of the robot with a number of point stands out as an approach that simplifies the complexity of trajectory generation. This book addresses different methods in the design of desired trajectories in serial robots as well as the time optimal trajectory generation, taking into account the achievable speed and acceleration constraints in actuators. By using these motion trajectories, the simulation of robot motion in two modes, namely forward and inverse dynamics, is investigated. Subsequently, with the use of these simulation tools, robot calibration will be addressed in detail. The calibration process involves adjusting the kinematic and dynamic parameters of the robot to match with its actual values, ensuring accurate and reliable results in practical applications.

The dynamic model of a robot is inherently nonlinear and multivariable. However, it is known that linear controllers are often used in the control systems of industrial robots. The reason for this can be found in the augmentation of the dynamics of actuators into the robot's dynamics. In this book, it will be demonstrated that for the robots using gearbox with high conversion ratios, the linear dynamics of the actuators dominate the nonlinear dynamics of the robot. In such cases, the multivariable and nonlinear system dynamics can be simplified into a number of single-input, single-output linear systems. This allows for leveraging linear control design methods for these robots. In light of this explanation, the book discusses and examines linear controller design methods for serial robots. To achieve a comprehensive linear controller design for serial robots, an efficient method for identifying the linear model of the robot and dynamically calibrating it through experiments is introduced. This novel method, not previously addressed in any other robotics book, provides a path for designing various linear controllers for the specific robot in hand.

Building on the foundation of linear controllers, this book studies advanced nonlinear controller design methods. The exploration begins with the introduction of a multivariable PD controller for the robot, enhanced by the incorporation of a feed-forward controller. Subsequently, a pivotal controller discussed is the "Inverse Dynamics Control," serving as a cornerstone for system linearization through feedback and is recognized as a fundamental nonlinear controller in robotics. The implementation of this method will be facilitated by studying partial feedback linearization method, specifically focusing on compensating for gravitational torques. Finally, robust and adaptive versions of inverse dynamics control will be discussed and examined in detail.

In scenarios where a robot engages with its surrounding environment, traditional motion control methods may prove ineffective. This book addresses the need for suitable control methods tailored for robot interaction scenarios. The exploration commences by developing the dynamic model of the robot in interaction scenarios, with a specific focus on addressing the compliance control of robots. Following this, the book elaborates on the implementation of direct force control in robots through a cascade control structure. Finally, a detailed exploration of impedance control methods in robots is given, offering insights into effectively managing the dynamic relation of robot motion and interaction forces.

This book stands out with two distinctive features: the comprehensive examination of multiple case studies and the practical use of engineering software such as MATLAB® and the Python programming language to work out these case studies. The first feature involves the thorough implementation of theoretical concepts discussed in the book on various industrial robots across its chapters. The case studies focus on the analysis and design of robots, including the planar 2$\underline{R}$ and 3$\underline{R}$ robots, 4-DOF SCARA robot, 6-DOF Stanford robot, 3-DOF and 6-DOF Elbow manipulators, and the Fanuc S-900w industrial robot. At the conclusion of each chapter, readers are encouraged to apply the theoretical concepts to four different planar robots, five spatial robots with up to four degrees of freedom, two robots with six degrees of freedom, and two Kuka industrial robots with six and seven degrees of freedom. This practical approach enables readers to gain proficiency in the design and analysis of serial robots by actively working on these problems.

The second distinguishing feature revolves around the utilization of MATLAB and Python programming for the analysis of the presented case studies. The book goes beyond theory by offering complete solutions and programming codes for the featured case studies in these software platforms. This approach empowers readers to acquire practical expertise in the analysis and design methods for serial robots. The inclusion of ready-to-use codes facilitates the seamless implementation of the discussed methods on the robots introduced in each chapter's problem section, enhancing accessibility for readers and promoting hands-on learning. The codes, and their updates could be accessed in the Author's Github: github.com/aras-labs/Fundamentals_of_Robotics.

This book has been typeset using LaTeX, with numerous figures and diagrams created using Inkscape software. Despite careful efforts to produce an error-free version, some imperfections may remain. The author invites readers to report any identified issues via email to hamid@cim.mcgill.ca. Your feedback is appreciated and will help to enhance the overall quality of the book. Moreover, all additional materials accompanying this book, such as MATLAB and Python codes, teaching slides, and any post-print corrections, will be made easily accessible to readers on the publisher's website.

For Instructors

Given its well-structured content, this book is anticipated to be valuable resource for two primary groups of readers. Firstly, it serves as a suitable reference for teaching robotics courses, either in the final year of undergraduate studies or the first year of graduate studies. Additionally, the content is suitable for an advanced course in dynamics and control of robots, specifically designed for senior graduate students. The interdisciplinary approach of this book enhances its accessibility for students in electrical engineering, mechanical engineering, and computer science. This approach allows students from diverse backgrounds to easily understand and acquire substantial knowledge in robotics engineering. The second category of readers includes industry professionals who use robotic technology in their career and are interested in gaining the necessary skills for the operation, design, and control of industrial robots for their specific applications.

The author has continuously taught the content of this book in robotics and robot dynamics and control courses at various universities for several years. The chapters' content and size have been tailored to meet the specific needs of these courses. Consequently, by selecting specific chapters from this book, it can be effectively used in the following courses.

1- Introduction to Robotics: 400 Level; Senior undergraduate students for a one semester course (16 weeks)

Chapter 1: Introduction

Chapter 2: Motion Description

Chapter 4: Forward Kinematics

Chapter 6: Jacobian Matrix

Chapter 7: Only "Singular Configuration" Section

Chapter 8: Dynamics Analysis

Chapter 11: Linear Motion Control

2- Robotics: 500 Level; Junior graduate students, for a one semester course (16 weeks)

Chapter 1: Introduction

Chapter 2: Motion Description

Chapter 3: Advanced Representations

Chapter 4: Forward Kinematics Analysis

Chapter 5: Review on Inverse Kinematics (selected section)

Chapter 6: Jacobian Matrix

Chapter 7: Singularity and Dexterity

Chapter 8: Dynamics Analysis

Chapter 11: Linear Motion Control

3- Robot Dynamics and Control: 600 Level; Senior graduate students for a one semester course (16 weeks)

Chapter 1-7: Review of Selected Sections

Chapter 8: Dynamics Analysis

Chapter 9: Properties and Formulations in Dynamics

Chapter 10: Simulation and Calibration

Chapter 11: Linear Motion Control

Chapter 12: Nonlinear Motion Control

Chapter 13: Force Control

Chapter 14: Impedance Control

Acknowledgments

The author expresses sincere gratitude to his esteemed colleague, Prof. Mahdi Tavakoli Afshari, a full professor at the Faculty of Electrical Engineering at the University of Alberta, Canada. His support during the author's visiting research period played a crucial role in laying the foundation for the main structure of this book. The project, which had been delayed for more than four years due to executive commitments, became operational with his invaluable support. Special thanks are extended to all those who contributed to the compilation and revision of this book, as well as to the students whose feedback has been valuable throughout the years of teaching this content.

The author acknowledges the thorough review and correction process undertaken by numerous colleagues and students for this book. Special appreciation is extended to Dr. Mohammad A. Khosravi, who played a meticulous and strategic role in reviewing and editing the text. The valuable contributions of Dr. Mohammad Motaharifar, and Dr. Seyed Ahmad Khalilpour are also acknowledged for their assistance in completing this edition with constructive feedback. Additionally, the author expresses appreciation to Mr. Mohammad Javad Ahmadi and Mr. Mohammad Mehdi Nazari Ardakani for their significant efforts in completing the codes, and uploading them to the book's GitHub repository.

In conclusion, the author expresses gratitude for the constructive feedback received from the readers of the book. This feedback will be duly considered in future editions of the book. The author hopes that this collection contributes significantly to the expansion of robotic knowledge and technology for the benefit of humanity.

Hamid D. Taghirad

Acknowledgements

About the Author

Hamid D. Taghirad received his M.Sc. in mechanical engineering (mechatronics) in 1993 and his Ph.D. in electrical engineering (control–robotics) in 1997, both from McGill University, Montreal, Canada. He is currently a Professor and the Director of the Applied Robotics and AI Solutions (ARAS). Additionally, he has served as a Visiting Professor at McGill University, ETS, Concordia University, and most recently at the University of Alberta in Canada. As a senior member of IEEE, Hamid D. Taghirad contributes to the field of Robotics as an associate editor of IEEE Transactions on Medical Robotics and Bionics (IEEE – TMRB) and Frontiers in Robotics and AI – Biomedical Robotics. He is also part of the editorial board of the International Journal of Robotics: Theory and Applications. His research interests primarily revolve around robust and nonlinear control applied to robotics, with a focus on medical robots, as well as the application of VR and AI technologies in medical applications. He boasts a substantial publication record, including seven books and over 300 papers in peer-reviewed international journals and conference proceedings.

Part I

Preliminary Studies

1

Introduction

The opening chapter of the book initiates with a comprehensive exploration of robotics. It begins by defining the essence of robotics and tracing its historical trajectory, offering readers a contextual understanding of its evolution over time. Following this conceptual groundwork, the chapter examines the fundamental components of robots, scrutinizing their anatomy with a focus on essential elements such as robot links and joints, including detailed explorations of primary and compound joints. Moving forward, the discussion shifts to the kinematic structures and notations of robots, categorizing them into serial, parallel, and hybrid structures, thereby providing insights into their diverse configurations and operational characteristics.

Expanding its scope, the chapter elucidates various applications of robotics across different domains, showcasing their roles in industrial applications, space exploration, surgery, exoskeleton technology, exploration, and humanoid robotics. This panoramic view aims to demonstrate the versatility of robotics across a spectrum of fields. Lastly, the chapter outlines the overarching goals and scope of the book, serving as a roadmap for readers and detailing what they can expect from the subsequent chapters. By providing clarity on the objectives, this chapter acts as a guide, ensuring readers are well-oriented for the more in-depth explorations into robot kinematics, dynamics, and control that will unfold in the ensuing chapters.

1.1 What Is Robotics?

In today's world, due to industrial advancements and the increasing demand for the production of various products on a large scale, the adoption of robotics technology has become an unavoidable necessity. Robotics is an intriguing and enigmatic field that involves creating intelligent machines capable of performing a wide array of tasks by imitating the designs found in nature. Robots move in a manner that closely resembles living creatures, possessing the ability to sense their surroundings and, much like humans, contribute to the production of essential products. Human interest in robotics can be traced back to their desire to create entities similar to themselves, a fascination that has been present since the early days of human existence. This interest is reflected in the enthusiastic engagement of children in playing with a variety of dolls. The ability to have a creature akin to themselves alongside them and under their control has been a significant driving force behind the advancement of robotic technology.

The introduction of robotic technology in the industry began with the emergence of robotic arms. During the industrial revolution and the expansion of mass production for various components and machines, there was a growing demand for skilled labor to carry out repetitive tasks in manufacturing. In 1954, the first programmable robot was invented and put to industrial use [5]. Subsequently, with the foundation of Unimation as the first company dedicated to manufacturing robots, robotic arms played a pivotal role in mass

DOI: 10.1201/9781003491415-1

production across various industries, particularly in the automotive sector. Thus, for many years, robots were primarily associated with robotic manipulators, drawing inspiration from the human arm [6].

However, it is important to note that robotics technology has extended far beyond just robotic manipulators. In the context of technology, innovations such as the design and construction of hands or end-effectors for grasping and manipulating objects, mobile and legged robots capable of navigating both on smooth and rugged terrains, machine vision systems for visual perception of the environment, sound processing systems that mimic auditory abilities, flying robots inspired by the flight of birds, humanoid robots, and the integration of artificial intelligence for perceiving and decision-making in robots have all been developed. This signifies the broad scope and evolution of robotics technology beyond its initial association with robotic manipulators [7].

While advancements in these domains owe their progress to multidisciplinary research, the contemporary landscape no longer categorizes these sciences and technologies as exclusively part of the context of robotic science. However, the remarkable progress in these technologies, particularly in recent years, has brought humanity one step closer to the aspiration of crafting a creature resembling itself. In this context, robotics knowledge represents an interdisciplinary domain within the fields of engineering and computer science, concerned with the creation, realization, application, and utilization of robots. Its ultimate objective is the development of machines that can serve and assist humans.

Even though humanoid robots, self-driving cars, and flying robots are fundamentally categorized as robots and share foundational principles of robot design, this book initiates its exploration by addressing the realm of conventional industrial robots, primarily focusing on robotic arms. To facilitate this, the book lays the groundwork for understanding this category of robots, encompassing two principal areas: mechanics and control. Within the scope of mechanical analysis, the book addresses topics such as the kinematics of motion and velocity, dynamics, and trajectory planning. In the control section, it delves into critical aspects, including robot identification or calibration, the development of linear and nonlinear motion and force controllers, and the practical implementation of these control systems. Please note that we will not directly discuss subjects like designing robot components or choosing sensors and actuators because these topics require a comprehensive review that is beyond the scope of this book.

1.1.1 Robot Definitions

Despite the integration of robots into human daily life and the recognition that each individual has their own interpretation of a robot, there is no universally accepted comprehensive definition of a robot. Typically, a robot is a mechanical or virtual entity that, in most cases, possesses electromechanical structures and is programmed by a computer in such a way that it appears to have independent intent or autonomy. While this description is comprehensive, it does not necessarily mean that a robot must have all of the following features, although it should have some of them:

- It is not natural and has been artificially created.
- It has the ability to perceive the environment.
- It can interact with its surrounding environment.
- It is programmable and to some extent intelligent.
- It moves in three-dimensional space.
- It has a certain level of autonomy and intent.

The last feature is perhaps the most distinguishing factor between a robot and an automated machine. In general, the closer the motion of a robot resembles human-like movement, the more it solidifies this characteristic in the viewer's mind, making it closer to a real robot.

As mentioned, a unified definition for a robot cannot yet be found. The International Organization for Standardization (ISO) defines a robot as follows:

> "A typical industrial robot is automatically controlled, programmable for various purposes, capable of movement in three or more degrees of freedom, and can be either stationary or mobile on the ground. It is primarily utilized in industrial automation applications."

This way, it becomes clear that this definition underscores the key features of a robot for the purpose of offering a specific definition. Joseph Engelberger, a trailblazer in the field of robotics, once expressed that he might not be able to precisely define a robot, but when he sees one, he can identify it [8].

The definition of the Cambridge and Oxford dictionaries for a robot is as follows:

> "A device that is automatically controlled by a computer to perform tasks."
> "A device that is similar to a human and is programmed to perform specific movements automatically."

A more comprehensive definition of a robot is provided by the American Robotics Society, in which a robot is defined as follows [9]:

> "A robot is a multifunctional arm with programmable capabilities that can be used for the transportation of objects, parts, tools, and specific equipment through various programmed movements and employed to perform a wide range of tasks."

As previously noted, differentiating between a robot and an automated machine in real-world scenarios can be quite challenging. The primary difference between the two could be encapsulated by emphasizing the heightened complexity and increased programmability of robots. In essence, a robot, equipped with adaptable programming, autonomously executes pre-defined tasks, and in advanced iterations of intelligent robots, a part of perception and decision-making is autonomously handled through its artificial intelligence. In contrast, automatic machines are designed for repetitive tasks, relying on on/off sensors and actuators, or human command for operation.

1.1.2 The History of Robotics

The term "robot" was first introduced to the world by the Czech writer Karel Čapek in 1921 in a play titled "Rossum's Universal Robots" [10]. This play portrays a factory where artificial organic materials are used to create emotionless workers who tirelessly labor for their masters. Čapek coined the term "Robota" to describe these hardworking laborers. While the robots we know today differ significantly in terms of nature and application from those described in this play, the word *robot* has become a part of the vocabulary of various countries.

It is important to note that from a historical perspective, the emergence of this concept in the engineering community occurred much later. After World War II, with the development of mass production, automated machines with digital control were employed. During this period, automatic machines began to replace human users for material handling tasks in nuclear power plants. The development of digital control technology, coupled with the use of automated material handling machines, naturally led to the development of the first generation of industrial robots, which were controlled pneumatically. Gradually, with technological advancement, computer control systems replaced digital control and electro-mechanical actuators replaced the previous generation of pneumatic actuators. A summary of significant events and developments in the evolution of robotics technology worldwide is presented in Table 1.1.

1.2 Robot Components

In general, the structure of a robot is composed of the following components:

Base: In general, the robot's chassis, which serves as the main frame for the robot's mechanical components, is called the base of the robot. If the robot's base is installed on the ground, the robot's movement will be constrained to a limited space near its installation location. However, if the robot's body is mounted on wheels or has the ability to move using robotic legs, the robot has a mobile base, and depending on the type of movement, it is referred to as a "mobile" or "legged" robot.

Arm and Legs: The arms and legs of a robot are composed of a number of links and joints that are connected to each other with a suitable structure. Similar to the joints and bones in the human arm and leg, these robot joints and links provide the robot with access to the surrounding space or the ability to take steps and move. Despite the differences in the intended purposes of human and robot arms and legs, the structure of their joints and links is very similar. For this reason, in most modern robots, designs inspired by the human arm structure are used in robot design.

Hand or End-Effector: Many industrial robots typically employ a simple structure for the purpose of grasping objects, which is commonly known as an "end-effector." The end-effector of the robot is usually positioned at the wrist of the robotic arm, situated at the end of the arm's components. In industrial robots, the end-effector frequently possesses just a single degree of freedom, allowing it to be utilized for the grasping and releasing of objects, often using pneumatic grippers or on/off magnetic actuators. When the task involves the manipulation of objects with uncertain geometries, it may necessitate a greater number of degrees of freedom and designs that draw inspiration from the human hand. The human hand, meticulously crafted for performing precise and intricate tasks such as painting or surgery, boasts an exceedingly intricate structure with 27 degrees of freedom, in contrast to the more limited degrees of freedom typically found in advanced robotic hands.

Actuators: Actuators in a robot replace the muscles in the human body. The movement of the robot is generated through motion commands issued by the robot's controller to the actuators. Common actuators in today's robots include various types of servomotors. This is in contrast to the past when pneumatic and hydraulic actuators were more commonly used. With the advancement of technology, there is ongoing research and development into creating actuators with structures resembling human muscles. In these types of actuators, new materials like "Shape Memory Alloys" are often employed.

Sensors: Various sensors are employed in robots to gather information about their surroundings, joint and arm movements, and interactions with the environment. Sensors in robots serve a role similar to the five human senses in interacting with the environment.

TABLE 1.1
Timeline of Significant Developments in the Field of Robotics Over Time.

1921	The introduction of the term "Robot" for the first time by Karel Čapek.
1927	The inaugural humanoid robot made its cinematic debut in the German film "Metropolis."
1938	The first patent for a "Controlled–Position Device" was filed by Pallard.
1942	Renowned writer Isaac Asimov introduced three fundamental laws of robotics.
1943	The field of cybernetics was inaugurated through the examination of the human nervous system and its utilization in creating artificial systems.
1954	The establishment of the first industrial robotics manufacturing company.
1960	The first sale of an industrial robotics company to General Motors (GM) and the use of robots in the manufacturing industry.
1967	The creation of the first industrial robot in Japan.
1968	The installation of the first hydraulic robot, "Unimation," in heavy industries in Kawasaki, Japan.
1969	The development of the first computer–controlled robot arm called the Stanford Robotic Arm.
1970	Significant progress in the development of intelligent humanoid robots at Waseda University, Japan.
1972	The creation of the first six-degree-of-freedom robot with electric actuators by Kuka in Germany.
1979	The establishment of the first Robotics Institute at Carnegie Mellon University in the US.
1982	A collaboration agreement between GM and FANUC led to the creation of the GMFanuc robot.
1994	The production of the first humanoid robot by Honda.
1995	The first approval for a robotic surgical assistant system named "Cyberknife."
1996	The development of the "Robotuna," the first robot inspired by nature.
1997	The successful mission of the "Rover" on the surface of Mars, lasting 83 days.
2000	The unveiling of ASIMO, the most advanced humanoid robot by Honda.
2000	The deployment of over 74,000 industrial robots worldwide, as reported by the United Nations.
2001	FDA approval for the da Vinci robotic surgical system for medical application.
2001	The installation of the Canadarm2 robotic system at the International Space Station.
2002	The release of the Roomba robotic vacuum cleaner by iRobot.
2005	The creation of the first modular robot, with self–assembly capabilities.
2011	The installation of Robonaut2, the largest robotic assistant for astronauts, on the space shuttle.
2017	Sophia, a humanoid robot, received honorary citizenship from Saudi Arabia.
2020	A four-legged robot named "Spot," that was revealed in 2016, was commercialized by Boston Dynamics in June 2020.
Now	The manufacturing and deployment of a wide array of industrial, medical, entertainment, automotive, and autonomous flying robots.

The robot's controller utilizes this acquired information to generate the necessary commands to the actuators based on the robot's control logic. Considering the sensors used in the human body, they are highly diverse, collecting a wide range of information from joint positions to the forces applied to the skin in a comprehensive manner. Furthermore, vision, hearing, and other sensory perceptions contribute to human perception of the environment and interaction with their surroundings. In the development of industrial robots, initially, only motion sensors were used to recognize the extent of movement in joints and arms. However, today, given the diverse applications of robotics, machine vision, laser rangefinders, speech recognition and processing, tactile and haptic sensors are also utilized depending on the application.

Controller: The cerebellum in the human brain, despite not having the expanse and complexity of the entire brain, controls all body movements. The robot's controller, much like the cerebellum, receives sensor information from the robot's primary processor (similar to the human brain) and is responsible for guiding and controlling the robot's movements. The key factor in the robotic motion performance is the use of feedback loops and information from motion or tactile sensors in controlling the robot's position or its interaction with the surrounding environment. The feedback loop continuously compares the robot's current position with the desired one in real-time, generating the necessary control commands sent to the robot's actuators. The presence of the feedback loop ensures that, even in the presence of environmental disturbances, measurement noise, and the lack of precise knowledge of the robot's model, suitable tracking can be achieved on the desired trajectory.

Processor: The processor plays the role of the brain in the robot. In the processor, the raw information of the sensors becomes meaningful information about the location of the robot's joints and links. By comparing the state of the robot with the desired state and calculating the movement speed of each joint to reach the desired state, this information is sent to the robot controller. The common processor in robots can be of ordinary computing systems, which are selected and used according to the amount of necessary calculations. For this purpose, depending on the industrial requirements and considerations, commercial microprocessors, industrial computers and programmable logic controllers (PLC) are used depending on the desired application.

Software: Three types of software are used in a robot. The operating system of the robot's processor is the primary software, which is real-time as it's required for online control of the robot. The second type of software in robots is responsible for executing control commands to the actuators. Additionally, a set of task-oriented routines, such as returning home, assembly, machine loading, parts retrieval and delivery, are embedded in the robot's software for user usage. Operation planning in robots usually occurs in two modes: manual and automatic. In the manual mode, the robot's automatic control system is relinquished to the manual robot controller, also known as the "teaching pendant." Using the teaching pendant, each of the robot's joints can be manually moved, and their information can be stored at desired states. Subsequently, this stored information is utilized for smooth movement from one state to another, employing the aforementioned routines. In many modern robots, pre-programmed planning has been replaced with flexible programming, incorporating artificial intelligence. In this type of programming, the motion points of the robot are not predetermined and stored through teaching but are calculated during movement. This calculation takes place using machine vision, information related to the robot's interaction with the environment, and the utilization of various artificial intelligence techniques.

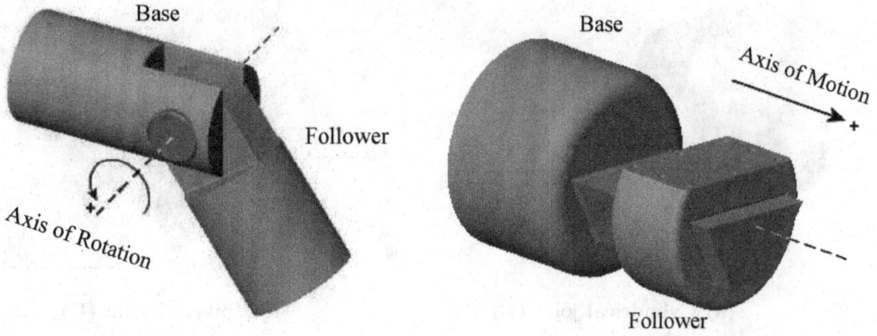

(a) Revolute joint (R) (b) Prismatic joint (P)

FIGURE 1.1
The schematics of primary joints of a robot.

1.2.1 Links

Robotic arms typically consist of a number of links and joints that, through their connections and a specific geometry, form the core structure of the robot. The degree of freedom in a robot's movement depends on the number, type, and geometry of the links and joints' connections. Over time, numerous creative designs have been developed to achieve the desired motion of a robot's end-effector. Among these designs, the links used in robotic arms define the robot's range of motion in its workspace. In industrial robots, rigid links are used to ensure precision and efficiency in performing motion maneuvers.

However, in applications such as space robots or cable-driven parallel robots, considerations related to weight limitations or practical constraints lead to the use of compliant links. These types of robots fall into the category of compliant robots, which, due to the complexity of their mechanical analysis and controller design, go beyond the scope of topics covered in this book. In this book, rigid links are assumed, and therefore, the analysis of their strain and deformation is neglected in comparison to the overall motion of the robot. With this assumption, the mechanical analysis and controller design for the robot, which will be discussed in the upcoming chapters, can be more easily accomplished.

From a kinematic perspective, a link of a robot is defined as a set of solid components connected to each other between two robot joints in such a way that it ideally eliminates relative motion between those two joints. In this manner, a simple metal rod with two rotational joints at its ends can be recognized as a robot link.

1.2.2 Joints

Robot links are connected to each other in pairs by joints. Joints enable relative motion between two links in one or more directions while constraining movement in other directions. Depending on the type and nature of the required motion between two links, various types of joints are used in robots, which are described as follows.

Revolute Joint (R)

The most commonly used joint in robotic arms is the "revolute" joint. As illustrated in the schematic of this joint in Figure 1.1(a), it allows for relative rotational motion between two links along the axis of the joint. Furthermore, this joint restricts rotation in directions

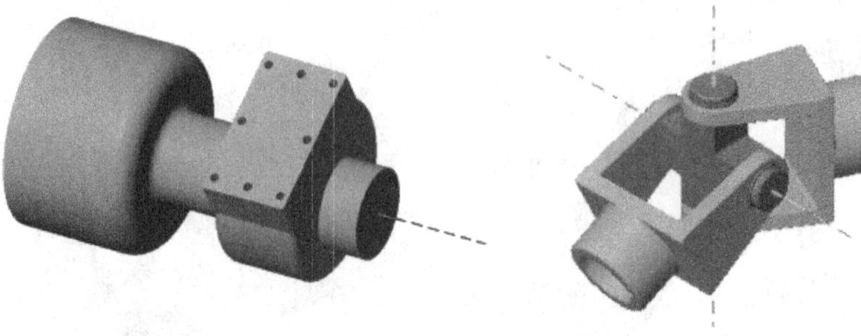

(a) Cylindrical joint (C) (b) Universal joint (U)

FIGURE 1.2
The schematics of robot's joints with two degrees of freedom.

perpendicular to the joint axis and translational motion in three directions. Thus, this joint provides one degree of freedom for rotational movement between two links and is classified as one of the *primary* types of joints. Many industrial robots are designed using solely this type of joint because it is simple to implement both the measurement of rotation using sensors such as encoders and the application of rotational torque commands using electric motors.

Prismatic Joint (P)

Another essential joint commonly used in robots is the "prismatic" joint. The schematic of this joint is depicted in Figure 1.1(b). This joint allows for translational motion between two links along the axis of the joint. Similarly, it restricts translational motion in directions perpendicular to the prismatic axis and rotational movements in three directions. Hence, this joint provides one degree of freedom for translational movement between two links and is categorized among the *primary joints*, as shown in Figure 1.1. In industrial robots that use hydraulic actuators, this type of joint is naturally employed based on the common cylinder-piston structure of such actuators. Furthermore, measuring the amount of translational movement can be easily implemented using sensors like the linear variable differential transformer (LVDT), in this joint.

Cylindrical Joint (C)

A cylindrical joint allows both rotational motion around an axis and simultaneous translational movement along the same axis between two links, as depicted in Figure 1.2(a). This type of joint provides two degrees of freedom for movement between two links, and therefore, it is not considered one of the primary joints in robots. While constructing this joint is not challenging, creating actuators that can achieve two simultaneous degrees of freedom is difficult. Consequently, the use of this joint is limited to specific cases in robots where passive joints are needed.

Universal Joint (U)

Similar to the cylindrical joint, the universal joint facilitates movement in two degrees of freedom between two interconnected components. Nevertheless, it differs from the cylindrical joint in that it creates two rotational degrees of freedom between the linked components, often oriented at right angles to each other. As depicted in Figure 1.2(b), this type of joint is also not commonly used in industrial robots, primarily due to the previously mentioned

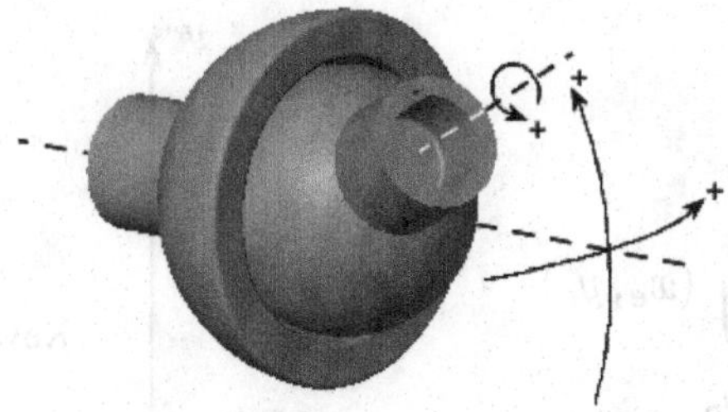

FIGURE 1.3
Schematics of a spherical joint with three rotational degrees of freedom.

reasons regarding two-degree-of-freedom joints. However, in some parallel robots, it serves as a passive intermediate joint. In automotive applications, this joint is used in power transmission systems in four-wheel-drive vehicles.

Spherical Joint (S)

A spherical joint allows the rotational movement of two links about all three principal axes. As shown in Figure 1.3, this type of joint restricts translational movements of the two links but enables rotational movement in all three directions, providing three degrees of freedom. A spherical joint is typically constructed by assembling a spherical ball into a precisely dimensioned spherical socket. This joint is commonly used in the structure of parallel robots. However, due to the complexity of constructing an active joint that allows rotational movement in three axes, spherical joints are often used in a passive manner, usually at the beginning and end of linear cylinder-piston actuators. In serial robots where each joint corresponds to a connected actuator, instead of using an active spherical joint, three revolute joints are often employed, with their rotational axes intersecting at a single point. This setup allows each joint to be actuated independently by separate motors.

1.3 Robot Kinematic Structures

The kinematic structure of robots is associated with the arrangement of their links and joints. Furthermore, whether a joint is active or actuated by an actuator and how it is actuated determine the robot's configuration. In the robotics literature, it is customary to represent a robot's abbreviated name by depicting the sequence of joint types. For instance, consider a simple planar robot consisting of two primary revolute joints denoted as R. The kinematic structure of this robot can be represented as RR or 2R. To clarify the type of actuation for the robot, the active joint, actuated by an actuator, is underlined. For example, if only the first joint of the two-jointed robot mentioned above is active, it is represented as $\underline{R}R$. If both joints are active, it can be represented as $\underline{RR}$ or simply as $2\underline{R}$.

1.3.1 Serial Robots

Within the robot's kinematic structure, some robots consist of a continuous sequence of interconnected joints in a serial configuration. In this case, by mounting the first joint on a base, the robot's end-effector will be suspended in space. This kinematic structure includes an open kinematic chain, which is known in robotics literature as a "serial robot structure."

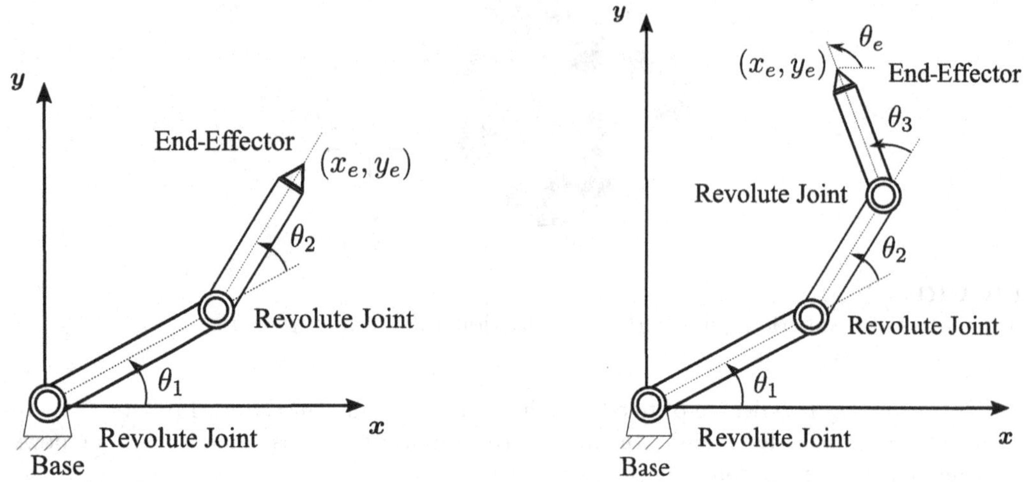

(a) Schematics of an R̲R̲ robot   (b) Schematics of an R̲R̲R̲ robot

FIGURE 1.4
Serial planar robots with revolute joints.

The simplest serial robot can be seen in Figure 1.4(a). This robot is planar and has two revolute joints, represented as R̲R̲ or 2R̲. The robot's base is positioned on the ground, and as illustrated in the figure, it can be designed to either face upward or be suspended. By controlling the two robot joints' motions with their actuators, the joint angles are adjusted in such a way that the robot's end-effector is located at the desired point with coordinates (x_e, y_e). If we want to control the orientation of the robot's end-effector on the plane as well, we can use an R̲R̲R̲ or 3R̲ planar robot. This robot, as shown in Figure 1.4(b), has three revolute joints, and uses three actuators to control the position and orientation of the robot's end-effector, represented as (x_e, y_e, θ_e).

The primary prismatic joint can also be employed in the kinematic structure of serial robots. This joint is typically actuated using hydraulic or pneumatic cylinder-piston actuators or linear motors and can be positioned either at the base or in a sequential arrangement of the robot's joints. Two examples of such robots used for rapid pick and place application on a plane are depicted in Figure 1.5. In the first sample, shown in Figure 1.5(a), the prismatic joint is mounted at the base and serves as the first joint of the robot. This planar robot, denoted as P̲R̲R̲, provides a greater linear displacement along the longitudinal axis compared to its revolute counterpart R̲R̲R̲. The robot illustrated in Figure 1.5(b) features two prismatic joints and is designed with a P̲R̲P̲ structure. The presence of a right angle in the design of the second arm showcases flexibility in the arrangement of the robot's joints. This flexibility aids the designer in creating an appropriate kinematic structure based on the intended application.

To facilitate the movement of objects in three-dimensional space, a common choice is to use a robotic arm with joints that allow movements beyond a single plane. The robot depicted in Figure 1.6 is the PUMA[1], designed by the Unimation company, that has found extensive use in various industries. As observed in this figure, the first joint of the PUMA robot depicted in this figure is oriented vertically on the base, and by rotating this joint, out-of-plane movements are generated in this robot. This robot is constructed with six revolute joints and follows a kinematic structure of 6R̲. The first three joints, along with their

---

[1] Programmable Universal Manipulation Arm

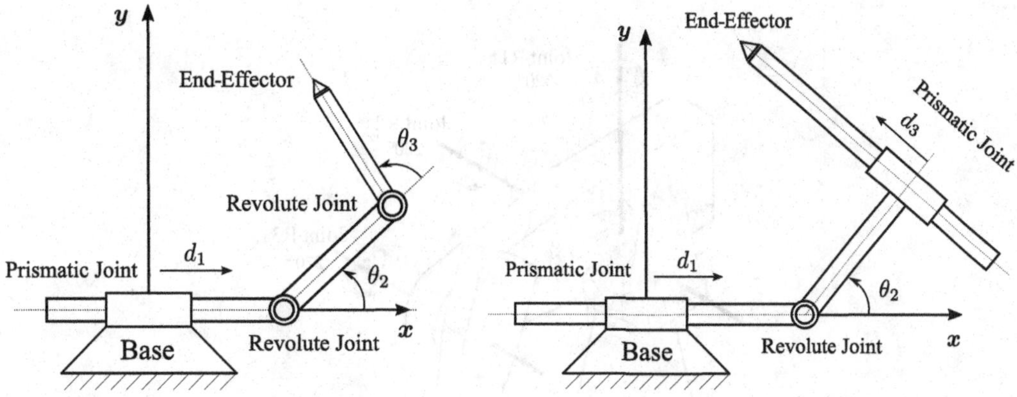

(a) Schematics of an PRR robot (b) Schematics of an PRP robot

FIGURE 1.5
Serial planar robots with revolute and prismatic joints.

corresponding links, form the robot's arm and provide access to the workspace. Meanwhile, the three end-effector joints, designed with intersecting joint axes at a single point, create an appropriate design for the robot's wrist with a spherical actuated joint.

Due to its six degrees of freedom in three-dimensional space, the PUMA robot is highly versatile and has been widely adopted in various industries for different applications. Unimation produced numerous PUMA models in the classes of 200, 500, and 700. The PUMA 560C model, the best-selling one, had a reach of 878 mm and a payload capacity ranging from 2.5 to 4 kg, making it suitable for a wide range of industrial applications. Unimation, which manufactured various types of these robots until the 1980s, was acquired by Westinghouse in 1980. In 1988, it continued its operations under the name Stäubli. Today, Stäubli is considered one of the leading manufacturers of industrial robots, offering new designs with diverse workspaces.

Another prominent serial robot widely used in the industries is renowned as the SCARA[1]. It is recognized as one of the most successful serial robot structures, particularly acclaimed for its effectiveness in rapid object handling. The kinematic structure of this robot is RRRP, where the first three joints are revolute joints with parallel rotation axes, and the fourth joint is prismatic. Due to the configuration of three parallel revolute joints, this robot exhibits a degree of compliance in the $x-y$ plane while maintaining complete rigidity along the vertical z-axis. This characteristic allows the robot to efficiently perform tasks involving object manipulation and part assembly. Moreover, the utilization of suitably designed belt-driven power transmission systems for the initial three joints allows for a substantial reduction in the weight of the moving components, and therefore, enhances the speed of the robot. With this structure, it is possible to achieve linear speeds of up to 7 m/s and perform 120 picks per minute. To observe the SCARA robot in action, performing swift pick-and-place maneuvers, just click on the designated link[2] to access the video.

1.3.2 Parallel Robots

A parallel robot also referred to as a parallel manipulator, is characterized by its structural arrangement, where multiple kinematic chains are linked to a shared mobile platform. These

[1]Selective Compliance Articulated Robot Arm
[2]https://youtu.be/-m1oKuFkSTE

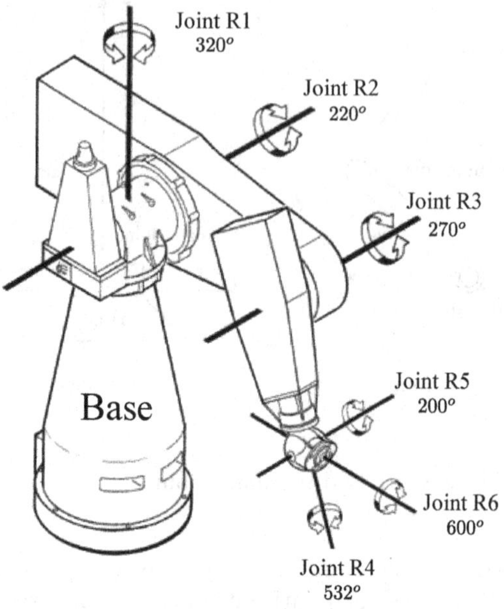

FIGURE 1.6
Schematics of a PUMA robot with a general 6R kinematic structure.

chains, commonly known as *legs*, with one end affixed to the base and the other end connected to the platform, establish a parallel configuration. In the context of parallel robots, the term "moving platform" is often used instead of end-effector, indirectly representing the multiple connections to the robot's base. The moving platform is usually employed for tasks like positioning, manipulating objects, or executing specific operations. The kinematic loops of the legs typically exhibit similar kinematic structures and consist of various joints and links, akin to a serial robot. The simplest type of these arms consists of a prismatic actuated joint connected to the base and the moving platform through two passive spherical or universal joints, denoted as S or U. This basic structure allows the use of hydraulic piston-cylinders to create robots with high payload capacities, such as those used in aircraft simulators.

The distinguishing feature of parallel robots is that all of their actuators are located on the base, while the moving platform moves in a coordinated manner due to the interconnected legs. This design provides several advantages, including high precision, stiffness, and the ability to handle heavy loads. Parallel robots are known for their use in applications that require accuracy, speed, and precision, such as flight simulators, medical surgery devices, and certain types of manufacturing processes [1].

In Figure 1.7, a schematic of a general-purpose parallel robot is depicted, where six closed kinematic chains connect the base to the moving platform. In this versatile setup, any kinematic structure can be used to link the base to the moving platform. Additionally, although this kinematic structure incorporates several revolute joints denoted as R and one prismatic joint denoted as P, typically only one of these joints is active, while the others remain passive. Consequently, with appropriate design, all the robot's actuators can be installed on the base, reducing the weight of the moving components. Furthermore, as evident in this figure, the weight of the mobile platform is distributed across multiple legs,

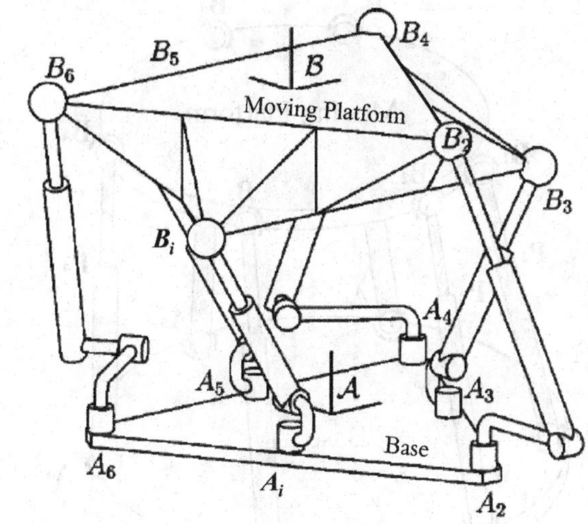

FIGURE 1.7
Schematics of a general-purpose parallel manipulator

enabling the use of lighter actuators in the robot. Given the inherent features of parallel robots, the payload-to-weight ratio and the robot's rigidity significantly increase, leading to improved performance accuracy.

Nevertheless, parallel robots have a notable limitation concerning their confined workspace. The close proximity of neighboring arms and joints introduces the possibility of collisions between various components, including arm components to the base, or the moving platform, consequently restricting the robot's workspace. Furthermore, in parallel robots, at singular points and under specific conditions, the robot may lose its stability and become uncontrollable [1].

The Stewart-Gough Platform (SGP) and the Delta robot stand out as two highly utilized parallel robots in the industry. Particularly, the Stewart-Gough platform, often referred to as the hexapod, holds significant prominence as an industrial example. It finds extensive applications in various domains, including aircraft, helicopter, and automotive simulators [11]. As shown in Figure 1.8, this robot employs a structure consisting of six similar arms with a kinematic structure of S$\underline{P}$S or U$\underline{P}$S, providing six degrees of freedom in three-dimensional space. Hence, due to the presence of six similar arms in this robot's structure, it is commonly denoted as 6S$\underline{P}$S or 6U$\underline{P}$S. The utilization of the universal joint (U) on the robot's base is preferred for design and ease of kinematic analysis, although the construction of the robot with a spherical joint (S) is simpler. Hydraulic piston-cylinders or linear electric motors are used as actuators in the prismatic joints of the robot. This enables the robot to provide the necessary six degrees of freedom in three-dimensional space, making it a suitable choice in applications that require a high payload capacity and a smaller workspace compared to serial robots with the structure 6$\underline{R}$. You can watch a video clip featuring the 6D motion of an SGP by following this link[1].

In applications where six degrees of freedom are not required, simpler structures can be employed in parallel robots. Another highly successful example of these robots, commercially produced by various manufacturers, is the Delta robot. The Delta robot is an excellent replacement for the SCARA in pick and place applications and is widely used in

[1] https://youtu.be/xiECumcaEx0

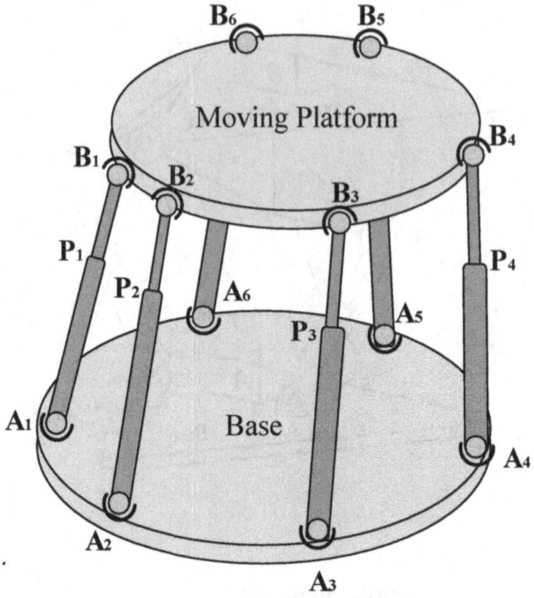

FIGURE 1.8
Schematics of a Stewart-Gauph platform with 6S$\underline{\text{P}}$S structure

the packaging industry. This robot is composed of three similar arms, attached to the base with revolute joints (R). The key idea behind this robot is the parallelogram structure in the second link, which keeps the orientation of the moving platform fixed. With this structure, the links of each of the three arms can be designed with minimal weight, enabling very fast movements. Due to these characteristics, commercial versions of these robots have been produced with a rapid displacement speed of over 200 picks per minute [12]. To witness the delta robot in action, executing rapid pick-and-place movements, simply click on the provided link[1] to access the video.

The robot's kinematic configuration is designated as 3$\underline{\text{R}}$RR, where the actuator motors are affixed to the base and control the initial revolute joint. The robot's second link features a parallelogram structure that provides the mobile platform with three translational degrees of freedom. If an additional rotational degree of freedom is required for the platform, a connecting rod can be employed to convey rotational motion to the mobile platform.

1.3.3 Hybrid Structures

The structure of the human arm and hand, as well as the motion system of some animals and even insects, is much more complex than the simple serial or parallel structures used in industrial robots. For instance, consider the structure of the human arm, in which the shoulder joint, with its unique configuration, provides three rotational degrees of freedom for motion. Structurally, this joint can be likened to a ball joint, in which the use of four muscle groups yield three rotational degrees of freedom.

On the other hand, the elbow and wrist of the hand are connected to each other by two bones, similar to the parallelogram structure of a delta robot. However, none of the elbow and wrist joints function like a simple hinge joint, and each of these joints individually creates two rotational degrees of freedom. The fascinating arrangement of muscles and

[1] https://youtu.be/6JIjCpW35CM

tendons in the forearm results in the combination of these angular movements, allowing us to consider a total of four independent degrees of freedom for them. Thus, the human arm utilizes seven actuated degrees of freedom to control the six degrees of freedom in the wrist.

This remarkable design, which combines both serial and parallel structures, exhibits redundancy to the actuation, a unique feature that bestows extraordinary motor skills upon the human arm. We will address the mathematical analysis of this subject in the upcoming chapters. Such astonishing designs can be found in many of Nature's creations.

Another example is the human wrist, which is composed of eight bone segments and six muscle groups, relying on a parallel structure to achieve additional dexterity for the three degrees of freedom wrist motion. On the other hand, each of the fingers of the hand can be categorized as a serial $3\underline{R}$ arm. However, the manner in which they are attached to the palm and the arrangement of muscles and tendons in them allow for out-of-plane movements for each finger.

Until now, due to the constraints of actuation technology, humans have not been able to create robotic technology that achieves performance similar to these astonishing designs. Nevertheless, to meet specific performance requirements and harness the advantages of both serial and parallel structures, engineers have developed robotic designs for specialized applications. In these designs, some components incorporate serial structures, while others employ parallel mechanisms.

Hybrid robots represent robotic systems that amalgamate elements and attributes from both serial and parallel robot designs within a unified structure. They typically harness the benefits of serial robots, which offer expansive workspace and flexibility, along with the advantages of parallel robots, which provide heightened stability and load-bearing capacity. By merging these diverse design approaches, hybrid robots can attain distinctive capabilities and adaptability for particular tasks, making them exceptionally suitable for applications where a combination of characteristics from both serial and parallel robots is essential for optimal performance. These robots frequently present innovative solutions by capitalizing on the array of features offered by each design type.

To illustrate the concept of a hybrid design, consider the case of the quadruped robot "Spot" accommodated with a robotic arm explicitly tailored for handling objects within its proximity. While each of Spot's legs could be engineered with a serial structure, the way these four legs connect to a mobile body, serving as a mobile platform, exhibits traits typically associated with parallel robots. Conversely, the robotic arm affixed to the robot's body exhibits characteristics more commonly found in serial robots, and it is integrated with the mobile platform. This particular instance effectively showcases the integration of diverse robotic structures, particularly suited for applications related to exploration and rescue [13].

As another example, consider a cable-driven "SpiderCam" robot, equipped with a three-degree-of-freedom stabilizing gimbal designed to effectively eradicate vibrations and ensure the stability of a camera for filming purposes. The cable-driven robot structure falls under the category of suspended parallel robots, while the stabilizing gimbal utilizes a three-degrees-of-freedom serial structure [14]. When it comes to the creation of hybrid robots, it is a conventional approach to incorporate both serial and parallel designs, with each design type having already proven its operational effectiveness. This fusion allows for the crafting of a robot equipped with unique features tailored to a specific task. Consequently, a comprehensive grasp of the design principles of both serial and parallel robots is a fundamental requirement for developing such innovative products. Nevertheless, it's important to note that human-made examples of these robots still have a significant journey ahead to replicate the intricacies found in natural structures.

1.4 Robotic Applications

Robots are excellent replacements for humans in applications that require higher capabilities and precision or involve hazardous environments. Currently, robots are extensively used in various industries and for a wide range of tasks. Where greater capability and precision are needed, robots have outperformed humans, while in cases requiring more flexibility and intelligence, robots have excelled. Nevertheless, the integration of robots into human daily life is such that it's challenging to imagine a world without them. Within the scope of this book, we only explore a very limited number of robotic applications and the technologies employed in them. However, even this brief overview provides a clear glimpse of the future of this technology and its impact on human life.

1.4.1 Industrial Robots

Industrial robots find applications across a broad spectrum. The advent of versatile robots has further expanded their widespread utilization in various industrial sectors. These multipurpose robots represent an advanced generation, such as the well-known "PUMA" robot, which often boasts six degrees of freedom and can adapt to small, medium, and large workspaces. By equipping these robots with the suitable end-effectors, they can perform various industrial tasks. Nevertheless, it's crucial to recognize that the general design of these robots may limit optimal performance in certain operations. For example, multipurpose robots can handle material movement, but for operations demanding optimized speed or quantity, specialized robots like "SCARA" or "Delta" robots, purpose-built and fine-tuned for specific tasks, offer better solutions. These applications encompass a wide range of industrial functions, including manufacturing and assembly, welding, painting and coating, material handling, palletizing and packaging, CNC machine tending, inspection and testing, pick and place operations, and 3D printing, among others. You can watch an impressive video featuring a Kuka robot performing precise arc welding by following this link[1].

1.4.2 Space Robots

Just as the development of space science and technology has led to significant advancements in other scientific fields, it has also made a notable contribution to the development of robotics technology. International space research, pursued by numerous countries, reached its pinnacle with the establishment of the International Space Station (ISS) in 2010 and human presence in space. But how was such a massive space station constructed in low Earth orbit? This feat was accomplished with the help of a space robot known as the "Space Station Remote Manipulator System (SSRMS)," developed as the primary robotic arm for handling the transportation of components and equipment as part of the space station's assembly program.

The first generation of this space robot, also known as the Canadarm, has six revolute joint actuators, measures 15.1 meters in length, and has a weight of 450 kilograms. It is remotely controlled by astronauts stationed in the space shuttle. The long arms of this robot are designed to be so lightweight that if they were to open on Earth, they wouldn't support their own weight. By its installation on the space shuttle in 1981, the Canadarm has been used in various shuttle missions, including deploying spacecraft into orbit, capturing and

[1] https://youtu.be/NJIgQjKDVUg

repairing satellites in orbit, assisting and guiding astronauts during spacewalks, repairing the Hubble Space Telescope, and most importantly, aiding in the assembly of the International Space Station [15].

The success of this robot during its over twenty years of missions in space led to the development of a second-generation robot inspired by the experience of designing and using the first version. In 2001, the Canadarm2 was installed on the International Space Station. Canadarm2 has seven revolute joint actuators, measures 17.6 meters in length, and has a weight of 1497 kilograms. It can be remotely controlled by astronauts both onboard the station and from Earth. This robot is capable of maneuvering payloads of up to 116,000 kilograms and has been instrumental in connecting the space shuttle to the International Space Station. With the help of these two robotic arms and the presence of astronauts in space, large and heavy components of the space station were delivered one after the other to orbit and installed on the space station. These two robots have also been used to repair and replace their own parts. In 2002, for instance, the wrist joint of the robot was replaced by spacewalking astronauts. In 2017 and 2018, two developed versions of the robot's end-effector were replaced by the other robot [16].

The most intricate space robot installed on the space station is the Dextre robot. This robot features two remotely controlled arms, designed to reduce astronauts' spacewalks. It was installed on the space station in 2008 and can also be automatically positioned on the wrist of the Canadarm2 robot. Dextre comprises two agile robotic arms, each measuring 3.5 meters in length and weighing 1662 kilograms, connected to its central pedestal. The kinematic structure and dimensions of each of Dextre's arms are scaled-down versions of the Canadarm2 robot. The motion control of each arm is performed independently to facilitate remote control, prevent arm collision, and maintain robot stability. A three-fingered hand for each arm has been designed and built, although it hasn't been sent to the space station yet. The development, installation, and use of these space robots brilliantly demonstrate human ingenuity and power in conquering space [17]. You can view a splendid animation of this prowess by following this link[1] where the Dextre robot is stationed on the Canadarm2 robot and remotely controlled by astronauts.

1.4.3 Surgical Robots

The use of robotic technology in various medical applications has significantly expanded. One of the most renowned commercial surgical robots is the da Vinci Surgical System, often referred to as the da Vinci robot. This robot is a cutting-edge robotic platform used to assist surgeons in performing minimally invasive surgeries. Developed by Intuitive Surgical and introduced in 2000, this system is operated by skilled surgeons to achieve highly precise and controlled movements during surgical procedures. It features multiple robotic arms, each equipped with specialized surgical instruments, which offer enhanced dexterity and flexibility. The da Vinci system provides surgeons with a 3D high-definition view of the surgical site, improving visualization and depth perception. It is widely used across various surgical specialties, such as urology, gynecology, general surgery, cardiovascular surgery, and more.

The kinematic configuration of this robotic system incorporates a Remote Center of Motion (RCM), which proves highly beneficial in the context of minimally invasive surgery. Minimally invasive procedures involve making small incisions, resulting in advantages such as reduced patient discomfort, shorter recovery periods, and minimized scarring. Surgeons oversee the system's operations from a control console located within the operating room, and the robot translates the instructions into precise movements executed by its robotic

[1] https://youtu.be/AXQ6iSFVwrk

arms. With an impressive track record, involving millions of procedures performed successfully, the da Vinci Surgical System has played a pivotal role in pushing the boundaries of surgical practices, ushering in a revolution in contemporary surgery. Furthermore, this robotic platform continually adapts and advances alongside ongoing developments in both robotics and medical technology.

An integral feature of this surgical robot is its incorporation of multiple robotic arms, which can be operated simultaneously by the surgeon and their assisting team. This collaborative control not only facilitates the management of surgical instruments but also allows for the adjustment of endoscopic cameras, controlled irrigation during surgical procedures, and the regulation of lighting within the surgical field, all conducted robotically [18]. You can view a descriptive movie clip on da Vinci robotic platform by following this link[1].

Given the precision of a human hand's movements and resulting tremors are on the order of about two hundred micrometers, it becomes evident that even the most skilled human surgeons lack the capability to perform highly accurate surgical procedures manually in certain surgeries. The integration of robotic technology is imperative in such cases, where levels of precision, particularly in eye surgeries involving the retina, demand working with dimensions of veins that are practically unattainable by even the most skilled surgeons without the assistance of robots. Extensive research in this field over the past two decades emphasizes the need for robotics in medical procedures.

The Preceyes robotic system, specifically designed for delicate surgical procedures, offers an exceptionally high level of precision. This precision is vital, especially in the field of eye surgery. While the exact values may vary depending on specific applications and versions of the Preceyes system, the typical precision falls within the range of micrometers, often going below 20 micrometers. This level of accuracy enables intricate and minimally invasive procedures, making it particularly well-suited for tasks involving the retina and other highly sensitive areas of the eye. Manual surgery struggles to achieve such accuracy, underscoring the crucial role of robotic assistance in these medical applications.

The development of this robotic system was initiated at the Eindhoven University of Technology in 2007, and the first commercial version was sold to a British biopharmaceutical company in 2016. This system comprises two similar robots designed for left and right-handed use, controlled remotely by two surgeons through a haptic console. The robot's kinematic structure involves a single closed kinematic loop with three revolute joints, classifying it among the serial robots. The utilization of a parallel-linkage structure in the design of the first and second links of this robot has enabled not only the fulfillment of the required Remote Center of Motion (RCM) characteristics for minimally invasive retina surgery but also the attachment of the robot's actuation motors to its base. Consequently, the weight of the mobile parts has been reduced, and the counterweights for the arms are positioned on the base joint. In this way, the system effectively eliminates hand tremors of the surgeon and allows for precise scaling of their hand movements, in order to perform delicate retina surgery [19]. You can view an animation of this system following this link[2].

1.4.4 Exoskeleton Robots

Exoskeleton robots are wearable mobile robots attached to the user's body, allowing users to achieve augmented strength or endurance. These devices may also include symbiotic robots, where the exoskeleton and humans become one unit. In these robots, which are worn by the user and act like their external skeleton, using electric motors and interacting with their limbs, more power or more endurance is created in their limbs. The most common

[1] https://youtu.be/HHKEHTr7o3Y
[2] https://youtu.be/a0v7vpRSNu0

application of exoskeleton robots is for medical purposes. Exoskeletons for medical use assists those with spinal cord injuries or other mobility impairments. Examples include devices such as powered leg braces to augment walking and stair climbing or special suits to help patients with mobility impairments perform physical therapy. Medical exoskeletons can be classified as therapeutic, rehabilitative, or assistive devices. Among the applications, one can mention two main applications of these robots in enhancing the soldiers' ability to carry more load and prolong their movement time on the one hand, and industrial applications of robots helping workers carry heavy loads.

You can see a movie clip of an exoskeleton robot designed for the rehabilitation of patients with spinal cord injuries by clicking on the following link[1]. This robot assist the patient to walk alone with the help of two crutches. The excessive happiness of patients who have been able to stand on their own feet and walk for the first time with the help of this robot is a gift that robotics technology has given to humanity [20]. Exoskeleton robots with a wide range of motion diversity have been developed for rehabilitation of the upper body, lower body, wrist, ankle, and similar limbs.

As another example, you may watch the video of a wearable power-enhancing exoskeleton worn by a soldier, by clicking on the following link[2]. In this clip, it is shown how the robot's external skeleton is adjusted to the user's body dimensions. Although the soldier needs to bear the weight of this wearable robot and its components, with the help of electric motors, he can carry much heavier weights on his back with more ease and can walk continuously for a longer time.

1.4.5 Exploration Robots

Exploration in unknown, dangerous, and inaccessible environments can only be achieved by exploration robots. Human presence in these types of environments is often very dangerous and unattainable, while robots can enter these environments without risk and explore them by walking or flying. The ability of robotic technology to perform remote operations in these environments has been demonstrated by the cleanup of nuclear reactors such as the one that occurred in the Fukushima tsunami in 2011. Exploration in space, underwater, and deep mines is also carried out by these types of robots. In the next subsections, we will review examples of these types of robots in two categories: mobile rovers and flying robots.

Mobile Robots

The most famous mobile robots built by humans are probably the Mars rovers. So far, six Mars rovers have been sent to explore the planet. Five of them were developed by NASA: Sojourner in 1997, Opportunity from 2004 to 2018, Spirit from 2004 to 2010, Curiosity from 2012 to present, and Perseverance from 2021 to present. The sixth rover, named Zhurong, has been on a mission to explore Mars since 2021 under the auspices of the China National Space Administration.

Among them, Spirit and Opportunity were two identical Mars rovers that were planned to explore the planet for only 90 Martian days. However, after the successful completion of this mission, the Spirit rover continued its activities until 2010, and the Opportunity rover continued its activities until 2018 on the surface of Mars. One of the extraordinary discoveries of this rover was that there was water on the planet Mars at some point in the past. Inspired by the successful mission of the initial Mars rovers, the Perseverance rover was designed with the dimensions of a car [21] and traveled for seven months after launch to reach Mars. The rover carried a flying robot, such as a helicopter, to Mars, and the first

[1] https://youtu.be/EhnMZLCQahI
[2] https://youtu.be/q5NjhGztlao

robotic flight outside the Earth's orbit took place on Mars on April 19, 2021. The mission of these robots is to explore Mars to find signs of microbiological life on the planet.

In addition to space exploration robots, many other mobile robots have also been developed. These types of robots are known as Unmanned Ground Vehicles (UGVs). A UGV is a vehicle that operates while in contact with the ground and without an onboard human presence. UGVs can be used for many applications where it may be inconvenient, dangerous, or impossible to have a human operator present. They are used in military applications, surveillance and exploration, search and rescue, agriculture, and similar applications. Tracked robots are used to move on rough terrain, while wheeled robots are used to move on flat terrain. Generally, the movement of these robots is controlled remotely by a human operator, but various cameras and sensors are used to perceive the environment and plan autonomous movements.

Robotic Drones

In addition to ground rovers, unmanned aerial vehicles (UAVs) also play an important role in surveillance and exploration operations. Initially developed only for military applications, drones have recently found their place in everyday human life with the expansion of multi-rotor designs in non-military applications such as aerial filming, postal delivery, agriculture, and entertainment. The maneuverability of this category of drones is such that the idea of replacing ground vehicles with personal flying vehicles has become closer to reality. An example of an electric flying car, which is a joint product of Italdesign and Audi automakers and Airbus, has been designed and prototyped in 2021.This self-driving electric car is made up of three independent modular parts. The passenger compartment has a lightweight body and is made of carbon fiber. The ground section is a self-driving electric car designed by automakers. The flying section of the car is also a quadcopter drone designed by Airbus.

The travel plan from origin to destination is done autonomously, and the passenger travels part of the route that is free of traffic by electric car. In case of traffic or passenger request, the flying section of the car will arrive at the location autonomously and lift the passenger compartment off the ground with a lock [22]. Although the idea of such a car has not yet been implemented commercially, this flying car has kept the hope of using a flying car alive in humans more than other alternatives made so far. You can watch an intriguing introductory video about this product by clicking on the following link[1].

1.4.6 Humanoid Robots

The expansion of robotics technology has led to humanoid robots that not only resemble humans in appearance but also interact like them. About thirty years ago, when robotics research began on mobile robots that used legs to move, it was never predicted that after this short period, a large number of humanoid robots would be developed by research centers and industrial companies. Today, there are so many of these robots that it is beyond the scope of this book to provide a brief overview of them, and a few examples of them are selected to be introduced in this section. Selecting these examples is also very difficult because each of these robots has a lot of similarities in shape and behavior with humans. Among the top ten humanoid robots selected by media and scientific organizations such as the Institute of Electrical and Electronics Engineers (IEEE) and the American Society of Mechanical Engineers (ASME), we will introduce three robots with fascinating designs and functions.

Perhaps the robot named Atlas developed by Boston Dynamics is the easiest selection to be introduce in this section. This robot has two legs and two arms and an advanced

[1] https://youtu.be/ZaKD6LYj5eE

programming and control system that has brought extraordinary motion dynamics to this robot. The latest version of this robot is 1.5 m tall and has a movement speed of 2.5 m/s. This robot has 28 degrees of freedom, all of which are controlled by powerful hydraulic actuators. Inspired by human body movements, agile and professional motion have been programmed into this robot, such that it can easily run at high speeds, cross obstacles and slopes, climb stairs, make stable jumps over obstacles, and even perform acrobatic movements [23]. You can watch an exhilarating video featuring the Atlas robot executing acrobatic maneuvers by clicking on the following link[1].

Another well-known humanoid robot to be introduced here is the ASIMO robot. Its name pays homage to the renowned writer Isaac Asimov, who introduced the famous "Three Laws of Robotics" in 1943. Honda first unveiled the initial version of this robot in Japan in 2000, with the latest iteration released in 2011. The research and development of this robot reached its culmination in 2018. This version of ASIMO stands at a height of 130 centimeters and has a weight of 48 kilograms. It can achieve a running speed of $1.7 m/s$. This robot boasts 57 degrees of freedom, with the majority of them located in its four-fingered hands. Furthermore, ASIMO can maintain continuous movement for up to an hour during both walking and running. It is equipped with the ability to detect and autonomously avoid obstacles in its path, as well as recognize facial images and expressions. It can also comprehend sound and environmental cues, allowing it to engage in friendly interactions with humans. One distinctive feature of this robot is its capacity to track moving objects in front of it and respond to commands and requests in both English and Japanese [24]. This technological advancement at Honda has also led to the development of exoskeleton robots designed to assist elderly individuals in climbing stairs and slopes. These exoskeletons are now available for commercial use, enhancing the mobility of elderly individuals. You can watch an introductory video showcasing the ASIMO robot engaging in various aspects of human daily life by clicking on the following link[2].

Among the various humanoid robots developed worldwide, we would like to introduce Sophia as the most recent addition in this section. Sophia made her debut in 2016 and has garnered widespread admiration from the global community for her striking facial resemblance and human-like facial movements. Leveraging artificial intelligence technology, Sophia is capable of mirroring the facial expressions of those she interacts with and providing responses to their inquiries. While her conversations are initially limited to predefined topics, her AI system connected to the global internet, enables her to offer acceptable answers to a wide array of audience questions. Furthermore, Sophia's intelligence continues to evolve, thanks to advancements in speech recognition technology developed by Google. Sophia's eyes employ stereo-machine vision technology, allowing her to establish and maintain eye contact with those she interacts with. Originally designed to interact with humans in a seated position, she was equipped with two robotic legs in 2018 to enable movement among humans. What sets Sophia apart is her ability to convey over 60 different facial expressions in response to her interlocutors. Notably, Sophia is the only robot to have received honorary citizenship in Saudi Arabia and the sole robot recognized as the first non-human innovation champion by the United Nations [25]. Explore a captivating video featuring Sophia's interview with Tony Robbins, where they discuss the similarities and differences between robots and human beings. Click on the following link[3] to watch it.

[1] https://youtu.be/tF4DML7FIWk
[2] https://youtu.be/CS23BQiF4h0
[3] https://youtu.be/Sq36J9pNaEo

1.5 The Aims and Scope of the Book

As elaborated within this chapter, the realization of our aspiration to create a being resembling ourselves necessitates extensive research and the generation of knowledge and technology across diverse scientific domains. Robotics, in its entirety, represents a multifaceted research domain that cannot be succinctly encapsulated within the confines of a single book. To provide clarity regarding the subjects addressed within this book, it might be advantageous to initially outline the scientific fields that fall outside the purview of this book's topics.

The design of humanoid robots encompasses a broad spectrum of topics, many of which extend beyond the boundaries of this book. For instance, fields such as machine vision [26], speech processing [27], artificial intelligence [28], and programming and decision-making [29], are each well-structured scientific domains explored in various dedicated publications. Moreover, the subjects of mobile and flying robots also draw from related scientific disciplines. In the field of mobile robots, topics like path planning and obstacle avoidance [30], kinematics, dynamics, and control of these types of robots [31], self-driving mobile robots [32], probabilistic robotics [33], and the sensors and actuators employed in these robots [34] constitute vital areas of study that fall outside the scope of this book. Additionally, the intricate design of complex structures such as multi-fingered hands [35] and the tactile sensors integrated within them [36] necessitates comprehensive exploration, which will not be addressed within this book.

Based on the earlier discussion concerning the classification of robotic structures into serial and parallel robots, it is evident that the analysis and design of parallel robots pose greater complexity compared to serial robots. Additionally, expertise in the analysis and design of serial robots is foundational for understanding. The underlying concepts of parallel robots. Consequently, this book will concentrate on the analysis and design of serial robots. Readers with an interest in parallel robots are encouraged to explore another book authored by the same writer as referenced in [1].

Within the context of designing serial robots, we will not address the mechanical design of such robots, the selection of suitable actuators and sensors, the implementation of teaching pendants, or the incorporation of data-driven techniques and artificial intelligence for their control. The primary objective of this book is to provide readers with the essential knowledge and skills required for the analysis and design of serial robots in two crucial domains, namely mechanics and control. Through a comprehensive study of the subjects presented in this book, readers will acquire the necessary expertise to analyze both simple and intricate structures within an open kinematic loop, possessing multiple degrees of freedom, which are commonly referred to as serial robots. By honing these skills, individuals will be equipped to evaluate and analyze a diverse array of robots employed in numerous applications, including but not limited to industrial, exoskeleton, space, and surgical robotics.

Considering the book's primary focus on serial robots, it will elaborate on the mathematical representation of motion within three-dimensional space, offering precise definitions for both location and orientation. Various methods for representing these concepts will be thoroughly examined. Furthermore, the kinematic analysis of serial robots will be explored through two distinct methods: the conventional approach and the utilization of screw theory. The representation of three-dimensional motion through screw theory holds remarkable advantages when analyzing the kinematics of robot motion. Notably, it seamlessly aligns with the fundamental principles of physical motion. It's worth mentioning that in the field of kinematic analysis of robots, two distinct strategies have been developed. The first approach, emphasizing numerical and recursive methods (as outlined in [2]), is closely aligned with computer science, while the second approach, which emphasizes geometric and analytical methods (as detailed in [3]), is more in harmony with engineering and analytical perspectives. This book will primarily focus on the second approach, incorporating mathematical methods that are closely linked to the relevant physical concepts.

The Jacobian matrix plays a pivotal role in the field of robotics, particularly in the context of differential kinematic analysis, and this book provides an in-depth exploration of this fundamental concept. Many crucial aspects involved in the design of serial robots necessitate the analysis of velocities and the Jacobian matrix of the robot. The true beauty of this knowledge emerges when one employs analytical methods to establish connections between the mathematical principles governing the kinematic analysis of robots and their physical characteristics. This process, inspired by the optimal design seen in the nature, fosters a deeper appreciation for the contemplative aspects of this knowledge.

This book meticulously explores the dynamic analysis of robots, with special attention given to striking the right balance between the level of computational complexity and the depth of physical understanding required to formulate the governing dynamic equations of a robot. This aspect is a distinctive feature often underrepresented in similar references [2, 4]. In this book, three distinct methods namely, Newton-Euler, Euler-Lagrange, and virtual work are introduced, with a primary emphasis on the Euler-Lagrange method. This book elaborates on the derivation of the robot's dynamic equations in a closed-form, along with an exploration of the associated properties. It is interesting to note that the governing dynamic formulation of all robots, regardless of their structure and degrees of freedom, can be expressed in a general form. This form includes a mass matrix that generalizes the mass of a rigid body into high-dimensional space, a gravity vector that encompasses the reacting forces on the robot links due to gravity, and a Centrifugal and Coriolis vector that includes all the interacting forces in the joints due to the multi-degree of freedom motion of the links. Although these vectors and matrices differ in size and shape for different robots, they share certain properties in all cases. These properties enable the computer simulation and model-based controller design of robots to be done systematically.

One of the key aspects of robot design is visualizing the motion characteristics and analyzing the controller performance of robots. Computer simulations can help designers achieve these objectives. The general formulation of robot dynamics, makes it feasible to perform computer simulation for different robots, by using a systematic approach into a standard simulation software. In this book MATLAB® software and Python programming language are effectively harnessed to perform the simulation and do the analysis. Furthermore, the generation of desired trajectories and methods for crafting optimal time paths in robots, while taking into account the speed and acceleration limitations inherent to the actuators are extensively discussed.

The standard model for a robot's dynamics is generally nonlinear and involves multiple variables. Despite this complexity, industrial robots often utilize linear controllers to accommodate the characteristics of their actuators. This book demonstrates that when robots are designed with gearboxes featuring high reduction ratios, the linear dynamics of the actuators can effectively counteract the influence of the robot's nonlinear dynamics. Consequently, the intricate multivariable system can be simplified into a series of more manageable single-variable linear systems. This simplification facilitates the application of linear control design methods in constructing the robot's control system.

Designing linear control systems that can achieve multiple objectives has been a significant challenge for control system designers. However, the use of feedback loops, regardless of the type of controller, can help achieve these objectives in a control system. This is perhaps the most important reason for the success of control systems in practice. By observing the system status continuously, it is possible to achieve the desired tracking in the presence of disturbances and uncertainties in the system model. This concept is called "the magic of Feedback," which is extensively discussed in the book. Using feedback loops, the design procedure of PD and PID controller for robots are discussed in details.

In robotic applications where high reduction ratio gearboxes are not used, the governing dynamic formulation remains nonlinear and multivariable. In such cases, a wide range of nonlinear controllers are developed for robotic applications. One of the bases of such controller design is full or partial feedback linearization, in which the nonlinear dynamics are compensated by an inverse dynamics control scheme. Since this method is the basis for many other developments, it is extensively elaborated in this book. Despite the popularity of inverse dynamic control in various applications, this control structure suffers from two major limitations. The efficiency of this control structure depends on the complete knowledge of the dynamic model of the system and the calibration of its parameters. In practice, no matter how much effort is made to determine the exact model of the system, the existence of uncertainty in the model is inevitable.

Two general approaches of robust and adaptive control have been proposed in the literature to deal with the uncertain effects of the model in the transient and steady-state dynamics. In robust control methods, the uncertainty of the model in the robot workspace is quantified and its upper bound is determined. A corrective term is then added to the inverse dynamic control to compensate for the uncertain effects in the worst possible case. On the other hand, in adaptive control methods, the nominal model of the system is initially used in the inverse dynamic loop. By observing the tracking error dynamics, the model parameters are updated based on the linear regression form of the model. In this way, the parameters of the model are gradually adjusted to obtain the desired tracking.

In numerous robot applications, especially in human-robot interactions, effective force control is crucial for safety and alignment with Azimov's three laws of robotics. This book elaborates on control frameworks tailored for managing interaction forces between robots and their surroundings. It explores specific control structures like compliance, direct force, hybrid, and impedance control, offering readers a comprehensive understanding of their application in various robotic scenarios. This in-depth exploration not only enhances theoretical knowledge but also equips readers to apply these control strategies in their own robotic projects.

Within this book, readers will find a wealth of illustrative examples and case studies, thoughtfully curated and strategically placed alongside specific applications. This deliberate arrangement ensures a continuous narrative throughout the chapters, allowing readers to seamlessly follow the theoretical underpinnings. The intention is to empower readers to not only grasp the conceptual framework but also to adeptly apply the discussed concepts and methodologies to their unique robotic projects. At the conclusion of each chapter, readers are encouraged to apply the theoretical concepts to a number of other robots. This practical approach enables readers to gain proficiency in the design and analysis of serial robots by actively solving these problems. To augment practical implementation, the book goes a step further by generously sharing the accompanying codes. These codes, developed in both MATLAB and Python, are made readily accessible to the audience, enriching the learning journey and encouraging hands-on exploration. This open accessibility, available on the book's GitHob repository, serves as a valuable resource for readers looking to focus deeper into the practical aspects of robotics through real-world examples and applicable code snippets.

Exercises

1.1 Create a timeline outlining significant developments in robotics research over the last five years, highlighting the pivotal role of AI technology in shaping the future of robotics research.

1.2 What are Isaac Asimov's three laws of robotics, and how have they been incorporated into robotics literature? To what extent have these laws influenced the development of new robots worldwide?

1.3 Share your perspective on the distinctions between serial and parallel robots, outlining the merits and drawbacks of each. Additionally, discuss the challenges associated with designing hybrid robots.

1.4 Explore recent applications of robotics beyond what has been discussed, and provide your findings. Include details about the nature of the application and the types of robotic systems employed in these instances.

1.5 Investigate the following concepts in robotics and provide brief statements conveying your understanding of each item:

- Degrees of freedom.
- Kinematics: Forward verses inverse kinematics.
- Workspace: Reachable and dexterous workspaces.
- Path and trajectory.
- Differential kinematics: Jacobian matrix and its properties.
- Dynamics: Newton-Euler versus Euler-Lagrange Method.
- Dynamics: Closed-form formulation.
- Control: Motion versus force control.
- Control: Robust versus adaptive control.

2

Motion Description

This chapter focuses deeply on the complexities of describing motion, with a particular focus on representing rigid body motion in three-dimensional space. It begins with an introduction then progresses to discuss the representation of rigid body motion in three dimensions, covering how to determine the position of a point on a rigid body and providing fundamental insights into describing motion. Furthermore, it elaborates on the orientation of rigid bodies, introducing concepts and methods such as rotation matrices and Euler angles for this purpose. The discussion on rotation matrices highlights their role in expressing orientation and outlines their associated properties. In contrast, Euler angles offer an alternative approach, using fewer parameters for orientation mapping.

There is also a notable emphasis on defining the pose of a rigid body in three-dimensional space, emphasizing the integration of both position and orientation. The chapter provides a thorough examination of homogeneous transformations, a mathematical framework that unifies translation and rotation. This examination includes introducing homogeneous coordinates for different types of vectors and defining the homogeneous transformation matrix, which is essential for expressing transformations. Additionally, it covers arithmetic operations involved in homogeneous transformations, such as consecutive transformations and the inverse of such transformation, in detail. Finally, the chapter presents various problems, allowing readers to practice diverse representations of rigid body motion and become adept at transitioning between them.

2.1 Introduction

In the examination of robotic motion, it becomes imperative to meticulously articulate the position and orientation of diverse components within the robotic structure. This necessity arises from the frequent execution of motion maneuvers by robots in three-dimensional space, making it essential to describe their motion within the spatial dimension. This representation holds significant relevance across a spectrum of robotic applications, including mobile and aerial robotics, humanoid robots, and various engineering domains like aerospace and underwater applications, where entities navigate freely in three-dimensional space. As a consequence, an array of methods has been devised and refined over time, originating from the early contemplation of flight by humans [37]. This section elaborates an examination of some of these methods, which find application in the analysis of robot motion.

For a meticulous definition of the motion of a specific point on an object in space, the introduction of a reference frame for motion measurement becomes imperative. Conventionally, a coordinate system affixed to the ground serves as the foundation for establishing this reference frame. However, to address the analysis of speed and acceleration, it is important that this reference frame remain fixed. Acknowledging that no point within space remains devoid of motion, in classical mechanics the effect of Earth's motion in scrutinizing

DOI: 10.1201/9781003491415-2

the movements of objects positioned on it is neglected, and thus, the ground-installed coordinate system is selected as the reference frame for motion measurement. While various coordinate systems have been proposed for motion measurement, the Cartesian coordinate system stands out as more prevalent in application compared to alternatives such as cylindrical and spherical coordinate systems.

2.2 Motion Representation in 3D Space

If one can ascertain the position of every point on a rigid body with respect to a reference coordinate system, the complete spatial location, or "pose," of the body is determined. Here, the term "pose" is employed as a concise representation of both the position and orientation of the rigid body. When the motion of the body is confined to a plane, a set of three variables suffices to articulate the body's pose: two variables to detail the location of any arbitrary point on the rigid body in relation to the reference coordinate system, and an additional variable to characterize the orientation of the body with respect to the reference frame.

In order to quantify these variables and achieve a precise mathematical interpretation of the body's pose, it is adequate to affix a movable Cartesian coordinate system onto the body. Considering the rigidity of the body, all points on it appear stationary to an observer on this coordinate system. Consequently, it is sufficient to express the position of the origin of this movable coordinate system relative to the reference coordinate system through two translational variables, and furthermore, the orientation of this coordinate system with respect to the reference frame using one rotational variable. This method necessitates three variables to depict the pose of a rigid body confined to a plane.

Now, if the objective is to depict the pose of a rigid body in three-dimensional space, it suffices to affix a mobile coordinate system to the moving body in a similar manner. The location of the origin of this mobile coordinate system with respect to the reference frame can be expressed through three translational variables, while the orientation of this coordinate system with respect to the reference frame can be represented using three supplementary rotational variables. Consequently, to represent the dynamic motion of a rigid body in space, a total of six variables will be requisite.

With this introduction, consider the spatial motion of a rigid body as depicted in Figure 2.1. Envision the Cartesian coordinate system $\{\mathcal{A}\}$ at the origin O_A, illustrated with three orthogonal axes (x, y, z), as the reference coordinate system for representing the motion of the moving body. Furthermore, to describe the rotational motion of the rigid body in three-dimensional space, consider another Cartesian coordinate system $\{\mathcal{B}\}$ installed on the moving body at point O_B. This coordinate system is represented with orthogonal axes (u, v, w), indicating its mobility.

In this manner, the pose of the rigid body can be expressed mathematically by describing the position and orientation of the mobile coordinate system $\{\mathcal{B}\}$ relative to the reference frame $\{\mathcal{A}\}$. This description involves six variables, three for the translational motion of the mobile coordinate system relative to the reference system and three for the rotational motion of these two coordinate systems relative to each other. The position can be easily displayed with vector interpretation, while the description of orientation is slightly more involved, with various methods available for representation. With a proper description of translational and rotational motion of the mobile coordinate system relative to the reference system, one can easily determine the absolute position of any arbitrary point P on the rigid body relative to the reference coordinate system.

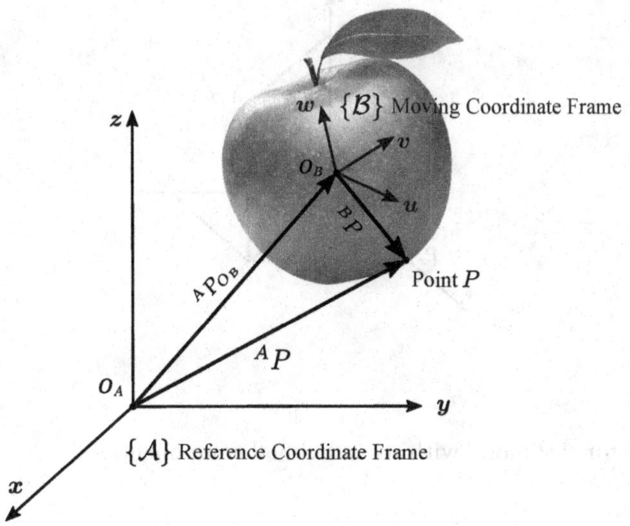

FIGURE 2.1
Motion representation of a rigid body in three dimensional Space

2.3 Position of a Point

The position of a point P relative to the coordinate system $\{\mathcal{A}\}$ can be represented by a vector containing three components along the three orthogonal axes (x, y, z). If this vector is expressed relative to a fixed coordinate system, it represents the absolute position of point P. However, if it is specified with respect to a moving coordinate system, it describes the relative position of this point. Considering the significance of the chosen coordinate system in representing the location of a point, the notation $^A\boldsymbol{P}$ is used to symbolize the position vector of point P relative to the coordinate system $\{\mathcal{A}\}$. This is read as the vector P with respect to A. With this description, the position of point P relative to the coordinate system A is represented as a vector with three components, as follows.

$$^A\boldsymbol{P} = \begin{bmatrix} P_x \\ P_y \\ P_z \end{bmatrix} = [P_x,\ P_y,\ P_z]^T \tag{2.1}$$

In this representation, as shown in Figure 2.2, the elements x, y, z of the vector $\boldsymbol{P}$ determine the components of the vector in the Cartesian coordinates of $\{\mathcal{A}\}$. Similarly, the vector representing the position of P with respect to another coordinate system, such as $\{\mathcal{B}\}$, can be derived. In this case, as seen in Figure 2.1, this vector is denoted by $^B\boldsymbol{P}$. The primary advantage of using a vector representation for the position of a point in space is the ability to leverage vector algebra properties. These vectors form a vector space where vector addition, scalar multiplication, vector multiplication, linearity, and translational invariance properties, among others, can be utilized.

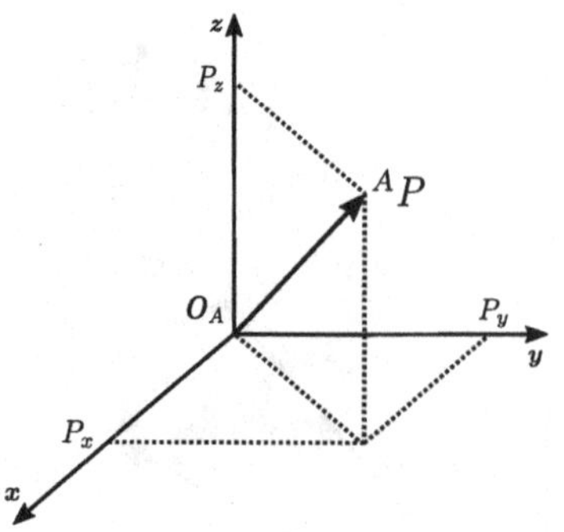

FIGURE 2.2
Displaying the vector $^A\boldsymbol{P}$ along with its individual components.

2.4 Orientation of a Rigid Body

The definition of orientation fundamentally differs from the definition of position. Each point on an object in space has its unique position vector, whereas the orientation of a rigid body is uniform for all its points. Hence, orientation is defined as a representation for the entire body. In the previous section, we saw that, to express the positions of points on a rigid body, a movable Cartesian coordinate system attached to the body can be used. In defining the orientation of the body, we will similarly utilize the amount of rotation of this coordinate system with respect to the reference frame. Unlike position, which can be easily defined using a vector with three components in space, various representations exist for describing the orientation of a body, developed from very past to the present.

The most common method for representing the orientation of a rigid body involves using a *rotation matrix*. In this representation, a 3×3 matrix is employed, with its columns derived from the representation of unit vectors of the coordinate axes attached to the body with respect to the reference coordinate frame. Although the rotation matrix is widely used for representing the orientation of a rigid body in space, describing orientation using this method involves six parameters, which is an additional requirement. This has led to the utilization of alternative representations, despite the excellent properties of the rotation matrix, that require fewer parameters. These include the representation through screw axes, different Euler angles, and quaternions, all of which will be explored in more detail in this chapter and the following one.

2.4.1 Rotation Matrix

One of the most common methods for representing the orientation of a body is using a rotation matrix. For a better understanding of this representation, let's assume that a rigid body only undergoes rotational motion and has no translational movement. Given this assumption, to represent the orientation of the body, consider a Cartesian coordinate system attached to the body, as shown in Figure 2.3, rotating freely in space. In this figure, the body transitions from its initial state, represented as a dotted body with the coordinate system $\{\mathcal{A}\}$, to the final state depicted with a solid line and the moving coordinate system $\{\mathcal{B}\}$. Assuming that the origins of the reference and moving coordinate systems coincide, the

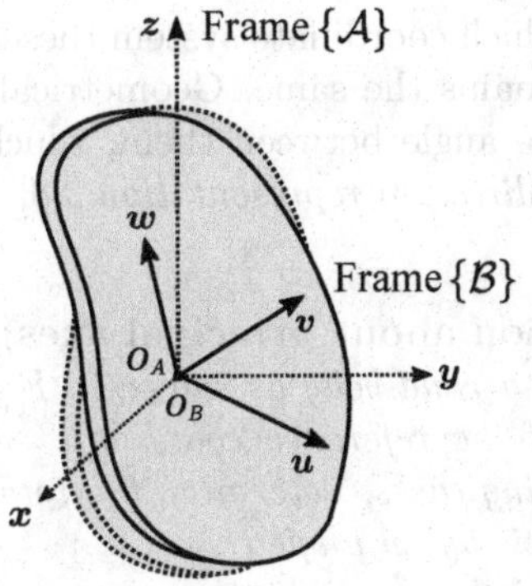

FIGURE 2.3
Representation of the pure rotation of a body along with the rotation-oriented coordinate axes within it.

unit vector directions $(\hat{\boldsymbol{x}}, \hat{\boldsymbol{y}}, \hat{\boldsymbol{z}})$ will not align with the unit vector directions $(\hat{\boldsymbol{u}}, \hat{\boldsymbol{v}}, \hat{\boldsymbol{w}})$. Note that in this book, the symbol $(\hat{\cdot})$ is used to represent unit vectors. With this assumption, we define the 3×3 rotation matrix as the operator representing this rotational motion as follows.

$$^{A}\boldsymbol{R}_B := \left[\ {}^{A}\hat{\boldsymbol{x}}_B \mid {}^{A}\hat{\boldsymbol{y}}_B \mid {}^{A}\hat{\boldsymbol{z}}_B\ \right] = \begin{bmatrix} r_{11} & r_{12} & r_{13} \\ r_{21} & r_{22} & r_{23} \\ r_{31} & r_{32} & r_{33} \end{bmatrix} \tag{2.2}$$

In this definition, ${}^{A}\hat{\boldsymbol{x}}_B$, ${}^{A}\hat{\boldsymbol{y}}_B$, ${}^{A}\hat{\boldsymbol{z}}_B$ are unit vectors of the coordinate system $\{\mathcal{B}\}$ represented in the reference coordinate system $\{\mathcal{A}\}$. Furthermore, since these vectors belong to the Cartesian coordinate system $\{\mathcal{B}\}$, they are perpendicular to each other. This implies that the elements forming the rotation matrix, denoted by r_{ij} in the right side of this equation, all have magnitudes less than or equal to one. Now, if we represent the unit vectors of the coordinate system $\{\mathcal{B}\}$ by their components, we have:

$$\begin{aligned} {}^{A}\hat{\boldsymbol{x}}_B &= {}^{A}\hat{\boldsymbol{u}} = u_x\hat{i} + u_y\hat{j} + u_z\hat{k} \\ {}^{A}\hat{\boldsymbol{y}}_B &= {}^{A}\hat{\boldsymbol{v}} = v_x\hat{i} + v_y\hat{j} + v_z\hat{k} \\ {}^{A}\hat{\boldsymbol{z}}_B &= {}^{A}\hat{\boldsymbol{w}} = w_x\hat{i} + w_y\hat{j} + w_z\hat{k} \end{aligned} \tag{2.3}$$

Thus, the rotation matrix can be expressed as follows:

$$^{A}\boldsymbol{R}_B = \left[\ {}^{A}\hat{\boldsymbol{u}} \mid {}^{A}\hat{\boldsymbol{v}} \mid {}^{A}\hat{\boldsymbol{w}}\ \right] = \begin{bmatrix} u_x & v_x & w_x \\ u_y & v_y & w_y \\ u_z & v_z & w_z \end{bmatrix} \tag{2.4}$$

The components of the unit vectors constituting the rotation matrix can be obtained by taking the dot product of the unit vectors of the coordinate system $\{\mathcal{B}\}$ with the unit vectors of the coordinate system $\{\mathcal{A}\}$ as follows:

$$^{A}\boldsymbol{R}_B = \left[\ {}^{A}\hat{\boldsymbol{x}}_B \mid {}^{A}\hat{\boldsymbol{y}}_B \mid {}^{A}\hat{\boldsymbol{z}}_B\ \right] = \begin{bmatrix} \hat{x}_B \cdot \hat{x}_A & \hat{y}_B \cdot \hat{x}_A & \hat{z}_B \cdot \hat{x}_A \\ \hat{x}_B \cdot \hat{y}_A & \hat{y}_B \cdot \hat{y}_A & \hat{z}_B \cdot \hat{y}_A \\ \hat{x}_B \cdot \hat{z}_A & \hat{y}_B \cdot \hat{z}_A & \hat{z}_B \cdot \hat{z}_A \end{bmatrix} \tag{2.5}$$

Note that in representing this vector product, superscripts of ${}^{A}(\cdot)$ have been omitted intentionally. This is because the result of the dot product of two numerical vectors results into a scalar, and regardless of in which coordinate system these two vectors are interpreted, the result of their dot product remains the same. Geometrically, the dot product of two unit vectors yields the cosine of the angle between them, which is why the rotation matrix is also referred to as the *Cosine direction representation* [38].

Example 2.1 (Pure rotation about principal axes) *Consider a moving coordinate system $\{\mathcal{B}\}$ attached to a rigid body as shown in Figure 2.4, rotating only around one of the principal axes of the reference coordinate system $\{\mathcal{A}\}$ with the following angles. In each of the following cases, determine the rotation matrix.*
A) Rotation around the x-axis by an angle α.
B) Rotation around the y-axis by an angle β.
C) Rotation around the z-axis by an angle γ.

Solution: The rotation matrix in these three cases can be obtained by specifying the components of the unit vectors of the coordinate axes $\{\mathcal{B}\}$ with respect to the reference coordinate system $\{\mathcal{A}\}$, as:

$$\textbf{A)} \quad {}^{A}\boldsymbol{R}_{B} = \boldsymbol{R}_{\boldsymbol{x}}(\alpha) = \begin{bmatrix} 1 & 0 & 0 \\ 0 & \cos\alpha & -\sin\alpha \\ 0 & \sin\alpha & \cos\alpha \end{bmatrix}$$

$$\textbf{B)} \quad {}^{A}\boldsymbol{R}_{B} = \boldsymbol{R}_{\boldsymbol{y}}(\beta) = \begin{bmatrix} \cos\beta & 0 & \sin\beta \\ 0 & 1 & 0 \\ -\sin\beta & 0 & \cos\beta \end{bmatrix}$$

$$\textbf{C)} \quad {}^{A}\boldsymbol{R}_{B} = \boldsymbol{R}_{\boldsymbol{z}}(\gamma) = \begin{bmatrix} \cos\gamma & -\sin\gamma & 0 \\ \sin\gamma & \cos\gamma & 0 \\ 0 & 0 & 1 \end{bmatrix}$$

As seen in these expressions, the columns of the rotation matrices are formed from the representation of unit vectors along the coordinate axes of $\{\mathcal{B}\}$ in the reference coordinate system $\{\mathcal{A}\}$. These basis rotation matrices will be used in subsequent sections to calculate rotation matrices in more complex rotations.

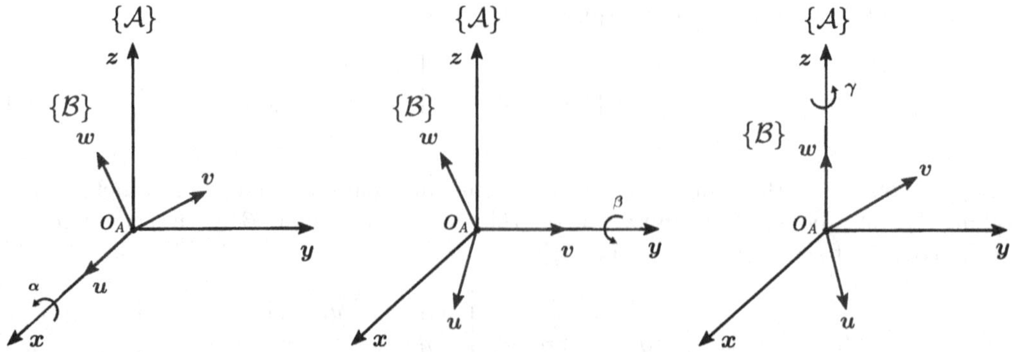

(a) α rotation around the $\boldsymbol{x}$-axis (b) β rotation around the $\boldsymbol{y}$-axis (c) γ rotation around the $\boldsymbol{z}$-axis

FIGURE 2.4
Pure rotation about principal axes.

Rotation Matrix Properties

Property I) Orthonormality

The rotation matrix is an orthonormal matrix; in other words, its columns are formed by vectors with unit magnitudes:

$$\boldsymbol{u}\cdot\boldsymbol{u}=\boldsymbol{v}\cdot\boldsymbol{v}=\boldsymbol{w}\cdot\boldsymbol{w}=1 \tag{2.6}$$

which are mutually perpendicular:

$$\boldsymbol{u}\cdot\boldsymbol{v}=\boldsymbol{v}\cdot\boldsymbol{w}=\boldsymbol{w}\cdot\boldsymbol{u}=0 \tag{2.7}$$

Furthermore, considering the orthogonality property of these vectors, the following relationships hold:

$$\boldsymbol{u}\times\boldsymbol{v}=\boldsymbol{w},\quad \boldsymbol{v}\times\boldsymbol{w}=\boldsymbol{u},\quad \boldsymbol{w}\times\boldsymbol{u}=\boldsymbol{v} \tag{2.8}$$

Property II) Transpose

Considering the definition of the rotation matrix and revisiting Equation 2.5, where the components are derived from the inner product of orthonormal axes within each coordinate system, one can infer that the rows of the rotation matrix can similarly be derived from the components of orthonormal vectors of the coordinate axes $\{\mathcal{A}\}$ with respect to the reference coordinate system $\{\mathcal{B}\}$.

$$^{A}\boldsymbol{R}_{B}=\left[\begin{array}{c|c|c} ^{A}\hat{\boldsymbol{x}}_{B} & ^{A}\hat{\boldsymbol{y}}_{B} & ^{A}\hat{\boldsymbol{z}}_{B}\end{array}\right]=\begin{bmatrix} ^{B}\hat{\boldsymbol{x}}_{A}^{T}\\ ^{B}\hat{\boldsymbol{y}}_{A}^{T}\\ ^{B}\hat{\boldsymbol{z}}_{A}^{T}\end{bmatrix} \tag{2.9}$$

This means that the rotation matrix of the reference coordinate system $\{\mathcal{A}\}$ relative to the coordinate system $\{\mathcal{B}\}$, denoted by $^{B}\boldsymbol{R}_{A}$, is the transpose of the rotation matrix $^{A}\boldsymbol{R}_{B}$.

$$^{B}\boldsymbol{R}_{A}={^{A}\boldsymbol{R}_{B}}^{T} \tag{2.10}$$

Property III) Inverse

From property II, it can be inferred that the inverse of a rotation matrix is equal to its transpose. This property is proven using the rules of linear algebra for orthonormal matrices (for more details on linear algebra, refer to Appendix A). However, considering the physical concept of the rotation of one coordinate system relative to another embedded in the rotation matrix, this property can also be logically deduced from a physical perspective. To mathematically prove this claim, multiply the rotation matrix by its transpose:

$${^{A}\boldsymbol{R}_{B}}^{T}\,{^{A}\boldsymbol{R}_{B}}=\begin{bmatrix} {^{A}\hat{\boldsymbol{x}}_{B}}^{T}\\ {^{A}\hat{\boldsymbol{y}}_{B}}^{T}\\ {^{A}\hat{\boldsymbol{z}}_{B}}^{T}\end{bmatrix}\left[\begin{array}{c|c|c} ^{A}\hat{\boldsymbol{x}}_{B} & ^{A}\hat{\boldsymbol{y}}_{B} & ^{A}\hat{\boldsymbol{z}}_{B}\end{array}\right]=\boldsymbol{I}_{3\times 3} \tag{2.11}$$

As you can see, considering property I, this product results in the identity matrix. Therefore,

$$^{B}\boldsymbol{R}_{A}={^{A}\boldsymbol{R}_{B}}^{-1}={^{A}\boldsymbol{R}_{B}}^{T} \tag{2.12}$$

Property IV) Pure Rotation Mapping

In robotics, the concept of rotation can be utilized to interpret the position vectors of a point on a body in different coordinate systems. As mentioned in the previous properties, the rotation matrix inherently encapsulates the concept of rotation about a specific coordinate system. This concept can be employed as a matrix mapping to represent a vector from one coordinate system to another that has pure rotations relative to each other. To express this mapping mathematically, consider the position vector of a point on the body with respect to the coordinate system attached to it, denoted by $\{\mathcal{B}\}$ as ${}^B\boldsymbol{P}$. The goal is to find the position vector with respect to the reference coordinate system $\{\mathcal{A}\}$, represented by ${}^A\boldsymbol{P}$. To determine the components of this vector, one can use its dot product with the unit vectors of the coordinate system in the reference frame, as follows.

$$\begin{aligned} {}^AP_x &= {}^A\hat{\boldsymbol{x}}_A \cdot {}^A\boldsymbol{P} \\ {}^AP_y &= {}^A\hat{\boldsymbol{y}}_A \cdot {}^A\boldsymbol{P} \\ {}^AP_z &= {}^A\hat{\boldsymbol{z}}_A \cdot {}^A\boldsymbol{P} \end{aligned} \tag{2.13}$$

Given that these components are scalar values, the dot product can be obtained in any other coordinate frame. Thus, by representing these vectors in the $\{\mathcal{B}\}$ frame and calculating their dot product, one can write:

$$\begin{aligned} {}^AP_x &= {}^B\hat{\boldsymbol{x}}_A \cdot {}^B\boldsymbol{P} = {}^B\hat{\boldsymbol{x}}_A^T\ {}^B\boldsymbol{P} \\ {}^AP_y &= {}^B\hat{\boldsymbol{y}}_A \cdot {}^B\boldsymbol{P} = {}^B\hat{\boldsymbol{y}}_A^T\ {}^B\boldsymbol{P} \\ {}^AP_z &= {}^B\hat{\boldsymbol{z}}_A \cdot {}^B\boldsymbol{P} = {}^B\hat{\boldsymbol{z}}_A^T\ {}^B\boldsymbol{P} \end{aligned} \tag{2.14}$$

Rewrite Equation 2.14 in matrix form.

$${}^A\boldsymbol{P} = \begin{bmatrix} {}^B\hat{\boldsymbol{x}}_A^T \\ {}^B\hat{\boldsymbol{y}}_A^T \\ {}^B\hat{\boldsymbol{z}}_A^T \end{bmatrix} {}^B\boldsymbol{P} \tag{2.15}$$

In view of Equation 2.9, this relationship can be succinctly expressed in matrix form using the rotation matrix:

$${}^A\boldsymbol{P} = {}^A\boldsymbol{R}_B{}^B\boldsymbol{P} \tag{2.16}$$

Thus, mathematically, the rotation matrix can be interpreted as a mapping that represents the transformation of a vector from one coordinate system to another. Moreover, considering the appropriate notation used for vector representations relative to different coordinate frames, and also the notation used to represent the rotation matrix, this mapping can be easily interpreted by canceling the left superscript B from the position vector ${}^B\boldsymbol{P}$ with the right subscript of the rotation matrix ${}^A\boldsymbol{R}_B$, resulting into the position vector of point P in the desired coordinate system $\{\mathcal{A}\}$ denoted as ${}^A\boldsymbol{P}$.

Example 2.2 (Pure rotational mapping) *Assume that the rotation of the coordinate system attached to a rigid body $\{\mathcal{B}\}$ with respect to the reference coordinate system $\{\mathcal{A}\}$ is given by the following rotation matrix.*

$${}^A\boldsymbol{R}_B = \begin{bmatrix} 0.933 & -0.067 & -0.354 \\ -0.067 & 0.933 & -0.354 \\ 0.354 & 0.354 & 0.866 \end{bmatrix}$$

Furthermore, assume that the position vector of point P on this body with respect to the frame $\{\mathcal{B}\}$ is given as ${}^B\boldsymbol{P} = \begin{bmatrix} 2 & -1 & 0 \end{bmatrix}^T$. In this case, determine the representation of this vector with respect to the reference coordinate system $\{\mathcal{A}\}$, denoted by ${}^A\boldsymbol{P}$.

Solution: By using the mapping relation 2.16, the vector ${}^A\boldsymbol{P}$ in the $\{\mathcal{A}\}$ coordinate system can be obtained as follows.

$$
\begin{aligned}
{}^A\boldsymbol{P} &= {}^A\boldsymbol{R}_B\,{}^B\boldsymbol{P} \\
&= \begin{bmatrix} 0.933 & -0.067 & -0.354 \\ -0.067 & 0.933 & -0.354 \\ 0.354 & 0.354 & 0.866 \end{bmatrix} \begin{bmatrix} 2 \\ -1 \\ 0 \end{bmatrix} = \begin{bmatrix} 1.93 \\ -1.07 \\ 0.35 \end{bmatrix}
\end{aligned}
$$

If we pay attention to the obtained result, it can be observed that the magnitude of the vector $\boldsymbol{P}$ in both coordinate systems, which is determined by the Euclidean norm $\|\boldsymbol{P}\|_2$, are equal.

$$
\|\boldsymbol{P}\|_2 = \sqrt{(2^2 + (-1)^2)} = \sqrt{(1.93^2 + (-1.07)^2 + 0.35^2)} = \sqrt{5}
$$

This occurs because, in a pure rotation mapping the magnitude or length of vectors does not change, and only their components change with respect to the coordinate system under consideration.

Property V) Determinant

Determinant of a rotation matrix is equal to one.

$$
\det\left({}^A\boldsymbol{R}_B\right) = 1 \tag{2.17}
$$

This property can also be proven both mathematically and interpreted physically. To mathematically prove this property, by expanding the determinant of the rotation matrix in terms of its components, we can write:

$$
\begin{aligned}
\det\left({}^A\boldsymbol{R}_B\right) &= w_x(u_y v_z - v_y u_z) - w_y(u_x v_z - v_x u_z) + w_x(u_x v_y - v_x u_y) \\
&= \boldsymbol{w} \cdot (\boldsymbol{u} \times \boldsymbol{v}) = \boldsymbol{w} \cdot \boldsymbol{w} = 1
\end{aligned}
$$

The physical interpretation of this relationship is consistent with the previous property and, in a sense, expresses the constancy of vector magnitudes in this transformation.

Property VI) Eigenvalues

The eigenvalues of the rotation matrix are equal to 1, $e^{i\theta}$, and $e^{-i\theta}$, where θ is calculated from the following relationship.

$$
\theta = \cos^{-1} \frac{\operatorname{tr}\left({}^A\boldsymbol{R}_B\right) - 1}{2} \tag{2.18}
$$

In this relation, $\cos^{-1}$ represents the inverse cosine function, and $\operatorname{tr}\left({}^A\boldsymbol{R}_B\right)$ denotes the trace of the rotation matrix ${}^A\boldsymbol{R}_B$.

To prove this property, calculating the eigenvalues of the rotation matrix (λ) using the components provided in Equation 2.4, one can write:

$$
\det\left({}^A\boldsymbol{R}_B - \lambda\boldsymbol{I}\right) = \begin{vmatrix} u_x - \lambda & v_x & w_x \\ u_y & v_y - \lambda & w_y \\ u_z & v_z & w_z - \lambda \end{vmatrix} = 0 \tag{2.19}
$$

By taking the determinant with respect to the first column and using the definition of the trace of a matrix, along with Equations 2.8 and 2.17, simplify this determinant as follows.

$$\begin{aligned}
\det\left({}^A\boldsymbol{R}_B - \lambda \boldsymbol{I}\right) &= (u_x - \lambda)\left[(v_y - \lambda)(w_z - \lambda) - v_z w_y\right] \\
&\quad - u_y\left[v_x(w_z - \lambda) - v_z w_x\right] + u_z\left[v_x w_y - w_x(v_y - \lambda)\right] \\
&= -\lambda^3 + (u_x + v_y + w_z)\lambda^2 \\
&\quad - \left[(v_y w_z - v_z w_y) + (u_x w_z - u_z w_x) + (u_x v_y - u_y v_x)\right]\lambda \\
&\quad + \det\left({}^A\boldsymbol{R}_B\right) \\
&= -\lambda^3 + \operatorname{tr}\left({}^A\boldsymbol{R}_B\right)\lambda^2 \\
&\quad - \left[(\boldsymbol{v}\times\boldsymbol{w})_x + (\boldsymbol{w}\times\boldsymbol{u})_y + (\boldsymbol{u}\times\boldsymbol{v})_z\right]\lambda + 1 \\
&= -\lambda^3 + \operatorname{tr}\left({}^A\boldsymbol{R}_B\right)\lambda^2 - (u_x + v_y + w_z)\lambda + 1
\end{aligned}$$

As a result, the characteristic equation of the rotation matrix simplifies to:

$$\lambda^3 - \operatorname{tr}\left({}^A\boldsymbol{R}_B\right)\lambda^2 + \operatorname{tr}\left({}^A\boldsymbol{R}_B\right)\lambda - 1 = 0 \tag{2.20}$$

or,

$$(\lambda - 1)\left(\lambda^2 + \left(1 - \operatorname{tr}\left({}^A\boldsymbol{R}_B\right)\right)\lambda + 1\right) = 0 \tag{2.21}$$

With the help of this equation, the eigenvalues of the rotation matrix can be expressed as:

$$\lambda = 1, e^{i\theta},\ e^{-i\theta} \tag{2.22}$$

where $i = \sqrt{-1}$ and $e^{\pm i\theta} \triangleq \cos\theta \pm i\sin\theta$. Furthermore, by using equation 2.21, the angle θ is calculated from the following relation, and the proof is completed.

$$\cos\theta = \frac{\operatorname{tr}\left({}^A\boldsymbol{R}_B\right) - 1}{2}$$

Example 2.3 (Rotation matrix eigenvalues) *Find the eigenvalues and eigenvectors of the rotation matrix provided in Example 2.2 and verify Property IV.*

$${}^A\boldsymbol{R}_B = \begin{bmatrix} 0.933 & -0.067 & -0.354 \\ -0.067 & 0.933 & -0.354 \\ 0.354 & 0.354 & 0.866 \end{bmatrix}$$

Solution: Eigenvalues and eigenvectors of the rotation matrix are determined as follows:

$$\lambda_1 = 1\ ;\ v_1 = \begin{bmatrix} -0.707 \\ +0.707 \\ 0 \end{bmatrix}$$

$$\lambda_{2,3} = 0.866 \pm 0.5i = e^{\pm\pi/6}\ ;\ v_{2,3} = \begin{bmatrix} \pm 0.5i \\ \pm 0.5i \\ 0.707 \end{bmatrix}$$

It can easily be shown that the angle θ, using Equation 2.18, is exactly equal to $\pi/6$.

The angle θ is one of the intrinsic characteristics of the rotation matrix, indicating the rotation angle in this representation. Furthermore, the eigenvector corresponding to the real eigenvalue 1 in the rotation matrix defines a direction in space around which the rotation takes place. This representation of rotation, encompassing its invariant physical characteristics, will be further discussed in the forthcoming chapter, when describing the orientation and pose of a rigid body by the use of "screw theory."

2.4.2 Euler Angles

As observed earlier, expressing the orientation of a rigid body in space through a rotation matrix involves nine independent parameters. However, considering the intuitive nature of rotational motion in three-dimensional space, it seems more appropriate to utilize only three independent parameters. Numerous approaches outlined in existing literature aim to characterize orientation in space with just three independent parameters. These methodologies typically hinge on the orientation measurement type employed by rotation instrumentation systems. Among these representations, the most prevalent ones involve the use of Euler angles in the form of two fixed-axes and moving axes conventions, a topic we will elaborate on into more depth.

Euler angles consist of an array of angles that elucidate the orientation of a rigid body within three-dimensional space. They articulate a sequence of three fundamental rotations, each executed around one of the axes with respect to a coordinate system. The specific Euler angle convention is determined by the order of these rotations and whether the reference frame for the rotation is fixed or moving, therefore, two distinct sets of Euler angles are delineated in the literature. In the first set, rotations transpire around the axes of a stationary coordinate system. Conversely, in the second set of Euler angles, the rotation is conceptualized around the axes of a moving frame. In both representations, any desired orientation in three-dimensional space can be achieved by amalgamating the pure rotations around the principal axes introduced in Example 2.1.

Euler angles provide an intuitive approach to visualize and understand rotations, yet they come with challenges such as gimbal lock, which can limit their applicability in certain scenarios. Gimbal lock occurs when one rotational axis aligns with another, resulting in the loss of one degree of freedom and potential complications in representing specific orientations. Despite these challenges, Euler angles continue to be widely used in various fields, including aerospace, robotics, and computer graphics. In the following sections, we will introduce a fixed-axis Euler angles representation, specifically with respect to the $\boldsymbol{x}-\boldsymbol{y}-\boldsymbol{z}$ axes, and explore three additional representations of Euler angles with respect to different combinations of axes in the moving frame. We will also address the characteristics of gimbal lock for each representation.

- Fixed-Axes Euler Angles

In this particular orientation representation of a rigid body in space, rotations are performed around the axes of the fixed coordinate system. As depicted in Figure 2.5, out of twelve available combinations [2, Appendix B], rotations around the lateral, longitudinal, and vertical axes of the fixed coordinate system are employed to depict the orientation of the rigid body. As shown in Figure 2.5, the rotation around the lateral axis is denoted as *pitch*, the rotation around the longitudinal axis is termed *roll*, and the rotation around the vertical axis is labeled *yaw*. Hence, this spatial orientation representation is commonly referred to as the "pitch–roll–yaw Euler angles."

To precisely define the fixed-axis rotation angles in this case and establish their relationship with the rotation matrix, consider that the orientation of the rigid body in space involves three consecutive rotations around the fixed axes $\boldsymbol{x}, \boldsymbol{y}, \boldsymbol{z}$ of the fixed coordinate system $\{\mathcal{A}\}$, as depicted in Figure 2.6. The sequence of rotations is outlined in detail as

FIGURE 2.5
Fixed-axis pitch–roll–yaw representation in the motion of a vehicle

follows.

> First, consider the moving coordinate system aligned with the fixed coordinate system $\{\mathcal{A}\}$. Then, sequentially rotate the moving coordinate system about the $\boldsymbol{x}$-axis by the angle α to reach the intermediate coordinate system $\{\mathcal{B}'\}$. Next, rotate this intermediate coordinate system about the $\boldsymbol{y}$ axis of the reference coordinate system $\{\mathcal{A}\}$ by the angle β to reach the second intermediate coordinate system $\{\mathcal{B}''\}$. Finally, rotate this second intermediate coordinate system about the $\boldsymbol{z}$ axis of the reference coordinate system $\{\mathcal{A}\}$ by the angle γ to reach the final coordinate system $\{\mathcal{B}\}$.

In this representation, an attempt has been made to decompose the final orientation of the rigid body, interpreted with the coordinate system $\{\mathcal{B}\}$ attached to it, relative to the reference coordinate system $\{\mathcal{A}\}$ through three sequential rotations as elucidated earlier. From

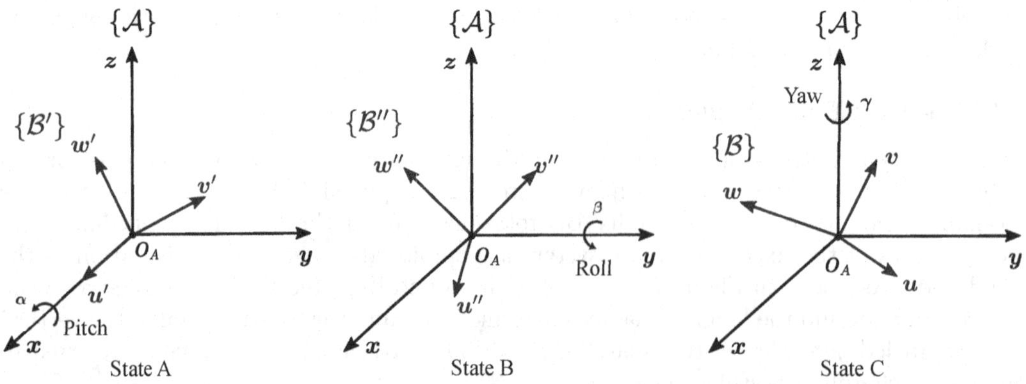

FIGURE 2.6
Sequential rotation about the fixed coordinate axes by the pitch angle α around the $\boldsymbol{x}$ axis, then by the roll angle β around the $\boldsymbol{y}$ axis, and finally by the yaw angle γ around the $\boldsymbol{z}$ axis.

a mathematical standpoint, understanding the compound rotation matrix as the result of combining three successive rotational mappings renders this concept easily comprehensible. In this example, the rotational mapping is applied first about the $\boldsymbol{x}$-axis, then about $\boldsymbol{y}$ axis, and finally about $\boldsymbol{z}$-axis. In this manner, calculating the rotation matrix derived from the composition of three transformations becomes straightforward through the successive pre-multiplication of three fundamental rotation matrices about the $\boldsymbol{x}, \boldsymbol{y}, \boldsymbol{z}$ axes.

$$\begin{aligned} \boldsymbol{R}_{pry}(\alpha, \beta, \gamma) &= \boldsymbol{R}_z(\gamma)\boldsymbol{R}_y(\beta)\boldsymbol{R}_x(\alpha) \\ &= \begin{bmatrix} c\gamma & -s\gamma & 0 \\ s\gamma & c\gamma & 0 \\ 0 & 0 & 1 \end{bmatrix} \begin{bmatrix} c\beta & 0 & s\beta \\ 0 & 1 & 0 \\ -s\beta & 0 & c\beta \end{bmatrix} \begin{bmatrix} 1 & 0 & 0 \\ 0 & c\alpha & -s\alpha \\ 0 & s\alpha & c\alpha \end{bmatrix} \end{aligned} \tag{2.23}$$

In this expression, the notation $c\alpha$ is used to represent $\cos\alpha$, and $s\alpha$ is used for $\sin\alpha$ and so forth. Furthermore, note that the result of rotations about fixed axes leads to the pre-multiplication of rotation matrices on each other. Considering that matrix multiplication is not commutative, the order of rotation angles becomes important. Mathematically, pre-multiplication refers to the multiplication of rotation matrices from the left. Thus, the first rotation, corresponding to the primary rotation matrix $\mathbf{R}_x(\alpha)$, is placed on the right, and the last rotation, corresponding to the primary rotation matrix $\mathbf{R}_z(\gamma)$, is placed on the left. By performing the matrix multiplication in this order and according to equation 2.23, the rotation matrix resulting from the pitch–roll–yaw Euler angles, $\mathbf{R}_{pry}$, is obtained as follows.

$$\boldsymbol{R}_{pry}(\alpha, \beta, \gamma) = \begin{bmatrix} c\beta c\gamma & s\alpha s\beta c\gamma - c\alpha s\gamma & c\alpha s\beta c\gamma + s\alpha s\gamma \\ c\beta s\gamma & s\alpha s\beta s\gamma + c\alpha c\gamma & c\alpha s\beta s\gamma - s\alpha c\gamma \\ -s\beta & s\alpha c\beta & c\alpha c\beta \end{bmatrix} \tag{2.24}$$

The inverse solution for this representation becomes crucial when analyzing the rotational motion of a rigid body. Given the rotation matrix representing the orientation of a rigid body in space, the pitch–roll–yaw Euler angles can be computed by the inverse solution. In this inverse solution, we encounter not only three trigonometric equations with three unknowns but also six implicit algebraic constraints derived from the intrinsic properties of the rotation matrix. As a result, the inverse problem may yield multiple solutions.

Here, we present one of the simplest solutions in which, instead of using all nine elements of the rotation matrix, only elements with less complexity are utilized. Consider the first column of the rotation matrix 2.24. By summing the squares of elements r_{11} and r_{21}, $c^2\beta$ can be obtained. In this way, two acceptable solutions for β can be found from $\pm\sqrt{r_{11}^2 + r_{21}^2}$ and r_{31}. Subsequently, γ can be obtained from elements r_{11} and r_{21}, and α from elements r_{32} and r_{33} of the rotation matrix, with the condition that $c\beta \neq 0$. Provided that this condition is satisfied, the solution can be expressed as follows.

$$\begin{aligned} \beta &= \text{atan2}(-r_{31}, \pm\sqrt{r_{11}^2 + r_{21}^2}) \\ \gamma &= \text{atan2}(r_{21}/c\beta, r_{11}/c\beta) \\ \alpha &= \text{atan2}(r_{32}/c\beta, r_{33}/c\beta) \end{aligned} \tag{2.25}$$

In these equations, the four-quadrant inverse tangent function, denoted as $\text{atan2}(y, x)$, is employed instead of the standard inverse tangent. In this function, akin to the regular inverse tangent, the relationship $\text{atan}(y/x)$ is utilized. However, the crucial distinction lies in how the sign of both arguments, namely x and y, impacts the sign of the resulting angle. For instance, while $\text{atan}(-1/-1) = \pi/4$, the expression $\text{atan2}(-1, -1)$ yields $-3\pi/4$. Utilizing this function enables the determination of rotation angles in each quadrant of the trigonometric circle, confined to the interval $[-\pi, \pi]$. Notably, this function adeptly handles scenarios where x is zero, avoiding division by zero issues encountered in the atan function, while the result is contingent on the sign of y. Consequently, it offers numerous advantages over the atan function for performing calculations.

Equation 2.25 provides two acceptable solutions for each rotation matrix. Furthermore, note that this solution is valid under the assumption that $c\beta \neq 0$, and it cannot be used in the case of $c\beta = 0$ or $\beta = \pm\pi/2$. In such cases, the rotation matrix simplifies, and for instance, in the case of $\beta = \pi/2$, it can be represented as follows:

$$\boldsymbol{R}_{pry}(\alpha, \gamma) = \begin{bmatrix} 0 & s(\alpha-\gamma) & c(\alpha-\gamma) \\ 0 & c(\alpha-\gamma) & -s(\alpha-\gamma) \\ -1 & 0 & 0 \end{bmatrix} \tag{2.26}$$

Keep in mind that this case constitute the gimbal lock state associated with this Euler angle representation. When $\beta = \pm\pi/2$, s rotated $\boldsymbol{x}$ axis will be aligned with the $\boldsymbol{z}$ axis. Consequently, the inverse formulation degenerates, and only the algebraic sum of the two angles α and γ can be calculated as follows:

$$\begin{cases} \beta = \pi/2 \\ \alpha - \gamma = \text{atan2}(r_{12}, r_{22}) \end{cases} \quad \text{or} \quad \begin{cases} \beta = -\pi/2 \\ \alpha + \gamma = \text{atan2}(-r_{12}, r_{22}) \end{cases} \tag{2.27}$$

Example 2.4 (Pitch-Roll-Yaw Euler Angles) *Consider the rotation of a rigid body represented using the rotation matrix given in Example 2.2. Represent this rotation using the fixed-angle conventions of pitch-roll-yaw Euler angles.*

Solution: For this rotation matrix, the pitch–roll–yaw Euler angles are calculated from equation 2.25 as follows.

$$\alpha = 22.23^\circ \;\; ; \;\; \beta = -20.73^\circ \;\; ; \;\; \gamma = -4.11^\circ$$

Example 2.5 *Represent the rotation of a rigid body, given by the following rotation matrix, using pitch–roll–yaw Euler angles.*

$$\boldsymbol{R} = \begin{bmatrix} 0 & 0.5 & 0.866 \\ 0 & 0.866 & -0.5 \\ -1 & 0 & 0 \end{bmatrix}$$

Solution: For this rotation matrix, computing the pitch–roll–yaw angles using the relation provided in Equation 2.25 is not feasible. This is due to the fact that $r_{11} = r_{21} = 0$, indicating that the angle β can only be either positive or negative 90 degrees. In these specific scenarios, only the algebraic sum of the two angles α and γ can be determined using Equation 2.27.

$$\begin{cases} \beta = \pi/2 \\ \alpha - \gamma = \pi/6 \end{cases} \quad \text{or} \quad \begin{cases} \beta = -\pi/2 \\ \alpha + \gamma = -\pi/6 \end{cases}$$

- The $\{u - v - w\}$ Euler Angles

As mentioned earlier, another common method for representing the rotation of a rigid body involves rotating around the axes of a moving coordinate system, known as the moving-axes

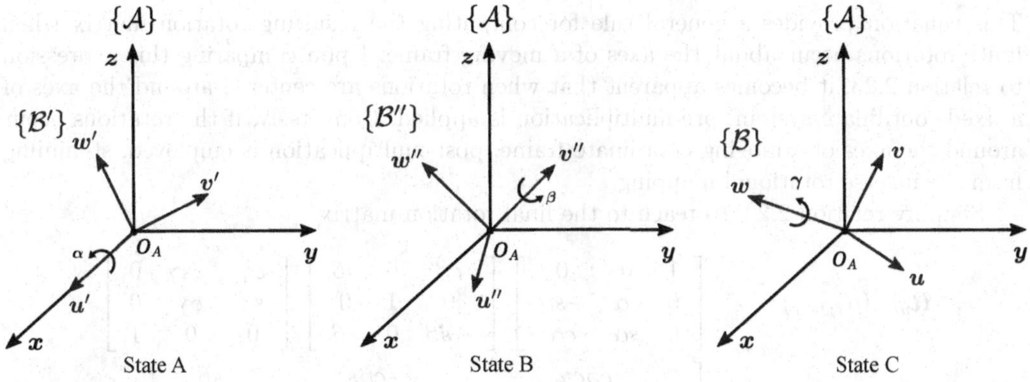

FIGURE 2.7
Sequential rotation around the moving coordinate axes by an angle α around the $\boldsymbol{u}$ axis, then by an angle β around the $\boldsymbol{v}'$ axis, and finally by an angle γ around the $\boldsymbol{w}''$ axis.

Euler angles representation. Consider, for example, one of the twelve possible configurations of this representation [2, Appendix B], such as the $\boldsymbol{u}-\boldsymbol{v}-\boldsymbol{w}$ representation:

> As shown in Figure 2.7, consider initially aligning the moving coordinate system with the fixed frame attached to the body $\{\mathcal{A}\}$. Rotate this moving coordinate system first around the $\boldsymbol{u}$ axis by an angle α to reach the intermediate coordinate system $\{\mathcal{B}'\}$. Then, rotate this intermediate coordinate system around the $\boldsymbol{v}'$ axis of the moving frame $\{\mathcal{B}'\}$ by an angle β to reach the second intermediate coordinate system $\{\mathcal{B}''\}$. Finally, rotate this intermediate coordinate system around the $\boldsymbol{w}''$ axis of the moving coordinate system $\{\mathcal{B}''\}$ by an angle γ to reach the final coordinate system $\{\mathcal{B}\}$.

In this representation, the final orientation of the rigid body is interpreted with the coordinate system attached to it, through three consecutive rotations in the specified order relative to the reference coordinate system $\{\mathcal{A}\}$. Note that in this representation, rotations occur relative to the axes of the moving coordinate systems. Therefore, the pre-multiplication rule to calculate the final rotation matrix is not valid. By using the inverse rotational mapping, it can be shown that in this case, the rotation matrix resulting from the post-multiplication of three consecutive rotation matrices around the axes $\boldsymbol{u}, \boldsymbol{v}, \boldsymbol{w}$ shall be used [1].

This mapping could be interpreted as the inverse rotational map of frame $\{\mathcal{A}\}$ with respect to $\{\mathcal{B}\}$. For this means, consider frame $\{\mathcal{A}\}$ to be fixed and the rotation is performed about its axes. The frame $\{\mathcal{A}\}$ is first rotated about $\boldsymbol{u}$ with the angle of $-\alpha$, then it will rotate about $\boldsymbol{v}$ with the angle of $-\beta$, and finally rotate about $\boldsymbol{z}$ with the angle of $-\gamma$. In such case one may write:

$$ {}^{B}\boldsymbol{R}_{A}(-\alpha,-\beta,-\gamma) = \boldsymbol{R}_{w}(-\gamma)\boldsymbol{R}_{v}(-\beta)\boldsymbol{R}_{u}(-\alpha) \tag{2.28} $$

Equation 2.28 can be written as:

$$ \begin{aligned} {}^{A}\boldsymbol{R}_{B}(\alpha,\beta,\gamma) &= {}^{B}\boldsymbol{R}_{A}^{-1}(-\alpha,-\beta,-\gamma) \\ &= [\boldsymbol{R}_{w}(-\gamma)\boldsymbol{R}_{v}(-\beta)\boldsymbol{R}_{u}(-\alpha)]^{-1} \\ &= \boldsymbol{R}_{u}^{-1}(-\alpha)\boldsymbol{R}_{v}^{-1}(-\beta)\boldsymbol{R}_{w}^{-1}(-\gamma) \end{aligned} $$

and finally as:

$$ {}^{A}\boldsymbol{R}_{B}(\alpha,\beta,\gamma) = \boldsymbol{R}_{u}(\alpha)\boldsymbol{R}_{v}(\beta)\boldsymbol{R}_{w}(\gamma) \tag{2.29} $$

This equation provides a general rule for computing the resulting rotation matrix when finite rotations occur about the axes of a moving frame. Upon comparing this expression to relation 2.23, it becomes apparent that when rotations are centered around the axes of a fixed coordinate system, pre-multiplication is applied. Conversely, if the rotations occur around the axes of a moving coordinate frame, post-multiplication is employed, stemming from the inverse rotational mapping.

Simplify relation 2.29, to reach to the final rotation matrix.

$$\begin{aligned} \boldsymbol{R}_{uvw}(\alpha,\beta,\gamma) &= \begin{bmatrix} 1 & 0 & 0 \\ 0 & c\alpha & -s\alpha \\ 0 & s\alpha & c\alpha \end{bmatrix} \begin{bmatrix} c\beta & 0 & s\beta \\ 0 & 1 & 0 \\ -s\beta & 0 & c\beta \end{bmatrix} \begin{bmatrix} c\gamma & -s\gamma & 0 \\ s\gamma & c\gamma & 0 \\ 0 & 0 & 1 \end{bmatrix} \\ &= \begin{bmatrix} c\beta c\gamma & -c\beta s\gamma & s\beta \\ s\alpha s\beta c\gamma + c\alpha s\gamma & -s\alpha s\beta s\gamma + c\alpha c\gamma & -s\alpha c\beta \\ -c\alpha s\beta c\gamma + s\alpha s\gamma & c\alpha s\beta s\gamma + s\alpha c\gamma & c\alpha c\beta \end{bmatrix} \end{aligned} \tag{2.30}$$

Similar to the previous case, the inverse problem can be solved by following relations, provided that $c\beta \neq 0$.

$$\begin{aligned} \beta &= \text{atan2}(r_{13}, \pm\sqrt{r_{11}^2 + r_{12}^2}) \\ \alpha &= \text{atan2}(-r_{23}/c\beta, r_{33}/c\beta) \\ \gamma &= \text{atan2}(-r_{12}/c\beta, r_{11}/c\beta) \end{aligned} \tag{2.31}$$

Otherwise, if $c\beta = 0$ this case constitute the gimbal lock state associated with this type of Euler angle representation. When $\beta = \pm\pi/2$, the rotated $\boldsymbol{u}$ axis will be aligned with the $\boldsymbol{w}$ axis. Consequently, the inverse formulation degenerates, and only the algebraic sum of the two angles α and γ can be calculated as follows:

$$\begin{cases} \beta = \pi/2 \\ \alpha + \gamma = \text{atan2}(r_{32}, r_{22}) \end{cases} \quad \text{or} \quad \begin{cases} \beta = -\pi/2 \\ \alpha - \gamma = \text{atan2}(r_{32}, r_{22}) \end{cases} \tag{2.32}$$

Example 2.6 (The $\{\boldsymbol{u}-\boldsymbol{v}-\boldsymbol{w}\}$ Euler Angles) *Consider the rotation matrix introduced in Example 2.2, and analyzed in Examples 2.3 and 2.4. Represent this rotation with the $\{\boldsymbol{u}-\boldsymbol{v}-\boldsymbol{w}\}$ Euler angles.*

Solution: Considering this rotation matrix, $\{\boldsymbol{u}-\boldsymbol{v}-\boldsymbol{w}\}$ Euler angles can be derived by using Equation 2.31, as follows.

$$\alpha = 22.23^\circ \;\; ; \;\; \beta = -20.73^\circ \;\; ; \;\; \gamma = 4.11^\circ$$

Example 2.7 *Consider rotation of a rigid body, which is represented by the following rotation matrix.*

$$\boldsymbol{R} = \begin{bmatrix} 0 & 0 & 1 \\ 0.966 & -0.259 & 0 \\ 0.259 & 0.966 & 0 \end{bmatrix}$$

Represent this rotation with the $\{\boldsymbol{u}-\boldsymbol{v}-\boldsymbol{w}\}$ Euler angles.

Solution: For this rotation matrix, the $\{\boldsymbol{u}-\boldsymbol{v}-\boldsymbol{w}\}$ Euler angles can not be calculated by using Equation 2.31, since $r_{11} = r_{12} = 0$ and this will result into gimbal lock of this representation. Consequently, the inverse formulation degenerates, and only the algebraic sum of the two angles α and γ can be calculated from Equation 2.32:

$$\begin{cases} \beta = \pi/2 \\ \alpha + \gamma = 5\pi/12 \end{cases} \quad \text{or} \quad \begin{cases} \beta = -\pi/2 \\ \alpha - \gamma = 5\pi/12 \end{cases}$$

- The $\{\boldsymbol{w}-\boldsymbol{v}-\boldsymbol{w}\}$ Euler Angles

Another representation of Euler angles may be represented by a rotation of the angle α around the axis $\boldsymbol{w}$, then the rotation of the angle β around the axis $\boldsymbol{v}$ and finally the rotation with the angle γ around the axis $\boldsymbol{w}$. In this case, by post-multiplying the individual rotation matrices, the final rotation matrix is obtained as follows.

$$\begin{aligned} \boldsymbol{R}_{wvw}(\alpha, \beta, \gamma) &= \boldsymbol{R}_w(\alpha)\boldsymbol{R}_v(\beta)\boldsymbol{R}_w(\gamma) \\ &= \begin{bmatrix} c\alpha c\beta c\gamma - s\alpha s\gamma & -c\alpha c\beta s\gamma - s\alpha c\gamma & c\alpha s\beta \\ s\alpha c\beta c\gamma + c\alpha s\gamma & -s\alpha c\beta s\gamma + c\alpha c\gamma & s\alpha s\beta \\ -s\beta c\gamma & s\beta s\gamma & c\beta \end{bmatrix} \end{aligned} \tag{2.33}$$

In the same way as the previous methods, in the case that the angle $s\beta \neq 0$, the inverse problem solution may be obtained from the following relations.

$$\begin{aligned} \beta &= \text{atan2}(\pm\sqrt{r_{31}^2 + r_{32}^2}, r_{33}) \\ \alpha &= \text{atan2}(r_{23}/s\beta, r_{13}/s\beta) \\ \gamma &= \text{atan2}(r_{32}/s\beta, -r_{31}/s\beta) \end{aligned} \tag{2.34}$$

Note that, if $s\beta = 0$, this case constitute the gimbal lock state associated with this type of Euler angle representation. When $\beta = 0$ or π, the inverse formulation degenerates, and only the algebraic sum of the two angles α and γ can be calculated as follows:

$$\begin{cases} \beta = 0 \\ \gamma + \alpha = \text{atan2}(-r_{12}, r_{11}) \end{cases} \quad \text{or} \quad \begin{cases} \beta = \pi \\ \gamma - \alpha = \text{atan2}(r_{12}, -r_{11}) \end{cases} \tag{2.35}$$

Example 2.8 (The $\{\boldsymbol{w}-\boldsymbol{v}-\boldsymbol{w}\}$ Euler Angles) *Consider the rotation of a rigid body with the rotation matrix introduced in the Example 2.2. Represent this rotational mapping with Euler angles $\{\boldsymbol{w}-\boldsymbol{v}-\boldsymbol{w}\}$.*

Solution: For this rotation matrix, the $\{\boldsymbol{w}-\boldsymbol{v}-\boldsymbol{w}\}$ Euler angles can be calculated from Equation 2.34 as follows.

$$\alpha = -135^\circ \ ; \ \beta = 30^\circ \ ; \ \gamma = 135^\circ$$

As can be seen in this representation, the Euler angles obtained are much more regular than the previous angles, and they can be visualized with a better understanding of the physical concept of rotation.

- The $\{w - u - w\}$ Euler Angles

Another representation of Euler angles involves rotation by an angle α about the axis $\boldsymbol{w}$, followed by a rotation of angle β about the axis $\boldsymbol{u}$ (replacing the $\boldsymbol{v}$ frame in the previous case), and finally, a rotation with an angle γ about the axis $\boldsymbol{w}$. In this case, the final rotation matrix is obtained by multiplying the corresponding rotation matrices, as follows.

$$\begin{aligned} \boldsymbol{R}_{wuw}(\alpha,\beta,\gamma) &= \boldsymbol{R}_w(\alpha)\boldsymbol{R}_u(\beta)\boldsymbol{R}_w(\gamma) \\ &= \begin{bmatrix} c\alpha c\gamma - s\alpha c\beta s\gamma & -c\alpha s\gamma - s\alpha c\beta c\gamma & s\alpha s\beta \\ s\alpha c\gamma + c\alpha c\beta s\gamma & -s\alpha s\gamma + c\alpha c\beta c\gamma & -c\alpha s\beta \\ s\beta s\gamma & s\beta c\gamma & c\beta \end{bmatrix} \end{aligned} \tag{2.36}$$

The solution to the inverse problem in this representation, in the case where the angle $s\beta \neq 0$, is obtained from the following equations.

$$\begin{aligned} \beta &= \text{atan2}(\pm\sqrt{r_{31}^2 + r_{32}^2}, r_{33}) \\ \alpha &= \text{atan2}(r_{13}/s\beta, -r_{23}/s\beta) \\ \gamma &= \text{atan2}(r_{31}/s\beta, r_{32}/s\beta) \end{aligned} \tag{2.37}$$

Furthermore, if the angle $s\beta = 0$, the inverse problem equations will degenerate and this will result into a gimbal lock of this representation. Consequently, only the algebraic sum of the two angles α and γ can be calculated from Equation 2.38 as follows:

$$\begin{cases} \beta = 0 \\ \alpha + \gamma = \text{atan2}(-r_{12}, r_{11}) \end{cases} \quad \text{or} \quad \begin{cases} \beta = \pi \\ \alpha - \gamma = \text{atan2}(r_{12}, r_{11}) \end{cases} \tag{2.38}$$

Example 2.9 (The $\{w - u - w\}$ Euler Angles) *Consider the rotation of a rigid body using the rotation matrix introduced in Example 2.2. Represent this rotation using the $\{w - u - w\}$ Euler angles.*

Solution: For this rotation matrix, the $\{w - u - w\}$ Euler angles are calculated as follows using the relation in Equation 2.37:

$$\alpha = -45^\circ \ ; \ \beta = -30^\circ \ ; \ \gamma = 45^\circ$$

It is apparent that the Euler angles obtained in this representation, similar to the previous example, exhibit a notably more regular pattern. These angles facilitate a clearer visualization, offering a more intuitive understanding of the fundamental physical concept of rotation.

Example 2.10 *Consider the rotation of a rigid body with the depicted configuration in Figure 2.8. Suppose the body is initially in the shown position with the coordinate system $\{\mathcal{I}\}$ and ends up in the final state illustrated with the coordinate system $\{\mathcal{F}\}$. Obtain the rotation matrix and $\{w - u - w\}$ Euler angles representing the rotation of this rigid body using the geometric aspects of the motion.*

Solution: The rotation matrix can be easily derived by using its definition given in the relation 2.2.

$${}^{I}\boldsymbol{R}_F = \begin{bmatrix} -1 & 0 & 0 \\ 0 & 0 & 1 \\ 0 & 1 & 0 \end{bmatrix}$$

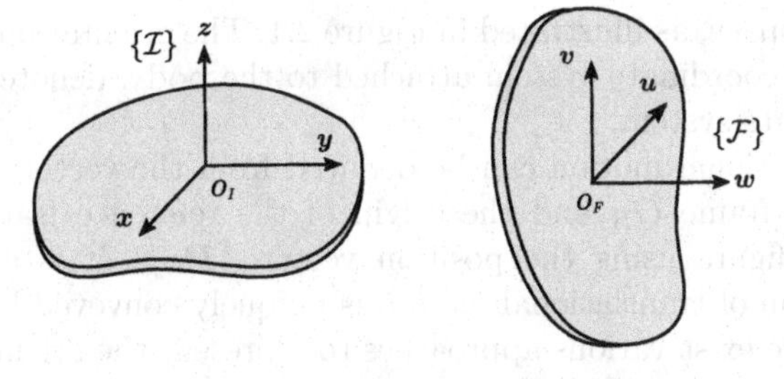

FIGURE 2.8
An example of the rotation of a rigid body with specified rotation angles.

Based on the motion geometry as depicted in Figure 2.8, this rotation can be represented with two rotations. First, a rotation around the moving $\boldsymbol{w}$ axis with an angle of 180 degrees, followed by a rotation around the moving $\boldsymbol{u}$ axis with an angle of 90 degrees. Thus, the $\{\boldsymbol{w}-\boldsymbol{u}-\boldsymbol{w}\}$ Euler angles may be determined as:

$$\alpha = \pi \;\; ; \;\; \beta = \pi/2 \;\; ; \;\; \gamma = 0$$

These angles may be derived from the inverse problem solution given in Equation 2.37, as well.

It's important to note that, despite describing the rotation of a body with different Euler angles, a unique rotation matrix is obtained corresponding to all representation. Furthermore, the inverse problem does not have a unique solution, leading to at least two sets of corresponding angles for its representation. In gimbal lock situations, where the inverse problem equation degenerates, the choice of Euler angles representing the rotation of the body has an infinite number of solutions. In such cases, it becomes impossible to determine independent Euler angles in all three directions. Therefore, when employing Euler angles in the motion analysis of robotic arm movements, special attention must be given to avoiding gimbal lock.

To tackle this issue, it is customary to employ motion continuity constraints to prevent abrupt transitions between potential solutions. Furthermore, to avoid gimbal lock in representing rotations with Euler angles, two sets of these angles are concurrently utilized in the representation of rotational motion. This enables the use of the second representation to articulate the motion when the first one encounters gimbal lock situation. Despite the widespread use of Euler angles in representing the motion of objects in three-dimensional space, there is a preference for using rotation matrices as a non-singular foundation for representation. In the upcoming chapter, we will explore alternative approaches for orientation representations.

2.5 Rigid Body's Pose

In the preceding section, the orientation of a rigid body in pure rotational motion was studied using two methods. If this rigid body undergoes translational motion in addition to rotational motion, the analysis of its motion requires considering the combination of translational and rotational motion. For this means, consider the general motion of a body in three-dimensional space, as illustrated in Figure 2.1. The orientation of the body in space is represented by the coordinate system attached to the body, denoted as $\{\mathcal{B}\}$, relative to the reference coordinate system.

The body's translational motion can be deduced from the vector distance between the origin of the moving frame O_B and the origin of the reference frame O_A. This motion is illustrated in the figure using the position vector ${}^A\boldsymbol{P}_{O_B}$. It is essential to recognize that the representation of translational motion is uniquely conveyed by the position vector ${}^A\boldsymbol{P}_{O_B}$. However, there exist various approaches to represent the orientation of the body in space. Fortunately, as illustrated in the preceding section, it was demonstrated that diverse representations of a rigid body's orientation in space can all be conveyed using a unique rotation matrix denoted as ${}^A\boldsymbol{R}_B$. Consequently, the *pose* of a rigid body in space can be precisely determined by knowing the following details.

1) The position vector from the origin of the moving coordinate system O_B to the origin of the reference coordinate system $\{\mathcal{A}\}$, represented by the vector ${}^A\boldsymbol{P}_{O_B}$.

2) The orientation of a rigid body or a coordinate system attached to it, denoted as $\{\mathcal{B}\}$, relative to the reference coordinate system $\{\mathcal{A}\}$ is represented by the rotation matrix ${}^A\boldsymbol{R}_B$. This rotation matrix can be obtained from any type of orientation representation of the rigid body in space.

In this manner, the absolute position vector of any arbitrary point P on the body, represented with respect to the reference coordinate system $\{\mathcal{A}\}$ by the vector ${}^A\boldsymbol{P}$, can be determined using the translational motion of the body represented by the vector ${}^A\boldsymbol{P}_{O_B}$, along with the rotational motion of the body and its orientation represented by the rotation matrix ${}^A\boldsymbol{R}_B$. This can be achieved using the following relation:

$$ {}^A\boldsymbol{P} = {}^A\boldsymbol{P}_{O_B} + {}^A\boldsymbol{R}_B\, {}^B\boldsymbol{P} \tag{2.39} $$

Here, ${}^A\boldsymbol{P}_{O_B}$ represents the translational motion vector, ${}^A\boldsymbol{R}_B$ is the rotation matrix representing the orientation, and ${}^B\boldsymbol{P}$ is the position vector of point P with respect to the body-fixed coordinate system $\{\mathcal{B}\}$. With this interpretation of the body's pose in space, it can be envisioned that initially, the rigid body is in its initial state, where the coordinate system attached to the body, $\{\mathcal{B}\}$, aligns with the reference coordinate system, $\{\mathcal{A}\}$. Then, it undergoes translational motion along the vector ${}^A\boldsymbol{P}_{O_B}$ without any rotational movement, positioning the coordinate system attached to the body at a distance ${}^A\boldsymbol{P}_{O_B}$ parallel to $\{\mathcal{A}\}$. Subsequently, it performs its rotational motion, determined by the rotation matrix ${}^A\boldsymbol{R}_B$, to reach the final state.

The decomposition of general motion in space into translational and rotational components has been a central focus in classical mechanics and is mathematically formalized in the Mozzi-Chasles' Theorem [39]. Particularly, this theorem, which deals with the decomposition of general motion into translational and rotational elements, highlights the instantaneous axis of rotation or the screw axis. Additional insights into this topic will be provided in the upcoming chapter and in Section 3.3.

Example 2.11 (Pose of a rigid body) *Consider the motion of a rigid body as illustrated in Figure 2.9. In this motion, the rigid body has transitioned from the depicted state with the coordinate system* $\{0\}$ *to the shown state with the coordinate system* $\{1\}$*, undergoing both translational and rotational movements. Given the provided values in the figure, determine the absolute position vector of point* P *with respect to the reference frame* $\{0\}$.

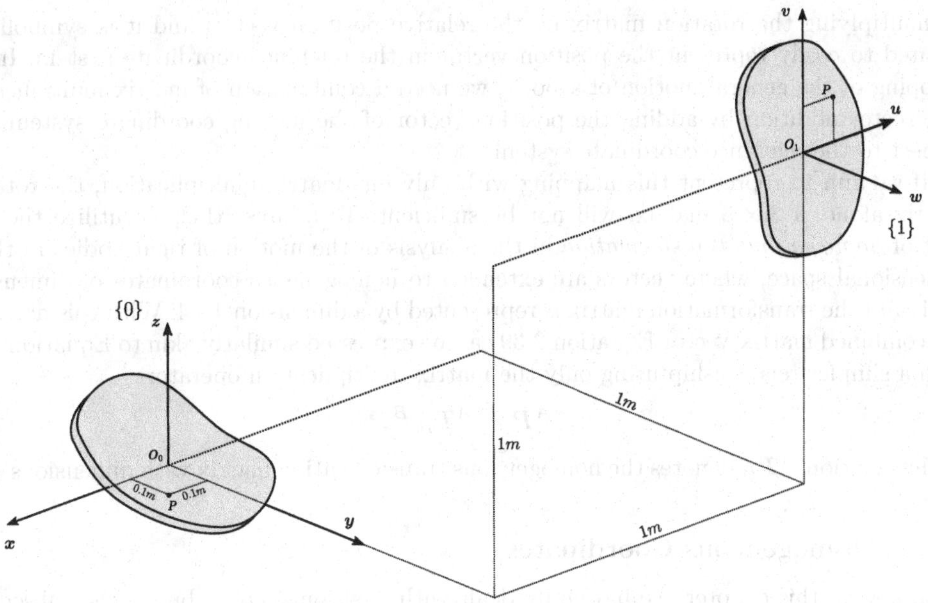

FIGURE 2.9
An example of general motion of a rigid body in 3-dimensional space.

Solution: Considering Figure 2.9, the relative position of point P with respect to the coordinate system attached to the body can be represented by the vector ${}^1\boldsymbol{P} = \begin{bmatrix} 0.1 & 0.1 & 0 \end{bmatrix}^T$. Additionally, the translational motion of the body is determined by the vector ${}^0\boldsymbol{P}_{o_1}$, and its rotational motion is characterized by the rotation matrix ${}^0\boldsymbol{R}_1$ as follows:

$$
{}^0\boldsymbol{P}_{o_1} = \begin{bmatrix} -1 \\ 1 \\ 1 \end{bmatrix} \quad ; \quad {}^0\boldsymbol{R}_1 = \begin{bmatrix} -1 & 0 & 0 \\ 0 & 0 & 1 \\ 0 & 1 & 0 \end{bmatrix}
$$

By this means, the absolute position vector of point P with respect to the reference coordinate system $\{0\}$ is obtained from the following relationship:

$$
\begin{aligned}
{}^0\boldsymbol{P} &= {}^0\boldsymbol{R}_1\,{}^1\boldsymbol{P} + {}^0\boldsymbol{P}_{o_1} \\
&= \begin{bmatrix} -1 & 0 & 0 \\ 0 & 0 & 1 \\ 0 & 1 & 0 \end{bmatrix} \begin{bmatrix} 0.1 \\ 0.1 \\ 0 \end{bmatrix} + \begin{bmatrix} -1 \\ 1 \\ 1 \end{bmatrix} = \begin{bmatrix} -1.1 \\ 1.0 \\ 1.1 \end{bmatrix}
\end{aligned}
$$

The accuracy of these calculations can be easily verified using the provided figure.

2.6 Homogeneous Transformation

In the prelude sections, we introduced pure rotational motion in Equation 2.16 and the mapping of general motion in Equation 2.39. The mapping of rotational motion is defined by multiplying the rotation matrix by the relative position vector, and it is symbolically notated to easily represent the position vector in the reference coordinate system. In the mapping of the general motion of a body, we need a combination of matrix multiplication and vector addition by adding the position vector of the moving coordinate system with respect to the reference coordinate system.

If we aim to represent this mapping with only one matrix multiplication, the rotation matrix alone, a 3×3 matrix, will not be sufficient. To address this, we utilize the concept of *homogeneous transformation* in the analysis of the motion of rigid bodies in three-dimensional space, where vectors are extended to homogeneous coordinates of dimensions 4×1, and the transformation matrix is represented by a dimension 4×4. With this notation, the combined matrix-vector Equation 2.39 can be expressed similarly akin to Equation 2.16, with a simpler relationship using only the matrix multiplication operator.

$$ {}^{A}\boldsymbol{P} = {}^{A}\boldsymbol{T}_{B}\, {}^{B}\boldsymbol{P} \tag{2.40} $$

In this relation ${}^{A}\boldsymbol{T}_{B}$ denotes the homogeneous transformation matrix with dimensions 4×4.

2.6.1 Homogeneous Coordinates

Until now in this chapter, we have only dealt with position vectors, but in the subsequent chapters, other vectors such as linear and angular velocity and acceleration, as well as force and torque, will be defined and used in the robotics analysis. All these vectors can be categorized into two groups, namely line vectors and free vectors. Line vectors are associated with physical quantities where the point of application is crucial for their analysis. The position vector of a point on a rigid body falls into the category of line vectors because this vector depends on the specific point, and with even a small displacement of that point, the representing vector will change. Other quantities such as linear velocity of a point on the body, as well as the force acting on the body, also fall into this category of vectors.

On the other hand, there are vectors that are not dependent on the point of application and represent the overall physical quantity of the entire body. The orientation of a body is one such physical quantity that belongs to this category. Vectors representing the angular velocity and acceleration of the body, as well as the torque applied to a rigid body, also fall into the category of free vectors because they are not dependent on the point of application.

In the transformation of coordinates for line vectors, both translation and rotation of the coordinate system significantly affect the calculation of their values. Therefore, to use homogeneous transformation and the relation in Equation 2.40 for their representation, we use homogeneous coordinate vectors of dimensions 4×1 by adding one to the fourth component of the regular vector as follows:

$$ \text{For line vectors:} \quad \boldsymbol{P} = \left[\begin{array}{c} p_x \\ p_y \\ p_z \\ \hline 1 \end{array}\right] \tag{2.41} $$

This adjustment allows us to use the homogeneous transformation matrix to represent both translation and rotation in a unified manner.

In free vectors, only the orientation will be affected by the coordinate transformation. Therefore, by adding zero to the fourth component of the regular vector, homogeneous coordinates for these vectors are defined as follows:

$$ \text{For free vectors:} \quad \boldsymbol{\omega} = \left[\begin{array}{c} \omega_x \\ \omega_y \\ \omega_z \\ \hline 0 \end{array}\right] \tag{2.42} $$

Upon the introduction of homogeneous coordinates, it is important to emphasize that this representation is employed specifically when utilizing homogeneous transformations to alter the coordinate system for a vector quantity.

2.6.2 Homogeneous Transformation Matrix

The *homogeneous transformation matrix* is a 4×4 matrix used to transform a vector quantity expressed in homogeneous coordinates from one coordinate system to another. This transformation is realized in a simple matrix form, as shown in Equation 2.40. Considering the relation in Equation 2.39 and the simplified matrix form in Equation 2.40, the homogeneous transformation matrix is composed of the rotation matrix ${}^{A}\boldsymbol{R}_{B}$ and the translation vector ${}^{A}\boldsymbol{P}_{O_B}$, as illustrated below.

$$
{}^{A}\boldsymbol{T}_{B} = \left[\begin{array}{ccc|c} & {}^{A}\boldsymbol{R}_{B} & & {}^{A}\boldsymbol{P}_{O_B} \\ \hline 0 & 0 & 0 & 1 \end{array}\right] \tag{2.43}
$$

The transformation matrix ${}^{A}\boldsymbol{T}_{B}$ describes the mapping for line vectors represented in homogeneous coordinates, which are shown as 4×1 matrices. In this mapping, both translational and rotational motions have an effect.

$$
\left[\begin{array}{c} {}^{A}\boldsymbol{P} \\ \hline 1 \end{array}\right] = \left[\begin{array}{ccc|c} & {}^{A}\boldsymbol{R}_{B} & & {}^{A}\boldsymbol{P}_{O_B} \\ \hline 0 & 0 & 0 & 1 \end{array}\right] \left[\begin{array}{c} {}^{B}\boldsymbol{P} \\ \hline 1 \end{array}\right] \tag{2.44}
$$

$$
\begin{aligned} {}^{A}\boldsymbol{P} &= {}^{A}\boldsymbol{R}_{B}\,{}^{B}\boldsymbol{P} + {}^{A}\boldsymbol{P}_{O_B} \\ 1 &= 1 \end{aligned} \tag{2.45}
$$

For the transformation of free vectors, where only the rotational matrix is influential, the homogeneous transformation 2.40 correctly performs the coordinate transformation, as well.

$$
\left[\begin{array}{c} {}^{A}\boldsymbol{\omega} \\ \hline 0 \end{array}\right] = \left[\begin{array}{ccc|c} & {}^{A}\boldsymbol{R}_{B} & & {}^{A}\boldsymbol{P}_{O_B} \\ \hline 0 & 0 & 0 & 1 \end{array}\right] \left[\begin{array}{c} {}^{B}\boldsymbol{\omega} \\ \hline 0 \end{array}\right] \tag{2.46}
$$

$$
\begin{aligned} {}^{A}\boldsymbol{\omega} &= {}^{A}\boldsymbol{R}_{B}\,{}^{B}\boldsymbol{\omega} \\ 0 &= 0 \end{aligned} \tag{2.47}
$$

Example 2.12 (Homogeneaous Transforamations) *Consider the general motion of a rigid body in space, as shown in Figure 2.9. Assume that the relative position vector of point P, denoted by* ${}^{1}\boldsymbol{P}$*, and the angular velocity vector of the body represented in the moving coordinate system {1}, denoted by* ${}^{1}\boldsymbol{\omega}$*, are given as follows:*

$$
{}^{1}\boldsymbol{P} = \begin{bmatrix} 0.1 \\ 0.1 \\ 0 \end{bmatrix} \quad ; \quad {}^{1}\boldsymbol{\omega} = \begin{bmatrix} 1 \\ 0 \\ 1.2 \end{bmatrix}
$$

Obtain the absolute position vector of point P and the angular velocity of the body with respect to the reference coordinate system $\{0\}$.

Solution: To determine the absolute position vector of point P, the homogeneous transformation and the relation 2.40 can be used. In this example, the homogeneous transformation matrix can be written in the following form.

$$ {}^0\boldsymbol{T}_1 = \left[\begin{array}{ccc|c} & {}^0\boldsymbol{R}_1 & & {}^0\boldsymbol{P}_{O_1} \\ \hline 0 & 0 & 0 & 1 \end{array}\right] = \left[\begin{array}{ccc|c} -1 & 0 & 0 & -1 \\ 0 & 0 & 1 & 1 \\ 0 & 1 & 0 & 1 \\ \hline 0 & 0 & 0 & 1 \end{array}\right] $$

Furthermore, since the position vector of point P is a line vector and the angular velocity of the body $\boldsymbol{\omega}$ is a free vector, the following homogeneous coordinates should be used to transform them.

$$ {}^1\boldsymbol{P} = \left[\begin{array}{c} 0.1 \\ 0.1 \\ 0 \\ \hline 1 \end{array}\right] \quad ; \quad {}^1\boldsymbol{\omega} = \left[\begin{array}{c} 1 \\ 0 \\ 1.2 \\ \hline 0 \end{array}\right] $$

In this way, one can easily calculate the homogeneous transformation of ${}^1\boldsymbol{P}$ and ${}^1\boldsymbol{\omega}$ with respect to the reference coordinate system as follows.

$$ {}^0\boldsymbol{P} = {}^0\boldsymbol{T}_1\,{}^1\boldsymbol{P} = \left[\begin{array}{c} -1.1 \\ 1.0 \\ 1.1 \\ \hline 1 \end{array}\right] \quad ; \quad {}^0\boldsymbol{\omega} = {}^0\boldsymbol{T}_1\,{}^1\boldsymbol{\omega} = \left[\begin{array}{c} -1 \\ 1.2 \\ 0 \\ \hline 0 \end{array}\right] $$

2.6.3 Homogeneous Transformation Arithmetics

In this section, the necessary mathematical operations for homogeneous transformations will be presented. In this regard, we will review how to perform successive coordinate transformations and obtain the inverse of a homogeneous transformation.

Consecutive Transformations

Consider the successive motion of a rigid body from the depicted state in the coordinate system $\{\mathcal{A}\}$ to $\{\mathcal{B}\}$ and finally to the coordinate system $\{\mathcal{C}\}$ as shown in Figure 2.10. Suppose the position vector of point P in the final state of the body is given by ${}^C\boldsymbol{P}$ in the $\{\mathcal{C}\}$ coordinate system, and the objective is to find this vector relative to the reference coordinate system $\{\mathcal{A}\}$. Given that the state of the rigid body in two successive states is known, the vector ${}^C\boldsymbol{P}$ can be obtained first in the intermediate coordinate system $\{\mathcal{B}\}$ and then in the reference coordinate system using the transformation relation 2.40. For this purpose, by using the transformation matrix ${}^B\boldsymbol{T}_C$, the position vector with respect to the $\{\mathcal{B}\}$ coordinate system is determined.

$$ {}^B\boldsymbol{P} = {}^B\boldsymbol{T}_C\,{}^C\boldsymbol{P} \tag{2.48} $$

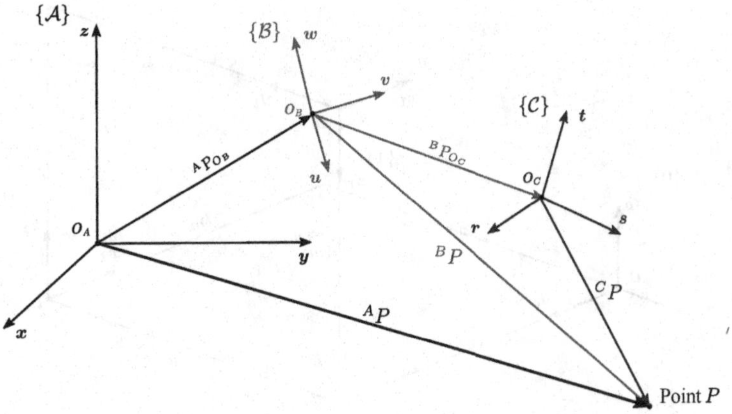

FIGURE 2.10
Successive motion of a rigid body from the depicted state in coordinate systems $\{\mathcal{A}\}$, $\{\mathcal{B}\}$, and finally in coordinate system $\{\mathcal{C}\}$.

Furthermore, the transformation matrix $^A\boldsymbol{T}_B$ is then used to obtain the position vector relative to the reference coordinate system $\{\mathcal{A}\}$ as follows.

$$^A\boldsymbol{P} = {}^A\boldsymbol{T}_B\, {}^B\boldsymbol{P} \tag{2.49}$$

By combining the equations 2.48 and 2.49, one may write:

$$^A\boldsymbol{P} = {}^A\boldsymbol{T}_B\, {}^B\boldsymbol{T}_C\, {}^C\boldsymbol{P} \tag{2.50}$$

The combined transformation matrix, obtained as the product of two homogeneous transformation matrices in a sequential motion, is given by:

$$^A\boldsymbol{T}_C = {}^A\boldsymbol{T}_B\, {}^B\boldsymbol{T}_C \tag{2.51}$$

If we want to analytically express the transformation matrix $^A\boldsymbol{T}_C$ in terms of its constituent transformation matrices, i.e., $^A\boldsymbol{T}_B$ and $^B\boldsymbol{T}_C$, we can use the general form of the homogeneous transformation matrix 2.43, which can be written as:

$$^A\boldsymbol{T}_C = \left[\begin{array}{ccc|c} & {}^A\boldsymbol{R}_B\, {}^B\boldsymbol{R}_C & & {}^A\boldsymbol{R}_B\, {}^B\boldsymbol{P}_{O_C} + {}^A\boldsymbol{P}_{O_B} \\ \hline 0 & 0 & 0 & 1 \end{array}\right] \tag{2.52}$$

Now, if a composite transformation involves more than two consecutive transformations, we can express the relationship 2.51 in a more general form, as shown in the following equation. Here, a sequence of movements in different states is considered:

$$^A\boldsymbol{T}_Z = {}^A\boldsymbol{T}_B\, {}^B\boldsymbol{T}_C\, {}^C\boldsymbol{T}_D \cdots {}^Y\boldsymbol{T}_Z \tag{2.53}$$

Example 2.13 (Consecutive Homogeneous Transformation) *Consider the consecutive motion of a rigid body in three different positions shown with coordinate systems $\{0\}$, $\{1\}$, and $\{2\}$ in Figure 2.11. Calculate the homogeneous transformation matrices $^0\boldsymbol{T}_1$, $^1\boldsymbol{T}_2$, and the composite transformation matrix $^0\boldsymbol{T}_2$. Verify the result based on the definition of homogeneous matrices.*

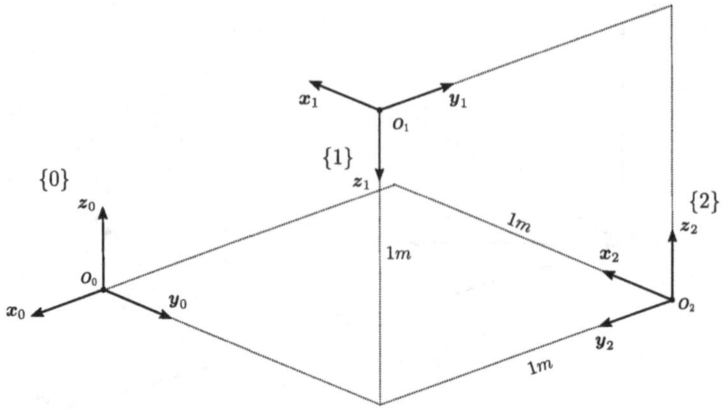

FIGURE 2.11
An example of the consecutive motion of a rigid body in three different states is illustrated with coordinate systems $\{0\}, \{1\}, \{2\}$.

Solution: With the definition of rotation matrices and homogeneous transformation matrices, and based on the depicted motion of the rigid body in space in Figure 2.11, the transformation matrices ${}^0\boldsymbol{T}_1$ and ${}^0\boldsymbol{T}_2$ can be obtained in the following forms:

$$
{}^0\boldsymbol{T}_1 = \left[\begin{array}{ccc|c} 0 & -1 & 0 & 0 \\ -1 & 0 & 0 & 1 \\ 0 & 0 & -1 & 1 \\ \hline 0 & 0 & 0 & 1 \end{array}\right], \quad {}^1\boldsymbol{T}_2 = \left[\begin{array}{ccc|c} 1 & 0 & 0 & 0 \\ 0 & -1 & 0 & 1 \\ 0 & 0 & -1 & 1 \\ \hline 0 & 0 & 0 & 1 \end{array}\right]
$$

With the direct multiplication of the two matrices above in accordance with Equation 2.53, or using the analytical relationship 2.43, the transformation matrix ${}^0\boldsymbol{T}_2$ can be obtained as follows:

$$
{}^0\boldsymbol{T}_2 = \left[\begin{array}{ccc|c} 0 & 1 & 0 & -1 \\ -1 & 0 & 0 & 1 \\ 0 & 0 & 1 & 0 \\ \hline 0 & 0 & 0 & 1 \end{array}\right]
$$

With the help of Figure 2.11 and the definition of the homogeneous transformation matrix, the accuracy of the above calculations can be easily verified.

Example 2.14 (Consecutive rotation with a finite angle) *In this example, it is demonstrated that consecutive rotations with finite angles are not commutative. However, if the rotation angles are infinitesimally small, the transformation becomes commutative. To illustrate this, two consecutive rotations of a rigid body are considered: first around the $\hat{x}$ axis with a finite rotation angle θ_1, and then around the $\hat{z}$ axis with a finite rotation angle θ_2. Find the resulting rotation from the first and second rotations. Subsequently, calculate these consecutive rotations with infinitesimally small angles $\delta\theta_1$ and $\delta\theta_2$, and compare the results.*

Solution: First, consider infinitesimal rotations around two axes $\hat{x}$ and $\hat{z}$.

$$\boldsymbol{R}_x(\theta_1) = \begin{bmatrix} 1 & 0 & 0 \\ 0 & c_1 & -s_1 \\ 0 & s_1 & c_1 \end{bmatrix}, \quad \boldsymbol{R}_z(\theta_2) = \begin{bmatrix} c_2 & -s_2 & 0 \\ s_2 & c_2 & 0 \\ 0 & 0 & 1 \end{bmatrix}$$

Given the matrix multiplication expression presented in Equation 2.53, the combination of rotations in the first scenario is obtained as follows.

$$\boldsymbol{R}_1 = \boldsymbol{R}_x(\theta_1)\,\boldsymbol{R}_z(\theta_2) = \begin{bmatrix} c_2 & -s_2 & 0 \\ c_1 s_2 & c_1 c_2 & -s_1 \\ s_1 s_2 & s_1 c_2 & c_1 \end{bmatrix}$$

Here, short hand notations $c_1 = \cos(\theta_1)$, $s_1 = \sin(\theta_1)$, $c_2 = \cos(\theta_2)$, and $s_2 = \sin(\theta_2)$ are used. Considering that finite rotations are interpreted by multiplying rotation matrices, and matrix multiplication is not commutative, if we change the order of rotation sequences, we will obtain a completely different representation as follows.

$$\boldsymbol{R}_2 = \boldsymbol{R}_z(\theta_2)\,\boldsymbol{R}_x(\theta_1) = \begin{bmatrix} c_2 & -c_1 s_2 & s_1 s_2 \\ s_2 & c_1 c_2 & -s_1 c_2 \\ 0 & s_1 & c_1 \end{bmatrix}$$

This will result in rotations that are not equal, i.e., $\boldsymbol{R}_1 \neq \boldsymbol{R}_2$. In other words, if the order of rotations in the sequence changes, the resulting rotation will be completely different.

However, if we consider the consecutive rotations to be infinitesimally small, the situation will be different. By substituting $\theta_1 = \delta\theta_1$ and $\theta_2 = \delta\theta_2$ into the above equations and assuming the angles to be infinitesimally small, we can use the following simplifications:

$$s_1 \simeq \delta\theta_1, \quad c_1 \simeq 1, \quad s_2 \simeq \delta\theta_2, \quad c_2 \simeq 1$$

By this means:

$$\boldsymbol{R}_1 = \boldsymbol{R}_x(\theta_1)\,\boldsymbol{R}_z(\theta_2) = \begin{bmatrix} 1 & -\delta\theta_2 & 0 \\ \delta\theta_2 & 1 & -\delta\theta_1 \\ \delta\theta_1\delta\theta_2 & \delta\theta_1 & 1 \end{bmatrix}$$

$$\boldsymbol{R}_2 = \boldsymbol{R}_z(\theta_2)\,\boldsymbol{R}_x(\theta_1) = \begin{bmatrix} 1 & -\delta\theta_2 & \delta\theta_1\delta\theta_2 \\ \delta\theta_2 & 1 & -\delta\theta_1 \\ 0 & \delta\theta_1 & 1 \end{bmatrix}$$

Given that the rotation angles $\delta\theta_i$ are infinitesimally small, their product is of the second order and can be neglected: $\delta\theta_1\delta\theta_2 \simeq 0$. Thus, the result of two consecutive rotations will be the same in both cases when considering infinitesimally small rotation angles.

$$\boldsymbol{R}_1 = \boldsymbol{R}_2 = \begin{bmatrix} 1 & -\delta\theta_2 & 0 \\ \delta\theta_2 & 1 & -\delta\theta_1 \\ 0 & \delta\theta_1 & 1 \end{bmatrix}$$

This reasoning demonstrates that rotations with infinitesimally small angles exhibit the property of commutativity, and they can also be represented using vectors. This will be addressed in the upcoming chapter and in Section 3.3. However, this argument is primarily employed to prove the vector representation of angular velocity and its derivatives.

Inverse of a Homogeneous Transformation

Direct calculating the inverse of a 4×4 matrix may not be computationally efficient. Therefore, in this section, we obtain the inverse using the concept of mapping in homogeneous transformation. Considering that part of the homogeneous transformation matrix is composed of the rotation matrix, and its inverse can be easily calculated based on its transpose, computing the inverse in this way involves less computation cost. On the other hand, another part of the homogeneous transformation matrix is derived from its translation vector, representing the displacement of the moving coordinate system's origin relative to the reference coordinate system $\{\mathcal{A}\}$. In the inverse mapping, we need to use a vector of the same magnitude but opposite direction. This vector can be denoted as ${}^B\boldsymbol{P}_{OA}$. Consequently, the inverse of the homogeneous transformation matrix ${}^A\boldsymbol{T}_B$, denoted as ${}^B\boldsymbol{T}_A$, can be determined from these two components as follows.

$$ {}^B\boldsymbol{T}_A = \left[\begin{array}{ccc|c} & {}^B\boldsymbol{R}_A & & {}^B\boldsymbol{P}_{O_A} \\ \hline 0 & 0 & 0 & 1 \end{array}\right] \tag{2.54} $$

However, as previously mentioned, ${}^B\boldsymbol{R}_A = {}^A\boldsymbol{R}_B^T$, and to calculate the vector ${}^B\boldsymbol{P}_{O_A}$ based on the discussion, one can write:

$$ {}^B\boldsymbol{P}_{OB} = {}^B\boldsymbol{R}_A \, {}^A\boldsymbol{P}_{OB} + {}^B\boldsymbol{P}_{OA} $$

And since the left side of the above equation is zero, we will have:

$$ {}^B\boldsymbol{P}_{OA} = -{}^A\boldsymbol{R}_B^T \, {}^A\boldsymbol{P}_{OB} \tag{2.55} $$

In this way, the inverse of the homogeneous transformation matrix is obtained as follows.

$$ {}^B\boldsymbol{T}_A = {}^A\boldsymbol{T}_B^{-1} = \left[\begin{array}{ccc|c} & {}^A\boldsymbol{R}_B^T & & -{}^A\boldsymbol{R}_B^T \, {}^A\boldsymbol{P}_{OB} \\ \hline 0 & 0 & 0 & 1 \end{array}\right] \tag{2.56} $$

The equation 2.56 calculates the inverse matrix for any transformation matrix with optimized computation.

Example 2.15 *Assume the homogeneous transformation matrix of a body is represented by*

$$ {}^0\boldsymbol{T}_1 = \begin{bmatrix} 0.933 & 0.167 & 0.354 & 2 \\ 0.067 & 0.933 & -0.354 & 1 \\ -0.354 & 0.354 & 0.866 & 0 \\ 0 & 0 & 0 & 1 \end{bmatrix} $$

It seems that there might be some mistakes in reporting certain elements of its rotation matrix. Obtain the correct representation of the homogeneous transformation matrix and then calculate its inverse. Furthermore, if the position vector of point P with respect to the $\{0\}$ coordinate system is given as ${}^0\boldsymbol{P} = \begin{bmatrix} 0.25 & 0.43 & 0.86 \end{bmatrix}^T$ Find the representation of this vector in frame $\{1\}$.

Solution: By examining the columns of the rotation matrix, it is easily understood that the Euclidean norm of the second column is not equal to one. Furthermore, by inspecting the row norms of the rotation matrix, it becomes evident that the element r_{12} was mistakenly written as 0.167 and should be 0.067. The inverse of the corrected transformation matrix can be obtained using equation 2.56 with the following calculations.

$$^1\boldsymbol{T}_0 = {}^0\boldsymbol{T}_1^{-1} = \left[\begin{array}{ccc|c} & {}^0\boldsymbol{R}_1^T & & -{}^0\boldsymbol{R}_1^T\,{}^0\boldsymbol{P}_{O_1} \\ \hline 0 & 0 & 0 & 1 \end{array}\right] = \left[\begin{array}{ccc|c} 0.933 & 0.067 & -0.354 & -1.80 \\ 0.067 & 0.933 & 0.354 & 0.80 \\ 0.354 & -0.354 & 0.866 & -1.06 \\ \hline 0 & 0 & 0 & 1 \end{array}\right]$$

Using this transformation matrix, the vector $^1\boldsymbol{P}$ can be calculated from the following relation.

$$^1\boldsymbol{P} = {}^1\boldsymbol{T}_0\,{}^0\boldsymbol{P} = \left[\begin{array}{ccc} -1.84 & 1.52 & -0.38 \end{array}\right]^T$$

Problems

2.1 Consider coordinate systems $\{\mathcal{A}\}, \{\mathcal{B}\}, \{\mathcal{C}\}$ in the depicted configuration in Figure 2.12. Obtain the rotation matrices $^A\boldsymbol{R}_B$, $^A\boldsymbol{R}_C$, and $^B\boldsymbol{R}_C$, by using the definition.

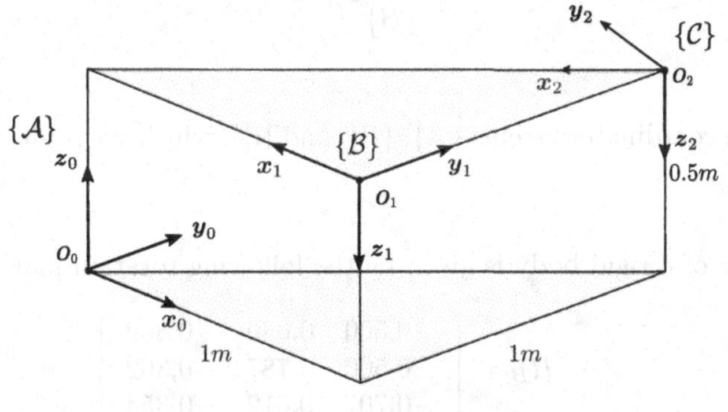

FIGURE 2.12
Illustration of coordinate systems $\{\mathcal{A}\}$, $\{\mathcal{B}\}$, and $\{\mathcal{C}\}$ placed at the corners of a wedge.

2.2 For the coordinate systems shown in the Figure 2.12, obtain the rotation matrices $^B\boldsymbol{R}_A$, $^C\boldsymbol{R}_A$ and $^C\boldsymbol{R}_B$ by using the definition. Furthermore, calculate the inverse rotation matrix obtained in Problem 2.1, and verify your results.

2.3 Repeat Problem 2.1 for the coordinate systems shown in Figure 2.13.

2.4 Reiterate Problem 2.2 for the coordinate systems depicted in Figure 2.13.

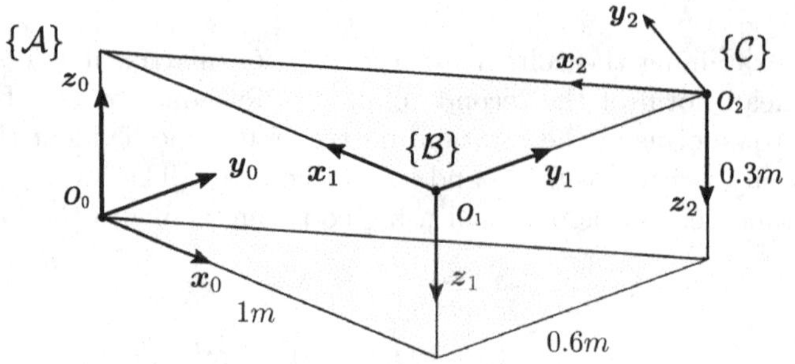

FIGURE 2.13
Representation of coordinate systems $\{\mathcal{A}\}$, $\{\mathcal{B}\}$, and $\{\mathcal{C}\}$, which are positioned at the corners of a wedge.

2.5 Duplicate the exercise outlined in Problem 2.1 for the coordinate systems illustrated in Figure 2.14.

2.6 Redo the task presented in Problem 2.2 for the coordinate systems depicted in Figure 2.14.

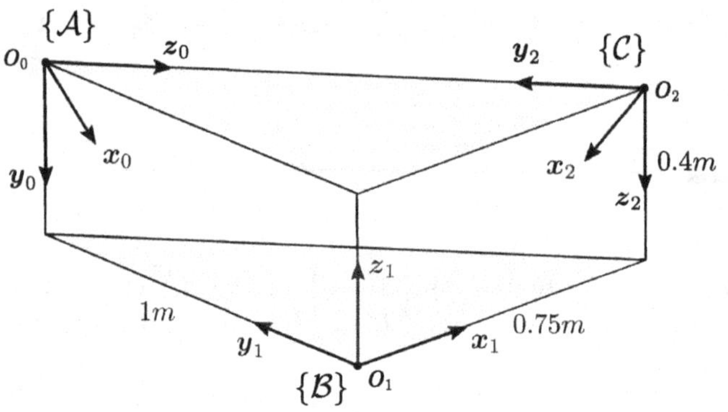

FIGURE 2.14
Representation of coordinate systems $\{\mathcal{A}\}$, $\{\mathcal{B}\}$, and $\{\mathcal{C}\}$, which are positioned at the corners of a wedge.

2.7 The rotation of a rigid body is given by the following rotation matrix:

$$^{A}\boldsymbol{R}_{B} = \begin{bmatrix} 0.500 & 0.080 & 0.862 \\ 0.500 & 0.787 & -0.362 \\ -0.707 & 0.612 & 0.354 \end{bmatrix}$$

a) If the relative position vector of point P is given as: $^{B}\boldsymbol{P} = [\ 0.707 \quad -0.707 \quad 0\ ]^{T}$ find the absolute position vector of this point, denoted by $^{A}\boldsymbol{P}$.

b) If the absolute position vector of point P is given as: $^{A}\boldsymbol{P} = [\ 0.577 \quad -0.577 \quad -0.577\ ]^{T}$ find the relative position vector of point P, denoted by $^{B}\boldsymbol{P}$.

c) Find the eigenvalues and eigenvectors of the rotation matrix $^{A}\boldsymbol{R}_{B}$.

d) Calculate the pitch-roll-yaw Euler angles of this rotation.

e) Calculate the $\boldsymbol{w}-\boldsymbol{v}-\boldsymbol{u}$ Euler angles of this rotation.

2.8 The rotation of a rigid body is given by the following rotation matrix:

$$^A\boldsymbol{R}_B = \begin{bmatrix} -0.146 & -0.854 & 0.500 \\ 0.854 & 0.146 & 0.500 \\ -0.500 & 0.500 & 0.707 \end{bmatrix}$$

a) Find the eigenvalues and eigenvectors of the rotation matrix $^A\boldsymbol{R}_B$.

b) Calculate the pitch-roll-yaw Euler angles of this rotation.

c) Calculate the $\boldsymbol{w} - \boldsymbol{v} - \boldsymbol{u}$ Euler angles of this rotation.

d) Calculate the $\boldsymbol{w} - \boldsymbol{v} - \boldsymbol{w}$ Euler angles of this rotation.

e) Calculate the $\boldsymbol{w} - \boldsymbol{u} - \boldsymbol{w}$ Euler angles of this rotation.

2.9 The rotational motion of a rigid body is represented by the following rotation matrix. It seems that there are errors in recording the elements of this matrix. Correct it using the properties of rotation matrices. Then, repeat Problem 2.7 for the corrected rotation matrix.

$$^A\boldsymbol{R}_B = \begin{bmatrix} 0.099 & -0.370 & 0.924 \\ 0.984 & -0.112 & -0.146 \\ 0.149 & 0.924 & 0.354 \end{bmatrix}$$

2.10 Repeat Problem 2.7 for the following rotation matrix.

$$^A\boldsymbol{R}_B = \begin{bmatrix} 0.933 & 0.067 & 0.354 \\ 0.067 & 0.933 & -0.354 \\ -0.354 & 0.354 & 0.866 \end{bmatrix}$$

2.11 Reiterate the exercise in Problem 2.8 using the rotation matrix from Problem 2.10.

2.12 Duplicate the task outlined in Problem 2.7 for the given rotation matrix.

$$^A\boldsymbol{R}_B = \begin{bmatrix} 0 & 0.259 & 0.966 \\ 0 & 0.966 & -0.250 \\ -1 & 0 & 0 \end{bmatrix}$$

2.13 Redo the exercise in Problem 2.8 using the rotation matrix provided in Problem 2.12.

2.14 Reiterate the task described in Problem 2.7 for the provided rotation matrix.

$$^A\boldsymbol{R}_B = \begin{bmatrix} -0.800 & -0.498 & 0.335 \\ 0.462 & 0.155 & 0.873 \\ -0.383 & 0.654 & 0.354 \end{bmatrix}$$

2.15 Repeat the activity described in Problem 2.8 with the rotation matrix given in Problem 2.14.

2.16 In this exercise, we want to prove that the rotation matrix is an orthogonal matrix, and the inverse of such matrix is equal to its transpose. For this purpose, consider two unit vectors, $\hat{\boldsymbol{s}}$ and $\hat{\boldsymbol{r}}$, located on a rigid body. Since the angle between these two vectors remains constant for any arbitrary rotation, provide a proof for the above mentioned properties of the rotation matrix.

2.17 Assume a rotational motion of a rigid body is represented by the following pitch-roll-yaw Euler angles.

$$\boldsymbol{R}_x(\pi/6), \boldsymbol{R}_y(\pi/4), \boldsymbol{R}_z(\pi/4)$$

a) Find the rotation matrix ${}^A\boldsymbol{R}_B$ representing the motion of this body.

b) Find the eigenvalues and eigenvectors of the rotation matrix ${}^A\boldsymbol{R}_B$.

c) Calculate the $\boldsymbol{w}-\boldsymbol{v}-\boldsymbol{u}$ Euler angles of this rotation.

d) Find the $\boldsymbol{w}-\boldsymbol{v}-\boldsymbol{w}$ Euler angles of this rotation.

e) Calculate the $\boldsymbol{w}-\boldsymbol{u}-\boldsymbol{w}$ Euler angles of this rotation.

2.18 Assume that the rotation of a rigid body is given with the following angles relative to the following fixed coordinate system:

$$\boldsymbol{R}_z(\alpha), \boldsymbol{R}_y(\beta), \boldsymbol{R}_z(\gamma).$$

a) Find the rotation matrix ${}^A\boldsymbol{R}_B$ representing the motion of this body.

b) Evaluate the rotation matrix for $\alpha = \pi/2, \beta = -\pi/4, \gamma = \pi/6$.

c) Find the eigenvalues and eigenvectors of the rotation matrix ${}^A\boldsymbol{R}_B$.

d) Calculate the pitch-roll-yaw Euler angles of this rotation.

e) Calculate the $\boldsymbol{w}-\boldsymbol{u}-\boldsymbol{w}$ Euler angles of this rotation.

f) Find the $\boldsymbol{w}-\boldsymbol{v}-\boldsymbol{w}$ Euler angles of this rotation.

2.19 Assume that the rotation of a rigid body is given with constant angles relative to the following moving coordinate system:

$$\boldsymbol{R}_v(\alpha), \boldsymbol{R}_w(\beta), \boldsymbol{R}_u(\gamma).$$

a) Find the rotation matrix ${}^A\boldsymbol{R}_B$ representing the motion of this body.

b) Evaluate the rotation matrix for $\alpha = \pi/4, \beta = -\pi/2, \gamma = \pi/3$.

c) Find the eigenvalues and eigenvectors of the rotation matrix ${}^A\boldsymbol{R}_B$.

d) Calculate the pitch-roll-yaw Euler angles of this rotation.

e) Find the $\boldsymbol{w}-\boldsymbol{v}-\boldsymbol{u}$ Euler angles of this rotation.

f) Calculate the $\boldsymbol{w}-\boldsymbol{v}-\boldsymbol{w}$ Euler angles of this rotation.

2.20 Consider the coordinate systems shown in Figure 2.12 and obtain the homogeneous transformation matrices ${}^A\boldsymbol{T}_B$, ${}^A\boldsymbol{T}_C$, ${}^B\boldsymbol{T}_C$ for this configuration.

2.21 Consider the coordinate systems shown in Figure 2.12 and obtain the transformation matrices ${}^B\boldsymbol{T}_A$, ${}^C\boldsymbol{T}_A$, ${}^C\boldsymbol{T}_B$ using the definition. Then, validate your calculations using the inverse transformation matrix relationships obtained in Problem 2.20 and with the help of Equation 2.56.

2.22 Repeat Problem 2.20 for the coordinate systems shown in Figure 2.13.

2.23 Redo the task described in Problem 2.21 for the coordinate systems illustrated in Figure 2.13.

2.24 Duplicate the exercise outlined in Problem 2.20 for the coordinate systems depicted in Figure 2.14.

2.25 Reiterate the task presented in Problem 2.21 for the coordinate systems shown in Figure 2.14.

2.26 If the homogeneous transformation matrix ${}^A\boldsymbol{T}_B$ is given, obtain the matrix ${}^B\boldsymbol{T}_A$.

$$
{}^A\boldsymbol{T}_B = \left[\begin{array}{ccc|c} 0.250 & 0.612 & 0.750 & 0.422 \\ 0.612 & 0.500 & -0.612 & 1.673 \\ -0.750 & 0.612 & -0.250 & -0.172 \\ \hline 0 & 0 & 0 & 1 \end{array}\right]
$$

2.27 Consider the coordinate systems shown in Figure 2.12 and assume that the vectors ${}^C\boldsymbol{P}$ and ${}^C\boldsymbol{\omega}$ are represented as follows:

$$
{}^C\boldsymbol{P} = \left[\begin{array}{ccc} 1 & -1 & 1 \end{array}\right]^T \; ; \; {}^C\boldsymbol{\omega} = \left[\begin{array}{ccc} 1 & 1.75 & -1.2 \end{array}\right]^T
$$

Using the consecutive homogeneous transformation matrices, obtain the values of ${}^B\boldsymbol{P}, {}^B\boldsymbol{\omega}, {}^A\boldsymbol{P}, {}^A\boldsymbol{\omega}$. Then, directly calculate the values of ${}^A\boldsymbol{P}, {}^A\boldsymbol{\omega}$ from the homogeneous matrix ${}^A\boldsymbol{T}_C$ and confirm the accuracy of your calculations by comparing the results of the two methods.

2.28 Repeat Problem 2.27 for the coordinate systems shown in Figure 2.13.

2.29 Redo the task outlined in Problem 2.27 for the coordinate systems depicted in Figure 2.14.

2.30 Consider the coordinate systems shown in Figure 2.12 and assume that the vectors ${}^A\boldsymbol{P}$ and ${}^A\boldsymbol{\omega}$ are represented as follows:

$$
{}^A\boldsymbol{P} = \left[\begin{array}{ccc} 1 & 0 & -1 \end{array}\right]^T \; ; \; {}^A\boldsymbol{\omega} = \left[\begin{array}{ccc} 1 & -1.25 & -2.15 \end{array}\right]^T
$$

Using the consecutive homogeneous transformation matrices, obtain the values of ${}^B\boldsymbol{P}, {}^B\boldsymbol{\omega}, {}^C\boldsymbol{P}, {}^C\boldsymbol{\omega}$. Subsequently, compute the values of ${}^C\boldsymbol{P}$ and ${}^C\boldsymbol{\omega}$ directly from the homogeneous matrix ${}^C\boldsymbol{T}_A$. Validate your calculations by comparing the results obtained through the two methods.

2.31 Repeat Problem 2.30 for the coordinate systems shown in Figure 2.13.

2.32 Reiterate the task presented in Problem 2.30 for the coordinate systems depicted in Figure 2.14.

3

Advanced Representations

The chapter explores the advanced methods for representing spatial motion through the lens of screw theory. The chapter covers an introduction to "screw theory," providing a mathematical framework for representing rigid body motions. This theory proves beneficial in articulating displacements, velocities, forces, and torques in three-dimensional space, rooted in the idea that rigid body motion combines rotation about a specific axis and translation along that axis.

Furthermore, this chapter addresses the representation of orientation, including screw and quaternion, and the application of screw representation in describing the motion of rigid bodies. A detailed examination of linear and angular velocities is presented next, covering aspects such as angular velocity of a rigid body, angular velocity and its relation to the rate of rotation matrix, rate of Euler angles, quaternion rates, and furthermore, linear velocity of a point, and screw coordinates.

The discussion extends to linear and angular accelerations, encompassing topics like angular acceleration of a rigid body and linear acceleration of a point. In essence, this chapter provides a comprehensive foundation in advanced representations, shedding light on the intricacies of motion, orientation, velocity, and acceleration within the context of rigid body dynamics.

3.1 Introduction to Screw Theory

The exploration of the kinematics and dynamics of a rigid body takes on a new dimension when approached through the lens of screw theory. Within this framework, the principle of transference emerges as a conspicuous outcome, as highlighted by McCarthy and Soh in [40]. This principle yields notable advantages, particularly in the establishment of an analogy between geometric computations for points utilizing vectors and the analogous calculations for lines achieved through the substitution of vectors with screws.

An inherent advantage of screw theory lies in its capacity to articulate displacements, velocities, forces, and torques in three-dimensional space. This capability is rooted in the foundational idea that any motion of a rigid body can be elegantly expressed as the inseparable fusion of a about a specific axis and a translation along that very axis. This axis coincides with the object or particle undergoing displacement, providing a unifying and intuitive framework for understanding rigid body general motion. The following results in different wordings are among the highlights of such representation of rigid body motion.

> Any motion of a rigid body can be achieved by combining a rotation around a specific axis with a parallel translation along that same axis.
>
> The most general motion of a rigid body can be accomplished through a translation along a line, either followed or preceded by a rotation around that particular line.

DOI: 10.1201/9781003491415-3

In the Euclidean space, the displacement of a rigid body can be effectively described as a rotation around a distinct axis coupled with a parallel translation along that axis.

While Mozzi and Cauchy receive credit for attaining comparable results, the recognition of the renowned theorem is specifically attributed to Michel Chasles in numerous books. To appropriately acknowledge both scientists, we denote the theorem as Mozzi–Chasles' theorem in this book.

Theorem 3.1 (Mozzi–Chasles' Theorem (1830)) *The most general displacement for a rigid body can be generated by executing a translation along a line, known as its screw axis or Mozzi axis, either followed or preceded by a rotation about an axis parallel to that specific line.*

The decomposition of a general motion into a pure rotation and a translation predates Mozzi and Chasles. However, it seems that Mozzi was the first to conceptualize that both the rotation and the translation occurring along the same line [41]. This line, initially referred to as "asse di Mozzi" in Italian, or the Mozzi axis in early references, is now adopted as the "screw axis" in this book, aligning with the terminology widely recognized in the literature. Even today, after about two hundred years later, the notion that the translation occurs along a line while the rotation is about the same line remains somewhat challenging to grasp.

In his contribution Mozzi certainly describes the general motion of a rigid body based on the concept of screw axis. Several decades later, Giorgini proved that although the results reported by Mozzi were in general accurate, the procedures Mozzi used had been partially incorrect. On the contrary, Chasles accurately states the theorem and provided a rigorous proof for it in [42]. Another important contribution to this field was given by Plücker in 1866. Plücker or line coordinates, are a way to assign six homogeneous coordinates to each line in projective 3D-space denoted by $\mathbb{P}^3$. This representation proves to be very useful for the screws, twists and wrenches in the theory of kinematics used for robotics. However, despite these pioneering contributions the framework of the theory of screws is credited to Ball based on his book "A treatise on the theory of screws" [43].

As briefly outlined, screw theory boasts a rich history spanning more than two centuries. Presently, it stands as an integral component of the theoretical framework employed in robotics research, playing a pivotal role in the evolution of sophisticated robotic systems capable of intricate and precise movements. This book adeptly utilizes this compelling concept for the representation of motion, kinematics analysis, and the control of robot manipulators. However, a comprehensive exploration of the concepts and methodologies within screw theory necessitates a dedicated investment of time and energy, surpassing the scope of this book [44, 45, 46]. In this book, fundamental concepts and crucial methodologies essential for the analysis and design of serial robots are presented.

3.2 Orientation Representation

In the preceding chapter, the definition and characteristics of the orientation of a rigid body is introduced. The representation of orientation using a rotation matrix and Euler angles are also discussed. Although the rotation matrix is preferred for its unique features, it requires nine parameters to represent the orientation. On the other hand, Euler angles only need

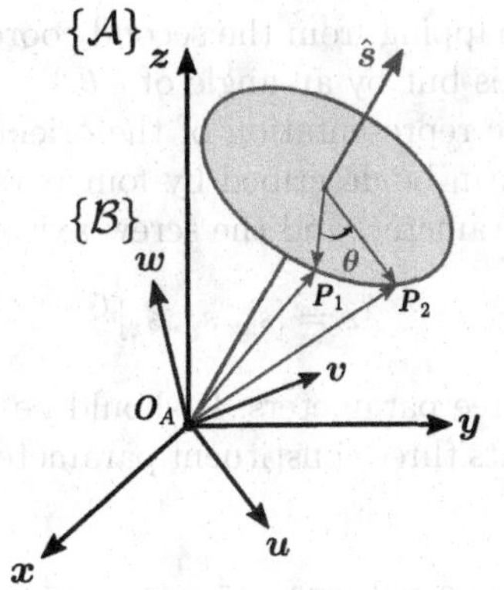

FIGURE 3.1
The pure rotational motion of a rigid body represented about a screw axis.

three parameters, but they are susceptible to gimbal lock situations and suffer from the regeneracy of their inverse problem solution.

In this chapter, we will explore two other well-known representations for orientation. The first representation employs the theory of screws to uniquely represent the orientation of a rigid body with only four parameters. The second representation also utilizes four parameters but employs the concept of unit quaternions to represent the motion. Subsequent sections will provide detailed explanations of these representations.

3.2.1 Screw Representation

As expressed in the properties of the rotation matrix, the eigenvalues of the rotation matrix are 1 and $e^{\pm i\theta}$, where θ is one of the intrinsic properties of the rotation matrix that indirectly represents the amount of rotation of the rigid body. Moreover, the eigenvector corresponding to the real eigenvalue $\lambda = 1$ of the rotation matrix is also a vector with real coefficients that specifies a defined direction in three-dimensional space. This direction can be introduced as the instantaneous axis of rotation of the rigid body, as shown in Figure 3.1.

Consequently, instead of interpreting the rotation of a rigid body by a matrix or about three (fixed or moving) axes, the rotation of the body at any moment can be interpreted about an instantaneous axis with an equivalent rotation angle. This representation of rotation is referred to in the robotics literature as the *equivalent angle and axis of rotation*, while the instantaneous axis of rotation is recognized as the *screw axis*. The use of the screw axis to represent the pure rotation of a rigid body has the feature that it can be generalized to the general motion of a rigid body, which includes translational and rotational motion, using the theory of screws. This will be discussed later in Section 3.3.

The description of rotation using the equivalent angle and axes of rotation takes advantage of the physical concept of the rotation of a rigid body through a rotation matrix. If we consider the rotation matrix as a mapping of a vector from the reference coordinate system $\{\mathcal{B}\}$ to the moving coordinate system $\{\mathcal{A}\}$, rotation only involves a change in the viewing angle in this coordinate transformation and does not introduce a change in its magnitude. This property is manifested in the unit magnitude of all eigenvectors of the rotation matrix and its unit determinant. It can be demonstrated that the instantaneous axis of rotation or the screw axis is another intrinsic characteristic of rotation. For example, if we consider a rotation from one coordinate system to another around this instantaneous axis, and by an angle θ, the inverse rotation mapping from the second coordinate system to the first is also expressible along this same axis but by an angle of $-\theta$.

With this introduction, the representation of the orientation of a rigid body using the rotation angle the screw axis can be described by four parameters. In this description, the rotation angle θ is the first parameter, and the screw axis, represented by the unit vector

$$^{A}\hat{\boldsymbol{s}} = [s_x, s_y, s_z]^T$$

encompasses the remaining three parameters. It should be noted, however, that due to the unit magnitude of this vector, its three constituent parameters satisfy the following algebraic constraint:

$$\hat{\boldsymbol{s}} \cdot \hat{\boldsymbol{s}} = 1 \quad \text{or} \quad s_x^2 + s_y^2 + s_z^2 = 1 \tag{3.1}$$

This constraint arises from the unit length requirement of the screw axis vector. Therefore, this representation can be expressed with the following four parameters placed in an array.

$$\{\theta, s_x, s_y, s_z\} \quad \text{subject to} \quad s_x^2 + s_y^2 + s_z^2 = 1 \tag{3.2}$$

Now, if the vector symbol $\boldsymbol{\theta}$ is used as the "screw vector" in the following form, this representation can also be interpreted with the following three parameters.

$$\boldsymbol{\theta} = \begin{bmatrix} \theta_x \\ \theta_y \\ \theta_z \end{bmatrix}, \quad \theta_x = \theta s_x, \quad \theta_y = \theta s_y, \quad \theta_z = \theta s_z \tag{3.3}$$

If a rotation matrix describing the orientation of a rigid body is given, it is straightforward to obtain the four parameters of the screw representation by calculating its eigenvalues and the corresponding unit eigenvector. To calculate the rotation angle θ, it is sufficient to use the relationship given in Equation 2.18,

Although calculating the screw axis can be easily done by finding the eigenvector of the rotation matrix, we will use the "Rodrigues' Rotation Formula" here to obtain the analytical expression for the screw axis. In this formula, the relationship between the vector $^{A}\boldsymbol{P}$ in the reference coordinate system $\{\mathcal{A}\}$ and the vector $^{B}\boldsymbol{P}$ in the rotated coordinate system $\{\mathcal{B}\}$ is found by multiplying it to the vector of the equivalent rotation axis $\hat{\boldsymbol{s}}$ and the trigonometric functions of the rotation angle θ as follows [47].

$$^{A}\boldsymbol{P} = (\cos\theta)\,^{B}\boldsymbol{P} + (\sin\theta)(\hat{\boldsymbol{s}} \times {}^{B}\boldsymbol{P}) + (1 - \cos\theta)(^{B}\boldsymbol{P} \cdot \hat{\boldsymbol{s}})\hat{\boldsymbol{s}} \tag{3.4}$$

If we represent this relationship in a conventional matrix form,

$$^{A}\boldsymbol{P} = {}^{A}\boldsymbol{R}_B\,^{B}\boldsymbol{P} \tag{3.5}$$

The rotation matrix in terms of the rotation angle θ and the screw axis $^{A}\hat{\boldsymbol{s}} = [s_x, s_y, s_z]^T$ is determined by the following equation [48].

$$^{A}\boldsymbol{R}_B = \begin{bmatrix} s_x^2 v\theta + c\theta & s_x s_y v\theta - s_z s\theta & s_x s_z v\theta + s_y s\theta \\ s_y s_x v\theta + s_z s\theta & s_y^2 v\theta + c\theta & s_y s_z v\theta - s_x s\theta \\ s_z s_x v\theta - s_y s\theta & s_z s_y v\theta + s_x s\theta & s_z^2 v\theta + c\theta \end{bmatrix} \tag{3.6}$$

In which notations $s\theta = \sin\theta$, $c\theta = \cos\theta$, and $v\theta = 1 - \cos\theta$ have been used. Equation 3.6 interprets the rotation of a rigid body with the parameters of the rotation angle and the screw axis. If the rotation angle and the screw axis are given, the rotation matrix is easily and uniquely calculated from this equation. The inverse problem is also not as challenging as the Euler angles. If the rotation matrix is given, the rotation angle is determined using Equation 2.18, which is restated here for reference.

$$\theta = \cos^{-1} \frac{\operatorname{tr}\left({}^{A}\boldsymbol{R}_{B}\right) - 1}{2} \tag{3.7}$$

If the rotation angle is not zero, the components of the screw axis can also be easily calculated using the following equation.

$$\begin{aligned} s_x &= \frac{r_{32} - r_{23}}{2s\theta} \\ s_y &= \frac{r_{13} - r_{31}}{2s\theta} \\ s_z &= \frac{r_{21} - r_{12}}{2s\theta} \end{aligned} \tag{3.8}$$

From Equations 3.7 and 3.8, it can be understood that the rotation of a body can be represented by two sets, namely $\theta, \hat{s}$ and $-\theta, -\hat{s}$, which have the same physical nature represented by the unique screw vector $\boldsymbol{\theta}$. Furthermore, if the rotation angle becomes zero, it is natural that there is no rotation axis, and Equation 3.8 cannot be used. However, if the rotation angle becomes $\pm\pi$ radians, a singular case occurs, and it is not possible to obtain the rotation axis from Equation 3.8. Similar to gimbal lock situation in Euler angle representation, in this case, the rotation matrix simplifies, and the components of the screw vector can be obtained using the diagonal elements of the matrix with the following equation.

$$s_x^2 = 1 + r_{11} \ , \quad s_y^2 = 1 + r_{22} \ , \quad s_z^2 = 1 + r_{33} \tag{3.9}$$

Example 3.1 (Screw Representation) *Assume, similar to what we had in Example 2.1, the moving coordinate system $\{\mathcal{B}\}$ attached to a rigid body rotates only around one of the principal axes of the reference coordinate system $\{\mathcal{A}\}$ with the following angles, as illustrated in Figure 2.4:*

A) Rotation around the x-axis by an angle α.

B) Rotation around the y-axis by an angle β.

C) Rotation around the z-axis by an angle γ.

Represent these rotations using the screw vector $\boldsymbol{\theta}$.

Solution: The screw parameters in these three cases can be obtained using Equation 3.7 and 3.8 in the following form.

A) For the first case, we have:

$$\begin{aligned} \theta &= \cos^{-1}((1 + 2\cos\alpha - 1)/2) = \alpha \\ s_x &= (s\alpha - (-s\alpha))/2s\theta = 1 \\ s_y &= s_z = 0 \end{aligned}$$

Therefore, as expected,

$$\boldsymbol{\theta}_A = \alpha \begin{bmatrix} 1 & 0 & 0 \end{bmatrix}^T$$

B) Similarly, for the second case, we can derive:

$$\boldsymbol{\theta}_B = \beta \begin{bmatrix} 0 & 1 & 0 \end{bmatrix}^T$$

C) And for the third case:

$$\boldsymbol{\theta}_C = \gamma \begin{bmatrix} 0 & 0 & 1 \end{bmatrix}^T$$

The correctness of the calculated results in this example can easily be verified and confirmed with the intuitive understanding of the rotation angle and screw axis.

Example 3.2 *Consider the rotation of a rigid body around the axis* $\hat{\boldsymbol{s}} = \frac{1}{\sqrt{3}}\begin{bmatrix} 1 & 1 & 1 \end{bmatrix}^T$ *with an angle* $\theta = \pi/6$. *Find the rotation matrix in this case.*

Solution: The rotation matrix in this case is obtained using Equation 3.6 as follows.

$$^{A}\boldsymbol{R}_B = \begin{bmatrix} 0.911 & -0.244 & 0.333 \\ 0.333 & 0.911 & -0.244 \\ -0.244 & 0.333 & 0.911 \end{bmatrix}$$

To validate the calculations, we utilize the eigenvalues and real eigenvectors of the rotation matrix. The eigenvalues of the rotation matrix $^{A}\boldsymbol{R}_B$ are $\lambda_{1,2,3} = 1, e^{\pm i\pi/6}$, and its real eigenvector is $\boldsymbol{v}_1 = \begin{bmatrix} 0.577 & 0.577 & 0.577 \end{bmatrix}^T$ which precisely corresponds to the screw axis. Thus, the accuracy of the calculations is verified.

Example 3.3 *Consider the rotation matrix of Example 2.2, which is in the form of*

$$^{A}\boldsymbol{R}_B = \begin{bmatrix} 0.933 & -0.067 & -0.354 \\ -0.067 & 0.933 & -0.354 \\ 0.354 & 0.354 & 0.866 \end{bmatrix}$$

Obtain the screw representation of this rotation.

Solution: The rotation of this matrix, using Equation 3.7, is equal to $\theta = \pi/6$. Additionally, the components of the equivalent rotation axis, using Equation 3.8, are equal to:

$$\hat{\boldsymbol{s}} = \begin{bmatrix} 0.707 & -0.707 & 0 \end{bmatrix}^T$$

Thus, the screw vector in this rotation is equal to:

$$\boldsymbol{\theta} = \begin{bmatrix} 0.37 & -0.37 & 0 \end{bmatrix}^T.$$

3.2.2 Quaternion

One of the applications of complex numbers is the representation of planar rotations on a unit circle. The unit quaternion, by extending this concept and using complex numbers, can similarly represent rotations in three-dimensional space. Quaternions were first introduced by Hamilton in 1843. Hamilton's primary motivation was to find an algebraic representation for points (x, y, z) in three-dimensional space through extending complex numbers. Initially, he sought a three-dimensional hyper-complex system with suitable properties [49]. For over a decade, Hamilton attempted to extend this concept by defining a triad consisting of a real part and two imaginary parts. Eventually, he realized that, for this purpose, he needed three imaginary units with the following properties instead of two [50].

$$i^2 = j^2 = k^2 = -1 \tag{3.10}$$

$$ij = k \qquad ji = -k \tag{3.11}$$

$$i \to j \to k \to i \tag{3.12}$$

In which Equation (3.12) describes a rotational cyclic permutation. Hamilton referred to the resulting four-element structure as a Quaternion.

$$Q = q_1 i + q_2 j + q_3 k + q_4 \tag{3.13}$$

In the representation of Quaternion, unlike Euler angle representation, four numbers are used instead of three to represent rotation. The use of Quaternion in this context is one of the most common methods for representing the orientation of a rigid body in space.

Assume that the rotation of a rigid body is described by four parameters introduced in the screw representation by the rotation angle and screw axis $\{s_x,\ s_y,\ s_z,\ \theta\}$. With this assumption, by using the quaternion definition given in Equation 3.13, the *Euler Parameters* are defined by the following four parameters.

$$\epsilon_1 = s_x \sin(\theta/2); \quad \epsilon_2 = s_y \sin(\theta/2); \quad \epsilon_3 = s_z \sin(\theta/2); \quad \epsilon_4 = \cos(\theta/2) \tag{3.14}$$

If the Euler parameters are represented in a 4×1 array as follows,

$$\boldsymbol{\epsilon} = \begin{bmatrix} \epsilon_1 & \epsilon_2 & \epsilon_3 & \epsilon_4 \end{bmatrix}^T \tag{3.15}$$

This array forms a unit quaternion because its Euclidean norm is equal to one.

$$\|\boldsymbol{\epsilon}\|_2^2 = \epsilon_1^2 + \epsilon_2^2 + \epsilon_3^2 + \epsilon_4^2 = 1 \tag{3.16}$$

The rotation matrix can be calculated using the components of the unit quaternion as follows [51].

$$^{A}\boldsymbol{R}_B = \begin{bmatrix} 1 - 2\epsilon_2^2 - 2\epsilon_3^2 & 2(\epsilon_1\epsilon_2 - \epsilon_3\epsilon_4) & 2(\epsilon_1\epsilon_3 + \epsilon_2\epsilon_4) \\ 2(\epsilon_1\epsilon_2 + \epsilon_3\epsilon_4) & 1 - 2\epsilon_1^2 - 2\epsilon_3^2 & 2(\epsilon_2\epsilon_3 - \epsilon_1\epsilon_4) \\ 2(\epsilon_1\epsilon_3 - \epsilon_2\epsilon_4) & 2(\epsilon_2\epsilon_3 + \epsilon_1\epsilon_4) & 1 - 2\epsilon_1^2 - 2\epsilon_2^2 \end{bmatrix} \tag{3.17}$$

Furthermore, it can be shown that by solving the inverse problem, if the rotation matrix is given, the components of the unit quaternion can be determined as follows.

$$\begin{aligned} \epsilon_1 &= \frac{(r_{32} - r_{23})}{4\epsilon_4} \\ \epsilon_2 &= \frac{(r_{13} - r_{31})}{4\epsilon_4} \\ \epsilon_3 &= \frac{(r_{21} - r_{12})}{4\epsilon_4} \\ \epsilon_4 &= \frac{1}{2}\sqrt{1 + r_{11} + r_{22} + r_{33}} \end{aligned} \tag{3.18}$$

Although the unit quaternion representation does not directly inherit the physical concept of rotation as seen in the screw representation, it is less singular compared to it. In a sense, if the rotation angle is zero, the unit quaternion will be as expected, equal to $\boldsymbol{\epsilon} = \begin{bmatrix} 0 & 0 & 0 & 1 \end{bmatrix}^T$. The singular case in quaternion only occurs when the rotation angle is $\pm\pi$, a situation that is avoided in quaternion-based measurement systems.

Example 3.4 (Rotation around principal axes) *Represent the rotation around the principal axes, as introduced in Example 2.1, and further analyzed in Example 3.1, using quaternions.*

Solution: The representation of quaternions in three rotation cases around the principal axes can be obtained using Equation 3.14 as follows.

a) For the first case, we have,

$$\epsilon = \begin{bmatrix} \sin(\alpha/2) & 0 & 0 & \cos(\alpha/2) \end{bmatrix}^T \tag{3.19}$$

b) Similarly, for the second case, we can write,

$$\epsilon = \begin{bmatrix} 0 & \sin(\beta/2) & 0 & \cos(\beta/2) \end{bmatrix}^T \tag{3.20}$$

c) Finally, for the third case:

$$\epsilon = \begin{bmatrix} 0 & 0 & \sin(\gamma/2) & \cos(\gamma/2) \end{bmatrix}^T \tag{3.21}$$

Example 3.5 *Consider the rotation of a rigid body using the quaternion representation below.*

$$\epsilon = \begin{bmatrix} 0.149 & 0.149 & 0.149 & 0.966 \end{bmatrix}^T \tag{3.22}$$

Find the rotation matrix in this case.

Solution: The rotation matrix in this case is calculated as:

$$^A\boldsymbol{R}_B = \begin{bmatrix} 0.911 & -0.244 & 0.333 \\ 0.333 & 0.911 & -0.244 \\ -0.244 & 0.333 & 0.911 \end{bmatrix}$$

It is observed that this rotation matrix is equal to what has been calculated in Example 3.2. To validate the calculations, compare the quaternion in this example with the screw representation in Example 3.2. Evaluating these two representations reveals that both depict the same rotation angle and the direction of the screw axis, confirming the accuracy of the calculations.

Example 3.6 *Consider the rotation matrix given in Example 2.2. Find the quaternion representation of this rotation.*

$$^A\boldsymbol{R}_B = \begin{bmatrix} 0.933 & -0.067 & -0.354 \\ -0.067 & 0.933 & -0.354 \\ 0.354 & 0.354 & 0.866 \end{bmatrix}$$

Solution: The quaternion corresponding to this matrix can be obtained using the Equation 3.18 as follows.

$$\epsilon = \begin{bmatrix} -0.183 & 0.183 & 0 & 0.966 \end{bmatrix}^T$$

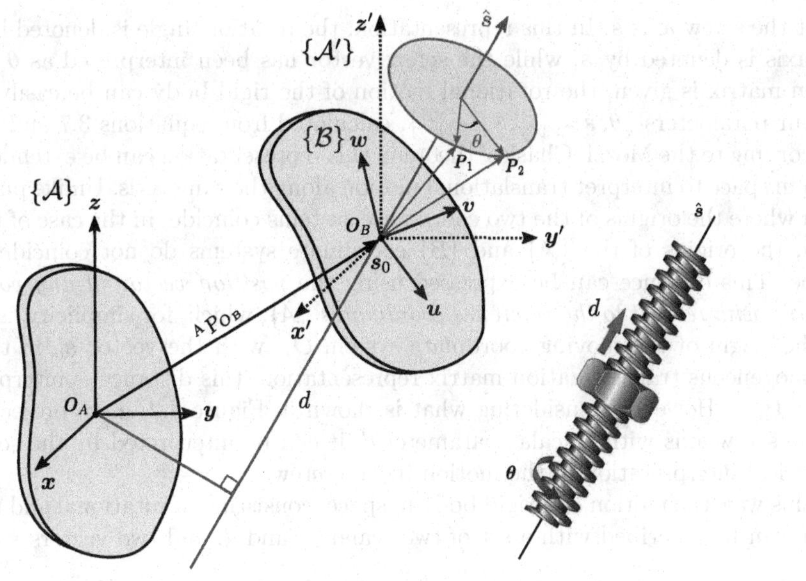

FIGURE 3.2
The general motion of a rigid body represented by a screw.

3.3 Screw Representation for Motion

As explained earlier, the Mozzi–Chasles's theorem states that the general motion of a rigid body can be obtained by combining its instantaneous rotational motion about an axis and its translational motion along the same axis. If we accept this claim, given that the direction of translational motion is aligned with the rotational axis, the use of the term "Screw" becomes more natural. This is because, as seen in Figure 3.2, considering the motion of a screw, depending on the pitch of the screw, the rotational motion of the screw around its axis induces translational motion along the same axis on the nut.

The first part of this theorem, which deals with the pure rotational motion of a body, is discussed in Section 3.2.1 with the introduction of the rotation angle and the screw axis. However, the second part of this theorem, which interprets translational motion along the same axis, is by no means intuitive and requires further investigation, which we address in this section.

Consider the general motion of a rigid body in space as shown in Figure 3.2. As observed in this figure, the rigid body has undergone a translational motion represented by the vector ${}^{A}\boldsymbol{P}_{O_B}$ and a pure rotational motion described by the angle θ and the screw axis $\hat{\boldsymbol{s}}$, starting from its initial state represented by the coordinate system $\{\mathcal{A}\}$. The final state of the rigid body is depicted with the coordinate system $\{\mathcal{B}\}$, and an intermediate coordinate system $\{\mathcal{A}'\}$, shown with the dashed line, is used to separate translational and rotational motions from each other. As seen in the figure, the origin of the intermediate coordinate system which is parallel to $\{\mathcal{A}\}$, coincides with the origin of the $\{\mathcal{B}\}$ coordinate system. Thus, this coordinate system only represents the translational motion of the body along the vector ${}^{A}\boldsymbol{P}_{O_B}$.

First, consider only the rotational motion of the rigid body, which is described by two coordinate systems $\{\mathcal{A}'\}$ and $\{\mathcal{B}\}$, respectively. As detailed in Section 3.2.1 and depicted in Figure 3.1, pure rotational motion of a rigid body can be represented by the rotation angle

θ about the screw axis $\hat{\boldsymbol{s}}$. In this representation, the rotation angle is denoted by θ, and the screw axis is denoted by $\hat{\boldsymbol{s}}$, while the screw vector has been interpreted as $\boldsymbol{\theta} = \theta\hat{\boldsymbol{s}}$. If the rotation matrix is given, the rotational motion of the rigid body can be easily represented with four parameters $\{\theta, \hat{\boldsymbol{s}} = [s_x, s_y, s_z]^T\}$, calculated from equations 3.7 and 3.8.

According to the Mozzi–Chasles' theorem, this representation can be extended to general motion in space to interpret translational motion along the same axis. Unlike pure rotational motion where the origins of the two coordinate systems coincide, in the case of translational motion, the origins of the $\{\mathcal{A}\}$ and $\{\mathcal{B}\}$ coordinate systems do not coincide and have a distance. This distance can be expressed using the *position vector of any point from the screw axis with respect to the reference coordinates* $\{\mathcal{A}\}$, which, for simplicity, is represented from the origin of the moving coordinate system O_B with the vector $\boldsymbol{s}_o$ in the figure. In the homogeneous transformation matrix representation, this distance is interpreted by the vector ${}^A\boldsymbol{P}_{O_B}$. However, considering what is shown in Figure 3.2, if we project this vector onto the screw axis with a scalar parameter d, it can be interpreted by the screw pitch as the physical interpretation of the motion by the screw.

In this way, the motion of a rigid body in space, consisting of rotational and translational motion, can be described with a set of two scalars θ and d, and two vectors $\hat{\boldsymbol{s}}$ and $\boldsymbol{s}_o$.

$$\{\theta\ ,\ \hat{\boldsymbol{s}} = [s_x, s_y, s_z]^T \quad ; \quad d\ ,\ \boldsymbol{s}_o = [s_{o_x}, s_{o_y}, s_{o_z}]^T\} \tag{3.23}$$

Despite the use of eight parameters in this representation, considering the two algebraic constraints:

$$\hat{\boldsymbol{s}}^T\hat{\boldsymbol{s}} = 1 \quad ; \quad \boldsymbol{s}_o^T\hat{\boldsymbol{s}} = d \tag{3.24}$$

as expected, six independent parameters are sufficient to represent motion in three-dimensional space.

If the homogeneous transformation matrix is given, one can easily obtain all eight parameters of the screw representation. The rotation parameters θ and $\hat{\boldsymbol{s}}$ can be calculated using the relationships 3.7 and 3.8, which are reiterated here for reference.

$$\begin{aligned} \theta &= \cos^{-1}\frac{\operatorname{tr}\left({}^A\boldsymbol{R}_B\right) - 1}{2} \\ s_x &= \frac{r_{32} - r_{23}}{2s\theta} \\ s_y &= \frac{r_{13} - r_{31}}{2s\theta} \\ s_z &= \frac{r_{21} - r_{12}}{2s\theta} \end{aligned} \tag{3.25}$$

The translational motion parameters can also be easily obtained using the following relationship.

$$\boldsymbol{s}_o = {}^A\boldsymbol{P}_{O_B} \quad ; \quad d = \boldsymbol{s}_o^T\hat{\boldsymbol{s}} \tag{3.26}$$

The solution to the inverse problem can also be presented by performing the following calculations [3]. If the eight parameters represent the screw representation are given,

$$\{\theta, \hat{\boldsymbol{s}} = [s_x, s_y, s_z]^T \quad ; \quad d, \boldsymbol{s}_o = [s_{o_x}, s_{o_y}, s_{o_z}]^T\} \tag{3.27}$$

then the homogeneous transformation matrix is obtained using the following relationship.

$${}^A\boldsymbol{T}_B = \left[\begin{array}{ccc|c} s_x^2 v\theta + c\theta & s_x s_y v\theta - s_z s\theta & s_x s_z v\theta + s_y s\theta & p_x \\ s_y s_x v\theta + s_z s\theta & s_y^2 v\theta + c\theta & s_y s_z v\theta - s_x s\theta & p_y \\ s_z s_x v\theta - s_y s\theta & s_z s_y v\theta + s_x s\theta & s_z^2 v\theta + c\theta & p_z \\ \hline 0 & 0 & 0 & 1 \end{array}\right] \tag{3.28}$$

which in this case

$$\begin{aligned}
p_x &= d\, s_x - s_{o_x}(s_x^2 - 1)v\theta - s_{o_y}(s_x s_y v\theta - s_z s\theta) - s_{o_z}(s_x s_z v\theta + s_y s\theta) \\
p_y &= d\, s_y - s_{o_x}(s_y s_x v\theta + s_z s\theta) - s_{o_y}(s_y^2 - 1)v\theta - s_{o_z}(s_y s_z v\theta - s_x s\theta) \\
p_z &= d\, s_z - s_{o_x}(s_z s_x v\theta - s_y s\theta) - s_{o_y}(s_z s_y v\theta + s_x s\theta) - s_{o_z}(s_z^2 - 1)v\theta
\end{aligned} \tag{3.29}$$

Similarly, here, the shorthand notations $s\theta = \sin\theta$, $c\theta = \cos\theta$, and $v\theta = 1 - \cos\theta$ have been used.

Example 3.7 (Motion Representation with Screw) *Assume that the motion of a rigid body is represented by the following homogeneous transformation matrix. Obtain the parameters of this motion based on the screw representation.*

$$^{A}\boldsymbol{T}_B = \left[\begin{array}{ccc|c} 0.750 & 0.612 & -0.250 & 1.414 \\ -0.612 & 0.500 & -0.612 & -1.576 \\ -0.250 & 0.612 & 0.750 & -1.414 \\ \hline 0 & 0 & 0 & 1 \end{array}\right]$$

Solution: The parameters of rotational motion, θ and $\hat{\mathbf{s}}$, in the screw representation are obtained using the relationship 3.25.

$$\left\{\theta = \frac{\pi}{3} \quad ; \quad \hat{\mathbf{s}} = \frac{1}{\sqrt{2}}[1, 0, -1]^T\right\}$$

The translational motion parameters of the screw, represented by the symbols d and $\boldsymbol{s}_o$, can also be calculated using the relationship 3.26 as follows.

$$\{d = 2 \quad ; \quad \boldsymbol{s}_o = [1.414, -1.576, -1.414]^T\}$$

Example 3.8 *Suppose the general motion of a rigid body is represented by the screw representation with the following parameters. In this representation, the point* $\boldsymbol{s}_o$ *is an arbitrary point chosen on the screw axis.*

$$\left\{\theta = \frac{\pi}{4}, \hat{\mathbf{s}} = \frac{1}{\sqrt{6}}[1, -2, -1]^T \quad ; \quad d = 0.5, \boldsymbol{s}_o = [0, 0, 1]^T\right\}$$

Represent this motion in space using the homogeneous transformation matrix.

Solution: In this example, equation 3.28 is used for determining the homogeneous transformation matrix.

$$^{A}\boldsymbol{T}_B = \left[\begin{array}{ccc|c} 0.756 & 0.191 & -0.626 & 0.830 \\ -0.386 & 0.902 & -0.191 & -0.217 \\ 0.529 & 0.386 & 0.756 & 0.040 \\ \hline 0 & 0 & 0 & 1 \end{array}\right]$$

3.4 Linear and Angular Velocities

In this section, we elaborate the linear and angular velocity quantities in the motion of a rigid body. The linear velocity vector of a point on a rigid body is determined by the derivative of its position vector with respect to time. If this differentiation is performed with respect to the reference coordinate system, the absolute velocity of that point is obtained. This is in contrast to deriving the relative position vector of a point on the body, which is measured with respect to the moving coordinate system, to determine its relative velocity. In robotics, establishing the relationship between the absolute velocity of a point on a body in space and its relative velocity holds special significance, and this is the focus of this section.

On the other hand, the angular velocity quantity falls into the category of free vectors, which is defined for the entire body. Angular velocity represents the rate of change of the rotation vector with respect to time. Considering that the orientation of a body is represented in various ways, determining the angular velocity vector of a body and its relationship with the time derivatives of variables defining the orientation in a rigid body, such as the rotation matrix, Euler angles, screw axis, and quaternion, is a topic that we will explore further. It is natural that representing the angular velocity vector of a body in certain representations offers more convenience, and using them in the analysis of robotic arm movements is preferable.

3.4.1 Angular Velocity of a Rigid Body

In order to define the angular velocity of a rigid body, as before, we use a coordinate system attached to it, denoted as the moving frame $\{\mathcal{B}\}$. The analysis of the body's angular velocity, as shown in Figure 3.3, is performed by examining the relative angular motion of this moving coordinate system with respect to the reference coordinate system $\{\mathcal{A}\}$. The angular velocity, denoted by $\boldsymbol{\Omega}$, is a quantity related to the entire rigid body, and its magnitude is the same for different points on the body, making it a free vector. The direction of the angular velocity vector at any time indicates the instantaneous axis of rotation of the moving coordinate system relative to the fixed coordinate system.

If you recall the definition of the screw axis, it is evident that the instantaneous rotation of the rigid body at each instance occurs around the screw axis. Additionally, the magnitude of the angular velocity vector at any moment could be determined by the time rate of change of the rotation angle θ. With this explanation, the angular velocity vector of a rigid body can be defined as follows.

$$\boldsymbol{\Omega} \triangleq \dot{\theta}\, \hat{\boldsymbol{s}} \tag{3.30}$$

Given this definition, it becomes clear that employing the screw vector $\boldsymbol{\theta}$ in the orientation representation results in the body's angular velocity precisely corresponding to the time derivative of this vector, expressed as $\boldsymbol{\Omega} = \dot{\boldsymbol{\theta}}$. This is the reason why utilizing the screw representation in the analysis of robotic motions is more straightforward, and therefore preferable, compared to alternative approaches.

Note that the angular velocity is a vector that can be represented in any coordinate frame. As before, the superscript before the vector is used to indicate the coordinate system in which the angular velocity is represented. For example, the vector ${}^{B}\boldsymbol{\Omega}$ represents the angular velocity of the body represented in the $\{\mathcal{B}\}$ coordinate system. The angular velocity vector can be represented in any coordinate system with its components as follows.

$$\begin{aligned} {}^{A}\boldsymbol{\Omega} &= \Omega_x \hat{\boldsymbol{x}} + \Omega_y \hat{\boldsymbol{y}} + \Omega_z \hat{\boldsymbol{z}} \\ &= \dot{\theta}\left(s_x \hat{\boldsymbol{x}} + s_y \hat{\boldsymbol{y}} + s_z \hat{\boldsymbol{z}}\right) \end{aligned} \tag{3.31}$$

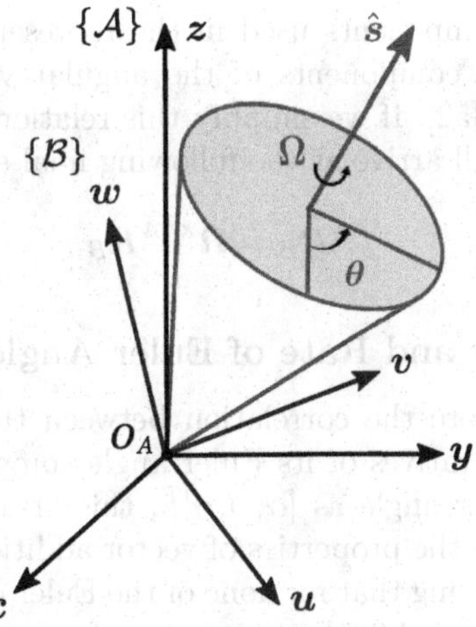

FIGURE 3.3
The representation of the instantaneous axis of rotation of a rigid body and the angular velocity vector

In this representation, $\Omega_x, \Omega_y, \Omega_z$ are the respective components of the angular velocity vector of the body, expressed in the $\{\mathcal{A}\}$ coordinate system and along the axes $\hat{\boldsymbol{x}}, \hat{\boldsymbol{y}}, \hat{\boldsymbol{z}}$.

3.4.2 Angular Velocity and the Rate of Rotation Matrix

As previously discussed, the rotation matrix serves as one of the most versatile approaches to represent the orientation of a body, facilitating the establishment of connections between diverse representations of the body's orientation. In this section, we will explore the connection between the derivative of the rotation matrix and the angular velocity vector of the body. Unfortunately, unlike the screw representation, where rotational motion is depicted with the vector $\boldsymbol{\theta}$, the rotation matrix representation conveys this concept using a matrix. It is evident that its time derivative will not be directly equivalent to the angular velocity of the body. Within this section, we will derive the mathematical relationship between the derivative of the rotation matrix and the angular velocity vector of the body.

The rotation matrix is an orthogonal matrix, meaning its transpose is equal to its inverse. Therefore,

$$ {}^{A}\boldsymbol{R}_{B}\,{}^{A}\boldsymbol{R}_{B}^{T} = \boldsymbol{I} \tag{3.32}$$

in which, $\boldsymbol{I}$ is the 3×3 identity matrix. By differentiating both sides of this equation with respect to time,

$$ {}^{A}\dot{\boldsymbol{R}}_{B}\,{}^{A}\boldsymbol{R}_{B}^{T} + {}^{A}\boldsymbol{R}_{B}\,{}^{A}\dot{\boldsymbol{R}}_{B}^{T} = \boldsymbol{0} \tag{3.33}$$

and by substituting ${}^{A}\boldsymbol{R}_{B}^{T} = {}^{A}\boldsymbol{R}_{B}^{-1}$ and ${}^{A}\boldsymbol{R}_{B} = ({}^{A}\boldsymbol{R}_{B}^{-1})^{T}$, we can arrive at the following relationship.

$$ \left({}^{A}\dot{\boldsymbol{R}}_{B}\,{}^{A}\boldsymbol{R}_{B}^{-1}\right) + \left({}^{A}\dot{\boldsymbol{R}}_{B}\,{}^{A}\boldsymbol{R}_{B}^{-1}\right)^{T} = \boldsymbol{0} \tag{3.34}$$

From this relationship, it can be concluded that the matrix ${}^{A}\dot{\boldsymbol{R}}_{B}\,{}^{A}\boldsymbol{R}_{B}^{-1}$ is a 3×3 skew-symmetric matrix. Therefore, it can be represented in the form of a general skew-symmetric matrix denoted as $\boldsymbol{\Omega}^{\times}$ as follows.

$$ \boldsymbol{\Omega}^{\times} \equiv {}^{A}\dot{\boldsymbol{R}}_{B}\,{}^{A}\boldsymbol{R}_{B}^{-1} = \begin{bmatrix} 0 & -\Omega_z & \Omega_y \\ \Omega_z & 0 & -\Omega_x \\ -\Omega_y & \Omega_x & 0 \end{bmatrix} \tag{3.35}$$

It can be shown that the components used in the representation of this skew-symmetric matrix are the same as the components of the angular velocity vector of the body, as introduced in equation 3.31 [52]. If we simplify this relationship in terms of the derivative of the rotation matrix, we will arrive at the following final equation.

$$ {}^{A}\dot{\boldsymbol{R}}_{B} = \boldsymbol{\Omega}^{\times}\ {}^{A}\boldsymbol{R}_{B} \tag{3.36}$$

3.4.3 Angular Velocity and Rate of Euler Angles

In this section, we will explore the correlation between the angular velocity vector of a rigid body and the time derivatives of its Euler angle components. If we express an array containing each type of Euler angle as $[\alpha, \beta, \gamma]^T$, this array does not qualify as a vector due to its failure to adhere to the properties of vector addition and commutativity on finite rotation angles. It is worth noting that for none of the Euler angle representations, including the ones introduced in Section 2.4.2, the time rates of change of these angles are not equal to the angular velocity vector.

$$\boldsymbol{\Omega} \neq \begin{bmatrix} \dot{\alpha} \\ \dot{\beta} \\ \dot{\gamma} \end{bmatrix} \tag{3.37}$$

It can be demonstrated that the relationship between the time derivative of this array and the angular velocity vector of the body is determined by the matrix $\boldsymbol{E}(\alpha, \beta, \gamma)$, which needs to be specified in the motion analysis for each type of Euler angle.

$$\boldsymbol{\Omega} = \boldsymbol{E}(\alpha, \beta, \gamma) \begin{bmatrix} \dot{\alpha} \\ \dot{\beta} \\ \dot{\gamma} \end{bmatrix} \tag{3.38}$$

To determine this matrix, one can use the relationship given in Equation 3.35. For this purpose, it is sufficient to obtain the rotation matrix associated with the desired Euler angles and differentiate it with respect to time. Subsequently, equation 3.35 can be simplified using the components of the rotation matrix and its derivatives.

$$\begin{aligned} \Omega_x &= \dot{r}_{31} r_{21} + \dot{r}_{32} r_{22} + \dot{r}_{33} r_{23} \\ \Omega_y &= \dot{r}_{11} r_{31} + \dot{r}_{12} r_{32} + \dot{r}_{13} r_{33} \\ \Omega_z &= \dot{r}_{21} r_{11} + \dot{r}_{22} r_{12} + \dot{r}_{23} r_{13} \end{aligned} \tag{3.39}$$

Here, r_{ij} represents the (i, j) component of the rotation matrix associated with the Euler angles of interest. By performing these calculations and simplifying them, the matrix $\boldsymbol{E}(\alpha, \beta, \gamma)$ can be determined. As an example, consider the $\boldsymbol{w} - \boldsymbol{v} - \boldsymbol{w}$ Euler angles, whose rotation matrix is given in Equation 2.33 as follows.

$$\boldsymbol{R}_{wvw}(\alpha, \beta, \gamma) = \begin{bmatrix} c\alpha c\beta c\gamma - s\alpha s\gamma & -c\alpha c\beta s\gamma - s\alpha c\gamma & c\alpha s\beta \\ s\alpha c\beta c\gamma + c\alpha s\gamma & -s\alpha c\beta s\gamma + c\alpha c\gamma & s\alpha s\beta \\ -s\beta c\gamma & s\beta s\gamma & c\beta \end{bmatrix}$$

Taking the derivative of this matrix with respect to time, employing the relationship in 3.39, and simplifying and categorizing the result in terms of the variables $\dot{\alpha}, \dot{\beta}, \dot{\gamma}$, we obtain the following result.

$$\boldsymbol{\Omega} = \boldsymbol{E}_{wvw} \begin{bmatrix} \dot{\alpha} \\ \dot{\beta} \\ \dot{\gamma} \end{bmatrix} \quad ; \quad \boldsymbol{E}_{wvw} = \begin{bmatrix} 0 & -s\alpha & c\alpha s\beta \\ 0 & c\alpha & s\alpha s\beta \\ 1 & 0 & c\beta \end{bmatrix} \tag{3.40}$$

3.4.4 Angular Velocity and Rate of Quaternion

In this section, we will explore the connection between the angular velocity vector of a rigid body and the time derivatives of its unit quaternion components. While quaternions utilize parameters associated with the screw vector, they do not directly incorporate the physical concept of the rotation vector. This is due to the specific arrangement chosen in their definition, as introduced in Equation 3.14. Consequently, the relationship between the angular velocity and the time derivatives of the quaternion components, represented by the four-dimensional array $\dot{\boldsymbol{\epsilon}} = \left[\dot{\epsilon}_1, \dot{\epsilon}_2, \dot{\epsilon}_3, \dot{\epsilon}_4\right]^T$, is not as straightforward as the screw representation.

For an in-depth examination of this relationship, quaternion algebra must be employed. Here, we introduce some key relationships commonly used in the literature on this subject. However, for further details, it is recommended to review specialized references in this field such as [53]. The relationship between the body's angular velocity and the derivative of its quaternion components can be obtained as follows [54]:

$$\boldsymbol{\Omega} = 2\boldsymbol{\epsilon}^{\dagger} \odot \dot{\boldsymbol{\epsilon}} \tag{3.41}$$

In this relation homogeneous coordinates $\boldsymbol{\Omega} = \begin{bmatrix}\Omega_x & \Omega_y & \Omega_z & 0\end{bmatrix}^T$ are used to represent the vector of angular velocity. Furthermore, the symbol $\odot$ is employed in this equation to denote quaternion multiplication. To define this multiplication, consider two quaternions $\boldsymbol{a}, \boldsymbol{b}$ decomposed into its vector and scalar components as follows:

$$\boldsymbol{a} = \begin{bmatrix}\mathbf{a}\\ a\end{bmatrix}, \quad \boldsymbol{b} = \begin{bmatrix}\mathbf{b}\\ b\end{bmatrix} \tag{3.42}$$

In this representation $\mathbf{a}, \mathbf{b}$ are the first three-dimensional arrays of unit quaternions, while a, b correspond to the last scalar components. Then, according to the definition, the product of these two quaternions can be obtained from the following equation [54]:

$$\boldsymbol{a} \odot \boldsymbol{b} = \begin{bmatrix} a\boldsymbol{I} + \mathbf{a}^{\times} & \mathbf{a}\\ -\mathbf{a}^T & a\end{bmatrix}\begin{bmatrix}\mathbf{b}\\ b\end{bmatrix} \tag{3.43}$$

Furthermore, in Equation 3.41, the notation $(\cdot)^{\dagger}$ is also used to represent the quaternion conjugate, which is defined by the following equation

$$\boldsymbol{a}^{\dagger} = \begin{bmatrix}\mathbf{a}\\ a\end{bmatrix}^{\dagger} = \begin{bmatrix}-\mathbf{a}\\ a\end{bmatrix} \tag{3.44}$$

When utilizing unit quaternions to denote the rotation of a body, the quaternion conjugate signifies a rotation around the principal quaternion axis (or the screw axis) with a negative rotation angle $(-\theta)$. Consequently, it can serve as the inverse quaternion in the representation of the pure rotation of a body.

With this concise overview of quaternion algebra, the connection between the angular velocity of a body and the time derivative of quaternion components, as defined by Equation 3.41, can be streamlined through some calculations as follows.

$$\begin{bmatrix}\Omega_x\\ \Omega_y\\ \Omega_z\end{bmatrix} = 2\,\boldsymbol{E}(\boldsymbol{\epsilon})\,\dot{\boldsymbol{\epsilon}} \tag{3.45}$$

in which,

$$\boldsymbol{E}(\boldsymbol{\epsilon}) = \begin{bmatrix}\epsilon_4 & \epsilon_3 & -\epsilon_2 & -\epsilon_1\\ -\epsilon_3 & \epsilon_4 & \epsilon_1 & -\epsilon_2\\ \epsilon_2 & -\epsilon_1 & \epsilon_4 & -\epsilon_3\end{bmatrix} \tag{3.46}$$

In this equation, $\boldsymbol{\epsilon} = \left[\epsilon_1, \epsilon_2, \epsilon_3, \epsilon_4\right]^T$ denotes the quaternion for rotational motion, and $\dot{\boldsymbol{\epsilon}}$ represents its time derivative. As evident, the connection between the angular velocity of the body and the time derivative of the quaternion is established using a 3×4 matrix. Although this introduces additional complexity compared to the screw representation, it is considerably less complex than the rates in Euler angles.

The solution to the inverse of this representation can be achieved using the following equation.

$$\dot{\boldsymbol{\epsilon}} = \frac{1}{2}\boldsymbol{\epsilon} \odot \boldsymbol{\Omega} \tag{3.47}$$

In this equation, $\boldsymbol{\Omega} = \left[\Omega_x \quad \Omega_y \quad \Omega_z \quad 0\right]^T$ signifies the angular velocity represented in homogeneous coordinates. Utilizing quaternion algebra, the equation can be alternatively expressed in terms of the components of the quaternion array as follows.

$$\dot{\boldsymbol{\epsilon}} = \frac{1}{2}\,\boldsymbol{E}^T(\boldsymbol{\epsilon}) \begin{bmatrix} \Omega_x \\ \Omega_y \\ \Omega_z \end{bmatrix} \tag{3.48}$$

In this equation, the relationship between the rate of change of quaternions and the angular velocity of the body is established using a 4×3 matrix, namely $\boldsymbol{E}^T(\boldsymbol{\epsilon})$. Remarkably, this matrix is the transpose of the one introduced in Equation 3.46.

3.4.5 Linear Velocity of a Point

The linear velocity vector of point P is easily obtained from the time derivative of the position vector $\boldsymbol{P}$. A noteworthy point in this equation is that if the representation of the position vector is in the fixed reference frame, then its time derivative determines the absolute linear velocity vector of that point.

$$\boldsymbol{v}_p = \dot{\boldsymbol{P}} = \left(\frac{d\boldsymbol{P}}{dt}\right)_{fix} \tag{3.49}$$

To emphasize the differentiation of the position vector with respect to the fixed coordinate system, the notation $(d(\cdot)/dt)_{fix}$ is used in this equation. If the changes in the position vector are measured with respect to the moving coordinate system, then the relative linear velocity vector will be obtained. To differentiate this type of derivation, the notation $(\partial(\cdot)/\partial t)_{mov}$ is used intentionally to differentiate it from absolute derivatives.

$$\boldsymbol{v}_{rel} = \left(\frac{\partial \boldsymbol{P}}{\partial t}\right)_{mov} \tag{3.50}$$

In this equation, $\boldsymbol{v}_{rel}$ represents the relative velocity of point P as observed from the moving coordinate system $\{\mathcal{B}\}$.

In classical mechanics, the relationship between the derivative of a quantity in the fixed coordinate system $\{\mathcal{A}\}$, denoted by $(d(\cdot)/dt)_{fix}$, and its derivative in the moving coordinate system $\{\mathcal{B}\}$, denoted by $(\partial(\cdot)/\partial t)_{mov}$, is given by the following equation [55].

$$\left(\frac{d(\cdot)}{dt}\right)_{fix} = \left(\frac{\partial(\cdot)}{\partial t}\right)_{mov} + \boldsymbol{\Omega} \times (\cdot) \tag{3.51}$$

The link between these two types of vector derivatives is provided by the vector $\boldsymbol{\Omega}$. This vector represents the angular velocity of the moving coordinate system $\{\mathcal{B}\}$ with respect to the fixed coordinate system $\{\mathcal{A}\}$. Additionally, the notation $\boldsymbol{\Omega} \times (\cdot)$ denotes the cross product of the angular velocity vector $\boldsymbol{\Omega}$ with the desired quantity $(\cdot)$. Since matrix multiplication is commonly used in robotics literature, the cross product term in Equation 3.51 can be expressed using the skew-symmetric matrix $\boldsymbol{\Omega}^{\times}$.

$$\boldsymbol{\Omega}^{\times} = \begin{bmatrix} 0 & -\Omega_z & \Omega_y \\ \Omega_z & 0 & -\Omega_x \\ -\Omega_y & \Omega_x & 0 \end{bmatrix} \tag{3.52}$$

This notation was previously defined in Equation 3.35. By substituting it, the matrix form of Equation 3.51 can be written as follows:

$$\left(\frac{d(\cdot)}{dt}\right)_{fix} = \left(\frac{\partial(\cdot)}{\partial t}\right)_{mov} + \boldsymbol{\Omega}^{\times}(\cdot) \tag{3.53}$$

Using this notation, it is straightforward to obtain the relationship between the absolute and relative derivatives of matrix quantities as well. For example, the derivative of the rotation matrix can be easily expressed using Equation 3.53 as follows:

$$\left(\frac{d({}^{A}\boldsymbol{R}_B)}{dt}\right)_{fix} = \left(\frac{\partial({}^{A}\boldsymbol{R}_B)}{\partial t}\right)_{mov} + \boldsymbol{\Omega}^{\times}({}^{A}\boldsymbol{R}_B) \tag{3.54}$$

From the perspective of the moving coordinate system, no change is observed in the rotation matrix ${}^{A}\boldsymbol{R}_B$, Hence

$$\left(\frac{\partial({}^{A}\boldsymbol{R}_B)}{\partial t}\right)_{mov} = 0$$

Therefore,

$${}^{A}\dot{\boldsymbol{R}}_B = \boldsymbol{\Omega}^{\times}({}^{A}\boldsymbol{R}_B) \tag{3.55}$$

This approach, in addition to providing a quick proof for the relation 3.35, demonstrates that the skew-symmetric matrix used in this equation is derived from the components of the angular velocity vector $\boldsymbol{\Omega}$.

Now, consider the general motion of a rigid body, including both translational and rotational motion, as shown in the Figure 3.4. In this figure, by attaching the coordinate system $\{\mathcal{B}\}$ to the rigid body, we aim to find the absolute velocity vector of point P with respect to the reference coordinate system $\{\mathcal{A}\}$. In this scenario, the general motion of the rigid body is described by the combination of its translational motion, represented by the vector ${}^{A}\boldsymbol{P}_{O_B}$, and its rotational motion, represented by the screw vector of rotation.

In order to determine the absolute linear velocity vector of point P, rewrite the vector describing the absolute position of this point provided in Equation 2.39.

$${}^{A}\boldsymbol{P} = {}^{A}\boldsymbol{P}_{O_B} + {}^{A}\boldsymbol{R}_B\, {}^{B}\boldsymbol{P} \tag{3.56}$$

To establish the relationship between the absolute and relative linear velocities of point P, differentiate both sides of the equation 3.56 with respect to time.

$$\begin{aligned} {}^{A}\dot{\boldsymbol{P}} &= {}^{A}\dot{\boldsymbol{P}}_{O_B} + {}^{A}\dot{\boldsymbol{R}}_B\, {}^{B}\boldsymbol{P} + {}^{A}\boldsymbol{R}_B\, {}^{B}\dot{\boldsymbol{P}} \\ {}^{A}\boldsymbol{v}_p &= {}^{A}\boldsymbol{v}_{O_B} + {}^{A}\dot{\boldsymbol{R}}_B\, {}^{B}\boldsymbol{P} + {}^{A}\boldsymbol{R}_B\, {}^{B}\boldsymbol{v}_p \end{aligned} \tag{3.57}$$

In this equation, ${}^{B}\boldsymbol{v}_p = v_{rel}$ represents the relative velocity vector of point P as observed from the perspective of the moving coordinate system $\{\mathcal{B}\}$. The derivative of the rotation matrix ${}^{A}\dot{\boldsymbol{R}}_B$ can also be obtained using the verified equation 3.55 as follows.

$${}^{A}\dot{\boldsymbol{R}}_B = {}^{A}\boldsymbol{\Omega}^{\times}\, {}^{A}\boldsymbol{R}_B \tag{3.58}$$

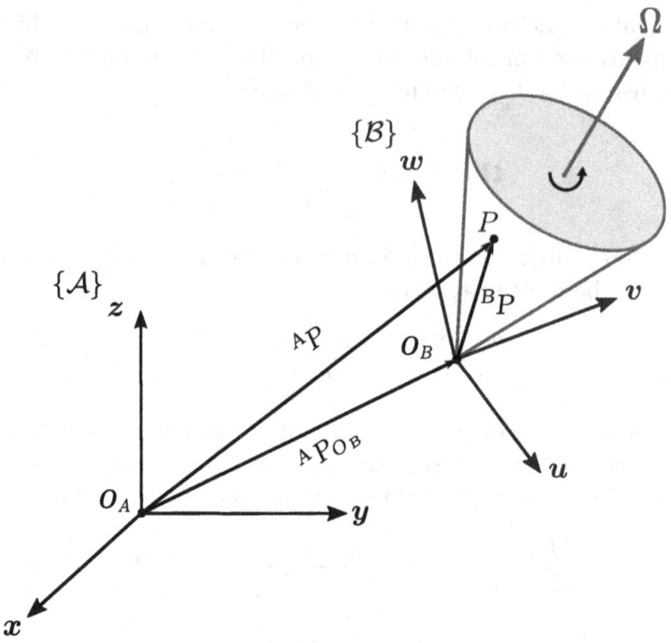

FIGURE 3.4
Instantaneous velocity of point P relative to the reference coordinate system $\{\mathcal{A}\}$.

Hence,

$$ {}^A\boldsymbol{v}_p = {}^A\boldsymbol{v}_{O_B} + {}^A\boldsymbol{R}_B\, {}^B\boldsymbol{v}_p + {}^A\boldsymbol{\Omega}^\times\, {}^A\boldsymbol{R}_B\, {}^B\boldsymbol{P} \tag{3.59} $$

In the case where P is a point on a rigid body, this point does not have relative motion with respect to the body, and ${}^B\boldsymbol{v}_p$ will be equal to zero. In this situation,

$$ {}^A\boldsymbol{v}_p = {}^A\boldsymbol{v}_{O_B} + {}^A\boldsymbol{\Omega}^\times\, {}^A\boldsymbol{R}_B\, {}^B\boldsymbol{P} \tag{3.60} $$

This relationship will be used in Chapter 6, in the analysis of the end-effector velocity in terms of its joint variables for the robot.

3.5 Screw Coordinates

We have seen that finite rotation of a rigid body can be described by the angle of rotation θ about the screw axis $\hat{\boldsymbol{s}}$. Additionally, we mentioned that the angular velocity vector of the body can also be represented along this axis with a magnitude of $\dot{\theta}$. In Section 3.3, using the Mozzi–Chasles' theorem, it was demonstrated that translational motion of a body can be represented along the screw axis with a pitch parameter d. Considering that the general motion of a rigid body is described by decomposing it into two parts: rotation about the screw axis and translational motion along that axis, the screw representation bears a resemblance to the motion of a screw and nut, justifying the use of the term "screw" for this representation.

In this section, we introduce "Screw Coordinates" that effectively describe simultaneous rotational and translational motion. For this purpose, consider the general motion of a rigid body, a combination of rotational motion around the screw axis and translational motion along this axis. Furthermore, similar to the screw pitch in the motion of a screw and nut, define the screw pitch λ as the ratio of translational motion to rotational motion, given by $\lambda = d/\theta$, where the unit of pitch measurement is m/rad. The ratio of translational velocity to angular velocity can also be used to define the screw pitch using the following relationship.

$$\lambda = \frac{\dot{d}}{\dot{\theta}} \tag{3.61}$$

The general motion of a rigid body, which consists of a combination of rotational and translational motion, can be represented by a 6×1 array composed of two corresponding vectors representing rotational and translational motion. This array, known by various names in robotics literature, is introduced in this book as *Screw Coordinates* [43, 56]. Thus, the unit screw coordinates, denoted by $\hat{\$}$, are defined as a 6×1 array, combining the following two vectors:

$$\hat{\$} = \begin{bmatrix} \hat{\boldsymbol{s}} \\ \boldsymbol{s}_o \times \hat{\boldsymbol{s}} + \lambda \hat{\boldsymbol{s}} \end{bmatrix} = [\$_1, \$_2, \$_3, \$_4, \$_5, \$_6]^T \tag{3.62}$$

In this definition, the unit vector $\hat{\boldsymbol{s}}$ indicates the axis of rotation or the screw axis, and the parameter λ, from the relationship 3.61, represents the ratio of translational to rotational motion. Additionally, $\boldsymbol{s}_o$ is a vector specifying the distance from the point of interest for velocity analysis, to the screw axis. To determine the characteristics of the rotational and translational motion of a body, it is sufficient to multiply the unit screw by the motion intensity $\dot{q}$. The motion intensity in rotational motion is the angular velocity $\dot{q} = \dot{\theta}$, and in translational motion, it is the linear translational velocity $\dot{q} = \dot{d}$. This multiplication results in a unique quantity called the *twist*.

$$\$ = \dot{q}\,\hat{\$} \tag{3.63}$$

In order to grasp the physical concept of the "twist," let us consider two specific cases initially. Later, we will examine the general motion of a rigid body by referring to these two specific cases. To this end, suppose that, according to Figure 3.5(a), the rigid body has only rotational motion, and we want to examine the characteristics of both rotational and translational motion of point P. In this case, $\lambda = 0$, and the screw coordinates in this specific case simplify to the following form.

$$\hat{\$} = \begin{bmatrix} \hat{\boldsymbol{s}} \\ \boldsymbol{s}_o \times \hat{\boldsymbol{s}} \end{bmatrix} \tag{3.64}$$

The six-dimensional quantity of twist is obtained by multiplying the rotational rate $\dot{\theta}$ by the screw of unit pitch, as follows:

$$\$ = \hat{\$}\,\dot{\theta} = \begin{bmatrix} \hat{\boldsymbol{s}} \\ \boldsymbol{s}_o \times \hat{\boldsymbol{s}} \end{bmatrix} \dot{\theta} \tag{3.65}$$

If we examine the vectors forming the twist in this case, we will find that the first vector determines the angular velocity of the body.

$$\hat{\boldsymbol{s}}\dot{\theta} = \boldsymbol{\Omega} \tag{3.66}$$

If we consider the vector $\boldsymbol{s}_o$ according to Figure 3.5(a) as the vector from the point P to the screw axis with the direction indicated in the figure, then the second vector forming the twist determines the linear velocity of this point.

$$(\boldsymbol{s}_o \times \hat{\boldsymbol{s}})\,\dot{\theta} = \boldsymbol{s}_o \times \boldsymbol{\Omega} = \boldsymbol{\Omega} \times (-\boldsymbol{s}_o) = \boldsymbol{v}_p \tag{3.67}$$

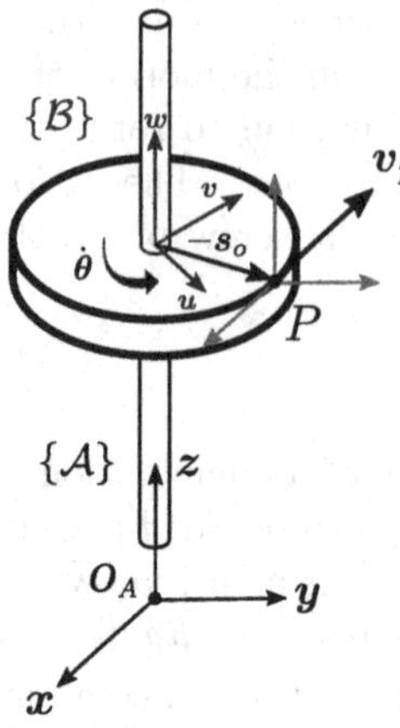

(a) Pure Rotational Motion

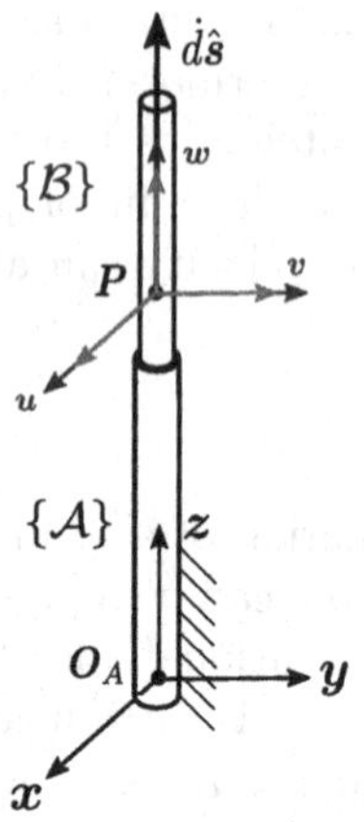

(b) Pure Translational Motion

FIGURE 3.5
Representation of screw coordinates in rotational and translational motions

According to the twist definition, the vector $\boldsymbol{s}_o$ is equal to the vectorial distance from the origin of the moving coordinate system to the fixed coordinate system, i.e., ${}^A\boldsymbol{P}_{O_B} = \boldsymbol{s}_o$. However, this interpretation cannot be directly used to determine the translational motion characteristics of point P that is moving with a rigid body. To determine the translational motion characteristics of point P, assume that the coordinate system $\{\mathcal{A}\}$ is instantly placed on the point of interest without changing the orientation of its principal axes. In this way, the vector $-\boldsymbol{s}_o$ can be determined by using the inverse of this vector, i.e., ${}^B\boldsymbol{P}_{O_A}$, and with this arrangement, the translational motion characteristics of any point on the body can be easily obtained by simply placing the origin of the $\{\mathcal{A}\}$ coordinate system instantaneously on it.

Similarly, pure translational motion in a prismatic joint can be examined as another special case of this representation. As shown in Figure 3.5(b), consider a point P on a rigid body with only translational motion. In this case, the pitch of motion is calculated asymptotically, since $\dot{\theta}$ is infinitesimally small, therefore, $\lambda = \infty$, and the velocity of motion is considered to be $\dot{q} = \dot{d}$ in the limit. The screw coordinates in this special case simplify to the following form.

$$\$\,\dot{d} = \begin{bmatrix} \boldsymbol{0} \\ \hat{\boldsymbol{s}} \end{bmatrix} \dot{d} \tag{3.68}$$

In this case, the first vector, which is zero, indicates that the body does not have any rotational motion. However, the second vector precisely determines the translational velocity of point P, if the origin of the $\{\mathcal{A}\}$ coordinate system is instantaneously placed on the point P according to the convention:

$$\boldsymbol{v}_p = \hat{\boldsymbol{s}}\,\dot{d} \tag{3.69}$$

This condition similarly holds for the general motion of a rigid body, as illustrated in Figure 3.6. In this case, the six-dimensional screw coordinate is written as follows.

$$\$ = \begin{bmatrix} \hat{\boldsymbol{s}} \\ \boldsymbol{s}_o \times \hat{\boldsymbol{s}} + \lambda\hat{\boldsymbol{s}} \end{bmatrix} \dot{\theta} \tag{3.70}$$

Note that the first three components of this array determine the angular velocity of the body.

$$\hat{\boldsymbol{s}}\,\dot{\theta} = \boldsymbol{\Omega} \tag{3.71}$$

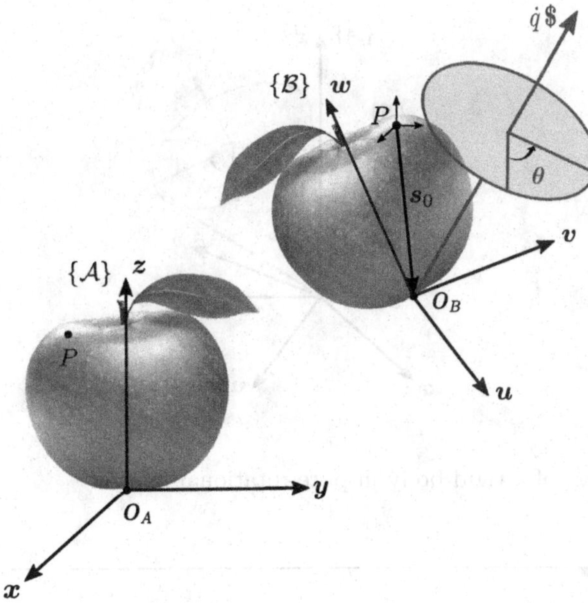

FIGURE 3.6
Instantaneous placement of the origin of the $\{\mathcal{A}\}$ coordinate system on point P to determine the screw coordinates.

The other three components of this array also determine the velocity vector of point P on the rigid body if we place the origin of coordinate system $\{\mathcal{A}\}$ instantly on point P according to the convention.

$$\begin{aligned}(\boldsymbol{s}_o \times \hat{\boldsymbol{s}} + \lambda \hat{\boldsymbol{s}})\,\dot{\theta} &= \boldsymbol{s}_o \times \dot{\theta}\hat{\boldsymbol{s}} + \lambda\,\dot{\theta}\,\hat{\boldsymbol{s}} \\ &= \boldsymbol{s}_o \times \boldsymbol{\Omega} + \lambda\,\dot{\theta}\,\hat{\boldsymbol{s}} \\ &= \boldsymbol{\Omega} \times (-\boldsymbol{s}_o) + \dot{d}\,\hat{\boldsymbol{s}} \\ &= \boldsymbol{\Omega} \times {}^{B}\boldsymbol{P}_{O_A} + \dot{d}\,\hat{\boldsymbol{s}} \end{aligned} \tag{3.72}$$

Note that the final relation in this equation, is obtained by placing the origin of coordinate system $\{\mathcal{A}\}$ instantly on the point P according to the convention. This results in ${}^{A}\boldsymbol{P}_{O_B} = \boldsymbol{s}_o$, and for this reason, ${}^{B}\boldsymbol{P}_{O_A} = -\boldsymbol{s}_o$.

By comparing this result with the linear velocity vector representation of point P, as expressed in equation 3.60, it becomes clear that the second term in equation 3.72 is a result of the translational motion of the body, corresponding to ${}^{A}\boldsymbol{v}_{O_B}$ in equation 3.60. The first term in equation 3.72 is a result of the rotational motion of the rigid body, contributing to a portion of the linear velocity vector of point P.

In the field of robotics, the use of the six-dimensional array in screw coordinates is not only used for interpreting the general motion of a point on a rigid body but can also be employed to represent the six dimensional force-torque applied to a rigid body at a specific point. The term "wrench" is used to describe this six-dimensional array, and we will elaborate on the algebra governing it and its relationship with screw coordinates in the subsequent chapters.

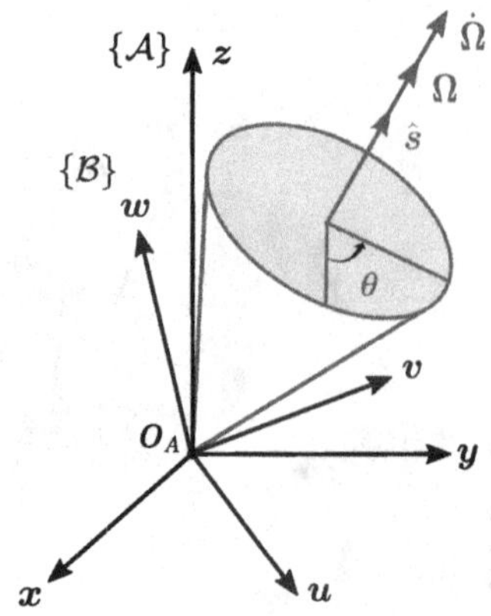

FIGURE 3.7
Angular acceleration of a rigid body in pure rotational motion

3.6 Linear and Angular Accelerations

In this section, we elaborate on the definition of linear and angular accelerations in the motion of a rigid body. The acceleration vector of a point on the body is determined by the derivative of its velocity vector with respect to time. If the differentiation is performed with respect to a fixed coordinate system, the absolute acceleration of that point is obtained. This is in contrast to differentiating the relative velocity vector of a point on the body with respect to the moving coordinate system, which determines the relative acceleration of that point.

On the other hand, the quantity of angular acceleration falls into the category of free vectors, defined for the entire rigid body. Angular acceleration determines the rate of change of angular velocity of the body with respect to time. By using the method introduced in the section 3.4.5 , which relates the absolute derivative of a quantity to its relative derivative, one can easily obtain the vectors of angular acceleration of the body and the linear acceleration of a point.

3.6.1 Angular Acceleration of a Rigid Body

To define the angular acceleration of a rigid body, as illustrated in Figure 3.7, we examine its motion by installing a moving coordinate system $\{\mathcal{B}\}$ relative to the reference coordinate system $\{\mathcal{A}\}$, similar to before. The absolute angular acceleration of the body is a characteristic of its overall motion, obtained by differentiating the angular velocity vector with respect to the reference coordinate system. Thus, the *angular acceleration vector*, denoted by $\dot{\boldsymbol{\Omega}}$, is obtained by differentiating the relation in 3.30 as follows:

$$\begin{aligned} \dot{\boldsymbol{\Omega}} = \frac{d\boldsymbol{\Omega}}{dt} &= \ddot{\theta}\,\hat{\boldsymbol{s}} + \dot{\theta}\dot{\hat{\boldsymbol{s}}} \\ &= \ddot{\theta}\,\hat{\boldsymbol{s}} + \dot{\theta}\,(\boldsymbol{\Omega} \times \boldsymbol{s}) \\ &= \ddot{\theta}\,\hat{\boldsymbol{s}} \end{aligned} \tag{3.73}$$

Note that in this relation, since the angular velocity vector of the body lies along the screw axis, the cross product $\boldsymbol{\Omega} \times \boldsymbol{s}$ equals zero. Consequently, it is demonstrated that not only is the instantaneous angular velocity of the body aligned with the screw axis at any given moment, but its angular acceleration also remains in the same direction as shown in Figure 3.7. This crucial characteristic only arises when representing the rotational motion of the body using a screw representation.

3.6.2 Linear Acceleration of a Point

Linear acceleration of point P can easily be obtained from the derivative of its linear velocity vector $\boldsymbol{v}_p$ and is given by:

$$\boldsymbol{a}_p = \dot{\boldsymbol{v}}_p = \left(\frac{d\boldsymbol{v}_p}{dt}\right)_{fix} \tag{3.74}$$

Note that this differentiation must be taken with respect to the fixed coordinate system to accurately calculate the absolute acceleration of point P. This importance is emphasized in this equation by representing absolute differentiation with the notation $(d(\cdot)/dt)_{fix}$.

If the linear velocity vector of point P is differentiated with respect to the moving coordinate system, the result will be the calculation of the *relative acceleration* of that point, denoted as $\boldsymbol{a}_{rel}$ and calculated as follows:

$$\boldsymbol{a}_{rel} = \left(\frac{\partial \boldsymbol{v}_p}{\partial t}\right)_{mov} \tag{3.75}$$

In this equation, differentiation with respect to the moving coordinate system, as before, is represented by $(\partial(\cdot)/\partial t)_{mov}$. Fortunately, the relationship between absolute and relative differentiation of a physical quantity can be determined using equations 3.51 or 3.53.

Now, consider the general motion of a rigid body in space, as shown in Figure 3.4. Let's analyze the linear acceleration of point P by attaching the moving coordinate system $\{\mathcal{B}\}$ to the body, as before, to find the linear acceleration of point P in the reference coordinate system $\{\mathcal{A}\}$. For this purpose, by rewriting the relation 3.59 in the interpretation of absolute and relative velocities of point P, we have:

$${}^A\boldsymbol{v}_p = {}^A\boldsymbol{v}_{O_B} + {}^A\boldsymbol{R}_B\,{}^B\boldsymbol{v}_p + {}^A\boldsymbol{\Omega}^\times\,{}^A\boldsymbol{R}_B\,{}^B\boldsymbol{P}$$

Differentiate this relation with respect to time to determine the linear acceleration of point P:

$$\begin{aligned}
{}^A\dot{\boldsymbol{v}}_p &= {}^A\dot{\boldsymbol{v}}_{O_B} + \frac{d}{dt}\left({}^A\boldsymbol{R}_B\,{}^B\boldsymbol{v}_p\right) + \frac{d}{dt}\left({}^A\boldsymbol{\Omega}^\times\,{}^A\boldsymbol{R}_B\,{}^B\boldsymbol{P}\right)\\
{}^A\boldsymbol{a}_p &= {}^A\boldsymbol{a}_{O_B} + \left({}^A\boldsymbol{R}_B\,{}^B\boldsymbol{a}_p + {}^A\boldsymbol{\Omega}^\times\,{}^A\boldsymbol{R}_B\,{}^B\boldsymbol{v}_P\right)\\
&+ {}^A\dot{\boldsymbol{\Omega}}^\times\,{}^A\boldsymbol{R}_B\,{}^B\boldsymbol{P} + {}^A\boldsymbol{\Omega}^\times\left({}^A\boldsymbol{\Omega}^\times\,{}^A\boldsymbol{R}_B\,{}^B\boldsymbol{P}\right) + {}^A\boldsymbol{\Omega}^\times\,{}^A\boldsymbol{R}_B\,{}^B\boldsymbol{v}_P
\end{aligned} \tag{3.76}$$

By simplifying this equation, the absolute acceleration of point P is expressed in terms of five terms as follows:

$$\begin{aligned}
{}^A\boldsymbol{a}_p &= {}^A\boldsymbol{a}_{O_B} + {}^A\boldsymbol{R}_B\,{}^B\boldsymbol{a}_p +\\
&\quad {}^A\dot{\boldsymbol{\Omega}}^\times\,{}^A\boldsymbol{R}_B\,{}^B\boldsymbol{P} + {}^A\boldsymbol{\Omega}^\times\left({}^A\boldsymbol{\Omega}^\times\,{}^A\boldsymbol{R}_B\,{}^B\boldsymbol{P}\right) + 2\,{}^A\boldsymbol{\Omega}^\times\,{}^A\boldsymbol{R}_B\,{}^B\boldsymbol{v}_P
\end{aligned} \tag{3.77}$$

In this equation, ${}^A\boldsymbol{a}_{O_B}$ represents the linear acceleration of the origin of the moving coordinate system $\{\mathcal{B}\}$. Furthermore, ${}^A\boldsymbol{R}_B\,{}^B\boldsymbol{a}_p$ represents the relative acceleration of point P with respect to the moving coordinate system $\{\mathcal{B}\}$. The term ${}^A\dot{\boldsymbol{\Omega}}^\times\,{}^A\boldsymbol{R}_B\,{}^B\boldsymbol{P}$ indicates the contribution of the angular acceleration of the moving coordinate system $\{\mathcal{B}\}$ to the linear acceleration of point P, and the two other terms determine the centrifugal and Coriolis accelerations of point P, respectively.

If point P belongs to the rigid body, its relative velocity and acceleration with respect to the moving coordinate system will be zero: ${}^{B}\boldsymbol{v}_p = {}^{B}\boldsymbol{a}_p = 0$. Consequently, the expression for the linear acceleration of point P simplifies to:

$$
{}^{A}\boldsymbol{a}_p = {}^{A}\boldsymbol{a}_{O_B} + {}^{A}\dot{\boldsymbol{\Omega}}^{\times}\ {}^{A}\boldsymbol{R}_B\,{}^{B}\boldsymbol{P} + {}^{A}\boldsymbol{\Omega}^{\times}\ ({}^{A}\boldsymbol{\Omega}^{\times}\ {}^{A}\boldsymbol{R}_B\,{}^{B}\boldsymbol{P}) \tag{3.78}
$$

This equation will be used in Chapter 8 for the dynamic analysis of robot motion.

Problems

3.1 Rotation of a rigid body is given by the following rotation matrix:

$$
{}^{A}\boldsymbol{R}_B = \begin{bmatrix} 0.500 & 0.080 & 0.862 \\ 0.500 & 0.787 & -0.362 \\ -0.707 & 0.612 & 0.354 \end{bmatrix}
$$

a) Find the eigenvalues and eigenvectors of the rotation matrix ${}^{A}\boldsymbol{R}_B$.

b) Obtain the representation of this rotation with a screw axis, determine the rotation angle θ, and find the axis of rotation ${}^{A}\hat{\boldsymbol{s}} = [\ s_x \quad s_y \quad s_z\ ]^T$. Verify your calculations with the eigenvalues and eigenvectors obtained in the previous part.

c) Represent this rotational motion with its corresponding quaternion.

3.2 Revisit Problem 3.1 using the given rotation matrix.

$$
{}^{A}\boldsymbol{R}_B = \begin{bmatrix} -0.146 & -0.854 & 0.500 \\ 0.854 & 0.146 & 0.500 \\ -0.500 & 0.500 & 0.707 \end{bmatrix}.
$$

3.3 Reiterate Problem 3.1 with the provided rotation matrix.

$$
{}^{A}\boldsymbol{R}_B = \begin{bmatrix} 0.0990 & -0.3696 & 0.9239 \\ 0.9839 & -0.1024 & -0.1464 \\ 0.1487 & 0.9235 & 0.3536 \end{bmatrix}
$$

3.4 Reattempt Problem 3.1 using the given rotation matrix.

$$
{}^{A}\boldsymbol{R}_B = \begin{bmatrix} 0.933 & 0.067 & 0.354 \\ 0.067 & 0.933 & -0.354 \\ -0.354 & 0.354 & 0.866 \end{bmatrix}.
$$

3.5 Repeat Problem 3.1 with the provided rotation matrix.

$$
{}^{A}\boldsymbol{R}_B = \begin{bmatrix} 0 & 0.259 & 0.966 \\ 0 & 0.966 & -0.250 \\ -1 & 0 & 0 \end{bmatrix}.
$$

3.6 Redo Problem 3.1 using the given rotation matrix.

$$
{}^{A}\boldsymbol{R}_B = \begin{bmatrix} -0.8001 & -0.4975 & 0.3351 \\ 0.4619 & -0.1546 & 0.8733 \\ -0.3827 & 0.8536 & 0.3536 \end{bmatrix}
$$

3.7 Assume that the rotation of a rigid body is given with constant angles relative to the following fixed coordinate systems:

$$\boldsymbol{R}_z(\pi/2), \boldsymbol{R}_y(-\pi/4), \boldsymbol{R}_z(\pi/6).$$

a) Obtain the representation of this rotation with a screw axis, determine the rotation angle θ, and find the axis of rotation ${}^A\hat{\boldsymbol{s}} = \begin{bmatrix} s_x & s_y & s_z \end{bmatrix}^T$.

b) Represent this rotational motion with its corresponding quaternion.

3.8 Revisit Problem 3.7 for the rotation of a rigid body with the the following pitch-roll-yaw Euler angles.

$$\boldsymbol{R}_x(\pi/6), \boldsymbol{R}_y(\pi/4), \boldsymbol{R}_z(\pi/4).$$

3.9 Reiterate Problem 3.7 for the rotation of a rigid body relative to the following moving coordinate systems:

$$\boldsymbol{R}_v(\pi/4), \boldsymbol{R}_w(-\pi/2), \boldsymbol{R}_u(\pi/3).$$

3.10 Redo Problem 3.7 for the rotation of a rigid body relative to the following moving coordinate systems:

$$\boldsymbol{R}_v(\pi/4), \boldsymbol{R}_w(-\pi/2), \boldsymbol{R}_u(\pi/3).$$

3.11 Assume that the rotation of a rigid body is represented by a screw axis with the following parameters:

$$\left\{\theta = \pi/2, \hat{\boldsymbol{s}} = \begin{bmatrix} -0.577 & 0.577 & 0.577 \end{bmatrix}^T\right\}.$$

a) Determine the rotation matrix ${}^A\boldsymbol{R}_B$ for this body.

b) Calculate the pitch-roll-yaw Euler angles in this rotation.

c) Calculate the $\boldsymbol{w}-\boldsymbol{v}-\boldsymbol{w}$ Euler angles in this rotation.

d) Calculate the $\boldsymbol{w}-\boldsymbol{u}-\boldsymbol{w}$ Euler angles in this rotation.

e) Represent this rotational motion with its corresponding quaternion.

3.12 Repeat Problem 3.11 for the rotation of a rigid body represented by a screw axis with the following parameters:

$$\left\{\theta = \pi/4, \hat{\boldsymbol{s}} = \begin{bmatrix} 0.707 & -0.707 & 0 \end{bmatrix}^T\right\}.$$

3.13 Revisit Problem 3.11 for the rotation of a rigid body represented by the following screw vector:

$$\boldsymbol{\theta} = \frac{\pi}{6}\begin{bmatrix} 0.447 & 0 & -0.894 \end{bmatrix}^T.$$

3.14 Reiterate Problem 3.11 for the rotation of a rigid body represented by a screw vector:

$$\boldsymbol{\theta} = \frac{\pi}{12}\begin{bmatrix} 2 & -1 & -2 \end{bmatrix}^T.$$

3.15 Assume that the rotation of a rigid body is represented by the following quaternion:

$$\epsilon = \begin{bmatrix} 0.707 & -0.707 & 0 & 0 \end{bmatrix}^T$$

a) Determine the rotation matrix ${}^A\boldsymbol{R}_B$ for this body.

b) Calculate the pitch-roll-yaw Euler angles in this rotation.

c) Calculate the $\boldsymbol{w}-\boldsymbol{v}-\boldsymbol{w}$ Euler angles in this rotation.

d) Calculate the $\boldsymbol{w}-\boldsymbol{u}-\boldsymbol{w}$ Euler angles in this rotation.

e) Represent this rotational motion with a screw vector $\boldsymbol{\theta}$.

3.16 Repeat Problem 3.15 for the rotation of a rigid body represented by the following quaternion:

$$\epsilon = \begin{bmatrix} 0.408 & 0.408 & 0.408 & 0.707 \end{bmatrix}^T$$

3.17 Reiterate Problem 3.15 for the rotation of a rigid body represented by the following quaternion:

$$\epsilon = \begin{bmatrix} 0.204 & 0.408 & -0.204 & 0.866 \end{bmatrix}^T$$

3.18 Revisit Problem 3.15 for the rotation of a rigid body represented by the following quaternion:

$$\epsilon = \begin{bmatrix} -0.204 & 0.204 & 0.408 & 0.866 \end{bmatrix}^T$$

3.19 Redo Problem 3.15 for the rotation of a rigid body represented by the following quaternion:

$$\epsilon = \begin{bmatrix} 0.289 & 0.577 & 0.289 & 0.707 \end{bmatrix}^T$$

3.20 Assume that the rotation of a rigid body is represented by a screw axis with the following parameters:

$$\left\{\theta, \hat{\boldsymbol{s}} = \begin{bmatrix} s_x & s_y & s_z \end{bmatrix}^T\right\}.$$

Show that the rotation matrix can be obtained using the exponential function as follows:

$$^{A}\boldsymbol{R}_B = e^{\hat{\boldsymbol{s}}^{\times}\theta}, \tag{3.79}$$

where the matrix $\hat{\boldsymbol{s}}^{\times}$ is defined as:

$$\hat{\boldsymbol{s}}^{\times} = \begin{bmatrix} 0 & -s_z & s_y \\ s_z & 0 & -s_x \\ -s_y & s_x & 0 \end{bmatrix}. \tag{3.80}$$

3.21 Verify that if the motion of a rigid body is described by a quaternion, the rotation matrix can be determined from the following equation:

$$^{A}\boldsymbol{R}_B = \begin{bmatrix} 1-2\epsilon_2^2-2\epsilon_3^2 & 2(\epsilon_1\epsilon_2-\epsilon_3\epsilon_4) & 2(\epsilon_1\epsilon_3+\epsilon_2\epsilon_4) \\ 2(\epsilon_1\epsilon_2+\epsilon_3\epsilon_4) & 1-2\epsilon_1^2-2\epsilon_3^2 & 2(\epsilon_2\epsilon_3-\epsilon_1\epsilon_4) \\ 2(\epsilon_1\epsilon_3-\epsilon_2\epsilon_4) & 2(\epsilon_2\epsilon_3+\epsilon_1\epsilon_4) & 1-2\epsilon_1^2-2\epsilon_2^2 \end{bmatrix}.$$

3.22 For a general motion in space that includes rotational and translational motion, derive the Rodrigues formula as given in Equation 3.4:

$$^{A}\boldsymbol{P} = \boldsymbol{s}_o + d\,\hat{\boldsymbol{s}} + (^{B}\boldsymbol{P} - \boldsymbol{s}_o)c\theta + \hat{\boldsymbol{s}} \times (^{B}\boldsymbol{P} - \boldsymbol{s}_o)s\theta + \left[(^{B}\boldsymbol{P} - \boldsymbol{s}_o)^T\hat{\boldsymbol{s}}\right] v\theta\,\hat{\boldsymbol{s}},$$

where notations $c\theta = \cos\theta, s\theta = \sin\theta, v\theta = 1 - \cos\theta$ are used. Furthermore, prove the relationship 3.29.

3.23 Show how to determine the rotation matrix of a rigid body, as given in equation 3.6, using the Rodrigues' formula.

3.24 Assume that the motion of a rigid body is represented by the following homogeneous transformation matrix. Obtain the parameters of this motion based on the screw representation.

$$^{A}\boldsymbol{T}_{B} = \left[\begin{array}{ccc|c} 0.5 & -0.75 & 0.433 & 0.5 \\ 0.866 & 0.433 & -0.25 & 0.866 \\ 0 & 0.5 & 0.866 & 0.1 \\ \hline 0 & 0 & 0 & 1 \end{array}\right]$$

3.25 Repeat Problem 3.24 for the motion represented by the following homogeneous transformation matrix.

$$^{A}\boldsymbol{T}_{B} = \left[\begin{array}{ccc|c} +0.808 & -0.575 & -0.130 & 0.991 \\ +0.534 & +0.620 & +0.575 & 1.083 \\ -0.250 & -0.534 & +0.808 & 0.148 \\ \hline 0 & 0 & 0 & 1 \end{array}\right]$$

3.26 Revisit Problem 3.24 for the motion represented by the following homogeneous transformation matrix.

$$^{A}\boldsymbol{T}_{B} = \left[\begin{array}{ccc|c} +0.250 & -0.967 & -0.058 & 0.916 \\ +0.433 & +0.058 & +0.900 & 0.587 \\ -0.866 & -0.250 & +0.433 & 0.527 \\ \hline 0 & 0 & 0 & 1 \end{array}\right]$$

3.27 Reiterate Problem 3.24 for the motion represented by the following homogeneous transformation matrix.

$$^{A}\boldsymbol{T}_{B} = \left[\begin{array}{ccc|c} -0.575 & -0.635 & -0.516 & 0.746 \\ +0.621 & -0.750 & +0.231 & 1.782 \\ -0.534 & -0.188 & +0.826 & 0.850 \\ \hline 0 & 0 & 0 & 1 \end{array}\right]$$

3.28 Suppose the general motion of a rigid body is represented by the screw representation with the following parameters. In this representation, the point $\boldsymbol{s}_o$ is an arbitrary point chosen on the screw axis.

$$\left\{\theta = \frac{\pi}{4}, \hat{\boldsymbol{s}} = \frac{1}{\sqrt{6}}[1, -2, -1]^T \quad ; \quad d = 0.5, \boldsymbol{s}_o = [0, 0, 1]^T\right\}$$

Represent this motion in space using the homogeneous transformation matrix.

3.29 Repeat Problem 3.28 for the motion described by the following Screw representation.

$$\left\{\theta = \frac{-\pi}{4}, \hat{\boldsymbol{s}} = \frac{1}{\sqrt{3}}[1, -1, -1]^T \quad ; \quad d = 0.75, \boldsymbol{s}_o = [0, -1, 1]^T\right\}$$

3.30 Revisit Problem 3.28 for the motion described by the subsequent Screw representation.

$$\left\{\theta = \frac{\pi}{2}, \hat{\boldsymbol{s}} = \frac{1}{\sqrt{2}}[1, 0, -1]^T \quad ; \quad d = 1.5, \boldsymbol{s}_o = [-1, 0, -1]^T\right\}$$

3.31 Reiterate Problem 3.28 for the motion described by the following Screw representation.

$$\left\{\theta = \frac{-\pi}{6}, \hat{\boldsymbol{s}} = \frac{1}{3}[-2, -2, -1]^T \quad ; \quad d = 1, \boldsymbol{s}_o = [1, 1, 1]^T\right\}$$

3.32 Redo Problem 3.28 for the motion described by the subsequent Screw representation.

$$\left\{\theta = \frac{\pi}{3}, \hat{\boldsymbol{s}} = \frac{1}{\sqrt{3}}[1, -1, 1]^T \quad ; \quad d = 2, \boldsymbol{s}_o = [1, -1, 1]^T\right\}$$

Part II

Kinematics

4

Forward Kinematics

This chapter commences with a comprehensive introduction aimed at providing readers with a contextual understanding of the forward kinematic analysis. The chapter then progresses to clarify the loop-closure method, elucidating its significance and application in the context of forward kinematic analysis. Loop closure is the fundamental method to formulate forward kinematic problem for robots.

Subsequently, the Denavit-Hartenberg convention takes center stage in the chapter. The convention is thoroughly explored, encompassing not only a detailed description but also an examination of the coordinate system and DH parameters associated with this widely-used method. Further, the chapter elaborates on DH homogeneous transformations, offering readers a deeper insight into the practical implementation of the convention. By using this convention, the loop-closure method is revisited, followed by an in-depth exploration of the Denavit-Hartenberg solution. Readers are likely guided through the steps and intricacies of employing these methods in several case studies.

The chapter proceeds to introduce the screw-based method, a more specialized approach to forward kinematic analysis. This method is elucidated through discussions on screw homogeneous transformation, consecutive screws, forward kinematics formulation, and the screw-based solution. This section is likely to cater to readers interested in a more advanced and specialized understanding of forward kinematics. Lastly, the chapter concludes with a section on problems, serving as an interactive component for readers to apply and test their understanding of the concepts covered in the chapter. These problems range from theoretical exercises to practical applications, fostering a hands-on approach to learning and reinforcing the knowledge gained throughout the chapter.

4.1 Introduction

In the examination of robot kinematics, the focus is on studying the geometry of motion without considering the factors causing that motion. Kinematic analysis operates under the assumption that the robot is in a specific static position; the robot's end-effector possesses a specific position and orientation, and the robot's links and joints are individually in specific poses within the space. The investigation into determining the relationship between the position and orientation of the robot's end-effector in relation to its joint motions formulates the robot forward kinematic analysis. If, along with the robot's motion state, its speed and acceleration are considered, it leads to the exploration of differential kinematics. The analysis of speed in robots will be covered in Chapter 6, and the analysis of acceleration and the dynamics of motion in Chapter 8.

As previously mentioned, serial robots consist of a sequential arrangement of joints and links with regular geometry. This type of robot is characterized by an open kinematic chain, where the initial joint is positioned on the ground or a mobile platform, and the robot's end-

DOI: 10.1201/9781003491415-4

effector moves to execute the desired tasks in a plane or within space. Typically, revolute or prismatic joints are prevalent in the structure of robots. In specific cases where joints with two or three degrees of freedom are employed, they can be substituted by equivalent sets of primary joints. Consequently, the ultimate motion of the robot's end-effector is a result of the amalgamation of movements from successive joints in the robot. This description underscores that the kinematic analysis of the robot centers on investigating the correlation between the position of the end-effector and the movements of its joints.

Consider the 3R robot shown in Figure 4.1(a) that moves in a plane. As observed in this figure, three independent motion variables are used to describe the end-effector's motion in the "task space." If we denote the position of the end-effector relative to the reference coordinate system with the vector $\boldsymbol{x}_e$ and with two independent variables (x_e, y_e) and a third variable θ_e for its orientation, the description of the end-effector's pose in space can be represented by the array $\boldsymbol{\mathcal{X}} = [x_e, y_e, \theta_e]^T$. On the other hand, the motion variables in the robot's joints can also be represented by another three-dimensional array denoted as $\boldsymbol{q} = [\theta_1, \theta_2, \theta_3]^T$, describing the angular movements of the joints. Thus, three independent variables are sufficient to describe the motion of this robot in the plane. The relationship between the "joint space" variables $\boldsymbol{q}$ and the "task space" variables $\boldsymbol{\mathcal{X}}$ is determined in the analysis of the robot's kinematics.

This same principle applies to robots with six degrees of freedom. Consider the 6R robot shown in Figure 4.1(b), which moves in three-dimensional space. To describe the end-effector's motion in the task space, six independent motion variables are utilized. If we represent the end-effector's position relative to the reference coordinate system with the vector $\boldsymbol{x}_e$ and with three independent variables (x_e, y_e, z_e) for the end-effector's position and a screw vector $\boldsymbol{\theta}_e$, which is composed of three independent variables $(\theta_{e_x}, \theta_{e_y}, \theta_{e_z})$ representing the end-effector's orientation in space, the description of the end-effector's pose in space can be expressed by the six-dimensional twist array $\boldsymbol{\mathcal{X}} = [x_e, y_e, z_e, \theta_{e_x}, \theta_{e_y}, \theta_{e_z}]^T$. In this robot, the motion variables in the joints can also be represented by another

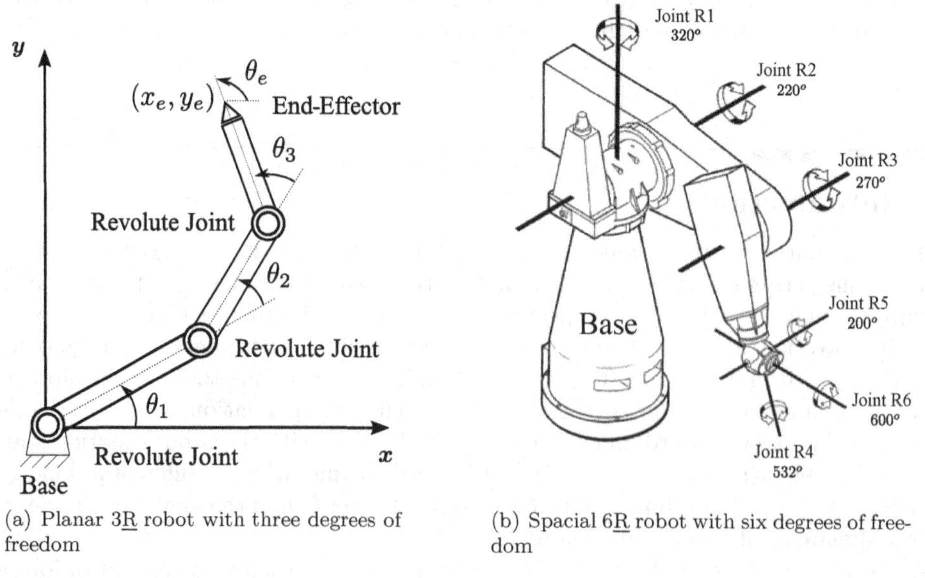

(a) Planar 3R robot with three degrees of freedom

(b) Spacial 6R robot with six degrees of freedom

FIGURE 4.1
An example of typical planar and spatial robots

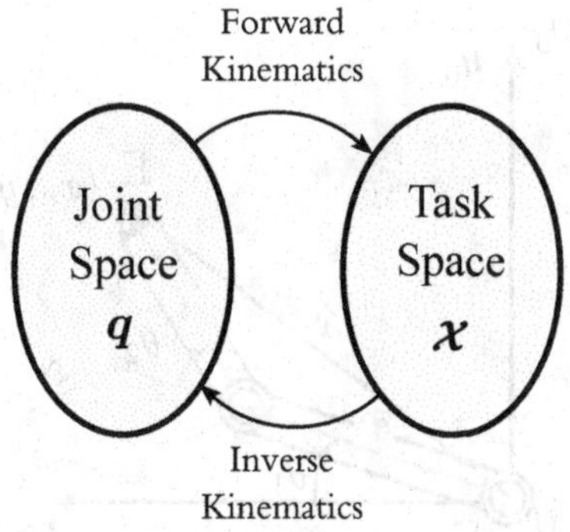

FIGURE 4.2
Representation of the joint and task space of the robot and forward and inverse kinematics

six-dimensional array denoted as $\boldsymbol{q} = [\theta_1, \theta_2, \theta_3, \theta_4, \theta_5, \theta_6]^T$, describing the angular movements of the joints. Thus, six independent variables are sufficient to describe the motion of this robot in three-dimensional space. Here again, the relationship between the joint space variables $\boldsymbol{q}$ and the task space variables $\boldsymbol{\mathcal{X}}$ is explored in the analysis of the robot's kinematics.

It should be noted that here the screw vector $\boldsymbol{\theta}_e$ is used to describe the rotational motion of the end-effector. As detailed in the previous chapter, the use of the screw vector has the unique advantage that its first derivative directly yields the angular velocity of the body. Given this feature, in this book, the screw vector is frequently used to describe the orientation of the end-effector. This should not be construed as an exclusion of alternative methods for representing rotational motion, such as Euler angles and quaternions, when describing variables in the task space. It is paramount to comprehend the accurate interpretation of the rate of these representations related to the angular velocity of the end-effector, particularly in the analysis of differential kinematics.

The two robots depicted in Figure 4.1 incorporate revolute joints, although many robots also integrate prismatic joints. In such instances, the set of joint space variables for the robot, delineating motion in prismatic joints, is complemented using translational variables as follows:

$$q_i = \begin{cases} \theta_i & \text{For Revolute Joints} \\ d_i & \text{For Prismatic Joints} \end{cases} \tag{4.1}$$

With this description of joint and task space, forward and inverse kinematic analyses can be defined as follows, which is illustrated in Figure 4.2:

Forward kinematics refers to an analysis of the geometric motion of the robot where the joint variables $\boldsymbol{q}$ are given, and the goal is to determine the variables in the task space $\boldsymbol{\mathcal{X}}$. In serial robots, this analysis typically leads to a straightforward and unique solution without much complexity. On the other hand, inverse kinematics in serial robots refers to an analysis of the robot's motion where the motion variables in the task space are given, and the objective is to find the joint variables. Solving inverse kinematics in serial robots is often more complex; it may not lead to a solution or may result in multiple valid solutions.

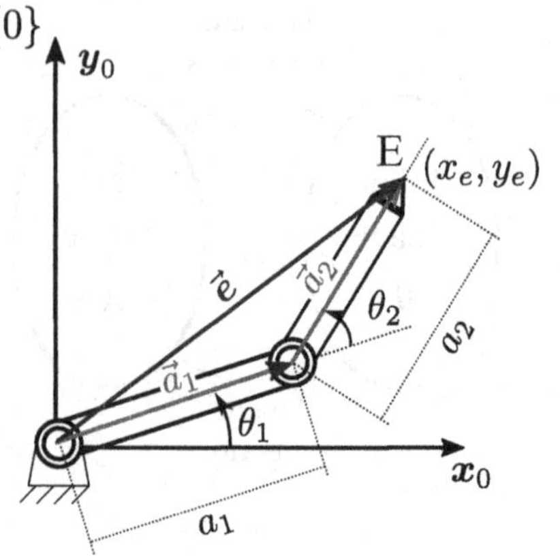

FIGURE 4.3
Representation of the loop-closure method in the kinematic analysis of a planar 2R robot

In this chapter, we will elaborate on the analysis of forward kinematics in serial robots, reserving the examination of inverse kinematics for the subsequent chapter.

4.2 Loop-Closure Method

In this section, we will analyze the kinematics of serial robots by presenting a conceptual example. To this end, consider a two-degree-of-freedom robot 2R as depicted in Figure 4.3, where the lengths of the robot's links are denoted by a_1 and a_2, and the angles of the first and second joints are denoted by θ_1 and θ_2, respectively. The robot's end-effector is located at the position indicated in the figure with the position vector $\boldsymbol{e} = [x_e, y_e]^T$. The reference coordinate system $\{0\}$ is considered at the fixed base of the robot and at the center of the first revolute joint.

Furthermore, contemplate the vectors $\vec{a}_1$ and $\vec{a}_2$ aligned with the robot's links. Through the loop-closure technique, the summation of these vectors and equating it to the end-effector vector $\boldsymbol{e}$, allow us to establish the correlation between the task space variables x_e and y_e with the joint variables θ_1 and θ_2. In this example, the loop-closure equation in vector form is written as follows:

$$\vec{a}_1 + \vec{a}_2 = \boldsymbol{e} \tag{4.2}$$

If this equation is expressed based on its components in the reference coordinate system $\{0\}$, we obtain the following two equations:

$$a_1 \cos\theta_1 + a_2 \cos(\theta_1 + \theta_2) = x_e \tag{4.3}$$

$$a_1 \sin\theta_1 + a_2 \sin(\theta_1 + \theta_2) = y_e \tag{4.4}$$

As indicated in these equations, the task space variables of the robot, denoted as $\mathcal{X} = [x_e, y_e]^T$, are represented as trigonometric functions of the robot's joint variables $\boldsymbol{q} = [\theta_1, \theta_2]^T$ and the lengths of the robot's links a_1 and a_2. Consequently, if the joint variables $\boldsymbol{q}$ are known and the task space variables are sought, the forward kinematics of the robot can be readily solved. This will result in a unique solution. Conversely, when the task space variables of the robot are provided, and the goal is to determine the joint variables, these nonlinear equations must be solved in the analysis of inverse kinematics, inherently involving greater complexity. It can be demonstrated that these equations are only solvable within the robot's workspace, and for each desired position of the robot's end-effector, two valid solutions can be obtained.

The use of the loop-closure method, in which the end-effector's position is related to the sum of its link vectors, proves to be a highly suitable approach for formulating solutions to both forward and inverse kinematic problems in robotics. However, as the number of links increases or for robots with intricate geometric structures, the simplicity of this method diminishes. This necessitates the development of approaches capable of accurately defining and solving all potential configurations. In this context, diverse methods have been developed in the robotics literature. In the following sections, we will specifically focus on the Denavit-Hartenberg method and the screw-based method as two effective approaches in the kinematic analysis of serial robots.

In the upcoming sections, the problem of the forward kinematics of serial robots in the most general case is defined, and two common methods for solving it are discussed. In the first method, the connection between two consecutive links of the robot is considered in its most general form. Through a suitable convention, coordinate systems are attached to the main joints of each connected link. In this scenario, it is assumed that the robot is composed of either revolute or prismatic joints. Then, using the geometric parameters of each link, such as length, offset, and twist, along with the joint motion variable, the corresponding homogeneous transformation matrices are determined. Since these transformation matrices are defined for the most general structure of two consecutive links in the robot, the loop-closure method can be formulated for all serial robots in matrix representation. The forward solution of this problem is discussed in Section 4.3, and its inverse solution is detailed in next chapter.

In the second approach, the application of screw theory is employed to formulate and address the kinematic challenges in robotics. The use of screws for motion representation offers the benefit of employing six-dimensional arrays and the associated algebra in the kinematic analysis of robots. As a result, the required calculations become more straightforward, and it offers a clearer representation of the fundamental physical concepts involved. The intricacies of this method will be thoroughly examined in Section 4.5.

4.3 The Denavit-Hartenberg Convention

As we have seen, a serial robot consists of a base and n consecutive links connected by n joints, forming an open kinematic chain. The relative motion of each joint is controlled by its actuator, and due to technological constraints in the development of actuators, primary revolute or prismatic joints are typically used in the kinematic structure of serial robots.

To describe the kinematic structure of a robot, consider a link as a rigid body that describes the relative motion between two adjacent joints. In this representation, the joint axis is defined in space, indicating either rotational motion around this axis or translational motion along it. Thus, a revolute joint is characterized by its rotational axis in space, while a prismatic joint considers the axis of translational motion. In cases where joints with more degrees of freedom are employed, they are first decomposed into two or three primary revolute or prismatic joints. Subsequently, each joint is represented by its own axis.

Typically, the links and joints of a robot are numbered from the base to the end-effector with indices 0 to n. Accordingly, the first link of the robot is connected to the base by the first joint, and the subsequent links and joints are numbered from 1 to n. With this representation, the i-th link of the robot extends from its initial joint, denoted by index i, to the joint $i + 1$ positioned at the end of the link.

In the robotics literature, two different notations have been proposed. The numbering order mentioned above, introduced by Paul in [57], is in accordance with the common notations used in most references [3, 4, 58]. Due to the generality of this representation, this notation will be used for the kinematic analysis of robots in this book. The alternative representation, although very similar to the common one, differs slightly in defining link parameters. Given the prevalence of the common representation, this book employs that, and for studying the alternative representation, you may refer to [2].

4.3.1 Convention Description

To describe the kinematic features of each link, defining only two parameters is sufficient. Since each link connects two adjacent joints, and each joint is defined by an axis in space, the distance between these two joint axes can be defined as the link length. For this means, the geometric distance between the axes of joints i and $i + 1$, denoted as the "link length a_i," is considered.

The distance of two lines in space is precisely defined based on the length of their common perpendicular. In this context, when considering two skew lines (neither intersecting nor parallel), a unique common perpendicular line to both is identified, and the length of that serves as the link length. In cases where the axes of adjacent joints are parallel, numerous common perpendiculars can be drawn between them, all with a unit length and in parallel. This is in contrast to the situation where the two axes intersect, making it impossible to draw a common perpendicular, and resulting in a zero length for this case.

The second quantity required for link description is the "link twist angle." To precisely define the link twist angle, visualize two axes representing the initial joint i and the final joint $i + 1$ on a plane. Measure the angle between them from the i axis toward the $i + 1$ axis, this angle, denoted as α_i, is known as the link twist angle. Usually, the initial and final joints of each link are parallel or perpendicular. In the first case, the twist angle is zero, and in the second case, it is $\pm\pi/2$. However, special cases may use twist angles other than zero or ninety degrees in robot design. In such cases, the sign of the twist angle is considered from the initial axis toward the final axis.

Now, consider the kinematic structure of two adjacent links. It can be demonstrated that the kinematic structure of two adjacent links can generally be described with only two other parameters. To achieve this, consider two intermediate links of the robot. If we focus on the relationship between these two links, we find that they share a common joint, and their motion is described by a common axis in space. On this common axis, two mutually perpendicular axes for the two links are obtained, determining the lengths of the two links. The distance between these two mutually perpendicular axes along the common joint axis introduces a third parameter describing the relative positioning of two adjacent links. This distance, termed the "link offset," is denoted as d_i.

Additionally, the angle between these two mutually perpendicular axes about the common joint axis, is called the "joint angle," and is denoted as θ_i. Note that if the common joint is a prismatic joint, the link offset d_i is a variable that changes with translational joint motion. Conversely, if it is a revolute joint, the joint angle θ_i is a variable that changes with rotational joint motion. Thus, as mentioned in Equation 4.1, if this variable is identified as a link joint variable, it is denoted as q_i. With this explanation, it can be said that to describe the motion of each link and its relationship with the adjacent link, four main parameters can be defined, known as Denavit-Hartenberg parameters [59].

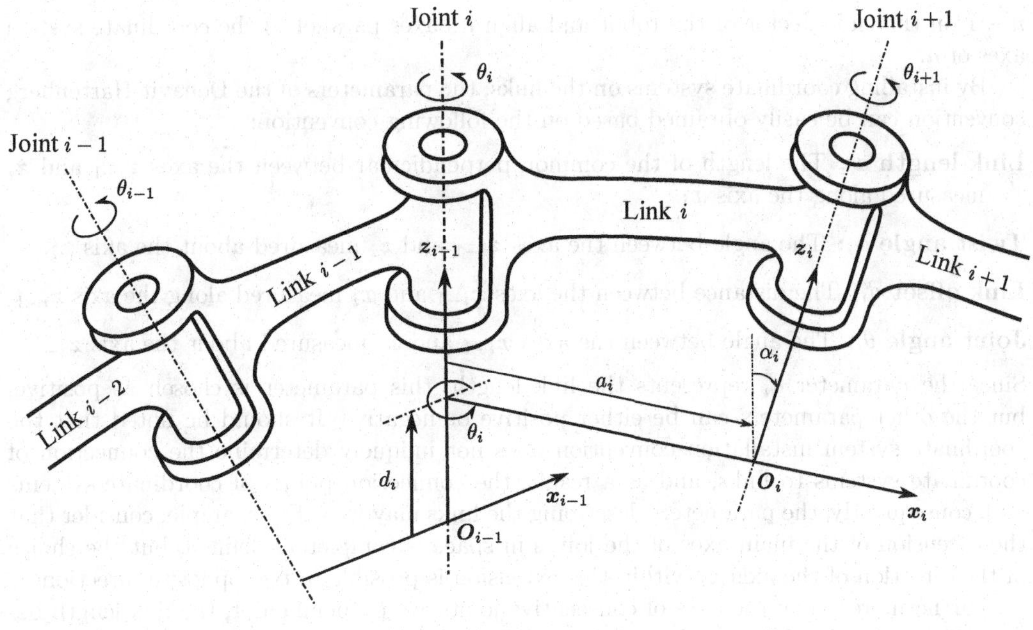

FIGURE 4.4
Numbering and parameters of a sample link in the serial robots.

4.3.2 Coordinate System and DH Parameters

To precisely define the four main parameters introduced in describing each link and its relationship with its adjacent link, it is necessary to first attach the coordinate frame that describes motion on each of the links. To do this, consider the i-th link in the structure of a serial robot, as shown in Figure 4.4. For this purpose, and drawing inspiration from [57], consider the i-th link of the robot extending from its initial joint, denoted by the index i, to joint $i+1$ located at the end of the link.

To precisely define the parameters of this link, install the coordinate system $i-1$ on the i-th link in such a way that the axis z_{i-1} lies along the extension of the initial joint of the link. Note that in this representation, the coordinate system i is placed at the end of the i-th link. Thus, the axis z_i represents the axis of rotation in the revolute joint or the translational motion axis in the prismatic joint.

Perform the same procedure for the other joints of the robot and then determine the common perpendicular between the joint axes. As seen in Figure 4.4, represent it with the notation x_i. The origin of the coordinate system i is determined at the intersection of the z_i and x_i axes, and the coordinate frame at this point could be established using right-hand rule.

To perform a similar procedure for the remaining links of the robot, defining all the coordinate systems describing the rotational and/or translational motion of the links is necessary. The installation of coordinate systems on the base and end-effector of the robot is an exception to the above rule because they do not have a joint sequence. For this purpose, place the reference coordinate system 0 on the base, and align the axis z_0 along the axis z_1. Furthermore, for the revolute joint, place its origin on the origin of the first coordinate system and choose the axis x_0 in a way that its rotational motion is measured by the appropriate angle θ_1. For the prismatic joint, align the axis x_0 parallel to x_1 and choose the coordinate system origin in a way that its translational motion is measured by the appropriate displacement variable d_1. Similarly, place the origin of the coordinate system

$n+1$ on the end-effector of the robot and align its axes parallel to the coordinate system axes of n.

By installing coordinate systems on the links, the parameters of the Denavit-Hartenberg convention can be easily obtained based on the following convention:

Link length a_i: The length of the common perpendicular between the axes $\boldsymbol{z}_{i-1}$ and $\boldsymbol{z}_i$ measured along the axis $\boldsymbol{x}_i$.

Twist angle α_i: The angle between the axes $\boldsymbol{z}_{i-1}$ and $\boldsymbol{z}_i$ measured about the axis $\boldsymbol{x}_i$.

Link offset d_i: The distance between the axes $\boldsymbol{x}_{i-1}$ and $\boldsymbol{x}_i$ measured along the axis $\boldsymbol{z}_{i-1}$.

Joint angle θ_i: The angle between the axes $\boldsymbol{x}_{i-1}$ and $\boldsymbol{x}_i$ measured about the axis $\boldsymbol{z}_{i-1}$.

Since the parameter a_i represents the link length, this parameter is chosen as positive, but the other parameters can be either positive or negative. It should be noted that the coordinate system installation convention does not uniquely determine the connection of coordinate systems to links, and as a result, the connection points of coordinate systems and, consequently, the parameters describing the links may vary. For example, consider that the extension of the main axes of the joints in space is completely defined, but the choice of the direction of the axis $\boldsymbol{z}_i$ within this extension is possible in two opposite directions.

Furthermore, when the axes of consecutive joints are perpendicular, the link length becomes zero, but the choice of the axis $\boldsymbol{x}_i$ is still possible in two opposite directions. On the other hand, if the axes of consecutive joints are parallel, the choice of the common perpendicular between them and the origin of the coordinate system on these axes is arbitrary. However, in these cases, an attempt is made to make the selection as simple as possible to zero out the corresponding parameters of adjacent links if possible. Additionally, by utilizing prismatic joints in the robot structure, there is freedom in choosing the location and extension of the installation of coordinate systems for adjacent links.

4.3.3 DH Homogeneous Transformations

By installing coordinate systems on the links of the robot and determining the Denavit-Hartenberg parameters, the homogeneous transformation between two consecutive links, which is generally a function of four link parameters, can be obtained. As observed in Figure 4.5, the coordinate system $\{i-1\}$ will reach the coordinate system $\{i\}$ through four consecutive motions. To describe these four consecutive motions and obtain the final homogeneous transformation matrix, intermediate coordinate systems $\{\mathcal{P}\}$, $\{\mathcal{Q}\}$, and $\{\mathcal{R}\}$, differing from each other in only one link parameter, are used. Finally, the final homogeneous transformation matrix is obtained by combining the homogeneous transformations of these intermediate coordinate systems, as follows.

$$ {}^{i-1}\boldsymbol{T}_i = {}^{i-1}\boldsymbol{T}_P \, {}^{P}\boldsymbol{T}_Q \, {}^{Q}\boldsymbol{T}_R \, {}^{R}\boldsymbol{T}_i \tag{4.5} $$

According to this relationship, the homogeneous transformation matrix ${}^{i-1}\boldsymbol{T}_i$ can be obtained by combining four transformation matrices, each representing a pure translational or rotational motion.

1) The coordinate system $\{i-1\}$ undergoes a translational motion by d_i along $\boldsymbol{z}_{i-1}$ to align with the coordinate system $\{\mathcal{P}\}$. The homogeneous transformation matrix for this translation is as follows,

$$ {}^{i-1}\boldsymbol{T}_P = \boldsymbol{T}(z, d_i) = \left[\begin{array}{ccc|c} 1 & 0 & 0 & 0 \\ 0 & 1 & 0 & 0 \\ 0 & 0 & 1 & d_i \\ \hline 0 & 0 & 0 & 1 \end{array}\right] $$

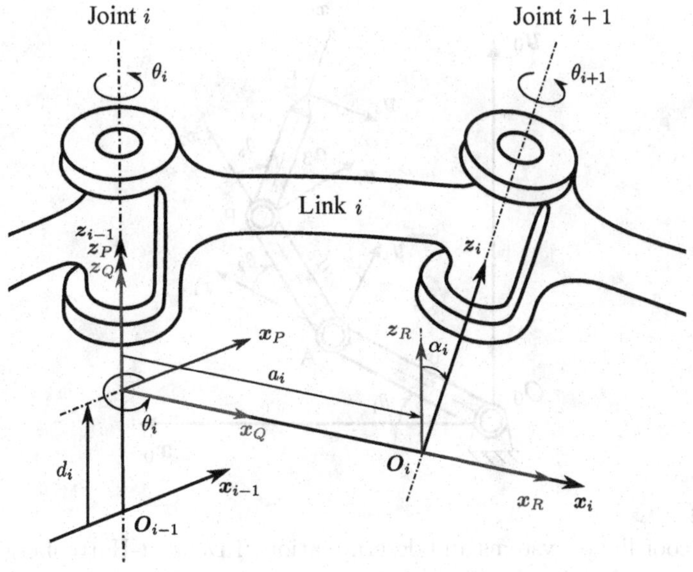

FIGURE 4.5
Intermediate coordinate systems for determining the homogeneous transformation of consecutive links

2) Subsequently, the coordinate system $\{\mathcal{P}\}$ undergoes a rotational motion with an angle θ_i around the $\boldsymbol{z}_P$ axis to align with the coordinate system $\{\mathcal{Q}\}$. In this case, the homogeneous transformation matrix for this rotation is as follows.

$$ {}^{P}\boldsymbol{T}_Q = \boldsymbol{T}(z, \theta_i) = \left[\begin{array}{ccc|c} c\theta_i & -s\theta_i & 0 & 0 \\ s\theta_i & c\theta_i & 0 & 0 \\ 0 & 0 & 1 & 0 \\ \hline 0 & 0 & 0 & 1 \end{array}\right] $$

3) Now, the coordinate system $\{\mathcal{Q}\}$ undergoes a translational motion by a_i along $\boldsymbol{x}_Q$ to align with the coordinate system $\{\mathcal{R}\}$. The homogeneous transformation matrix for this translation is given by

$$ {}^{Q}\boldsymbol{T}_R = \boldsymbol{T}(x, a_i) = \left[\begin{array}{ccc|c} 1 & 0 & 0 & a_i \\ 0 & 1 & 0 & 0 \\ 0 & 0 & 1 & 0 \\ \hline 0 & 0 & 0 & 1 \end{array}\right] $$

4) Finally, the coordinate system $\{\mathcal{R}\}$ undergoes a rotational motion with an angle α_i around the $\boldsymbol{x}_R$ axis to align with the coordinate system $\{i\}$. The homogeneous transformation matrix for this rotation is as follows.

$$ {}^{R}\boldsymbol{T}_i = \boldsymbol{T}(x, \alpha_i) = \left[\begin{array}{ccc|c} 1 & 0 & 0 & 0 \\ 0 & c\alpha_i & -s\alpha_i & 0 \\ 0 & s\alpha_i & c\alpha_i & 0 \\ \hline 0 & 0 & 0 & 1 \end{array}\right] $$

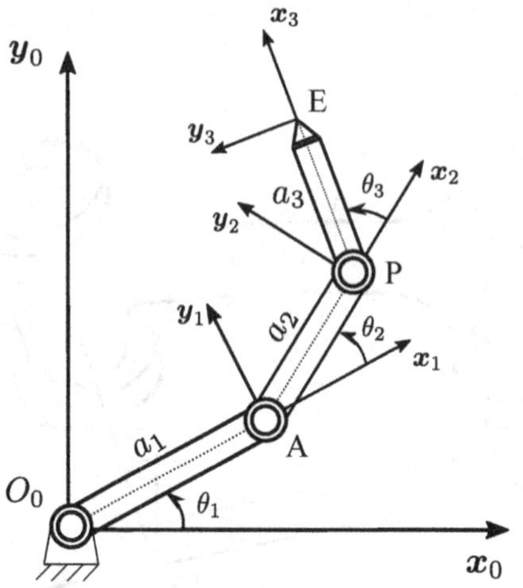

FIGURE 4.6
Installation of coordinate systems and determination of Denavit-Hartenberg parameters for the planar robot in Case Study 4.1.

In this way, the final homogeneous transformation matrix, obtained by the right multiplication of the four transformation matrices above, can be written as follows:

$$\begin{aligned} {}^{i-1}\boldsymbol{T}_i &= {}^{i-1}\boldsymbol{T}_P \cdot {}^{P}\boldsymbol{T}_Q \cdot {}^{Q}\boldsymbol{T}_R \cdot {}^{R}\boldsymbol{T}_i \\ &= \left[\begin{array}{ccc|c} c\theta_i & -c\alpha_i s\theta_i & s\alpha_i s\theta_i & a_i c\theta_i \\ s\theta_i & c\alpha_i c\theta_i & -s\alpha_i c\theta_i & a_i s\theta_i \\ 0 & s\alpha_i & c\alpha_i & d_i \\ \hline 0 & 0 & 0 & 1 \end{array}\right] \end{aligned} \tag{4.6}$$

In this representation, the following notations $c\theta_i = \cos(\theta_i), s\theta_i = \sin(\theta_i), c\alpha_i = \cos(\alpha_i), s\alpha_i = \sin(\alpha_i)$ are used. The inverse of this transformation matrix is calculated using the physical concept of transformation and the Equation 2.56 as follows:

$${}^{i}\boldsymbol{T}_{i-1} = \left[\begin{array}{ccc|c} c\theta_i & s\theta_i & 0 & -a_i \\ -c\alpha_i s\theta_i & c\alpha_i c\theta_i & s\alpha_i & -d_i s\alpha_i \\ s\alpha_i s\theta_i & -s\alpha_i c\theta_i & c\alpha_i & -d_i c\alpha_i \\ \hline 0 & 0 & 0 & 1 \end{array}\right] \tag{4.7}$$

Case Study 4.1 (Planar 3<u>R</u> Robot) *Consider the planar 3<u>R</u> robot, as shown in Figure 4.6. By installing appropriate coordinate systems and defining the Denavit-Hartenberg parameters for this robot, obtain the transformation matrix for the robot's end-effector at point E with respect to the fixed coordinate system attached to the robot's base.*

Solution: First, consider the rotational axes of the robot's joints. Since all revolute joints lie in the plane, their rotational axes are all considered perpendicular and outward to the plane. Thus, by drawing the coordinate system $\{0\}$ on the robot's base, three other coordinate systems $\{1\}, \{2\}, \{3\}$ can be considered along the lengths of the links at their respective endpoints A, P, E.

TABLE 4.1
Denavit-Hartenberg parameters for the Case Study 4.1

i	a_i	α_i	d_i	θ_i
1	a_1	0	0	θ_1
2	a_2	0	0	θ_2
3	a_3	0	0	θ_3

The Denavit-Hartenberg parameters for this robot are determined according to Table 4.1. As observed in this table, the lengths of the three robot arms are denoted by a_1, a_2, a_3, and the joint variables (colored in blue) are $\theta_1, \theta_2, \theta_3$. Due to the geometry of the robot, the joints' twist angles and link offsets are all equal to zero. Using the relationship 4.6, the homogeneous transformation matrix for three consecutive links can be obtained as follows:

$$
{}^0\boldsymbol{T}_1 = \left[\begin{array}{ccc|c} c_1 & -s_1 & 0 & a_1c_1 \\ s_1 & c_1 & 0 & a_1s_1 \\ 0 & 0 & 1 & 0 \\ \hline 0 & 0 & 0 & 1 \end{array}\right], \quad {}^1\boldsymbol{T}_2 = \left[\begin{array}{ccc|c} c_2 & -s_2 & 0 & a_2c_2 \\ s_2 & c_2 & 0 & a_2s_2 \\ 0 & 0 & 1 & 0 \\ \hline 0 & 0 & 0 & 1 \end{array}\right]
$$

$$
{}^2\boldsymbol{T}_3 = \left[\begin{array}{ccc|c} c_3 & -s_3 & 0 & a_3c_3 \\ s_3 & c_3 & 0 & a_3s_3 \\ 0 & 0 & 1 & 0 \\ \hline 0 & 0 & 0 & 1 \end{array}\right]
$$

where $c_i = \cos(\theta_i)$ and $s_i = \sin(\theta_i)$. By multiplying the transformation matrices of consecutive links, the homogeneous transformation matrix for the robot's end-effector point E, with respect to the coordinate system fixed to the robot's base, can be obtained as follows:

$$
\begin{aligned}
{}^0\boldsymbol{T}_3 &= {}^0\boldsymbol{T}_1 \cdot {}^1\boldsymbol{T}_2 \cdot {}^2\boldsymbol{T}_3 \\
&= \left[\begin{array}{ccc|c} c_{123} & -s_{123} & 0 & a_1c_1 + a_2c_{12} + a_3c_{123} \\ s_{123} & c_{123} & 0 & a_1s_1 + a_2s_{12} + a_3s_{123} \\ 0 & 0 & 1 & 0 \\ \hline 0 & 0 & 0 & 1 \end{array}\right]
\end{aligned} \tag{4.8}
$$

In which, the notations $c_{12} = \cos(\theta_1 + \theta_2)$, $s_{12} = \sin(\theta_1 + \theta_2)$, $c_{123} = \cos(\theta_1 + \theta_2 + \theta_3)$, and $s_{123} = \sin(\theta_1 + \theta_2 + \theta_3)$ are used.

Case Study 4.2 (3DOF Elbow Manipulator) *Consider the robot shown in Figure 4.7. By installing appropriate coordinate systems and defining Denavit-Hartenberg parameters for this robot, obtain the end-effector transformation matrix for the robot's end-effector with respect to the coordinate system fixed to the robot's base.*

Solution: First, consider the rotation axes of the robot's joints. Given the geometry of the robot, it can be observed that the rotation axis of the first revolute joint is along the vertical axis, producing out of the plane motion for the robot's end-effector. The other two revolute joints have rotation axes parallel to each other and perpendicular

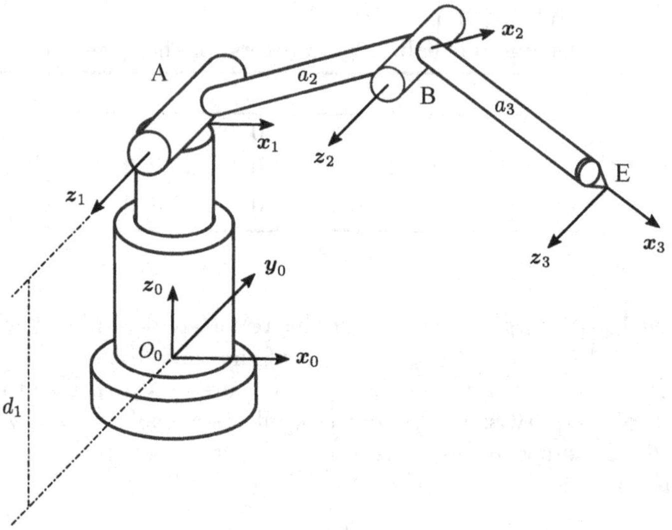

FIGURE 4.7
Installation of coordinate systems and determination of Denavit-Hartenberg parameters for the three-degree-of-freedom Elbow manipulator in Case Study 4.2

to the vertical plane. Place the reference coordinate system {0} on the robot's base, as shown in the figure. Install the coordinate system for the first link {1} at the end of the link in a way that its z_1 axis aligns with the rotation axis, and its x_1 axis lies in the plane passing through the second link. In this configuration, the length of the first link is zero, its offset is d_1, and the joint angle is θ_1, forming the first row of Table 4.2.

Place the other two coordinate systems parallel to {1}, at the ends of the second and third links of the robot, respectively. By this means, the lengths of the second and third links are a_2 and a_3, respectively, and the joint angles are θ_2 and θ_3. With this representation Table 4.2 is completed (The robot's joint variables are highlighted in blue color).

Now, using the equation 4.6, obtain the homogeneous transformation matrix for three consecutive links as follows.

$$
{}^0\boldsymbol{T}_1 = \left[\begin{array}{ccc|c} c_1 & 0 & s_1 & 0 \\ s_1 & 0 & -c_1 & 0 \\ 0 & 1 & 0 & d_1 \\ \hline 0 & 0 & 0 & 1 \end{array}\right], \quad {}^1\boldsymbol{T}_2 = \left[\begin{array}{ccc|c} c_2 & -s_2 & 0 & a_2c_2 \\ s_2 & c_2 & 0 & a_2s_2 \\ 0 & 0 & 1 & 0 \\ \hline 0 & 0 & 0 & 1 \end{array}\right]
$$

$$
{}^2\boldsymbol{T}_3 = \left[\begin{array}{ccc|c} c_3 & -s_3 & 0 & a_3c_3 \\ s_3 & c_3 & 0 & a_3s_3 \\ 0 & 0 & 1 & 0 \\ \hline 0 & 0 & 0 & 1 \end{array}\right]
$$

where notations $c_i = \cos(\theta_i)$ and $s_i = \sin(\theta_i)$ are used. By multiplying consecutive link transformation matrices together, the end-effector homogeneous transformation matrix,

TABLE 4.2
Denavit-Hartenberg parameters for the Case Study 4.2

i	a_i	α_i	d_i	θ_i
1	0	$\pi/2$	d_1	θ_1
2	a_2	0	0	θ_2
3	a_3	0	0	θ_3

with respect to the coordinate system fixed to the robot's base is obtained as follows.

$$
\begin{aligned}
{}^0T_3 &= {}^0T_1 \cdot {}^1T_2 \cdot {}^2T_3 \\
&= \left[\begin{array}{ccc|c} c_1c_{23} & -c_1s_{23} & s_1 & c_1\,(a_2c_2 + a_3c_{23}) \\ s_1c_{23} & -s_1s_{23} & -c_1 & s_1\,(a_2c_2 + a_3c_{23}) \\ s_{23} & c_{23} & 0 & d_1 + a_2s_2 + a_3s_{23} \\ \hline 0 & 0 & 0 & 1 \end{array}\right]
\end{aligned} \tag{4.9}
$$

where shorthand notations $c_{23} = \cos(\theta_2 + \theta_3)$ and $s_{23} = \sin(\theta_2 + \theta_3)$ are used.

Case Study 4.3 (SCARA Robot) *Consider the SCARA robot shown in Figure 4.8. By installing appropriate coordinate systems and defining Denavit-Hartenberg parameters for this robot, obtain the end-effector transformation matrix relative to the coordinate system fixed to the robot's base.*

Solution: The rotation axes of the three revolute joints of the robot are all aligned with the vertical axis in the plane. The third prismatic joint also moves in this direction,

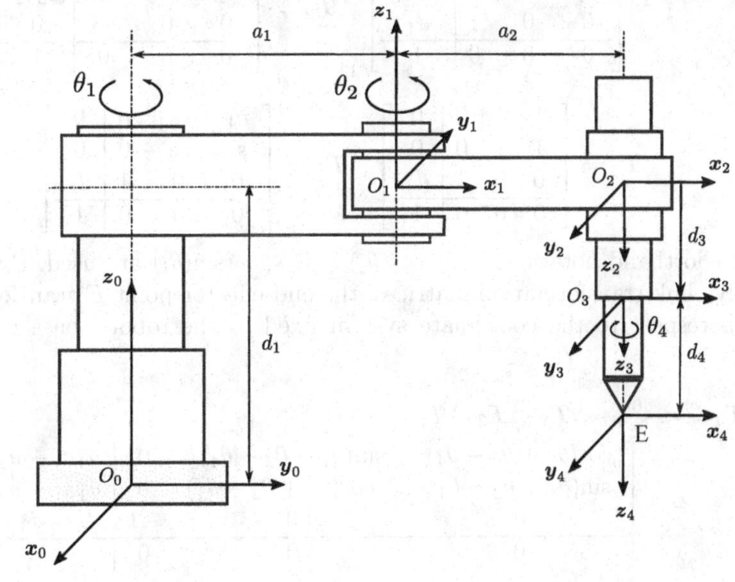

FIGURE 4.8
Installing coordinate systems and determining Denavit-Hartenberg parameters for the SCARA robot of Case Study 4.3

therefore, its axis is parallel to the rotation axes of the other joints. Install the reference coordinate system $\{0\}$ according to the figure on the robot's base, and furthermore, install the coordinate system of the first link $\{1\}$ at the end of the link in such a way that its $\boldsymbol{z}_1$ axis is along the vertical axis and its $\boldsymbol{x}_1$ axis is along the length of the second link. In this way, the length of the first link is a_1, its offset is d_1, and the joint angle is θ_1, completing the first row of Table 4.3.

Similarly, the coordinate system of the second link can be installed at the end of that link. The direction of the $\boldsymbol{z}_2$ axis can be chosen upwards or downwards. In this case study downward direction is used for this axis, to represent the downward motion of the robot's end-effector. Thus, the length of the second link is a_2, its twist angle is $\alpha_2 = \pi$, and the joint angle is θ_2, constituting the second row of Table 4.3.

The $\boldsymbol{z}_3$ axis is in the direction of the prismatic joint's motion and parallel to the $\boldsymbol{z}_2$ axis. This is while, considering the alignment of these two axes, the origin of the coordinate system can be placed on any point on these two axes. To obtain a proper representation of the prismatic joint's motion, place the origin of the third coordinate system $\{3\}$ at the end of this link. Thus, the length of this link is zero, its translational motion variable is d_3, and these parameters are reported in the third row of Table 4.3.

Finally, place the last coordinate system $\{4\}$ at the end of the fourth link and on the desired point of the robot's end-effector, denoted by E. In this way, the length of the link is zero, its offset is d_4, and the joint angle is θ_4, completing the fourth row of Table 4.3. Thus, the table of Denavit-Hartenberg parameters for the robot is completed (with joint variables highlighted in blue). It is clear that the choice of the installation location of the coordinate-systems is not unique, and other representations can be considered for this robot.

Using Equation 4.6, the homogeneous transformation matrix for four consecutive links can be easily obtained as follows.

$$
{}^0\boldsymbol{T}_1 = \left[\begin{array}{ccc|c} c_1 & -s_1 & 0 & a_1c_1 \\ s_1 & c_1 & 0 & a_1s_1 \\ 0 & 0 & 1 & d_1 \\ \hline 0 & 0 & 0 & 1 \end{array}\right], \quad {}^1\boldsymbol{T}_2 = \left[\begin{array}{ccc|c} c_2 & s_2 & 0 & a_2c_2 \\ s_2 & -c_2 & 0 & a_2s_2 \\ 0 & 0 & -1 & 0 \\ \hline 0 & 0 & 0 & 1 \end{array}\right]
$$

$$
{}^2\boldsymbol{T}_3 = \left[\begin{array}{ccc|c} 1 & 1 & 0 & 0 \\ 1 & 1 & 0 & 0 \\ 0 & 0 & 1 & d_3 \\ \hline 0 & 0 & 0 & 1 \end{array}\right], \quad {}^3\boldsymbol{T}_4 = \left[\begin{array}{ccc|c} c_4 & -s_4 & 0 & 0 \\ s_4 & c_4 & 0 & 0 \\ 0 & 0 & 1 & d_4 \\ \hline 0 & 0 & 0 & 1 \end{array}\right]
$$

In which shorthand notations $c_i = \cos(\theta_i)$ and $s_i = \sin(\theta_i)$ are used. By multiplying consecutive link transformation matrices, the end-effector point E transformation matrix, with respect to the coordinate system fixed to the robot's base is obtained as follows.

$$
\begin{aligned}
{}^0\boldsymbol{T}_4 &= {}^0\boldsymbol{T}_1 \cdot {}^1\boldsymbol{T}_2 \cdot {}^2\boldsymbol{T}_3 \cdot {}^3\boldsymbol{T}_4 \\
&= \left[\begin{array}{ccc|c} \cos(\theta_1+\theta_2-\theta_4) & \sin(\theta_1+\theta_2-\theta_4) & 0 & a_1c_1 + a_2c_{12} \\ \sin(\theta_1+\theta_2-\theta_4) & -\cos(\theta_1+\theta_2-\theta_4) & 0 & a_1s_1 + a_2s_{12} \\ 0 & 0 & -1 & d_1 - d_3 - d_4 \\ \hline 0 & 0 & 0 & 1 \end{array}\right]
\end{aligned} \tag{4.10}
$$

where notations $c_{12} = \cos(\theta_1 + \theta_2)$, $s_{12} = \sin(\theta_1 + \theta_2)$ are used.

TABLE 4.3
Denavit-Hartenberg parameters for the Case Study 4.3

i	a_i	α_i	d_i	θ_i
1	a_1	0	d_1	θ_1
2	a_2	π	0	θ_2
3	0	0	d_3	0
4	0	0	d_4	θ_4

Case Study 4.4 (Stanford Robot) *Consider the Stanford robot shown in Figure 4.9. By installing appropriate coordinate systems and defining Denavit-Hartenberg parameters for this robot, derive the transformation matrix of the end-effector with respect to the coordinate system affixed to the base of the robot.*

Solution: The Stanford robot shown in Figure 4.9 has six degrees of freedom, consisting of five revolute joints and one prismatic joint. To analyze the kinematics of this robot, first, install the coordinate axes according to the figure on the robot's joints. Given the motion of the robot's links, it can be inferred that the rotation axis of the first revolute joint is along the vertical axis which produces out-of-plane motion for the end-effector. The second revolute joint has a rotation axis perpendicular to the vertical axis. Meanwhile, the prismatic joint's translational motion axis in this robot is aligned with the horizontal axis. The three revolute joint axes forming the robot's wrist intersect

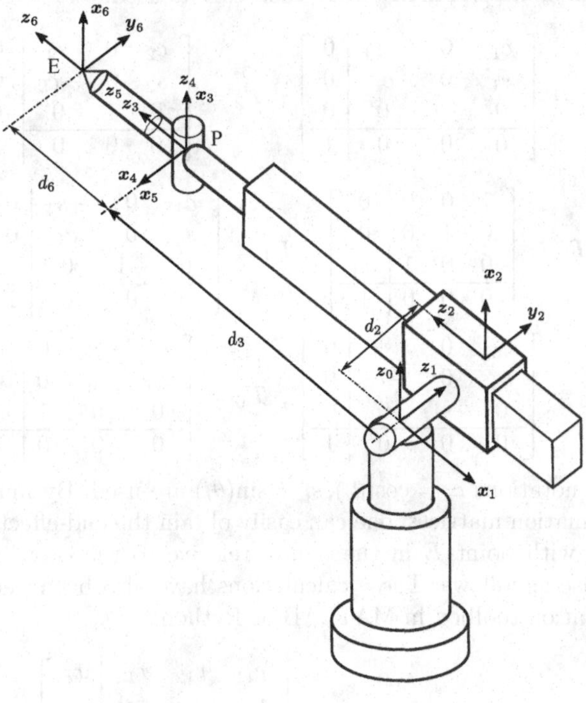

FIGURE 4.9
Installation of coordinate systems and determination of Denavit-Hartenberg parameters for the Stanford robot of Case Study 4.4

at a single point, resembling the Euler angles $\boldsymbol{w}-\boldsymbol{v}-\boldsymbol{w}$ structure, providing suitable orientation control for the robot's end-effector.

Install the reference coordinate system $\{0\}$ according to the figure on the robot's base. In this example, contrary to the two previous cases and for simplicity in calculations, the origin of this coordinate system is placed at the end of the first link to set the offset of the first link d_1 to zero. Install the coordinate system of link one $\{1\}$, on the end of the link such that its axis $\boldsymbol{z}_1$ is along the rotation axis, and its axis $\boldsymbol{x}_1$ is along the length of the first link. In this case, the length of the first link is also zero, its offset is d_1, and the joint angle is θ_1, completing the first row of Table 4.4.

Install the coordinate system of link two according to Figure 4.9, at the end of the second joint in such a way that its axis $\boldsymbol{z}_2$ is along the prismatic joint's translational motion, and its axis $\boldsymbol{x}_2$ is parallel to $\boldsymbol{x}_1$. This choice is made to simplify the kinematic analysis of the robot. Place the origins of the coordinate systems from the third to the fifth on the wrist point P with a distance of d_3 from the origin of the second coordinate system. Furthermore, place the origin of the sixth coordinate system on the end-effector point E of the robot, which is at a distance of d_6 from the wrist point. Consider the axes $\boldsymbol{z}_i$ in coordinate systems three to six along the rotation axes of the respective joints, completing the installation of the coordinate systems on the robot.

In this way, the Denavit-Hartenberg parameters table for the robot is completed (where the robot's joint variables are shown in blue). It is clear that the choice of the installation points of the coordinate systems is not unique, and other representations can also be considered for the same robot.

Now, using equation 4.6, the homogeneous transformation matrix for six consecutive links can be obtained as follows. To calculate this transformation, the symbolic computation toolbox of MATLAB® or Python can be used.

$$
{}^0\boldsymbol{T}_1 = \left[\begin{array}{ccc|c} c_1 & 0 & -s_1 & 0 \\ s_1 & 0 & c_1 & 0 \\ 0 & -1 & 0 & 0 \\ \hline 0 & 0 & 0 & 1 \end{array}\right], \; {}^1\boldsymbol{T}_2 = \left[\begin{array}{ccc|c} c_2 & 0 & s_2 & 0 \\ s_2 & 0 & -c_2 & 0 \\ 0 & 1 & 0 & d_2 \\ \hline 0 & 0 & 0 & 1 \end{array}\right]
$$

$$
{}^2\boldsymbol{T}_3 = \left[\begin{array}{ccc|c} 1 & 0 & 0 & 0 \\ 0 & 1 & 0 & 0 \\ 0 & 0 & 1 & d_3 \\ \hline 0 & 0 & 0 & 1 \end{array}\right], \; {}^3\boldsymbol{T}_4 = \left[\begin{array}{ccc|c} c_4 & 0 & -s_4 & 0 \\ s_4 & 0 & c_4 & 0 \\ 0 & -1 & 0 & 0 \\ \hline 0 & 0 & 0 & 1 \end{array}\right]
$$

$$
{}^4\boldsymbol{T}_5 = \left[\begin{array}{ccc|c} c_5 & 0 & s_5 & 0 \\ s_5 & 0 & -c_5 & 0 \\ 0 & -1 & 0 & 0 \\ \hline 0 & 0 & 0 & 1 \end{array}\right], \; {}^5\boldsymbol{T}_6 = \left[\begin{array}{ccc|c} c_6 & -s_6 & 0 & 0 \\ s_6 & c_6 & 0 & 0 \\ 0 & 0 & 1 & d_6 \\ \hline 0 & 0 & 0 & 1 \end{array}\right]
$$

where shorthand notations $c_i = \cos(\theta_i), s_i = \sin(\theta_i)$ are used. By multiplying consecutive link transformation matrices, one can easily obtain the end-effector transformation matrix, depicted with point E in the figure, relative to the coordinate system fixed to the robot's base as follows. These calculations have also been performed using the symbolic computation toolbox in MATLAB or Python.

$$
{}^0\boldsymbol{T}_6 = {}^0\boldsymbol{T}_1 \cdot {}^1\boldsymbol{T}_2 \cdots {}^5\boldsymbol{T}_6 = \left[\begin{array}{ccc|c} r_{11} & r_{12} & r_{13} & x_e \\ r_{21} & r_{22} & r_{23} & y_e \\ r_{31} & r_{32} & r_{33} & z_e \\ \hline 0 & 0 & 0 & 1 \end{array}\right] \tag{4.11}
$$

TABLE 4.4
Denavit-Hartenberg parameters for the Case Study 4.4

i	a_i	α_i	d_i	θ_i
1	0	$-\pi/2$	0	θ_1
2	0	$+\pi/2$	d_2	θ_2
3	0	0	d_3	0
4	0	$-\pi/2$	0	θ_4
5	0	$+\pi/2$	0	θ_5
6	0	0	d_6	θ_6

and the components of the homogeneous matrix are obtained using the notations $c_i = \cos(\theta_i)$, $s_i = \sin(\theta_i)$ as follows.

$$
\begin{aligned}
r_{11} &= c_1\left[c_2(c_4c_5c_6 - s_4s_6) - s_2s_5c_6\right] - s_1(s_4c_5c_6 + c_4s_6) \\
r_{21} &= s_1\left[c_2(c_4c_5c_6 - s_4s_6) - s_2s_5c_6\right] + c_1(s_4c_5c_6 + c_4s_6) \\
r_{31} &= s_2(s_4s_6 - c_4c_5c_6) - c_2s_5c_6, \\
r_{12} &= c_1\left[-c_2(c_4c_5c_6 + s_4c_6) + s_2s_5s_6\right] + s_1(s_4c_5s_6 - c_4c_6) \\
r_{22} &= s_1\left[-c_2(c_4c_5s_6 + s_4c_6) + s_2s_5s_6\right] + c_1(-s_4c_5s_6 + c_4c_6) \\
r_{32} &= s_2(c_4c_5s_6 + s_4c_6) + c_2s_5c_6 \\
r_{13} &= c_1(c_2c_4s_5 + s_2c_5) - s_1s_4s_5 \\
r_{23} &= s_1(c_2c_4s_5 + s_2c_5) + c_1s_4s_5 \\
r_{33} &= -s_2c_4s_5 + c_2c_5 \\
x_e &= -s_1d_2 - d_3(s_1c_4 + c_1c_2s_4) + d_6(c_1s_2c_5 - s_1s_4s_5 + c_1c_2c_4s_5) \\
y_e &= c_1d_2 + d_3(c_1c_4 - s_1c_2s_4) + d_6(s_1s_2c_5 + c_1s_4s_5 + s_1c_2c_4s_5) \\
z_e &= s_2s_4d_3 + d_6(c_2c_5 - c_4s_2s_5)
\end{aligned}
$$

Case Study 4.5 (6DOF Elbow Manipulator) *Consider the 6DOF Elbow Manipulator shown in Figure 4.10. This robot is created by adding a 3DOF wrist to the 3DOF Elbow manipulator introduced in case study 4.2. By installing appropriate coordinate systems and defining Denavit-Hartenberg parameters for this robot, obtain the end-effector transformation matrix relative to the coordinate system fixed to the robot's base.*

Solution: First, obtain the Denavit-Hartenberg parameters for the six-degree-of-freedom elbow manipulator using the coordinate systems installed on the robot joints, as shown in Figure 4.10. The three last coordinate systems are installed on the wrist of the manipulator, aligning the z axes along the corresponding axis of rotations. The Denavit-Hartenberg parameters of this robot can be easily extended from the 3DOF case, using this figure. The final parameters are reported in Table 4.5 (where the joint variables are highlighted in blue). The homogeneous transformation matrices for the

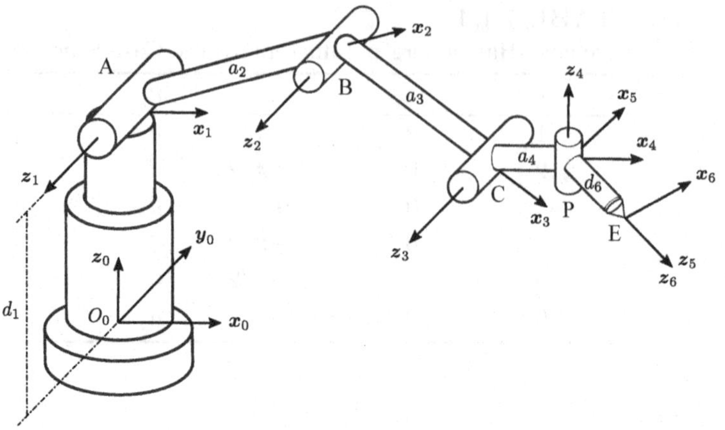

FIGURE 4.10
Schematic of the 6DOF Elbow Manipulator, created by adding a 3DOF wrist to the 3DOF Elbow Manipulator

robot may be easily obtained by using theses Denavit-Hartenberg parameters.

$$^0\boldsymbol{T}_1 = \left[\begin{array}{ccc|c} c_1 & 0 & -s_1 & 0 \\ s_1 & 0 & c_1 & 0 \\ 0 & 1 & 0 & 0 \\ \hline 0 & 0 & 0 & 1 \end{array}\right], {}^1\boldsymbol{T}_2 = \left[\begin{array}{ccc|c} c_2 & -s_2 & 0 & a_2c_2 \\ s_2 & c_2 & 0 & a_2s_2 \\ 0 & 0 & 1 & 0 \\ \hline 0 & 0 & 0 & 1 \end{array}\right]$$

$$^2\boldsymbol{T}_3 = \left[\begin{array}{ccc|c} c_3 & -s_3 & 0 & a_3c_3 \\ s_3 & c_3 & 0 & a_3s_3 \\ 0 & 0 & 1 & 0 \\ \hline 0 & 0 & 0 & 1 \end{array}\right], {}^3\boldsymbol{T}_4 = \left[\begin{array}{ccc|c} c_4 & 0 & -s_4 & a_4c_4 \\ s_4 & 0 & c_4 & a_4s_4 \\ 0 & -1 & 0 & 0 \\ \hline 0 & 0 & 0 & 1 \end{array}\right]$$

$$^4\boldsymbol{T}_5 = \left[\begin{array}{ccc|c} c_5 & 0 & s_5 & 0 \\ s_5 & 0 & -c_5 & 0 \\ 0 & 1 & 0 & 0 \\ \hline 0 & 0 & 0 & 1 \end{array}\right], {}^5\boldsymbol{T}_6 = \left[\begin{array}{ccc|c} c_6 & -s_6 & 0 & 0 \\ s_6 & c_6 & 0 & 0 \\ 0 & 0 & 1 & d_6 \\ \hline 0 & 0 & 0 & 1 \end{array}\right]$$

TABLE 4.5
Denavit-Hartenberg parameters for the Case Study 4.5

i	a_i	α_i	d_i	θ_i
1	0	$\pi/2$	0	θ_1
2	a_2	0	0	θ_2
3	a_3	0	0	θ_3
4	a_4	$-\pi/2$	0	θ_4
5	0	$\pi/2$	0	θ_5
6	0	0	d_6	θ_6

4.4 Forward Kinematic Analysis Methods

The objective of forward kinematic analysis in robotics is to establish an analytical relationship between the joint variables $\boldsymbol{q}$ and the task variables within the robot's workspace $\boldsymbol{\mathcal{X}}$, and subsequently solve it. The loop closure method can generally formulate this relationship in vector form. In robots with simple kinematic structures, often planar in nature, this vector-based formulation can be elucidated through analytical geometry and trigonometric relationships. This method entails expressing the formulation with algebraic equations that govern both the joint variables $\boldsymbol{q}$ and the workspace variables $\boldsymbol{\mathcal{X}}$. This intuitive approach provides a guide for solving the problem in more general cases and in robots with more complex kinematic structures. The geometric principles used in this method will also prove insightful in the subsequent analysis of inverse kinematics for serial robots, as discussed in Chapter 5.

On the other hand, for solving the forward kinematic problem of serial robots in a general sense, homogeneous transformation matrices can also be employed. As previously explained, in the most general case, serial robots and the sequential connection between robot links can be effectively described using Denavit-Hartenberg parameters through the homogeneous transformation matrix of each link. By this means, the closed-loop relationship can be rewritten using matrix operations and then solved. This method is detailed in Section 4.4.2 and is applied to various case studies.

Other methods have been proposed for forward kinematic analysis of serial robots in robotics literature [44, 45]. Among these methods, the use of screw algebra is another efficient approach that is less commonly mentioned in related books [3, 58]. In this method, the closed-loop relationship is significantly simplified using six-dimensional screw arrays, and solving it becomes more straightforward by utilizing screw algebra. A thorough exposition of this approach will be delineated in Section 4.5.

It is worth noting that, despite the differences in the formulation of forward kinematics solutions using these methods, they all lead to a unified result. In practice, the appropriate method should be chosen based on the complexity of the robot's structure and the computational cost, as determined by the user. Despite the differences in the formulation of the problem, the geometric foundations in solving the closed-loop method are very insightful. Therefore, in the following section, we continue the discussion using this method.

4.4.1 Loop Closure Method

In order to provide a more detailed description of the loop closure method, consider the 3$\underline{R}$ robot depicted in Figure 4.11. In the forward kinematic problem, the assumption is that the joint space variables $\boldsymbol{q}$ are given, and the objective is to find the necessary algebraic relationships to determine the task space variables $\boldsymbol{\mathcal{X}}$. The 3$\underline{R}$ robot consists of three revolute joints, providing three degrees of freedom. Thus, in this robot, the joint space variables are the joint angles, represented in the vector form $\boldsymbol{q} = [\theta_1, \theta_2, \theta_3]^T$. On the other hand, the end-effector of the robot, denoted by point E in the figure, has three degrees of freedom in the plane. Its configuration is fully described by the position vector with two variables (x_e, y_e) and the end-effector orientation with an angle θ_e. Therefore, the task space variables of the robot form a three-dimensional array $\boldsymbol{\mathcal{X}} = [x_e, y_e, \theta_e]^T$, as well. Thus, in solving the forward kinematic problem, we are looking for three independent equations that relate the three components of the task space vector $\boldsymbol{\mathcal{X}}$ to the joint space vector $\boldsymbol{q}$.

In the loop closure method, we describe the end-effector's position using two vector paths. The first path is formed by adding the vectors describing the positions of the robot

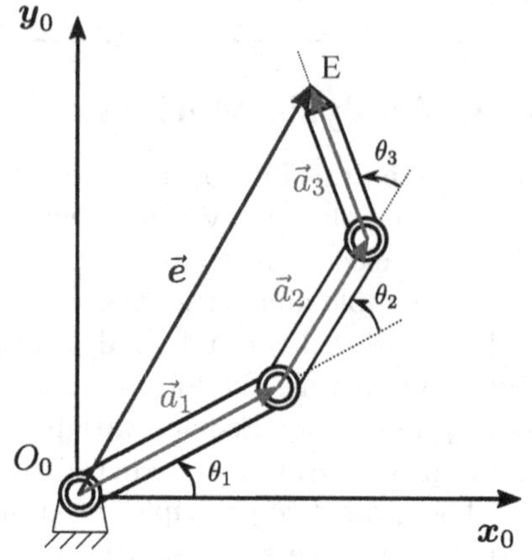

FIGURE 4.11
Vector description of the loop closure method in the planar 3<u>R</u> Robot

links, represented in the figure by the vectors $\vec{a}_1, \vec{a}_2, \vec{a}_3$ (in green). The second path directly targets the end-effector point vector $\vec{e}$ (depicted in blue) in this figure. The loop closure method implies that, regardless of the path taken to reach point E, the sum of the traversed vectors is equal. The vector relation is given by:

$$\vec{e} = \vec{a}_1 + \vec{a}_2 + \vec{a}_3 \tag{4.12}$$

This vector equation only describes the location of the end-effector point E, and it is necessary to determine the orientation of the end-effector from two paths. Fortunately, the orientation of the robot's end-effector in the plane is described by a single angle, denoted here as θ_e. Therefore, we need to find how to generate this angle from the vector of joint angles. In this robot, using geometric relations, it can be easily seen that this relation is also a function of the algebraic sum of the joint angles.

$$\theta_e = \theta_1 + \theta_2 + \theta_3 \tag{4.13}$$

Now, if we express the vector equation 4.12 in terms of its components along $\boldsymbol{x}$ and $\boldsymbol{y}$ axes and combine it with equation 4.13, we will obtain the forward kinematics equations of the 3<u>R</u> robot, as follows.

$$\begin{aligned} x_e &= a_1 c_1 + a_2 c_{12} + a_3 c_{123} \\ y_e &= a_1 s_1 + a_2 s_{12} + a_3 s_{123} \\ \theta_e &= \theta_1 + \theta_2 + \theta_3 \end{aligned} \tag{4.14}$$

In these relations, shorthand notations $c_{12} = \cos(\theta_1 + \theta_2), s_{12} = \sin(\theta_1 + \theta_2), c_{123} = \cos(\theta_1 + \theta_2 + \theta_3), s_{123} = \sin(\theta_1 + \theta_2 + \theta_3)$, are used. In this way, if the components of the vector of joint space variables $\boldsymbol{q} = [\theta_1, \theta_2, \theta_3]^T$ are given, the components of the task space variables $\boldsymbol{\mathcal{X}} = [x_e, y_e, \theta_e]^T$ are easily and uniquely calculated using this relation, and the problem of forward kinematics for the robot is solved.

4.4.2 Denavit-Hartenberg Method

The loop closure method forms the main framework for solving the forward kinematics of robots. When the kinematic structure of a robot becomes more complex or the robot moves in three-dimensional space, the vector relationship of the loop-closure method, where we describe the two paths of the robot's end-effector, can still be easily extended to three-dimensional space. However, describing the orientation of the end-effector is not straightforward, and different methods for representing the end-effector orientation in three-dimensional space need to be utilized.

In this section, we describe the position and orientation of the end-effector using homogeneous transformation matrices. Similar to what we had in the loop closure method, we perform these descriptions through two paths and equate them. Using this approach, we can easily represent the position and orientation of the end-effector in the form of a homogeneous transformation matrix. To do this, let's assume that the vector representing the position of the desired point on the robot's end-effector $\vec{e}$ is given by $\vec{e} = [x_e, y_e, z_e]^T$. Furthermore, assume that the orientation of the end-effector is represented by the screw vector $\boldsymbol{\theta}_e$. In this case, the homogeneous matrix describing the position and orientation of the robot's end-effector in the reference frame $\{0\}$ is as follows.

$$ {}^0\boldsymbol{T}_e = \left[\begin{array}{ccc|c} & & & {}^0x_e \\ & {}^0\boldsymbol{R}_e(\boldsymbol{\theta}_e) & & {}^0y_e \\ & & & {}^0z_e \\ \hline 0 & 0 & 0 & 1 \end{array}\right] \tag{4.15} $$

In which ${}^0\boldsymbol{R}_e(\boldsymbol{\theta}_e)$ is the rotation matrix describing the orientation of the robot's end-effector with respect to the reference frame $\{0\}$. Given the general nature of the rotation matrix in representing the orientation of the robot's end-effector, any type of Euler angle or quaternion representation can be used to describe the end-effector's orientation.

Now, consider another path for representing the position and orientation of the robot's end-effector through the links, using their homogeneous transformation matrices. This path is formed by multiplying the transformation matrices of the consecutive links in the kinematic chain. If the robot in question has n consecutive links, this path can be obtained from the following relationship.

$$ {}^0\boldsymbol{T}_n = {}^0\boldsymbol{T}_1 \cdot {}^1\boldsymbol{T}_2 \cdots {}^{n-1}\boldsymbol{T}_n \tag{4.16} $$

By equating the two paths in the loop closure method, the problem of the robot's forward kinematics can be formulated and examined using the following relationship.

$$ {}^0\boldsymbol{T}_e = {}^0\boldsymbol{T}_n \tag{4.17} $$

If we express this relationship in terms of the components of the two sides, we will obtain the representation of the forward kinematics matrix for the robot in the general case.

$$ \left[\begin{array}{ccc|c} & & & {}^0x_e \\ & {}^0\boldsymbol{R}(\boldsymbol{\theta}_e) & & {}^0y_e \\ & & & {}^0z_e \\ \hline 0 & 0 & 0 & 1 \end{array}\right] = {}^0\boldsymbol{T}_1(q_1) \cdot {}^1\boldsymbol{T}_2(q_2) \cdots {}^{n-1}\boldsymbol{T}_n(q_n) \tag{4.18} $$

Note that in this matrix representation, if the joint space variables $\boldsymbol{q} = [q_1, q_2, \ldots, q_n]^T$ are given, using the Denavit-Hartenberg parameters, we can obtain the transformation matrices of the right-hand side of the equation. In this way, the components of the robot's workspace variables $\boldsymbol{\mathcal{X}} = [x_e, y_e, z_e, \theta_{e_x}, \theta_{e_y}, \theta_{e_z}]^T$ are obtained by equating the obtained transformation matrix with the robot's end-effector transformation matrix.

The three components of the position of the robot's end-effector point E are analytically determined by equating x_e, y_e, z_e components with their corresponding vector on the right-hand side of Equation 4.18. Meanwhile, the orientation components of the robot are implicitly obtained through the equality of the rotation matrices on both sides. Moreover, if numerical values for the joint variables $\boldsymbol{q} = [q_1, q_2, \dots, q_n]^T$ are provided, the end-effector rotation matrix ${}^0\boldsymbol{R}(\boldsymbol{\theta}_e)$ is also numerically obtained, and one can obtain the representation of screw, quaternion, or any desired Euler angle for it.

Case Study 4.6 (Planar 3<u>R</u> Robot) *Consider the planar 3<u>R</u> robot introduced in Case Study 4.1 and illustrated in Figure 4.6. Utilizing homogeneous transformation matrices, analyze the forward kinematics of the end-effector of the robot. Compare the results with the expressions given in Equation 4.14, obtained using the loop closure method.*

Solution: The task space variables for this robot consist of two position components and the end-effector orientation angle, represented as a three-dimensional array: $[x_e, y_e, \theta_e]$. Considering that the robot motion is confined to the $x-y$ plane with $z_e = 0$, and the rotational angle with respect to the $x-y$ axes equal to zero, it follows that $\theta_{e_x} = \theta_{e_y} = 0$ and $\theta_{e_z} = \theta_e$. Consequently, the six-dimensional array representing the task space variables for planar robots in general and specifically for this robot is $\boldsymbol{\mathcal{X}} = [x_e, y_e, 0, 0, 0, \theta_e]^T$. Thus, the left-hand side of Equation 4.18 can be summarized as follows:

$$
{}^0\boldsymbol{T}_e = \left[\begin{array}{ccc|c} c\theta_e & -s\theta_e & 0 & x_e \\ s\theta_e & c\theta_e & 0 & y_e \\ 0 & 0 & 1 & 0 \\ \hline 0 & 0 & 0 & 1 \end{array}\right]
$$

On the other hand, the homogeneous transformation matrix ${}^0\boldsymbol{T}_3$ for the robot is obtained in Case study 4.1, and the successive multiplication of the homogeneous matrices of the links, is reported in Equation 4.8, and reiterated here.

$$
{}^0\boldsymbol{T}_3 = \left[\begin{array}{ccc|c} c_{123} & -s_{123} & 0 & a_1c_1 + a_2c_{12} + a_3c_{123} \\ s_{123} & c_{123} & 0 & a_1s_1 + a_2s_{12} + a_3s_{123} \\ 0 & 0 & 1 & 0 \\ \hline 0 & 0 & 0 & 1 \end{array}\right]
$$

By equating these two matrices, three independent equations are derived as the solution to the forward kinematics of this robot:

$$
\begin{aligned} x_e &= a_1c_1 + a_2c_{12} + a_3c_{123} \\ y_e &= a_1s_1 + a_2s_{12} + a_3s_{123} \\ \theta_e &= \theta_1 + \theta_2 + \theta_3 \end{aligned} \tag{4.19}
$$

This is exactly equivalent to the geometric solution of the forward kinematics of the robot using the loop closure method, as reported in Equation 4.14.

Case Study 4.7 (3DOF Elbow Manipulator) *Consider the 3DOF Elbow Manipulator introduced in Case Study 4.2 and depicted in Figure 4.7. Analyze the forward kinematics of the end-effector of the robot using homogeneous transformation matrices.*

Solution: This robot has motion in three-dimensional space. Therefore, its task variables are expressed as $\boldsymbol{\mathcal{X}} = [x_e, y_e, z_e, \theta_{e_x}, \theta_{e_y}, \theta_{e_z}']^T$, and its comprehensive homogeneous transformation matrix is given by Equation 4.18.

On the other hand, the homogeneous transformation matrices for the links of the robot have been obtained in Case Study 4.2, and the homogeneous transformation matrix ${}^{0}\boldsymbol{T}_3$ is calculated from the successive multiplication of the homogeneous matrices of the links, as reported in Equation 4.9. By equating these two homogeneous transformation matrices, we arrive at the following equation.

$$\left[\begin{array}{ccc|c} & {}^{0}\boldsymbol{R}(\boldsymbol{\theta}_e) & & \begin{array}{c}{}^{0}x_e\\ {}^{0}y_e\\ {}^{0}z_e\end{array} \\ \hline 0 & 0 & 0 & 1 \end{array}\right] = \left[\begin{array}{ccc|c} c_1c_{23} & -c_1s_{23} & s_1 & c_1\left(a_2c_2+a_3c_{23}\right) \\ s_1c_{23} & -s_1s_{23} & -c_1 & s_1\left(a_2c_2+a_3c_{23}\right) \\ s_{23} & c_{23} & 0 & d_1+a_2s_2+a_3s_{23} \\ \hline 0 & 0 & 0 & 1 \end{array}\right]$$

Thus, the forward kinematics of the robot end-effector position is obtained from the following equation.

$$\begin{aligned} x_e &= c_1\left(a_2c_2+a_3c_{23}\right) \\ y_e &= s_1\left(a_2c_2+a_3c_{23}\right) \\ z_e &= d_1+a_2s_2+a_3s_{23} \end{aligned} \tag{4.20}$$

The orientation of the robot end-effector is also obtained from the following rotation matrix.

$${}^{0}\boldsymbol{R}(\boldsymbol{\theta}_e) = \begin{bmatrix} c_1c_{23} & -c_1s_{23} & s_1 \\ s_1c_{23} & -s_1s_{23} & -c_1 \\ s_{23} & c_{23} & 0 \end{bmatrix} \tag{4.21}$$

If we consider Equation 4.20, it can be observed that the translational motions along the $\boldsymbol{x}, \boldsymbol{y}, \boldsymbol{z}$ axes are not independent of each other. Furthermore, by using Equation 4.21, the orientation of the robot end-effector can be calculated based on the screw representation, unit quaternion, or any desired Euler angles. However, it should be noted that the obtained angles in this case are also not independent. This is aligned with the fact that this robot has only three degrees of freedom, meaning only three variables can be freely chosen from the six task space variables in any given robot configuration.

Case Study 4.8 (SCARA Robot) *Consider the SCARA robot introduced in Case Study 4.3 and illustrated in Figure 4.8. Analyze the forward kinematics of the end-effector of the robot using homogeneous transformation matrices.*

Solution: The task space variables for this robot include three position components and one orientation angle of the robot end-effector, represented as a four-dimensional array: $[x_e, y_e, z_e, \theta_e]$. Considering that the rotation of the robot end-effector is only around the $\boldsymbol{z}$ axis, the six-dimensional array representing the task space variables for this robot is reduced to $\boldsymbol{\mathcal{X}} = [x_e, y_e, z_e, 0, 0, \theta_e]^T$. Thus, the left-hand side of Equation 4.18 can be summarized as follows:

$${}^{0}\boldsymbol{T}_e = \left[\begin{array}{ccc|c} c_{\theta_e} & -s_{\theta_e} & 0 & x_e \\ s_{\theta_e} & c_{\theta_e} & 0 & y_e \\ 0 & 0 & 1 & z_e \\ \hline 0 & 0 & 0 & 1 \end{array}\right]$$

On the other hand, the homogeneous transformation matrices for the links of the robot have been obtained in Case Study 4.3, and the homogeneous transformation matrix

${}^0\boldsymbol{T}_4$ is calculated from the successive multiplication of the homogeneous matrices of the links, which is reported in Equation 4.10, and reiterated here.

$$
{}^0\boldsymbol{T}_4 = \left[\begin{array}{ccc|c} \cos(\theta_1+\theta_2-\theta_4) & \sin(\theta_1+\theta_2-\theta_4) & 0 & a_1c_1+a_2c_{12} \\ \sin(\theta_1+\theta_2-\theta_4) & -\cos(\theta_1+\theta_2-\theta_4) & 0 & a_1s_1+a_2s_{12} \\ 0 & 0 & -1 & d_1-d_3-d_4 \\ \hline 0 & 0 & 0 & 1 \end{array}\right]
$$

By equating these two homogeneous transformation matrices and comparing the resulting rotation matrices, the analytical solution for the forward kinematics of the robot can be obtained as follows.

$$
\begin{aligned} x_e &= a_1c_1+a_2c_{12} \\ y_e &= a_1s_1+a_2s_{12} \\ z_e &= d_1-d_3-d_4 \\ \theta_e &= -\theta_1-\theta_2+\theta_4 \end{aligned} \tag{4.22}
$$

Case Study 4.9 (Stanford Robot) *Consider the Stanford robot introduced in Case Study 4.4 and illustrated in Figure 4.9. Analyze the forward kinematics of the end-effector of the robot using homogeneous transformation matrices.*

Solution: This robot has six degrees of freedom; therefore, the motion variables in its task space are represented by a complete six-dimensional array: $\boldsymbol{\mathcal{X}} = [x_e, y_e, z_e, \theta_{e_x}, \theta_{e_y}, \theta_{e_z}]^T$. Consider Equation 4.18 in its complete form for $n = 6$. On the other hand, the homogeneous transformation matrices for the links of the robot have been obtained in Case Study 4.4, and the homogeneous transformation matrix ${}^0\boldsymbol{T}_6$, calculated from the successive multiplication of the homogeneous matrices of the links, is reported in Equation 4.11 and reiterated here.

$$
{}^0\boldsymbol{T}_6 = {}^0\boldsymbol{T}_1 \cdot {}^1\boldsymbol{T}_2 \cdots {}^5\boldsymbol{T}_6 = \left[\begin{array}{ccc|c} r_{11} & r_{12} & r_{13} & x_e \\ r_{21} & r_{22} & r_{23} & y_e \\ r_{31} & r_{32} & r_{33} & z_e \\ \hline 0 & 0 & 0 & 1 \end{array}\right] \tag{4.23}
$$

By equating these two homogeneous transformation matrices, the analytical solution for the forward kinematics of the robot can be calculated as follows.

$$
\begin{aligned} x_e &= -s_1d_2 - d_3(s_1c_4 + c_1c_2s_4) + d_6(c_1s_2c_5 - s_1s_4s_5 + c_1c_2c_4s_5) \\ y_e &= c_1d_2 + d_3(c_1c_4 - s_1c_2s_4) + d_6(s_1s_2c_5 + c_1s_4s_5 + s_1c_2c_4s_5) \\ z_e &= s_2s_4d_3 + d_6(c_2c_5 - c_4s_2s_5) \end{aligned} \tag{4.24}
$$

and,

$$
{}^0\boldsymbol{R}(\boldsymbol{\theta}_e) = \begin{bmatrix} r_{11} & r_{12} & r_{13} \\ r_{21} & r_{22} & r_{23} \\ r_{31} & r_{32} & r_{33} \end{bmatrix} \tag{4.25}
$$

where the rotation matrix components are given in Equation 4.11. Using Equation 4.25, the orientation of the robot end-effector can be calculated based on the screw representation, unit quaternion, or any desired Euler angles. Furthermore, unlike the two previous case studies, in this robot, all six task space variables are obtained independently provided that the joint space variables are given.

4.5 Screw-Based Method

As mentioned earlier, in the forward kinematic analysis of robots, an algebraic relationship between the joint space variables $\boldsymbol{q}$, and the task space variables $\boldsymbol{\mathcal{X}}$, is derived. Methods such as loop closure and homogeneous transformation matrices can generally formulate this in a vector or matrix form. In this section, the screw-based method for defining and solving the forward kinematic problem of serial robots is studied. In this method, the loop closure relationship is rewritten using the six-dimensional arrays of joint screws in a way that a new formulation of the forward kinematic problem for serial robots is obtained. Subsequently, the solution to this problem is addressed using the screw algebra.

In the screw-based method, transformation matrices corresponding to the screws are also used to describe the overall motion of a rigid body. This description provides more freedom in representing the rotational and translational motions of revolute and prismatic joints commonly found in serial robots. Additionally, the end-effector position vector of the robot can be transformed from one configuration to another using the corresponding transformation matrices, and the forward kinematic problem can be analyzed accordingly. Although the choice of these configurations is entirely arbitrary, selecting an initial end-effector placement configuration that easily determines its position and orientation makes solving the forward kinematic problem very straightforward.

It should be noted that the transformation matrices used in this section are entirely different from the matrices corresponding to the Denavit-Hartenberg parameters used in the previous section, and they should not be confused with each other. In the following, by reviewing the general motion description of a rigid body in space using screw representation, the motion of consecutive screws, defining the forward kinematic problem, and finally, the solution method will be elaborated and examined through several case studies.

4.5.1 Screw-Based Homogeneous Transformation

In classical mechanics, the Mozzi–Chasles' theorem states that the general motion of a rigid body can be represented by combining its rotational motion about its instantaneous rotation axis and translational motion along the same axis. Thus, as demonstrated in Section 3.3, a screw can be represented with the following eight parameters:

$$\{\theta, \hat{\boldsymbol{s}} = [s_x, s_y, s_z]^T \quad ; \quad d, \boldsymbol{s}_o = [s_{o_x}, s_{o_y}, s_{o_z}]^T\} \tag{4.26}$$

Here, $\hat{\boldsymbol{s}}$ is the direction of screw axis, and $\boldsymbol{s}_o$ is the distance from this axis to the origin of the reference coordinate system. We also observed that if these eight parameters are given, the homogeneous transformation matrix can be obtained from the relation 3.28, which is reiterated here.

$$\boldsymbol{T} = \left[\begin{array}{ccc|c} s_x^2 v\theta + c\theta & s_x s_y v\theta - s_z s\theta & s_x s_z v\theta + s_y s\theta & d_x \\ s_y s_x v\theta + s_z s\theta & s_y^2 v\theta + c\theta & s_y s_z v\theta - s_x s\theta & d_y \\ s_z s_x v\theta - s_y s\theta & s_z s_y v\theta + s_x s\theta & s_z^2 v\theta + c\theta & d_z \\ \hline 0 & 0 & 0 & 1 \end{array}\right] \tag{4.27}$$

in which,

$$\begin{aligned} d_x &= d\, s_x - s_{o_x}(s_x^2 - 1)v\theta - s_{o_y}(s_x s_y v\theta - s_z s\theta) - s_{o_z}(s_x s_z v\theta + s_y s\theta) \\ d_y &= d\, s_y - s_{o_x}(s_y s_x v\theta + s_z s\theta) - s_{o_y}(s_y^2 - 1)v\theta - s_{o_z}(s_y s_z v\theta - s_x s\theta) \\ d_z &= d\, s_z - s_{o_x}(s_z s_x v\theta - s_y s\theta) - s_{o_y}(s_z s_y v\theta + s_x s\theta) - s_{o_z}(s_z^2 - 1)v\theta \end{aligned} \tag{4.28}$$

and the notations $s\theta = \sin\theta$, $c\theta = \cos\theta$, and $v\theta = 1 - \cos\theta$ have been used as before.

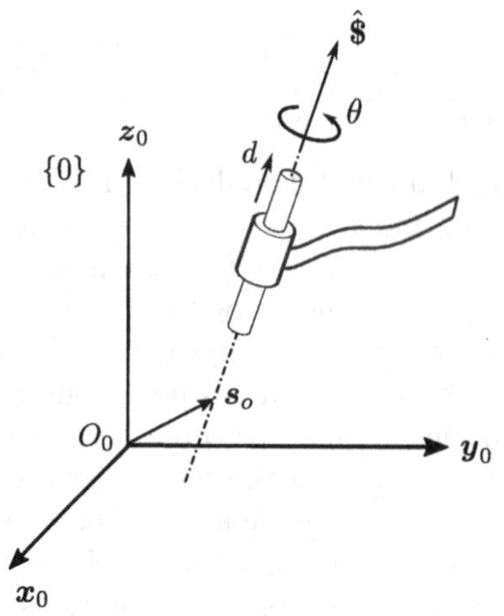

FIGURE 4.12
Representation of the general motion of a rigid body in space by a screw.

The representation of a screw in space is also expressed in Section 3.5, with the screw coordinates denoted by $\$$ and using the six-dimensional array named as "twist." To utilize this representation in defining the forward kinematic problem for robots, it should be noted that since the kinematic structure of serial robots consists of primary revolute and prismatic joints, determining the principal axis of motion or the screw axis $\hat{s}$ can easily be done along the rotation axis in revolute joints and along the translational motion axis in prismatic joints.

Furthermore, considering the joint's location in the robot's kinematic structure, its distance from the reference coordinate system can also be determined, as shown in Figure 4.12, with the vector $\boldsymbol{s}_o$. For the revolute joint, the parameter θ is a joint variable, and d is zero. This is in contrast to the prismatic joint, where the parameter d is a joint variable, and the parameter θ is zero. With this description, all eight parameters describing the motion of the screw in each robot joint can be easily determined.

Hence, as shown in Figure 4.12, we consider the symbol $\$$ along the principal axis of the screw. This representation describes the rotational or translational motion of the joint, while the kinematic arrangement of the robot can also be easily obtained by determining its distance from the origin of the reference coordinate system. By determining the screws in the kinematic structure of each robot, the homogeneous transformation matrix for each link can also be obtained using the relation 4.27. Here, the notation $\boldsymbol{T}_i$ is used for simplicity in interpreting the homogeneous transformation matrices corresponding to the screw definition.

Contrary to the Denavit-Hartenberg method, which requires defining a coordinate system for each joint, in screw algebra, it is sufficient to consider a fixed reference coordinate system to describe the robot's motion. This coordinate system can be arbitrarily placed at any point. However, it is usually placed at the robot's base for convenient representation of the robot's motion with respect to it.

In accordance with Figure 4.12, name this coordinate system as $\{0\}$. Thus, if we represent the first joint of the robot with index 0, and if this joint is revolute, by substituting $\theta = \theta_1, d = 0$, the transformation matrix for this joint, denoted as $\boldsymbol{T}_1$, and the screw associated with this joint, denoted as $\$_1$, are obtained. Now, if this joint is prismatic, then by substituting $\theta = 0, d = d_1$, the transformation matrix for this joint and the corresponding screw will be calculated. Similarly, the screw associated with the i-th joint of the robot,

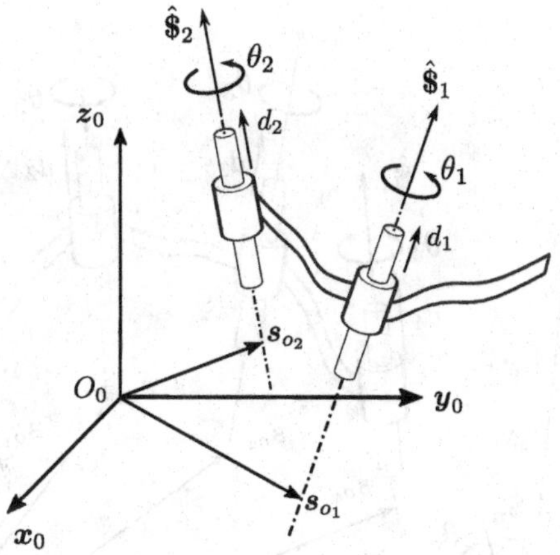

FIGURE 4.13
Representation of the motion of robot links by two consecutive screws.

denoted as $\$_i$, and the corresponding transformation matrix $\boldsymbol{T}_i$ can be obtained in a similar manner.

4.5.2 Consecutive Screws

Now consider the motion of two consecutive joints, represented by two screws $\$_1$ and $\$_2$, as shown in Figure 4.13. As observed in this figure, the screw $\$_1$ is used to represent the motion of the first joint of the robot, and the screw $\$_2$ is used for the second joint. The first link of the robot moves, based on the motion described by the screw $\$_1$, with respect to the fixed robot base. Similarly, the second link of the robot moves based on the motion described by the screw $\$_2$, with respect to the first link of the robot.

The result of the motion in the second link of the robot can be obtained as a sequential motion of rotational (or translational) motion of the first link around (or along) the axis of the first screw and then rotational (or translational) motion of the second link around (or along) the axis of the second moving screw. If we consider a rigid point in the second link and measure its position with respect to the fixed reference frame through consecutive motions, it can be shown that the result of the motion is obtained by the right multiplication of the corresponding transformation matrices as follows [48, 3]:

$$\boldsymbol{T}_r = \boldsymbol{T}_1 \cdot \boldsymbol{T}_2 \tag{4.29}$$

Here, $\boldsymbol{T}_1$ is the homogeneous transformation matrix related to the screw $\$_1$, and $\boldsymbol{T}_2$ is the homogeneous transformation matrix related to the screw $\$_2$. It should be noted that the resulting transformation from this motion in space, $\boldsymbol{T}_r$, can be interpreted as the screw $\$_r$, which contains its instantaneous axis for the rotational motion of the rigid body in space. Furthermore, considering that the multiplication of homogeneous transformation matrices is not commutative, the sequence of robot link motions is effective in the order of multiplying these matrices, while right multiplication in these matrices should be used.

With this introduction, the motion around consecutive screws can be generalized for an n degrees-of-freedom robot. To do this, consider the n degrees-of-freedom robot shown in Figure 4.14, where the joint motion is represented by consecutive screws $\$_1, \$_2, \ldots, \$_n$. The robot's end-effector is represented by point E at the end of the nth link. Furthermore, in this figure, the reference coordinate system $\{0\}$ is fixed on the ground, and the coordinate

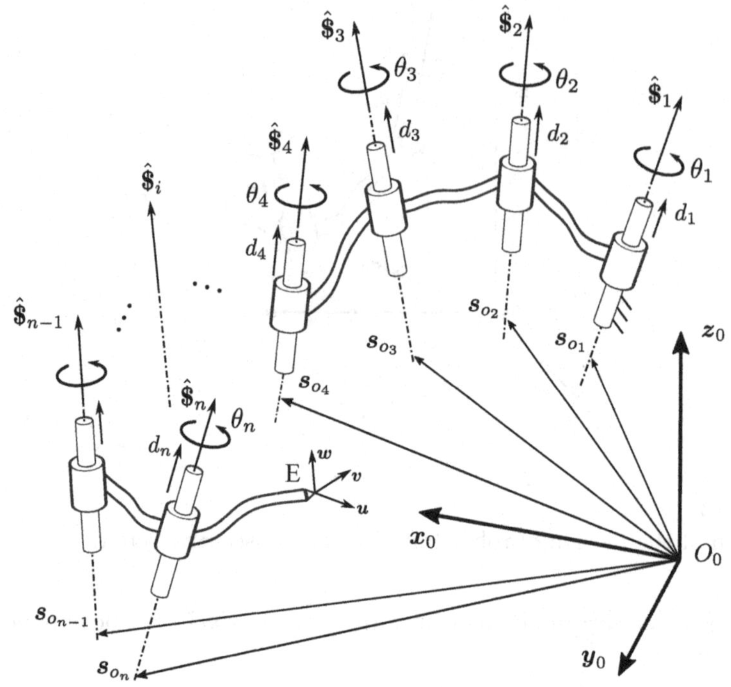

FIGURE 4.14
Representation of the motion of n degree-of-freedom robot links by consecutive screws.

system attached to the robot's end-effector is shown with $\{\boldsymbol{E}\}$. Extend the relation 4.29 for this case as:

$$\boldsymbol{T}_r = \boldsymbol{T}_1 \cdot \boldsymbol{T}_2 \cdots \boldsymbol{T}_n \tag{4.30}$$

The resulting transformation matrix $\boldsymbol{T}_r$ can be utilized to associate the robot's end-effector position in different motion states. A detailed description of this method will be provided in the upcoming section.

4.5.3 Forward Kinematic Formulation

The forward kinematic problem of serial robots can be defined using consecutive screws. In this case, we use the relation 4.30 to describe the position and orientation of the robot. To do this, consider the robot in two different states. Let us denote the position vector of the robot's end-effector in the initial state with respect to the reference coordinate system $\{0\}$ as $\boldsymbol{x}_e^0$. Furthermore, represent the position vector of the robot's end-effector in the final state with respect to the reference coordinate system as $\boldsymbol{x}_e^f$. The relationship between these two vectors can be written using the relation 4.30 as follows:

$$\begin{aligned} \boldsymbol{x}_e^f &= \boldsymbol{T}_r \cdot \boldsymbol{x}_e^0 \\ &= \boldsymbol{T}_1 \cdot \boldsymbol{T}_2 \cdots \boldsymbol{T}_n \cdot \boldsymbol{x}_e^0 \end{aligned} \tag{4.31}$$

Note that this relation encompasses the concept of rigid body motion by consecutive screws; meaning that the position vector of a point on the rigid body experiencing the motion of consecutive screws $\$_1, \$_2, \ldots, \$_n$ is transferred from one state to another in space, and the result of the motion can be obtained from the product of the transformation matrices of these screws. The vector $\boldsymbol{x}_e$ in this representation is arbitrary, and any other vector that has undergone a change from the initial state to the final state can be obtained using a similar relationship.

With this explanation and to describe the orientation of the robot's end-effector, consider the rotation matrix of the end-effector coordinate system $\{\boldsymbol{E}\}$ with respect to the reference coordinate system $\{0\}$ by the unit vectors constituting it. Assume that this rotation matrix in the initial state is represented by $\boldsymbol{R}_e^0 = [\boldsymbol{u}^0, \boldsymbol{v}^0, \boldsymbol{w}^0]$, where the columns of this matrix are interpreted with the unit vectors of the end-effector coordinate system $\{\boldsymbol{E}\}$ with respect to the reference coordinate system $\{0\}$. Now, by representing the orientation of the end-effector in the final state with $\boldsymbol{R}_e^f = [\boldsymbol{u}^f, \boldsymbol{v}^f, \boldsymbol{w}^f]$, the consecutive screws relation for the vectors constituting these rotation matrices can be written as follows:

$$\begin{aligned} \boldsymbol{u}^f &= \boldsymbol{T}_r \cdot \boldsymbol{u}^0 \\ \boldsymbol{v}^f &= \boldsymbol{T}_r \cdot \boldsymbol{v}^0 \quad \text{where,} \ \boldsymbol{T}_r = \boldsymbol{T}_1 \cdot \boldsymbol{T}_2 \cdots \boldsymbol{T}_n \\ \boldsymbol{w}^f &= \boldsymbol{T}_r \cdot \boldsymbol{w}^0 \end{aligned} \tag{4.32}$$

Next, let us demonstrate how to fully define the forward kinematic problem of serial robots using the relationships 4.31 and 4.32. To do this, first consider the robot in its initial state. Try to choose this state in such a way that the end-effector's position vector $\boldsymbol{x}_e^0$ and orientation described by the matrix $\boldsymbol{R}_e^0$ can be easily obtained.

Then, consider the robot in its final state, where all robot joints have moved according to joint variables $\boldsymbol{q} = [q_1, q_2, \ldots, q_n]^T$, and the end-effector is in the state represented by the robot task space variables array $\boldsymbol{\mathcal{X}} = [x_e, y_e, z_e, \theta_{e_x}, \theta_{e_y}, \theta_{e_z}]$.

In this way, Equation 4.31 expresses the algebraic relation between joint variables and the position of the robot end-effector, and the three equations 4.32 define this relationship for the orientation variables of the robot. If the robot joint variables $\boldsymbol{q}$ are given, these equations formulate the forward kinematic problem of the robot, by which the state of the robot's end-effector $\boldsymbol{\mathcal{X}}$ can be easily determined. The inverse solution of this problem is more complex and will be discussed and examined in the next chapter.

Before further proceeding to the details, let us examine the advantages of this method over the Denavit-Hartenberg method. The first advantage is that in this method, there is no need to define Denavit-Hartenberg parameters and coordinate systems for joints. Only one reference coordinate system $\{0\}$ and one attached coordinate system to the end-effector $\{\boldsymbol{E}\}$ are used to apply this method. The second advantage of this method is that, considering the robot's kinematic structure and the concept of screws, the eight parameters of the robot joint screws can be easily obtained by intuition. This geometric interpretation will be very useful in the analysis of robot differential kinematics which will be elaborated on in chapter 6. Another advantage of this method is the freedom to choose the initial state of the robot. Although the choice of the initial and final states of the robot is entirely arbitrary, by appropriately selecting the initial state of the end-effector placement in a way that the end-effector's position and orientation can be easily determined, solving the forward kinematic problem becomes very easy. Usually, this state is chosen in such a way that all joint variables are zero, and the robot arm is fully stretched and lies in a plane. Thus, the position vector $\boldsymbol{x}_e^0$ and the matrix describing the orientation of the robot $\boldsymbol{R}_e^0$ in the initial state are also easily calculated, and the formulation of the forward kinematic problem of the robot becomes more straightforward.

4.5.4 Screw-Based Analysis

In this section, we will first summarize the forward kinematics solution method for rigid robots and then apply it to several robots. To solve the forward kinematic problem of serial robots, follow these steps:

Step 1: Consider the reference coordinate system $\{0\}$ at a suitable location on the fixed base of the robot. Furthermore, attach a moving coordinate system $\{\boldsymbol{E}\}$ on the robot's end-effector to determine the position and orientation of the robot's end-effector.

Step 2: Determine the corresponding screws for the robot's joint motions based on the robot structure. In general, each screw is determined by eight parameters introduced in Equation 4.26. Consider the i-th joint screw and denote it as $\$_i$. If this joint is revolute, the vector $\boldsymbol{s}_i$ is in the direction of the rotation axis with $\theta = \theta_i$, and $d = 0$. The vector $\boldsymbol{s}_{o_i}$ determines the distance of a point from this axis relative to the reference coordinates. If this joint is prismatic, the vector $\boldsymbol{s}_i$ indicates the direction of translational motion, while $\theta = 0$, and $d = d_i$, while the vector $\boldsymbol{s}_{o_i}$ is determined as described before.

Step 3: Choose the initial configuration of the robot such that all joint variables are zero. The robot's arm is fully extended, and if possible, it lies in a plane. By drawing a schematic of the robot in this state, the successive screws $\$_i$ for each joint are indicated in this state. Then, obtain the homogeneous transformation matrix corresponding to the i-th joint using its parameters in this configuration, denoted as $\boldsymbol{T}_i$. Furthermore, represents the end-effector position vector relative to the reference coordinate system $\{0\}$ with $\boldsymbol{x}_e^0$. Determine the orientation of the coordinate system $\{\boldsymbol{E}\}$ attached to the end-effector with the rotation matrix $\boldsymbol{R}_e^0 = [\boldsymbol{u}^0, \boldsymbol{v}^0, \boldsymbol{w}^0]$.

Step 4: Consider the final configuration of the robot, where all joints have finite rotational or translational motions. Specify the position of the robot's end-effector with the position vector $\boldsymbol{x}_e^f$ and the orientation matrix $\boldsymbol{R}_e^f = [\boldsymbol{u}^f, \boldsymbol{v}^f, \boldsymbol{w}^f]$.

Step 5: Reformulate the loop closure equations for the position vector $\boldsymbol{x}_e$ and the vectors constituting the rotation matrix $\boldsymbol{R}_e$ in the initial and final states based on the successive screw relations 4.31 and 4.32. Assume that the joint variables of the robot $\boldsymbol{q}$ are given in these equations. By solving these equations for the robot's task space variables in the arbitrary final state $\boldsymbol{\mathcal{X}}$, the forward kinematic analysis of the robot is accomplished.

Case Study 4.10 (Planar 3R Robot) *Consider the planar 3R robot introduced in Case Study 4.1, as shown in Figure 4.6. Analyze the forward kinematics of the robot's end-effector using the successive screws method and compare the results with the expressions 4.19 obtained from the Denavit-Hartenberg method.*

Solution: The initial configuration of the robot, where all joint variables are set to zero rotations, is shown in Figure 4.15. As depicted in this figure, the reference coordinate

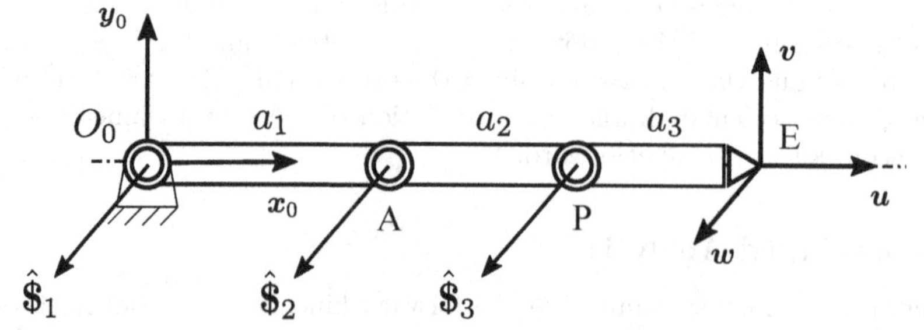

FIGURE 4.15
Schematic representation of the successive screws in the initial configuration of the planar robot in Case Study 4.10.

system $\{0\}$ is on the base, and the moving coordinate system $\{\boldsymbol{E}\}$ is attached parallel to the reference coordinate system on the robot's end-effector. Consequently, successive screws for the joint motions are all parallel to each other and along the z_0 axis. Additionally, the position and orientation of the end-effector in this configuration can be easily determined as follows.

$$\boldsymbol{x}_e^0 = \begin{bmatrix} a_1 + a_2 + a_3 \\ 0 \\ 0 \end{bmatrix}, \quad \boldsymbol{R}_e^0 = [\boldsymbol{u}^0, \boldsymbol{v}^0, \boldsymbol{w}^0] = \begin{bmatrix} 1 & 0 & 0 \\ 0 & 1 & 0 \\ 0 & 0 & 1 \end{bmatrix}$$

The parameters of the successive screws for the robot joints can be easily obtained based on what is reported in Table 4.6. Furthermore, by using Equation 4.27, the corresponding transformation matrices for these screws can be determined as follows.

$$\boldsymbol{T}_1 = \left[\begin{array}{ccc|c} c_1 & -s_1 & 0 & 0 \\ s_1 & c_1 & 0 & 0 \\ 0 & 0 & 1 & 0 \\ \hline 0 & 0 & 0 & 1 \end{array}\right], \boldsymbol{T}_2 = \left[\begin{array}{ccc|c} c_2 & -s_2 & 0 & a_1 v_2 \\ s_2 & c_2 & 0 & -a_1 s_2 \\ 0 & 0 & 1 & 0 \\ \hline 0 & 0 & 0 & 1 \end{array}\right]$$

$$\boldsymbol{T}_3 = \left[\begin{array}{ccc|c} c_3 & -s_3 & 0 & (a_1 + a_2) v_3 \\ s_3 & c_3 & 0 & -(a_1 + a_2) s_3 \\ 0 & 0 & 1 & 0 \\ \hline 0 & 0 & 0 & 1 \end{array}\right]$$

where notations $s_i = \sin\theta_i, c_i = \cos\theta_i, v_i = 1 - \cos\theta_i$ have been used. As a result, the resulting motion transformation matrix from these successive screws, based on Equation 4.30, is obtained as follows.

$$\boldsymbol{T}_r = \left[\begin{array}{ccc|c} c_{123} & -s_{123} & 0 & a_1 c_1 + a_2 c_{12} - (a_1 + a_2) c_{123} \\ s_{123} & c_{123} & 0 & a_1 s_1 + a_2 s_{12} - (a_1 + a_2) s_{123} \\ 0 & 0 & 1 & 0 \\ \hline 0 & 0 & 0 & 1 \end{array}\right]$$

Now consider the final state of the robot in a configuration where all the joints have finite motion, and the robot's end-effector is represented by the task space variables in the final state $\boldsymbol{\mathcal{X}} = [x_e, y_e, 0, 0, 0, \theta_e]^T$. In this state, the position vector and orientation matrix are given as follows.

$$\boldsymbol{x}_e^f = \begin{bmatrix} x_e \\ y_e \\ 0 \end{bmatrix}, \quad \boldsymbol{R}_e^0 = [\boldsymbol{u}^f, \boldsymbol{v}^f, \boldsymbol{w}^f] = \begin{bmatrix} c\theta_e & -s\theta_e & 0 \\ s\theta_e & c\theta_e & 0 \\ 0 & 0 & 1 \end{bmatrix}$$

where $c\theta_e = \cos\theta_e$ and $s\theta_e = \sin\theta_e$.

The loop closure equation for the position vector $\boldsymbol{x}_e$ in the initial and final states, based on the successive screw equation 4.31, can be obtained as follows.

$$\begin{aligned} \boldsymbol{x}_e^f &= \boldsymbol{T}_r \cdot \boldsymbol{x}_e^0 \\ x_e &= a_1 c_1 + a_2 c_{12} + a_3 c_{123} \\ y_e &= a_1 s_1 + a_2 s_{12} + a_3 s_{123} \\ z_e &= 0 \end{aligned}$$

TABLE 4.6
Parameters of the consecutive screws in the initial state of the planar 3R robot of Case Study 4.10

i	θ_i	$\hat{\boldsymbol{s}}_i$	d_i	$\boldsymbol{s}_{O_i}$
1	θ_1	$[0,0,1]^T$	0	$[0,0,0]^T$
2	θ_2	$[0,0,1]^T$	0	$[a_1,0,0]^T$
3	θ_3	$[0,0,1]^T$	0	$[a_1+a_2,0,0]^T$

Now, rewrite the rotation matrix $\boldsymbol{R}_e$ in the initial and final states based on the successive screw equation 4.32. Assuming that the joint variables of the robot $\boldsymbol{q}$ are given, in this case:

$$\boldsymbol{R}_e^f = \begin{bmatrix} c\theta_e & -s\theta_e & 0 \\ s\theta_e & c\theta_e & 0 \\ 0 & 0 & 1 \end{bmatrix} = \begin{bmatrix} c_{123} & -s_{123} & 0 \\ s_{123} & c_{123} & 0 \\ 0 & 0 & 1 \end{bmatrix}$$

which easily leads to the following equation.

$$\theta_e = \theta_1 + \theta_2 + \theta_3$$

If we write the non-zero algebraic equations in conjunction with the loop closure equations for the robot's position and orientation, we will arrive at the following equations:

$$\begin{aligned} x_e &= a_1c_1 + a_2c_{12} + a_3c_{123} \\ y_e &= a_1s_1 + a_2s_{12} + a_3s_{123} \\ \theta_e &= \theta_1 + \theta_2 + \theta_3 \end{aligned} \tag{4.33}$$

which exactly matching the forward kinematic solution of the robot using the Denavit-Hartenberg method, as reported in Equation 4.19.

Case Study 4.11 (3DOF Elbow Manipulator) *Consider the 3DOF Elbow manipulator introduced in Case Study 4.2, and shown in Figure 4.7. Using the consecutive screws method, analyze the forward kinematics of the robot's end-effector and compare the results with what was obtained in Case Study 4.2 using the Denavit-Hartenberg method.*

Solution: The initial state of the robot when all joint variables are set to zero is shown in Figure 4.16. As seen in this figure, the reference coordinate system $\{0\}$ is located on the base, and the moving coordinate system $\{\boldsymbol{E}\}$ is installed on the robot's end-effector. In this state, the screws of the first joint are aligned with the $\boldsymbol{z}_0$ axis, and the screws of the second and third joints are aligned with the $-\boldsymbol{y}_0$ axis. In addition, the position vector and orientation of the end-effector can be easily determined in this state.

$$\boldsymbol{x}_e^0 = \begin{bmatrix} a_1 + a_2 \\ 0 \\ d_1 \end{bmatrix}, \quad \boldsymbol{R}_e^0 = [\boldsymbol{u}^0, \boldsymbol{v}^0, \boldsymbol{w}^0] = \begin{bmatrix} 1 & 0 & 0 \\ 0 & 0 & -1 \\ 0 & 1 & 0 \end{bmatrix}$$

The parameters of the consecutive screws for the joints can be easily obtained based on what is reported in Table 4.7. By using Equation 4.27, the transformation matrices

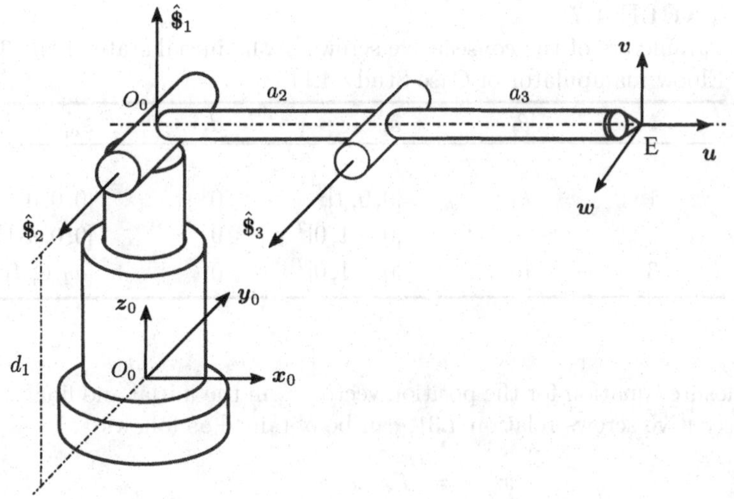

FIGURE 4.16
Representation of consecutive screws in the initial state of the 3DOF Elbow manipulator of Case Study 4.11.

corresponding to these screws can be determined as follows.

$$T_1 = \left[\begin{array}{ccc|c} c_1 & -s_1 & 0 & 0 \\ s_1 & c_1 & 0 & 0 \\ 0 & 0 & 1 & 0 \\ \hline 0 & 0 & 0 & 1 \end{array}\right], T_2 = \left[\begin{array}{ccc|c} c_2 & 0 & -s_2 & d_1 s_2 \\ 0 & 1 & 0 & 0 \\ s_2 & c_2 & 1 & d_1 v_1 \\ \hline 0 & 0 & 0 & 1 \end{array}\right]$$

$$T_3 = \left[\begin{array}{ccc|c} c_3 & 0 & -s_3 & d_1 s_3 + a_2 v_3 \\ 0 & 1 & 0 & 0 \\ s_3 & c_3 & 1 & d_1 v_1 - a_2 s_3 \\ \hline 0 & 0 & 0 & 1 \end{array}\right]$$

Where notations $s_i = \sin\theta_1, c_i = \cos\theta_i, v_i = 1 - \cos\theta_i$ have been used. As a result, the motion transformation matrix derived from these consecutive screws is obtained based on Equation 4.30, as follows.

$$T_r = \left[\begin{array}{ccc|c} c_1 c_{23} & -s_1 & -c_1 s_{23} & c_1 \left(d_1 s_{23} + a_2(c_2 - c_{23})\right) \\ s_1 c_{23} & c_1 & -s_1 s_{23} & s_1 \left(d_1 s_{23} + a_2(c_2 - c_{23})\right) \\ s_{23} & 0 & c_{23} & d_1 v_{23} + a_2(s_2 - s_{23}) \\ \hline 0 & 0 & 0 & 1 \end{array}\right]$$

Consider the final state of the robot in which all joints have finite motion, and the robot's end-effector is represented in the final state by the task space variables $\mathcal{X} = [x_e, y_e, z_e, \theta_{e_x}, \theta_{e_y}, \theta_{e_z}]$. The position vector and orientation matrix are represented as follows:

$$x_e^f = \begin{bmatrix} x_e \\ y_e \\ z_e \end{bmatrix}, \quad R_e^0 = [u^f, v^f, w^f]$$

TABLE 4.7
Parameters of the consecutive screws in the initial state of the 3DOF Elbow manipulator of Case Study 4.11

i	θ_i	$\hat{\boldsymbol{s}}_i$	d_i	$\boldsymbol{s}_{O_i}$
1	θ_1	$[0, 0, 1]^T$	0	$[0, 0, 0]^T$
2	θ_2	$[0, -1, 0]^T$	0	$[0, 0, d_1]^T$
3	θ_3	$[0, -1, 0]^T$	0	$[a_2, 0, d_1]^T$

The loop closure equation for the position vector $\boldsymbol{x}_e$ in the initial and final states, based on the consecutive screws relation 4.31, can be obtained as follows.

$$\begin{aligned} \boldsymbol{x}_e^f &= \boldsymbol{T}_r \cdot \boldsymbol{x}_e^0 \\ x_e &= c_1(a_2 c_2 + a_3 c_{23}) \\ y_e &= s_1(a_2 c_2 + a_3 c_{23}) \\ z_e &= d_1 + a_2 s_2 + a_3 s_{23} \end{aligned} \tag{4.34}$$

Now, rewrite the rotation matrix $\boldsymbol{R}_e$ in the initial and final states based on the consecutive screws relation 4.32. Assume that the joint variables of the robot are given by $\boldsymbol{q}$. In this case,

$$\boldsymbol{R}_e^f = \begin{bmatrix} c_1 c_{23} & -c_1 s_{23} & s_1 \\ s_1 c_{23} & -s_1 s_{23} & -c_1 \\ s_{23} & c_{23} & 0 \end{bmatrix} \tag{4.35}$$

The resulting equations 4.34 and 4.35 are equivalent to what was obtained in Case Study 4.2 using the Denavit-Hartenberg method, as reported in equations 4.20 and 4.21.

Case Study 4.12 (6DOF Elbow Manipulator) *Consider the 6DOF Elbow manipulator introduced in Case Study 4.5, and shown in Figure 4.10. Analyze the forward kinematics of the robot's end-effector using the method of consecutive screws.*

Solution: Consider the initial state of the robot, assuming all joint variables are zero as shown in Figure 4.17. As depicted in this figure, the reference coordinate system $\{0\}$ is located on the base and at the center of the second joint, while the moving coordinate system $\{\boldsymbol{E}\}$ is mounted on the robot's end-effector. Consequently, the first joint screw is aligned with the $\boldsymbol{z}_0$ axis, and the screws of the second, third, and fourth joints are aligned with the $-\boldsymbol{y}_0$ axis. Given the kinematic structure of the robot's joint, the fifth joint screw is perpendicular to the fourth joint screw and aligned with the $\boldsymbol{z}_0$ axis, and the sixth joint screw is aligned with the $\boldsymbol{x}_0$ axis. With this description, the position and orientation vector of the robot's end-effector in this situation can be easily determined.

$$\boldsymbol{x}_e^0 = \begin{bmatrix} a_2 + a_3 + a_4 + d_6 \\ 0 \\ 0 \end{bmatrix}, \quad \boldsymbol{R}_e^0 = [\boldsymbol{u}^0, \boldsymbol{v}^0, \boldsymbol{w}^0] = \begin{bmatrix} 0 & 0 & 1 \\ 0 & -1 & 0 \\ 1 & 0 & 0 \end{bmatrix}$$

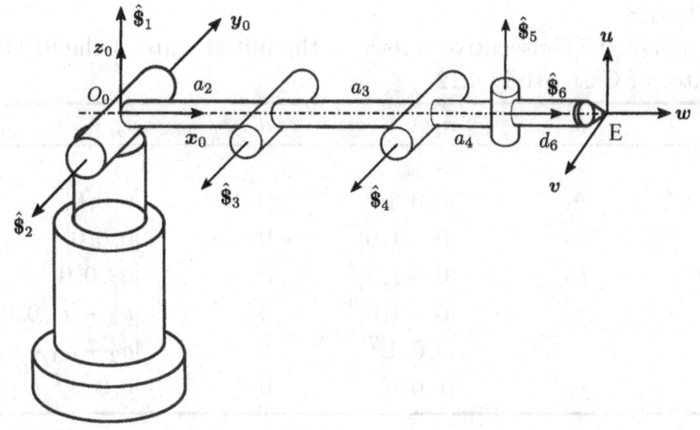

FIGURE 4.17
Representation of the consecutive screws in the initial state of the 6DOF Elbow Manipulator of Case Study 4.12.

The parameters of the consecutive joint screws can easily be obtained based on what is reported in Table 4.8, and using Equation 4.27, the transformation matrices corresponding to these screws can be determined as follows.

$$T_1 = \left[\begin{array}{ccc|c} c_1 & -s_1 & 0 & 0 \\ s_1 & c_1 & 0 & 0 \\ 0 & 0 & 1 & 0 \\ \hline 0 & 0 & 0 & 1 \end{array}\right], T_2 = \left[\begin{array}{ccc|c} c_2 & 0 & -s_2 & 0 \\ 0 & 1 & 0 & 0 \\ s_2 & 0 & c_2 & 0 \\ \hline 0 & 0 & 0 & 1 \end{array}\right]$$

$$T_3 = \left[\begin{array}{ccc|c} c_3 & 0 & -s_3 & a_2 v_3 \\ 0 & 1 & 0 & 0 \\ s_3 & 0 & c_3 & -a_2 s_3 \\ \hline 0 & 0 & 0 & 1 \end{array}\right], T_4 = \left[\begin{array}{ccc|c} c_4 & 0 & -s_4 & (a_2+a_3)v_4 \\ 0 & 1 & 0 & 0 \\ s_4 & 0 & c_4 & -(a_2+a_3)s_4 \\ \hline 0 & 0 & 0 & 1 \end{array}\right]$$

$$T_5 = \left[\begin{array}{ccc|c} c_5 & -s_5 & 0 & (a_2+a_3+a_4)v_4 \\ s_5 & c_5 & 0 & 0 \\ 0 & 0 & 1 & -(a_2+a_3+a_4)s_4 \\ \hline 0 & 0 & 0 & 1 \end{array}\right], T_6 = \left[\begin{array}{ccc|c} 1 & 0 & 0 & 0 \\ 0 & c_6 & -s_6 & 0 \\ 0 & s_6 & c_6 & 0 \\ \hline 0 & 0 & 0 & 1 \end{array}\right]$$

Now consider the final state of the robot in a situation where all joints have finite motion, and the robot's end-effector is represented by the variables $\mathcal{X} = [x_e, y_e, z_e, \theta_{e_x}, \theta_{e_y}, \theta_{e_z}]$ with the following position vector and orientation matrix.

$$x_e^f = \begin{bmatrix} x_e \\ y_e \\ z_e \end{bmatrix}, \quad R_e^0 = [u^f, v^f, w^f]$$

The loop closure equation for the position vector x_e in both initial and final states, based on the successive screw equation 4.31, is obtained as follows.

$$\begin{aligned} x_e^f &= T_r \cdot x_e^0 \\ x_e &= (a_2+a_3+a_4)s_1 s_5 \\ &+ c_1 c_2[a_2 v_3 + (a_2+a_3)(s_3 s_4 + c_3 v_4) - (a_2+a_3+a_4)c_{34} v_5] \\ &+ c_1 s_2[a_2 s_3 + (a_2+a_3)(c_3 s_4 - s_3 v_4) - (a_2+a_3+a_4)s_{34} v_5] \\ &- (a_2+a_3+a_4+d_6)[s_1 s_5 - c_1 c_2 c_{34} c_5 + c_1 s_2 s_{34} c_5] \end{aligned} \quad (4.36)$$

TABLE 4.8
Parameters of the consecutive screws in the initial state of the 6DOF Elbow manipulator of Case Study 4.12

i	θ_i	$\hat{\boldsymbol{s}}_i$	d_i	$\boldsymbol{s}_{o_i}$
1	θ_1	$[0, 0, 1]^T$	0	$[0, 0, 0]^T$
2	θ_2	$[0, -1, 0]^T$	0	$[0, 0, 0]^T$
3	θ_3	$[0, -1, 0]^T$	0	$[a_2, 0, 0]^T$
4	θ_4	$[0, -1, 0]^T$	0	$[a_2 + a_3, 0, 0]^T$
5	θ_5	$[0, 0, 1]^T$	0	$[a_2 + a_3 + a_4, 0, 0]^T$
6	θ_6	$[1, 0, 0]^T$	0	$[0, 0, 0]^T$

$$\begin{aligned} y_e &= -(a_2 + a_3 + a_4)c_1 s_5 \\ &+ s_1 c_2[a_2 v_3 + (a_2 + a_3)(s_3 s_4 + c_3 v_4) + (a_2 + a_3 + a_4)c_{34} v_5] \\ &+ s_1 s_2[a_2 s_3 + (a_2 + a_3)(c_3 s_4 - s_3 v_4) - (a_2 + a_3 + a_4)s_{34} v_5] \\ &+ (a_2 + a_3 + a_4 + d_6)[c_1 s_5 + s_1 c_2 c_{34} c_5 - s_1 s_2 s_{34} c_5] \\ z_e &= a_2 s_2 + a_3 s_{23} + a_4 s_{234} + \tfrac{1}{2} d_6 (s_{2345} + \sin(\theta_2 + \theta_3 + \theta_4 - \theta_5)) \end{aligned} \tag{4.37}$$

Finally, for the rotation matrix $\boldsymbol{R}_e$ based on the successive screw equation 4.32, we have:

$$\boldsymbol{R}_e = \begin{bmatrix} r_{11} & r_{12} & r_{13} \\ r_{21} & r_{22} & r_{23} \\ r_{31} & r_{32} & r_{33} \end{bmatrix} \tag{4.38}$$

where,

$$\begin{aligned} r_{11} &= s_1 c_5 s_6 - c_1 c_2[s_{34} c_6 - c_{34} s_5 s_6] - c_1 s_2[c_{34} c_6 + s_{34} s_5 s_6] \\ r_{12} &= s_1 c_5 c_6 - c_1 c_2[s_{34} s_6 + c_{34} s_5 c_6] + c_1 s_2[c_{34} s_6 - s_{34} s_5 c_6] \\ r_{13} &= -s_1 s_5 + c_1 c_2 c_{34} c_5 - c_1 s_2 s_{34} c_5 \\ r_{21} &= -c_1 c_5 s_6 - s_1 c_2[s_{34} c_6 - c_{34} s_5 s_6] - s_1 s_2[c_{34} c_6 + s_{34} s_5 s_6] \\ r_{22} &= -c_1 c_5 c_6 + s_1 c_2[s_{34} s_6 + c_{34} s_5 c_6] + s_1 s_2[c_{34} s_6 - s_{34} s_5 c_6] \\ r_{23} &= c_1 s_5 + s_1 c_2 c_{34} c_5 - s c_1 s_2 s_{34} c_5 \\ r_{31} &= c_2[c_{34} c_6 + s_{34} s_5 s_6] - s_2[s_{34} c_6 - c_{34} s_5 s_6] \\ r_{32} &= s_2[s_{34} s_6 + c_{34} s_5 c_6] - c_2[c_{34} s_6 - s_{34} s_5 c_6] \\ r_{33} &= s_{234} c_5 \end{aligned} \tag{4.39}$$

Case Study 4.13 (Stanford Robot) *Analyze the forward kinematics of the end-effector of the Stanford robot, introduced in Case Study 4.13, using the method of successive screws.*

Solution: The initial configuration of the Stanford robot, assuming all joint variables are zero ($d_3 = 0$), is illustrated in Figure 4.18. As depicted in this figure, the reference coordinate system $\{0\}$ is located on the base, and the moving coordinate system $\{\boldsymbol{E}\}$ is installed on the robot's end-effector. The screws of the robot's joints are represented in

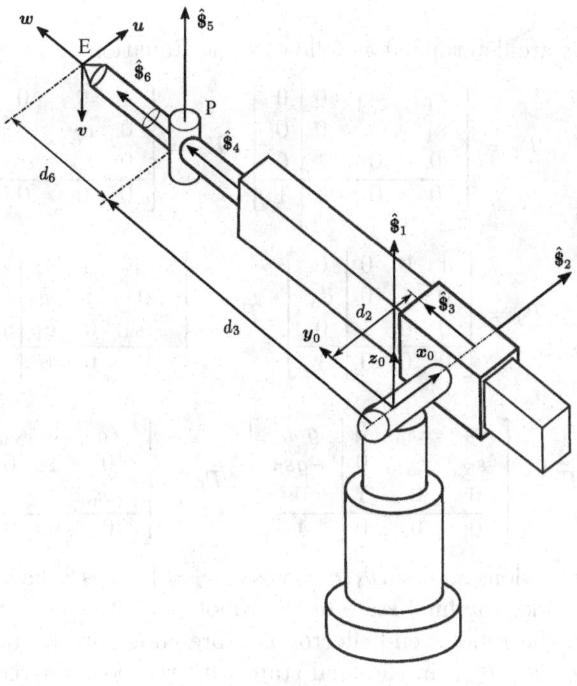

FIGURE 4.18
Display of the successive screws in the initial state of the Stanford robot of Case Study 4.13.

this configuration. The position and orientation vector of the end-effector can be easily determined in this situation.

$$\boldsymbol{x}_e^0 = \begin{bmatrix} d_2 \\ d_6 \\ 0 \end{bmatrix}, \quad \boldsymbol{R}_e^0 = [\boldsymbol{u}^0, \boldsymbol{v}^0, \boldsymbol{w}^0] = \begin{bmatrix} 1 & 0 & 0 \\ 0 & 0 & 1 \\ 0 & -1 & 0 \end{bmatrix}$$

The parameters of the consecutive screws for the joints, based on the information provided in Table 4.9, are obtained, and the transformation matrices corresponding to

TABLE 4.9
Parameters of the consecutive screws in the initial state of the Stanford robot of Case Study 4.13

i	θ_i	$\hat{\boldsymbol{s}}_i$	d_i	$\boldsymbol{s}_{O_i}$
1	θ_1	$[0,0,1]^T$	0	$[0,0,0]^T$
2	θ_2	$[1,0,0]^T$	0	$[0,0,0]^T$
3	0	$[0,1,0]^T$	d_3	$[d_2,0,0]^T$
4	θ_4	$[0,1,0]^T$	0	$[d_2,0,0]^T$
5	θ_5	$[0,0,1]^T$	0	$[d_2,0,0]^T$
6	θ_6	$[0,1,0]^T$	0	$[d_2,0,0]^T$

these screws are determined as follows using Equation 4.27.

$$\boldsymbol{T}_1 = \left[\begin{array}{ccc|c} c_1 & -s_1 & 0 & 0 \\ s_1 & c_1 & 0 & 0 \\ 0 & 0 & 1 & 0 \\ \hline 0 & 0 & 0 & 1 \end{array}\right], \boldsymbol{T}_2 = \left[\begin{array}{ccc|c} 1 & 0 & 0 & 0 \\ 0 & c_2 & -s_2 & 0 \\ 0 & s_2 & c_2 & 0 \\ \hline 0 & 0 & 0 & 1 \end{array}\right]$$

$$\boldsymbol{T}_3 = \left[\begin{array}{ccc|c} 1 & 0 & 0 & 0 \\ 0 & 1 & 0 & d_3 \\ 0 & 0 & 1 & 0 \\ \hline 0 & 0 & 0 & 1 \end{array}\right], \boldsymbol{T}_4 = \left[\begin{array}{ccc|c} c_4 & 0 & s_4 & d_2 v_4 \\ 0 & 1 & 0 & 0 \\ -s_4 & 0 & c_4 & d_2 s_4 \\ \hline 0 & 0 & 0 & 1 \end{array}\right]$$

$$\boldsymbol{T}_5 = \left[\begin{array}{ccc|c} c_5 & -s_5 & 0 & g v_5 \\ s_5 & c_5 & 0 & -g s_5 \\ 0 & 0 & 1 & 0 \\ \hline 0 & 0 & 0 & 1 \end{array}\right], \boldsymbol{T}_6 = \left[\begin{array}{ccc|c} c_6 & 0 & s_6 & d_2 v_6 \\ 0 & 1 & 0 & 0 \\ -s_6 & 0 & c_6 & d_2 s_6 \\ \hline 0 & 0 & 0 & 1 \end{array}\right]$$

where the notations $s_i = \sin\theta_1, c_i = \cos\theta_i, v_i = 1 - \cos\theta_i$ have been used.

Now consider the final state of the robot in a situation where all joints have finite motion and the robot's end-effector is represented by the task space variables $\boldsymbol{\mathcal{X}} = [x_e, y_e, z_e, \theta_{e_x}, \theta_{e_y}, \theta_{e_z}]$, in the final state with the position vector $\boldsymbol{x}_e^f = [x_e, y_e, z_e]$ and the orientation matrix $\boldsymbol{R}_e^0 = [\boldsymbol{u}^f, \boldsymbol{v}^f, \boldsymbol{w}^f]$.

$$\boldsymbol{x}_e^f = \begin{bmatrix} x_e \\ y_e \\ z_e \end{bmatrix}, \quad \boldsymbol{R}_e^0 = [\boldsymbol{u}^f, \boldsymbol{v}^f, \boldsymbol{w}^f]$$

By employing the loop closure equation for the position vector $\boldsymbol{x}_e$ in the initial and final states, based on Eq. 4.31, one can express,

$$\begin{aligned} \boldsymbol{x}_e^f &= \boldsymbol{T}_r \cdot \boldsymbol{x}_e^0 \\ x_e &= d_2 c_1 - d_3 s_1 c_2 - d_6(s_1 c_2 c_5 + c_1 c_4 s_5 - s_1 s_2 s_4 s_5) \\ y_e &= d_2 s_1 + d_3 c_1 c_2 + d_6(c_1 c_2 c_5 - s_1 c_4 s_5 - c_1 s_2 s_4 s_5) \\ z_e &= d_3 s_2 + d_6(s_2 c_5 + c_2 s_4 s_5) \end{aligned} \tag{4.40}$$

Finally, for the rotation matrix $\boldsymbol{R}_e$ based on Eq. 4.32, we have,

$$\boldsymbol{R}_e = \begin{bmatrix} r_{11} & r_{12} & r_{13} \\ r_{21} & r_{22} & r_{23} \\ r_{31} & r_{32} & r_{33} \end{bmatrix} \tag{4.41}$$

where,

$$\begin{aligned} r_{11} &= -s_1 c_2 s_5 c_6 - c_1(s_4 s_6 - c_4 c_5 c_6) - s_1 s_2(c_4 s_6 + s_4 c_5 c_6) \\ r_{12} &= +s_1 c_2 s_5 s_6 - c_1(s_4 c_6 + c_4 c_5 s_6) - s_1 s_2(c_4 c_6 - s_4 c_5 s_6) \\ r_{13} &= s_1 s_2 s_4 s_5 - c_1 c_4 s_5 - s_1 c_2 c_5 \\ r_{21} &= +c_1 c_2 s_5 c_6 - s_1(s_4 s_6 - c_4 c_5 c_6) + c_1 s_2(c_4 s_6 + s_4 c_5 c_6) \\ r_{22} &= -c_1 c_2 s_5 c_6 - s_1(s_4 c_6 + c_4 c_5 s_6) + c_1 s_2(c_4 c_6 - s_4 c_5 s_6) \\ r_{23} &= c_1 s_2 s_4 s_5 - s_1 c_4 s_5 + c_1 c_2 c_5 \\ r_{31} &= s_2 s_5 c_6 - c_2(c_4 s_6 + s_4 c_5 c_6) \\ r_{32} &= -s_2 s_5 s_6 - c_2(c_4 c_6 - s_4 c_5 s_6) \\ r_{33} &= s_2 c_5 + c_2 s_4 s_5 \end{aligned} \tag{4.42}$$

Problems

4.1 Consider the two-degrees-of-freedom RP robot shown in Figure 4.19. Analyze the forward kinematics of this robot using the loop closure method.

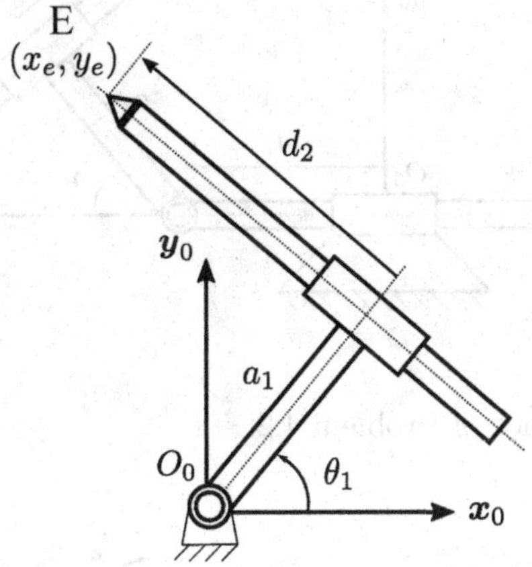

FIGURE 4.19
Schematics of the 2DOF robot of Problem 4.1.

4.2 Examine the forward kinematics of the three-degree-of-freedom PRP robot depicted in Figure 4.20 utilizing the loop closure technique.

4.3 Investigate the forward kinematics of the robot outlined in Problem 4.2 utilizing the Denavit-Hartenberg approach.

4.4 Analyze the forward kinematics of the robot described in Problem 4.2 using the successive screw method.

4.5 Examine the forward kinematics of the three-degrees-of-freedom PRR robot depicted in Figure 4.21 using the loop closure technique.

4.6 Explore the forward kinematics of the robot detailed in Problem 4.5 using the Denavit-Hartenberg method.

4.7 Examine the forward kinematics of the robot specified in Problem 4.5 employing the method of successive screws.

4.8 Analyze the forward kinematics of the three-degrees-of-freedom RPR robot depicted in Figure 4.22 by using the loop closure method.

4.9 Investigate the forward kinematics of the robot outlined in Problem 4.8 employing the Denavit-Hartenberg method.

4.10 Examine the forward kinematics of the robot specified in Problem 4.8 by applying the method of successive screws.

4.11 Analyze the forward kinematics of the two-degrees-of-freedom RR robot depicted in Figure 4.23 utilizing the Denavit-Hartenberg method.

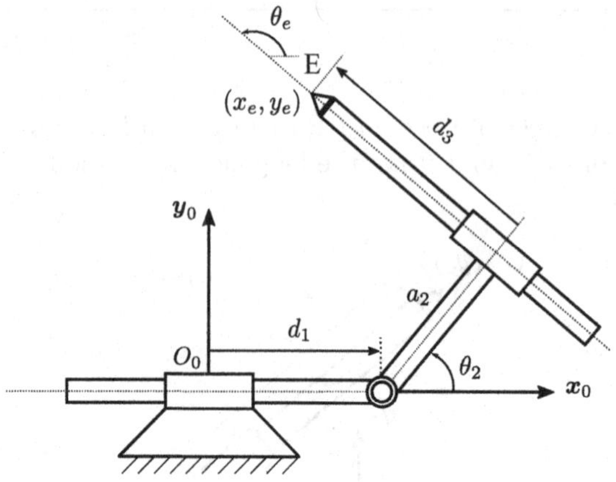

FIGURE 4.20
Schematics of the 3DOF robot of Problem 4.2.

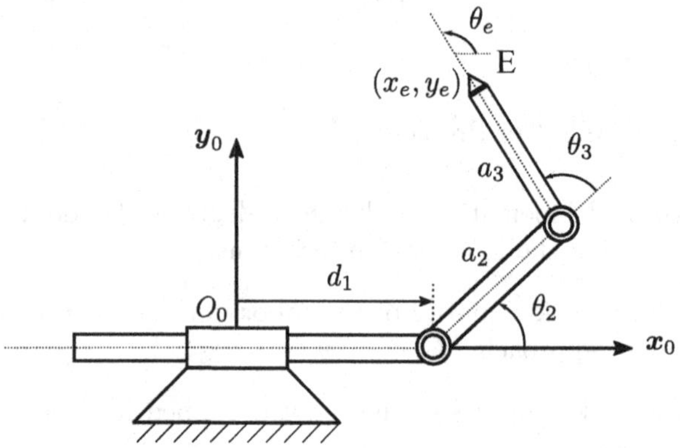

FIGURE 4.21
Schematics of the 3DOF robot of Problem 4.5.

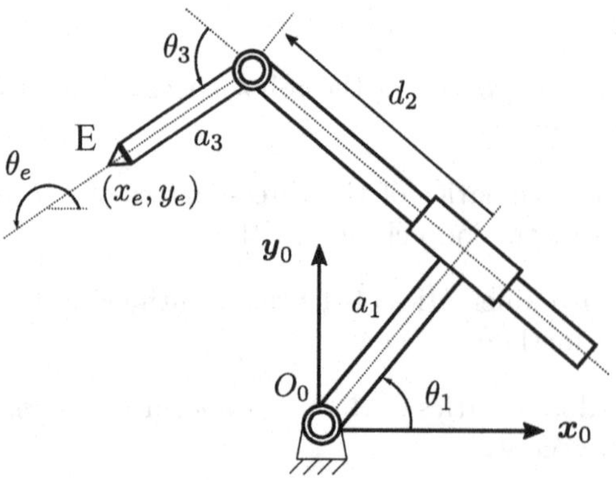

FIGURE 4.22
Schematics of the 3DOF robot of Problem 4.8.

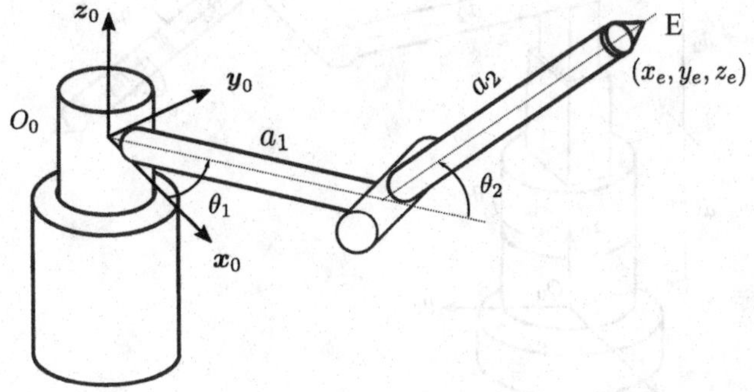

FIGURE 4.23
Schematics of the 2DOF robot of Problem 4.11.

4.12 Investigate the forward kinematics of the robot outlined in Problem 4.11 by applying the method of successive screws.

4.13 Consider the four-degrees-of-freedom RRRP robot shown in Figure 4.24. Analyze the forward kinematics of this robot using the Denavit-Hartenberg method.

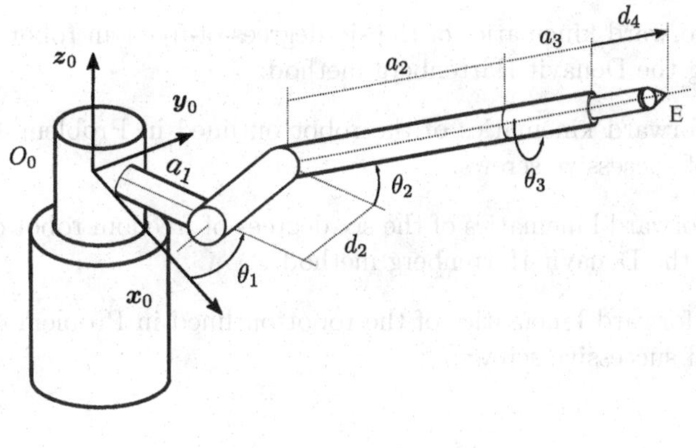

FIGURE 4.24
Schematics of the 4DOF robot of Problem 4.13.

4.14 Analyze the forward kinematics of the robot described in Problem 4.13 using the method of successive screws.

4.15 Examine the forward kinematics of the four-degrees-of-freedom RPRR robot illustrated in Figure 4.25 employing the Denavit-Hartenberg approach.

4.16 Explore the forward kinematics of the robot outlined in Problem 4.15 by employing the method of successive screws.

4.17 Examine the forward kinematics of the four-degrees-of-freedom 4R robot depicted in Figure 4.26 utilizing the Denavit-Hartenberg method.

4.18 Investigate the forward kinematics of the robot described in Problem 4.17 using the method of successive screws.

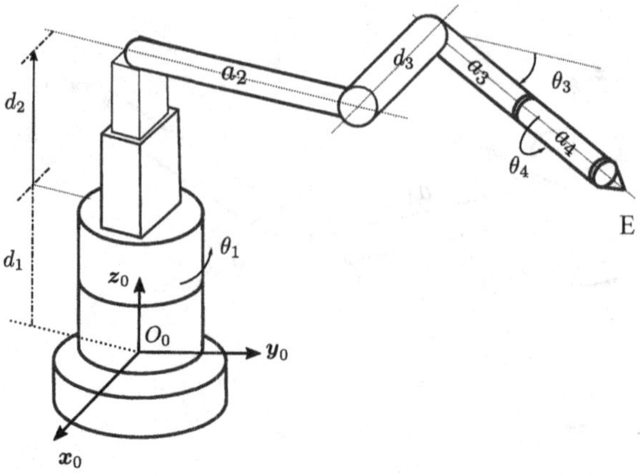

FIGURE 4.25
Schematics of the 4DOF robot of Problem 4.15.

4.19 Analyze the forward kinematics of the four-degrees-of-freedom 4R robot depicted in Figure 4.27 utilizing the Denavit-Hartenberg method.

4.20 Examine the forward kinematics of the robot outlined in Problem 4.19 by employing the method of successive screws.

4.21 Analyze the forward kinematics of the six-degrees-of-freedom robot illustrated in Figure 4.28 using the Denavit-Hartenberg method.

4.22 Explore the forward kinematics of the robot outlined in Problem 4.21 by employing the method of successive screws.

4.23 Analyze the forward kinematics of the six-degrees-of-freedom robot depicted in Figure 4.29 utilizing the Denavit-Hartenberg method.

4.24 Examine the forward kinematics of the robot outlined in Problem 4.23 by employing the method of successive screws.

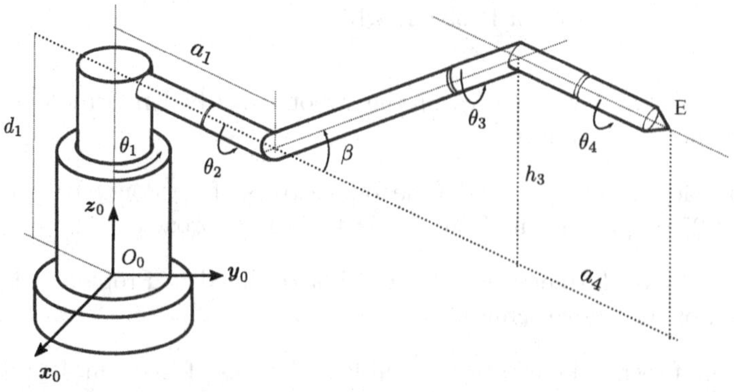

FIGURE 4.26
Schematics of the 4DOF robot of Problem 4.17.

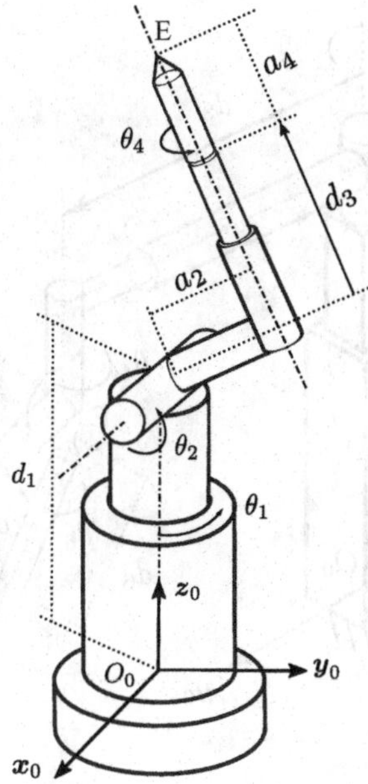

FIGURE 4.27
Schematics of the 4DOF robot of Problem 4.19.

4.25 Consider the industrial 6-DOF "Kuka KR6 Agilus" robot shown in Figure 4.30. Specify the Denavit-Hartenberg parameters of this robot and analyze its forward kinematics by the Denavit-Hartenberg method.

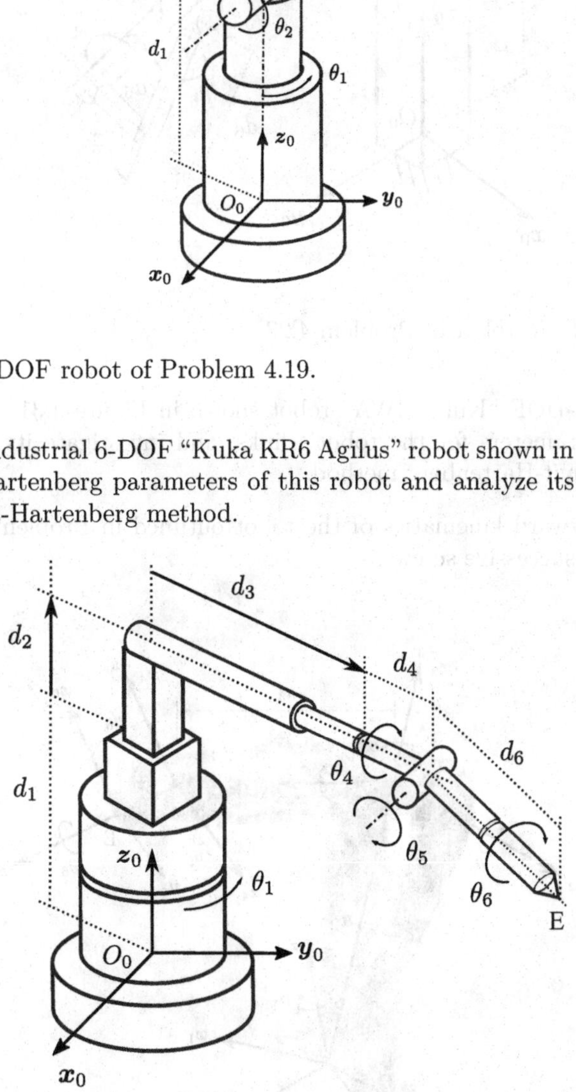

FIGURE 4.28
Schematics of the 6DOF robot of Problem 4.21.

4.26 Examine the forward kinematics of the robot outlined in Problem 4.25 by employing the method of successive screws.

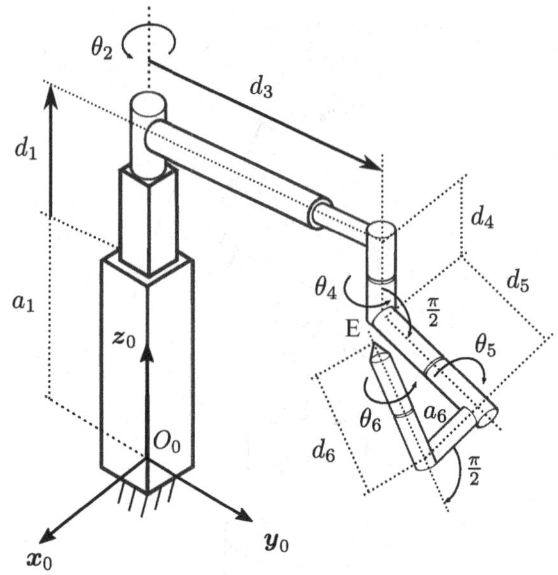

FIGURE 4.29
Schematics of the 6DOF robot of Problem 4.23.

4.27 Consider the 7-DOF "Kuka IIWA" robot shown in Figure 4.31. Define the Denavit-Hartenberg parameters for the robot's links, and investigate its forward kinematics using the Denavit-Hartenberg method.

4.28 Explore the forward kinematics of the robot outlined in Problem 4.27 by employing the method of successive screws.

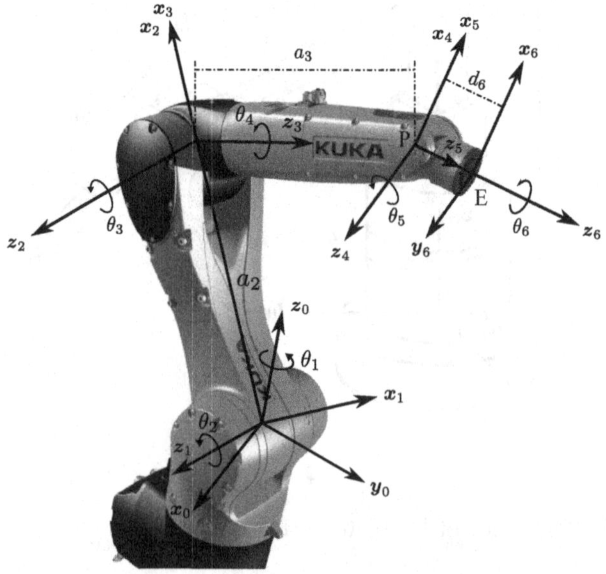

FIGURE 4.30
Schematics of the 6DOF "KUKA KR6 Agilus" robot of Problem 4.25

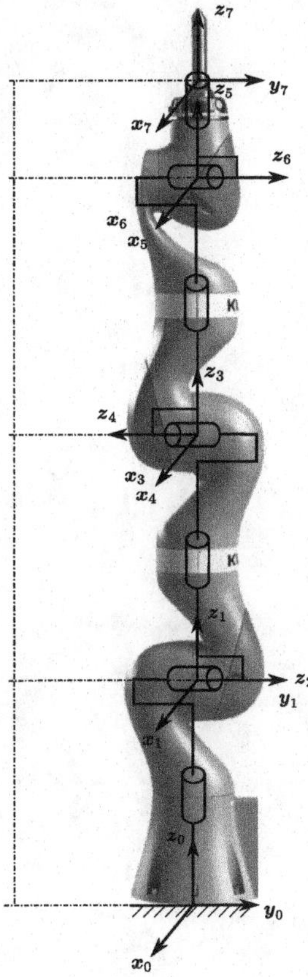

FIGURE 4.31
Schematics of the industrial 7-DOF "KUKA IIWA" robot of Problem 4.27

5

Inverse Kinematics

This chapter commences with an introduction, providing a comprehensive overview and the stage for understanding the complexities and intricacies involved in inverse kinematics. Following this, a conceptual example is presented to offer a practical illustration of the theoretical concepts discussed. The chapter then progresses to explore solvability and the robot workspace, emphasizing the significance of determining feasible solutions and understanding the spatial boundaries in which the robot can operate effectively. Subsequently, the discussion focuses on solution methods, covering a range of techniques. These include the algebraic method, which involves solving equations algebraically, the geometric approach, which considers the geometric relationships between links, and the reduction to polynomial method, simplifying complex expressions to polynomial equations. Furthermore, the chapter addresses the numerical method, utilizing numerical techniques for solving inverse kinematics problems, and introduces the concept of simplification through the use of the wrist point. This technique involves streamlining calculations by strategically selecting a point on the robot, typically the wrist, for simplifying computations.

The chapter provides case studies to exemplify the application of these solution methods. Notable case studies include the SCARA robot and the Fanuc S-900w robot analyzed employing the Denavit-Hartenberg method, the Stanford robot and the 6DOF Elbow Manipulator employing the method of consecutive screws. The chapter concludes with a section on accuracy and repeatability, emphasizing the importance of precision in the execution of robotic movements and the ability to replicate them consistently. Finally, a set of problems designed to reinforce the understanding of the material covered are given, providing an opportunity for readers to apply the concepts in a practical context.

5.1 Introduction

In the previous chapter, we addressed the forward kinematic problem of the robot using the Denavit-Hartenberg method and the method of consecutive screws. In this chapter, we focus on the inverse of this problem, examining various methods that lead to a solution. As mentioned in Chapter 4 and illustrated in Figure 4.2, the inverse kinematic problem in robots refers to the analytical analysis of the robot's motion geometry, in which, the variables describing the robot's end-effector in the task space $\boldsymbol{\mathcal{X}}$ are known, and we aim to determine the corresponding joint variables $\boldsymbol{q}$, for the robot's motion. Solving the inverse kinematic problem in serial robots is often more complex than that of forward kinematics. This challenge has been addressed through various methods for simplification and transformation of the problem into multiple lower-dimensional problems, depending on the type of robot.

It should be noted that the inverse kinematic problem may not always have a solution. This can occur when the desired position or orientation of the robot's end-effector is outside

DOI: 10.1201/9781003491415-5

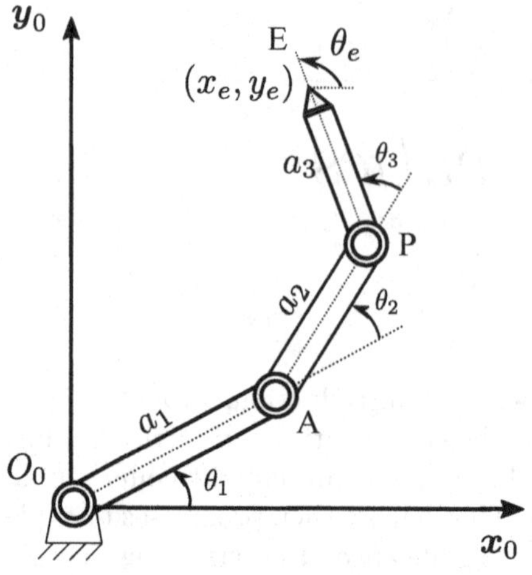

FIGURE 5.1
Representation of the algebraic method in the analysis of the inverse kinematics of a planar robot

the robot's reachable workspace. Furthermore, there might be multiple different solutions to this problem, a consequence of the nonlinear relationships describing the kinematics of the robot's motion. In this section, we first address the inverse kinematics of a simple robot using a conceptual example. Subsequently, we elaborate on topics such as manipulability and dexterity, different robot workspace definitions, and various methods for simplification and solving the inverse kinematic problem.

5.2 Conceptual Example

First, let us analyze the inverse kinematics of serial robots by presenting a conceptual example. For this purpose, consider the planar 3$\underline{\text{R}}$ robot depicted in Figure 5.1. The lengths of the robot's links are denoted as a_1, a_2, a_3, and the end-effector of the robot is in the desired position, given by the task space variables $\boldsymbol{\mathcal{X}} = [x_e, y_e, \theta_e]^T$. In inverse kinematics, we aim to determine the joint variables of the robot, $\boldsymbol{q} = [\theta_1, \theta_2, \theta_3]^T$, in a way that the robot's end-effector is in the desired configuration.

As shown in the figure, place the reference coordinate system $\{0\}$ for kinematic analysis on the fixed base of the robot and at the center of the first revolute joint. Furthermore, consider point E on the end-effector and point P on the robot's wrist, located at the center of the third revolute joint. By using the loop closure equations, the kinematics formulation of this robot is derived and expressed in Equation 4.14 and reiterated here.

$$x_e = a_1 c_1 + a_2 c_{12} + a_3 c_{123} \tag{5.1}$$

$$y_e = a_1 s_1 + a_2 s_{12} + a_3 s_{123} \tag{5.2}$$

$$\theta_e = \theta_1 + \theta_2 + \theta_3 \tag{5.3}$$

In the inverse kinematic problem, the motion variables of the workspace, seen on the left side of these equations, are given, and the goal is to calculate the joint variables $\theta_1, \theta_2, \theta_3$ that generate this motion. As observed, these variables appear nonlinearly and in the form of trigonometric functions in these equations, making the solution of this problem challenging.

To simplify the solution of the inverse kinematic problem in this example, we separate the position and orientation formulation of the robot's end-effector from each other and try to solve each independently. This approach which is applicable to many robots including this one, involves considering the wrist point P and redefining the position formulation for this point. For this purpose, the position vector of this point, considering its longitudinal and transverse components as $\boldsymbol{P} = [x_p,\ y_p]^T$, is defined, and is determined based on the robot's task space variables.

$$x_p = x_e - a_3 \cos\theta_e \ , \ y_p = y_e - a_3 \sin\theta_e \tag{5.4}$$

These equations are easily derived based on the geometry of the robot's end-effector with respect to its wrist. By defining these intermediate variables, the components of this vector can be obtained based on the robot's joint variables. Note that the position vector of the wrist joint depends only on the motion of the first and second robot links, and the third joint variable θ_3, does not appear in this relation.

$$\begin{aligned} x_p &= a_1 c_1 + a_2 c_{12} \\ y_p &= a_1 s_1 + a_2 s_{12} \end{aligned} \tag{5.5}$$

Compare this equation with Equation 5.1; it is observed that despite their similar structure, the nonlinear inverse kinematics equations of the robot have reduced from three equations with three unknowns to two equations with two unknowns. This reduction in dimension has been achieved by considering the position vector of the wrist joint. However, the trigonometric equations obtained are still nonlinear and coupled, making them not easily solvable.

Here, we use another trick to further simplify the equations. Utilizing trigonometric identities in this example will be instrumental. For any angle θ_i, the trigonometric identity $c_i^2 + s_i^2 = 1$ holds true. Square and sum the two sides of Equation 5.5,

$$x_p^2 + y_p^2 = a_1^2 + a_2^2 + 2a_1a_2c_2 \tag{5.6}$$

This simplification is achieved by using the following trigonometric angle summation identities.

$$\begin{aligned} c_{12} &= c_1c_2 - s_1s_2 \\ s_{12} &= c_1s_2 + s_1c_2 \end{aligned}$$

Thus, angle c_2 can be obtained from Equation 5.6.

$$c_2 = \frac{x_p^2 + y_p^2 - a_1^2 - a_2^2}{2a_1a_2} \tag{5.7}$$

This relation is only valid if its right-hand side is a number between 1 and -1. If we geometrically examine the condition for the solvability of this relationship, we find that if the point (x_p, y_p) is at a distance greater than the sum of the link lengths $a_1 + a_2$, this point is not reachable, and the relationship will not have a solution. In this way, by examining the feasibility of solving the inverse kinematics of robot manipulators, the reachable workspace can be determined. This concept will be studied in more detail in section 5.3.

Now, assuming that the wrist point is in the robot's workspace, by having c_2 and using the following relationship for s_2 the angle θ_2 can be obtained by four-quadrant inverse tangent.

$$s_2 = \pm\sqrt{1 - c_2^2} \ , \ \theta_2 = \text{atan2}(s_2, c_2) \tag{5.8}$$

As mentioned in the previous chapter, using the atan2 function has the advantage over the regular inverse tangent function, as it yields solutions in the range of $[-\pi, \pi]$.

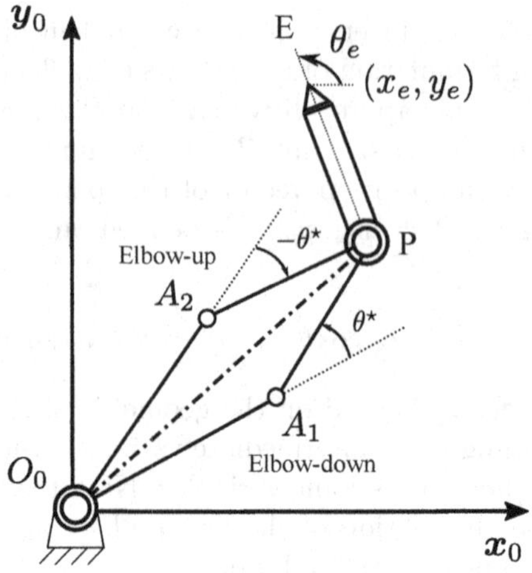

FIGURE 5.2
Representation of the two valid solutions of the inverse kinematics of a planar robot in the "elbow-up" and "elbow-down" configurations.

On the other hand, it is observed that due to the $\pm$ sign, two different solutions are obtained in the inverse kinematics of the robot, and both of them are valid. The geometric interpretation of these two solutions is shown in Figure 5.2. As can be seen, the "elbow-up" and "elbow-down" configurations interpret the correct solutions of the inverse kinematic problems, physically.

Upon determining the angle θ_2 and in order to determine the angle θ_1, rewrite the equation 5.5 in terms of these variables:

$$\begin{aligned} x_p &= b_1 c_1 - b_2 s_1 \\ y_p &= b_1 s_1 + b_2 c_1 \end{aligned} \tag{5.9}$$

where,

$$\begin{aligned} b_1 &= a_1 + a_2 c_2 \\ b_2 &= a_2 s_2 \end{aligned}$$

Equation 5.9 is a system of two equations with two unknowns expressed in terms of its variables c_1, s_1, which can be solved using matrix methods or other techniques. It can be shown that the angle θ_1 can be obtained from the following equation [2].

$$\theta_1 = \text{atan2}(y_p, x_p) - \text{atan2}(b_2, b_1) \tag{5.10}$$

Finally, by having the angles θ_1, θ_2, the angle θ_3 can be obtained from the end-effector orientation equation in 5.1.

$$\theta_3 = \theta_e - (\theta_1 + \theta_2) \tag{5.11}$$

The analysis of robot inverse kinematics begins with the kinematic equations of the robot, which in this example have been obtained using the loop closure method. It is natural that for robots with more complex kinematic structures, methods such as the Denavit-Hartenberg or consecutive screws can be employed to determine these equations. As observed in this example, solving the inverse kinematics even for the simplest planar robot requires a high level of knowledge in using algebraic, geometric, and trigonometric equations, indicating a certain complexity in solving this problem.

On the other hand, many robotic concepts, such as workspace, the existence of multiple solutions, and similar issues, are topics that receive attention in solving the inverse kinematics of robots. A deeper examination of these topics and a description of practical methods for solving inverse kinematic problems will be addressed in the upcoming sections.

5.3 Solvability and the Robot Workspace

As observed in the previous section, in planar robots, writing kinematic equations leads to a minimum of three nonlinear coupled equations. However, a robotic end-effector in three-dimensional space experiences three translational degrees of freedom and three rotational degrees of freedom. Thus, we may use the notation of $n = 3$ to describe the spatial dimension in planar robots and $n = 6$ in spatial robots.

Consider a robot with six degrees-of-freedom that moves in three-dimensional space. In this case, the kinematic relations of the robot have six inputs represented by the components of $\boldsymbol{\mathcal{X}}$ and correspondingly have six unknowns represented by the components of the joint variable vector $\boldsymbol{q}$. The description of the end-effector position in this scenario involves three equations. However, considering the orientation description of the robot's end-effector with rotation matrices, an additional 9 relations are generated for the robot's orientation. These 12 relations are not all independent, and only 6 independent relations can be obtained among them. Thus, the inverse kinematic relations are not only nonlinear and coupled but also have a direct dependence on the representation of the robot's end-effector orientation in space. It is obvious that these relations are very complex, and a general method cannot be used to solve them.

Considering the intricate nature of the inverse kinematic relations, finding a solution might prove challenging or even infeasible. The attainment of an analytical solution for these relations adds another layer of complexity. Nevertheless, exploring the solvability of kinematic relations, a question that pertains solely to the existence of a solution, can be approached through various methods. If, for a given point in the six-dimensional space defined by the task space variables $\boldsymbol{\mathcal{X}}$, the robot's inverse kinematic relations can be solved, then that point lies within the reachable workspace (RW) of the robot. The collection of points encompassing these solvable points is referred to as the reachable workspace or simply the robot's workspace. The boundary of this workspace possesses distinctive characteristics, which will be elaborated upon later.

If a robot has m actuated joints, where the number of joints is less than the degrees of freedom in the workspace ($n = 3$ for planar robots and $n = 6$ for spatial robots), then $m < n$ and the robot is referred to as an "underactuated robot." An example of such robots is examined in Example 4.2 for a three-degrees-of-freedom elbow manipulator. On the other hand, if the number of actuated joints in a robot exceeds its workspace dimension $m > n$, it is termed a "redundant robot." Naturally, in underactuated robots, the number of joint variables is less than the number of kinematic relations, and thus, the number of workspace variables of the robot. In this type of robot, not every arbitrary point in the workspace can be reached with a desired orientation. In other words, the inverse kinematic problem for such robots is solvable only at specific configurations. On the contrary, in redundant robots, the number of joint variables is greater than the number of task space variables. This suggests that reaching a point in the workspace with a desired orientation may yield more than one attainable solution. Consequently, the incorporation of redundancy in serial robots enables the achievement of a particular point in the workspace through various orientations. This adaptability in redundancy proves immensely advantageous in robot design, offering benefits in obstacle avoidance and structural optimization. With this explanation, it should be noted

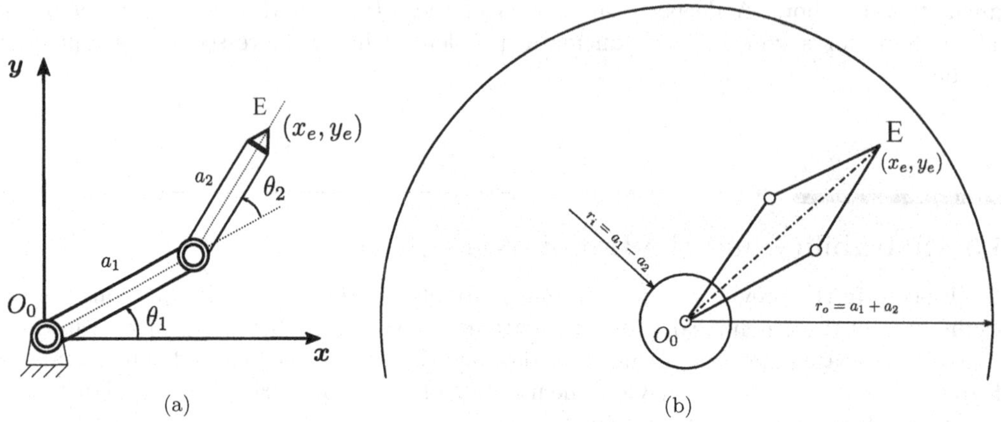

FIGURE 5.3
(a) Schematic of a 2R robotic arm. (b) A section of its reachable workspace.

that the solvability of inverse kinematic relations in a robot and the number of its solutions lead to important concepts in the design and analysis of robots.

Now, we will introduce two additional categories of workspaces in robots. The dexterous workspace encompasses a collection of points within the robot's workspace, enabling access to *any* desired orientation of the robot's end-effector. This workspace naturally falls within the reachable workspace, where the robot's end-effector maneuvers with increased dexterity and agility. Leveraging redundancy in robot actuators facilitates the establishment or enlargement of this specific workspace in robots.

In six-degrees-of-freedom robots, achieving end-effector access with any desired orientation is challenging. For this reason, a workspace with constant orientation, known as the *Constant Orientation Workspace*, has been defined in robotics literature. This workspace includes a set of points surrounding the robot that can be accessed with a fixed orientation in space. Here, the constancy of the robot's orientation means considering a fixed orientation vector in three-dimensional space, denoted as $\boldsymbol{\theta} = cte$.

Example 5.1 (Planar 2R Robot) *For the 2R robot shown in Figure 5.3(a), assuming no constraints on the joint angles, and obtain the reachable and dexterous workspaces for two cases: $a_1 > a_2$ and $a_1 = a_2$.*

Solution: In the case of $a_1 > a_2$, the reachable workspace of the robot, as illustrated in Figure 5.3(b), consists of a set of points inside a ring with an inner radius $r_i = a_1 - a_2$ and an outer radius $r_o = a_1 + a_2$. It is evident that end-effector access with any desired orientation is not feasible at any point within this space, and thus, the dexterous workspace is an empty set.

In the case of $a_1 = a_2$, this ring expands into a circle with a radius of $r = 2a_1$. The center point of this circle is the only point where the robot's end-effector can be reached with any desired orientation. Therefore, the dexterous workspace of the robot in this case is solely comprised of this central point.

In solving the inverse kinematic problem for a robot, points within the workspace differ fundamentally from points located on the boundary of the workspace. For instance, as seen in Example 5.1 and illustrated in Figure 5.3(a), within the workspace, there are two distinct solutions to the inverse kinematic problem of a 2R robot, representing the upper- and lower-elbow configuration. However, if you focus on the points forming the boundary of the robot's workspace, at these points these two solutions converge resulting in a repeated solution in solving the inverse kinematic problem of the robot. Points creating the outer circle indicate a state where the robot arm is fully extended, while the inner circle represents the arm in a fully retracted state.

The connection of boundary points to the solutions of the robot's inverse kinematics is not limited to solving this problem and the existence of repeated solutions; it is also related to the dexterity of motion at these points. In Chapter 7, we will elaborate more precisely on this relationship by defining singular configurations in serial robots.

5.4 Solution Methods

Since solving the inverse kinematic equations of serial robots involves solving a set of coupled nonlinear equations, unlike linear equations, there is no general method for determining the analytical solutions. Nevertheless, we consider the inverse kinematics as solvable if, by any possible method, we can obtain *all possible solutions* for the joint variables of the robot that bring the robot's end-effector to the desired position and orientation [60]. The key point in this definition is obtaining all possible solutions. For this reason, in many robotics reference books, numerical solutions to the inverse kinematic equations are not discussed. This stems from the fact that numerical methods inherently rely on iterative techniques in solving nonlinear equations, making it challenging to obtain all possible solutions. In this book, we will briefly explore numerical methods for solving inverse kinematics, but a detailed examination of these methods requires a comprehensive description [61, 62, 63], that goes beyond the scope of this book.

If we dedicate our study to the analytical solution of inverse kinematics, the closed-form solution of kinematic equations is only achievable when we transform them into linear equations or polynomials of degree less than four. This may be accomplished by using methods similar to what was presented in the conceptual example of this section. In this case, the possibility of obtaining the solution analytically and in a closed-form will be feasible. To achieve this, three general approaches for simplifying kinematic equations are discussed in the literature, namely, algebraic methods, geometric methods, and polynomial transformations. Although this classification is not very precise, as these methods are often mixed, a separate study of these approaches will familiarize us with the fundamental tools for simplifying complex equations into more easily solvable forms.

Research on the kinematics of serial robots indicates that for a general six degrees-of-freedom serial robots with revolute and prismatic joints, inverse kinematics is solvable [64]. However, obtaining this analytical solution for a robot in the general case has not been reported so far. This has only been accomplished for specific cases where all robot joints are revolute while the twist angle parameters of the links, denoted as α_i, are restricted to zero or $\pm\pi/2$ [65, 66]. Furthermore, if the axes of three neighboring joints in a robot are intersecting, a situation arises where, similar to the conceptual example explained in this section, it is possible to separate the kinematics of position from that of the orientation of the robot's end-effector and obtain a closed-form inverse kinematic solution [67]. This topic will also be studied further in section 5.4.5.

5.4.1 Algebraic Method

In this section, we introduce the algebraic method for solving the inverse kinematics of robots by applying it to the 3R robot, which was discussed in the conceptual example of this section. Consider the planar robot 3R depicted in Figure 5.1. The lengths of the robot's links are denoted by a_1, a_2, a_3, and the desired end-effector pose is given by the task space variables $\boldsymbol{\mathcal{X}} = [x_e, y_e, \theta_e]^T$. The forward kinematics equations for this robot are provided in Equation 5.1. Initially, we separate the inverse kinematic problem into the position and orientation of the robot's end-effector by considering the wrist point P and redefining the position problem of this point. The two components of the position vector of this point, based on the robot's joint variables, are expressed in Equation 5.5, which is reiterated here.

$$\begin{aligned} x_p &= a_1 c_1 + a_2 c_{12} \\ y_p &= a_1 s_1 + a_2 s_{12} \end{aligned} \tag{5.12}$$

Square the two sides of this equation and then sum them up:

$$x_p^2 + y_p^2 = a_1^2 + a_2^2 + 2a_1 a_2 c_2 \tag{5.13}$$

Thus, the angle θ_2 is obtained from the following equation:

$$\theta_2 = \cos^{-1}\kappa \ ; \ \kappa = \frac{x_p^2 + y_p^2 - a_1^2 - a_2^2}{2a_1 a_2} \tag{5.14}$$

In this equation,

1) If $|\kappa| < 1$, then this equation will have two distinct solutions. In this case, if one of the solutions is $\theta_2 = \theta_2^\star$, then the second valid solution is $\theta_2 = -\theta_2^\star$, where $-\pi \le \theta_2 \le 0$. These two solutions are illustrated in Figure 5.2. As observed, the solution $\theta_2 = \theta_2^\star$ represents the elbow-up configuration, and the other solution represents the elbow-down configuration, both being valid inverse kinematic solutions.

2) If $|\kappa| = 1$, then this equation has only one solution. This solution can be considered as a limiting case of the two valid solutions obtained in the previous case, where they have converged to a repeated solution. From a kinematic perspective, in this case, the robot's end-effector is positioned on the boundary of the workspace, and its motion toward the outside of the workspace is infeasible.

3) If $|\kappa| > 1$, then this equation has no solution. In this case, the desired point $\boldsymbol{\mathcal{X}}$ is not within the robot's workspace.

By specifying the angle θ_2, the equation 5.12 can be rewritten as a matrix form in terms of the angle θ_1 as follows:

$$\begin{bmatrix} x_p \\ y_p \end{bmatrix} = \begin{bmatrix} b_1 & -b_2 \\ b_2 & b_1 \end{bmatrix} \begin{bmatrix} c_1 \\ s_1 \end{bmatrix}$$

where,

$$\begin{aligned} b_1 &= a_1 + a_2 c_2 \\ b_2 &= a_2 s_2 \end{aligned}$$

The solution to this system of linear equations in terms of the variables c_1, s_1 is expressed by solving the matrix equation as follows:

$$\begin{bmatrix} c_1 \\ s_1 \end{bmatrix} = \frac{1}{a_1^2 + a_2^2 + 2a_1 a_2 c_2} \begin{bmatrix} b_1 & b_2 \\ -b_2 & b_1 \end{bmatrix} \begin{bmatrix} x_p \\ y_p \end{bmatrix}$$

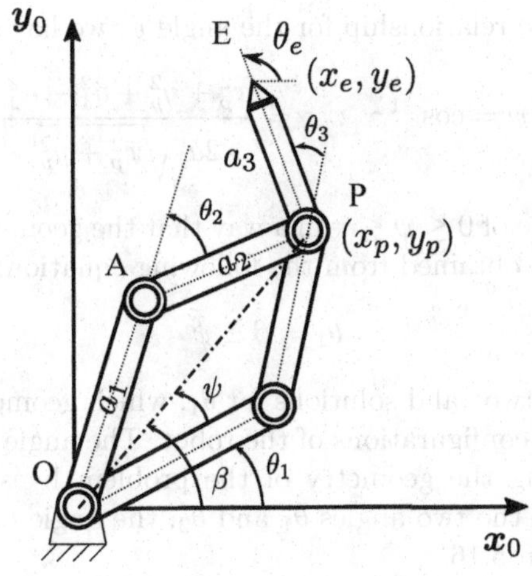

FIGURE 5.4
Geometric approach used in the inverse kinematic analysis of a planar robot.

In this way, for each valid solution of θ_2, a valid solution for θ_1 is obtained.

$$\theta_1 = \text{atan2}(s_1, c_1) \tag{5.15}$$

Finally, with the angles θ_1, θ_2, the angle θ_3 can easily be obtained from the end-effector orientation equation:

$$\theta_3 = \theta_e - (\theta_1 + \theta_2) \tag{5.16}$$

In this way, solving the inverse kinematics for this robot yields two valid solutions, as depicted in Figure 5.2, representing the reflective orientation of the robot arms relative to the *OP* line.

5.4.2 Geometric Approach

In solving the inverse kinematic problem for serial robots by using the geometric approach, an attempt is made to constrain the configuration of the robot arms in 3D space, to a few limited cases in the plane. This assumption is not applicable to all robots, but it is feasible when the twist angle α_i of the links is zero or $\pm\pi/2$. In such cases, joint angles can be determined using geometric relationships.

To illustrate how this method is applied, consider the 3$\underline{\text{R}}$ robot. As shown in Figure 5.4, geometric relationships can be employed in the triangle OAP. The length of the hypotenuse of this triangle can be obtained as a function of the opposite angle using the following equation.

$$x_p^2 + y_p^2 = a_1^2 + a_2^2 - 2a_1a_2\cos(\pi - \theta_2) \tag{5.17}$$

Considering that $\cos(\pi - \theta_2) = -\cos\theta_2$, this geometric relationship simplifies to its counterpart in the algebraic method 5.13, and the angle θ_2 is obtained similar to the previous case from equation 5.14.

To determine the angle θ_1, geometric relationships can further be used. For this purpose, consider two angles β, ψ as shown in Figure 5.4. The angle β can be easily obtained using the following equation.

$$\beta = \text{atan2}(y_p, x_p) \tag{5.18}$$

Next, by writing the cosine relationship for the angle ψ, we have:

$$\psi = \cos^{-1}\gamma \ ; \ \gamma = \frac{x_p^2 + y_p^2 + a_1^2 - a_2^2}{2a_1\sqrt{x_p^2 + y_p^2}} \tag{5.19}$$

By obtaining ψ in the range of $0 \leq \psi \leq \pi$ in a way that the geometry of the triangle remains unchanged, the angle θ_1 is obtained from the following equation:

$$\theta_1 = \beta \pm \psi \tag{5.20}$$

Here, the $\pm$ sign provides two valid solutions for θ_1, which geometrically correspond to the elbow-up and elbow-down configurations of the robot. The angle of the robot's end-effector can also be obtained using the geometry of the problem by summing the joint angles. Therefore, by determining the two angles θ_1 and θ_2, the angle of the third joint θ_3 can be obtained from the equation 5.16.

$$\theta_3 = \theta_e - (\theta_1 + \theta_2)$$

5.4.3 Reduction to Polynomial

Solving non-algebraic equations is usually complex and challenging because algebraic relations and analytical methods cannot be effectively used in determining the solutions of non-polynomial equations. To simplify the solution of these equations, the inverse kinematics equations of serial robots, which typically involve trigonometric functions, can be transformed into corresponding algebraic equations using half-angle tangent relations. For this purpose, if we use the variable u to represent the tangent of half the angle,

$$u = \tan\frac{\theta}{2}$$

then the sine and cosine functions of that angle can be obtained using the following relations.

$$\cos\theta = \frac{1-u^2}{1+u^2} \ ; \ \sin\theta = \frac{2u}{1+u^2} \tag{5.21}$$

When these values are substituted into the kinematic equations of the robot, if there is only one unknown in these equations, the trigonometric relation undergoes a transformation into an algebraic equation. Upon simplification, we reach a polynomial of only one unknown. Ultimately, if the polynomial's degree is less than or equal to four, a closed-form solution can be determined.

Example 5.2 *Solve the following trigonometric equation by transforming it into an algebraic equation and determining all roots of the polynomial.*

$$a\cos\theta + b\sin\theta = c \tag{5.22}$$

Solution: By using the relation 5.21, trigonometric functions can be reduced to algebraic functions.

$$a(1-u^2) + 2bu = c(1+u^2)$$

By simplifying this equation, we will arrive at a second-degree polynomial.

$$(a+c)u^2 - 2bu + (c-a) = 0$$

The roots of it are analytically obtained as follows.

$$u = \frac{b \pm \sqrt{b^2 + a^2 - c^2}}{a + c}$$

Therefore, the angle θ can be obtained from the following equation.

$$\theta = \text{atan}\left(\frac{b \pm \sqrt{b^2 + a^2 - c^2}}{a + c}\right)$$

This equation only has a solution when the expression under the square root remains positive; this condition determines the solvability of the equations. In the presence of a solution, if the expression under the square root is non-zero, two valid solutions will be obtained, and if it becomes zero, we will encounter a repeated solution at the boundary of the workspace. Furthermore, if the denominator in this equation becomes zero, the argument of the tangent inverse becomes infinite, resulting in $\theta = \pm\pi$. These considerations should be taken into account in the practical implementation of these equations.

5.4.4 Numerical Method

With the increasing processing power of computers, numerical solutions for inverse kinematics in robots have gained more attention than ever before. In cases where an analytical solution for inverse kinematics is not reported for a robot, or in the case of redundant robots, numerical methods may be the only option for solving the inverse kinematic problem.

To define the problem and shed light on the subject, consider that in general, the forward kinematic relationship between the joint variables of a robot, denoted by $\boldsymbol{q}$, and the position and orientation of the end-effector, represented by the task variable $\boldsymbol{\mathcal{X}}$, is given by a multi-variable function $\boldsymbol{f}$.

$$\boldsymbol{\mathcal{X}} = \boldsymbol{f}(\boldsymbol{q}) \tag{5.23}$$

Furthermore, suppose that for a given specific configuration of the robot's end-effector $\boldsymbol{\mathcal{X}}$ or a time-varying trajectory of the robot's task space variables $\boldsymbol{\mathcal{X}}(t)$, we aim to find the robot's joint configuration by specifying its joint variables. In numerical methods for solving nonlinear equations, we begin with an initial guess and iteratively refine it to obtain the residual or computational error as follows.

$$\mathbf{Res}(\boldsymbol{q}) = \boldsymbol{\mathcal{X}} - \boldsymbol{f}(\boldsymbol{q}) \tag{5.24}$$

In general, this residual is a vector whose components determine the error in different directions of the workspace. In a situation where we have reached the solution of the inverse kinematics, all components of this residual vector will be zero. To achieve this solution, we consider the Euclidean norm of the residual vector $e(\boldsymbol{q}) = \|\mathbf{Res}(\boldsymbol{q})\|_2$ and, through an iterative method, update the components of the vector $\boldsymbol{q}$ in such a way that the overall magnitude of all components of the residual vector, and consequently its norm, becomes small. By repeating this process a sufficient number of times, the computational error $e(\boldsymbol{q})$ can be minimized with the desired accuracy.

Various methods have been developed to update the joint variable $\boldsymbol{q}$ appropriately. In these methods, the direction of changing the components of the solution is determined in a way that all components of the residual vector decrease. Some of these methods are generalizations of the conventional Newton-Raphson method, in which the appropriate direction for changing the joint variables is determined using the analytical or numerical solution of the Jacobian matrix of the function $\boldsymbol{f}$. This method uses the inverse of the Jacobian matrix, but due to the computational complexity required to determine the inverse of the Jacobian matrix, the transpose of the Jacobian matrix is used as an alternative in the "Steepest Descent" method [4]. Another efficient method used for this purpose is the Levenberg-Marquardt method [68, 69, 70]. This method provides a compromise between the above two methods, using an interpolation of the results of these two methods at each iteration.

As mentioned earlier, numerical methods involve many details that cannot be covered within the scope of this book. Fortunately, engineering software has implemented suitable algorithms for this purpose. For instance, the `fsolve` command in MATLAB® easily solves the inverse of nonlinear multivariable equations. In this function, by introducing the residual function, various numerical methods can be selected to iteratively obtain the solution to the inverse kinematics. If the initial guess provided by this algorithm is not too far from the desired result, this function can generally handle the numerical solution of complex equations. Although the use of this command does not require simplification of kinematic equations, its recursive nature in solving equations introduces complexity, making the calculations needed to reach the desired solution time-consuming. More important than the required computation time is its variability based on the given initial conditions, making it challenging to directly use these recursive methods in real-time applications.

To avoid this issue and speed up reaching the solution, in many cases analytical methods discussed in previous sections are used to simplify and reduce the dimension of kinematic equations, followed by the use of numerical methods. Using these combined methods significantly increases the computation speed and alleviates the difficulty of analytically solving all details. The goal of these hybrid computations is to obtain a nonlinear relationship with only one unknown such that the solution can be found more quickly. In such cases, the `fzero` command in MATLAB and similar functions in other optimization routines can be used to find the root of a single-variable nonlinear equation.

It should be noted that solving the inverse kinematic problem using numerical methods will lead to finding only one solution. In the implementation of these methods, if determining the inverse kinematic solution for a time trajectory of the robot task space variable $\boldsymbol{\mathcal{X}}(t)$, is considered, usually after determining the solution at time zero, this solution is considered as the initial guess for the next step. This way, in addition to obtaining the solution in a shorter time, by considering the continuity of the robot's motion, the occurrence of jumps in the response is prevented. This method is very effective in points inside the workspace, however, as we approach the boundary of the workspace, two or more different and valid solutions of inverse kinematics get close to each other, and selecting from them with a fixed threshold becomes challenging. It should be noted that, selecting from different valid solutions in analytical methods also faces the same challenge. In such cases, constraints such as determining the upper- or lower-elbow configuration and selecting a state based on the robot's working conditions should be considered.

5.4.5 Simplification by Using Wrist Point

As mentioned earlier, although finding a comprehensive solution to inverse kinematics in six-degrees-of-freedom robots is challenging, it becomes somewhat simpler when specific conditions are considered in robot design. One of the initial applied research works in this area is reported in Donald Lee Pieper's doctoral dissertation at Stanford University in 1969, which forms the basis for the design of many industrial robots to date [67]. The dissertation demonstrates that the inverse kinematics of serial six-degrees-of-freedom robots with revolute or prismatic joints, having at least three consecutive joint axes intersecting at a common point, has an analytical solution. In most industrial robots, the final three joints that form the wrist of the robot are considered to be revolute, with their joint axes intersecting at a common point. Due to this consideration, most robots employ a spherical structure at their wrist to control the orientation of the end-effector.

If we designate the intersection point of the joint axes as the wrist point P, its position can be determined using the joint variables of the first three links of the robot, without needing to involve all the joint variables in this problem. By solving the position equations of the inverse kinematic problem, it is possible to analyze the orientation of the robot's wrist in a separate problem. By this means the dimensions of the inverse kinematic problem are reduced to two sets of three degrees of freedom. The general method of this separation, using the Denavit-Hartenberg method and successive screws, will be discussed further in the upcoming subsections, and its application in solving the inverse kinematics of multiple industrial robots will be addressed in Section 5.5.

Denavit-Hartenberg Method

In this method, we first establish the relationship between the end-effector point and the wrist point of the robot. To achieve this, we can determine the position vector of the wrist point relative to the robot's end-effector coordinate system, denoted as $\{6\}$, which is mounted on its end-effector. Considering that the linear distance from the robot's end-effector point E to the wrist point is specified by the parameter d_6, this can be expressed as follows.

$$^{6}\boldsymbol{P} = [0, 0, -d_6]^T \tag{5.25}$$

On the other hand, since the three axes of the robot's end-effector joints are usually orthogonal, the relationship between this vector and the position vector of the robot's end-effector point can often be obtained without the need for matrix calculations. Using the geometry of the problem, the general relationship between this vector and the position vector of the robot's end-effector point can be expressed as follows.

$$^{0}\boldsymbol{P} = \begin{bmatrix} x_p \\ y_p \\ z_p \end{bmatrix} = \begin{bmatrix} e_x \\ e_y \\ e_z \end{bmatrix} + \boldsymbol{R}(\boldsymbol{\theta}_e) \begin{bmatrix} 0 \\ 0 \\ -d_6 \end{bmatrix} \tag{5.26}$$

where the task space pose of the robot's given by the components of the variable $\boldsymbol{\mathcal{X}} = [e_x, e_y, e_x, \theta_{e_x}, \theta_{e_y}, \theta_{e_z}]^T$, and $\boldsymbol{R}(\boldsymbol{\theta}_e)$ is the orientation matrix of the end-effector with respect to the fixed coordinate system. By the use of the components of the position vector of point P, the kinematic position of this point can be written in terms of the joint variables of the first three joints of the robot using the following relationship.

$$^{0}\boldsymbol{P} = {}^{0}\boldsymbol{T}_3 \, {}^{3}\boldsymbol{P} \tag{5.27}$$

The solution to this problem is pursued using the methods described in this chapter. Typically, by utilizing the inverse transformation matrix $^{0}\boldsymbol{T}_1$, this relationship can be factored in terms of the first joint variable as follows, and then its sequential solution can be followed.

$$^{0}\boldsymbol{T}_1^{-1} \, {}^{0}\boldsymbol{P} = {}^{1}\boldsymbol{T}_3 \, {}^{3}\boldsymbol{P} \tag{5.28}$$

By solving the inverse kinematics of the position of the robot's wrist point, the joint variables $\theta_1, \theta_2, \theta_3$ are obtained, and the kinematics of the end-effector's orientation can be written using the following equation.

$$^{3}\boldsymbol{R}_6 = {}^{0}\boldsymbol{R}_3^{-1} \, {}^{0}\boldsymbol{R}_6(\boldsymbol{\theta}) \tag{5.29}$$

Note that the right side of this equation is known, and the left side depends only on the joint variables of the robot's wrist, i.e., the fourth to sixth variables. Using the general methods outlined in this chapter, the inverse kinematics of the wrist orientation can be solved. Practical applications of this method in solving the inverse kinematics of multiple industrial robots are elaborated in Section 5.5.

Method of Succesive Screws

If we use the representation of screws in the forward kinematics of the robot instead of the Denavit-Hartenberg parameters, a similar approach allows separating the kinematics of the wrist position from the orientation of the end-effector. In this method, similar to the Denavit-Hartenberg method, consider the wrist position vector, but based on the screw theory, relate the wrist position vector in the initial configuration to its final state by 5.26:

$$\boldsymbol{x}_p^f = \boldsymbol{T}_1\, \boldsymbol{T}_2\, \boldsymbol{T}_3\, \boldsymbol{x}_p^0 \tag{5.30}$$

This relationship can be rewritten step by step by multiplying the inverse of the screw-based homogeneous transformation matrix from left, while it can be solved using the basic methods mentioned in this section.

$$\boldsymbol{T}_1^{-1}\, \boldsymbol{x}_p^f = \boldsymbol{T}_2\, \boldsymbol{T}_3\, \boldsymbol{x}_p^0 \tag{5.31}$$

By solving the inverse kinematics of the wrist point, the joint variables $\theta_1, \theta_2, \theta_3$ are obtained, and the kinematics of end-effector orientation can be rewritten in terms of the screw-based coordinates using the constituent vectors of the end-effector rotation matrix.

$$\begin{aligned} \boldsymbol{u}^f &= \boldsymbol{T}_r\, \boldsymbol{u}^0 \\ \boldsymbol{v}^f &= \boldsymbol{T}_r\, \boldsymbol{v}^0 \quad \text{where, } \boldsymbol{T}_r = \boldsymbol{T}_1\, \boldsymbol{T}_2 \cdots \boldsymbol{T}_6 \\ \boldsymbol{w}^f &= \boldsymbol{T}_r\, \boldsymbol{w}^0 \end{aligned} \tag{5.32}$$

Given the knowledge of the joint variables $\theta_1, \theta_2, \theta_3$ obtained from solving the inverse kinematics of the wrist point, these relationships can be rewritten in terms of the three end-effector joint variables $\theta_4, \theta_5, \theta_6$, as follows.

$$\begin{aligned} \boldsymbol{T}_3^{-1}\, \boldsymbol{T}_2^{-1}\, \boldsymbol{T}_1^{-1} \boldsymbol{u}^f &= \boldsymbol{T}_4\, \boldsymbol{T}_5\, \boldsymbol{T}_6\, \boldsymbol{u}^0 \\ \boldsymbol{T}_3^{-1}\, \boldsymbol{T}_2^{-1}\, \boldsymbol{T}_1^{-1} \boldsymbol{v}^f &= \boldsymbol{T}_4\, \boldsymbol{T}_5\, \boldsymbol{T}_6\, \boldsymbol{v}^0 \\ \boldsymbol{T}_3^{-1}\, \boldsymbol{T}_2^{-1}\, \boldsymbol{T}_1^{-1} \boldsymbol{w}^f &= \boldsymbol{T}_4\, \boldsymbol{T}_5\, \boldsymbol{T}_6\, \boldsymbol{w}^0 \end{aligned} \tag{5.33}$$

Writing down these equations and determining one of them that exhibits less coupling between the unknowns of the inverse kinematic problem, an attempt could be made to solve them using the methods described in this chapter. Practical applications of this method in solving the inverse kinematics of multiple industrial robots are further discussed in the following section.

5.5 Case Studies

5.5.1 SCARA Robot: Denavit Hartenberg Method

Consider the SCARA robot shown in Figure 4.8. The homogeneous transformation matrices of this robot, based on the Denavit-Hartenberg method, are given in Case Study 4.3, and the forward kinematic equations for this robot are derived in Case Study 4.8. Given that this robot has only four degrees of freedom, the end-effector of the robot cannot move to any arbitrary position and orientation in space. Therefore, the inverse kinematics of the robot

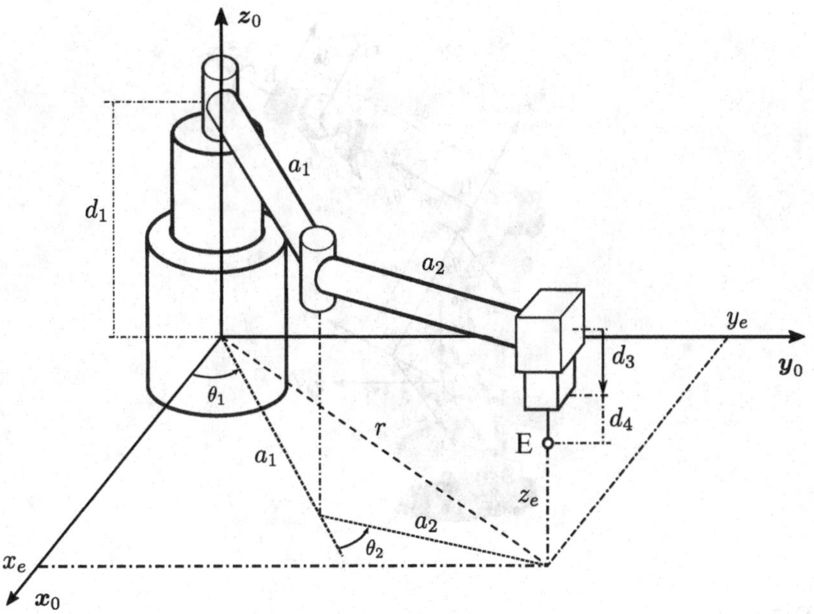

FIGURE 5.5
Schematics of the SCARA robot arms on the horizontal plane.

is only solvable if the orientation rotation matrix of the robot is confined to the following form.

$$\boldsymbol{R}(\theta_e) = \begin{bmatrix} \cos\theta_e & \sin\theta_e & 0 \\ \sin\theta_e & -\cos\theta_e & 0 \\ 0 & 0 & 1 \end{bmatrix} \tag{5.34}$$

In this case,

$$\theta_e = -\theta_1 - \theta_2 + \theta_4 \tag{5.35}$$

to use the geometric method is used for solving the inverse kinematic equations, project the robot configuration on the horizontal $x - y$ plane as depicted in Figure 5.5, and apply cosine relations to find c_2 as follows.

$$c_2 = \frac{e_x^2 + e_y^2 - a_1^2 - a_2^2}{2a_1a_2} \tag{5.36}$$

In this way, the angle θ_2 can be obtained from the following relation:

$$\theta_2 = \text{atan2}(c_2, \pm\sqrt{1 - c_2^2}) \tag{5.37}$$

As observed, this equation provides two valid solutions for each point inside the robot's workspace, corresponding to the elbow-left and elbow-right configurations in the robot's configuration. To determine the angle θ_1, similar to the inverse kinematics of the 3$\underline{\text{R}}$ robot, we will follow the calculations, and reach to similar result given in Equation 5.10, as follows.

$$\theta_1 = \text{atan2}(y_p, x_p) - \text{atan2}(a_2s_2, a_1 + a_2c_2) \tag{5.38}$$

In this way, two valid solutions for θ_1 are obtained, corresponding to the two solutions obtained for θ_2. Now, if the angles of the first and second joints are known, the angle θ_4 can be easily obtained from Equation 5.35.

$$\theta_4 = \theta_e + \theta_1 + \theta_2 \tag{5.39}$$

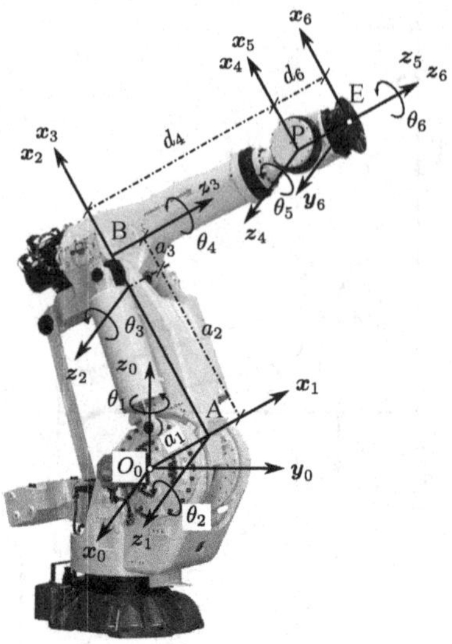

FIGURE 5.6
A representation of the Fanuc S-900W robot and its joint rotation axes.

Finally, the value of d_3 is obtained based on Equation 4.22, and the inverse kinematic analysis of the robot is completed.

$$d_3 = d_1 - z_e - d_4 \tag{5.40}$$

5.5.2 Fanuc S-900W Robot: Denavit-Hartenberg Method

Consider a six-degrees-of-freedom Fanuc S-900W robot with articulated joints, shown in Figure 5.6. As observed in this figure, the axis of rotation for the first joint of the robot is along the vertical axis $\boldsymbol{z}_0$, and the second joint is perpendicular to this axis with an offset a_1. The third joint axis is parallel to the second joint axis with a link length of a_2, and the fourth joint axis is perpendicular to the third joint axis with an offset a_3. Additionally, the three final joint axes intersect at point P and are mutually orthogonal. The distance from the wrist to the second joint axis, representing the offset of the fourth link, is denoted as d_4. The robot's end-effector is also considered at a distance of d_6 from the wrist.

In this robot, a four-bar parallel mechanism is used to transfer power from the robot's base-mounted motor to the third joint axis. This mechanism contributes highly favorable characteristics to the robot's design. However, this structure does not introduce any changes in the kinematic analysis of the robot.

Using coordinate systems installed on the robot, as illustrated in Figure 5.6, the Denavit-Hartenberg parameters for the robot can be determined according to Table 5.1. Now, using the transformation matrix given by Equation 4.6, the homogeneous transformation matrix for six consecutive links is obtained as follows. Symbolic calculations have been performed using the Symbolic Toolbox in MATLAB or Python:

$$^0\boldsymbol{T}_1 = \left[\begin{array}{ccc|c} c_1 & 0 & s_1 & a_1 c_1 \\ s_1 & 0 & -c_1 & a_1 s_1 \\ 0 & 1 & 0 & 0 \\ \hline 0 & 0 & 0 & 1 \end{array}\right], {}^1\boldsymbol{T}_2 = \left[\begin{array}{ccc|c} c_2 & -s_2 & 0 & a_2 c_2 \\ s_2 & c_2 & 0 & a_2 s_2 \\ 0 & 0 & 1 & 0 \\ \hline 0 & 0 & 0 & 1 \end{array}\right],$$

TABLE 5.1
Denavit-Hartenberg parameters for the Fanuc S-900W Robot

i	a_i	α_i	d_i	θ_i
1	a_1	$\pi/2$	0	θ_1
2	a_2	0	0	θ_2
3	a_3	$\pi/2$	0	θ_3
4	0	$-\pi/2$	d_4	θ_4
5	0	$\pi/2$	0	θ_5
6	0	0	d_6	θ_6

$$
{}^2\boldsymbol{T}_3 = \left[\begin{array}{ccc|c} c_3 & 0 & s_3 & a_3c_3 \\ s_3 & 0 & c_3 & a_3s_3 \\ 0 & 1 & 0 & 0 \\ \hline 0 & 0 & 0 & 1 \end{array}\right], {}^3\boldsymbol{T}_4 = \left[\begin{array}{ccc|c} c_4 & 0 & -s_4 & 0 \\ s_4 & 0 & c_4 & 0 \\ 0 & -1 & 0 & d_4 \\ \hline 0 & 0 & 0 & 1 \end{array}\right],
$$

$$
{}^4\boldsymbol{T}_5 = \left[\begin{array}{ccc|c} c_5 & 0 & s_5 & 0 \\ s_5 & 0 & -c_5 & 0 \\ 0 & 1 & 0 & 0 \\ \hline 0 & 0 & 0 & 1 \end{array}\right], {}^5\boldsymbol{T}_6 = \left[\begin{array}{ccc|c} c_6 & -s_6 & 0 & 0 \\ s_6 & c_6 & 0 & 0 \\ 0 & 0 & 1 & d_6 \\ \hline 0 & 0 & 0 & 1 \end{array}\right],
$$

where, as before, notations $c_i = \cos(\theta_i)$ and $s_i = \sin(\theta_i)$ are used in here. By multiplying the transformation matrices of consecutive links together, the end-effector transformation matrix, denoted by point E in the figure, with respect to the fixed coordinate system attached to the robot's base is obtained as follows.

$$
{}^0\boldsymbol{T}_6 = {}^0\boldsymbol{T}_1 \cdot {}^1\boldsymbol{T}_2 \cdots {}^5\boldsymbol{T}_6 = \left[\begin{array}{ccc|c} u_x & v_x & w_x & e_x \\ u_y & v_y & w_y & e_y \\ u_z & v_z & w_z & e_z \\ \hline 0 & 0 & 0 & 1 \end{array}\right]. \tag{5.41}
$$

These calculations have also been performed in two steps using the symbolic computation toolbox in MATLAB and Python. For this purpose, first, the homogeneous matrix for the initial three joints of the robot is calculated as follows.

$$
{}^0\boldsymbol{T}_3 = {}^0\boldsymbol{T}_1 \cdot {}^1\boldsymbol{T}_2 \cdot {}^2\boldsymbol{T}_3 \tag{5.42}
$$

$$
= \left[\begin{array}{ccc|c} c_1c_{23} & s_1 & c_1s_{23} & c_1(a_1 + a_2c_2 + a_3c_{23}) \\ s_1c_{23} & -c_1 & s_1s_{23} & s_1(a_1 + a_2c_2 + a_3c_{23}) \\ s_{23} & 0 & -c_{23} & a_2s_2 + a_3s_{23} \\ \hline 0 & 0 & 0 & 1 \end{array}\right]. \tag{5.43}
$$

Then, by combining the homogeneous transformations, the homogeneous matrix for the final three joints of the robot can be obtained.

$$
{}^3\boldsymbol{T}_6 = {}^3\boldsymbol{T}_4 \cdot {}^4\boldsymbol{T}_5 \cdot {}^5\boldsymbol{T}_6
$$

$$
= \left[\begin{array}{ccc|c} c_4c_5c_6 - s_4s_6 & -c_4c_5s_6 - s_4c_6 & c_4s_5 & d_6c_4s_5 \\ s_4c_5c_6 + c_4s_6 & -s_4c_5s_6 + c_4c_6 & s_4s_5 & d_6s_4s_5 \\ -s_5c_6 & s_5s_6 & c_5 & d_4 + d_6c_5 \\ \hline 0 & 0 & 0 & 1 \end{array}\right]. \tag{5.44}
$$

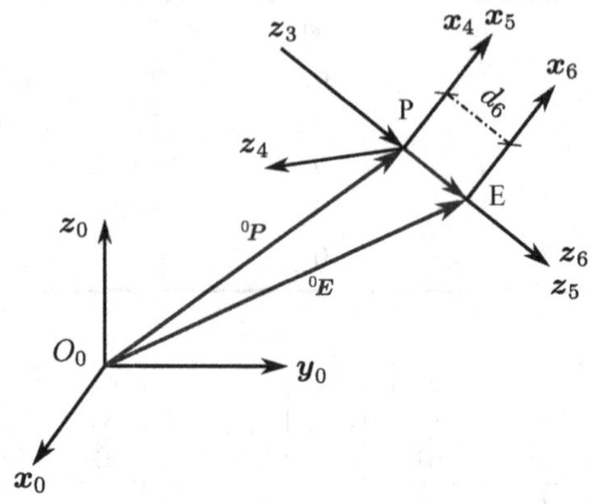

FIGURE 5.7
Relationship of the wrist point with the end-effector of the Fanuc robot in the coordinate system $\{6\}$.

By multiplying these two homogeneous matrices, the overall homogeneous matrix for the robot is obtained, and the forward kinematic problem of the robot is completed.

For the purpose of solving the inverse kinematics, we will follow the simplification method using the wrist point. To do this, first, the kinematic relationship of the robot's wrist point P, with the end-effector of the robot is determined. Note that the distance from the wrist point to the robot's end-effector, as shown in the figure, is along the $-\boldsymbol{z}_6$ axis with a length of d_6. Thus,

$$^6\boldsymbol{P} = [0, 0, -d_6, 1]^T. \tag{5.45}$$

The position and orientation of the wrist relative to the reference coordinate system $\{0\}$ are given in Equation 5.41. Therefore, the absolute position vector of the wrist point can be calculated using the following equation.

$$^0\boldsymbol{P} = {}^0\boldsymbol{T}_6 \cdot {}^6\boldsymbol{P} = \begin{bmatrix} e_x - d_6 w_x \\ e_y - d_6 w_y \\ e_z - d_6 w_z \\ 1 \end{bmatrix}. \tag{5.46}$$

In this equation, the position vector of the robot's wrist point is obtained based on the robot's task space variables $\boldsymbol{\mathcal{X}}$. In order to obtain this point using joint variables, note that only the first three joint variables $\theta_1, \theta_2, \theta_3$ are required. Obtain the position vector of the wrist point in the coordinate system $\{3\}$, taking into account that the fourth link length d_4 is along the $\boldsymbol{z}_4$ axis:

$$^3\boldsymbol{P} = [0, 0, d_4, 1]^T. \tag{5.47}$$

Next, you can use the loop closure method and obtain the equivalence of these vectors using homogeneous transformation matrices:

$$^0\boldsymbol{P} = {}^0\boldsymbol{T}_3 \cdot {}^3\boldsymbol{P} \tag{5.48}$$

In this equation, the position kinematics of the wrist point is written based on the first three joint variables of the robot, $\theta_1, \theta_2, \theta_3$, completely independent of the end-effector orientation relations.

To solve the position kinematics, first multiply the inverse transformation matrix ${}^0\boldsymbol{T}_1$ from the left on both sides of Equation 5.48.

$$({}^0\boldsymbol{T}_1)^{-1} \cdot {}^0\boldsymbol{P} = {}^1\boldsymbol{T}_3 \cdot {}^3\boldsymbol{P}. \tag{5.49}$$

Next, obtain the inverse of ${}^0\boldsymbol{T}_1$, and by using it find ${}^1\boldsymbol{T}_3$.

$$({}^0\boldsymbol{T}_1)^{-1} = \left[\begin{array}{ccc|c} c_1 & s_1 & 0 & -a_1 \\ 0 & 0 & 1 & 0 \\ s_1 & -c_1 & 0 & 0 \\ \hline 0 & 0 & 0 & 1 \end{array}\right], \quad {}^1\boldsymbol{T}_3 = \left[\begin{array}{ccc|c} c_{23} & 0 & s_{23} & a_2c_2 + a_3c_{23} \\ s_{23} & 0 & -c_{23} & a_2s_2 + a_3s_{23} \\ 0 & 1 & 0 & \\ \hline 0 & 0 & 0 & 1 \end{array}\right] \tag{5.50}$$

Then simplify Equation 5.49 by substituting the matrices from Equation 5.50.

$$x_pc_1 + y_ps_1 - a_1 = a_2c_2 + a_3c_{23} + d_4s_{23}, \tag{5.51}$$

$$z_p = a_2s_2 + a_3s_{23} - d_4c_{23}, \tag{5.52}$$

$$x_ps_1 - y_pc_1 = 0. \tag{5.53}$$

In this equation, x_p, y_p, z_p are given in 5.46, and using Equation 5.53, you can directly obtain the angle θ_1 as follows.

$$\theta_1 = \text{atan}\frac{y_p}{x_p} \tag{5.54}$$

In this way, two valid solutions for θ_1 are obtained. If $\theta_1 = \theta_1^\star$ is a solution for Equation 5.54, then $\theta_1 = \theta_1^\star + \pi$ will also be a valid solution of the inverse kinematic problem of this robot, where the first solution corresponds to the robot's in-front workspace and the second one to the robot the in-back workspace. However, due to the rotational joint constraints, only the in-front workspace is accessible. We use this solution to solve the other joint variables of the robot.

If you carefully examine the kinematic structure of the robot, you will see that the distance from the wrist point to the center of the robot's base is only a function of the third joint angle θ_3. This means that the angles θ_1, θ_2 can be eliminated from these equations. To do this, you can square and sum the three equations 5.52, 5.53, and 5.53 and simplify them as follows:

$$b_1s_3 + b_2c_3 = b_3 \tag{5.55}$$

where,

$$b_1 = 2a_2d_4 \;, \quad b_2 = 2a_2a_3$$

$$b_3 = x_p^2 + y_p^2 + z_p^2 - 2x_pa_1c_1 - 2y_pa_1s_1 + a_1^2 - a_2^2 - a_3^2 - d_4^2$$

By using the tangent of half-angles, Equation 5.55 can be transformed into the following second-order polynomial:

$$(b_2 + b_3)u_3^2 - 2b_1u_3 + (b_3 - b_2) = 0,$$

and using that, two valid solutions can be obtained.

$$\theta_3 = \text{atan}\frac{b_1 \pm \sqrt{b_1^2 + b_2^2 - b_3^2}}{b_2 + b_3}. \tag{5.56}$$

In this equation:

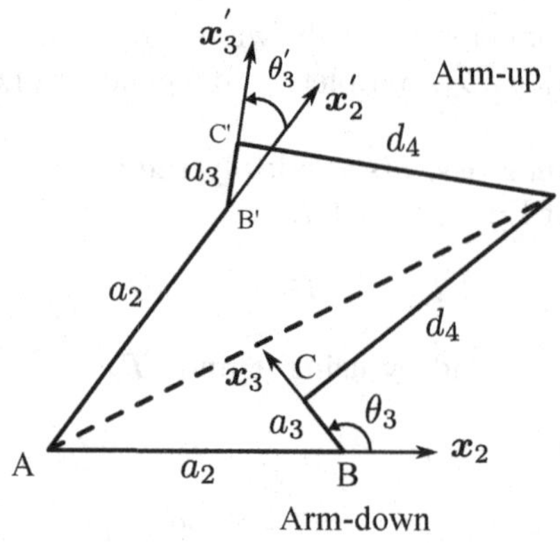

FIGURE 5.8
Two valid solutions for the angle θ_3 in the arm-up and arm-down configurations of the Fanuc robot.

1) If $b_1^2+b_2^2-b_3^2 > 0$, then this equation has two valid solutions. One corresponds to the arm-up configuration of the robot, and the other corresponds to the arm-down configuration. These two configurations are illustrated in Figure 5.8.

2) If $b_1^2 + b_2^2 - b_3^2 = 0$, then this equation has only one solution. This solution can be considered as a limiting case of the two valid solutions obtained in the previous case, where they coincide and result into a repeated solution. In this case, the robot's end-effector is located at the boundary of the workspace, and its movement toward the outside of the workspace is infeasible.

3) If $b_1^2 + b_2^2 - b_3^2 < 0$, then this equation has no solution. In this case, the point $\boldsymbol{\mathcal{X}}$ is not within the robot's workspace.

Once the angles θ_1 and θ_3 are known and substituted into Equations 5.52 and 5.53, we arrive at the following system of linear equations:

$$\begin{bmatrix} \alpha_1 & \beta_1 \\ \alpha_2 & \beta_2 \end{bmatrix} \cdot \begin{bmatrix} c_2 \\ s_2 \end{bmatrix} = \begin{bmatrix} \gamma_1 \\ \gamma_2 \end{bmatrix}$$

where,

$$\begin{aligned} \alpha_1 &= a_2 + a_3 c_3 + d_4 s_3, & \beta_1 &= -a_3 s_3 + d_4 c_3, & \gamma_1 &= x_p c_1 + y_p s_1 - a_1, \\ \alpha_2 &= a_3 s_3 - d_4 c_3, & \beta_2 &= a_2 + a_3 c_3 + d_4 s_3, & \gamma_2 &= z_p. \end{aligned}$$

By using the matrix solution method, we will arrive at the following solution:

$$\begin{bmatrix} c_2 \\ s_2 \end{bmatrix} = \frac{1}{\alpha_1\beta_2 - \alpha_2\beta_1} \begin{bmatrix} \beta_2 & -\beta_1 \\ -\alpha_2 & \alpha_1 \end{bmatrix} \cdot \begin{bmatrix} \gamma_1 \\ \gamma_2 \end{bmatrix}$$

Thus, for each valid solution for θ_3, we obtain a valid solution for θ_2 as follows:

$$\theta_2 = \text{atan2}(s_2, c_2). \tag{5.57}$$

In summary, the solution to the inverse kinematics for the wrist point of the robot results in four potential solutions in general. However, due to the constraint on the motion of the first joint of the robot, only two of them are valid.

By determining the joint angles $\theta_1, \theta_2, \theta_3$, the orientation kinematics equations for the robot's end-effector can be written as follows:

$$^3\boldsymbol{T}_6 = (^0\boldsymbol{T}_3)^{-1} \cdot {}^0\boldsymbol{T}_6. \tag{5.58}$$

The right-hand side of this equation is known, and we want to find the joint angles $\theta_4, \theta_5, \theta_6$. The transformation matrix on the right-hand side of this equation is also given in Equation 5.44. If you focus on this matrix, you can easily obtain the angle θ_5 using the $(3,3)$ element of the matrix with the following equation:

$$\theta_5 = \cos^{-1}(r_{33}), \tag{5.59}$$

where,

$$r_{33} = w_x c_1 s_2 3 + w_y s_1 s_{23} - w_z c_{23}.$$

In this way, for each valid solution of the first three joint angles $\theta_1, \theta_2, \theta_3$:

1) If $|r_{33}| < 1$, then this equation has two valid solutions.

2) If $|r_{33}| = 1$, then this equation has only one solution, where $\theta_5 = 0$ or $\theta_5 = \pi$. In this case, the axis $\boldsymbol{z}_5$ aligns with the axis $\boldsymbol{z}_3$, which is referred to as a wrist singularity.

3) Physically, it is not possible for $|r_{33}|$ to be greater than one.

Assuming the robot is in the first state and $s_5 \neq 0$, in this case, one can obtain the angle θ_4 using the $(1,3)$ and $(2,3)$ elements of the transformation matrix in Equation 5.44 as follows:

$$\theta_4 = \text{atan2}(s_4, c_4), \tag{5.60}$$

where,

$$\begin{aligned} c_4 &= \frac{1}{s_5}(w_x c_1 c_{23} + w_y s_1 c_{23} + w_z s_{23}), \\ s_4 &= \frac{1}{s_5}(w_x s_1 - w_y c_1). \end{aligned}$$

Similarly, using the $(3,1)$ and $(3,2)$ elements at positions of the transformation matrix in Equation 5.44, you can obtain the angle θ_6 as follows:

$$\theta_6 = \text{atan2}(s_6, c_6), \tag{5.61}$$

in which,

$$\begin{aligned} c_6 &= \frac{-1}{s_5}(u_x c_1 s_{23} + u_y s_1 s_{23} - u_z c_{23}), \\ s_6 &= \frac{1}{s_5}(v_x c_1 s_{23} + v_y s_1 s_{23} - v_z c_{23}). \end{aligned}$$

If $s_5 = 0$, the robot is in a singular configuration, and you can only obtain the algebraic sum of the joint angles θ_4 and θ_6. In the end, for each valid solution of the joint angles from the first to the third, two valid configurations for the robot's wrist are obtained. Therefore, in general, there are eight possible solutions for the inverse kinematics of the robot. However, due to the constraint on the motion of the first joint, only four valid solutions remain feasible.

5.5.3 Stanford Robot: Consecutive Screws

Consider the six-degrees-of-freedom Stanford robot shown in Figure 4.9. The forward kinematics of this robot have been analyzed based on the successive screw methods in Case Study 4.13, and the homogeneous transformation matrices corresponding to the joint screws have been obtained. To analyze the inverse kinematics of this robot, consider the robot's wrist point P, at the intersection of screws $\$_4$, $\$_5$, and $\$_6$. Given that in this robot the three final screws intersect each other at a point, the position kinematics of the wrist point can be separated from the orientation kinematics of the robot's end-effector. To accomplish this, the position and orientation vectors of the wrist point can be determined in the initial configuration, based on Figure 4.18, as follows.

$$\boldsymbol{p}^0 = \begin{bmatrix} d_2 \\ 0 \\ 0 \end{bmatrix}, \quad \boldsymbol{R}_p^0 = [\boldsymbol{u}^0, \boldsymbol{v}^0, \boldsymbol{w}^0] = \begin{bmatrix} 1 & 0 & 0 \\ 0 & 0 & 1 \\ 0 & -1 & 0 \end{bmatrix}$$

Now, consider the final configuration of the robot as a state where all joints have finite motions. In the final state, the wrist of the robot is represented by its position vector and the orientation matrix with respect to the robot's base, in terms of the robot's task space variables, as given below.

$$\boldsymbol{p}^f = \begin{bmatrix} x_p \\ y_p \\ z_p \end{bmatrix}, \quad \begin{aligned} \boldsymbol{u}^f &= [u_x, u_y, u_z]^T \\ \boldsymbol{v}^f &= [v_x, v_y, v_z]^T \\ \boldsymbol{w}^f &= [w_x, w_y, w_z]^T \end{aligned}$$

In which, the position vector of the wrist point in terms of the robot's task space variables, $\boldsymbol{\mathcal{X}} = [e_x, e_y, e_z, \theta_{e_x}, \theta_{e_y}, \theta_{e_z}]^T$, is given by:

$$\begin{bmatrix} x_p \\ y_p \\ z_p \end{bmatrix} = \begin{bmatrix} e_x \\ e_y \\ e_z \end{bmatrix} - d_6 \begin{bmatrix} w_x \\ w_y \\ w_z \end{bmatrix} \tag{5.62}$$

In this robot, the initial and final states of the wrist point's position vector can be related to each other using the following equation through the use of successive screws.

$$\boldsymbol{x}_p^f = \boldsymbol{T}_1 \boldsymbol{T}_2 \boldsymbol{T}_3 \, \boldsymbol{x}_p^0 \tag{5.63}$$

where,

$$\boldsymbol{T}_1 \boldsymbol{T}_2 \boldsymbol{T}_3 = \left[\begin{array}{ccc|c} c_1 & -s_1 c_2 & s_1 s_2 & -d_3 s_1 c_2 \\ s_1 & c_1 c_2 & -c_1 s_2 & d_3 c_1 c_2 \\ 0 & s_2 & c_2 & d_3 s_2 \\ \hline 0 & 0 & 0 & 1 \end{array}\right]$$

By substituting this transformation matrix into equation 5.63 and simplifying, we obtain the following equations.

$$x_p = d_2 c_1 - d_3 s_1 c_2 \tag{5.64}$$
$$y_p = d_2 s_1 + d_3 c_1 c_2 \tag{5.65}$$
$$z_p = d_3 s_2 \tag{5.66}$$

By summing the squares of the three equations and simplifying, we can write:

$$x_p^2 + y_p^2 + z_p^2 = d_2^2 + d_3^2 \tag{5.67}$$

In this way, the variable d_3 is obtained using the following equation.

$$d_3 = \pm\sqrt{x_p^2 + y_p^2 + z_p^2 - d_2^2} \tag{5.68}$$

If the following expression is positive, this relation has two valid solutions. However, considering the length of the prismatic joint d_3 as positive, the only valid solution with a positive sign is considered in the inverse kinematics of the robot. If $d_3 = 0$, the robot's wrist point lies along the axis of the first revolute joint, referred to as a singular configuration; this situation does not practically occur for a real robot, due to the limitation in the retraction of the prismatic actuator. Furthermore, increasing the length of the prismatic link has practical limitations $d_3 \leq d_{max}$. This constraint also restricts the accessible workspace of the robot, although the inverse kinematics of the robot are solvable in this scenario.

Assuming that the value of d_3 is found from Equation 5.66, the following equation can be used to determine the angle θ_2.

$$\theta_2 = \sin^{-1}(z_p/d_3) \tag{5.69}$$

For each value of d_3, two corresponding solutions for θ_2 can be obtained from equation 5.69. In this case, if $\theta_2 = \theta_2^\star$ is one solution, then $\theta_2 = \pi - \theta_2^\star$ will be another valid solution in the inverse kinematics of this robot.

By determining the angle θ_2 and the length of d_3, the equations 5.65 and 5.66 can be rewritten as follows.

$$c_1 = (d_2x_p + d_3y_pc_2)/(d_2^2 + d_3^2c_2^2) \tag{5.70}$$

$$s_1 = (d_2y_p - d_3x_pc_2)/(d_2^2 + d_3^2c_2^2) \tag{5.71}$$

Finally, the angle θ_1 is obtained using the following equation.

$$\theta_1 = \text{atan2}(s_1, c_1) \tag{5.72}$$

Thus, for each point in the robot's workspace, four valid solutions are obtained.

Now, let's examine the orientation kinematics of the robot's end-effector. For this purpose, by multiplying the rotation matrices of the first three joints together, we will have:

$$\boldsymbol{R}_1\boldsymbol{R}_2\boldsymbol{R}_3 = \begin{bmatrix} c_1 & -s_1c_2 & s_1s_2 \\ s_1 & c_1c_2 & -c_1s_2 \\ 0 & s_2 & c_2 \end{bmatrix} \tag{5.73}$$

Next, by multiplying the rotation matrices of the final three joints with the mutually perpendicular joint axes, we will have:

$$\boldsymbol{R}_4\boldsymbol{R}_5\boldsymbol{R}_6 = \begin{bmatrix} c_4c_5c_6 - s_4s_6 & -c_4s_5 & c_4c_5s_6 + s_4c_6 \\ s_5c_6 & c_5 & s_5s_6 \\ -s_4c_5c_6 - c_4s_6 & s_4s_5 & -s_4c_5s_6 + c_4c_6 \end{bmatrix} \tag{5.74}$$

Now, it is possible to express the relationship for consecutive joints of the vector $\boldsymbol{w}$ in the initial and final states, and simplify it as follows.

$$(\boldsymbol{R}_1\boldsymbol{R}_2\boldsymbol{R}_3)^T\boldsymbol{w}^f = \boldsymbol{R}_4\boldsymbol{R}_5\boldsymbol{R}_6\boldsymbol{w}^0 \tag{5.75}$$

If we represent the right-hand side of the above equation with the vector ${}^3\boldsymbol{w}$, we will have:

$${}^3w_x = w_xc_1 + w_ys_1 = -c_4s_5 \tag{5.76}$$

$${}^3w_y = (-w_xs_1 + w_yc_1)c_2 + w_zs_2 = c_5 \tag{5.77}$$

$${}^3w_z = (w_xs_1 - w_yc_1)s_2 + w_zc_2 = s_4s_5 \tag{5.78}$$

As seen in these equations, the vector $\boldsymbol{w}$ is not a function of the sixth joint variable. By using the equation 5.77, the angle θ_5 can be determined.

$$\theta_5 = \cos^{-1}[(-w_x s_1 + w_y c_1)c_2 + w_z s_2] \tag{5.79}$$

In this way, for each valid solution, the wrist joint angles will be derived by:

1) If $|{}^3w_y| < 1$, then this equation will have two valid solutions.

2) If $|{}^3w_y| = 1$, then this equation has only one solution, in which $\theta_5 = 0$ or $\theta_5 = \pi$. In this case, the axis $\boldsymbol{z}_6$ aligns with the axis $\boldsymbol{z}_4$, and this is referred to as a singular configuration of the robot.

3) Physically, the case where $|{}^3w_y| > 1$ cannot happen.

In the scenario where the first case is true, by using the equations 5.76 and 5.78, the angle θ_4 can be uniquely determined.

$$\theta_4 = \text{atan2}[((w_x s_1 - w_y c_1)s_2 + w_z c_2)/s_5, -(w_x c_1 + w_y s_1)/s_5] \tag{5.80}$$

Finally, to obtain the angle θ_6, the relationship for consecutive joints for the vector $\boldsymbol{u}$ in the initial and final states can be written as follows.

$$(\boldsymbol{R}_1\boldsymbol{R}_2\boldsymbol{R}_3)^T\boldsymbol{u}^f = \boldsymbol{R}_4\boldsymbol{R}_5\boldsymbol{R}_6\boldsymbol{u}^0 \tag{5.81}$$

If we represent the right-hand side of the above equation with the vector ${}^3\boldsymbol{u}$, we will have:

$$ {}^3u_x = u_x c_1 + u_y s_1 = c_4 c_5 c_6 - s_4 s_6 \tag{5.82}$$
$$ {}^3u_y = (-u_x s_1 + u_y c_1)c_2 + u_z s_2 = s_5 c_6 \tag{5.83}$$
$$ {}^3u_z = (u_x s_1 - u_y c_1)s_2 + u_z c_2 = -s_4 c_5 c_6 - c_4 s_6 \tag{5.84}$$

Now, by multiplying the right-hand side of the equation 5.82 by s_4, multiplying the right-hand side of the equation 5.84 by c_4, and adding the results, we will have:

$${}^3u_x s_4 + {}^3u_z c_4 = -s_6 \tag{5.85}$$

By using this equation and the right-hand side of the equation 5.83, the angle θ_6 can be determined.

$$\theta_6 = \text{atan2}[-({}^3u_x s_4 + {}^3u_z c_4), {}^3u_y/s_5] \tag{5.86}$$

In this way, the inverse kinematic analysis of the robot is complete, and eight solutions are obtained. However, only four of these solutions are valid considering the motion constraints in the robot's joints.

5.5.4 6DOF Elbow Manipulator: Consecutive Screws

Consider the six-degrees-of-freedom Elbow manipulator shown in Figure 4.10. The forward kinematics of this robot have been analyzed based on the successive screw methods in Case Study 4.12, where the transformation matrices corresponding to the joint screws are derived. To analyze the inverse kinematics of this robot, consider the wrist point P at the intersection of screws $\$_5$ and $\$_6$. Despite the fact that the three final screws of the robot do not intersect with each other in this robot, it is possible to analyze the inverse kinematics using this wrist point.

The position and orientation vector of the wrist in the initial configuration can easily be determined based on Figure 4.17 as follows.

$$p^0 = \begin{bmatrix} a_2 + a_3 + a_4 \\ 0 \\ 0 \end{bmatrix}, \quad R_p^0 = [u^0, v^0, w^0] = \begin{bmatrix} 0 & 0 & 1 \\ 0 & -1 & 0 \\ 1 & 0 & 0 \end{bmatrix}$$

Now, consider the final configuration of the robot as a state where all joints have finite movements, and the wrist point of the robot, as a function of task space variables, is represented by its position vector and orientation matrix below.

$$p^f = \begin{bmatrix} x_p \\ y_p \\ z_p \end{bmatrix}, \quad \begin{aligned} u^f &= [u_x, u_y, u_z]^T \\ v^f &= [v_x, v_y, v_z]^T \\ w^f &= [w_x, w_y, w_z]^T \end{aligned}$$

In this robot, using the successive screw method, the initial and final states of the wrist position vector can be related to each other through the following equation.

$$x_p^f = T_1 T_2 T_3 T_4 x_p^0 \tag{5.87}$$

By multiplying both sides of this equation by T_1^{-1}, we can write:

$$(T_1)^{-1} \begin{bmatrix} x_p \\ y_p \\ z_p \\ 1 \end{bmatrix} = T_2 T_3 T_4 \begin{bmatrix} a_2 + a_3 + a_4 \\ 0 \\ 0 \\ 1 \end{bmatrix} \tag{5.88}$$

where,

$$(T_1)^{-1} = \left[\begin{array}{ccc|c} c_1 & s_1 & 0 & 0 \\ -s_1 & c_1 & 0 & 0 \\ 0 & 0 & 0 & 1 \\ \hline 0 & 0 & 0 & 1 \end{array}\right]$$

$$T_2 T_3 T_4 = \left[\begin{array}{ccc|c} c_{234} & 0 & -s_{234} & a_2 c_2 + a_3 c_{23} - (a_2 + a_3) c_{234} \\ 0 & 1 & 0 & 0 \\ s_{234} & 0 & c_{234} & a_2 s_2 + a_3 s_{23} - (a_2 + a_3) s_{234} \\ \hline 0 & 0 & 0 & 1 \end{array}\right]$$

Therefore,

$$x_p c_1 + y_p s_1 = a_2 c_2 + a_3 c_{23} + a_4 s_{234} \tag{5.89}$$

$$-x_p s_1 + y_p c_1 = 0 \tag{5.90}$$

$$z_p = a_2 s_2 + a_3 s_{23} + a_4 s_{234} \tag{5.91}$$

By substituting the homogeneous transformation matrices into Equation 5.88 and simplifying, we will arrive at the following equations.

$$\theta_1 = \operatorname{atan} \frac{y_p}{x_p} \tag{5.92}$$

Unfortunately, in this problem, due to the non-intersecting nature of the three final screws, the solution for the position kinematics is not entirely decoupled from the orientation. For the orientation, the component w of the wrist can be expressed as follows.

$$w^f = T_1 T_2 T_3 T_4 T_5 w^0 \tag{5.93}$$

Now, consider only the rotation matrix components in this equation and multiply both sides by $\boldsymbol{R}_1^{-1}$.

$$(\boldsymbol{R}_1)^T \boldsymbol{w}^f = \boldsymbol{R}_2 \boldsymbol{R}_3 \boldsymbol{R}_4 \boldsymbol{R}_5 \boldsymbol{w}^0 \tag{5.94}$$

where the right-hand side rotation matrix is obtained from the following equation.

$$\boldsymbol{R}_2 \boldsymbol{R}_3 \boldsymbol{R}_4 \boldsymbol{R}_5 = \begin{bmatrix} c_5 c_{234} & -s_5 c_{234} & -s_{234} \\ s_5 & c_5 & 0 \\ c_5 s_{234} & -s_5 s_{234} & c_{234} \end{bmatrix}$$

By substituting the transformation matrices into Equation 5.94, we will obtain the following equations.

$$w_x c_1 + w_y s_1 = c_5 c_{234} \tag{5.95}$$

$$-w_x s_1 + w_y c_1 = s_5 \tag{5.96}$$

$$w_z = c_5 s_{234} \tag{5.97}$$

For each solution of θ_1, two corresponding solutions for θ_5 can be obtained from the equation 5.96.

$$\theta_5 = \sin^{-1}(-w_x s_1 + w_y c_1) \tag{5.98}$$

In this case, if $\theta_5 = \theta_5^\star$ is one solution, then $\theta_5 = \pi - \theta_5^\star$ will also be another valid solution in the inverse kinematics of this robot. By determining the angles θ_1 and θ_5, the angle θ_{234} can be obtained from equations 5.95 and 5.97 using the following equation.

$$\theta_{234} = \text{atan2}[w_z/c_5, (w_x c_1 + w_y s_1)/c_5] \tag{5.99}$$

Now, by using equations 5.89 and 5.91, we can determine the angles θ_2 and θ_3. Rewrite these equations for simplicity as follows.

$$a_2 c_2 + a_3 c_{23} = k_1 \tag{5.100}$$

$$a_2 s_2 + a_3 s_{23} = k_2 \tag{5.101}$$

where,

$$k_1 = x_p c_1 + y_p s_1 - a_4 c_{234} \ , \qquad k_2 = z_p - a_4 s_{234}$$

By summing the squares of the two above equations and simplifying, we will have:

$$k_1^2 + k_2^2 = a_2^2 + a_3^2 + 2a_2 a_3 c_3 \tag{5.102}$$

In this way, the angle θ_3 can be obtained using the following equation.

$$\theta_3 = \cos^{-1} \frac{k_1^2 + k_2^2 - a_2^2 - a_3^2}{2a_2 a_3} \tag{5.103}$$

This equation also creates two valid solutions for the angle θ_3. In this case, if $\theta_3 = \theta_3^\star$ is one solution, then $\theta_3 = -\theta_3^\star$ will also be another valid solution in the inverse kinematics of this robot. By determining the solution for the angle θ_3, one can easily obtain the angle θ_2 using equations 5.89 and 5.91.

$$\theta_2 = \text{atan2}[(k_2 - a_2 s_2)/a_3, (k_1 - a_2 c_2)/a_3] - \theta_3 \tag{5.104}$$

Next, the angle θ_4 can be easily calculated using the following equation.

$$\theta_4 = \theta_{234} - \theta_2 - \theta_3 \tag{5.105}$$

To determine the angle θ_6, the relationship for consecutive screws for the vector $\boldsymbol{u}$ in the initial and final states can be written as follows and simplified.

$$(\boldsymbol{R}_1\boldsymbol{R}_2\boldsymbol{R}_3\boldsymbol{R}_4)^T\boldsymbol{u}^f = \boldsymbol{R}_5\boldsymbol{R}_6\boldsymbol{u}^0 \tag{5.106}$$

By calculating the rotation matrices on both sides of the equation and substituting them, we will arrive at the following equations.

$$u_x c_1 c_{234} + u_y s_1 c_{234} + u_s s_{234} = s_5 s_6 \tag{5.107}$$

$$-u_x s_1 + u_y c_1 = -c_5 s_6 \tag{5.108}$$

$$-u_x c_1 s_{234} - u_y s_1 s_{234} + u_z c_{234} = c_6 \tag{5.109}$$

By multiplying Equation 5.108 by s_5 and subtracting it from the product of Equation 5.109 by c_5, we can obtain s_6 as follows.

$$s_6 = s_5(u_x c_1 c_{234} + u_y s_1 c_{234} + u_s s_{234}) - c_5(-u_x s_1 + u_y c_1) \tag{5.110}$$

Now, using Equation 5.109, one unique solution for the angle θ_6 can be obtained from the following equation.

$$\theta_6 = \text{atan2}(s6, (-u_x c_1 s_{234} - u_y s_1 s_{234} + u_z c_{234})) \tag{5.111}$$

In this way, the inverse kinematics of the robot is completed, and eight valid solutions are obtained.

5.6 Accuracy and Repeatability

In industrial robots, point-to-point motion is taught to the robot by a teaching pendant. By using this device, the motion of the robot's arms are manually controlled, the end-effector is positioned at the desired pose, and the information of the robot's joints in this pose is monitored and saved. The same process is repeated for other poses where the robot is intended to move. Subsequently, the robot's motion processor plans a direct path or appropriate motion according to the required performance, from one point to another. In this way and in a point-to-point motion using this method, workspace coordinates are not entered in the robot's motion planning, and there is no need to solve the inverse kinematics of the robot in the feedback loop.

With this method, robot programming does not require precision in reaching absolute robot coordinates; instead, the focus is on how the robot reaches the desired point in different motion instances. For this reason, in industrial robots, repeatability index is often reported instead of the robot's accuracy. To determine this index, the robot's motion is taught from one point to another at various points in the workspace, and the robot is programmed to perform this motion repeatedly. By accurately measuring the end-effector's position in these experiments, statistical characteristics of the relative motion error are reported as the repeatability index. The repeatability value is calculated by:

$$\text{Repeatibility Index} = \frac{\max(\text{Measured Value}) - \min(\text{Measured Value})}{\max(\text{Measured Value}) + \min(\text{Measured Value})} \tag{5.112}$$

A lower repeatability index indicates better repeatability, meaning the robot can consistently achieve the specified pose with less variation. Conversely, a higher repeatability index suggests greater variability in the robot's performance and may require adjustments or improvements to enhance precision. The repeatability index provides a quantitative measure of the robot's positional accuracy and consistency, helping to assess its suitability for various industrial or research applications.

If it is necessary for the end-effector's position and orientation in the workspace to reach desired points in the workspace represented by their absolute values, the inverse kinematics of the robot must be used. This is because the motion control of serial robots occurs usually in their joints, and through inverse kinematics, the mapping of motion from the task space to the joint space is performed. In robots where this type of motion planning is used, it is necessary to accurately identify kinematic information such as link lengths and joint offsets. This is typically done by performing various motion experiments with the robot in the workspace and precisely measuring the end-effector's position and orientation; this process in robotics is known as robot calibration. Efficient methods for robot calibration have been developed in the robotics literature [71, 72].

In robots where motion planning is not predetermined, for instance, when the robot tracks and locates the desired object on a conveyor belt using a camera, the latter method for motion planning needs to be employed, and such motion experiments in robots determine the accuracy of the robot's motion in the workspace. The accuracy of the end-effector's motion is a quantity that can be challenging to evaluate, and in industrial robots, it is usually limited to reporting repeatability. Although the accuracy of robot motion is related to the level of repeatability, it should be noted that the accuracy is also related to the accurate information on kinematic parameters of the robot. Therefore, the accuracy of robot motion is usually less than the reported repeatability given in the technical specifications of the robot.

Problems

5.1 For the two-degrees-of-freedom RP robot depicted in Figure 5.9, the forward kinematics were previously analyzed in Problem 4.1. Determine the inverse kinematic solutions for this robot and examine the number of solutions.

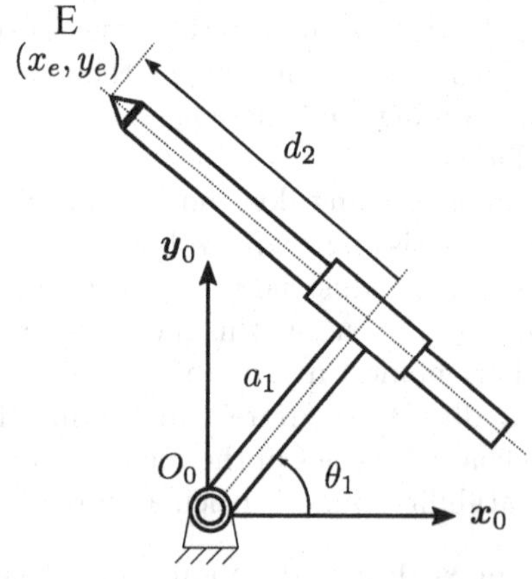

FIGURE 5.9
Schematics of the 2DOF robot of Problem 5.1.

5.2 For the PRP robot with three degrees of freedom shown in Figure 5.10, forward kinematics were previously investigated in Problem 4.2. Now, ascertain the inverse kinematic solutions for this robot and analyze the number of solutions.

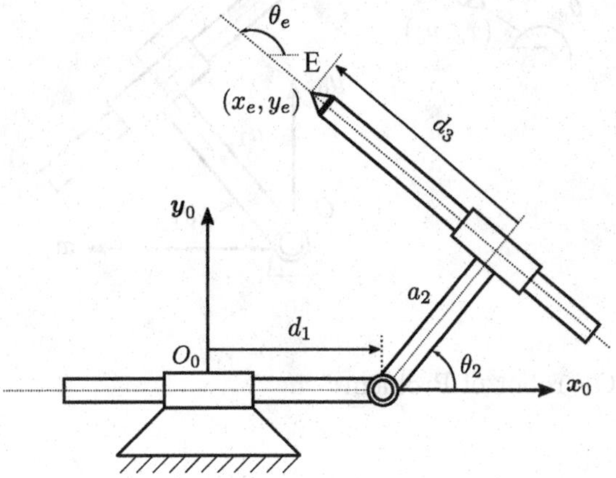

FIGURE 5.10
Schematics of the 3DOF robot of Problem 5.2.

5.3 For the three-degree-of-freedom PRR robot illustrated in Figure 5.11, the forward kinematics were previously examined in Problem 4.5. Now, determine the inverse kinematic solutions for this robot and assess the quantity of solutions.

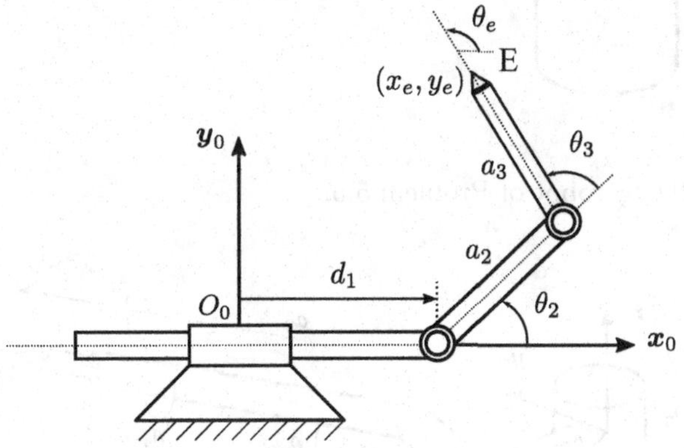

FIGURE 5.11
Schematics of the 3DOF robot of Problem 5.3.

5.4 For the three-degrees-of-freedom RPR robot depicted in Figure 5.12, the forward kinematics were previously analyzed in Problem 4.8. Determine the inverse kinematic solutions for this robot and examine the number of solutions.

5.5 For the RR robot with two degrees of freedom depicted in Figure 5.13, forward kinematics were previously explored in Problem 4.11. Establish the inverse kinematic solutions for this robot and evaluate the number of solutions.

5.6 For the four-degrees-of-freedom RRRP robot depicted in Figure 5.14, the forward kinematics were previously analyzed in Problem 4.13. Determine the inverse kinematic solutions for this robot and examine the number of solutions.

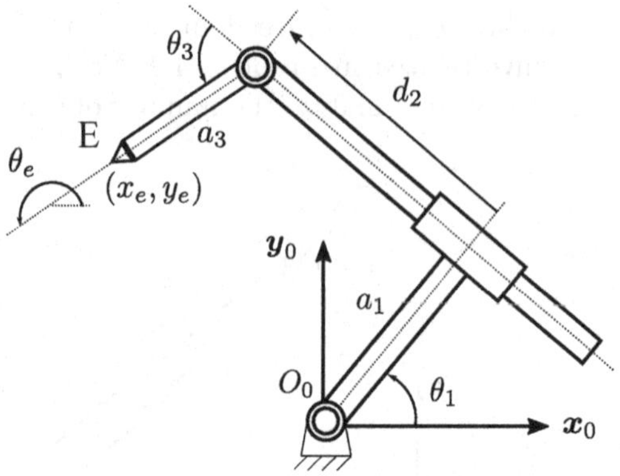

FIGURE 5.12
Schematics of the 3DOF robot of Problem 5.4.

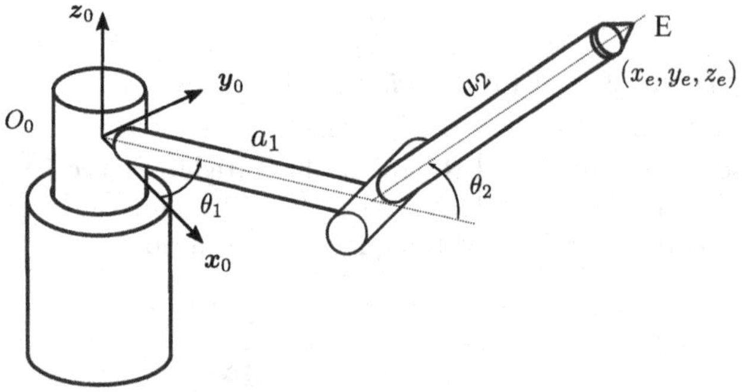

FIGURE 5.13
Schematics of the 2DOF robot of Problem 5.5.

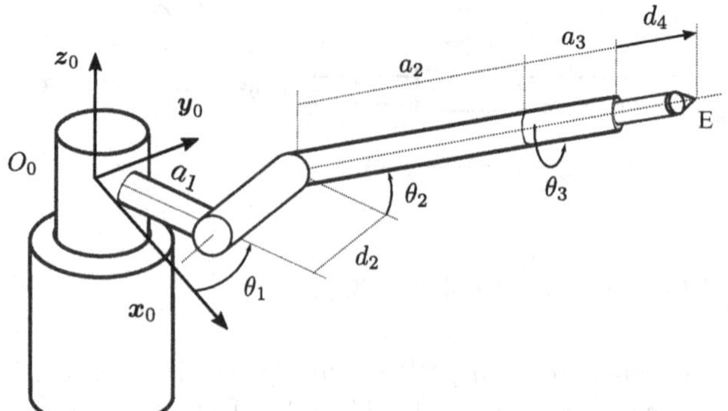

FIGURE 5.14
Schematics of the 4DOF robot of Problem 5.6.

5.7 Analyze the inverse kinematic solutions for the four-degree-of-freedom RPRR robot illustrated in Figure 5.15, following the examination of forward kinematics in Problem 4.15. Determine the inverse kinematic solutions for this robot and evaluate the

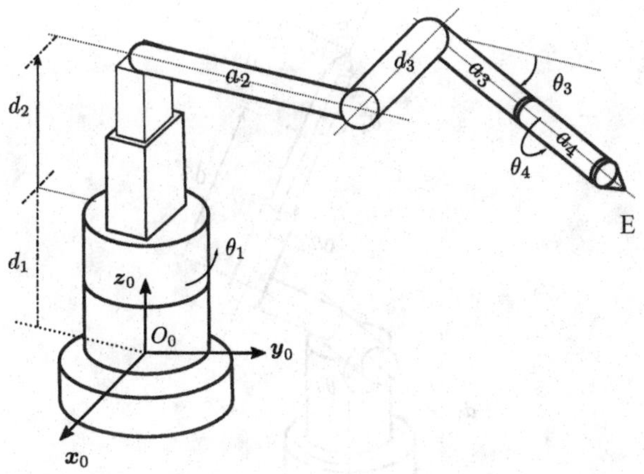

FIGURE 5.15
Schematics of the 4DOF robot of Problem 5.7.

number of solutions as well.

5.8 For the four-degrees-of-freedom 4R robot depicted in Figure 5.16, the forward kinematics were previously analyzed in Problem 4.17. Determine the inverse kinematic solutions for this robot and examine the number of solutions.

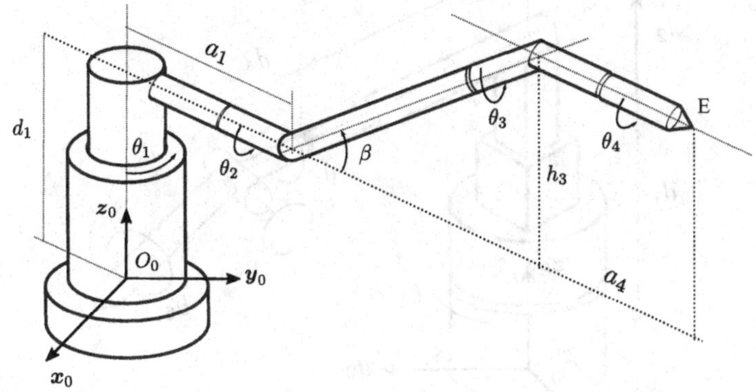

FIGURE 5.16
Schematics of the 4DOF robot of Problem 5.8.

5.9 Investigate the inverse kinematic solutions for the four-degree-of-freedom 4R robot illustrated in Figure 5.17, subsequent to the analysis of forward kinematics in Problem 4.19. Determine the inverse kinematic solutions for this robot, and furthermore, assess the number of solutions.

5.10 For the six-degrees-of-freedom robot depicted in Figure 5.18, the forward kinematics were previously analyzed in Problem 4.21. Determine the inverse kinematic solutions for this robot and examine the number of solutions.

5.11 For the six-degrees-of-freedom robot depicted in Figure 5.19, the forward kinematics were previously analyzed in Problem 4.23. Determine the inverse kinematic solutions for this robot and examine the number of solutions.

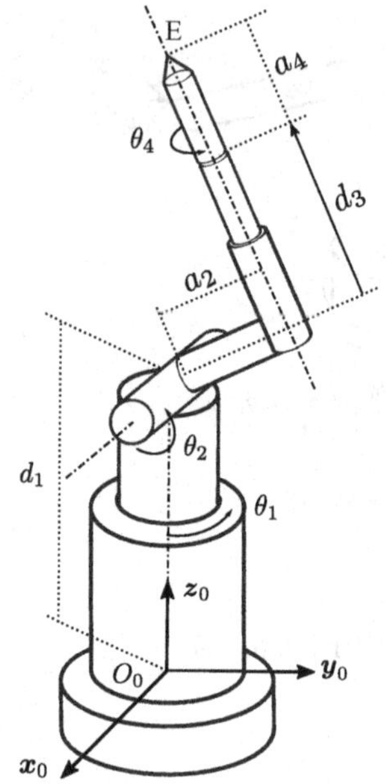

FIGURE 5.17
Schematics of the 4DOF robot of Problem 5.9.

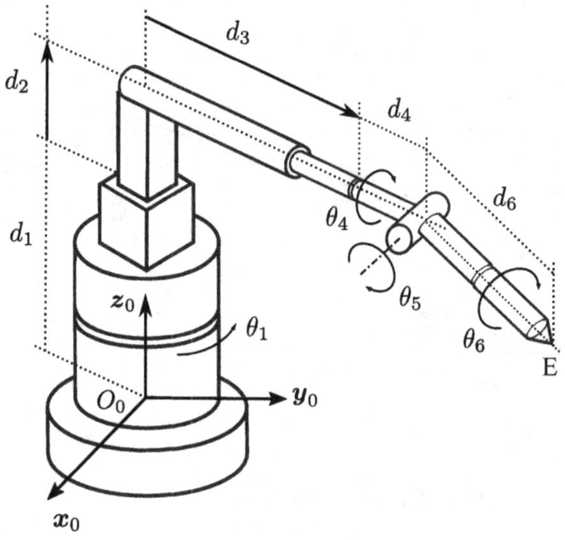

FIGURE 5.18
Schematics of the 6DOF robot of Problem 5.10.

5.12 For the six-degrees-of-freedom "Kuka KR6 Agilus" robot depicted in Figure 5.20, the forward kinematics were previously analyzed in Problem 4.25. Determine the inverse kinematic solutions for this robot and examine the number of solutions.

5.13 Consider the two-degrees-of-freedom <u>RP</u> robot in Problem 5.1 illustrated in Figure 5.9. For the given numerical values:

$$a_1 = 1, \ \ \theta_1 = [0 - 2\pi], \ \ d_2 = [0 - 1]$$

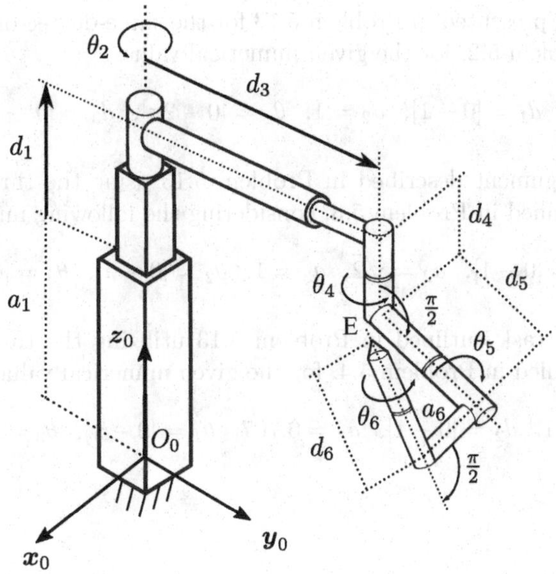

FIGURE 5.19
Schematics of the 6DOF robot of Problem 5.11.

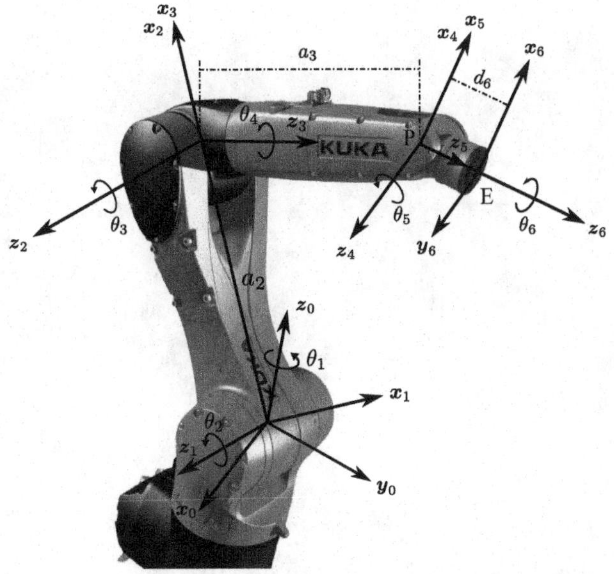

FIGURE 5.20
Schematics of the 6DOF "KUKA KR6 Agilus" Robot of Problem 5.12

identify the robot's workspace by assessing the feasibility of the inverse kinematics solution within a numerical mesh grid surrounding the robot, and visualize it in three-dimensional space. Furthermore, explore instances where the inverse kinematic problem yields repetitive solutions and verify the boundaries of the ascertained robot workspace accordingly.

5.14 Repeat the task presented in Problem 5.13 for the three-degree-of-freedom PRP robot outlined in Problem 5.2, for the given numerical values.

$$d_1 = [0 - 1], \;\; a_2 = 1, \;\; \theta_2 = [0 - 2\pi], \;\; d_3 = [0 - 1].$$

5.15 Revisit the assignment described in Problem 5.13 using the three-degree-of-freedom PRR robot specified in Problem 5.3, considering the following numerical values.

$$d_1 = [0 - 1], \;\; a_2 = \sqrt{2}, \;\; a_3 = 1, \;\; \theta_2 = [0 - \pi], \;\; \theta_3 = [0 - 2\pi].$$

5.16 Re-examine the task outlined in Problem 5.13 utilizing the three-degree-of-freedom RPR robot detailed in Problem 5.4, for the given numerical values.

$$a_1 = 1, \;\; d_2 = [0 - 1], \;\; a_3 = 0.707, \;\; \theta_1 = [0 - \pi], \;\; \theta_3 = [0 - 2\pi].$$

6

Jacobian Matrix

In preceding chapters, we explored the forward and inverse kinematics of robots. This chapter expands on robot kinematics, moving beyond the static state to focus on velocity analysis. The primary emphasis is on examining the linear and angular velocities of the robot's end-effector and their relationship with joint angular velocities. This exploration naturally leads to the introduction of the robot's Jacobian matrix. Referring to the definitions of linear velocity of a point and angular velocity of a rigid body, we introduce the Jacobian matrix and outline a general approach for its computation. Subsequently, we present an alternative method for determining the Jacobian matrix based on screw theory, incorporating the screw representation and its connection to physical intuition in motion. Finally, we illustrate how the relationship between static forces applied to the robot's end-effector and actuator torques is articulated through the transpose of the Jacobian matrix using the principle of virtual work. Throughout the chapter and at the end, various problems are provided to reinforce understanding and facilitate practice.

6.1 Definition of the Jacobian Matrix

As previously mentioned, to describe the end-effector's motion in the task space, six independent motion variables are generally used. Accordingly, the configuration of the robot's end-effector in space is represented by a six-dimensional array on the task space variables $\boldsymbol{\mathcal{X}} = [x_e, y_e, z_e, \theta_{e_x}, \theta_{e_y}, \theta_{e_z}]^T$, which is called the "twist." The movements of the robot's joints are also described by another n-dimensional array, representing angular or translational motions in the joints, denoted by the joint variables $\boldsymbol{q} = [q_1, q_2, \ldots, q_n]^T$. The joint variables of the robot's workspace are obtained based on the rotational motions of the revolute joints or translational motions of the prismatic joints, by using the following relationship.

$$q_i = \begin{cases} \theta_i & \text{For Revolute Joints} \\ d_i & \text{For Prismatic Joints} \end{cases} \tag{6.1}$$

With this description, the forward kinematics of the robot is interpreted as an analytical function of the task space variables $\boldsymbol{\mathcal{X}}$ with respect to the robot's joint variables $\boldsymbol{q}$. If we express the relationship for the forward kinematics of the robot in general form with a vector function $\boldsymbol{f}$, it can be formulated as follows:

$$\boldsymbol{\mathcal{X}} = \boldsymbol{f}(\boldsymbol{q}) \tag{6.2}$$

In the general case, this relationship is a multivariable and nonlinear function of the joint variables. If the workspace variables have a dimension of m and the robot has n degrees of freedom, this relationship can be rewritten in terms of its components using m components

DOI: 10.1201/9781003491415-6

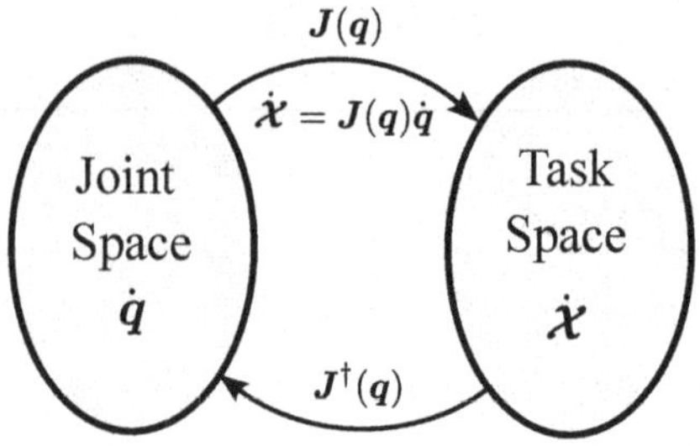

FIGURE 6.1
Differential kinematics and the relationship of the Jacobian matrix with respect to the velocity variables in joint and task spaces.

f_i of the vector function $\boldsymbol{f}$.

$$\mathcal{X}_i = f_i(q_1, q_2, \ldots, q_n) \ , \quad i = 1, 2, \ldots, m \tag{6.3}$$

To examine the differential kinematics of the robot, differentiate this relationship with respect to time.

$$\dot{\mathcal{X}}_i = \frac{\partial f_i}{\partial q_1}\dot{q}_1 + \frac{\partial f_i}{\partial q_2}\dot{q}_2 + \cdots \frac{\partial f_i}{\partial q_n}\dot{q}_n \ , \quad i = 1, 2, \ldots, m \tag{6.4}$$

and rewrite it in matrix form.

$$\dot{\boldsymbol{\mathcal{X}}} = \boldsymbol{J}(\boldsymbol{q})\dot{\boldsymbol{q}} \tag{6.5}$$

In this equation, $\boldsymbol{J}(\boldsymbol{q})$ is called the Jacobian matrix of the robot, defined as follows:

$$\boldsymbol{J}(\boldsymbol{q}) = \frac{\partial \boldsymbol{f}(\boldsymbol{q})}{\partial \boldsymbol{q}} = \begin{bmatrix} \frac{\partial f_1}{\partial q_1} & \frac{\partial f_1}{\partial q_2} & \cdots & \frac{\partial f_1}{\partial q_n} \\ \frac{\partial f_2}{\partial q_1} & \frac{\partial f_2}{\partial q_2} & \cdots & \frac{\partial f_2}{\partial q_n} \\ \vdots & & \ddots & \vdots \\ \frac{\partial f_m}{\partial q_1} & \frac{\partial f_m}{\partial q_2} & \cdots & \frac{\partial f_m}{\partial q_n} \end{bmatrix} \tag{6.6}$$

As inferred from this equation, the robot's Jacobian matrix $\boldsymbol{J}(\boldsymbol{q})$ is a function of the joint variables, and its value changes based on the configuration of the robot and the position of the robot's end-effector in the workspace. From this perspective, the Jacobian matrix can be considered as a mapping from joint space velocity vector to the end-effector velocity vector in the task space. This mapping is illustrated in Figure 6.1.

In general, if we express the six-dimensional twist rates of the robot's end-effector $\dot{\boldsymbol{\mathcal{X}}}$ in terms of linear velocity vectors $\boldsymbol{v}_e$ and angular velocity vectors $\boldsymbol{\omega}_e$, we can decompose the robot's Jacobian matrix into two submatrices, $\boldsymbol{J}_v$ and $\boldsymbol{J}_\omega$, as follows:

$$\begin{bmatrix} \boldsymbol{v}_e \\ \boldsymbol{\omega}_e \end{bmatrix} = \begin{bmatrix} \boldsymbol{J}_v \\ \boldsymbol{J}_\omega \end{bmatrix} \dot{\boldsymbol{q}} \tag{6.7}$$

In this equation, $\boldsymbol{J}_v$ is the submatrix of the Jacobian corresponding to linear velocities, and $\boldsymbol{J}_\omega$ is the submatrix of the Jacobian corresponding to angular velocities of the robot's end-effector. If the Jacobian matrix is given, then these two submatrices are determined by separating its upper and lower three rows, respectively.

$$\boldsymbol{J} = \begin{bmatrix} \boldsymbol{J}_v \\ \boldsymbol{J}_\omega \end{bmatrix} \tag{6.8}$$

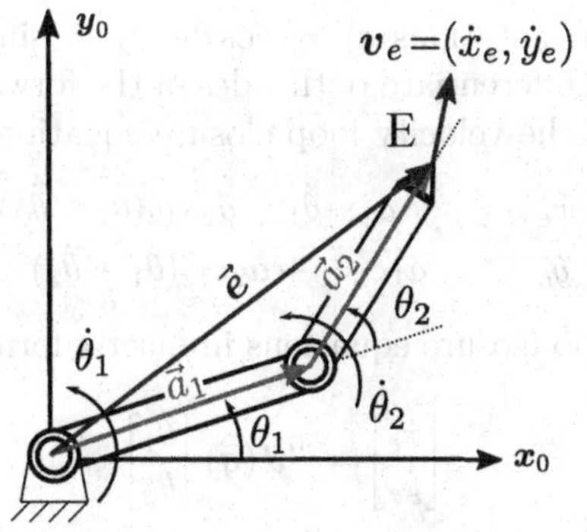

FIGURE 6.2
Velocity loop closure method for deriving the Jacobian matrix of a planar robot.

As seen in Equation 6.6, the Jacobian matrix can be defined with any dimensional matrix (even non-square). The number of rows in the Jacobian matrix indicates the degrees of freedom of the robot in the task space, while its number of columns is equal to the number of joint variables of the robotic manipulator. In planar robots where $n = 3$, the Jacobian matrix can be represented by three rows. However, for redundant manipulators where the number of actuated joints exceeds the number of degrees of freedom, the number of columns in the Jacobian matrix can surpass the number of rows.

6.2 Jacobian Matrix Derivation Methods

While relation 6.6 provides a comprehensive definition of the Jacobian matrix and its relationship with the robot's direct kinematics, it does not always offer a straightforward method for its determination. In the following, we will explore efficient methods for deriving the Jacobian matrix in serial robots.

6.2.1 Velocity Loop Closure Method

The velocity loop closure method is a fundamental approach for determining the Jacobian matrix and its relation to the robot's forward kinematics. In this method, the robot's forward kinematics is obtained using the loop closure technique, and by differentiating it with respect to time, the relationship between the linear and angular velocities of the robot's end-effector and the joint velocities is derived.

In this section, we present the velocity loop closure method through a conceptual example. For this purpose, consider a two-degrees-of-freedom $2\underline{R}$ robot as shown in Figure 6.2, where the lengths of the robot's links are denoted by a_1 and a_2, and the joint angles are θ_1 and θ_2. The end-effector of the robot, represented by the position vector $\boldsymbol{e} = [x_e, y_e]^T$, is located at the position shown in Figure 6.2. According to the figure, to analyze the kinematics of the robot, we place the reference coordinate system $\{0\}$ on the fixed base of the robot and at the center of the first revolute joint. By applying the loop closure method, the forward kinematics equations of the robot can be obtained as follows.

$$\begin{aligned} x_e &= a_1 c_1 + a_2 c_{12} \\ y_e &= a_1 s_1 + a_2 s_{12} \end{aligned}$$

In these equations, common notations $c_1 = \cos\theta_1$, $s_1 = \sin\theta_1$, $c_{12} = \cos(\theta_1 + \theta_2)$, and $s_{12} = \sin(\theta_1 + \theta_2)$ are used. Differentiate both sides of the forward kinematic equations with respect to time, to arrive at the velocity loop closure equations.

$$\dot{x}_e = -a_1 s_1 \dot{\theta}_1 - a_2 s_{12}(\dot{\theta}_1 + \dot{\theta}_2) \tag{6.9}$$

$$\dot{y}_e = a_1 c_1 \dot{\theta}_1 + a_2 c_{12}(\dot{\theta}_1 + \dot{\theta}_2) \tag{6.10}$$

By rewriting the velocity loop closure equations in matrix form, according to Equation 6.5, we have:

$$\begin{bmatrix} \dot{x}_e \\ \dot{y}_e \end{bmatrix} = {}^0\boldsymbol{J}(\boldsymbol{q}) \begin{bmatrix} \dot{\theta}_1 \\ \dot{\theta}_2 \end{bmatrix} \tag{6.11}$$

and the Jacobian matrix is easily obtained.

$${}^0\boldsymbol{J}(\boldsymbol{q}) = \begin{bmatrix} -(a_1 s_1 + a_2 s_{12}) & -a_2 s_{12} \\ a_1 c_1 + a_2 c_{12} & a_2 c_{12} \end{bmatrix} \tag{6.12}$$

In this equation, the superscript ${}^0(\cdot)$ is used to indicate the interpretation of the Jacobian matrix in the reference coordinate frame $\{0\}$. Thus, the relationship between the end-effector velocity in the $\{0\}$ coordinate system and the angular velocities of the two revolute joints is described by this matrix.

As evident from Equation 6.12, the Jacobian matrix of the robot is influenced by the robot's configuration in the workspace, and its properties vary across different points. Now, let's highlight a crucial aspect of this mapping by examining the determinant of the Jacobian matrix. If we view the Jacobian matrix as the mapping between the joint angular velocities of the robot and the end-effector velocity, as illustrated in Figure 6.1, its determinant serves as an indicator of the gain in this mapping. Consequently, a large determinant implies that small angular velocities in the robot's joints lead to significant linear velocities at the robot's end-effector. Conversely, a small determinant indicates that large angular velocities in the robot's joints result in minimal linear velocities at the robot's end-effector. If, in a specific robot configuration, the determinant of the Jacobian matrix equals zero, the robot's joints are incapable of generating motion at the end-effector at least in one specific direction in the workspace.

To investigate this feature in the planar robot under consideration, calculate the determinant of the Jacobian matrix using the trigonometric equality $c_1 s_{12} - s_1 c_{12} = s_2$.

$$\begin{aligned} \det({}^0\boldsymbol{J}(\boldsymbol{q})) &= -a_1 a_2 s_1 c_{12} - a_2^2 c_{12} s_{12} + a_1 a_2 c_1 s_{12} + a_2^2 c_{12} s_{12} \\ &= a_1 a_2 s_2 \end{aligned} \tag{6.13}$$

Note that if $s_2 = 0$, the determinant of the Jacobian matrix will be zero. This situation occurs when θ_2 is either zero or π. In such cases, the robot arm is either fully stretched or fully retracted. As explained before, this condition occurs when the robot's end-effector is located at a point on the boundary of its workspace. In this situation, intuitively, even with applying very large angular velocities to the robot's joints, it is not possible to create the smallest movement in the end-effector in the radial direction and perpendicular to the boundary of the workspace.

In this case, the usual two solutions to the inverse kinematics problem converges to one repeated solution, while the determinant of the Jacobian matrix becomes zero. It is said that the robot's Jacobian matrix is in a *singular configuration* and is not invertible. The convergence of two seemingly different features in the inverse kinematic analysis and the differential kinematics of the robot's end-effector in a singular configuration is highly intriguing. It reveals that by numerically analyzing the Jacobian matrix in different configurations, desirable features in robot design can be determined. This topic is elaborated in detail in next chapter and in Section 7.1.

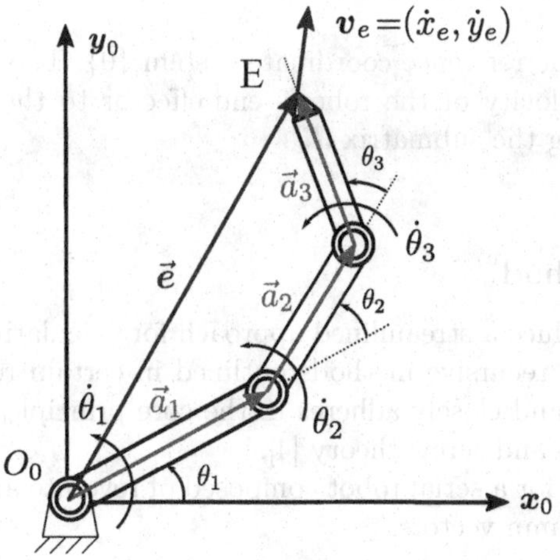

FIGURE 6.3
The velocity loop closure method for determining the Jacobian matrix of the planar 3R robot.

Case Study 6.1 (Planar 3R Robot) *Consider the planar 3R robot shown in Figure 6.3. Using the velocity loop closure method, obtain the Jacobian matrix of the robot.*

Solution:
In order to apply the velocity loop closure method, consider the forward kinematic equations for the 3R robot, given in Equation 4.14, which is reiterated here.

$$\begin{aligned} x_e &= a_1c_1 + a_2c_{12} + a_3c_{123} \\ y_e &= a_1s_1 + a_2s_{12} + a_3s_{123} \\ \theta_e &= \theta_1 + \theta_2 + \theta_3 \end{aligned} \tag{6.14}$$

By taking the derivative of both sides of this equation with respect to time, we can arrive at the velocity loop closure equation.

$$\begin{aligned} \dot{x}_e &= -a_1s_1\dot{\theta}_1 - a_2s_{12}(\dot{\theta}_1 + \dot{\theta}_2) - a_3s_{123}(\dot{\theta}_1 + \dot{\theta}_2 + \dot{\theta}_3) \\ \dot{y}_e &= a_1c_1\dot{\theta}_1 + a_2c_{12}(\dot{\theta}_1 + \dot{\theta}_2) + a_3c_{123}(\dot{\theta}_1 + \dot{\theta}_2 + \dot{\theta}_3) \\ \dot{\theta}_e &= \dot{\theta}_1 + \dot{\theta}_2 + \dot{\theta}_3 \end{aligned} \tag{6.15}$$

Rewrite this equation according to the convention given in Equation 6.5,

$$\begin{bmatrix} \dot{x}_e \\ \dot{y}_e \\ \dot{\theta}_e \end{bmatrix} = {}^0\boldsymbol{J}(\boldsymbol{q}) \begin{bmatrix} \dot{\theta}_1 \\ \dot{\theta}_2 \\ \dot{\theta}_3 \end{bmatrix} \tag{6.16}$$

The robot's Jacobian matrix is expressed as:

$${}^0\boldsymbol{J}(\boldsymbol{q}) = \begin{bmatrix} -(a_1s_1 + a_2s_{12} + a_3s_{123}) & -(a_2s_{12} + a_3s_{123}) & -a_3s_{123} \\ a_1c_1 + a_2c_{12} + a_3c_{123} & a_2c_{12} + a_3c_{123} & a_3c_{123} \\ 1 & 1 & 1 \end{bmatrix} \tag{6.17}$$

The first two rows of this matrix constitute the submatrix $\boldsymbol{J}_v$, representing the relationship between the linear velocity of the robot's end-effector and the angular velocity of its revolute joints in the reference coordinate system $\{0\}$. The third row of the matrix relates the angular velocity of the robot's end-effector to the angular velocity of its revolute joints, forming the submatrix $\boldsymbol{J}_\omega$.

6.2.2 General Method

In this section, we introduce a streamlined approach for calculating the Jacobian matrix of serial robots. Unlike the recursive methods outlined in certain references [2], this method is more straightforward and closely adheres to the core principles of rigid body motion, as explained through twists and screw theory [4].

The Jacobian matrix for a serial robot composed of revolute and prismatic joints can be determined using its column vectors.

$$\boldsymbol{J} = \begin{bmatrix} \boldsymbol{J}_v \\ \boldsymbol{J}_\omega \end{bmatrix} = \left[\ \boldsymbol{J}_1 \mid \boldsymbol{J}_2 \mid \cdots \mid \boldsymbol{J}_n \ \right]_{6\times n} \tag{6.18}$$

In which, $\boldsymbol{J}_i$'s are the columns of the Jacobian matrix, and it can be decomposed by,

$$\boldsymbol{J}_i = \begin{bmatrix} \boldsymbol{J}_{vi} \\ \boldsymbol{J}_{\omega i} \end{bmatrix}_{6\times 1} \tag{6.19}$$

In this representation $\boldsymbol{J}_v$ is the submatrix of the Jacobian matrix related to the linear velocities of the robot's end-effector, and $\boldsymbol{J}_\omega$ is the submatrix related to the angular velocities.

It can be shown that the columns of the Jacobian matrix in **revolute joints** are determined by the relation

$$\boldsymbol{J}_i = \begin{bmatrix} \boldsymbol{z}_{i-1} \times \boldsymbol{p}_e^{i-1} \\ \boldsymbol{z}_{i-1} \end{bmatrix}, \tag{6.20}$$

and in **prismatic joints** by the following relation,

$$\boldsymbol{J}_i = \begin{bmatrix} \boldsymbol{z}_{i-1} \\ \boldsymbol{0} \end{bmatrix}, \tag{6.21}$$

as depicted in Figure 6.4. Here, $\boldsymbol{z}_{i-1}$ is the axis of rotation (or translation) of the ith revolute (or prismatic) joint, and the vector $\boldsymbol{p}_e^{i-1}$ is the *position vector from the origin of the coordinate system* $i-1$ to the end-effector point E.

Note that in these equations, the Jacobian matrix is focused on determining the linear and angular velocity relation of the end-effector point E. If we want to obtain the Jacobian matrix for any other arbitrary point on the robot, we can simply consider that point as the end-effector point and use the same equations. Additionally, these vectors are usually represented with respect to the reference frame $\{0\}$, in which case the Jacobian matrix with respect to the reference coordinate system will be derived , which is denoted as ${}^0\boldsymbol{J}(\boldsymbol{q})$. However, all calculations can be performed in any other arbitrary coordinate system, and the Jacobian matrix can be determined with respect to that coordinate system.

In these calculations, determining the vectors $\boldsymbol{z}_i$ is usually straightforward. To obtain the vector $\boldsymbol{p}_e^{i-1}$, according to Figure 6.4, the following vector relation can also be used.

$${}^0\boldsymbol{p}_e^{i-1} = {}^0\boldsymbol{e} - {}^0\boldsymbol{p}_{i-1} \tag{6.22}$$

where ${}^0\boldsymbol{e}$ is the end-effector position vector, with respect to the reference coordinate system, and ${}^0\boldsymbol{p}_{i-1}$ is the position vector of the origin of the coordinate system $i-1$ with respect

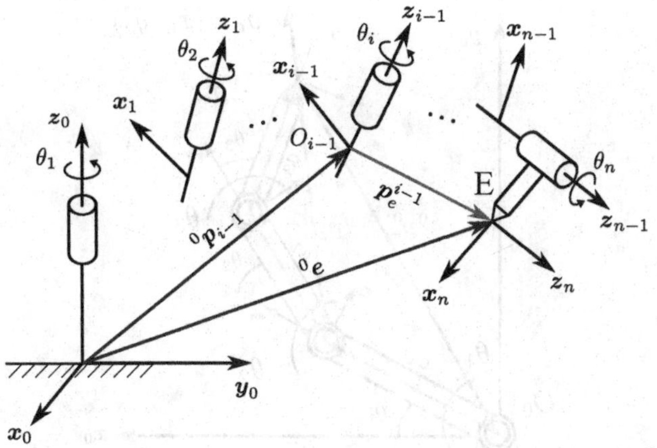

FIGURE 6.4
Representation of the vectors used in the general method for determining the Jacobian matrix.

to the same reference frame. In this way, in robots with simpler structural complexity, it is easy to obtain the vectors that make up the Jacobian matrix using this relation by intuition.

In the case of a more complex kinematic structure of the robot, one can obtain these vectors using Denavit-Hartenberg parameters. For this purpose, for $i = 1, 2, \ldots, n$, the joint axis vectors $\boldsymbol{z}_{i-1}$ can first be recursively obtained using the following relation.

$$ {}^{0}\boldsymbol{z}_{i-1} = {}^{0}\boldsymbol{R}_{i-1} \begin{bmatrix} 0 \\ 0 \\ 1 \end{bmatrix} \tag{6.23} $$

where ${}^{0}\boldsymbol{R}_{i-1}$, is the rotation matrix from the coordinate system $i - 1$ to the reference coordinate system $\{0\}$. Continuing for $i = n - 1, \ldots, 1, 0$, the position vector $\boldsymbol{p}_e^{i-1}$ can be back-propagated using the following relation.

$$ {}^{0}\boldsymbol{p}_e^{i-1} = {}^{0}\boldsymbol{R}_{i-1}\, {}^{i-1}\boldsymbol{p}_i^{i-1} + {}^{0}\boldsymbol{p}_e^{i} \tag{6.24} $$

In this relation, start the iteration with ${}^{0}\boldsymbol{p}_e^{n} = 0$ and use the following Denavit-Hartenberg parameters.

$$ {}^{i-1}\boldsymbol{p}_i^{i-1} = \begin{bmatrix} a_i c_i \\ a_i s_i \\ d_i \end{bmatrix} \tag{6.25} $$

In what follows, we will review how to perform these calculations in several Case Studies.

Case Study 6.2 (Planar 3R Robot) *Consider the planar 3R robot depicted in Figure 6.5. Utilize the general method to obtain the Jacobian matrix for this robot.*

Solution: To employ the general method for determining the Jacobian matrix of the 3R robot, we first obtain the vectors $\boldsymbol{z}_i$ according to Figure 6.5.

$$ \boldsymbol{z}_0 = \boldsymbol{z}_1 = \boldsymbol{z}_2 = \begin{bmatrix} 0 \\ 0 \\ 1 \end{bmatrix} $$

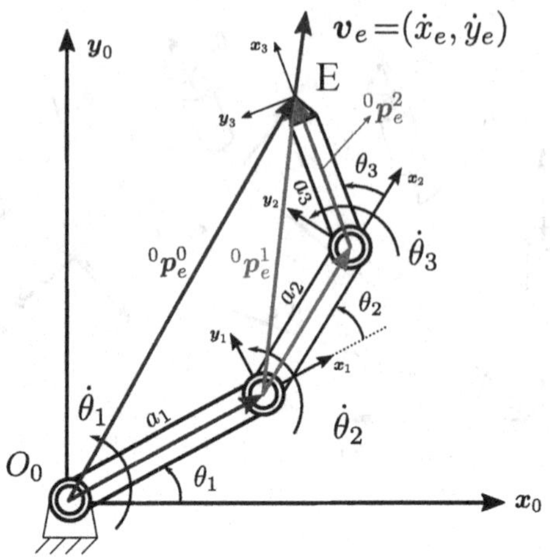

FIGURE 6.5
Visualization of vectors used in the general method for the Jacobian matrix derivation of a planar 3R robot.

Next, the vectors ${}^0\boldsymbol{p}_e^{i-1}$ are obtained based on the robot's geometry and in accordance with Figure 6.5. These vectors are represented in Figure 6.5 with different colors (red, green, and blue, respectively) and are annotated accordingly.

$$
{}^0\boldsymbol{p}_e^2 = \begin{bmatrix} a_3c_{123} \\ a_3s_{123} \\ 0 \end{bmatrix}, \quad {}^0\boldsymbol{p}_e^1 = \begin{bmatrix} a_2c_{12} + a_3c_{123} \\ a_2s_{12} + a_3s_{123} \\ 0 \end{bmatrix}, \quad {}^0\boldsymbol{p}_e^0 = \begin{bmatrix} a_1c_1 + a_2c_{12} + a_3c_{123} \\ a_1s_1 + a_2s_{12} + a_3s_{123} \\ 0 \end{bmatrix}
$$

By determining the necessary vectors, the columns of the Jacobian matrix are obtained according to the relation 6.20.

$$
\boldsymbol{J}_3 = \begin{bmatrix} -a_3s_{123} \\ a_3c_{123} \\ 0 \\ 0 \\ 0 \\ 1 \end{bmatrix}, \boldsymbol{J}_2 = \begin{bmatrix} -(a_2s_{12} + a_3s_{123}) \\ a_2c_{12} + a_3c_{123} \\ 0 \\ 0 \\ 0 \\ 1 \end{bmatrix}, \boldsymbol{J}_1 = \begin{bmatrix} -(a_1s_1 + a_2s_{12} + a_3s_{123}) \\ a_1c_1 + a_2c_{12} + a_3c_{123} \\ 0 \\ 0 \\ 0 \\ 1 \end{bmatrix}
$$

As observed, the columns of the determined Jacobian matrix using the general method are six-dimensional arrays. In planar robots, only the first, second, and sixth rows corresponding to linear velocities in the x, y directions and angular velocity about the z axis needed to be considered. Thus, if we rewrite the Jacobian matrix relation as follows,

$$
\begin{bmatrix} \dot{x}_e \\ \dot{y}_e \\ \dot{\theta}_e \end{bmatrix} = {}^0\boldsymbol{J}(\boldsymbol{q}) \begin{bmatrix} \dot{\theta}_1 \\ \dot{\theta}_2 \\ \dot{\theta}_3 \end{bmatrix}
$$

The Jacobian matrix, with the elimination of zero rows, is obtained as follows:

$$
{}^0\boldsymbol{J}(\boldsymbol{q}) = \begin{bmatrix} -(a_1s_1 + a_2s_{12} + a_3s_{123}) & -(a_2s_{12} + a_3s_{123}) & -a_3s_{123} \\ a_1c_1 + a_2c_{12} + a_3c_{123} & a_2c_{12} + a_3c_{123} & a_3c_{123} \\ 1 & 1 & 1 \end{bmatrix} \tag{6.26}
$$

This matrix exactly corresponds to what is obtained using the velocity loop closure method, reported in Equation 6.17.

Case Study 6.3 (SCARA Robot) *Consider the SCARA robot depicted in Figure 6.6. Utilize the general method to obtain the Jacobian matrix for this robot.*

Solution: To employ the general method for determining the Jacobian matrix of the robot, one can easily obtain the vectors $\boldsymbol{z}_i$ according to Figure 6.6.

$$\boldsymbol{z}_0 = \boldsymbol{z}_1 = \begin{bmatrix} 0 \\ 0 \\ 1 \end{bmatrix}, \boldsymbol{z}_2 = \boldsymbol{z}_3 = \begin{bmatrix} 0 \\ 0 \\ -1 \end{bmatrix}$$

Then, the vectors ${}^0\boldsymbol{p}_e^{i-1}$ are obtained based on the robot's geometry and according to Figure 6.6. To achieve this, we start from the end-effector.

$${}^0\boldsymbol{p}_e^3 = \begin{bmatrix} 0 \\ 0 \\ -d_4 \end{bmatrix}, \qquad {}^0\boldsymbol{p}_e^2 = \begin{bmatrix} 0 \\ 0 \\ -(d_3 + d_4) \end{bmatrix}$$

These two vectors are shown in Figure 6.6 (in purple and red, respectively). To determine the other two vectors, the recursive method introduced in Equation 6.24 can be utilized.

$${}^0\boldsymbol{p}_e^1 = {}^0\boldsymbol{R}_2 \begin{bmatrix} a_2 \\ 0 \\ 0 \end{bmatrix} + {}^0\boldsymbol{p}_e^2 = \begin{bmatrix} a_2 c_{12} \\ a_2 s_{12} \\ -(d_3 + d_4) \end{bmatrix}$$

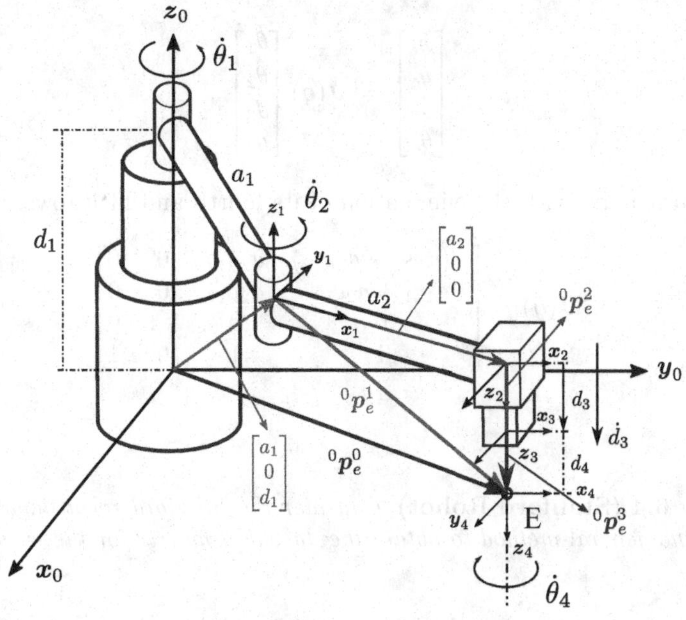

FIGURE 6.6
Visualization of vectors in the general method for determining the Jacobian matrix of a SCARA robot.

$$
{}^{0}\boldsymbol{p}_e^0 = {}^{0}\boldsymbol{R}_1 \begin{bmatrix} a_1 \\ 0 \\ d_1 \end{bmatrix} + {}^{0}\boldsymbol{p}_e^1 = \begin{bmatrix} a_1 c_1 + a_2 c_{12} \\ a_1 s_1 + a_2 s_{12} \\ d_1 - (d_3 + d_4) \end{bmatrix}
$$

By determining the necessary vectors, the columns of the Jacobian matrix are obtained based on Equation 6.20 for the first, second, and fourth revolute joints, and Equation 6.21 for the third prismatic joint.

$$
\boldsymbol{J}_4 = \begin{bmatrix} \boldsymbol{z}_3 \times \boldsymbol{p}_e^3 \\ \boldsymbol{z}_3 \end{bmatrix} = \begin{bmatrix} 0 \\ 0 \\ 0 \\ 0 \\ 0 \\ 1 \end{bmatrix}, \boldsymbol{J}_3 = \begin{bmatrix} \boldsymbol{z}_2 \\ \boldsymbol{0} \end{bmatrix} = \begin{bmatrix} 0 \\ 0 \\ -1 \\ 0 \\ 0 \\ 0 \end{bmatrix}
$$

$$
\boldsymbol{J}_2 = \begin{bmatrix} \boldsymbol{z}_1 \times \boldsymbol{p}_e^1 \\ \boldsymbol{z}_1 \end{bmatrix} = \begin{bmatrix} a_2 s_{12} \\ a_2 c_{12} \\ 0 \\ 0 \\ 1 \end{bmatrix}, \boldsymbol{J}_1 = \begin{bmatrix} \boldsymbol{z}_0 \times \boldsymbol{p}_e^0 \\ \boldsymbol{z}_0 \end{bmatrix} = \begin{bmatrix} -(a_1 s_1 + a_2 s_{12}) \\ a_1 c_1 + a_2 c_{12} \\ 0 \\ 0 \\ 0 \\ 1 \end{bmatrix}
$$

As observed in this example, the columns of the Jacobian matrix determined using the general method are six-dimensional arrays. Considering the degrees of freedom of the SCARA robot, only the first three rows plus the sixth row, corresponding to linear velocities in the x, y, z directions, and angular velocity about the z axis, are non-zero and should be taken into account in calculating the Jacobian matrix. Thus, if we rewrite the Jacobian matrix relation as follows,

$$
\begin{bmatrix} \dot{x}_e \\ \dot{y}_e \\ \dot{z}_e \\ \dot{\theta}_e \end{bmatrix} = {}^{0}\boldsymbol{J}(\boldsymbol{q}) \begin{bmatrix} \dot{\theta}_1 \\ \dot{\theta}_2 \\ \dot{d}_3 \\ \dot{\theta}_4 \end{bmatrix}
$$

The Jacobian matrix, with the elimination of its fourth and fifth rows, is obtained as follows:

$$
{}^{0}\boldsymbol{J}(\boldsymbol{q}) = \begin{bmatrix} -(a_1 s_1 + a_2 s_{12}) & a_2 s_{12} & 0 & 0 \\ a_1 c_1 + a_2 c_{12} & a_2 c_{12} & 0 & 0 \\ 0 & 0 & -1 & 0 \\ 1 & 1 & 0 & 1 \end{bmatrix} \tag{6.27}
$$

Case Study 6.4 (Stanford Robot) *Consider the Stanford robot depicted in Figure 6.7. Utilize the general method to obtain the Jacobian matrix for the wrist point of the robot.*

Solution: To use the general method in determining the Jacobian matrix for the robot's wrist point, first obtain the vectors $\boldsymbol{z}_i$ according to Figure 6.7 by using the recursive method. Begin with the vector $\boldsymbol{z}_0$, and using Equation 6.23, calculate the vectors $\boldsymbol{z}_i$ for

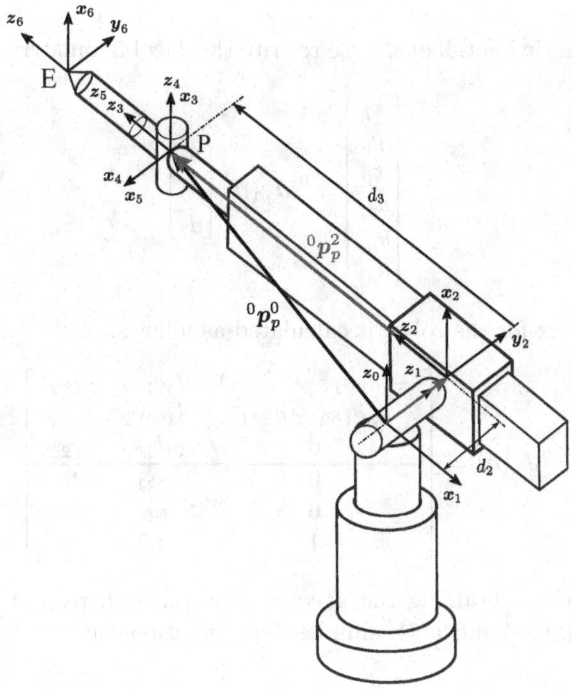

FIGURE 6.7
Visualization of vectors in the general method for determining the Jacobian matrix of the Stanford robot.

the first to third joints. Note that only the angular motions of these joints will affect the motion of the robot's wrist point.

$$z_0 = \begin{bmatrix}0\\0\\1\end{bmatrix}, \quad z_1 = {}^0R_1 z_0 = \begin{bmatrix}-s_1\\c_1\\0\end{bmatrix}, \quad z_2 = z_3 = {}^0R_2 z_0 = \begin{bmatrix}c_2 s_2\\s_1 s_2\\c_2\end{bmatrix}$$

Considering that the linear velocity of the robot's wrist point P is of interest, the vectors ${}^0p_p^{i-1}$ are obtained based on the robot's geometry and according to Figure 6.7 as follows. Since the angular velocities of the fourth to sixth joints have no effect on the linear velocity of the wrist point, start the back propagation method from the third joint using Equation 6.24.

$${}^0p_p^3 = \mathbf{0}, \quad {}^0p_p^2 = {}^0R_2 \begin{bmatrix}0\\0\\d_3\end{bmatrix} + {}^0p_p^3 = \begin{bmatrix}d_3 c_1 s_2\\d_3 s_1 s_2\\d_3 c_2\end{bmatrix}$$

$${}^0p_p^0 = {}^0p_p^1 = {}^0R_1 \begin{bmatrix}0\\0\\d_2\end{bmatrix} + {}^0p_p^2 = \begin{bmatrix}d_3 c_1 s_2 - d_2 s_2\\d_3 s_1 s_2 + d_2 c_1\\d_3 c_2\end{bmatrix}$$

These vectors are shown in Figure 6.7 (in green and black, respectively) and are annotated accordingly. By determining the necessary vectors, the columns of the Jacobian matrix are obtained based on Equation 6.20 for the revolute joints and Equation 6.21

for the third prismatic joint. Thus, if we rewrite the Jacobian matrix relation as follows,

$$\begin{bmatrix} \dot{x}_p \\ \dot{y}_p \\ \dot{z}_p \\ \dot{\theta}_{p_x} \\ \dot{\theta}_{p_y} \\ \dot{\theta}_{p_z} \end{bmatrix} = {}^0\boldsymbol{J}_p(\boldsymbol{q}) \begin{bmatrix} \dot{\theta}_1 \\ \dot{\theta}_2 \\ \dot{d}_3 \end{bmatrix}$$

The Jacobian matrix for the robot is calculated as follows:

$$ {}^0\boldsymbol{J}_p(\boldsymbol{q}) = \left[\begin{array}{ccc} -(d_3 s_1 s_2 + d_2 c_1) & d_3 c_1 c_2 & c_1 s_2 \\ d_3 c_1 s_2 - d_2 s_1 & d_3 s_1 c_2 & s_1 s_2 \\ 0 & -d_3 s_2 & c_2 \\ \hline 0 & -s_1 & 0 \\ 0 & c_1 & 0 \\ 1 & 0 & 0 \end{array}\right] \tag{6.28}$$

It can be shown that obtaining the Jacobian matrix with respect to the coordinate system $\{3\}$ results in a significantly simpler representation compared to what is obtained in Equation 6.28.

6.2.3 Screw-Based Jacobian Matrix

In this section, we will explore an alternative and efficient method based on screw theory for determining the Jacobian matrix of serial robots. Upon examining the columns of the Jacobian matrix calculated using the general method, a notable resemblance to screw coordinates becomes evident. The expressions provided for revolute joints in Equation 6.20 and for prismatic joints in Equation 6.21 closely align with the definition of screw coordinates for revolute and prismatic joints. Let's reiterate the definition of the six-dimensional screw coordinate here, as outlined in Equation 3.65, describing the pure rotational motion of a rigid body.

$$\$ = \hat{\$}\,\dot{\theta} = \begin{bmatrix} \hat{\boldsymbol{s}} \\ \boldsymbol{s}_o \times \hat{\boldsymbol{s}} \end{bmatrix} \dot{\theta} \tag{6.29}$$

Upon examining the vectors constituting the screw in this scenario, we find that the first vector corresponds to the angular velocity of the body, while the second vector describes the linear velocity of the specified point on the body. Interestingly, this aligns precisely with the two vectors in Equation 6.20, albeit with a different order; in a way that the first vector determines the linear velocity of the specified point on the body, and the second vector describes the angular velocity of the body. This exact correspondence is also observed in the definition of screw coordinates for pure translational motion, as given in Equation 3.68:

$$\$ = \hat{\$}\,\dot{d} = \begin{bmatrix} \boldsymbol{0} \\ \hat{\boldsymbol{s}} \end{bmatrix} \dot{d} \tag{6.30}$$

which is exactly analogous to Equation 6.21 for prismatic joints, where the vectors of linear and angular velocities of the translating body are interchanged.

Thus, if we redefine the Jacobian matrix in a way that angular velocities precede linear velocities, we can directly rewrite the Jacobian matrix columns based on the screw coordinates of the joints. To achieve this, we define the screw-based Jacobian matrix as follows.

$$\begin{bmatrix} \boldsymbol{\omega}_e \\ \boldsymbol{v}_e \end{bmatrix} = \boldsymbol{J}^s(\boldsymbol{q})\dot{\boldsymbol{q}} = \begin{bmatrix} \boldsymbol{J}_\omega \\ \boldsymbol{J}_v \end{bmatrix} \dot{\boldsymbol{q}} \tag{6.31}$$

In this equation as before, $\boldsymbol{J}_\omega$ represents the Jacobian matrix corresponding to angular velocities, and $\boldsymbol{J}_v$ represents the Jacobian matrix corresponding to linear velocities of the robot's end-effector. The screw-based Jacobian matrix, denoted as $\boldsymbol{J}^s$, is formed by combining these two matrices in the interchanged order compared to the conventional Jacobian matrix. To simultaneously compare the Jacobian matrix $\boldsymbol{J}$ and its screw-based version $\boldsymbol{J}^s$, consider the following relationships.

$$\boldsymbol{J} = \begin{bmatrix} \boldsymbol{J}_v \\ \boldsymbol{J}_\omega \end{bmatrix} \quad , \quad \boldsymbol{J}^s = \begin{bmatrix} \boldsymbol{J}_\omega \\ \boldsymbol{J}_v \end{bmatrix} \tag{6.32}$$

With this definition, it can be shown that the screw-based Jacobian matrix is formed from the unit screw coordinates of the robot's joints.

$$\boldsymbol{J}^s(\boldsymbol{q}) = \begin{bmatrix} \hat{\$}_1 \mid \hat{\$}_2 \mid \cdots \mid \hat{\$}_n \end{bmatrix}_{6\times n} \tag{6.33}$$

in which, unit screws for **Revolute Joints** are derived from:

$$\hat{\$}_i = \begin{bmatrix} \hat{\boldsymbol{s}}_i \\ \boldsymbol{s}_{O_i} \times \hat{\boldsymbol{s}}_i \end{bmatrix} \tag{6.34}$$

and unit screws for **Prismatic Joints**, from:

$$\hat{\$}_i = \begin{bmatrix} \boldsymbol{0} \\ \hat{\boldsymbol{s}}_i \end{bmatrix} \tag{6.35}$$

In these equations, according to Figure 6.8, the ith unit screw, denoted by $\hat{\$}_i$, is the rotational (or translational) axis of the ith revolute (or prismatic) joint, determined by the robot's geometry. The determination of $\boldsymbol{s}_{O_i}$ is also consistent with a convention established in the definition of screw coordinates for determining the linear velocity of the specified point.

According to this convention, place the coordinate system $\{0\}$ instantly on the specified point, without changing its principal axis directions. With this description, if the goal is to determine the Jacobian matrix for the robot's end-effector, place this coordinate system on the end-effector (and point E), and if you want to obtain the linear velocity of any other point, such as the robot's wrist P, place this coordinate system on the desired point P. Accordingly, as shown in Figure 6.8, the vector $\boldsymbol{s}_{O_i}$ represents the distance from the ith screw to this coordinate system.

If we want to obtain the Jacobian matrix for the robot's end-effector with respect to another coordinate system, it is sufficient to place a coordinate system, which is instantly located at point E, parallel to the desired coordinate system. Then, similarly, we determine the vectors $\boldsymbol{s}_{O_i}$ in this coordinate system. In Figure 6.8, a situation is illustrated where the linear velocity of the robot's end-effector is obtained with respect to the reference coordinate system $\{0\}$. In this case, the screw-based Jacobian matrix with respect to the reference coordinate system, denoted by ${}^0\boldsymbol{J}^s(\boldsymbol{q})$, is determined. In what follows, we will explain, through an illustrative Case Study, how to determine the vectors constituting the necessary screws in the Jacobian matrix.

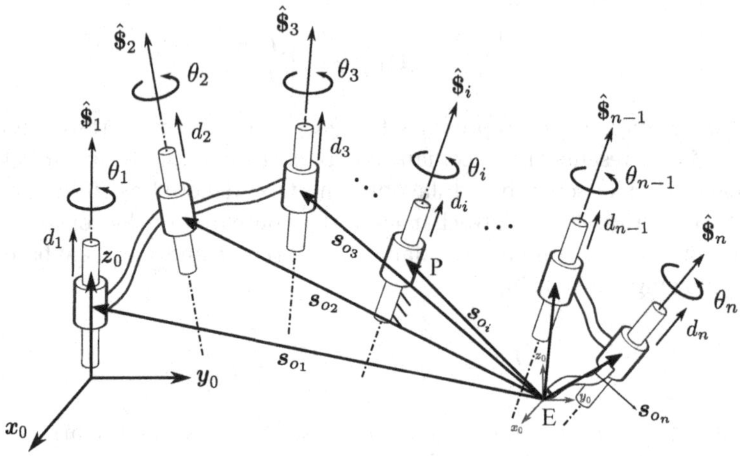

FIGURE 6.8
Definition of the necessary vectors in the method for determining the screw-based Jacobian matrix.

Case Study 6.5 (Planar 3R robot) *Consider the planar 3R robot shown in Figure 6.9. Determine the screw-based Jacobian matrix of its end-effector with respect to the reference frame* $\{0\}$.

Solution: In order to use the Screw-based method for determining the Jacobian matrix of the 3R robot, first, place a coordinate system instantaneously at the end-effector point E, parallel to the reference coordinate system. Subsequently, it is straightforward to determine the screw axis and vectors $\hat{\boldsymbol{s}}_i$ as shown in Figure 6.9.

$$\hat{\boldsymbol{s}}_1 = \hat{\boldsymbol{s}}_2 = \hat{\boldsymbol{s}}_3 = \begin{bmatrix} 0 \\ 0 \\ 1 \end{bmatrix}$$

Then, the vectors $\boldsymbol{s}_{o_i}$ are obtained from point E to the centers of the corresponding joints, considering the robot's geometry and according to Figure 6.9.

$$\boldsymbol{s}_{o_3} = {}^0\boldsymbol{R}_3 \begin{bmatrix} -a_3 \\ 0 \\ 0 \end{bmatrix} = \begin{bmatrix} -a_3 c_{123} \\ -a_3 s_{123} \\ 0 \end{bmatrix}$$

$$\boldsymbol{s}_{o_2} = \boldsymbol{s}_{o_3} + {}^0\boldsymbol{R}_2 \begin{bmatrix} -a_2 \\ 0 \\ 0 \end{bmatrix} = \begin{bmatrix} -a_2 c_{12} - a_3 c_{123} \\ -a_2 s_{12} - a_3 s_{123} \\ 0 \end{bmatrix}$$

$$\boldsymbol{s}_{o_1} = \boldsymbol{s}_{o_2} + {}^0\boldsymbol{R}_1 \begin{bmatrix} -a_1 \\ 0 \\ 0 \end{bmatrix} = \begin{bmatrix} -a_1 c_1 - a_2 c_{12} - a_3 c_{123} \\ -a_1 s_1 - a_2 s_{12} - a_3 s_{123} \\ 0 \end{bmatrix}$$

These vectors, shown in Figure 6.9 (in red, green, and blue, respectively), are accordingly annotated. By determining the necessary vectors, the joint screw coordinates and columns of the screw-based Jacobian matrix can be obtained according to the

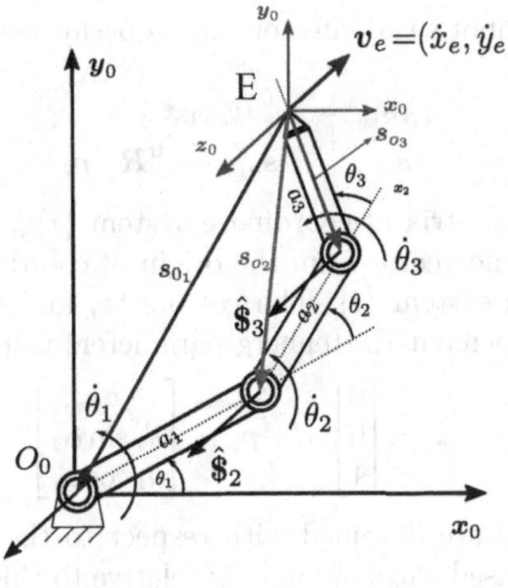

FIGURE 6.9
Vectors constituting the screw-based Jacobian matrix in the planar robot 3R̲

relationship in Equation 6.34 as follows.

$$\hat{\$}_3 = \begin{bmatrix} 0 \\ 0 \\ 1 \\ -a_3 s_{123} \\ a_3 c_{123} \\ 0 \end{bmatrix}, \hat{\$}_2 = \begin{bmatrix} 0 \\ 0 \\ 1 \\ -(a_2 s_{12} + a_3 s_{123}) \\ a_2 c_{12} + a_3 c_{123} \\ 0 \end{bmatrix}, \hat{\$}_1 = \begin{bmatrix} 0 \\ 0 \\ 1 \\ -(a_1 s_1 + a_2 s_{12} + a_3 s_{123}) \\ a_1 c_1 + a_2 c_{12} + a_3 c_{123} \\ 0 \end{bmatrix}$$

As observed, the columns of the Jacobian matrix determined using the screw coordinates are six-dimensional arrays. In planar robots, only the rows three to five are considered, which are corresponding to angular velocity around the z axis and linear velocities along the x and y directions. Thus, if we rewrite the screw-based Jacobian matrix as follows,

$$\begin{bmatrix} \dot{\theta}_e \\ \dot{x}_e \\ \dot{y}_e \end{bmatrix} = {}^0\boldsymbol{J}^s(\boldsymbol{q}) \begin{bmatrix} \dot{\theta}_1 \\ \dot{\theta}_2 \\ \dot{\theta}_3 \end{bmatrix}$$

The screw-based Jacobian matrix, with the removal of rows corresponding to zero components, is obtained as follows.

$$ {}^0\boldsymbol{J}^s(\boldsymbol{q}) = \begin{bmatrix} 1 & 1 & 1 \\ -(a_1 s_1 + a_2 s_{12} + a_3 s_{123}) & -(a_2 s_{12} + a_3 s_{123}) & -a_3 s_{123} \\ a_1 c_1 + a_2 c_{12} + a_3 c_{123} & a_2 c_{12} + a_3 c_{123} & a_3 c_{123} \end{bmatrix} \tag{6.36}$$

This matrix, by permuting its first row, aligns with what is obtained using the general method and reported in Equation 6.26.

If the robot under consideration has a complex structure, determining the necessary vectors for screw coordinates by intuition may not be straightforward. In this case, the vectors constituting the screw-based Jacobian matrix can be obtained recursively. To do so, assume we want to obtain the screw-based Jacobian matrix based of the end-effector E of the robot with respect to the reference frame. In this case, backward propagation for the screws preceding the robot's end-effector can be performed for $i = n, \dots, 2, 1$, using the following equations:

$$\boldsymbol{s}_i = {}^0\boldsymbol{R}_{i-1}\,\hat{\boldsymbol{z}} \tag{6.37}$$

$$\boldsymbol{s}_{o_i} = \boldsymbol{s}_{o_{i+1}} - {}^0\boldsymbol{R}_i\,{}^i\boldsymbol{p}_i \tag{6.38}$$

where ${}^0\boldsymbol{R}_i$ is the rotation matrix of coordinate system $\{i\}$ relative to the reference coordinate system, and ${}^i\boldsymbol{p}_i$ is the vector from the origin of coordinate system $\{i\}$ to the point E expressed in coordinate system $\{i\}$. The vectors ${}^i\boldsymbol{z}_i$ and ${}^i\boldsymbol{p}_i$ are determined either by intuition or based on the Denavit-Hartenberg parameters using the following Equation.

$$\hat{\boldsymbol{z}} = \begin{bmatrix} 0 \\ 0 \\ 1 \end{bmatrix}, \quad {}^i\boldsymbol{p}_i = \begin{bmatrix} a_i \\ d_i \sin\alpha_i \\ d_i \cos\alpha_i \end{bmatrix} \tag{6.39}$$

Note that all these vectors are obtained with respect to the reference coordinate system $\{0\}$, and thus, the screw-based Jacobian matrix relative to this coordinate system, denoted as ${}^0\boldsymbol{J}^s(\boldsymbol{q})$, is determined. In many cases, if we determine the Jacobian matrix relative to an intermediate coordinate system $\{j\}$, the results will be much simpler. To do this, it is sufficient to use the transformation matrix with respect to the desired coordinate system ${}^j\boldsymbol{R}_i$ instead of ${}^0\boldsymbol{R}_i$ in the above equations. Furthermore, if we want to obtain the linear and angular velocities of the robot's wrist point P, we only need to place the reference coordinate system at this point and determine the relevant vectors up to this point. This is because only the screws preceding this point in the robot's structure will affect the determination of the linear and angular velocities of the wrist.

Case Study 6.6 (Planar 3R robot) *Consider the planar 3R robot shown in Figure 6.9. Obtain the screw-based Jacobian matrix for the end-effector of this robot using the recursive method.*

Solution: To use the screw-based method for determining the Jacobian matrix of the 3R robot, first, place a coordinate system instantaneously at the end-effector point E in a way that is parallel to the reference coordinate system. Since the point of interest is located on the end-effector of the robot, recursive backward calculations must be performed to determine the screws of all joints. For this purpose, the vectors constituting the screws of the robot joints are obtained in the order given below, according to Figure 6.9.

For $i = 2, 1, 0$, and using Equations 6.37 and 6.38, the vectors constituting the screws of the joints are recursively determined.

$$\hat{\boldsymbol{s}}_3 = \begin{bmatrix} 0 \\ 0 \\ 1 \end{bmatrix}, \quad \boldsymbol{s}_{o_3} = -\,{}^0\boldsymbol{R}_3 \begin{bmatrix} a_3 \\ 0 \\ 0 \end{bmatrix} = \begin{bmatrix} -a_3 c_{123} \\ -a_3 s_{123} \\ 0 \end{bmatrix}$$

$$\hat{\boldsymbol{s}}_2 = \begin{bmatrix} 0 \\ 0 \\ 1 \end{bmatrix}, \quad \boldsymbol{s}_{o_2} = \boldsymbol{s}_{o_3} - {}^0\boldsymbol{R}_2 \begin{bmatrix} a_2 \\ 0 \\ 0 \end{bmatrix} = \begin{bmatrix} -a_2 c_{12} - a_3 c_{123} \\ -a_2 s_{12} - a_3 s_{123} \\ 0 \end{bmatrix}$$

$$\hat{\boldsymbol{s}}_1 = \begin{bmatrix} 0 \\ 0 \\ 1 \end{bmatrix}, \quad \boldsymbol{s}_{o_1} = \boldsymbol{s}_{o_2} - {}^0\boldsymbol{R}_1 \begin{bmatrix} a_1 \\ 0 \\ 0 \end{bmatrix} = \begin{bmatrix} -a_1 c_1 - a_2 c_{12} - a_3 c_{123} \\ -a_1 s_1 - a_2 s_{12} - a_3 s_{123} \\ 0 \end{bmatrix}$$

It can be observed that the obtained vectors are perfectly consistent with the vectors obtained from the conceptual method reported in Case Study 6.5. Accordingly, the screw-based Jacobian matrix that is obtained by this method will also be equal to the result reported in Equation 6.36.

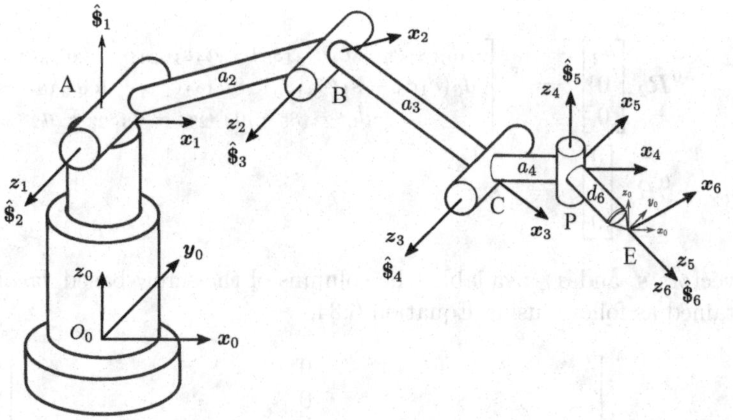

FIGURE 6.10
Vectors constituting the screw-based Jacobian matrix in the 6DOF Elbow Manipulator

Case Study 6.7 (6DOF Elbow Manipulator) *Obtain the screw-based Jacobian matrix for the end-effector of the six-degrees-of-freedom Elbow manipulator shown in Figure 6.10.*

Solution: The Denavit-Hartenberg parameters for this manipulator is derived in Case Study 4.5 and given in Table 4.5. Furthermore, the corresponding homogeneous transformations are also given in that case study. With the homogeneous transformations matrices of the robot available, using the recursive relationships in Equations 6.37 and 6.38, the vectors constituting the screw-based Jacobian matrix for the robot are determined as follows. To achieve this, first, calculate the vectors $\hat{\boldsymbol{s}}_i$.

$$\begin{aligned}
\hat{\boldsymbol{s}}_1 &= \begin{bmatrix} 0 \\ 0 \\ 1 \end{bmatrix}, \quad \hat{\boldsymbol{s}}_2 = \hat{\boldsymbol{s}}_3 = \hat{\boldsymbol{s}}_4 = {}^0\boldsymbol{R}_1 \begin{bmatrix} 0 \\ 0 \\ 1 \end{bmatrix} = \begin{bmatrix} s_1 \\ -c_1 \\ 0 \end{bmatrix} \\
\hat{\boldsymbol{s}}_5 &= {}^0\boldsymbol{R}_4 \begin{bmatrix} 0 \\ 0 \\ 1 \end{bmatrix} = \begin{bmatrix} -c_1 s_{234} \\ -s_1 s_{234} \\ c_{234} \end{bmatrix} \\
\hat{\boldsymbol{s}}_6 &= {}^0\boldsymbol{R}_6 \begin{bmatrix} 0 \\ 0 \\ 1 \end{bmatrix} = \begin{bmatrix} s_1 c_5 + c_1 c_{234} s_5 \\ -c_1 c_5 + s_1 c_{234} s_5 \\ s_{234} s_5 \end{bmatrix}
\end{aligned}$$

Subsequently, back propagate the vectors $\boldsymbol{s}_{o_i}$ as follows.

$$\begin{aligned}
\boldsymbol{s}_{o_4} = \boldsymbol{s}_{o_5} - {}^0\boldsymbol{R}_4 \begin{bmatrix} a_4 \\ 0 \\ 0 \end{bmatrix} &= \begin{bmatrix} -d_6(s_1c_5 + c_1c_{234}s_5) - a_4c_1c_{234} \\ d_6(c_1c_5 - s_1c_{234}s_5) - a_4s_1c_{234} \\ -d_6s_{234}s_5 - a_4s_{234} \end{bmatrix} \\
\boldsymbol{s}_{o_3} = \boldsymbol{s}_{o_4} - {}^0\boldsymbol{R}_3 \begin{bmatrix} a_3 \\ 0 \\ 0 \end{bmatrix} &= \begin{bmatrix} -d_6(s_1c_5 + c_1c_{234}s_5) - a_4c_1c_{234} - a_3c_{23}c_1 \\ d_6(c_1c_5 - s_1c_{234}s_5) - a_4s_1c_{234} - a_3c_{23}s_1 \\ -d_6s_{234}s_5 - a_4s_{234} - a_3s_{23} \end{bmatrix}
\end{aligned}$$

$$\boldsymbol{s}_{o_2} = \boldsymbol{s}_{o_3} - {}^0\boldsymbol{R}_2 \begin{bmatrix} a_2 \\ 0 \\ 0 \end{bmatrix} = \begin{bmatrix} -d_6(s_1c_5 + c_1c_{234}s_5) - a_4c_1c_{234} - a_3c_{23}c_1 - a_2c_1c_2 \\ d_6(c_1c_5 - s_1c_{234}s_5) - a_4s_1c_{234} - a_3c_{23}s_1 - a_2s_1c_2 \\ -d_6s_{234}s_5 - a_4s_{234} - a_3s_{23} - a_2s_2 \end{bmatrix}$$

$$\boldsymbol{s}_{o_1} = \boldsymbol{s}_{o_2} - {}^0\boldsymbol{R}_1 \begin{bmatrix} 0 \\ 0 \\ 0 \end{bmatrix} = \boldsymbol{s}_{o_2}$$

With the vectors $\hat{\boldsymbol{s}}_i$ and $\boldsymbol{s}_{o_i}$ available, the columns of the screw-based Jacobian matrix will be obtained as follows using Equation 6.34.

$$\hat{\$}_1 = \begin{bmatrix} 0 \\ 0 \\ 1 \\ -a_2s_1c_2 - a_3s_1c_{23} - a_4s_1c_{234} + d_6(c_1c_5 - s_1c_{234}s_5) \\ -a_2c_1c_2 - a_3c_1c_{23} - a_4c_1c_{234} + d_6(s_1c_5 + c_1c_{234}s_5) \\ 0 \end{bmatrix}$$

$$\hat{\$}_2 = \begin{bmatrix} s_1 \\ -c_1 \\ 0 \\ -c_1(a_2s_2 + a_3s_{23} + a_4s_{234} + d_6s_{234}s_5) \\ -s_1(a_2s_2 + a_3s_{23} + a_4s_{234} + d_6s_{234}s_5) \\ a_2c_2 + a_3c_{23} + a_4c_{234} + \frac{1}{2}d_6(s_{2345} - \sin(\theta_{234} - \theta_5)) \end{bmatrix}$$

$$\hat{\$}_3 = \begin{bmatrix} s_1 \\ -c_1 \\ 0 \\ -c_1(a_3s_{23} + a_4s_{234} + d_6s_{234}s_5) \\ -s_1(a_3s_{23} + a_4s_{234} + d_6s_{234}s_5) \\ a_3c_{23} + a_4c_{234} + \frac{1}{2}d_6(s_{2345} - \sin(\theta_{234} - \theta_5)) \end{bmatrix}$$

$$\hat{\$}_4 = \begin{bmatrix} s_1 \\ -c_1 \\ 0 \\ -c_1s_{234}(a_4 + d_6s_5) \\ -s_1s_{234}(a_4 + d_6s_5) \\ c_{234}(a_4 + d_6s_5) \end{bmatrix}, \hat{\$}_5 = \begin{bmatrix} -c_1s_{234} \\ -s_1s_{234} \\ c_{234} \\ d_6(c_1c_{234}c_5 - s_1s_5) \\ d_6(s_1c_{234}c_5 + c_1s_5) \\ d_6s_{234}c_5 \end{bmatrix}, \hat{\$}_6 = \begin{bmatrix} c_1c_{234}s_5 + s_1c_5 \\ s_1c_{234}s_5 - c_1c_5 \\ s_{234}s_5 \\ 0 \\ 0 \\ 0 \end{bmatrix}$$

Finally, the screw-based Jacobian matrix for the end-effector of the 6DOF Elbow manipulator with respect to the reference coordinate system {0} is determined as follows.

$${}^0\boldsymbol{J}^s(\boldsymbol{q}) = \left[\ \hat{\$}_1 \mid \hat{\$}_2 \mid \hat{\$}_3 \mid \hat{\$}_4 \mid \hat{\$}_5 \mid \hat{\$}_6\ \right].$$

In this case study, a comprehensive case is considered where the screw-based Jacobian matrix for the end-effector of the robot is calculated with respect to the reference coordinate system {0}. As observed, the Jacobian matrix is highly complex, and determining its intrinsic properties would be significantly challenging. In the following, we calculate the Jacobian matrix for the wrist point P of the robot with respect to the coordinate system {4}, demonstrating that the result will be much simpler.

Case Study 6.8 (6DOF Elbow Manipulator) *Obtain the screw-based Jacobian matrix for the wrist point P of the six-degree-of-freedom Elbow manipulator shown in Figure 6.10, with respect to the coordinate frame $\{4\}$.*

Solution: To use the screw-based method for determining the Jacobian matrix of the robot's wrist, first, place a coordinate system instantaneously at point P that is parallel to the coordinate system $\{4\}$. Since the point of interest is located at the origin of the coordinate system $\{4\}$, recursive backward calculations must be performed to determine the screws of the first to fourth joints. For this purpose, calculate the vectors constituting the screws of the robot joints. First, obtain the necessary rotation matrices relative to the coordinate system $\{4\}$ using the transformation matrices provided in Case Study 4.5.

$$
{}^4\boldsymbol{R}_3 = \begin{bmatrix} c_4 & s_4 & 0 \\ 0 & 0 & -1 \\ -s_4 & c_4 & 0 \end{bmatrix} \quad {}^4\boldsymbol{R}_2 = \begin{bmatrix} c_{34} & s_{34} & 0 \\ 0 & 0 & -1 \\ -s_{34} & c_{34} & 0 \end{bmatrix}
$$

$$
{}^4\boldsymbol{R}_1 = \begin{bmatrix} c_{234} & s_{234} & 0 \\ 0 & 0 & -1 \\ -s_{234} & c_{234} & 0 \end{bmatrix}, \; {}^4\boldsymbol{R}_0 = \begin{bmatrix} c_1 c_{234} & s_1 c_{234} & s_{234} \\ s_1 & -c_1 & 0 \\ -c_1 s_{234} & -s_1 s_{234} & c_{234} \end{bmatrix}
$$

With the rotation matrices of the robot available, using the recursive relationships in Equations 6.37 and 6.38, calculate the vectors constituting the screw-based Jacobian matrix for the robot. First, calculate the vectors $\hat{\boldsymbol{s}}_i$ relative to the coordinate system $\{4\}$.

$$
\begin{aligned}
\hat{\boldsymbol{s}}_4 &= {}^4\boldsymbol{R}_3 \begin{bmatrix} 0 \\ 0 \\ 1 \end{bmatrix} = \begin{bmatrix} 0 \\ -1 \\ 0 \end{bmatrix} = \hat{\boldsymbol{s}}_3 = \hat{\boldsymbol{s}}_2 \\
\hat{\boldsymbol{s}}_1 &= {}^4\boldsymbol{R}_0 \begin{bmatrix} 0 \\ 0 \\ 1 \end{bmatrix} = \begin{bmatrix} s_{234} \\ 0 \\ c_{234} \end{bmatrix}
\end{aligned}
$$

Subsequently, calculate the vectors $\boldsymbol{s}_{o_i}$ relative to the coordinate system $\{4\}$.

$$
\begin{aligned}
\boldsymbol{s}_{o_4} = \begin{bmatrix} 0 \\ 0 \\ 0 \end{bmatrix} - {}^4\boldsymbol{R}_4 \begin{bmatrix} a_4 \\ 0 \\ 0 \end{bmatrix} &= \begin{bmatrix} -a_4 \\ 0 \\ 0 \end{bmatrix} \\
\boldsymbol{s}_{o_3} = \boldsymbol{s}_{o_4} - {}^4\boldsymbol{R}_3 \begin{bmatrix} a_3 \\ 0 \\ 0 \end{bmatrix} &= \begin{bmatrix} -(a_4 + a_3 c_4) \\ 0 \\ a_3 s_4 \end{bmatrix} \\
\boldsymbol{s}_{o_2} = \boldsymbol{s}_{o_3} - {}^4\boldsymbol{R}_2 \begin{bmatrix} a_2 \\ 0 \\ 0 \end{bmatrix} &= \begin{bmatrix} -(a_4 + a_3 c_4 + a_2 c_{34}) \\ 0 \\ a_3 s_4 + a_2 s_{34} \end{bmatrix} \\
\boldsymbol{s}_{o_1} = \boldsymbol{s}_{o_2} - {}^4\boldsymbol{R}_1 \begin{bmatrix} 0 \\ 0 \\ 0 \end{bmatrix} &= \boldsymbol{s}_{o_2}
\end{aligned}
$$

With the vectors $\hat{\boldsymbol{s}}_i$ and $\boldsymbol{s}_{o_i}$ available, calculate the joint screw coordinates using Equation 6.34, and based on that, obtain the screw-based Jacobian matrix for the robot's wrist with respect to the coordinate system {4}.

$$
{}^4\boldsymbol{J}_p^s(\boldsymbol{q}) = \begin{bmatrix} s_{234} & 0 & 0 & 0 \\ 0 & -1 & -1 & -1 \\ c_{234} & 0 & 0 & 0 \\ 0 & a_2 s_{34} + a_3 s_4 & a_3 s_4 & 0 \\ a_2 c_2 + a_3 c_{23} + a_4 c_{234} & 0 & 0 & 0 \\ 0 & a_2 c_{34} + a_3 c_4 + a_4 & a_3 c_4 + a_4 & a_4 \end{bmatrix}
$$

As observed, the Jacobian matrix for the wrist point P of the robot with respect to the coordinate system {4} is significantly simpler in representation.

Case Study 6.9 (Stanford Robot) *Consider the Stanford robot shown in Figure 6.11. Obtain the screw-based Jacobian matrix for the wrist point P of this robot with respect to the coordinate system {3}.*

Solution: To use the screw-based method for determining the Jacobian matrix of the robot's wrist, a coordinate system must first be placed instantaneously at point P that is parallel to the coordinate system {3}. Since the point of interest is located at the origin of the coordinate system {3}, recursive backward calculations are performed to

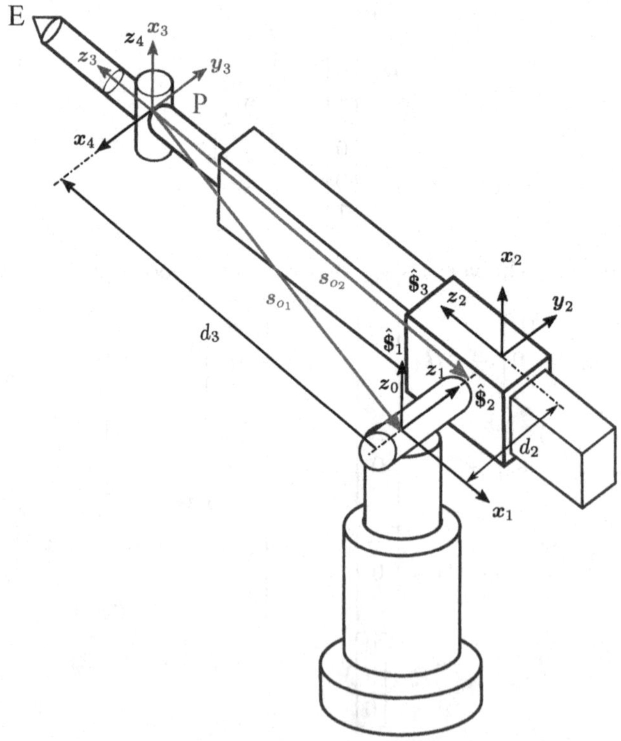

FIGURE 6.11
Vectors constituting the screw-based Jacobian matrix in the Stanford robot.

determine the screws of the first to third joints. For this purpose, obtain the necessary rotation matrices relative to the coordinate system {3} using the transformation matrices provided in Cace Study 4.4.

$$^3\boldsymbol{R}_1 = \begin{bmatrix} c_2 & s_2 & 0 \\ 0 & 0 & 1 \\ s_2 & -c_2 & 0 \end{bmatrix}, \ ^3\boldsymbol{R}_0 = \begin{bmatrix} s_1 & -c_1 & 0 \\ c_1c_2 & s_1c_2 & -s_2 \\ c_1s_2 & s_1s_2 & c_2 \end{bmatrix}$$

With the rotation matrices of the robot available, you can use the recursive relationships in Equations 6.37 and 6.38 to obtain the vectors constituting the screw-based Jacobian matrix for the robot. For this purpose, first determine the vectors $\hat{\boldsymbol{s}}_i$ relative to the coordinate system {3}.

$$\hat{\boldsymbol{s}}_3 = {}^3\boldsymbol{R}_2 \begin{bmatrix} 0 \\ 0 \\ 1 \end{bmatrix} = \begin{bmatrix} 0 \\ 0 \\ 1 \end{bmatrix}, \hat{\boldsymbol{s}}_2 = {}^3\boldsymbol{R}_1 \begin{bmatrix} 0 \\ 0 \\ 1 \end{bmatrix} = \begin{bmatrix} -1 \\ 0 \\ 0 \end{bmatrix}, \hat{\boldsymbol{s}}_1 = {}^3\boldsymbol{R}_0 \begin{bmatrix} 0 \\ 0 \\ 1 \end{bmatrix} = \begin{bmatrix} 0 \\ -s_2 \\ c_2 \end{bmatrix}$$

Next, obtain the vectors $\boldsymbol{s}_{o_i}$ relative to the coordinate system {3}.

$$\boldsymbol{s}_{o_3} = -{}^3\boldsymbol{R}_3 \begin{bmatrix} 0 \\ 0 \\ d_3 \end{bmatrix} = \begin{bmatrix} 0 \\ 0 \\ -d_3 \end{bmatrix}$$

$$\boldsymbol{s}_{o_2} = \boldsymbol{s}_{o_3} - {}^3\boldsymbol{R}_2 \begin{bmatrix} 0 \\ 0 \\ d_2 \end{bmatrix} = \begin{bmatrix} -d_2 \\ 0 \\ -d_3 \end{bmatrix} = \boldsymbol{s}_{o_1}$$

Note that these vectors can be intuitively determined based on the geometric structure of the robot and using Figure 6.11. With the vectors $\hat{\boldsymbol{s}}_i$ and $\boldsymbol{s}_{o_i}$ in hand, the columns of the screw-based Jacobian matrix of the Stanford robot relative to the coordinate system {3} can be obtained using Equation 6.34.

$$^3\boldsymbol{J}_p^s(\boldsymbol{q}) = \begin{bmatrix} 0 & -1 & 0 \\ -s_2 & 0 & 0 \\ c_2 & 0 & 0 \\ -d_3s_2 & 0 & 0 \\ -d_2c_2 & d_3 & 0 \\ -d_2s_2 & 0 & 1 \end{bmatrix}$$

As observed, the Jacobian matrix for the wrist point P of this robot is also represented in a simple form relative to the coordinate system {3}. Using this Jacobian matrix, the angular velocity of the robot wrist $\boldsymbol{\omega}_p$ and the linear velocity of the point P on the wrist $\boldsymbol{v}_p$ can be calculated relative to the coordinate system {3}.

$$\begin{aligned} {}^3\omega_{p_x} &= -\dot{\theta}_2, \ {}^3\omega_{p_y} = -s_2\dot{\theta}_1, \ {}^3\omega_{p_z} = c_2\dot{\theta}_1 \\ {}^3v_{p_x} &= -d_3s_2\dot{\theta}_1, \ {}^3v_{p_y} = -d_2c_2\dot{\theta}_1 + d_3\dot{\theta}_2, \ {}^3v_{p_z} = -d_2s_2\dot{\theta}_1 + \dot{d}_3 \end{aligned}$$

6.3 Coordinate Transformation in Jacobian Matrix

As observed in the previous section, the Jacobian matrix of a robot's end-effector or wrist point can be represented in various coordinate systems. If the Jacobian matrix of the robot is given in the coordinate system $\{\mathcal{A}\}$, it can be determined with respect to the coordinate system $\{\mathcal{B}\}$ based on the following relationship.

$$ {}^{A}\boldsymbol{J}(\boldsymbol{q}) = \left[\begin{array}{c|c} {}^{A}\boldsymbol{R}_B & \boldsymbol{0} \\ \hline \boldsymbol{0} & {}^{A}\boldsymbol{R}_B \end{array}\right] {}^{B}\boldsymbol{J}(\boldsymbol{q}) \tag{6.40} $$

In this equation, ${}^{A}\boldsymbol{R}_B$ is the rotation matrix of the coordinate system $\{\mathcal{B}\}$ with respect to the coordinate system $\{\mathcal{A}\}$, and the notation $[\boldsymbol{0}]$ denotes the 3×3 zero matrix. Since this relationship is derived from the coordinate transformation of linear and angular velocity vectors from one coordinate system to another, it can also be applied to the screw-based Jacobian matrices.

As it can be seen, the representation of the Jacobian matrix of a robot is not only a function of the robot's structure and its configuration but also on the designated coordinate system. However, it should be noted that the inherent properties of the Jacobian matrix, such as eigenvalues, determinant, and other features that will be introduced in next chapter, are not dependent on its representation in a particular coordinate system and do not change when moving from one coordinate system to another. With this notice, the choice of a suitable coordinate system, in which the Jacobian matrix can be written in the simplest possible form, is crucial for reducing computational burden in determining the inherent properties of this matrix. Usually, coordinate systems attached to the robot's base or end-effector are not appropriate choices because a chain of rotation matrix multiplications is required to determine the vectors constituting the Jacobian matrix. Typically, the robot's wrist, situated at the intermediate joints, would be a suitable choice for this purpose.

Case Study 6.10 (3R Robot) *Obtain the Jacobian matrix of the wrist point P in the 3R robot, as shown in Figure 6.12, with respect to the reference coordinate system*

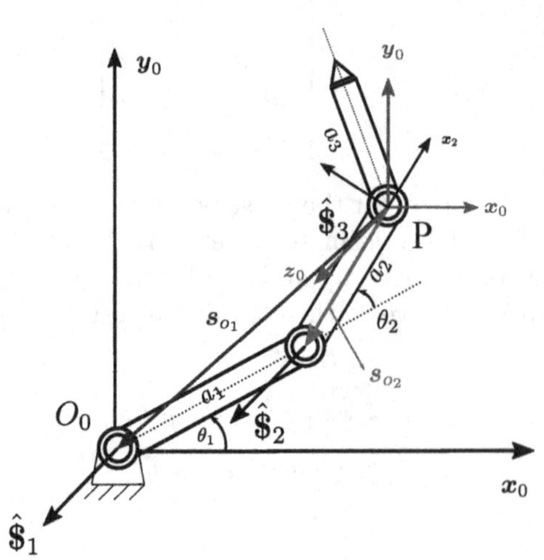

FIGURE 6.12
Vectors constituting the screw-based Jacobian matrix in the 3R robot.

{0} and the coordinate system {2}. Then, calculate the determinant of the Jacobian matrix in two cases and compare the results.

Solution: In order to determine the screw-based Jacobian matrix of the 3R robot with respect to the reference coordinate system 0, first, an instantaneous coordinate system is placed at the wrist point P parallel to the reference coordinate system. Subsequently, the vectors constituting the screws of the two joints of the robot are obtained in the following order, as depicted in Figure 6.12.

$$\hat{s}_1 = \hat{s}_2 = \begin{bmatrix} 0 \\ 0 \\ 1 \end{bmatrix}, \quad s_{o_2} = \begin{bmatrix} -a_2 c_{12} \\ -a_2 s_{12} \\ 0 \end{bmatrix}, \quad s_{o_1} = \begin{bmatrix} -a_1 c_1 - a_2 c_{12} \\ -a_1 s_1 - a_2 s_{12} \\ 0 \end{bmatrix}$$

By determining the necessary vectors, the joint screws' coordinates and the columns of the screw-based Jacobian matrix are obtained according to the relationship 6.34 as follows.

$$\hat{\$}_1 = \begin{bmatrix} 0 \\ 0 \\ 1 \\ -(a_1 s_1 + a_2 s_{12}) \\ a_1 c_1 + a_2 c_{12} \\ 0 \end{bmatrix}, \quad \hat{\$}_2 = \begin{bmatrix} 0 \\ 0 \\ 1 \\ -a_2 s_{12} \\ a_2 c_{12} \\ 0 \end{bmatrix}$$

As observed, the columns of the Jacobian matrix determined using the screw coordinates are six-dimensional arrays. To determine the submatrix for linear velocity, consider only the fourth and fifth rows corresponding to the linear velocities of the robot's joints along the x and y axes. Thus, if we rewrite the screw-based Jacobian matrix as follows,

$$\begin{bmatrix} \dot{x}_p \\ \dot{y}_p \end{bmatrix} = {}^0\boldsymbol{J}_p(\boldsymbol{q}) \begin{bmatrix} \dot{\theta}_1 \\ \dot{\theta}_2 \end{bmatrix}$$

The Jacobian matrix with respect to the reference coordinate system 0 is obtained.

$${}^0\boldsymbol{J}_p(\boldsymbol{q}) = \begin{bmatrix} -(a_1 s_1 + a_2 s_{12}) & -a_2 s_{12} \\ a_1 c_1 + a_2 c_{12} & a_2 c_{12} \end{bmatrix} \tag{6.41}$$

Now, let's calculate the Jacobian matrix of the robot's wrist point with respect to the coordinate system {2}.

$${}^2\boldsymbol{J}_p(\boldsymbol{q}) = {}^2\boldsymbol{R}_0 \, {}^0\boldsymbol{J}_p(\boldsymbol{q}) \tag{6.42}$$

where the rotation matrix ${}^2\boldsymbol{R}_0$ with dimensions 2×2 is calculated as follows.

$${}^2\boldsymbol{R}_0 = \begin{bmatrix} c_{12} & s_{12} \\ -s_{12} & c_{12} \end{bmatrix}$$

By substituting this transformation matrix into Equation 6.42 and some simplifications, the Jacobian matrix of the wrist point with respect to the coordinate system {2} is calculated as follows.

$${}^2\boldsymbol{J}_p(\boldsymbol{q}) = \begin{bmatrix} a_1 s_2 & 0 \\ a_1 c_2 + a_2 & a_2 \end{bmatrix} \tag{6.43}$$

As observed, the Jacobian matrix of the robot's wrist point with respect to the coordinate system {2} is significantly simpler than its representation in the reference coordinate system. Now, let's calculate the determinant of the Jacobian matrix in both representations.

$$\begin{aligned}\det({}^0\boldsymbol{J}_p(\boldsymbol{q})) &= -a_1a_2s_1c_{12} - a_2^2s_{12}c_{12} + a_1a_2c_1s_{12} + a_2^2c_{12}s_{12}, \\ &= a_1a_2s_2 \\ \det({}^2\boldsymbol{J}_p(\boldsymbol{q})) &= a_1a_2s_2\end{aligned}$$

In simplifying the first equation, the trigonometric identity $c_1s_{12} - s_1c_{12} = s_2$ has been utilized. Despite the fact that the results of both expressions are equal, performing calculations in the coordinate system 2 is much simpler. Considering the outcome of this case study, it is recommended that in cases where the characteristics of the Jacobian matrix are under study, the direct calculation of this matrix in an intermediate coordinate system such as the robot's wrist point be carried out to minimize the computational cost.

6.4 Static Force Analysis

For applications such as drilling and grinding, where the robot is in contact with the environment and applies forces or torques to it, force analysis in the open kinematic chain of the robot becomes essential. The forces and torques exerted by the robot's end-effector on the environment are generated by the actuator torques. The relationship between these forces and torques in the joint and task spaces is examined in this section under static conditions.

In this analysis, it is assumed that the robot is in a state of static equilibrium, with no consideration given to dynamic motion. Furthermore, to establish an accurate relationship between the forces and torques applied to the environment and the actuator torques, the weight of the robot is not considered. These simplifying assumptions will be relaxed in Chapter 8 in the dynamic motion analysis. To mathematically define the amount of force and torque applied to the environment, we use a six-dimensional array called a wrench, where the first three components represent force, and the last three components represent torque.

6.4.1 Methods Description

Two main methods have been developed for static force analysis in robots. The first method uses free-body diagrams of the forces and Newton-Euler laws in a static state. In this method, all robot links need to be separated from each other, and free-body diagrams for the forces need to be applied. The connections between the links are introduced by the active and support forces and torques in the joints, and then the Newton-Euler laws in the static state are applied to each robot link to determine the actuator torques by eliminating the support forces and torques.

In this method, all support forces and torques in the robot joints need to be considered, and then, they are determined or eliminated from the free-body diagram equilibrium equations. Therefore, this method involves significant computational costs. Moreover, these calculations are usually performed recursively, allowing for a more efficient determination of the actuator torques using a computer with less effort. Due to the extensive computations required, a detailed presentation of this method is avoided in this book, and readers are referred to other sources in this field [2, 3].

Another efficient method, which is discussed in this book, is the use of the principle of virtual work in static force analysis. In this method, support forces are not included in the calculations, leading to a substantial reduction in computational efforts. Furthermore, we will show that the relationship between the actuator torques in joint space, and the wrench applied to the environment in task space is determined using the transpose of the Jacobian matrix. With this representation, a conceptual description of this mapping is facilitated using the Jacobian matrix, and concepts related to the inherent characteristics of the Jacobian matrix will also be utilized in the static force analysis domain.

6.4.2 Virtual Work Principle Method

In this section, we will study the static force analysis in serial robots using the "Principle of Virtual Work." To introduce this principle, it is necessary to first define virtual displacement. Virtual displacement refers to an infinitely small change in position and orientation of the robot's end-effector caused by an infinitesimally small movement in the robot's joints. The term "virtual" is used in this definition to distinguish it from actual motion. During actual motion, the geometric structure of the robot, the magnitudes of forces and torques applied to the environment, and the torques exerted by the actuators will change. However, in virtual displacement, it is assumed that by performing this virtual displacement, the robot's geometry and the distribution of forces remain unchanged.

To distinguish virtual variables from real ones during this infinitely small movement, the notation δ is used. This is in contrast to real motions where the differential symbol d is used for this purpose. Thus, the virtual displacement of the robot's joints is represented by $\delta\boldsymbol{q}$, where $\boldsymbol{q}$ represents the joint variables.

$$\delta\boldsymbol{q} = \begin{bmatrix}\delta q_1 & \delta q_2 & \dots & \delta q_n\end{bmatrix}^T \tag{6.44}$$

Additionally, the virtual displacement of the robot's end-effector is represented by the notation $\delta\boldsymbol{\mathcal{X}}$, which, as before, is interpreted through a six-dimensional array by using screw coordinates.

$$\delta\boldsymbol{\mathcal{X}} = \begin{bmatrix}\delta x_e & \delta y_e & \delta z_e & \delta\theta_{e_x} & \delta\theta_{e_y} & \delta\theta_{e_z}\end{bmatrix}^T \tag{6.45}$$

Similarly, the vector of forces and torques exerted by the actuators is represented by an n-dimensional array $\boldsymbol{\tau} = \begin{bmatrix}\tau_1 & \tau_2 & \dots & \tau_n\end{bmatrix}^T$, where τ_i represents the torque at revolute joints or the force at prismatic joints generated by actuator i. It is important to note that the notation τ is used uniformly to represent both torques and forces. On the other hand, the forces and torques applied to the robot's end-effector due to interaction with the environment can be represented by a six-dimensional wrench.

$$\boldsymbol{\mathcal{F}} = \begin{bmatrix}\boldsymbol{f}_e \\ \boldsymbol{n}_e\end{bmatrix} \tag{6.46}$$

where $\boldsymbol{f}_e = \begin{bmatrix}f_{e_x} & f_{e_y} & f_{e_z}\end{bmatrix}^T$ is the force vector and $\boldsymbol{n}_e = \begin{bmatrix}n_{e_x} & n_{e_y} & n_{e_z}\end{bmatrix}^T$ is the torque vector exerted on the robot's end-effector at the point of contact with the environment.

The virtual work principle states that by neglecting gravitational and friction forces, the total virtual work done by the actuators and external interactive forces is equal to zero for virtual displacements in the end-effector and robot joints. If we denote the scalar virtual work by δW, the virtual work principle can be expressed by the following relationship:

$$\delta W = \boldsymbol{\tau}^T\delta\boldsymbol{q} - \boldsymbol{\mathcal{F}}^T\delta\boldsymbol{\mathcal{X}} = 0 \tag{6.47}$$

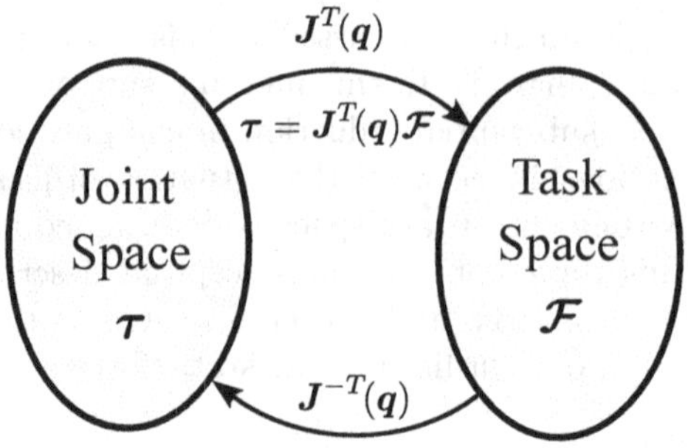

FIGURE 6.13
Transposed Jacobian matrix as a mapping of actuator torques in joint space and the wrench applied to the robot's end-effector in task space.

where the virtual work done by the actuators is directly determined by $\boldsymbol{\tau}^T \delta \boldsymbol{q}$. However, the virtual work on the environment is given by $-\boldsymbol{\mathcal{F}}^T \delta \boldsymbol{\mathcal{X}}$, and the negative sign is used because, according to the definition, $\boldsymbol{\mathcal{F}}$ describes the wrench applied to the robot's end-effector by the environment. Therefore, the virtual work on the environment is considered with a negative sign to account for this wrench applied to the end-effector in the calculations.

Since the Jacobian matrix represents the mapping between the derivatives of joint and task space variables according to the equation $\dot{\boldsymbol{\mathcal{X}}} = \boldsymbol{J}(\boldsymbol{q})\dot{\boldsymbol{q}}$, the relationship of virtual displacement between joint and task space variables can be obtained by using it, as well.

$$\delta \boldsymbol{\mathcal{X}} = \boldsymbol{J}(\boldsymbol{q})\delta \boldsymbol{q} \tag{6.48}$$

Substitute this mapping into the equation 6.47:

$$\left(\boldsymbol{\tau}^T - \boldsymbol{\mathcal{F}}^T \boldsymbol{J}(\boldsymbol{q})\right) \delta \boldsymbol{q} = 0 \tag{6.49}$$

This equation holds for any arbitrary virtual displacement $\delta \boldsymbol{q}$, thus:

$$\boldsymbol{\tau}^T - \boldsymbol{\mathcal{F}}^T \boldsymbol{J}(\boldsymbol{q}) = 0 \tag{6.50}$$

This equation identifies the relationship between the force/torque vector in joint space, and the resulting end-effector wrench in task space through the robot's Jacobian matrix transpose.

$$\boldsymbol{\tau} = \boldsymbol{J}^T(\boldsymbol{q})\boldsymbol{\mathcal{F}} \tag{6.51}$$

Equation 6.51 indicates that the Jacobian matrix not only maps between the velocities of joint and task variables but also, as depicted in Figure 6.13, its transpose describes the relationship between the actuator force/torques in joint space, and the resulting end-effector wrench in task space.

Case Study 6.11 (Planar 3R robot) *Consider the planar 3R robot shown in Figure 6.14. Using the Jacobian matrix mapping, determine the actuator torques $\boldsymbol{\tau}$ to apply forces $\boldsymbol{f}_e = \begin{bmatrix} f_{e_x} & f_{e_y} \end{bmatrix}^T$ and torque n_{e_z} on the robot's end-effector by the environment.*

Solution: We use the equation 6.51 for this purpose. By employing the transpose of the Jacobian matrix provided in equation 6.17, the actuator torques $\boldsymbol{\tau}$ for applying

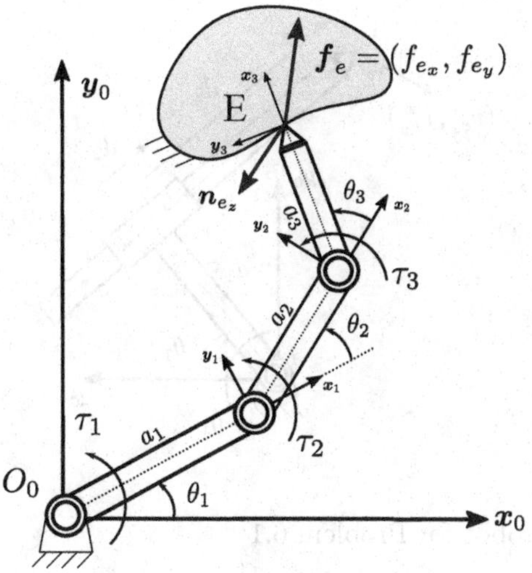

FIGURE 6.14
Actuator torques required to compensate the desired forces and torques applied to the end-effector of the 3R robot by the environment.

the desired forces and torques on the robot's end-effector can be obtained using the following relationships:

$$\begin{aligned}
\tau_1 &= -(a_1 s_1 + a_2 s_{12} + a_3 s_{123}) f_{e_x} + (a_1 c_1 + a_2 c_{12} + a_3 c_{123}) f_{e_y} + n_{e_z} \\
\tau_2 &= -(a_2 s_{12} + a_3 s_{123}) f_{e_x} + (a_2 c_{12} + a_3 c_{123}) f_{e_y} + n_{e_z} \\
\tau_3 &= -(a_3 s_{123}) f_{e_x} + (a_3 c_{123}) f_{e_y} + n_{e_z}
\end{aligned}$$

This relationship can also be derived using the free-body diagram method.

Problems

6.1 Consider the two-degrees-of-freedom RP robot shown in Figure 6.15, whose forward kinematics have been analyzed in Problem 4.1. Obtain the Jacobian matrix for the robot's end-effector using the velocity loop closure method.

6.2 Contemplate the three-degrees-of-freedom PRP robot depicted in Figure 6.16. Its forward kinematics have been examined in Problem 4.2. Determine the Jacobian matrix for the robot's end-effector using the velocity loop closure method.

6.3 Calculate the Jacobian matrix for the robot in Exercise 6.2 utilizing the general method.

6.4 Obtain the screw-based Jacobian matrix for the robot described in Exercise 6.2.

6.5 Consider the three-degrees-of-freedom PRR robot shown in Figure 6.17. The forward kinematics of this robot have been analyzed in Problem 4.5. Obtain the Jacobian

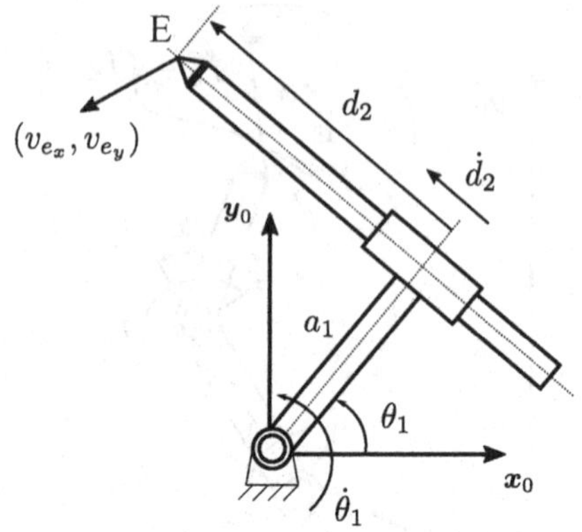

FIGURE 6.15
Two-degrees-of-freedom robot for Problem 6.1

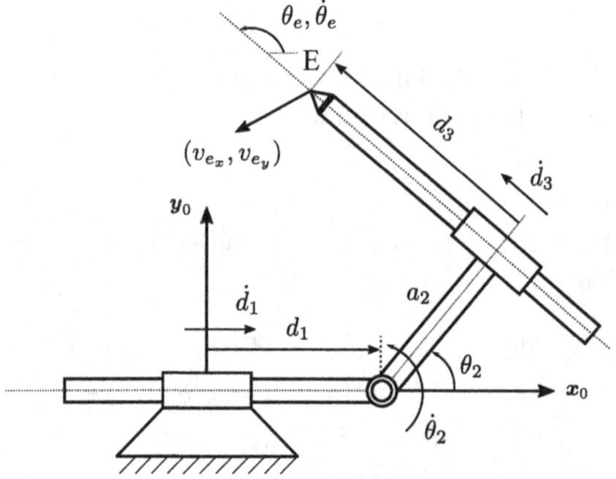

FIGURE 6.16
Three-degrees-of-freedom robot for Exercise 6.2

matrix for the robot's end-effector using the velocity loop closure method.

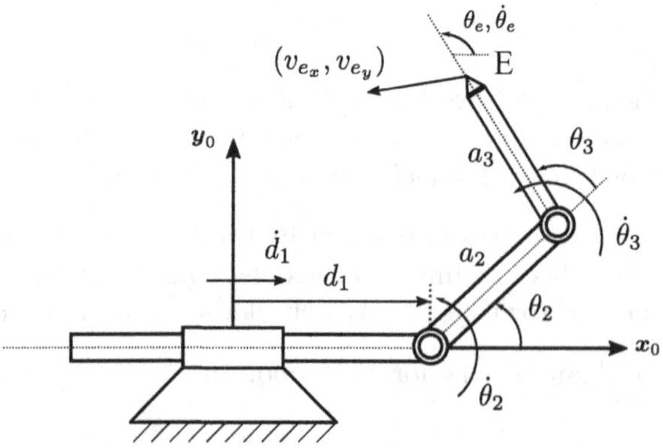

FIGURE 6.17
Three-degrees-of-freedom robot for Exercise 6.5

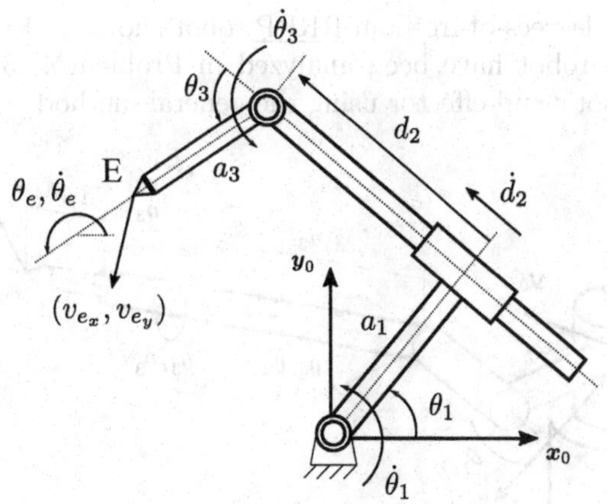

FIGURE 6.18
Three-degrees-of-freedom robot for Exercise 6.8

6.6 Obtain the Jacobian matrix for the robot in Exercise 6.5 utilizing the general method.

6.7 Compute the screw-based Jacobian matrix for the robot described in Exercise 6.5.

6.8 Calculate the Jacobian matrix for the end-effector of the three-degrees-of-freedom <u>RPR</u> robot illustrated in Figure 6.18 using the velocity loop closure method.

6.9 Calculate the Jacobian matrix for the robot described in Problem 6.8 using the general method.

6.10 Determine the screw-based Jacobian matrix for the robot presented in Problem 6.8.

6.11 Consider the two-degrees-of-freedom <u>RR</u> robot shown in Figure 6.19, whose forward kinematics have been analyzed in Problem 4.11. Obtain the Jacobian matrix for the robot's end-effector using the general method.

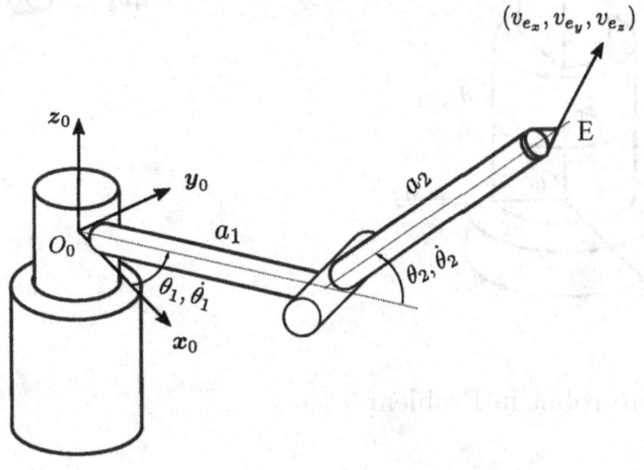

FIGURE 6.19
Two-degrees-of-freedom robot described in Problem 6.11

6.12 Derive the screw-based Jacobian matrix for the robot presented in Problem 6.11.

6.13 Consider the four-degrees-of-freedom RRRP robot shown in Figure 6.20. The forward kinematics of this robot have been analyzed in Problem 4.13. Obtain the Jacobian matrix for the robot's end-effector using the general method.

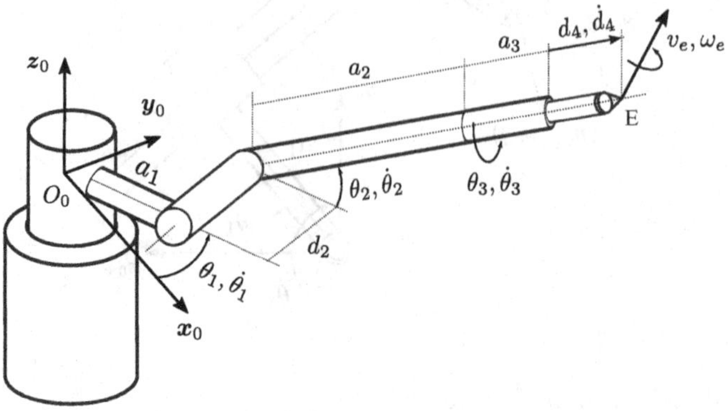

FIGURE 6.20
Four-degrees-of-freedom robot in Problem 6.13

6.14 Calculate the screw-based Jacobian matrix for the four-degrees-of-freedom RRRP robot described in Problem 6.13.

6.15 Contemplate the four-degrees-of-freedom RPRR robot shown in Figure 6.21. The forward kinematics of this robot have been analyzed in Problem 4.15. Obtain the Jacobian matrix for the robot's end-effector using the general method.

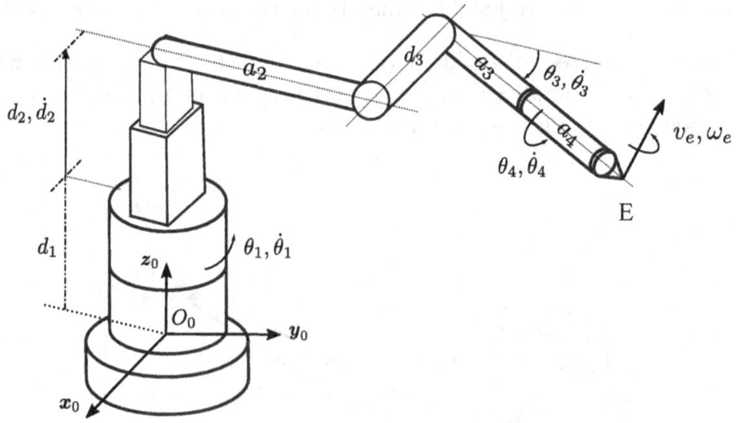

FIGURE 6.21
Four-degrees-of-freedom robot in Problem 6.15

6.16 Derive the screw-based Jacobian matrix for the robot presented in Problem 6.15.

6.17 Consider the four-degrees-of-freedom 4R robot shown in Figure 6.22. The forward kinematics of this robot have been analyzed in Problem 4.17. Obtain the Jacobian matrix for the robot's end-effector using the general approach.

6.18 Calculate the screw-based Jacobian matrix for the robot described in Problem 6.17.

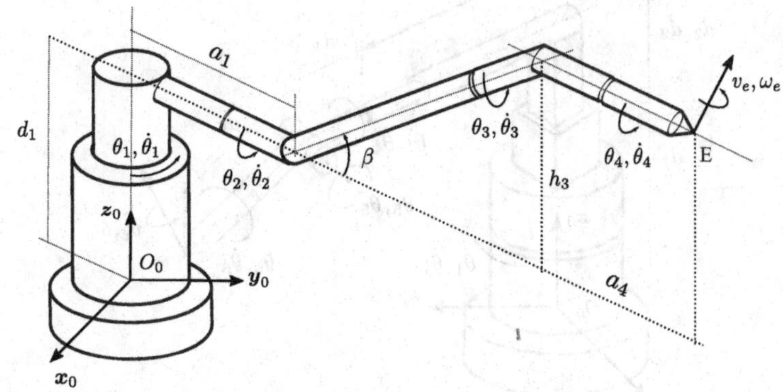

FIGURE 6.22
Four-degrees-of-freedom robot in Problem 6.17

6.19 Consider the four-degrees-of-freedom RRPR robot depicted in Figure 6.23, whose Forward kinematics have been examined in Problem 4.19. Determine the Jacobian matrix for the robot's end-effector using the general method.

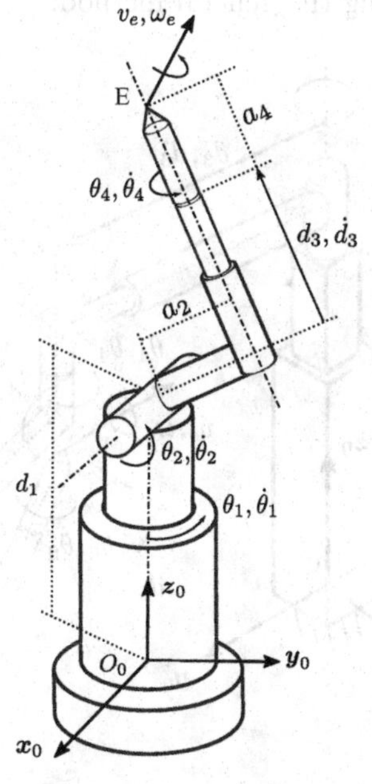

FIGURE 6.23
Four-degrees-of-freedom robot in Problem 6.19

6.20 Derive the screw-based Jacobian matrix for the robot presented in Problem 6.19.

6.21 Consider the six-degrees-of-freedom robot depicted in Figure 6.24. The forward kinematics of this robot have been analyzed in Problem 4.21. Obtain the Jacobian matrix for the robot's end-effector using the general method.

6.22 Derive the screw-based Jacobian matrix for the robot presented in Problem 6.21.

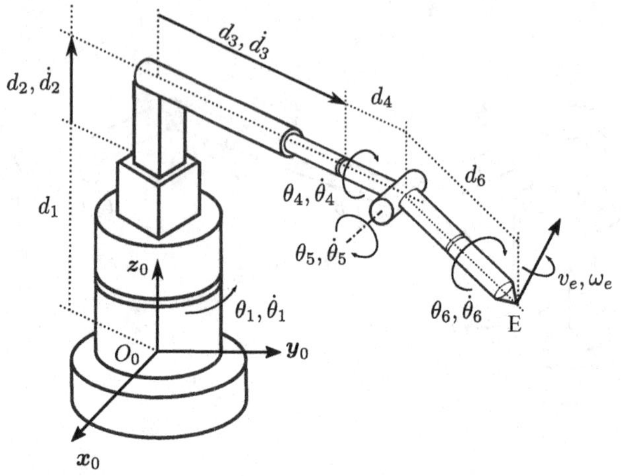

FIGURE 6.24
Six-degrees-of-freedom robot in Problem 6.21

6.23 Consider the six-degrees-of-freedom robot shown in Figure 6.25. The forward kinematics of this robot have been analyzed in Problem 4.23. Derive the Jacobian matrix for the robot's end-effector using the general method.

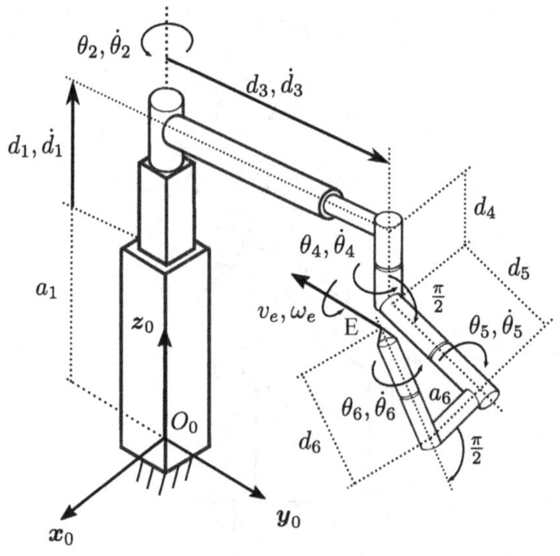

FIGURE 6.25
Six-degrees-of-freedom robot in Problem 6.23

6.24 Obtain the screw-based Jacobian matrix for the robot presented in Problem 6.23.

6.25 Consider the six-degrees-of-freedom Kuka KR6 Agilus robot shown in Figure 4.30, whose forward kinematics have been examined in Problem 4.25. Derive the Jacobian matrix for the robot's end-effector using the general approach.

6.26 Determine the screw-based Jacobian matrix for the Kuka KR6 Agilus robot presented in Problem 6.25.

6.27 Consider the seven-degrees-of-freedom Kuka IIWA robot shown in Figure 4.31, whose forward kinematics have been examined in Problem 4.27. Derive the Jacobian matrix for the robot's end-effector using the general approach.

6.28 Obtain the screw-based Jacobian matrix for the Kuka IIWA robot presented in Problem 6.27.

7

Singularity and Dexterity

The optimization of a robot's kinematic structure is a captivating subject within the field of robotics research and its technological advancements [73]. Within this domain, numerous studies have focused on refining criteria such as the dexterity and manipulability of robots, aspects intricately tied to the inherent properties of the Jacobian matrix [74, 75]. Given the pivotal role of the Jacobian matrix in robot design, this chapter explores the robot's dexterous motion analysis and the identification of singular configurations. To accomplish this, we will elaborate on the properties of the Jacobian matrix. Initially, we will scrutinize the robot's configurations in singular states, defining singular configurations and examining their correlation with the characteristics of the Jacobian matrix [76]. Subsequently, by considering a spectrum ranging from singularity to isotropy, we will define indices of dexterity and manipulability in the robot [77]. Additionally, within this context, an inquiry into ellipsoidal dexterity and its link to the distribution of the end-effector's velocity relative to the velocities of the actuators will be carried out [78].

Redundancy in actuation is a pivotal aspect influencing the performance of robots, akin to many biological musculoskeletal structures that leverage redundancy to enhance dexterity and avoid singularities in movement. This chapter introduces several biological examples illustrating the utilization of additional actuators to improve the kinematic conditions within musculoskeletal structures. The optimal utilization of redundant actuators in robot motion will be discussed in the context of redundancy resolution analysis, which forms a significant part of this chapter. Finally, the concept of robot compliance and stiffness will be established, and a comprehensive examination of the stiffness ellipsoid will be presented in detail.

7.1 Singular Configurations

In Chapter 6, it is demonstrated that the Jacobian matrix serves not only to establish the required mapping between velocities in joint and task spaces but also, its transpose translates actuator forces/torques in joint space into the corresponding wrench in task space. Moreover, it is demonstrated that, typically, this mapping stems from the robot's joint screws, which are represented by six-dimensional arrays.

Now consider a robot with n actuated joints. In general, the Jacobian matrix of the robot's end-effector, with dimensions $6 \times n$, describes the relationship between velocities in joint space $\dot{\boldsymbol{q}}$, and that in task space $\dot{\boldsymbol{\mathcal{X}}}$, as follows:

$$\dot{\boldsymbol{\mathcal{X}}} = \boldsymbol{J}(\boldsymbol{q})\dot{\boldsymbol{q}} = \boldsymbol{J} = \begin{bmatrix} \boldsymbol{J}_v \\ \boldsymbol{J}_\omega \end{bmatrix} = \begin{bmatrix} \boldsymbol{J}_1 \mid \boldsymbol{J}_2 \mid \cdots \mid \boldsymbol{J}_n \end{bmatrix} \dot{b}q \tag{7.1}$$

On the other hand, the transpose of the Jacobian matrix establishes the relationship between the wrench applied to the environment in the task space $\boldsymbol{\mathcal{F}}$, and the vector of actuator forces/torques in the joint space $\boldsymbol{\tau}$. For a more simplified depiction of this mapping, consider

DOI: 10.1201/9781003491415-7

the "Structural Matrix" of the robot denoted by $\boldsymbol{S}(\boldsymbol{q})$, defined as the transpose of the Jacobian matrix.

$$[\boldsymbol{S}(\boldsymbol{q})]_{n\times 6} = \boldsymbol{J}^T(\boldsymbol{q}) \tag{7.2}$$

With this definition, the relationship between the wrench applied to the environment in the task space $\boldsymbol{\mathcal{F}}$, and the vector of actuator forces/torques in the joint space $\boldsymbol{\tau}$, can be interpreted by the following equation, in which $\boldsymbol{J}_i$ is the ith column of the Jacobian matrix which represents the i-th row of the structural matrix of the robot, as well.

$$\boldsymbol{\tau} = \boldsymbol{S}(\boldsymbol{q})\boldsymbol{\mathcal{F}} = \boldsymbol{J}_1 f_{e_x} + \boldsymbol{J}_2 f_{e_y} + \boldsymbol{J}_3 f_{e_z} + \boldsymbol{J}_4 n_{e_x} + \boldsymbol{J}_5 n_{e_y} + \boldsymbol{J}_6 n_{e_z} \tag{7.3}$$

First, consider the mapping of velocities and the relationship given in equation 7.1, the end-effector velocity variable $\dot{\boldsymbol{\mathcal{X}}}$, in general, has six independent components. If we aim to generate any arbitrary velocity in the task space through the motion of the joints, this mapping must have full rank. This implies that the Jacobian matrix must have at least six independent columns. Considering that the columns of the Jacobian matrix are formed by the joint screws, the independence of at least six screws associated with the robot's joints is essential.

Naturally, if the robot is underactuated and the number of actuated joints is less than six, $n < 6$, this condition could not be satisfied, and it is not possible to achieve any arbitrary velocity in the task space for the robot's end-effector. On the other hand, if the robot has redundant actuators $n > 6$, the likelihood of having independence in at least six screws will be higher. In the usual case where the number of actuated joints is equal to the number of degrees of freedom of the robot, $n = 6$, arbitrary motion in any direction for the end-effector can only be produced if the Jacobian matrix has six independent columns.

With this explanation, if the Jacobian matrix has a rank less than 6, there are not six independent columns in the Jacobian matrix, and it is not possible to have the end-effector velocity in any arbitrary direction. This situation is referred to as a *singular configuration* in the robot. Since the Jacobian matrix $\boldsymbol{J}(\boldsymbol{q})$ is a function of the robot's configuration, it may become singular at certain poses in the workspace and non-singular at others. Analyzing singular configurations in robots, known as singularity analysis, is crucial in the design and motion analysis of robots.

Now, consider the force mapping and the relationship presented in equation 7.3, presuming the number of actuated joints aligns with the degrees of freedom of the robot's end-effector, i.e., $n = 6$. Recognizing that singularity in a matrix implies singularity in its transpose as well, the Jacobian singularity directly influences the analysis of robot forces. It's crucial to observe that in this scenario, the mapping direction has shifted from task variables to joint variables, leading to a modification in the physical interpretation of singularity.

If the structural matrix of the robot becomes singular, it is no longer invertible. Consequently, applying constrained torques $\boldsymbol{\tau}$ to the actuators can result in exceedingly large forces at the end-effector! Thus, positioning the robot in a singular configuration facilitates the lifting of heavy loads. Intuitively, when lifting a heavy suitcase, one often extends the arm fully. In such a scenario, the arm is in a singular configuration, causing the Jacobian matrix to lose its full rank, making it much easier to carry the heavy load. This case is further explored in the upcoming example.

Example 7.1 (Planar 2<u>R</u> Robot) *Consider the two-degrees-of-freedom planar 2<u>R</u> robot depicted in Figure 7.1. Determine the singularity configurations in this robot and study the characteristics of the velocity and force mapping in these configurations.*

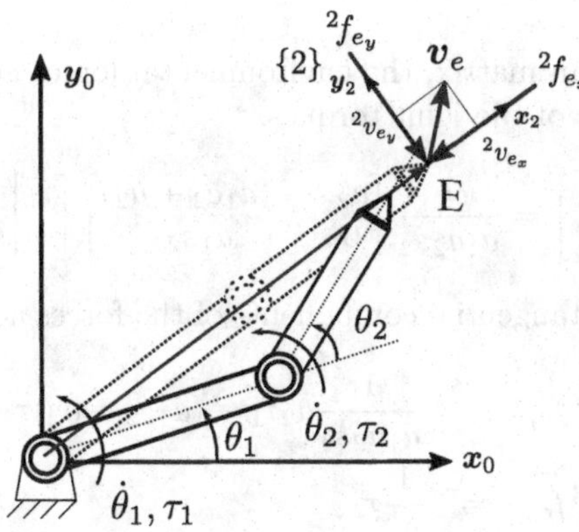

FIGURE 7.1
Representation of singularity and its characteristics in the planar 2R robot

Solution: The Jacobian matrix of the robot with respect to the reference coordinate system is given in Equation 6.12, and with respect to the coordinate system 2, is reported in Equation 6.43, which is reiterated here.

$$^2\boldsymbol{J}_p(\boldsymbol{q}) = \begin{bmatrix} a_1 s_2 & 0 \\ a_1 c_2 + a_2 & a_2 \end{bmatrix} \tag{7.4}$$

The mapping of velocities using this Jacobian matrix can be expressed as:

$$\begin{bmatrix} ^2v_{e_x} \\ ^2v_{e_y} \end{bmatrix} = \begin{bmatrix} a_1 s_2 & 0 \\ a_1 c_2 + a_2 & a_2 \end{bmatrix} \begin{bmatrix} \dot{\theta}_1 \\ \dot{\theta}_2 \end{bmatrix} \tag{7.5}$$

Singularity in this robot occurs when the Jacobian matrix loses its full rank; in this case, the determinant of the Jacobian matrix becomes zero.

$$\det(^2\boldsymbol{J}(\boldsymbol{q})) = a_1 a_2 s_2 = 0$$

This situation occurs when $s_2 = 0$, which means that θ_2 is either zero or π. Therefore, singularity occurs when the robot arm is either fully extended or fully retracted. Consider the extreme case shown in Figure 7.1, where θ_2 approaches zero, and the robot arms are fully extended. In this case, the first row of the Jacobian matrix becomes entirely zero. This means that, no matter how much you increase or decrease the joint velocities, the end-effector of the robot will have zero velocity along the radial direction and cannot be further extended in that direction. This situation occurs when the robot's end-effector is located on the boundary of its workspace, and intuitively, it can be understood that even with very large angular velocities in the robot joints, the robot cannot make the smallest movement in the perpendicular direction to the boundary of the workspace.

Now, let's consider the analysis of static forces in the robot in a singular configuration by determining the force mapping using the transpose of the Jacobian matrix in the coordinate system {2}.

$$\begin{bmatrix} \tau_1 \\ \tau_2 \end{bmatrix} = \begin{bmatrix} a_1 s_2 & a_1 c_2 + a_2 \\ 0 & a_2 \end{bmatrix} \begin{bmatrix} ^2f_{e_x} \\ ^2f_{e_y} \end{bmatrix} \tag{7.6}$$

Using the inverse Jacobian matrix, the environmental forces acting on the end-effector can be obtained in terms of the joint torques.

$$\begin{bmatrix} {}^2f_{e_x} \\ {}^2f_{e_y} \end{bmatrix} = \frac{1}{a_1 a_2 s_2} \begin{bmatrix} a_2 & -(a_1 c_2 + a_2) \\ 0 & a_1 s_2 \end{bmatrix} \begin{bmatrix} \tau_1 \\ \tau_2 \end{bmatrix} \tag{7.7}$$

Calculate the radial and tangential components of the forces acting on the end-effector,

$$ {}^2f_{e_x} = \frac{1}{a_1 a_2 s_2} a_2 \tau_1 - (a_1 c_2 + a_2)\tau_2 \tag{7.8}$$

$$ {}^2f_{e_y} = \frac{1}{a_1} \tau_2 \tag{7.9}$$

As observed in this equation, in a singular configuration, with the application of minimum torques in the joint actuators, the radial force on the end-effector, denoted as ${}^2f_{e_x}$, tends toward infinity. In fact, the radial force applied in a singular configuration acts as a supporting force to the robot's joints, and there is no need to apply actuator torques for their compensation. Therefore, it is recommended to keep the hands fully extended when lifting heavy loads. Unlike the velocity analysis, where singularity restricts the robot's motion along the radial direction, in the force analysis, the ability to carry heavier loads in a singular configuration becomes beneficial.

Finding singularity states in robots is crucial because, in a singular configuration:

- The end-effector's motion is not accessible in one or more directions. This typically occurs at the boundary of the robot's reachable workspace. An example of this situation

FIGURE 7.2
Utilizing the singular configuration or the extended neck posture of a giraffe for maximum workspace accessibility

FIGURE 7.3
Illustration of the desirable characteristic of heavy load lifting in singular configuration.

is observed in the giraffe's extended neck posture, as shown in Figure 7.2.

- The end-effector's velocities in one or more directions can only be achieved with infinite joint velocities, which is practically impossible.
- In one or more directions, significant force or torque can be exerted on the robot's end-effector without applying torques to the joints. This condition is desirable for lifting heavy loads, as illustrated in Figure 7.3.
- The end-effector is often located at the boundary of the workspace, allowing the determination of the reachable workspace. In this situation, the inverse kinematics has two repeated solutions, and a direct relationship between the robot's inverse kinematics and the singularity state of the Jacobian matrix is established.
- The end-effector loses access to points within the workspace with very small changes in its kinematic parameters.

Various methods have been developed to determine singularity configurations in robots [76, 79]. In the following section, the determination of singularity states in fully actuated robots with square Jacobian matrices will be studied. In this case, to identify singularity states, it is sufficient to find robot configurations where the determinant of the Jacobian matrix becomes zero. These calculations are not dependent on the coordinate system in which the Jacobian matrix is computed. Therefore, choosing a suitable coordinate system is crucial for minimizing computations. Additionally, for robots with six degrees of freedom, if the robot's wrist is composed of spherical joints, and their rotation axes intersect at a point, the Jacobian matrix can be decomposed into two submatrices for the arm and wrist. This decomposition facilitates the determination of singularity states.

7.2 Jacobian Matrix Decomposition

In Chapter 3, the kinematic analysis of serial robots was investigated using Denavit-Hartenberg parameters and screw coordinates. For this purpose, coordinate systems of robot joints were determined, and based on that, Denavit-Hartenberg parameters were defined. In addition, by using screw coordinates, the kinematic structure of the robot was determined, and forward and inverse kinematic analysis was performed. Although the forward kinematic equations, in this case, depend on the type and configuration of coordinate systems connected to the robot joints, we saw that determining repeated solutions and defining the workspace boundaries in the robot are not dependent on the coordinate systems used to represent the kinematic parameters but rather related to the main structure of the robot and its configuration. We also observed that in robots with spherical kinematic structures in their wrists composed of three intersecting revolute joints, the inverse kinematic analysis will be more easily performed with the decomposition of end-effector orientation from the wrist position.

Recognizing these facts helps to similarly analyze singularity configurations in serial robots with greater ease, by separating the Jacobian matrix into two submatrices, namely arm and wrist. With this separation, the singularity state of the robot can also be divided into two subcategories: singularity in the arm and singularity in the wrist. For this purpose, consider a six-degrees-of-freedom robot with a spherical wrist structure composed of three intersecting revolute joints. In this robot, the Jacobian matrix has dimensions of 6×6, and the singularity state occurs only in a configuration of the robot where the determinant of the Jacobian matrix becomes zero.

$$\det \boldsymbol{J}(\boldsymbol{q}) = 0 \tag{7.10}$$

In this context, we can use the conventional or screw-based Jacobian matrix. Let's follow the calculations in this section using the conventional Jacobian matrix. By rewriting the columns of the Jacobian matrix based on Equation 6.20 for revolute joints and Equation 6.21 for prismatic joints, and calculating its determinant, we will have:

$$\det \boldsymbol{J}(\boldsymbol{q}) = \begin{vmatrix} \boldsymbol{J}_1 & \boldsymbol{J}_2 & \boldsymbol{J}_3 & \boldsymbol{J}_4 & \boldsymbol{J}_5 & \boldsymbol{J}_6 \end{vmatrix} = 0 \tag{7.11}$$

In the event that the robot's wrist is formed by revolute joints with intersecting axes at point P, the Jacobian matrix can be decomposed into two submatrices for the arm and wrist. In this case, if we rewrite the conventional Jacobian matrix for point P, considering that $\boldsymbol{p}_e^{i-1}$ is zero in the last three rotation axes, the overall Jacobian matrix for the robot's wrist transforms into a lower triangular block matrix.

$$\boldsymbol{J}_p(\boldsymbol{q}) = \left[\begin{array}{c|c} \boldsymbol{J}_{11} & 0 \\ \hline \boldsymbol{J}_{12} & \boldsymbol{J}_{22} \end{array}\right] \tag{7.12}$$

where

$$\boldsymbol{J}_{11}(\boldsymbol{q}) = \begin{bmatrix} \boldsymbol{z}_0 \times \boldsymbol{p}_p^0 & \boldsymbol{z}_1 \times \boldsymbol{p}_p^1 & \boldsymbol{z}_2 \times \boldsymbol{p}_p^2 \end{bmatrix} \tag{7.13}$$

$$\boldsymbol{J}_{22}(\boldsymbol{q}) = \begin{bmatrix} \boldsymbol{z}_3 & \boldsymbol{z}_4 & \boldsymbol{z}_5 \end{bmatrix} \tag{7.14}$$

Hence, the determinant of the Jacobian matrix is obtained by multiplying the determinants of these two submatrices.

$$\det \boldsymbol{J}(\boldsymbol{q}) = \det \boldsymbol{J}_{11}(\boldsymbol{q}) \cdot \det \boldsymbol{J}_{22}(\boldsymbol{q}) \tag{7.15}$$

Therefore, singularity in the robot occurs when the submatrices for the arm and wrist, each with dimensions 3×3, lose linear independence and become singular. Analyzing this situation is more easily achievable compared to the analysis of the complete Jacobian matrix.

The same calculations can be performed using a screw-based matrix for point P, as reported in Equation 7.16.

$$\boldsymbol{J}_p^p(\boldsymbol{q}) = \left[\begin{array}{c|c} \boldsymbol{J}_{11}^s & \boldsymbol{J}_{22}^s \\ \hline \boldsymbol{J}_{12}^s & \boldsymbol{0} \end{array}\right] \tag{7.16}$$

where,

$$J_{12}^{s}(q) = \begin{bmatrix} s_{o_1} \times \hat{s}_1 & s_{o_2} \times \hat{s}_2 & s_{o_3} \times \hat{s}_3 \end{bmatrix} \tag{7.17}$$
$$J_{22}^{s}(q) = \begin{bmatrix} \hat{s}_4 & \hat{s}_5 & \hat{s}_6 \end{bmatrix} \tag{7.18}$$

In order to calculate the vectors s_{o_i}, according to the convention, the coordinate system is placed at the wrist point. Thus, the determinant of the Jacobian matrix is obtained by multiplying the determinants of these two submatrices.

$$\det J_p^s(q) = -\det J_{12}(q) \cdot \det J_{22}(q) \tag{7.19}$$

In this equation, the presence of a negative sign, in comparison to Equation 7.15, is because the vectors s_{o_i} are considered in the opposite direction to the vectors p_p^{i-1}; thus, the result of both equations is the same. A notable advantage of the screw-based Jacobian matrix is its better geometric intuition in visualizing the vectors that constitute the Jacobian matrix. Additionally, its index corresponds one-to-one with the joint number under study.

To enhance comprehension, we will separately examine the singularity analysis of typical arm and wrist structures in the upcoming sections.

7.2.1 Wrist Singularities

Determining the singularity status in the robot's wrist involves analyzing the submatrix of the wrist Jacobian matrix, as given in equations 7.14 or 7.18. Intuitively, it can be understood that the linear independence of the columns in these matrices is lost when the rotation axes or the screw axes of the three wrist joints become parallel to each other. Consider the spherical wrist structure in industrial robots, formed by three mutually perpendicular rotation about $w - u - w$ axes. An example of this structure, commonly used in the wrist of many industrial robots, is illustrated in Figure 7.4. As observed in this figure, the vectors z_3 and z_5 become parallel, or the screw axes $\hat{\$}_4$ and $\hat{\$}_6$ align only when the fifth joint angle θ_5, is either zero or π. This singularity occurs only in these specific joint configurations and is unavoidable unless the motion range of the fifth joint is constrained. It can be visually recognized from the robot's wrist structure that when two rotation axes become aligned, a singularity condition arises. This is because in this configuration, moving two joints with the same angle but in opposite directions results in a zero output motion.

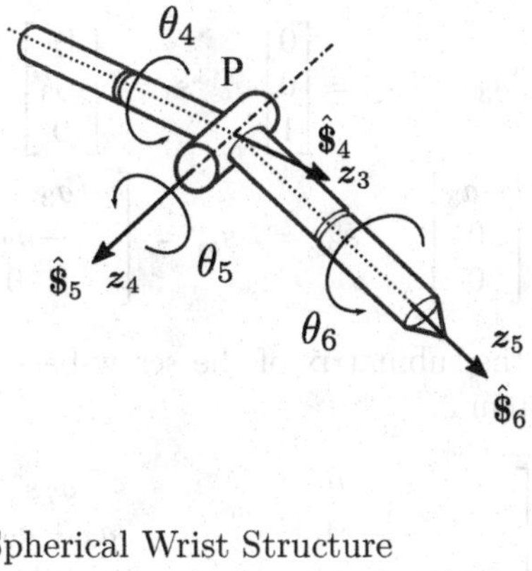

FIGURE 7.4
Singularity Analysis in the Spherical Wrist Structure

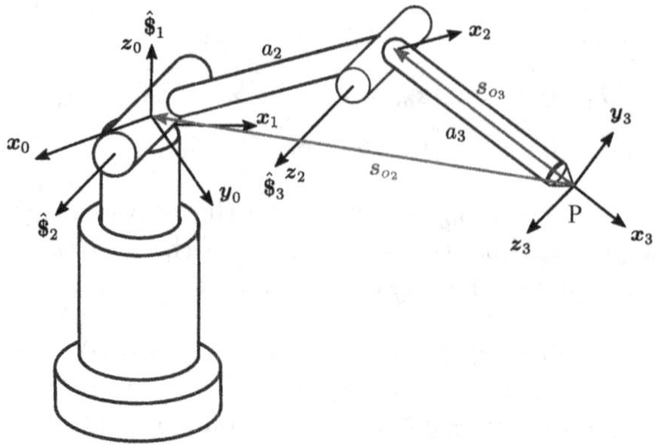

FIGURE 7.5
Analysis of singular configuration in a three-degrees-of-freedom Elbow manipulator

7.2.2 Arm Singularities

In order to examine the singular configuration in industrial robots, it is necessary to analyze the Jacobian submatrix $\boldsymbol{J}_{11}$ in a conventional representation or, equivalently, decompose and analyze the Jacobian submatrix $\boldsymbol{J}^s_{12}$ based on the screw representation. The singular configuration of these submatrices can be intuitively determined using the geometry of its constituent vector columns. Note that the columns of this submatrix are formed by vector cross products of the joint rotation axis with the distance vector from it to the wrist point P. Thus, in any situation where the cross-product of these two vectors in the first to third joints becomes collinear or when this cross-product becomes zero, the matrix will lose its full rank, and a singular configuration will occur in the robot. To investigate the singular configuration of the arm in several common industrial structures, we examine the determinant of the Jacobian matrix $\boldsymbol{J}_{11}$ and determine the singular configuration of the robot intuitively by analyzing the positioning of its constituent vectors.

Consider the kinematic structure of a three-degrees-of-freedom Elbow manipulator as a first case study. This structure, commonly used in many industrial robots, is illustrated in Figure 7.5. By obtaining the constituent vectors of the screw-based Jacobian matrix of the robot wrist point P relative to the coordinate system $\{3\}$, the Jacobian matrix can be expressed in its simplest form.

$$ {}^3\hat{\boldsymbol{s}}_3 = {}^3\hat{\boldsymbol{s}}_2 = \begin{bmatrix} 0 \\ 0 \\ 1 \end{bmatrix}, \quad {}^3\hat{\boldsymbol{s}}_1 = \begin{bmatrix} s_{23} \\ c_{23} \\ 0 \end{bmatrix} $$

$$ {}^3\hat{\boldsymbol{s}}_{o_3} = \begin{bmatrix} -a_3 \\ 0 \\ 0 \end{bmatrix}, \quad {}^3\hat{\boldsymbol{s}}_{o_2} = {}^3\hat{\boldsymbol{s}}_{o_1} = \begin{bmatrix} -(a_3 + a_2 c_3) \\ -a_2 s_3 \\ 0 \end{bmatrix} $$

Now, it is possible to derive the submatrix of the screw-based Jacobian of the robot with respect to the coordinate system 3.

$$ {}^3\boldsymbol{J}^s_{12} = \begin{bmatrix} 0 & a_2 s_3 & 0 \\ 0 & a_3 + a_2 c_3 & a_3 \\ -(a_3 + a_2 c_2) c_{23} - a_2 c_2 & 0 & 0 \end{bmatrix} \tag{7.20} $$

To determine the singular configuration of the arm in this robot, acquire configurations

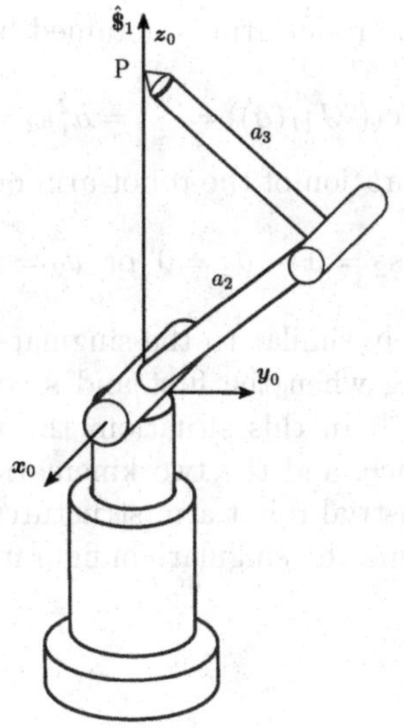

FIGURE 7.6
Geometric representation of one of the singular configurations in the 3DOF Elbow manipulator

where the determinant of this matrix becomes zero.

$$\det({}^3\boldsymbol{J}^s_{12}) = -a_2a_3s_3(a_2c_2 + a_3c_{23}) = 0 \tag{7.21}$$

In this way, the singular configuration of the robot arm occurs when,

$$s_3 = 0 \rightarrow \theta_3 = 0 \text{ or } \theta_3 = \pi \tag{7.22}$$

or,

$$a_2c_2 + a_3c_{23} = 0 \tag{7.23}$$

Both cases where the robot is in its singular configuration, have their own geometric intuitions. The first case, as expressed in equation 7.22, occurs when the robot's arm is fully extended or fully retracted. In this situation, the robot's wrist point is positioned on the boundary of its workspace, and the two kinematic inverse solutions converge to a repeated solution.

The second case, reported in equation 7.23, occurs when, according to Figure 7.6, the robot's wrist point is aligned with the vertical rotation axis z_0. This singular condition for the robot arm will occur in robots with a cylindrical structure like RRP robots, such as the Stanford robot and others when their second and third links have zero offsets. However, in the case of an elbow or cylindrical structure robot, if the second or third link has a non-zero offset $d_i \neq 0$, this singular condition will not occur.

Consider the SCARA robot's kinematic structure as a second example, depicted in Figure 6.6. The complete Jacobian matrix for this robot has been investigated in Case Study 6.3 and is reported in equation 6.27. Rewrite the 3×3 submatrix of the arm of this robot,

$${}^0\boldsymbol{J}_{11}(\boldsymbol{q}) = \begin{bmatrix} -(a_1s_1 + a_2s_{12}) & a_2s_{12} & 0 \\ a_1c_1 + a_2c_{12} & a_2c_{12} & 0 \\ 0 & 0 & -1 \end{bmatrix} \tag{7.24}$$

The singular configuration of the robot arm is obtained by setting the determinant of this matrix to zero.

$$\det({}^0\boldsymbol{J}_{11}(\boldsymbol{q})) = \cdots = a_1^2 s_2 = 0 \tag{7.25}$$

In this way, the singular configuration of the robot arm occurs when,

$$s_2 = 0 \rightarrow \theta_2 = 0 \text{ or } \theta_2 = \pi \tag{7.26}$$

This singular condition is entirely similar to the singular configuration in a three-degrees-of-freedom robot arm, occurring when the first and second links of the robot are either fully extended or fully retracted. In this situation, the robot's wrist point is positioned on the boundary of its workspace, and the two kinematic inverse solutions converge to a repeated solution. In other industrial robot arm structures, using what has been explained in this section, one can determine the singular configuration of the robot arm and obtain its geometric intuition.

7.3 Jacobian Matrix Inherent Features

Up to now, we have explored the connection between the Jacobian matrix and ideas such as poses in the workspace that are unreachable or the best configuration for lifting heavy objects. The significance of the inherent features of the Jacobian matrix in the optimal design of a robot's kinematic structure has garnered significant attention in diverse research studies. In the context of optimal design, researchers promptly focus on the biological skeletal structure of creatures in nature, with a particular emphasis on humans. Emulating these natural structures has led to the popularity of designing robots that inherit the extraordinary characteristics of these remarkable designs.

Let's elaborate on a few instances of the muscular-skeletal composition of certain animals to examine and enhance our mathematical comprehension of attributes like agility and dexterity in robots. As mentioned earlier, achieving full movement in three-dimensional space requires six degrees of freedom. Consequently, numerous industrial robots with full actuation offer these degrees of freedom. It raises the question: does nature adhere to a comparable pattern, and do the musculoskeletal structures of animals also employ six degrees of freedom?

Considering the skeletal structures of animals and humans, one notable example is the distinction in the number of vertebrae in the necks of mammals and birds. Specifically, the neck of all mammals typically comprises seven vertebrae, whereas the neck of all birds is structured by fourteen vertebrae. Undoubtedly, this variability in articulation serves to minimize or entirely eliminate singular configurations in the neck movements of mammals. In the case of birds, where survival, securing food, hunting prey, or escaping from predators hinge on rapid and agile neck movements, the number of neck vertebrae doubles, resulting in eight degrees of redundancy in neck articulation. Initially, it may appear astonishing that a giraffe, as illustrated in Figure 7.2, possesses only seven vertebrae on its neck, while a hummingbird, weighing just a few grams, boasts fourteen vertebrae on its neck.

This phenomenon is equally striking in the number of actuators present in the human shoulder, arm, and wrist structure, offering intriguing insights. The human shoulder, as well as that of many animals, facilitates complete rotational motion with three degrees of freedom for the arm. The shoulder joint is ingeniously designed with a ball-and-socket joint, providing these three degrees of freedom concurrently. However, this movement is induced by four muscle groups, introducing an additional degree of actuation redundancy.

FIGURE 7.7
Positioning of the king cobra before attacking its prey

One pivotal reason for incorporating actuation redundancy in this structure is to prevent singular configurations in the shoulder joint.

Upon examining the musculoskeletal structure of the human arm, in addition to the three degrees of freedom in the shoulder, two degrees of movement in the elbow, and two degrees of movement in the wrist are apparent. Consequently, the human arm utilizes seven actuators to achieve six degrees of freedom in motion, with an extra degree of actuation redundancy. The intricately intertwined musculoskeletal structure of the elbow and wrist involves a total of seven muscle groups contributing to the creation of four degrees of freedom motion. The design of the human wrist is also a masterpiece, utilizing eight bones and four muscle groups to generate rotational movements in the wrist.

As a concluding illustration in this section, take note of the king cobra's posture before launching an attack on its prey, illustrated in Figure 7.7. As evident in this depiction, the king cobra, before striking at its prey, raises its body from the ground to enhance accessibility and speed. Concurrently, it arches its neck backward, with its gaze fixed entirely on the prey. This arrangement maximizes the speed at the cobra's head, significantly reducing the likelihood of the prey escaping its venomous bite.

In these instances, as well as many others observed in nature, redundancy in actuation and appropriate configuration are utilized to attain maximum accessibility or achieve optimal end-effector speed. Translating these effective structures into inspiration, it becomes imperative to establish connections between these physical concepts and computable parameters for optimization in the design and implementation of humanoid robots. Unfortunately, formulating criteria that comprehensively encompass all desired features in the optimal design of a robot is not straightforward and necessitates extensive investigation. Consequently, researchers, drawing inspiration from nature and interpreting certain aspects, have introduced diverse criteria grounded in the singularity characteristics of the Jacobian matrix, each addressing a facet of reality. In the subsequent section, we will explore the mathematical definition of some of these criteria and metrics.

7.3.1 Dexterity Ellipsoid

Given the robot's structure and its specific configuration $\boldsymbol{q}$, we can construct the Jacobian matrix to linearly map the end-effector's velocity to the joint velocities of the robot, represented as $\dot{\boldsymbol{\mathcal{X}}} = \boldsymbol{J}(\boldsymbol{q})\dot{\boldsymbol{q}}$. By treating this mapping as a linear system with joint velocities as input and the end-effector's velocity as output, one can analyze its characteristics, such

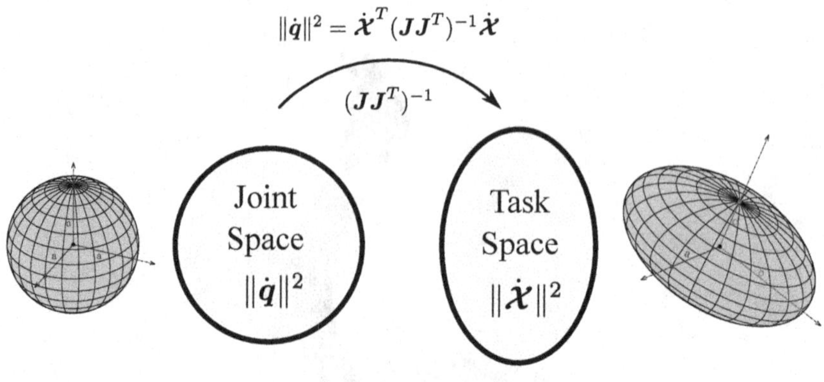

FIGURE 7.8
Robot's manipulability ellipsoid

as gain, to describe its behavior. This definition allows for the formulation of a strategy to either increase or decrease the magnitude of the end-effector's velocity with respect to the robot's joint velocities. It is important to acknowledge that this mapping involves multiple variables, and to define the system's gain, one can rely on the Euclidean norms of the input and output arrays.

To better illustrate the idea, let us consider a scenario where all the robot joints move uniformly with velocities less than or equal to one. In this case, the Euclidean norm of the n-dimensional vector $\dot{q}$ is constrained to be less than or equal to one.

$$\|\dot{q}\|_2^2 = \dot{q}_1^2 + \dot{q}_2^2 + \cdots + \dot{q}_n^2 \leq 1 \tag{7.27}$$

If the Jacobian matrix of the robot is square and the robot is not in a singular configuration, using the inverse of the Jacobian matrix relates the velocity of the joints to that of the end-effector.

$$\dot{q} = J^{-1}(q)\dot{\mathcal{X}} \tag{7.28}$$

In the case where the robot exhibits redundancy in its actuation, solving the velocity inverse problem can also be accomplished using the *pseudo-inverse* of the Jacobian matrix.

$$\dot{q} = J^{\dagger}(q)\dot{\mathcal{X}} \tag{7.29}$$

In the general form, the relationship between the Euclidean norms of joint velocities and task space velocities can be calculated as follows.

$$\begin{aligned} \|\dot{q}\|_2^2 = \dot{q}^T\dot{q} &= [J^{\dagger}\dot{\mathcal{X}}]^T[J^{\dagger}\dot{\mathcal{X}}] \\ &= \dot{\mathcal{X}}^T J^{\dagger^T} J^{\dagger}\dot{\mathcal{X}} \\ &= \dot{\mathcal{X}}^T (JJ^T)^{\dagger}\dot{\mathcal{X}} \end{aligned}$$

Considering that the matrix JJ^T is always square for any type of robot, regardless of being underactuated or redundant, this relationship simplifies to the following form.

$$\|\dot{q}\|_2^2 = \dot{\mathcal{X}}^T (JJ^T)^{-1}\dot{\mathcal{X}} \tag{7.30}$$

Equation 7.30 quantitatively illustrates the relationship between the end-effector velocity vector and the joint velocities of the robot. If the rank of the matrix JJ^T is full, and its inverse exists, the relationship between the Euclidean norms of joint velocities and the end-effector velocity array is given by a weighted norm along the ellipsoidal direction, as defined by this equation. This representation is referred to as the dexterity ellipsoid of the robot.

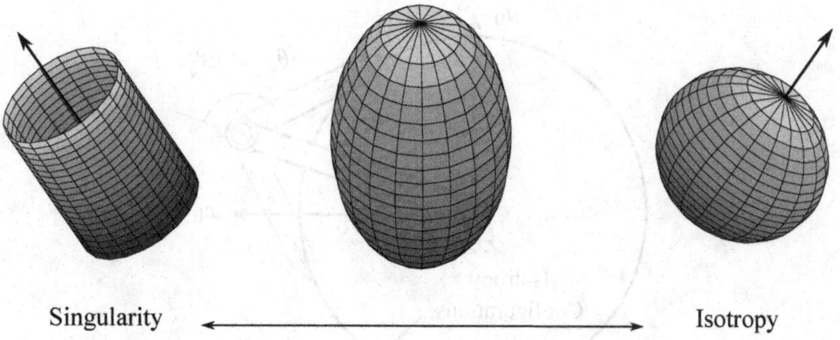

FIGURE 7.9
Dexterity ellipsoid of the robot from isotropy to singularity

According to Figure 7.8, in the case where the robot joint velocities are uniform and lie within an n-dimensional unit sphere ($|\dot{\boldsymbol{q}}|^2 = 1$), the six-dimensional array representing the end-effector velocities will be ellipsoidal. Along the ellipsoid longitudinal axis, this mapping has the maximum magnitude, while along its transverse axis, it has the minimum magnitude, all while maintaining a unit volume. In this way, the ellipsoidal dexterity not only determines the relationship between the maximum and minimum sizes of the input and output arrays of this mapping but also specifies the directions in which the end-effector velocities reach their maximum or minimum values.

7.3.2 From Singularity to Isotropy

The dexterity ellipsoid in serial robots delineates crucial features within the robot's Jacobian matrix. In contrast to the Jacobian matrix, which is limited to being square only for fully actuated robots, the dexterity matrix $\boldsymbol{J}\boldsymbol{J}^T$ for diverse robots—encompassing underactuated, fully actuated, and even redundantly actuated cases—is invariably square. Therefore, the attributes of the dexterity ellipsoid can be extended to various types of serial robots.

Now, let us explore the relationship between the determinant of the dexterity matrix and its eigenvalues. We know that the determinant of any square matrix is equal to the product of its eigenvalues. Therefore, in the dexterity matrix,

$$\det\left(\boldsymbol{J}(\boldsymbol{q})\boldsymbol{J}^T(\boldsymbol{q})\right) = \lambda_1 \cdot \lambda_2 \cdots \lambda_n \tag{7.31}$$

where, for planar robots, $n = 3$, and for robots with spatial motion in three-dimensional space, $n = 6$. One of the key features of the dexterity ellipsoid is described by the determinant of this matrix, and as observed, this characteristic is directly related to the product of the eigenvalues of the matrix.

For a deeper understanding, consider two limiting cases to define the spectrum influenced by this feature. In the first limit, assume that the determinant of the dexterity matrix $(\boldsymbol{J}(\boldsymbol{q})\boldsymbol{J}^T(\boldsymbol{q}))$ becomes zero in a specific configuration $\boldsymbol{q}$. In this case, the robot is in a singular configuration, and according to Equation 7.31, at least one of the eigenvalues of the dexterity matrix is zero. In this limiting case, the ellipsoidal dexterity is stretched along the eigenvector corresponding to the zero eigenvalues, turning into a cylinder, as depicted in Figure 7.9. The physical interpretation of this limit in the dexterity ellipsoid aligns directly with the characteristics of a singular configuration.

Considering that in Equation 7.30 the inverse of the dexterity matrix is multiplied by the norm of the task space variables, when the determinant of this matrix becomes zero,

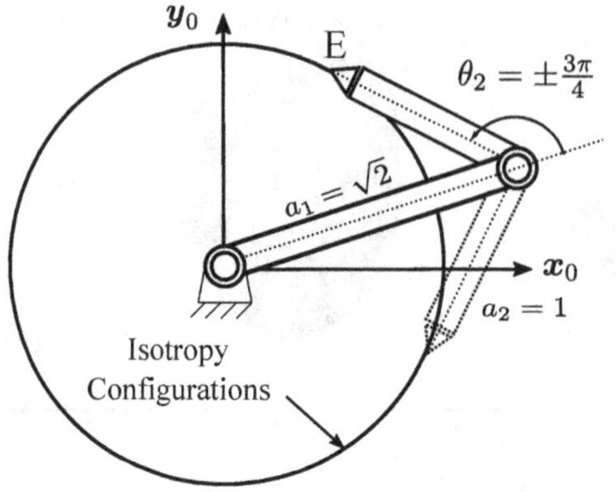

FIGURE 7.10
Investigation of isotropic condition in the 2R Robot

its inverse becomes infinite along the eigenvector corresponding to the zero eigenvalue. To maintain the product of the norm of task space variables finite and equal to one, it is necessary for these variables to approach zero in a specific direction in space. This situation occurs when the robot end-effector reaches the boundary of the workspace and cannot move along the direction perpendicular to it.

Now consider the second limiting case where all the eigenvalues of the dexterity matrix in a specific configuration $\boldsymbol{q}$ are equal to one, and therefore, according to Equation 7.31, the determinant of the dexterity matrix $(\boldsymbol{J}(\boldsymbol{q})\boldsymbol{J}^T(\boldsymbol{q}))$ will be equal to one. In this case, the dexterity ellipsoid transforms into an n-dimensional sphere with unit dimensions in all directions. This limiting case is also depicted in Figure 7.9 and is referred to as an *isotropic* configuration. The physical interpretation of this limiting case in dexterity ellipsoid is in contrast to the singularity configuration of the robot. In this limit, applying equal velocities to all joint variables results in symmetrical velocities in the end-effector task space variables. In an isotropic state, the motion of the robot's end-effector is uniformly distributed in all directions, and the dexterity of motion in the robot is maximized.

Note that the dexterity matrix is a function of the joint variable vector $\boldsymbol{q}$ and is dependent on the robot's configuration. Therefore, features like singularity may not occur at all or may only be observed in limited situations in the robot's workspace. However, a spectrum between singularity and isotropy will exist at different points in the workspace. In the optimal design of robots inspired by nature, the robot's state tends to be inclined toward isotropy, and a uniform distribution among the eigenvalues of the dexterity matrix and the end-effector velocities in different directions is sought. This way, the robot avoids singularities in its workspace and exhibits dexterous and agile maneuvers.

Example 7.2 (Planar 2R Robot) *Consider the planar 2R robot depicted in Figure 7.10. The lengths of the two arms of the robot are chosen to be proportional to human arm dimensions, with $a_1 = \sqrt{2}$ and $a_2 = 1$. Determine the eigenvalues of the dexterity matrix for various configurations of the robot and investigate the isotropic condition in the robot.*

Solution: The Jacobian matrix with respect to the coordinate system $\{2\}$ for the 2R robot was previously derived and reported in Equation 7.4. By substituting the link lengths into this Jacobian matrix, we have

$$^2\boldsymbol{J}(\boldsymbol{q}) = \begin{bmatrix} \sqrt{2}s_2 & 0 \\ \sqrt{2}c_2 + 1 & 1 \end{bmatrix}$$

The dexterity matrix for the robot is obtained as follows:

$$\boldsymbol{J}\boldsymbol{J}^T = \begin{bmatrix} 2s_2^2 & \sqrt{2}s_2(\sqrt{2}c_2 + 1) \\ \sqrt{2}s_2(\sqrt{2}c_2 + 1) & (\sqrt{2}c_2 + 1)^2 + 1 \end{bmatrix}$$

As observed, both the Jacobian matrix and the dexterity matrix in the robot are solely functions of the second joint angle θ_2, and their intrinsic characteristics change only by varying this joint angle. To determine the isotropic condition in the robot, it is necessary to find the eigenvalues of this matrix and set them equal to one. The characteristic equation of the robot's dexterity matrix can be formulated as follows:

$$\lambda^2 - 2\lambda(\sqrt{2}c_2 + 2) + 2s_2^2 = 0$$

The eigenvalues of the dexterity matrix become equal to one only if,

$$s_2^2 = 1/2 \quad \text{and} \quad c_2 = -1/\sqrt{2}$$

Given that the characteristic equation of the dexterity matrix is not a function of the angle θ_1, isotropic conditions in this robot whose dimensions are inspired from nature, occur for all θ_1 angles when $\theta_2 = \pm 3\pi/4$. The location of the isotropic state in the robot is depicted as a complete circle in Figure 7.10. Considering the conditions of isotropy and the positioning of the human arm, it becomes evident that in activities such as handwriting, painting, and surgery, where dexterity in hand movements is crucial, a person naturally places his/her arm close to this configuration based on experience. Additionally, note that if the ratio of the robot arm lengths is not mimicked from the biological designs and is not equal to $\sqrt{2}$, isotropy features are not found at any point within the workspace.

The singularity condition in the robot occurs when the robot arms are either fully extended or fully retracted, and the robot's end-effector is at the boundary of its workspace.

7.3.3 Dexterity Measure

As mentioned in the previous section, the characteristics of the dexterity matrix of a robot can vary across a wide spectrum of states, from isotropic conditions to singularities, in different configurations. Various metrics have been defined in the field of robotics to quantitatively express different aspects of the manipulability matrix. Among them, the "Dexterity Measure" has gained more popularity in robotics literature [75, 80]. In the following, we elaborate on an examination of this metric.

The determinant of the dexterity matrix, according to Equation 7.31, is obtained from the product of the eigenvalues of this matrix. However, the eigenvalues of the dexterity matrix $\boldsymbol{J}\boldsymbol{J}^T$ are directly related to the singular values of the Jacobian matrix, as expressed in the following equation.

$$\sigma_i(\boldsymbol{J}) = \sqrt{\lambda_i(\boldsymbol{J}\boldsymbol{J}^T)} \tag{7.32}$$

This relationship holds even for non-square Jacobian matrices. Thus, without the need to compute the product of the Jacobian matrix and its transpose $\boldsymbol{J}\boldsymbol{J}^T$, the dexterity measure can be determined by calculating the singular values of the Jacobian matrix.

With this introduction, the dexterity measure μ is defined, based on the singular values of the Jacobian matrix $\boldsymbol{J}(\boldsymbol{q})$, as follows.

$$\begin{aligned} \mu &= \sqrt{\det(\boldsymbol{J}\boldsymbol{J}^T)} = \sqrt{\lambda_1 \cdot \lambda_2 \cdots \lambda_n} \\ &= \sigma_1 \cdot \sigma_2 \cdots \sigma_n \end{aligned} \tag{7.33}$$

where λ_i are the eigenvalues of the dexterity matrix $\boldsymbol{J}\boldsymbol{J}^T$, and σ_i are the singular values of the Jacobian matrix $\boldsymbol{J}$. Since the Jacobian matrix depends on the robot's structure and configuration, the defined dexterity measure is a function of the robot's structure and configuration, as well. If the dexterity measure μ tends toward one, the robot moves toward its isotropic configuration.

On the other hand, when the dexterity measure μ tends toward zero, at least one of the singular values of the Jacobian matrix approaches zero. This occurs when the robot is close to its singular configuration. In this case, the rank of the Jacobian matrix becomes smaller than the number of degrees of freedom, i.e., $rank(\boldsymbol{J}(\boldsymbol{q})) < n$. Thus, avoiding singularity is manifested by keeping this measure away from zero. In redundantly actuated robots, due to the added column to the Jacobian matrix, the tendency toward singularity is reduced, and essentially, robots have a better dexterity measure. This has been qualitatively demonstrated in examples of musculoskeletal structures studied earlier in this chapter.

In addition to the eigenvalues of the dexterity matrix, the eigenvectors of this matrix also play an important role in the analysis of the dexterity ellipsoid of a robot. To analyze this feature in fully actuated robots, where the Jacobian matrix is square, it is not necessary to directly calculate the eigenvectors of the manipulability matrix $\boldsymbol{J}\boldsymbol{J}^T$. This is because the eigenvectors of the dexterity matrix can be obtained by decomposing the singular value matrix of the Jacobian.

Let us examine the norm relation between the joint velocities, and that of the end-effector, relating it to the manipulability measure.

$$\begin{aligned} \|\dot{\boldsymbol{q}}\|^2 &= \dot{\boldsymbol{\mathcal{X}}}^T(\boldsymbol{J}\boldsymbol{J}^T)^{-1}\dot{\boldsymbol{\mathcal{X}}} \\ &\le \frac{1}{\|\boldsymbol{J}\boldsymbol{J}^T\|}\|\dot{\boldsymbol{\mathcal{X}}}\|^2 \\ &\le \frac{1}{\mu^2}\|\dot{\boldsymbol{\mathcal{X}}}\|^2 \end{aligned} \tag{7.34}$$

Therefore, the bounds on end-effector velocities can be determined based on the manipulability measure as follows:

$$\|\dot{\boldsymbol{\mathcal{X}}}\| \ge \mu\|\dot{\boldsymbol{q}}\| \tag{7.35}$$

The physical interpretation of this relationship in singular and isotropic configurations can be elucidated as follows. In a singular configuration, where $\mu = 0$, it implies that in certain directions within the workspace, the end-effector velocity will be zero. In other words, movement in those directions is not accessible, and this condition usually occurs at the boundary of the workspace. In contrast, in an isotropic configuration where μ tends toward unity, applying uniform joint velocities results in uniform end-effector velocities in all directions. This signifies that the robot exhibits desirable dexterity, allowing for accessible motion in all directions in its workspace.

In addition to the magnitude of velocities, their directions, interpreted by the eigenvectors of the dexterity matrix, play a crucial role in the analysis of the dexterity ellipsoid.

For this purpose, consider the singular value decomposition of the Jacobian matrix $\boldsymbol{J}$ with dimensions $n \times m$ as follows:

$$\boldsymbol{J} = \boldsymbol{U}\,\boldsymbol{\Sigma}\,\boldsymbol{V}^T \tag{7.36}$$

where matrices $\boldsymbol{U}$ and $\boldsymbol{V}$ are given by

$$\boldsymbol{U} = \begin{bmatrix}\boldsymbol{u}_1 & \boldsymbol{u}_2 & \cdots & \boldsymbol{u}_n\end{bmatrix}_{n\times n}, \quad \boldsymbol{V} = \begin{bmatrix}\boldsymbol{v}_1 & \boldsymbol{v}_2 & \cdots & \boldsymbol{v}_m\end{bmatrix}_{m\times m} \tag{7.37}$$

are orthogonal matrices, and the matrix $\boldsymbol{\Sigma}$ can be expressed in terms of the Jacobian matrix singular values as:

$$\boldsymbol{\Sigma} = \left[\begin{array}{cccc|c} \sigma_1 & 0 & \cdots & 0 & \mathbf{0} \\ 0 & \sigma_2 & \cdots & 0 & \mathbf{0} \\ \vdots & & \ddots & \vdots & \vdots \\ 0 & 0 & \cdots & \sigma_n & \mathbf{0} \end{array}\right]_{n\times m} \tag{7.38}$$

Efficient numerical methods have been developed for the decomposition of singular value matrices and are employed in engineering software tools[1].

Suppose the maximum singular value of the Jacobian matrix in a specific configuration is denoted as $\sigma_{\max}$ along the direction $\boldsymbol{u}_{max}$, and similarly, the minimum singular value is denoted as $\sigma_{\min}$ along the direction $\boldsymbol{u}_{min}$. It can be demonstrated that these directions are aligned with the eigenvectors corresponding to the maximum and minimum singular values of the dexterity matrix.

$$\boldsymbol{u}_{max} = \boldsymbol{v}_{max}(\boldsymbol{J}\boldsymbol{J}^T) \;, \quad \boldsymbol{u}_{min} = \boldsymbol{v}_{min}(\boldsymbol{J}\boldsymbol{J}^T) \tag{7.39}$$

With this note, the maximum and minimum velocities of the robot's end-effector can be obtained along the directions of these eigenvectors.

$$\dot{\mathcal{X}}_{max} = |\sigma_{max}|\|\dot{\boldsymbol{q}}\| \rightarrow \text{ along } \boldsymbol{u}_{max} \tag{7.40}$$

$$\dot{\mathcal{X}}_{min} = |\sigma_{min}|\|\dot{\boldsymbol{q}}\| \rightarrow \text{ along } \boldsymbol{u}_{min} \tag{7.41}$$

In this way, not only the bounds of maximum and minimum velocities of the robot's end-effector can be determined, but also the directions of maximum and minimum dexterity ellipsoid axes can be obtained using these relationships. The importance of utilizing the dexterity measure in the quantitative and qualitative analysis of the features of the dexterity matrix has been emphasized in these equations. Note that for determining the maximum and minimum singular values and their corresponding directions, it is sufficient to perform the singular value decomposition of the Jacobian matrix, and there is no need to carry out these calculations on the more complex dexterity matrix $\boldsymbol{J}\boldsymbol{J}^T$.

Example 7.3 (Planar 2R Robot) *Consider the planar 2R robot shown in Figure 7.10. The lengths of its two links are chosen as inspired by the aspect ratio of human arms equal to $a_1 = \sqrt{2}, a_2 = 1$. For $\theta_2 = \pi/2$, find the bounds and directions of the maximum and minimum velocities of the robot's end-effector, and plot the dexterity ellipsoid in this configuration.*

Solution: Previously, the Jacobian matrix with respect to the coordinate system $\{2\}$ for the 2R robot was derived and reported in Equation 7.4. By substituting the lengths of the links and setting $\theta_2 = \pi/2$, the Jacobian matrix for the robot simplifies to:

$$^2\boldsymbol{J}(q) = \begin{bmatrix}\sqrt{2} & 0 \\ 1 & 1\end{bmatrix}$$

[1] Such as the `svd` command in `MATLAB`®

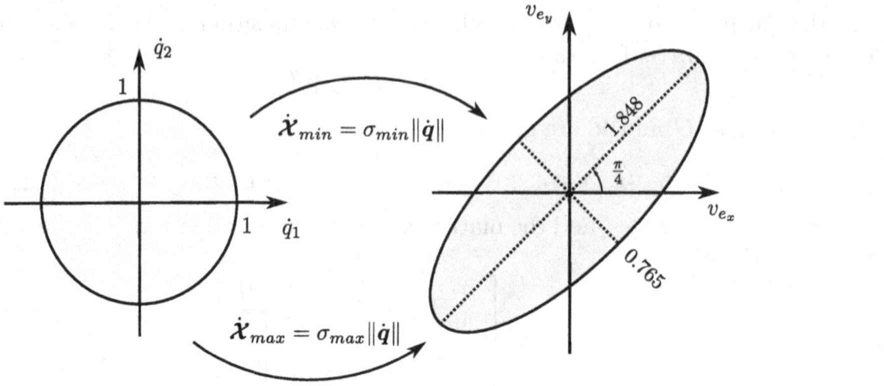

FIGURE 7.11
Characteristics of the dexterity ellipsoid in the 2$\underline{\text{R}}$ robot.

The singular value decomposition of this matrix is calculated as follows:

$$\boldsymbol{U} = \frac{1}{\sqrt{2}}\begin{bmatrix} -1 & -1 \\ -1 & 1 \end{bmatrix}, \quad \boldsymbol{\Sigma} = \begin{bmatrix} 1.848 & 0 \\ 0 & 0.765 \end{bmatrix}$$

As a result, the maximum velocity bound in this configuration is obtained as

$$\sigma_{max} = 1.848 \rightarrow \quad \text{along} \quad \begin{bmatrix} 1 \\ 1 \end{bmatrix}$$

while the minimum velocity bound in this mapping is,

$$\sigma_{min} = 0.765 \rightarrow \quad \text{along} \quad \begin{bmatrix} -1 \\ 1 \end{bmatrix}$$

and the dexterity ellipsoid in this configuration is shown in Figure 7.11. The dexterity measure in this configuration is equal to $\mu = \sigma_1 \cdot \sigma_2 = \sqrt{2}$.

7.3.4 Manipulability Measure

The dexterity measure encompasses all the singular values of the Jacobian matrix, which are associated with the dexterity matrix ellipsoid in the robot. However, as this measure is defined as the product of all the singular values of the Jacobian matrix, it is typically not dimensionless. Its value can exceed one, and at times, it could become significantly large. Additionally, assessing isotropy with this measure faces the challenge that, while individual singular values may not be unitary, their distribution might result in their product equaling one. In such a scenario, even though the dexterity measure is equal to one, as anticipated in an isotropic configuration, a uniform distribution in the robot's end-effector velocities may not be present.

To address these issues, a manipulability measure has been proposed in the robotic literature to assess the dexterity status of the robot. This measure is based on the condition number of the Jacobian matrix and can be utilized for underactuated, fully actuated, and redundant robots. The condition number of a matrix is defined as the ratio of its maximum singular value to its minimum singular value.

$$\text{Condition Number of } \boldsymbol{J}(\boldsymbol{q}) = \frac{\sigma_{max}(\boldsymbol{J}(\boldsymbol{q}))}{\sigma_{min}(\boldsymbol{J}(\boldsymbol{q}))} \tag{7.42}$$

In general, the condition number of a matrix can take values from 1 to infinity. If the condition number is close to 1, the singular values of the matrix are close to each other, indicating a near-isotropic condition. As the condition number increases, the ratio of the maximum singular value to the minimum singular value grows, and in the limit where the matrix becomes singular, the condition number approaches infinity. The use of the matrix condition number in linear algebra reduces numerical computation errors on matrices. When a matrix has a large condition number, appropriate numerical methods must be employed in calculations involving the matrix inverse and its counterparts to maintain accuracy [81].

With this explanation, to define a suitable dimensionless measure for assessing robot dexterity that ranges between zero and one, the reciprocal of the matrix condition number of the Jacobian matrix is utilized. This way, the manipulability measure in a robot is defined as the inverse of the matrix condition number of its Jacobian matrix at each configuration.

$$\kappa(\boldsymbol{J}(\boldsymbol{q})) = \frac{\sigma_{min}(\boldsymbol{J}(\boldsymbol{q}))}{\sigma_{max}(\boldsymbol{J}(\boldsymbol{q}))} \tag{7.43}$$

In this definition, $\sigma_{min}(\boldsymbol{J})$ and $\sigma_{max}(\boldsymbol{J})$ represent the minimum and maximum singular values of the Jacobian matrix of the robot, respectively. With this definition, the manipulability measure $\kappa(\boldsymbol{J})$ is a dimensionless quantity ranging between zero and one. A value close to zero indicates a singular configuration, while a value close to one signifies an isotropic condition in the robot.

The manipulability measure κ is, like other dexterity measures in a robot, a function of the robot's structure and its configuration $\boldsymbol{q}$, changing from one point to another in the workspace. Therefore, with knowledge of the robot's kinematic structure, the Jacobian matrix can be determined, and the dexterity measure can be obtained at any point in the workspace. It is evident that if the robot is designed properly, there will be a region in the workspace where the dexterity measure is suitably set close to one, while near the workspace boundaries, this measure tends to zero.

In addition to describing isotropy and singularity, the manipulability measure provides information about the directions of maximum and minimum dexterity through the vectors associated with the maximum and minimum singular values, denoted as $\boldsymbol{u}_{max}$ and $\boldsymbol{u}_{min}$. Although, unlike other dexterity measures, the definition of the manipulability measure does not explicitly consider other ellipsoidal axes, these directions can still be determined using the singular value decomposition of the Jacobian matrix.

7.3.5 Global Measures

As we have seen, dexterity measures in robots are functions of the robot's structure and its configuration $\boldsymbol{q}$, changing from one point to another in the workspace. If we intend to use these measures to optimize the kinematic structure of a robot and determine optimal kinematic parameters, we need to examine the condition of these measures across the entire workspace, instead of focusing on specific points. To address this, global performance measures have been introduced in the robotics literature [82], defined by averaging the desired measure over the entire robot's workspace.

In calculating global performance measures, the integral of the relevant measure is obtained over the entire workspace, while divided by the volume of the workspace. For example, if we consider the manipulability measure, the global manipulability measure $G\kappa$ is defined as follows.

$$G\kappa(\boldsymbol{J}) = \frac{A}{B} \tag{7.44}$$

where,

$$A = \int_W \kappa(\boldsymbol{q})\, dW, \;\; B = \int_W dW \tag{7.45}$$

In this definition, A is computed by integrating the desired metric over the entire robot workspace W, and B determines the volume of the robot's workspace. The average manipulability index is a measure of the robot's point-to-point manipulability throughout the entire workspace. Given that the manipulability index itself is dimensionless and lies in the range between zero and one, the global manipulability measure inherits these characteristics.

A similar definition can be applied to other global measures; it is sufficient to substitute the desired metric value for κ in Equation 7.45. If the metric of interest is dimensionless, its global measure will also be dimensionless. Additionally, in some applications, it is not necessary to compute the metric over the entire workspace; instead, if the goal is for the robot to exhibit dexterity in a specific region of the workspace, compute the metric of interest only in that particular region. In practical applications, when performing numerical computations to evaluate global measures, it is common to discretize the desired workspace into a uniform grid of points. The summation of the metric at each grid point can then be employed as an alternative to integration.

By defining the global measure, it can be utilized in optimizing the desired robot structure and determining optimal parameters. In simpler structures, finding the optimal point can be achieved by setting the derivative of the global measure with respect to the design parameters $\boldsymbol{h}$ to zero.

$$\frac{\partial G\kappa}{\partial \boldsymbol{h}} = 0 \tag{7.46}$$

In complex structures where the number of design parameters increases, numerical optimization methods can be employed. For a better understanding, let's consider an illustrative case of analytical optimal design taken from the reference [82], as presented in the following example.

Example 7.4 (Planar 2R Robot) *Consider the planar 2R robot depicted in Figure 7.10. Determine the optimal ratio of the two robot arm lengths, denoted by $\alpha = a_2/a_1$, over the working range $\theta_2 = [0, \pi]$, using the criterion of global manipulability. Compare this optimal ratio with a value inspired by nature, $\alpha = \sqrt{2}/2$, for reference.*

Solution: Previously, the Jacobian matrix for the planar 2R robot with respect to the coordinate system $\{2\}$ has been derived and reported in Equation 7.4. The determinant of the Jacobian matrix is also presented in Equation 6.13, which is given by $\det(\boldsymbol{J}) = a_1 a_2 s_2$, and it is a function of the robot link lengths and the angle θ_2. The manipulability criterion for the robot can be obtained based on the singular values of the Jacobian matrix [83].

$$\kappa = \frac{2a_1 a_2 s_2}{a_1^2 + a_2^2 + 2a_1 a_2 c_2}$$

Substitute the parameter representing the ratio of the lengths of the two links, $\alpha = a_1/a_2 > 0$, into this equation:

$$\kappa = \frac{2s_2}{1/\alpha + 2\alpha + 2c_2}$$

The value of the manipulability criterion for the robot over the range $\theta_2 = [0, \pi]$ and for various values of α is depicted in Figure 7.12. As observed, the criterion reaches unity only for $\alpha = \sqrt{2}/2$ in a portion of the workspace, indicating the robot's isotropic configuration. This scenario is explored in detail in Example 7.2.

Now, let's obtain the global manipulability measure in this workspace. For this purpose, the necessary integral can be computed with respect to the joint variables for $\theta_1 = [0, 2\pi]$ and $\theta_2 = [0, \pi]$, as follows.

$$B = \int_{\theta_1=0}^{2\pi} \int_{\theta_2=0}^{\pi} a_1 a_2 s_2 d\theta_2 d\theta_1 = 4\pi a_1 a_2$$

and

$$G\kappa = \frac{1}{4\pi a_1 a_2} \int_{\theta_1=0}^{2\pi} \int_{\theta_2=0}^{\pi} \left(\frac{s_2}{1/\alpha + 2\alpha + 2c_2} \right) a_1 a_2 s_2 d\theta_2 d\theta_1$$

After some simplification, we arrive at the following expression:

$$G\kappa = \int_{\theta_2=0}^{\pi} \left(\frac{2s_2^2}{1/\alpha + 2\alpha + 2c_2} \right) d\theta_2$$

To determine the optimal point, differentiate this expression with respect to α and set it equal to zero.

$$\frac{\partial G\kappa}{\partial \alpha} = (2 - 1/\alpha^2) \int_{\theta_2=0}^{\pi} \left(\frac{s_2^2}{(1/\alpha + 2\alpha + 2c_2)^2} \right) d\theta_2 = 0$$

Considering that the integral converges to a positive constant value, the expression is only equal to zero when:

$$2 - 1/\alpha^2 = 0 \rightarrow \alpha = \sqrt{2}/2$$

As observed, the result of optimizing the robot based on the manipulability criterion aligns with a ratio inspired by nature.

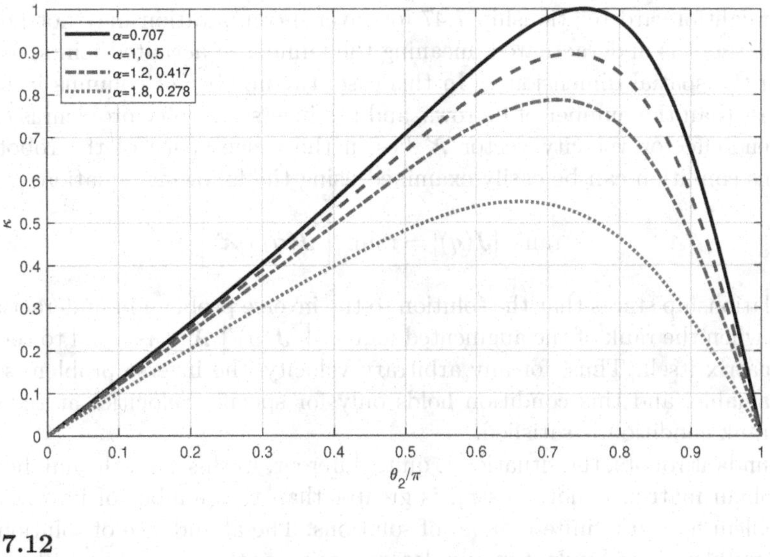

FIGURE 7.12
Manipulability measure values in the 2$\underline{\text{R}}$ robot for various α over the range of $\theta_2 = [0, \pi]$.

7.4 Inverse Problems

7.4.1 In Differential Kinematics

As seen the Jacobian matrix describes the mapping between the end-effector velocity and the joint velocities. Thus, if the joint velocities of the robot are given, the end-effector velocity can be obtained using the Jacobian matrix. Now, if the end-effector velocity of the robot is given and we want to find the joint velocities, we need to solve the inverse problem in differential kinematics. Unlike solving the inverse kinematics for serial robots, which can be quite complex, solving the inverse problem of differential kinematics for robots is straightforward and relatively easy.

In robots with full actuation, the Jacobian matrix is square, and solving the inverse problem in differential kinematics involves determining the inverse of the Jacobian matrix.

$$\dot{\boldsymbol{\mathcal{X}}} = \boldsymbol{J}(\boldsymbol{q})\dot{\boldsymbol{q}} \;\rightarrow\; \dot{\boldsymbol{q}} = \boldsymbol{J}^{-1}(\boldsymbol{q})\dot{\boldsymbol{\mathcal{X}}} \tag{7.47}$$

This problem will have a unique solution if the robot is not in a singular state. By examining the inverse velocity problem, we can also study the solution to the inverse acceleration problem. To do so, take derivatives on both sides of Equation 7.47:

$$\ddot{\boldsymbol{\mathcal{X}}} = \boldsymbol{J}(\boldsymbol{q})\ddot{\boldsymbol{q}} + \dot{\boldsymbol{J}}(\boldsymbol{q})\dot{\boldsymbol{q}} \tag{7.48}$$

In the inverse acceleration problem, the end-effector acceleration $\ddot{\boldsymbol{\mathcal{X}}}$ is given, and the goal is to determine $\ddot{\boldsymbol{q}}$ instantaneously. To achieve this, rewrite Equation 7.48 in terms of the unknown $\ddot{\boldsymbol{q}}$ and substitute $\dot{\boldsymbol{q}}$ from Equation 7.47 into it.

$$\ddot{\boldsymbol{q}} = \boldsymbol{J}^{-1}(\boldsymbol{q})\left[\ddot{\boldsymbol{\mathcal{X}}} - \dot{\boldsymbol{J}}(\boldsymbol{q})\boldsymbol{J}^{-1}(\boldsymbol{q})\dot{\boldsymbol{\mathcal{X}}}\right] \tag{7.49}$$

It is observed that the inverse acceleration relationship in robots is much more complex than the velocity counterpart, and determining it analytically using Equation 7.49 is only feasible for fully actuated robots.

Going back to the inverse velocity problem, it is noted that if the robot is not fully actuated or enters a singular configuration, the Jacobian matrix may not be invertible, and the straightforward relationship 7.47 for inverse computation may not hold. Initially, consider the robot is underactuated, meaning the number of actuated joints, denoted as m, is less than the spatial dimension n. In this case, the number of columns in the Jacobian matrix is less than the number of its rows, and the inverse velocity problem is only solvable when the end-effector velocity vector $\dot{\boldsymbol{\mathcal{X}}}$ lies in the range space of the robot's Jacobian matrix. This condition can be easily examined using the following equation.

$$\text{rank}\,[\boldsymbol{J}(\boldsymbol{q})] = \text{rank}\left[\;\boldsymbol{J}(\boldsymbol{q}) \mid \dot{\boldsymbol{\mathcal{X}}}\;\right] \tag{7.50}$$

This relationship states that the solution to the inverse problem in underactuated robots exists only when the rank of the augmented matrix $\left[\;\boldsymbol{J}(\boldsymbol{q}) \;\middle|\; \dot{\boldsymbol{\mathcal{X}}}\;\right]$ is equal to the rank of the Jacobian matrix itself. Thus, for any arbitrary velocity, the inverse problem solution may not be attainable, and this condition holds only for specific velocities at the end-effector where the rank condition is satisfied.

For redundant robots, the situation is quite different. In this case, the number of columns in the Jacobian matrix, denoted as m, is greater than the number of its rows n, and the inverse problem has an infinite number of solutions. The abundance of solutions in solving the inverse problem in redundant manipulators motivates the researchers to focus on finding optimal solutions. This means that among the numerous solutions, one can choose a solution that not only solves the inverse problem but also optimizes a desired criterion in the robot. This criterion could involve minimizing joint motion energy, avoiding obstacles, enhancing dexterity, and similar objectives. Given the practical importance of this topic in the design and implementation of redundant robots, detailed discussions on these matters will be provided in Section 7.5.

7.4.2 In Force Mapping

Similar to the inverse velocity problem in differential kinematics, the inverse problem in force mapping can also be explored. Note that the transpose of the Jacobian matrix, introduced as the robot's structural matrix, describes the mapping between the forces/torques vector in the joint space to the end-effector wrenches applied to the robot. Therefore, if the robot is fully actuated, one can also determine the joint forces/torques applied to the robot in terms of the wrenches exerted on the end-effector using the inverse of the structural matrix.

$$\boldsymbol{\tau} = \boldsymbol{S}(\boldsymbol{q})\boldsymbol{\mathcal{F}} \quad \rightarrow \quad \boldsymbol{\mathcal{F}} = \boldsymbol{S}^{-1}(\boldsymbol{q})\boldsymbol{\tau} \tag{7.51}$$

Nevertheless, if the robot is not fully actuated or encounters a singular configuration, the Jacobian matrix becomes uninvertible, rendering the direct relationship for inverse computation inapplicable. It's worth noting that, given the structural matrix of the robot is derived from the transpose of the Jacobian matrix, it typically exhibits a different dimension than the Jacobian matrix. Consequently, resolving the inverse problem for joint velocities may not align seamlessly with its counterpart in force space.

In the inverse force problem, considering that the transpose of the Jacobian matrix is under consideration, solving the inverse problem in redundant and underactuated cases switches from the previous scenario. Initially, assume that the robot is redundant, meaning the number of actuated joints, denoted as m, is greater than the spatial dimension n. In this case, the number of columns in the structural matrix is less than the number of its rows, and the inverse force problem is only solvable when the joint torque vector $\boldsymbol{\tau}$ lies in the range space of the structural matrix. This condition can be examined similarly to before using the following relationship:

$$\text{rank}\,[\boldsymbol{S}(\boldsymbol{q})] = \text{rank}\,[\,\boldsymbol{S}(\boldsymbol{q}) \mid \boldsymbol{\tau}\,] \tag{7.52}$$

Thus, for every wrench vector, it's not possible to obtain a solution to the inverse problem for arbitrary joint torques, and the problem is only solvable for specific torques that satisfy this condition.

In the case of underactuated robots, the scenario diverges. Here, the structural matrix has more rows than columns, resulting in an inverse problem with an infinite number of solutions. As with resolving the inverse velocity problem, one can select a solution from the myriad options that not only addresses the inverse problem but also optimizes various criteria. These criteria may include minimizing the energy needed in joint torques or achieving other specific objectives. Addressing the optimal force problem in this scenario parallels the process of tackling the optimal velocity problem in redundant robots, a discussion that will be thoroughly explored in the upcoming section.

7.5 Redundancy Resolution

Redundancy in actuation is a crucial feature in the performance of robots. In many industrial robots, the number of actuators used is equal to the number of degrees of freedom $n = 6$ to simplify robot design and reduce the number of actuators and the cost of manufacturing. However, this can limit the dexterity of the robot's end-effector and fail to provide optimal conditions. On the other hand, as illustrated in the examples in Section 7.3, many biological musculoskeletal structures utilize redundancy in actuation to perform their movements more dexterously and avoid singularities.

In general, for robots with redundant actuation, hereafter referred to as redundant robots, the number of actuated joints m exceeds the required degrees of freedom for motion n $(m > n)$. These additional actuators are employed to create better kinematic conditions in the robot's structure. Expected objectives in redundant robots include singularity avoidance [62], optimal posture [84], obstacle avoidance [85], and satisfaction of motion constraints in the joints [86], among others. The methods that make optimal use of additional actuation in the considered robot are known as "Redundancy Resolution."

Due to its practical applications, redundancy resolution has received significant attention from researchers in the past three decades. Many of these methods are based on the pseudo-inverse of the Jacobian matrix and least squares error minimization. This approach was first introduced by Whitney in 1969 [87] and further developed by Liégeois in 1977 [88]. In this method, the Euclidean norm of the optimization variable is minimized, which from a physical standpoint, is equivalent to minimizing the required energy. This method has proven to be highly applicable and has been utilized in various contexts, such as torque minimization in serial robot actuators [89], optimization based on the weighted pseudo-inverse of the Jacobian matrix [90], optimization of elliptical dexterity features [91, 92], and determining the optimal configuration of the robot in joint and task spaces [93, 86], among many other applications.

The redundancy resolution methods are intricately linked to optimization solvers in mathematics, both analytically and numerically. In analytical methods, Lagrange multipliers and KKT[1] conditions are used to formulate the optimization problem based on the first derivative of the optimization criterion under constraints. In numerical methods, local minimum of the optimization criterion is computed using Newton-Raphson or steepest descent methods in the presence of linear or nonlinear equality and inequality constraints. The details of these methods require further exploration beyond the scope of this book. Interested readers are recommended to study books on nonlinear programming [94]. In this section, we will provide a concise definition of the optimization problem and its numerical solution in the context of the discussed serial robots' applications.

7.5.1 Optimization Formulation

In the inverse problem in velocity space, suppose the robot's structure and its Jacobian matrix are given. Additionally, assume the desired end-effector twist rate $\dot{\mathcal{X}}$ is provided, and the goal is to determine the joint velocities of the robot. If the robot is redundant, with the number of actuators m greater than the degrees of freedom in the robot's space n, then the Jacobian matrix is not square. Consequently, solving the inverse problem cannot be easily obtained from the equation 7.47.

[1] Karush-Kuhn-Tucker

In this case, the number of unknowns in the problem is greater than the number of equations, resulting in countless solutions to the inverse problem. Among these numerous solutions, the pseudo-inverse of the Jacobian matrix yields a basic solution to this problem. In this pseudo-inverse solution, the redundancy problem is solved in a way that in addition to satisfying the relation $\dot{\boldsymbol{\mathcal{X}}} = \boldsymbol{J}(\boldsymbol{q})\dot{\boldsymbol{q}}$, the Euclidean norm of joint velocities $\|\dot{\boldsymbol{q}}\|_2$ is minimized. Thus, the solution to the redundancy problem in its basic form is defined as follows.

Basic Optimization Problem:

Find the optimal value of $\dot{\boldsymbol{q}}$ in a way that the velocity mapping relation $\dot{\boldsymbol{\mathcal{X}}} = \boldsymbol{J}(\boldsymbol{q})\dot{\boldsymbol{q}}$ holds while minimizing the Euclidean norm of joint velocities $\|\dot{\boldsymbol{q}}\|_2$.

The solution to this basic problem is obtained using the pseudo-inverse of the Jacobian matrix.

$$\dot{\boldsymbol{q}}_0 = \boldsymbol{J}(\boldsymbol{q})^{\dagger}\dot{\boldsymbol{\mathcal{X}}}, \tag{7.53}$$

where $\boldsymbol{J}(\boldsymbol{q})^{\dagger}$ is the right pseudo-inverse of the Jacobian matrix, obtained from the following relationship:

$$\boldsymbol{J}^{\dagger} = \boldsymbol{J}^T(\boldsymbol{J}\boldsymbol{J}^T)^{-1}. \tag{7.54}$$

Note that right pseudo-inverses are defined in rectangular matrices, where the number of columns is greater than the number of rows. Computing this inverse leads to determining the inverse of the dexterity matrix $\boldsymbol{J}\boldsymbol{J}^T$, which is square and invertible. It can also be easily verified that, by multiplying the right pseudo-inverse of the matrix $\boldsymbol{J}$ from the right, the identity matrix will be obtained.

$$\boldsymbol{J}\boldsymbol{J}^{\dagger} = \boldsymbol{J}\boldsymbol{J}^T(\boldsymbol{J}\boldsymbol{J}^T)^{-1} = \boldsymbol{I}. \tag{7.55}$$

By determining the base solution in this inverse problem, the set of all solutions can be obtained using the following relationship:

$$\dot{\boldsymbol{q}} = \dot{\boldsymbol{q}}_0 + \dot{\boldsymbol{q}}_n, \tag{7.56}$$

where,

$$\dot{\boldsymbol{q}}_n = (\boldsymbol{I} - \boldsymbol{J}^{\dagger}\boldsymbol{J})\boldsymbol{b}. \tag{7.57}$$

In this equation, $\boldsymbol{b} \in \mathbb{R}^n$ is an arbitrary array, and $(\boldsymbol{I} - \boldsymbol{J}^{\dagger}\boldsymbol{J})$ is nonzero. Using the fundamentals of linear algebra, it can be easily shown that $\dot{\boldsymbol{q}}_n$ lies in the null space of the Jacobian matrix $\dot{\boldsymbol{q}}_n \in \mathcal{N}(\boldsymbol{J})$. Conceptually, this means that there is a nonzero vector $\dot{\boldsymbol{q}}_n$ such that its product with the Jacobian matrix is equal to zero, and it has no effect on the end-effector velocity of the robot.

$$\boldsymbol{J}\dot{\boldsymbol{q}}_n = 0. \tag{7.58}$$

In this way, in redundant robots, some motions can be generated in the joints although the end-effector is motionless. This specific motion, lying in the null space of the Jacobian matrix, is referred to as "self-motion" in the robotics literature. This self-motion can be determined in a way that optimizes another criterion in the robot. For example, if obstacle avoidance is a concern, one can select a self-motion that maximizes the distance between the robot and obstacles. Similarly, if avoiding singularities is the goal, self-motion can be chosen to move the robot's configuration away from singularities and toward isotropy.

Determining optimal self-motion is a topic that can be addressed using analytical and numerical methods. Analytical methods are illuminating when the optimization criterion is relatively simple. Interested readers are recommended to study references [95, 96] for analytical methods. Here, we present a more general definition of the optimization problem, and its numerical solution will be discussed in subsequent sections.

To do this, first define the optimization criterion based on the requirements of the inverse problem. In numerical methods, there is no constraint on defining this criterion, and it can be a function of joint variables and their derivatives, $V(\boldsymbol{q}, \dot{\boldsymbol{q}})$, or workspace variables and their derivatives, $V(\boldsymbol{\mathcal{X}}, \dot{\boldsymbol{\mathcal{X}}})$. The only requirement is that this criterion aligns with the desired objective in the optimization problem. For example, if avoiding singularities is considered, the dexterity or manipulability measures, and even their weighted sum, can be used as the optimization criterion.

The relationship between joint and workspace variables can be introduced as an equality constraint in the optimization problem. For this purpose, the forward kinematic relationship of the robot is considered as the first equality constraint in the optimization problem, incorporating the robot's structure and its features into the formulation. Additionally, the Jacobian matrix relationship, as the mapping between the joint's and end-effector's velocities, is considered the second equality constraint in the optimization problem. If there are motion bounds in the joints or their velocities, or if there are bounds on the motion variables and their derivatives, they can be introduced as inequality constraints in the optimization problem. Thus, the general optimization problem for solving the inverse problem is defined as follows.

General Optimization Problem:

Find the optimal values for $\boldsymbol{q}$ and $\dot{\boldsymbol{q}}$ such that the criterion $V(\boldsymbol{q}, \dot{\boldsymbol{q}})$ in a robot with forward kinematics relation $\boldsymbol{\mathcal{X}} = \boldsymbol{f}_{fk}(\boldsymbol{q})$, and the velocity mapping relation of the Jacobian matrix $\dot{\boldsymbol{\mathcal{X}}} = \boldsymbol{J}(\boldsymbol{q})\dot{\boldsymbol{q}}$, is satisfied in a way that the joint variables and their derivatives are constrained within the specified upper and lower bounds.

$$\min_{\boldsymbol{q},\dot{\boldsymbol{q}}} V(\boldsymbol{q}, \dot{\boldsymbol{q}}), \quad \text{Subject to} \tag{7.59}$$

$$\begin{aligned} \boldsymbol{\mathcal{X}} &= \boldsymbol{f}_{fk}(\boldsymbol{q}) \\ \dot{\boldsymbol{\mathcal{X}}} &= \boldsymbol{J}(\boldsymbol{q})\dot{\boldsymbol{q}} \\ \boldsymbol{q}_{min} \leq\ & \boldsymbol{q} \leq \boldsymbol{q}_{max} \\ \dot{\boldsymbol{q}}_{min} \leq\ & \dot{\boldsymbol{q}} \leq \dot{\boldsymbol{q}}_{max} \\ & \vdots \end{aligned}$$

The numerical solution of this optimization problem has attracted the attention of researchers, and details can be found in many references such as [94, 97, 98]. In engineering software, one can easily obtain the solution to this optimization problem using classical methods and evolutionary algorithms[1]. In the following, we investigate the optimization problem for two different criteria in a redundant robot and will find their numerical solution.

[1]The `fmincon` command solves the classical problem, and the `ga` command solves it using evolutionary processing in the MATLAB software.

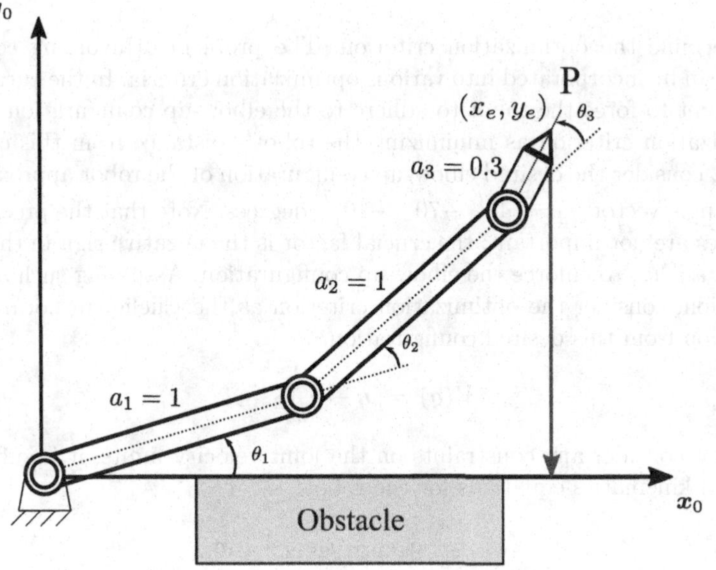

FIGURE 7.13
Redundancy resolution in the planar 3$\underline{\text{R}}$ robot for obstacle avoidance.

7.5.2 Obstacle Avoidance

Case Study 7.1 (Planar 3$\underline{\text{R}}$ Robot) *Consider the planar 3$\underline{R}$ robot shown in Figure 7.13, where the desired motion of its end-effector in the plane is along a vertical line. In this case, assume that the orientation of the robot's end-effector is not important, and only its motion along the desired vertical line matters. With this description, the robot has three actuated joints for maneuvering in the $x - y$ plane with two degrees of freedom and one degree of redundancy. Furthermore, assume that there is an obstacle in the robot's workspace arranged in such a way that if the robot is in the elbow-down configuration and attempts to perform this maneuver, it will collide with the obstacle. Using the redundant actuation, the goal is to find the joint variables such that the end-effector's motion follows the specified path without colliding with the obstacle.*

Solution: In this problem, the path of motion in the task space is given, and the motion path of the joint variables is unknown. Furthermore, the robot has one redundant degree of actuation that is intended to be used to avoid collision with the obstacle. Thus, this problem aligns with solving an inverse problem along with redundancy resolution and will lead to an obstacle avoidance solution. To solve this problem, we first formulate it as an optimization problem and then proceed to solve it using numerical methods.

Initially, determine the robot's trajectory in the workspace. Suppose the joint variables of the robot are initially set at angles $\boldsymbol{q}_0 = \begin{bmatrix} 20 & 30 & 20 \end{bmatrix}^T$ degrees. In this case, at the initial time, calculate the corresponding task space variables using the robot's forward kinematics as $\boldsymbol{\mathcal{X}}_0 = \begin{bmatrix} x_0 & y_0 \end{bmatrix}^T = \begin{bmatrix} 1.685 & 1.390 \end{bmatrix}^T$. Now, consider that within one second, the robot's end-effector reaches its final position with the values $\boldsymbol{\mathcal{X}}_f = \begin{bmatrix} 1.685 & 0 \end{bmatrix}^T$. This motion path can be described over time using a cubic polynomial as follows

$$\dot{x}(t) = 0 \quad ; \quad \dot{y}(t) = y_0(18t^2 - 6t). \tag{7.60}$$

Now, determine the optimization criterion. The problem of avoiding collisions with obstacles can be incorporated into various optimization criteria. In the current problem, it is sufficient to force the robot to adhere to the elbow-up configuration and consider the optimization criterion as minimizing the robot's distance from this configuration. To do this, consider the desired elbow-up configuration of the robot approximately with the joint angle vector $\boldsymbol{q}_d = \begin{bmatrix} 45 & -70 & -10 \end{bmatrix}^T$ degrees. Note that the precision of these angle values are not important; the crucial factor is the negative sign in the second and third joint angles, to enforce the elbow-up configuration. Assuming such a desired arm configuration, consider the optimization criterion as the Euclidean norm of the robot configuration from this desired configuration.

$$V(\boldsymbol{q}) = \|\boldsymbol{q} - \boldsymbol{q}_d\|_2. \tag{7.61}$$

If we do not consider any constraints on the joint velocity limits, it is sufficient to use the forward kinematics equations for the robot:

$$\begin{aligned} x &= a_1 c_1 + a_2 c_{12} + a_3 c_{123} \\ y &= a_1 s_1 + a_2 s_{12} + a_3 s_{123} \end{aligned}$$

and introduce its Jacobian matrix mapping as equality constraints in this optimization problem.

$$\begin{bmatrix} \dot{x} \\ \dot{y} \end{bmatrix} = \begin{bmatrix} -a_1 s_1 - a_2 s_{12} - a_3 s_{123} & -a_2 s_{12} - a_3 s_{123} & -a_3 s_{123} \\ a_1 c_1 + a_2 c_{12} + a_3 c_{123} & a_2 c_{12} + a_3 c_{123} & a_3 c_{123} \end{bmatrix} \begin{bmatrix} \dot{q}_1 \\ \dot{q}_2 \\ \dot{q}_3 \end{bmatrix}$$

Optimize the system at different positions along the trajectory. For example, discretize the robot's continuous motion path over one second into a hundred points with a sampling time of 0.01 seconds and solve the optimization problem numerically at each of these points. To compare optimal solutions with and without obstacle avoidance criteria, address the problem in two distinct scenarios.

In the initial scenario, neglect the requirement of obstacle avoidance and neglect positioning the robot in the elbow-up configuration within the optimization process. Instead, employ the pseudo-inverse Jacobian to directly solve the inverse differential kinematics problem. In this context, derive the joint velocities using the pseudo-inverse of the Jacobian and determine the angular motion of the joints accordingly. Subsequently, solve the optimization problem numerically using the criterion 7.61 at one hundred points within the robot trajectory, and compare the outcomes. To enhance result clarity, present only five instances among the hundred points, each within time intervals of 0.2 seconds.

The results of the base optimization solution are illustrated in Figure 7.14, and the optimized results with obstacle avoidance are shown in Figure 7.15. As observed in Figure 7.14, utilizing the Jacobian pseudo-inverse allows the robot's joints to reach the final configuration in the workspace with minimal motion from the initial state. Naturally, the robot will perform a maneuver in the elbow-down configuration, which involves collision with obstacles.

In the case where the obstacle avoidance optimization is solved considering the desired configuration of the arms in an elbow-up position, as depicted in Figure 7.15, the robot transitions from the elbow-down to elbow-up configuration in the initial moments and then follows the desired path without colliding with obstacles. It is evident that in this optimal response, the joint velocities are significantly larger than those in the base solution.

Now, let's assume that the joint velocities are constrained within the range $-1.2 \le \dot{q}_i \le 1.2$. If you introduce this limit as an inequality constraint in the optimization problem and resolve it, a different solution will be obtained, as shown in Figure 7.16. As observed in this figure, at the beginning of the motion and when the robot changes its state from elbow-down to elbow-up, high joint velocities are required, exceeding the speed limit. In the numerical solution of this problem, obstacle avoidance serves as the primary optimization criterion, and the velocities are constrained within the desired limit with a lower priority. Since Satisfying both objectives is not possible in this case, the robot's motion will slightly deviate from the desired vertical line, although it ensures the elbow-up configuration. Given that this robot has only one degree of redundancy, it is natural that multiple objectives cannot be satisfied in the presence of actuator joint speed constraints. If two degrees of redundancy are utilized, both criteria for optimal obstacle avoidance with bounded velocities could be achieved.

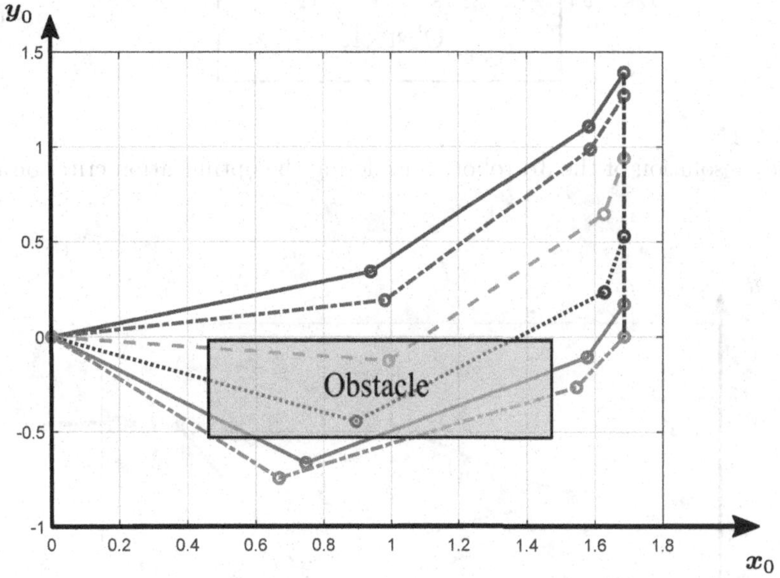

FIGURE 7.14
The base redundancy resolution of the 3R robot, obtained by pseudo-inverse of the Jacobian matrix.

7.5.3 Singularity Circumvention

Case Study 7.2 (Planar 3R Robot) *Now consider the 3R robot with the previous extension, with its end-effector moving in the plane along a vertical line. However, in this case, the initial configuration of the robot arms is chosen near a singularity, and the goal is to find joint variables that enable the end-effector's motion along the desired path with singularity circumvention.*

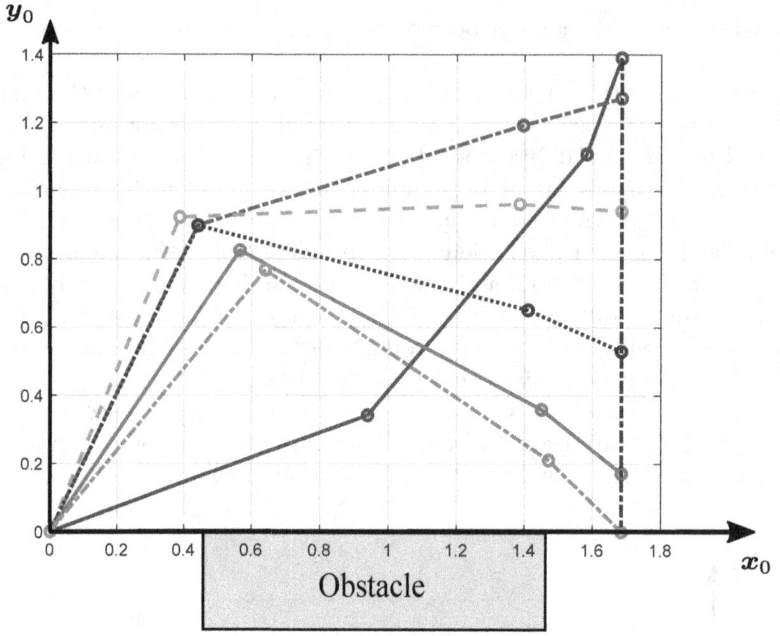

FIGURE 7.15
Redundancy resolution of the 3R̲ robot, considering the optimization criterion for obsyacle avoidance.

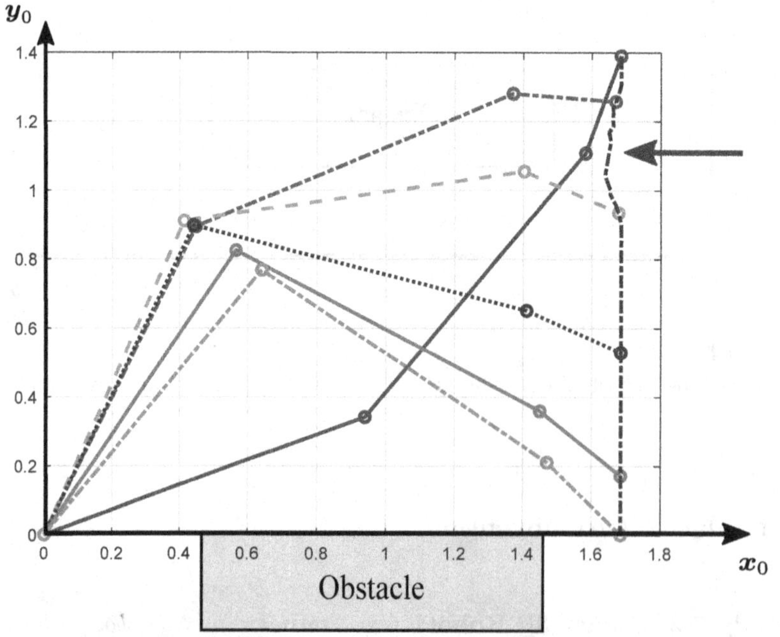

FIGURE 7.16
Solving the redundancy resolution problem in the 3R̲ robot, considering the obstacle avoidance optimization in the presence of joint speed constraints.

Solution: Firstly, determine the robot's motion path in the workspace. Assume that the robot's joint variables at the initial time are set at angles $\boldsymbol{q}_0 = \begin{bmatrix}-180 & -179 & 10\end{bmatrix}^T$ degrees. Note that the initial conditions place the robot arms in a configuration near a singularity, providing a test for the optimization method in this challenging setting. In this case, calculate its workspace variable at the initial time using the robot's direct kinematics as $\boldsymbol{\mathcal{X}}_0 = \begin{bmatrix}x_0 & y_0\end{bmatrix}^T = \begin{bmatrix}0.294 & 0.075\end{bmatrix}^T$. Now, consider that within one second, the robot's end-effector reaches its final position with the values $\boldsymbol{\mathcal{X}}_f = \begin{bmatrix}0.294 & 0\end{bmatrix}^T$. This motion path can be described over time using a cubic polynomial.

$$x(t) = x_0 \quad ; \quad y(t) = y_0(1 + 2t^3 - 3t^2), \tag{7.62}$$

Calculate the velocity of the robot's end-effector over time by taking the derivative of this equation.

$$\dot{x}(t) = 0 \quad ; \quad \dot{y}(t) = y_0(6t^2 - 6t). \tag{7.63}$$

Now, determine the optimization criterion. The singularity circumvention problem can be addressed by incorporating various measures of dexterity and manipulability in optimization. Here, let's assume the use of manipulability measures. Thus, consider the optimization criterion as follows:

$$V(\boldsymbol{q}) = -\kappa(\boldsymbol{J}) = -\frac{\sigma_{min}(\boldsymbol{J})}{\sigma_{max}(\boldsymbol{J})}. \tag{7.64}$$

Note that, considering the use of the `fmincon` function to determine the optimal response, the optimization criterion needs to be minimized. Therefore, $-\kappa$ is used. Finally, assume that the joint angles of the robot are constrained within the following limits into the optimization problem.

$$\begin{aligned} -190^o \leq \; & q_1, q_2 & \leq -170^o \\ -30^o \leq \; & q_3 & \leq 30^o \end{aligned}$$

Specify these constraints as inequality constraints in the optimization problem, and similar to before include the forward kinematics equations of the robot and its Jacobian matrix mapping as equality constraints. Solve this optimization problem at 100 different points along the path, and for better visualization of the results, report only five instances within time intervals of 0.2 seconds. The result of solving the basic optimization problem is shown in Figure 7.17, and the result of optimization for singularity circumvention is shown in Figure 7.18.

As seen in Figure 7.17, using the pseudo-inverse of the Jacobian matrix, the robot's joints move with the least possible motion from the initial state to the final state in the workspace. In this case, it is natural for the robot not to show significant changes in its configuration and remain close to the singularity.

In the case of optimizing the singularity circumvention problem using the manipulability criterion, as shown in Figure 7.18, the robot's movement quickly deviates from the singularity of the arms and then traverses the entire path with a more open angle between the two arms. By adjusting the constraints on the joint angles, a wider opening of the arms can be achieved in solving this constrained optimization problem.

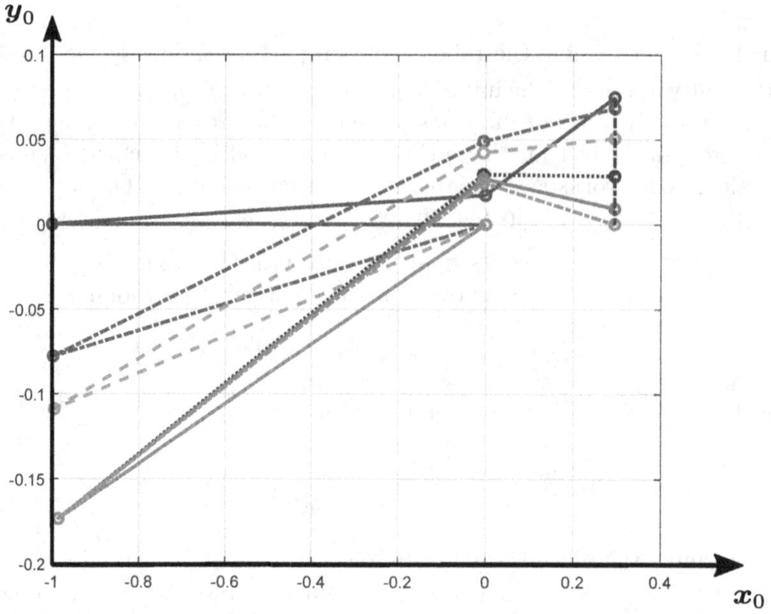

FIGURE 7.17
Solving the basic singularity circumvention problem in the planar 3$\underline{\text{R}}$ robot, considering the pseudo-inverse of Jacobian matrix.

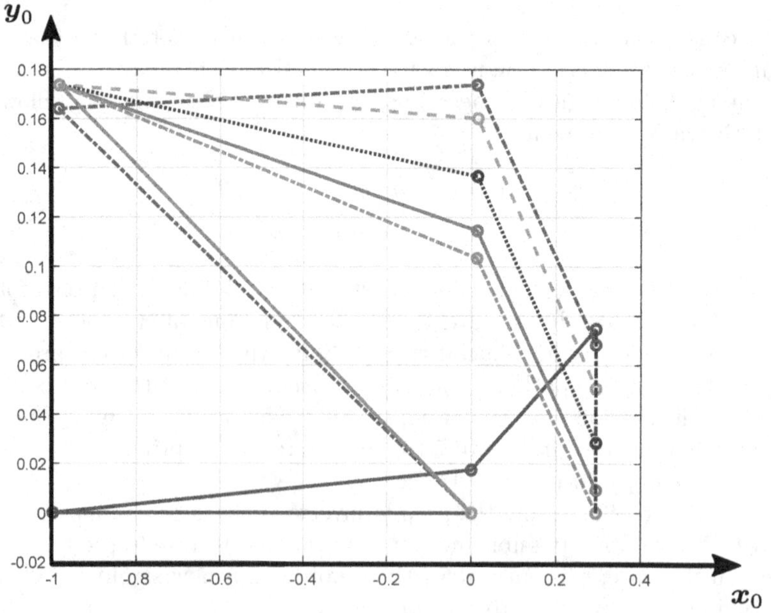

FIGURE 7.18
Optimizing the singularity circumvention problem in the planar 3$\underline{\text{R}}$ robot, considering singularity avoidance criteria in the presence of joint angle constraints.

7.6 Compliance and Stiffness in Robots

In the analysis of robot stiffness, we investigate the ratio of the applied forces at the robot's end-effector to the resulting displacement. To achieve this, consider that the robot's end-effector is in contact with the external environment and experiences deformation due to applied forces. If the robot is in a static state, the presence of flexibility in the robot's components and the application of forces lead to a small displacement in the robot's end-effector. This displacement is a function of the robot's structure and configuration, impacting the accuracy of the robot; therefore, its examination is crucial. However, it should be noted that the presence of compliance in robots is not only undesirable in some applications but also desirable, especially in scenarios where the robot interacts with humans.

The cause of compliance in robots are diverse. In the following, we will examine some of the main factors contributing to compliance in robots.

Robot Structure: The kinematic structure of a robot has a direct impact on its stiffness and compliance. Utilizing open kinematic chains in the robot's structure expands its workspace. Nevertheless, this advantage is accompanied by drawbacks such as heightened compliance and increased errors in tracking the end-effector's motion. Conversely, employing closed kinematic chains in the robot's structure boosts the robot's stiffness and enhances its capability to apply force to the moving platform. The analysis of the robot's structure's impact on stiffness characteristics can be analyzed through examination of the robot's Jacobian matrix.

Links: Industrial robots typically utilize rigid links to attain precise motion control. However, this comes at the cost of increased robot weight, which is undesirable in applications like space robotics. In such robots, lightweight links are commonly employed, inherently offering higher compliance and introducing greater flexibility into the robot's structure.

Joints: Robot joints are typically selected to be stiff enough to minimize unintended oscillations in the robot's motion. However, in applications such as robots that interact with humans, having compliance in robot joints is desirable, creating greater flexibility in interacting with humans. Therefore, in these types of robots, compliant joints are typically used.

Sensors and Actuators: A significant part of the robot components that cause compliance in the robot structure are the actuators and sensors of the robot. Hydraulic actuators and direct-drive systems inherently exhibit compliance. Various force sensors also have a compliant element that introduces compliance into the robot's structure.

Power Transmission One of the main factors leading to compliance in robot joints is the use of harmonic drive power transmission systems. Harmonic drive is a highly efficient type of gearbox that creates a very high transmission ratio in a compact volume. Many industrial robots use harmonic drives due to their efficiency. The harmonic drive employs a specific type of gear called "Flexspline," which possesses compliant properties. The high transmission ratio in this gearbox is achieved through the presence of this flexible element in the system. When using a harmonic drive, the robot joint becomes compliant, directly impacting the robot's overall compliance.

Control System: In addition to mechanical components, robot sensors, and actuators, the robot's control system also plays a significant role in determining the stiffness and compliance of the robot. Conventional control structures in robots utilize high control gains to ensure proper tracking and disturbance rejection in the system's output. The presence of high gains in the robot's control structure leads to an increase in the robot's stiffness. This may not be suitable when the robot interacts with humans, and reducing control gains can be employed to control the robot's compliance.

Due to the widespread adoption of harmonic drives in industrial robots, the joints of the robot exhibit compliance. This case is considered here for further exploration. If harmonic drives are used across all robot joints, applying a wrench to the robot's end-effector will result in minor displacements in the robot's end-effector. In this section, we will assess the stiffness or compliance of the robot and investigate the extent of displacement in the robot's end-effector when compliance is present in the robot joints.

7.6.1 Compliance and Stiffness Matrices

If we model the compliance in the i-th joint of the robot as a linear spring, we can use the following equation for modeling the compliance.

$$\tau_i = k_i \Delta q_i \qquad i = 1, 2, \ldots, m \tag{7.65}$$

where m is the number of actuated joints in the robot, and k_i is the stiffness coefficient of the i-th joint. This coefficient relates the applied torque in the robot joint τ_i to the small displacement generated in the robot joint Δq_i. This relationship can be rewritten in vector form for all joints.

$$\boldsymbol{\tau} = \boldsymbol{\mathcal{K}} \Delta \boldsymbol{q} \tag{7.66}$$

In which $\boldsymbol{\tau} = \begin{bmatrix} \tau_1 & \tau_2 & \ldots & \tau_m \end{bmatrix}^T$ represents the vector of motor torques and $\Delta \boldsymbol{q} = \begin{bmatrix} \Delta q_1 & \Delta q_2 & \ldots & \Delta q_m \end{bmatrix}^T$ is the small displacement vector in the robot's joints. Thus, the joint stiffness can be described with the following diagonal matrix:

$$\mathcal{K} = \begin{bmatrix} k_1 & & & 0 \\ & k_2 & & \\ & & \ddots & \\ 0 & & & k_m \end{bmatrix} \tag{7.67}$$

In order to analyze the stiffness of the robot, the relationship between the displacement of the robot's end-effector and the applied wrench is examined. For this purpose, the Jacobian matrix can be used as a mapping between the small joint displacements and the end-effector displacement of the robot. Additionally, the transpose of the Jacobian matrix serves as the mapping between the wrench applied to the end-effector and the joint torques.

$$\Delta \boldsymbol{\mathcal{X}} = \boldsymbol{J}(\boldsymbol{q}) \Delta \boldsymbol{q} \quad , \quad \boldsymbol{\tau} = \boldsymbol{J}^T(\boldsymbol{q}) \boldsymbol{\mathcal{F}} \tag{7.68}$$

By using these two mappings, the relationship between the displacement and the force of the robot's end-effector can be obtained.

$$\Delta \boldsymbol{\mathcal{X}} = \boldsymbol{J} \Delta \boldsymbol{q} = \boldsymbol{J} \boldsymbol{\mathcal{K}}^{-1} \boldsymbol{\tau} = \boldsymbol{J} \boldsymbol{\mathcal{K}}^{-1} \boldsymbol{J}^T \boldsymbol{\mathcal{F}}$$

Thus, the compliance matrix of the robot is defined as follows:

$$\Delta \boldsymbol{\mathcal{X}} = \boldsymbol{C} \cdot \boldsymbol{\mathcal{F}} \quad , \quad \boldsymbol{C}_{m \times m} = \boldsymbol{J} \boldsymbol{\mathcal{K}}^{-1} \boldsymbol{J}^T \tag{7.69}$$

The stiffness matrix of the robot is also obtained from the inverse of the compliance matrix.

$$\boldsymbol{\mathcal{F}} = \boldsymbol{K} \cdot \Delta \boldsymbol{\mathcal{X}} \quad , \quad \boldsymbol{K}_{m \times m} = \boldsymbol{J}^{-T} \boldsymbol{\mathcal{K}} \boldsymbol{J}^{-1} \tag{7.70}$$

Note that the stiffness and compliance matrices are functions of the robot's structure and configuration. Therefore, their values vary at different points in the workspace. Additionally, the above relation Equation (7.70) is defined only for nonsingular square Jacobian matrices, which are applicable in fully actuated robots with $m = n$.

7.6.2 Compliance and Stiffness Ellipsoids

If similar harmonic drive systems are used in all joints, the joint stiffness can be considered uniform,

$$k_1 = k_2 = \cdots = k_m = k$$

then,

$$\boldsymbol{C} = k^{-1}\boldsymbol{J}\boldsymbol{J}^T, \quad \boldsymbol{K} = k(\boldsymbol{J}\boldsymbol{J}^T)^{-1} \tag{7.71}$$

It's interesting to note that in this case, the calculation of the compliance matrix in the robot leads to the weighted dexterity matrix $\boldsymbol{J}\boldsymbol{J}^T$, where the weight of this matrix is equal to the stiffness level in the joint. Naturally, by weighting a matrix, its eigenvectors will not change, and only its eigenvalues will be weighted. Thus, the robot's compliance matrix will inherit all ellipsoidal dexterity features in a weighted manner.

Now, if the end-effector's wrench is considered to be uniform and with a unit norm, $\|\mathcal{F}\|_2 = 1$, the Euclidean norm of the end-effector's displacement due to the application of this wrench can be obtained from the compliance ellipsoid.

$$\Delta\boldsymbol{\mathcal{X}}^T\Delta\boldsymbol{\mathcal{X}} = \boldsymbol{\mathcal{F}}^T(\boldsymbol{C}^T\boldsymbol{C})\boldsymbol{\mathcal{F}} \tag{7.72}$$

Consider the singular value decomposition of the robot's compliance matrix,

$$\text{svd}(\boldsymbol{C}) = \boldsymbol{U}_c\,\boldsymbol{\Sigma}_c\,\boldsymbol{V}_c^T \tag{7.73}$$

and represent the maximum and minimum singular values of the compliance matrix as σ_{max} and σ_{min}, respectively. Let the corresponding singular vectors in the matrix $\boldsymbol{U}_c$ be denoted by $\boldsymbol{u}_{max}$ and $\boldsymbol{u}_{min}$. Then, the maximum and minimum displacements in the robot's end-effector can be obtained along these vectors.

$$\Delta\boldsymbol{\mathcal{X}}_{max} = |\sigma_{max}|\|\boldsymbol{\mathcal{F}}\| \rightarrow \text{along } \boldsymbol{u}_{max} \tag{7.74}$$

$$\Delta\boldsymbol{\mathcal{X}}_{min} = |\sigma_{min}|\|\boldsymbol{\mathcal{F}}\| \rightarrow \text{along } \boldsymbol{u}_{min} \tag{7.75}$$

Example 7.5 (Planar 2R Robot) *Consider the planar 2R robot shown in Figure 7.10, where the lengths of the two robot arms inspired by the human hand dimension ratio, are set as $a_1 = \sqrt{2}, a_2 = 1$. Furthermore, assume that the angle of the second joint is $\theta_2 = \pi/2$. Suppose the stiffness coefficient for both robot joints is $k = 10^3 N.m/rad$. If a uniform force with a unit Euclidean norm is applied to the robot's end-effector, calculate the maximum and minimum displacements of the robot's end-effector and plot the compliance ellipsoid for the robot in this configuration.*

Solution: Previously, the Jacobian matrix for the planar 2R robot in the coordinate system {2} was derived and reported in Equation 7.4. By substituting the link lengths, the angle $\theta_2 = \pi/2$, and the joint stiffness, the compliance matrix for the robot with respect to the coordinate system {2} is obtained as follows.

$${}^2\boldsymbol{C}(\boldsymbol{q}) = k^{-1}\boldsymbol{J}\boldsymbol{J}^T = 2 \times 10^{-3}\begin{bmatrix} 1 & \sqrt{2} \\ \sqrt{2} & 1 \end{bmatrix}$$

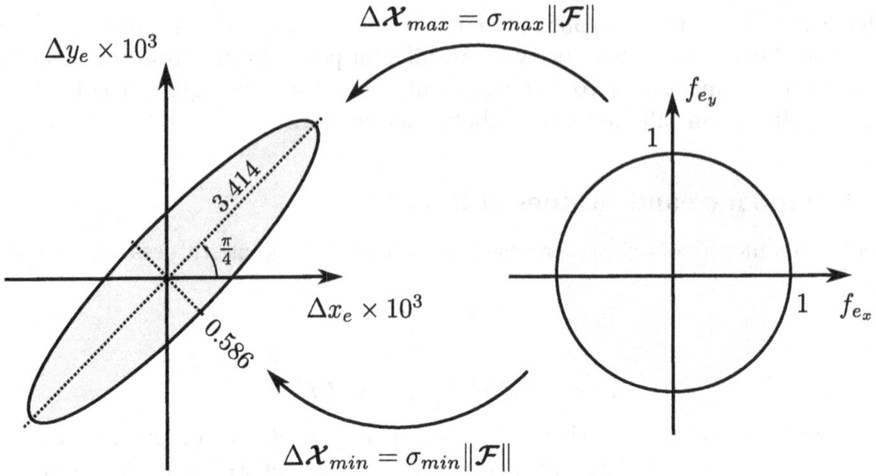

FIGURE 7.19
Magnitude of maximum and minimum displacements of the planar 2R robot's end-effector for the application of uniform force to its end-effector.

By singular value decomposing of this matrix, we can write,

$$\boldsymbol{U} = \frac{1}{\sqrt{2}}\begin{bmatrix} -1 & -1 \\ -1 & 1 \end{bmatrix}, \quad \boldsymbol{\Sigma} = \begin{bmatrix} 3.414 \times 10^{-3} & 0 \\ 0 & 0.586 \times 10^{-3} \end{bmatrix}$$

Thus, the maximum displacement limit in this mapping is

$$\sigma_{max} = 3.414 \times 10^{-3} \quad \text{along} \quad \begin{bmatrix} 1 \\ 1 \end{bmatrix}$$

and the minimum displacement limit in this mapping is

$$\sigma_{min} = 0.586 \times 10^{-3} \quad \text{along} \quad \begin{bmatrix} -1 \\ 1 \end{bmatrix}$$

The stiffness ellipsoid of the robot in this configuration can be plotted as shown in Figure 7.19. As seen in this figure, the displacement of the robot's end-effector is very small, on the order of millimeters. However, the displacement along the major axis of the ellipsoid is much larger than in the other direction. By changing the robot's configuration, different conditions can be achieved in the motion of the robot's end-effector.

Problems

7.1 Consider the two-degree-of-freedom RP robot in Problem 6.1, depicted in Figure 7.20. For numerical values:

$$a_1 = 1, \ \ \theta_1 = [0 - 2\pi], \ \ d_2 = [0 - 1],$$

calculate the dexterity and manipulability metrics in the discretized workspace of the robot and visualize them numerically using the `surf` command in MATLAB three-dimensional plots, or similar commands in other software. Then, compare these two metrics intuitively and examine their variations concerning the robot's singular configurations near the boundaries of its workspace.

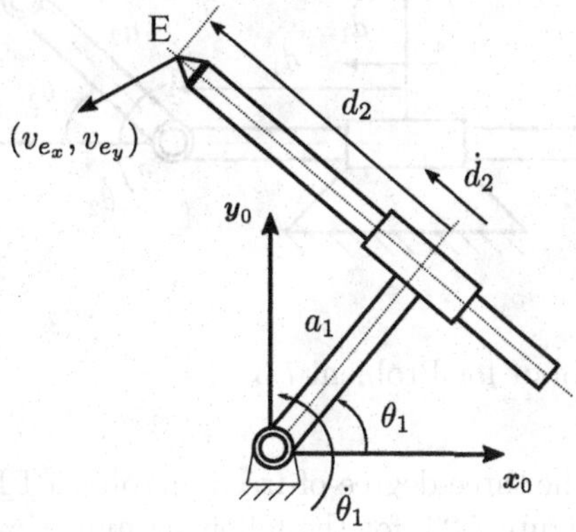

FIGURE 7.20
Two-degrees-of-freedom robot for Problem 7.1

7.2 Solve Problem 7.1 for the three-degree-of-freedom robot <u>PRP</u> introduced in Problem 6.2 and depicted in Figure 7.21, for the following numerical values.

$$d_1 = [0 - 1],\ a_2 = 1,\ \theta_2 = [0 - 2\pi],\ d_3 = [0 - 1].$$

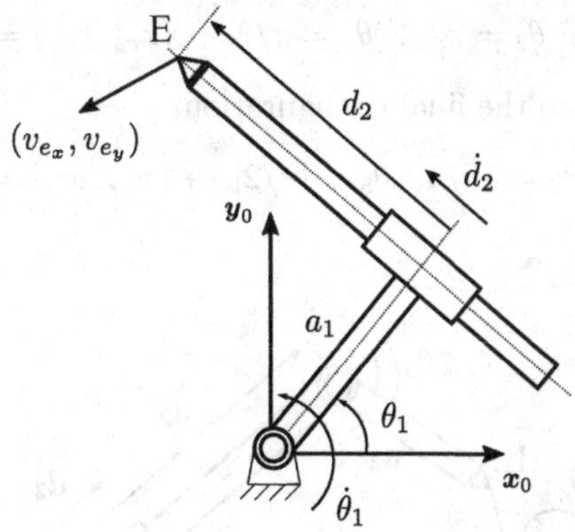

FIGURE 7.21
Two-degrees-of-freedom robot for Problem 7.2

7.3 Resolve Problem 7.1 for the three-degree-of-freedom <u>PRR</u> robot introduced in Problem 6.5 and illustrated in Figure 7.22, using the provided numerical values.

$$d_1 = [0 - 1],\ a_2 = \sqrt{2},\ a_3 = 1,\ \theta_2 = [0 - \pi],\ \theta_3 = [0 - 2\pi].$$

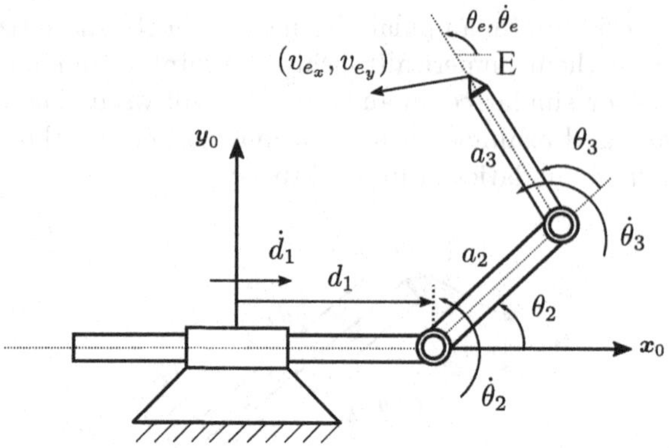

FIGURE 7.22
Three-degrees-of-freedom robot for Problem 7.3

7.4 Solve Problem 7.1 for the three-degree-of-freedom robot RPR introduced given in Problem 6.8, depicted in Figure 7.23, for the following numerical values.

$$a_1 = 1, \;\; d_2 = [0 - 1], \;\; a_3 = 0.707, \;\; \theta_1 = [0 - \pi], \;\; \theta_3 = [0 - 2\pi].$$

7.5 Consider the three-degree-of-freedom robot PRR from Problem 6.5, as illustrated in Figure 7.24.

Assume that the lengths of the robot's links are equal, $a_2 = a_3 = 1$, and there are no constraints on the translational and angular motions of the robot's joints. Furthermore, assume that during motion, the links will not collide with each other or the base. It is desired to move the robot from the initial configuration

$$d_1 = 1, \;\; \theta_2 = \pi/4, \;\; \theta_3 = \pi/2, \rightarrow (x_{e_0}, y_{e_0}) = (1, 1.414)$$

shown in Figure 7.24, to the final configuration

$$d_1 = 0.5, \;\; \theta_2 = 5\pi/4, \;\; \theta_3 = [\pi/2], \rightarrow (x_{e_f}, y_{e_f}) = (0.5, -1.414)$$

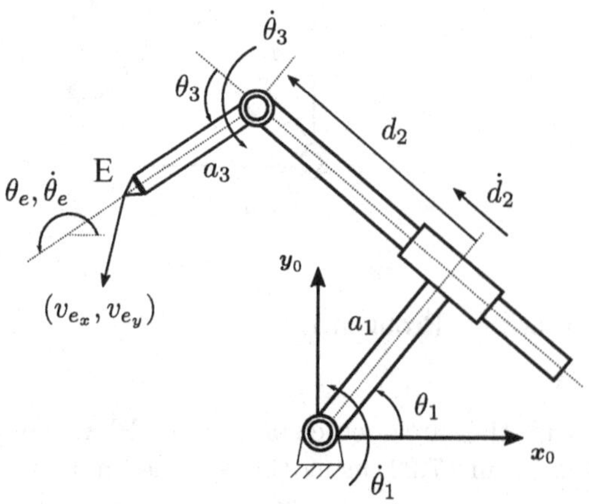

FIGURE 7.23
Three-degrees-of-freedom robot for Exercise 7.4

Move along a straight line so that the robot arms do not collide with the obstacle shown in the figure. Initially, use the pseudo-inverse of the Jacobian matrix in path planning and show that in this case, collision with the obstacle occurs. Then solve the redundancy resolution problem by performing suitable numerical optimization and display the robot's motion on the specified path. Finally, if the joint velocities have an upper limit,

$$-1.2 \leq \dot{d}_1 \leq 1.2, \quad -1.5 \leq \dot{\theta}_2 \leq 1.5, \quad -1.2 \leq \dot{\theta}_3 \leq 1.2,$$

solve the constrained optimization problem again and compare the final path with the previous case.

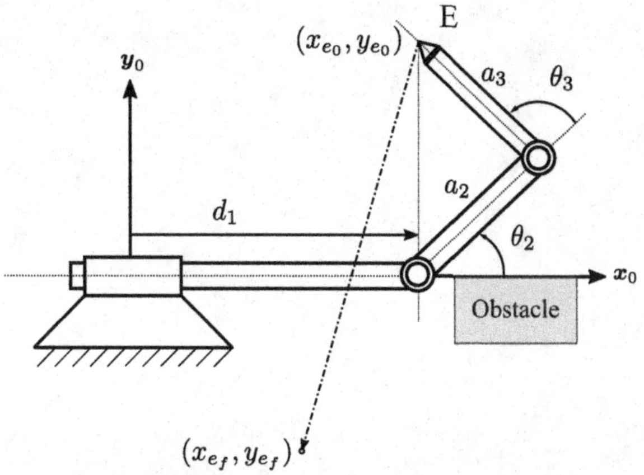

FIGURE 7.24
Solving the redundancy resolution problem for the three-degree-of-freedom robot in Problem 7.5 to avoid collision with the obstacle.

Part III

Dynamics

8

Dynamic Analysis

This chapter elaborates on the analysis of the dynamic formulation of serial robots. Unlike the previous chapters that primarily focused on kinematic analysis, which analyzed motion without considering the forces that cause the motion, this chapter places emphasis on the correlation between forces and torques applied to the robot and the resulting dynamic maneuvers. Dynamic analysis plays a pivotal role in designing robot components, selecting suitable sensors and actuators, conducting motion simulations, and analyzing motion maneuvers, particularly in the context of robot control.

This chapter provides a comprehensive review of rigid body dynamics for readers who may not be acquainted with dynamic analysis concepts. Special emphasis is placed on the Euler-Lagrange method for deriving the dynamic formulation of serial robots. Initially, the method is elucidated using a motivational example, followed by the introduction of a recursive approach for the dynamic formulation of general serial robots. Several case studies are presented to reinforce understanding and illustrate the practical application of the recursive method. Additionally, a series of problems are provided at the end of the chapter, offering readers an opportunity to practice the fundamental theories and methods covered throughout the chapter.

8.1 Introduction

Various methods have been studied for the dynamic analysis of systems, and among these approaches, the use of Newton-Euler equations as a fundamental method in this field has gained more popularity.

Newtonian mechanics is established based on three laws, first formulated by Sir Isaac Newton (1642-1727) in "The Mathematical Principles of Natural Sciences," published in 1687. Newton effectively built upon the findings of his predecessors, including renowned scientists like Galileo and Kepler. Although the first two laws were recognized by Galileo, Newton was the first to clearly articulate these three laws. These laws are formulated for a particle, requiring the existence of a reference coordinate system for their validity, known as an inertial frame or the Galilean reference system. The laws of motion introduced by Newton are as follows:

First Law: The Law of Inertia.

An object will stay at rest or keep moving at a constant velocity if there is no net force acting on it.

Second Law: The Law of Force and Acceleration.

Force is equal to the mass of an object times its acceleration ($F = ma$). The acceleration of an object is directly proportional to the force and inversely proportional to the mass.

DOI: 10.1201/9781003491415-8

Third Law: The Action and Reaction Law.

For every action there is an equal and opposite reaction. This law describes how forces always come in pairs.

In Newtonian mechanics, the dynamic formulation are directly derived from the application of Newton's second law, which describes dynamic systems in terms of forces and torques. This approach involves the use of free-body diagrams and Newton-Euler equations in a dynamic state. In this method, it is necessary to separate all the links of the robot from each other and study them in a free-body diagrams.

In the free-body diagram, the relationships between the links are studied through support forces/torques and actuator force/torque at the robot joints. By analyzing the acceleration of the center of mass of the links, the Newton-Euler equations in dynamic states are applied to each link of the robot to determine the relationship between these forces and torques in producing the motion of each link. However, it should be noted that ultimately, the support forces are eliminated from the equations to establish a direct relationship between actuator force/torque and the motion of robot links.

As noted, this method necessitates calculating both the support forces and torques and the acceleration of the center of mass of the robot links. Consequently, the computational demands associated with this approach are substantial, making its application reasonable only when intending to determine all support forces and torques in designing robot joints or selecting sensors and actuators. Due to the considerable computational requirements, this book refrains from providing an in-depth exposition of this method, directing readers to consult other sources in this field [2, 3].

An alternative effective approach to derive dynamic equations involves employing Euler-Lagrange relations. In this method, it suffices to analyze the velocities of the center of mass of the links, while eliminating the need to calculate support forces. The dynamic formulation is determined by differentiating the energy relations of the robot components, leading to a significant reduction in computational costs compared to the Newton-Euler method. The appeal of this approach extends across various engineering disciplines due to its lower prerequisite for prior mechanical knowledge. Another advantage of utilizing this method for dynamic motion equations is the ability to compute closed-form formulations for all serial robots, a crucial factor in implementing model-based control methods for robots. Given these considerations, the foundation for determining the dynamic motion equations of the robot in this book will be rooted in the Euler-Lagrange method.

The Virtual Work method, based on D'Alembert's Principle, represents an alternative approach discussed in robotics literature. In this method, it is essential to analyze the acceleration of the center of mass of the links. To achieve this, the derivative of the Jacobian matrices of these points is employed, reducing the required computational effort. Unlike the Newton-Euler method, this approach does not necessitate determining support forces, making it computationally more efficient. However, utilizing this method demands a more extensive background in classical mechanics compared to the Euler-Lagrange method. Due to the constraints on presenting various methods, this book refrains from elaboration on the details of this approach. Still, interested readers are encouraged to explore recent sources extended by the author in this field [99, 100, 101].

Several other methods have been developed for the dynamic analysis of systems, each offering advantages in specific cases over the Euler-Lagrange method. Readers of this book can broaden their understanding by exploring these various approaches. Among these, the Kane method [102] and the Gibbs-Appell method [103] are notable. It is noteworthy that the Gibbs-Appell method directly leads to the Euler-Lagrange formulation [104], and in additionally, this method can be advantageous for determining Kane equations [105]. The direct application of the Kane method in determining the motion equations of robots has also been directly worked out by researchers [106]. In this chapter, we will initially review the dynamics of a rigid body, followed by an exploration of only the Euler-Lagrange method.

8.2 Review on Rigid Body Dynamics

In this section, fundamental definitions and relationships for the dynamic analysis of rigid bodies will be examined. Considering that in Chapter 3 and in Section 3.6 methods for determining the linear acceleration of a point and the angular acceleration of a body have been discussed, the subsequent focus will be on the mass properties of the rigid body, linear and angular momentum, and the kinetic energy of the rigid body.

8.2.1 Center of Mass

Engineering students become familiar with the concept of mass in basic physics courses. Mass can be referred to as a property of a body that causes its weight in a gravitational field. As illustrated in Figure 8.1, the distribution of mass in a rigid body affects its size, shape, and weight. Considering the coordinate system $\{\mathcal{A}\}$, in which the mass properties of the body are measured, if we assume a uniform volumetric density ρ for the mass distribution within the body, the mass of a differential element of the body can be represented as ρdV, where dV is the infinitesimal volume of the element. If we denote the position vector of this element in the coordinate system $\{\mathcal{A}\}$ by $\boldsymbol{p}$, the total mass of the body is obtained by integrating the mass of its elements over the body's volume.

$$m = \int_V \rho dV \tag{8.1}$$

In that case, the position vector of the center of mass $\boldsymbol{C}$, denoted by $\boldsymbol{p}_c$ as depicted in the figure, can be determined based on the mass distribution within the body.

$$\boldsymbol{p}_c = \frac{1}{m}\int_V \boldsymbol{p}\,\rho dV \tag{8.2}$$

The center of mass is a theoretical point within an object where the entire mass of the object is concentrated, and it is evenly distributed around this point. Consequently, if the object is positioned on a sharp edge at its center of mass, the weight on both sides of the object is equal, leading to a state of equilibrium. In the dynamic analysis of rigid bodies, this point holds significant importance. Instead of scrutinizing the distributed mass within the body, one can simplify the dynamic equations by assuming the mass of the body is concentrated at this point. It is noteworthy that the mass properties of a rigid body are characterized by four parameters: the body's mass, denoted as m, and the three components of the position vector of its center of mass, $\boldsymbol{p}_c = \begin{bmatrix} p_{c_x} & p_{c_y} & p_{c_z} \end{bmatrix}^T$.

8.2.2 Moment of Inertia

Unlike the mass of a body that exhibits inertia against linear accelerations, the moment of inertia of a body manifests inertia against angular accelerations. In rotational motions, the mass distribution of a body with respect to the axis of rotation exhibits its moment of inertia. If we consider the rotational motion of a rigid body shown in Figure 8.1, the moment of inertia of the body with respect to the reference coordinate system $\{\mathcal{A}\}$ is represented

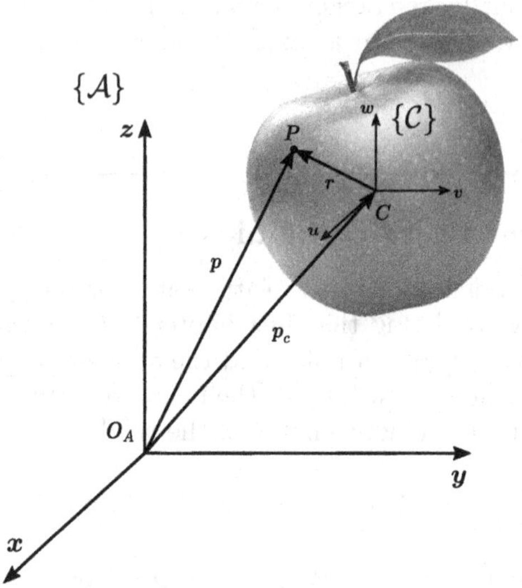

FIGURE 8.1
Mass properties of a rigid body.

by the inertia matrix ${}^{A}I$.

$$ {}^{A}\boldsymbol{I} = \begin{bmatrix} I_{xx} & I_{xy} & I_{xz} \\ I_{yx} & I_{yy} & I_{yz} \\ I_{zx} & I_{zy} & I_{zz} \end{bmatrix} \tag{8.3} $$

where:

$$ \begin{aligned} I_{xx} &= \int_V (y^2 + z^2)\rho dV & I_{xy} &= I_{yx} = -\int_V xy\ \rho dV \\ I_{yy} &= \int_V (x^2 + z^2)\rho dV & I_{yz} &= I_{zy} = -\int_V yz\ \rho dV \\ I_{zz} &= \int_V (x^2 + y^2)\rho dV & I_{xz} &= I_{zx} = -\int_V xz\ \rho dV \end{aligned} \tag{8.4} $$

What is observed in these equations is that the diagonal components of the inertia matrix are directly proportionally to the square of the radial distance of the particle mass to the axis of rotation. This is in contrast to the non-diagonal components, which are proportionally to the product of the distance from the mass particle to the other axes. Thus, the inertia matrix of a body is a symmetric matrix, requiring only six dynamic parameters for its description. Additionally, as evident from the equations, the body's inertia is dependent on the coordinate system in which these calculations are performed. Typically, these calculations are done with respect to the coordinate system attached to the center of mass of the body $\{\mathcal{C}\}$.

Calculating the inertia matrix for bodies is seldom done directly utilizing the definition provided in Equation 8.4. In contrast, for bodies with simple geometries, these computations are usually conducted using tables found in references [107, Appendix 5]. For bodies with intricate geometries, these computations are automated through computer-aided design (CAD) software. The only requisite step is to transform it from one coordinate system to another, a topic to be explored in the following sections.

8.2.3 Principal Axes

As mentioned, the inertia matrix of a body depends on the coordinate system with respect to which it is measured. For this reason, it is possible to find a coordinate system in space in which the inertia matrix becomes diagonal.

$$^{A}\boldsymbol{I} = \begin{bmatrix} I_{xx} & 0 & 0 \\ 0 & I_{yy} & 0 \\ 0 & 0 & I_{zz} \end{bmatrix} \tag{8.5}$$

As previously mentioned, coordinate axes in which the inertia matrix becomes diagonal are known as the principal axes of inertia. The diagonal components of this matrix in this configuration, denoted as I_{xx}, I_{yy}, I_{zz} are called the principal moments of inertia of the body. It can be demonstrated that the principal moments of inertia are among the fundamental mass properties of the body, easily obtainable through a transformation matrix consisting of the eigenvectors of the inertia matrix. Physically, the principal moments of inertia align with the symmetry axes of the body.

In this context, viewing the inertia matrix of a body as a mapping reveals that the principal components of the inertia matrix define the upper and lower bounds of this mapping. Furthermore, as the inertia matrix governs the mass properties of a body during rotational movements, the principal moments of inertia consistently remain positive. We will exploit these attributes when scrutinizing the characteristics of robot dynamic formulations.

8.2.4 Inertia Matrix Computations

As previously mentioned, the direct computation of body inertia matrices according to their definition is rarely performed. Typically, these values are derived from tables found in references or generated using CAD software. However, these inertia matrix values are commonly provided with respect to a coordinate system attached to the body at its center of mass or at its initial and final points. This stands in contrast to the dynamic analysis of bodies, where it becomes essential to ascertain inertia matrix values with respect to other coordinate systems. Consequently, comprehending the transformation of the inertia matrix from one coordinate system to another holds significant importance.

As illustrated in Figure 8.1, if the change in coordinates in the inertia matrix is limited to translational motion, these calculations can be made based on the parallel-axis theorem [108] and using the following equation

$$^{A}\boldsymbol{I} = {}^{C}\boldsymbol{I} + m\left(\boldsymbol{p}_c^T\boldsymbol{p}_c\boldsymbol{I}_{3\times3} - \boldsymbol{p}_c\boldsymbol{p}_c^T\right) \tag{8.6}$$

In this relation, $^{A}\boldsymbol{I}$ represents the inertia matrix relative to the coordinate system $\{\mathcal{A}\}$, and $^{C}\boldsymbol{I}$ is the inertia matrix relative to the coordinate system attached to the center of mass $\{\mathcal{C}\}$. Additionally, m denotes the mass of the body, the vector $\boldsymbol{p}_c = [x_c, y_c, z_c]^T$, as shown in Figure 8.1, represents the translation vector from the coordinate system $\{\mathcal{C}\}$ to the coordinate system $\{\mathcal{A}\}$, and $\boldsymbol{I}_{3\times3}$ is the identity matrix of size 3×3. This relation can be rewritten in terms of the components of the inertia matrix as follows.

$$\begin{aligned} ^{A}I_{xx} &= {}^{C}I_{xx} + m(y_c^2 + z_c^2) & \quad ^{A}I_{xy} &= {}^{C}I_{xy} + m\,x_c y_c \\ ^{A}I_{yy} &= {}^{C}I_{yy} + m(x_c^2 + z_c^2) & \quad ^{A}I_{yz} &= {}^{C}I_{yz} + m\,y_c z_c \\ ^{A}I_{zz} &= {}^{C}I_{zz} + m(x_c^2 + y_c^2) & \quad ^{A}I_{xz} &= {}^{C}I_{xz} + m\,x_c z_c \end{aligned} \tag{8.7}$$

As observed in this relation, the diagonal components of the inertia matrix are increased proportional to the square of the radial distance between the two coordinate systems, and the off-diagonal components are increased proportional to the product of the distances of the corresponding axes of the two coordinate systems.

If the coordinate transformation in determining the body's inertia matrix is limited to pure rotational motion, this transformation can be performed using the following equation.

$$^{A}\boldsymbol{I} = {}^{A}\boldsymbol{R}_C \, {}^{C}\boldsymbol{I} \, {}^{A}\boldsymbol{R}_C{}^{T} \tag{8.8}$$

In this equation, $^{A}\boldsymbol{R}_C$ is the rotation matrix from the coordinate system $\{\mathcal{C}\}$ to $\{\mathcal{A}\}$. In the dynamic analysis of serial robots, the inertia matrix of the links is often obtained relative to the coordinate system located at their own center of mass. Equation 8.8 is then often used to transform it into other coordinate systems that have pure rotational motion relative to it.

8.2.5 Linear and Angular Momentum

The Newton-Euler and Euler-Lagrange methods are based on principles related to linear and angular momentum, as well as kinetic energy. A thorough exploration of this subject in diverse coordinate systems demands a more extensive investigation, which falls outside the scope of this book. Enthusiastic readers are encouraged to read the references provided in this area [107, 103]. In this context, we will concisely outline the outcomes of these studies, directly relevant to the dynamic analysis of serial robots.

The linear momentum of a rigid body can be obtained by multiplying its mass by the velocity of its center of mass.

$$^{A}\boldsymbol{G} = m \, {}^{A}\boldsymbol{v}_c \tag{8.9}$$

In this context, it's important to emphasize that we consider the mass distribution around the center of mass as if it were concentrated solely at the center of mass itself. Consequently, we avoid computing individual particle momenta. This relationship is only valid using the velocity of the body's center of mass $\boldsymbol{v}_c$, and it cannot be determined using the velocities of other points on the body. Hence, the significance of precisely defining the center of mass and determining its velocity becomes evident in the dynamic analysis of rigid bodies.

As indicated by Equation 8.9, the linear momentum of a body, akin to its center of mass velocity, is a vector with interpretability across various coordinate systems. In this equation, the linear momentum of the body with respect to the coordinate system $\{\mathcal{A}\}$ is represented, and for this purpose, the velocity of the center of mass calculated in the coordinate system $\{\mathcal{A}\}$ is used. Considering the vector properties of linear momentum, the homogeneous transformation matrix can be used to transform it into another coordinate system as follows.

$$^{A}\boldsymbol{G} = {}^{A}\boldsymbol{T}_C \, {}^{C}\boldsymbol{G} = m \, {}^{A}\boldsymbol{T}_C \, {}^{C}\boldsymbol{v}_c \tag{8.10}$$

The angular momentum of a rigid body is also a vector, calculated with respect to the coordinate system attached to the body at its center of mass $\{\mathcal{C}\}$ as follows.

$$^{C}\boldsymbol{H} = {}^{C}\boldsymbol{I} \, \boldsymbol{\Omega} \tag{8.11}$$

With this definition, angular momentum extends the concept of linear momentum to rotational motions. Given that the angular velocity of the body $\boldsymbol{\Omega}$ is the same for all points on the body, calculating angular momentum only requires determining the inertia matrix of the body in the $\{\mathcal{C}\}$ coordinate system.

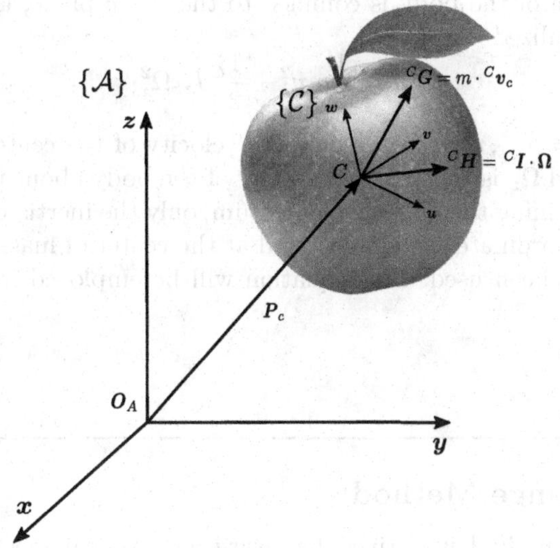

FIGURE 8.2
Linear and angular momentum of a rigid body about the $\{\mathcal{C}\}$ reference frame.

If we intend to calculate angular momentum with respect to another coordinate system, such as $\{\mathcal{A}\}$, it is no longer straightforward to accomplish this using a homogeneous transformation matrix. Instead, we need to use the following equation:

$$^A\boldsymbol{H} = \boldsymbol{p}_c \times \boldsymbol{G}_c + {}^C\boldsymbol{H} \tag{8.12}$$

As observed in this equation, to determine the angular momentum of the body with respect to the coordinate system $\{\mathcal{A}\}$, denoted by $^A\boldsymbol{H}$, it is necessary to sum the angular momentum of the body with respect to the center of mass coordinate system $\{\mathcal{C}\}$ with the cross product of the vector representing the distance from the center of mass to the origin of the coordinate system $\boldsymbol{p}_c$ and the linear momentum $\boldsymbol{G}_c$.

Note that in this equation, the mass distribution around the center of mass is also treated as a concentrated mass at the center of mass. Consequently, for calculating the angular momentum with respect to the coordinate system $\{\mathcal{A}\}$, it suffices to combine the angular momentum with respect to the center of mass coordinate system $\{\mathcal{C}\}$ and the effect of the linear momentum of the body about the $\{\mathcal{A}\}$ coordinate system.

8.2.6 Kinetic Energy

In the analysis of dynamics using the Euler-Lagrange method, the kinetic energy of a rigid body plays a crucial role. To calculate the kinetic energy of a rigid body, it is sufficient to obtain the total kinetic energy of the linear and angular motion of the body with respect to the coordinate system attached to the system's center of mass.

$$K = \frac{1}{2}\boldsymbol{v}_c \cdot {}^C\boldsymbol{G} + \frac{1}{2}\boldsymbol{\Omega} \cdot {}^C\boldsymbol{H} \tag{8.13}$$

In this equation, the kinetic energy of the body is a scalar quantity. Its linear motion component is calculated by the dot product of the linear velocity of the center of mass and its linear momentum, while the angular motion component is determined by the dot product of the angular velocity of the body and its angular momentum about the center of mass. If the motion of the body is confined to the $x - y$ plane, the following simplified relationship can be utilized.

$$K = \frac{1}{2} m \boldsymbol{v}_c^2 + \frac{1}{2} {}^C I_{zz} \Omega_z^2 \tag{8.14}$$

In this equation, $\boldsymbol{v}_c = [v_{c_x}, v_{c_y}]^T$ represents the velocity of the center of mass of the body in the $x - y$ plane, and Ω_z is the angular velocity of the body about the z axis. It is for this reason that, in determining the angular momentum, only the inertia component of the body with respect to the coordinate system centered at the center of mass and about the z axis, denoted by ${}^C I_{zz}$, has been used. This equation will be employed in dynamic formulation for planar robots.

8.3 Euler-Lagrange Method

The Euler-Lagrange method is derived by rewriting classical mechanics, combining the principles of conservation of linear and angular momentum with energy conservation. This method is particularly effective for dynamic analysis of multibody systems, offering efficiency by eliminating the need to account for support forces between system components in the equations of motion. As a result, the analysis is considerably simplified.

In the Euler-Lagrange method, the equations of motion for a system are determined using a set of *generalized coordinates.* These coordinates represent the minimal set of variables necessary to describe the system's motion, with the understanding that they are not uniquely defined. Consider the motion of a bead in a narrow groove, for example. If we attempt to derive the equations of motion for this system using the Newton-Euler method, we would need to determine the friction force between the bead and the groove. This force is often nonlinear and time-varying, making its determination challenging. However, without the need to consider this force and by solely considering the path of the bead's motion through generalized coordinates, we can easily derive its equations of motion. With this choice and without calculating support forces between different bodies, the system's dynamics can be analyzed.

By defining generalized coordinates, external forces and torques acting along these coordinates are determined as *generalized forces.* Subsequently, the kinetic and potential energy of the moving masses in the system are calculated in terms of generalized coordinates. Finally, the Euler-Lagrange equations are employed to determine the equations of motion. The details of this method will be thoroughly discussed in the following sections.

8.3.1 Generalized Coordinates and Forces

The connection between two masses in multi-body systems is described by kinematic constraints that restrict the relative motion between them in various directions. In serial robots, joints create these kinematic constraints, defining and limiting the system's degrees of freedom.

In a general classification, system constraints are termed "holonomic" if the relationship between position variables can be expressed by an algebraic equation between these variables and, if necessary, the time.

$$f(x_1, x_2, \ldots, t) = 0 \tag{8.15}$$

In this algebraic relationship, x_i represents position or orientation variables, while their derivatives are not seen in this equation. For example, consider the angular motion of a pendulum rotating about its joint. The coordinates x and y of each point on the pendulum with a radial distance of r from the center of rotation can be constrained by the algebraic relation $x^2 + y^2 = r^2$. Therefore, motion constrained by a revolute joint is holonomic. This property holds true for other types of joints such as prismatic, spherical, and others commonly used in robotic systems.

If kinematic constraints cannot be interpreted with an algebraic relation like 8.15, if derivatives of position and orientation variables appear in the kinematic constraint, if a differential relation in the position constraints appears, the system is called "non-holonomic." In these systems, the motion variables of the body depend on the path of motion, and interpreting algebraic constraints in them will not be possible. For example, consider a four-wheeled car. If the car departs from its initial parked state, maneuvers through the surrounding streets, and precisely returns to the initial point with the same parking configuration, it is reasonable to expect that the wheel positions in the initial and final states will differ, even though the car's overall state remains identical in both instances. This discrepancy points to the existence of non-holonomic constraints within the system. By simulating the motion of the wheels on the ground, it becomes evident that a corresponding algebraic relation, akin to 8.15, cannot be identified in such a system.

Consider a serial robot consisting of n links with primary joints. If revolute and prismatic joints are employed as the primary joints, its generalized coordinates can easily be expressed using the joint variables of the robot.

$$\boldsymbol{q} = [q_1, q_2, \ldots, q_n]^T \quad \rightarrow \quad q_i = \begin{cases} \theta_i & \text{For Revolute Joints} \\ d_i & \text{For Prismatic Joints} \end{cases} \tag{8.16}$$

With this description, the minimum number of independent variables that can describe the motion of the robot links are selected as generalized variables. It should be noted that there is no distinction between angular motion in a revolute joint θ_i and translational motion in a prismatic joint d_i when choosing them as generalized variables.

Likewise, it is straightforward to establish the definition of generalized forces. Let's begin by assuming that the only external forces acting on the system are their actuator force/torques. Under this circumstance, one can define generalized forces as follows.

$$\boldsymbol{Q} = [Q_1, Q_2, \ldots, Q_n]^T \quad \rightarrow \quad Q_i = \begin{cases} \tau_i & \text{For Revolute Joints} \\ f_i & \text{For Prismatic Joints} \end{cases} \tag{8.17}$$

In this relation, τ_i represents the actuator torque of the revolute joint, and f_i denotes the actuator force of the prismatic joint. If, in addition to the actuator forces, there is friction force at the robot joints, the generalized forces can be defined by subtracting the friction forces from the actuator forces.

$$\boldsymbol{Q} = [Q_1, Q_2, \ldots, Q_n]^T, \quad \rightarrow \quad Q_i = \begin{cases} \tau_i - \tau_{f_i} & \text{For Revolute Joints} \\ f_i - f_{f_i} & \text{For Prismatic Joints} \end{cases} \tag{8.18}$$

In this equation, τ_{f_i} represents the friction torque at the i-th revolute joint, and f_{f_i} denotes the friction force at the i-th prismatic joint. Now, let's assume that the robot's end-effector is exerting an external force $\boldsymbol{\mathcal{F}}_e$ on the environment. In this case, using the Jacobian transpose matrix $\boldsymbol{J}^T(\boldsymbol{q})$, one can map its effect in the joint space and add it to the generalized forces.

$$\boldsymbol{Q} = [Q_1, Q_2, \ldots, Q_n]^T + \boldsymbol{J}^T \boldsymbol{\mathcal{F}}_e \tag{8.19}$$

Now, if an external force is applied to the robot at a point other than the end-effector by using the Jacobian matrix of that point, one can easily map the force onto the robot's joints. Similar to Equation 8.19, this force can be added to the generalized forces. Note that in the Euler-Lagrange method, gravitational forces are not considered as generalized forces; instead, they are accounted for in the robot's potential energy.

8.3.2 Euler-Lagrange Formulation

By defining the generalized coordinates and forces within the robot, one only needs to compute the kinetic and potential energy of the robot's components to employ the Euler-Lagrange method. To accomplish this, the kinetic energy for all robot components can be determined by evaluating the angular velocity of the body and the linear velocity of its center of mass through Equation 8.13. Consequently, the kinetic energy of the i-th link of the robot can be computed as follows:

$$K_i = \frac{1}{2} m_i \boldsymbol{v}_{c_i}^T \boldsymbol{v}_{c_i} + \frac{1}{2} \boldsymbol{\omega}_i^T \, {}^C\boldsymbol{I}_i \boldsymbol{\omega}_i \tag{8.20}$$

In this equation, the vectors of linear and angular velocities with respect to any arbitrary coordinate system can be utilized since the result of their inner product is invariant with respect to that coordinate system. However, special attention must be paid to the link's moment of inertia matrix. As mentioned, this matrix can be readily obtained for each link with respect to its own center of mass coordinate system using tables or engineering CAD software. However, if we assume that the angular velocity calculations are performed in the reference coordinate system $\{0\}$, it is natural that the inertia matrix should be rewritten in this system, as well. Fortunately, this transformation can be easily achieved using the rotation matrix ${}^0\boldsymbol{R}_i$ and the relationship given in Equation 8.8. This means that if all calculations are performed in the reference coordinate system, the following equation can be employed:

$${}^0\boldsymbol{I}_i = {}^0\boldsymbol{R}_i \, {}^C\boldsymbol{I}_i \, {}^0\boldsymbol{R}_i^T \tag{8.21}$$

On the other hand, it should be noted that, depending on the complexity of the robot's kinematic structure, performing all calculations in the reference coordinate system $\{0\}$ may impose a significant computational burden. For this reason, in many cases, it is preferable to carry out these calculations in the coordinate system of each link. With these notes, ultimately, the total kinetic energy of the robot can be obtained by summing the kinetic energies of its components.

$$K = \sum_{i=1}^{n} K_i \tag{8.22}$$

Note that in this calculation, the accurate determination of the center of mass velocities of the robot's links is of particular importance. In robots with simple kinematic structures, the velocity of the center of mass of the robot's links can be intuitively obtained. However, in robots with complex kinematic structures, these calculations are performed with greater accuracy and ease, using recursive methods and the Jacobian matrix of the robot links. The details of this method will be elaborated in section 8.4.3.

To leverage the Euler-Lagrange equations, after calculating the system's kinetic energy, the potential energy of the robot components must be obtained in a similar manner. The potential energy of the robot components arises from gravity and, if present, the spring force of the link connectors.

$$P_i = -m_i \boldsymbol{g}^T \boldsymbol{p}_{c_i} + \frac{1}{2} k_i (\Delta q_i)^2 \tag{8.23}$$

In this equation, $\boldsymbol{g}$ is the gravitational acceleration vector, $\boldsymbol{p}_{c_i}$ is the vector from the center of mass of the robot's links to the reference potential level, and, if present, k_i is the stiffness coefficient of the link connectors at the robot's flexible joints. Thus, in the case of torsional compliance in revolute joints, this coefficient represents the torsional stiffness in the joint. Meanwhile, in the case of linear compliance in prismatic joints, this coefficient models the tensile stiffness in the joint. Ultimately, the total potential energy of the robot relative to the reference potential level is obtained from the sum of the potential energies of the robot components.

$$P = \sum_{i=1}^{n} P_i \tag{8.24}$$

By determining the potential and kinetic energies of the robot, it is sufficient to obtain the Lagrangian function, which is the difference between the kinetic and potential energies of the robot.

$$\mathcal{L} = K - P \tag{8.25}$$

In the following, by using the Lagrangian function, the dynamic equations of the robot are obtained using the Euler-Lagrange relationship.

$$\frac{d}{dt}\left(\frac{\partial \mathcal{L}}{\partial \dot{q}_i}\right) - \frac{\partial \mathcal{L}}{\partial q_i} = Q_i \qquad i = 1, 2, \ldots, n \tag{8.26}$$

This equation is used for each generalized coordinate. For this means, the partial derivative of the Lagrangian function is calculated once with respect to $\dot{q}_i$ while considering its subsequent derivative with respect to time. Furthermore, the partial derivative of the Lagrangian function with respect to q_i is calculated, and its result, along with the generalized force from the dynamic equation of motion, determines the dynamic formulation for that generalized coordinate.

8.4 Derivation Methods

The Euler-Lagrange method serves as a fundamental tool in deriving the equations of motion for dynamic systems. Two main approaches are commonly employed to derive these equations: the direct method and the recursive method. The direct method involves directly applying the Euler-Lagrange equations to the Lagrangian of the system. This method is straightforward and suitable for systems with a small number of degrees of freedom. It involves calculating the partial derivatives of the Lagrangian with respect to the generalized coordinates and their time derivatives and then substituting these derivatives into the Euler-Lagrange equations to obtain the dynamic formulation.

On the other hand, the recursive method is particularly useful for systems with a large number of degrees of freedom or complex structures. It involves decomposing the system into smaller sub-components, deriving the equations of motion for each sub-component separately, and then combining these equations to obtain the overall equations of motion for the entire system. This method is iterative and recursive in nature, allowing for the systematic analysis of complex systems by breaking them down into simpler parts.

Both methods have their advantages and limitations, and the choice between them depends on the specific characteristics of the dynamic system being analyzed. The direct method is more straightforward and suitable for simpler systems, while the recursive method offers a systematic approach for analyzing complex systems with numerous degrees of freedom. Let's begin the process of deriving the equation of motion for a robotic system by

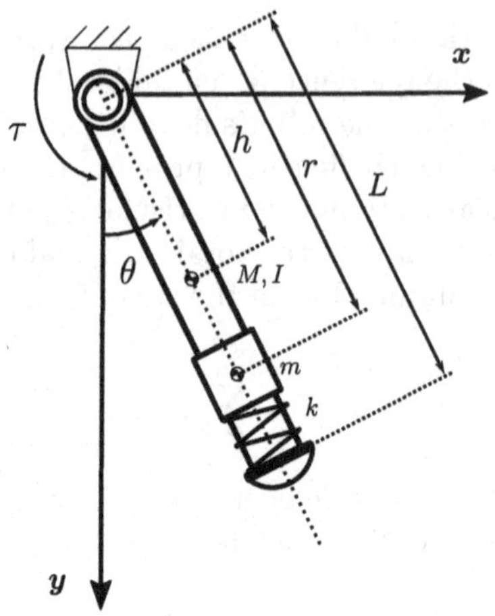

FIGURE 8.3
Schematic of a pendulum with suspended mass by a spring.

first introducing a motivational example. We will then proceed with the direct method and conclude with the recursive method.

8.4.1 Motivational Example

In order to describe how the Euler-Lagrange method is used to determine the dynamic formulation of multi-body systems, consider the pendulum and suspended mass system shown in Figure 8.3. In this system, the angular motion of the pendulum θ with length L and inertia parameters I, M is controlled by the applied torque τ. A weight with concentrated mass m is suspended from the pendulum by a linear spring with stiffness coefficient k. The distance from the pendulum's center of mass to the joint is denoted as h, and the longitudinal displacement of the mass along the pendulum axis is represented by the variable r. Friction in the revolute and prismatic motions is neglected.

To determine the dynamic formulation of this system, we first define generalized coordinates. Since the angular motion of the pendulum can be described by the variable $q_1 = \theta$ and the translational motion of the mass along the pendulum is described by the variable $q_2 = r$, these two variables serve as generalized coordinates providing a complete description of the system's motion.

$$\boldsymbol{q} = \begin{bmatrix} \theta & r \end{bmatrix}^T \tag{8.27}$$

Express the generalized forces along these variables in a corresponding manner. Since friction in the joints is neglected, and only the first joint is actuated, the generalized forces are easily determined.

$$\boldsymbol{Q} = \begin{bmatrix} \tau & 0 \end{bmatrix}^T \tag{8.28}$$

In order to determine the kinetic energy and potential energy of these two masses, it is first necessary to determine the linear velocities of the center of masses and their angular velocities. To calculate the linear velocity of the second mass $\boldsymbol{v}_{c_2}$ more accurately and avoid computational errors, it is advisable to first obtain the absolute position of this mass in the $x - y$ coordinates and then differentiate it.

$$\begin{aligned} x_{c_2} &= r \sin\theta \quad \rightarrow \quad \dot{x}_{c_2} = \dot{r}\sin\theta + r\dot{\theta}\cos\theta \\ y_{c_2} &= r\cos\theta \quad \rightarrow \quad \dot{y}_{c_2} = \dot{r}\cos\theta - r\dot{\theta}\sin\theta \end{aligned}$$

Hence,

$$v_{c_2}^2 = \dot{x}_{c_2}^2 + \dot{y}_{c_2}^2 = \dot{r}^2 + (r\dot{\theta})^2 + 2r\dot{r}\dot{\theta}\sin\theta\cos\theta - 2r\dot{r}\dot{\theta}\sin\theta\cos\theta$$

By this means, the linear and angular velocities for both masses are obtained as follows.

$$v_{c_1}^2 = (h\dot{\theta})^2, \;\; \omega_1 = \dot{\theta} \quad ; \quad v_{c_2}^2 = \dot{r}^2 + (h\dot{\theta})^2, \;\; \omega_2 = \dot{\theta}$$

By use of these quantities, it is straightforward to calculate the total kinetic energy of the system.

$$K = \frac{1}{2}M(h\dot{\theta})^2 + \frac{1}{2}I\dot{\theta}^2 + \frac{1}{2}m\left(\dot{r}^2 + (r\dot{\theta})^2\right) \tag{8.29}$$

Note that in this equation, the rotational kinetic energy of the second mass is considered zero because this body is modeled as a point mass, and its moment of inertia is assumed to be zero. Similarly, if we set the reference level for potential energy at the origin of the coordinate system, the potential energy of the system can be determined as follows.

$$P = -Mgh\cos\theta - mgr\cos\theta + \frac{1}{2}k(r - r_0)^2 \tag{8.30}$$

In this equation, r_0 represents the radial distance of the center of mass of the second mass when it is at the equilibrium state on the spring. As observed, the potential energy of the first mass is solely due to gravity, while the potential energy of the second mass is a combination of gravitational effects and the energy stored in the spring.

By determining the kinetic and potential energies of the system, the Lagrangian function for the system is expressed as follows:

$$\mathcal{L} = K - P = \frac{1}{2}(Mh^2 + mr^2 + I)\dot{\theta}^2 + \frac{1}{2}m\dot{r}^2 + (Mh + mr)g\cos\theta + \frac{1}{2}k(r - r_0)^2 \tag{8.31}$$

Now, for $q_1 = \theta$ and $Q_1 = \tau$, apply the Euler-Lagrange equation for the generalized variable q_1.

$$\frac{d}{dt}\left(\frac{\partial \mathcal{L}}{\partial \dot{\theta}}\right) - \frac{\partial \mathcal{L}}{\partial \theta} = \tau$$

For this purpose, the necessary differentiation can be performed in two steps, First find the partial derivative result,

$$\frac{d}{dt}\left((Mh^2 + mr^2 + I)\dot{\theta}\right) + (Mh + mr)g\sin\theta = \tau$$

Then, determine the equation of motion for the first generalized coordinate by obtaining the time derivative and simplifying the expression.

$$(Mh^2 + mr^2 + I)\ddot{\theta} + 2mr\dot{r}\dot{\theta} + (Mh + mr)g\sin\theta = \tau \tag{8.32}$$

Now, for $q_2 = r$ and $Q_2 = 0$, apply the Euler-Lagrange equation for the second generalized coordinate.

$$\frac{d}{dt}\left(\frac{\partial \mathcal{L}}{\partial \dot{r}}\right) - \frac{\partial \mathcal{L}}{\partial r} = 0$$

By differentiating and simplifying, the equation of motion for the second generalized coordinate is also obtained.

$$m\ddot{r} - mr\dot{\theta}^2 - mg\cos\theta + k(r - r_0) = 0 \tag{8.33}$$

Equations 8.32 and 8.33 describe the system's dynamic formulation with two coupled second-order nonlinear differential equations. As inferred from these equations, the inputs to these differential equations consist solely of the system's actuator torque, and there are no support forces playing a role in these equations.

8.4.2 Direct Method

In many robots with a relatively simple kinematic structure, the dynamic formulation can be directly obtained using the Euler-Lagrange method. Consider a serial robot with n actuated joints, where the generalized coordinates are $\boldsymbol{q} = [q_1, q_2, \ldots, q_n]^T$ and the generalized forces are $\boldsymbol{Q} = [Q_1, Q_2, \ldots, Q_n]^T$.

By defining the Lagrangian function as $\mathcal{L} = K(\boldsymbol{q}, \dot{\boldsymbol{q}}) - P(\boldsymbol{q})$, the Euler-Lagrange equations can be expressed in vector form.

$$\frac{d}{dt}\left(\frac{\partial \mathcal{L}}{\partial \dot{\boldsymbol{q}}}\right) - \frac{\partial \mathcal{L}}{\partial \boldsymbol{q}} = \boldsymbol{Q} \tag{8.34}$$

In the Lagrangian function, the kinetic energy is a function of the generalized variables and their derivatives, whereas the potential energy of the body is only a function of the generalized variables. By substituting the Lagrangian function into the equation 8.34, the equation can be expressed as:

$$\frac{d}{dt}\left(\frac{\partial K}{\partial \dot{\boldsymbol{q}}}\right) - \frac{\partial K}{\partial \boldsymbol{q}} + \frac{\partial P}{\partial \boldsymbol{q}} = \boldsymbol{Q} \tag{8.35}$$

To determine the closed-form of the dynamic formulation for the robot, the system's kinetic energy can be rewritten in terms of the system's mass matrix $\boldsymbol{M}(\boldsymbol{q})$ as follows:

$$K = \frac{1}{2}\dot{\boldsymbol{q}}^T \boldsymbol{M}(\boldsymbol{q})\dot{\boldsymbol{q}} \tag{8.36}$$

In addition, the gravitational vector $\mathfrak{g}$ can also be defined in terms of the system's potential energy.

$$\mathfrak{g}(\boldsymbol{q}) = \frac{\partial P(\boldsymbol{q})}{\partial \boldsymbol{q}} \tag{8.37}$$

By substituting these two quantities into 8.34 and simplifying, we will arrive at the closed-form dynamic formulation for the robot.

$$\boldsymbol{M}(\boldsymbol{q})\ddot{q} + \mathfrak{v}(\boldsymbol{q}, \dot{\boldsymbol{q}}) + \mathfrak{g}(\boldsymbol{q}) = \boldsymbol{Q} \tag{8.38}$$

Where $\mathfrak{v}(\boldsymbol{q}, \dot{\boldsymbol{q}})$ is the vector of Coriolis and centrifugal forces of the robot, which is derived from the following relation.

$$\mathfrak{v}(\boldsymbol{q}, \dot{\boldsymbol{q}}) = \left(\dot{\boldsymbol{M}}(\boldsymbol{q})\dot{\boldsymbol{q}} - \frac{\partial K}{\partial \boldsymbol{q}}\right) \tag{8.39}$$

In the following section, we will illustrate the process of performing these calculations and determining the mass matrix $\boldsymbol{M}(\boldsymbol{q})$, the gravitational vector $\mathfrak{g}(\boldsymbol{q})$, and the Coriolis and centrifugal vector $\mathfrak{v}(\boldsymbol{q}, \dot{\boldsymbol{q}})$ in the context of robot dynamic formulation, in a case study.

Case Study 8.1 (Planar 2R Robot) *Consider the planar 2R robot depicted in Figure 8.4. Use the Euler-Lagrange method to determine the dynamic formulation for the robot in closed-form and find the mass matrix, gravitational vector, Coriolis and centrifugal vector.*

Solution: As shown in Figure 8.4, the lengths of the robot's links are represented by a_i, and their mass and moment of inertia are denoted by m_i, and I_i, respectively. The actuator torques are given by τ_i. Furthermore, the centers of mass of the links are

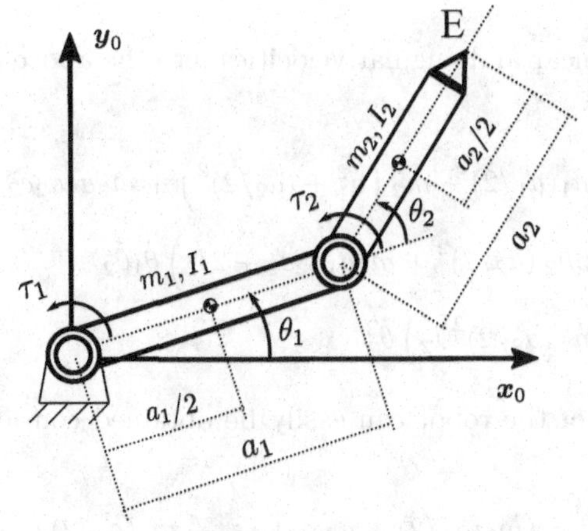

FIGURE 8.4
Schematic of the planar 2R robot in determining its dynamics formulations.

placed at the midpoint of their lengths, i.e., $a_i/2$. The vector of generalized coordinates is denoted by $\boldsymbol{q} = [\theta_1, \theta_2]^T$, and neglecting friction torques, the vector of generalized forces is given by $\boldsymbol{Q} = [\tau_1, \tau_2]^T$.

To determine the kinetic and potential energy of these two links, we first find the linear velocities of their center of masses and their angular velocities. To do this, we first obtain their absolute positions relative to the reference coordinate system $\{0\}$ and then differentiate is with respect to time to find the linear velocities.

$$
\begin{aligned}
x_{c_1} &= \frac{a_1}{2}c_1 && \rightarrow && \dot{x}_{c_1} = -\frac{a_1}{2}\dot{\theta}_1 s_1 \\
y_{c_1} &= \frac{a_1}{2}s_1 && \rightarrow && \dot{y}_{c_1} = \frac{a_1}{2}\dot{\theta}_1 c_1 \\
v_{c_1}^2 &= \dot{x}_{c_1}^2 + \dot{y}_{c_1}^2 && = && \left(\frac{a_1}{2}\dot{\theta}_1\right)^2
\end{aligned}
$$

Similarly,

$$
\begin{aligned}
x_{c_2} &= a_1 c_1 + \frac{a_2}{2}c_{12} && \rightarrow && \dot{x}_{c_2} = -a_1\dot{\theta}_1 s_1 - \frac{a_2}{2}(\dot{\theta}_1 + \dot{\theta}_2)s_{12} \\
y_{c_2} &= a_1 s_1 + \frac{a_2}{2}s_{12} && \rightarrow && \dot{x}_{c_2} = a_1\dot{\theta}_1 c_1 + \frac{a_2}{2}(\dot{\theta}_1 + \dot{\theta}_2)c_{12} \\
v_{c_2}^2 &= \dot{x}_{c_1}^2 + \dot{y}_{c_1}^2 && = && \left(a_1\dot{\theta}_1\right)^2 + \left(\frac{a_2}{2}(\dot{\theta}_1 + \dot{\theta}_2)\right)^2 + a_1 a_2\dot{\theta}_1(\dot{\theta}_1 + \dot{\theta}_2)c_2
\end{aligned}
$$

Determine angular velocities for the two links.

$$
\omega_1 = \dot{\theta}_1 \quad ; \quad \omega_2 = (\dot{\theta}_1 + \dot{\theta}_2)
$$

Given the linear and angular velocities, the total kinetic energy of the system can be obtained using the following equation:

$$
K = \frac{1}{2}m_1 v_{c_1}^2 + \frac{1}{2}I_1\dot{\omega}_1^2 + \frac{1}{2}m_2 v_{c_2}^2 + \frac{1}{2}I_2\dot{\omega}_2^2
$$

By substituting the linear and angular velocities into the above equation and simplifying, we will have:

$$\begin{aligned} K &= \frac{1}{2}\left(m_1(a_1/2) + m_2\left(a_1^2 + (a_2/2)^2\right) + m_2 a_1 a_2 c_2 + I_1 + I_2\right)\dot{\theta}_1^2 \\ &+ \frac{1}{2}\left(2m_2\left(a_2/2\right)^2 + m_2 a_1 a_2 c_2 + 2I_2\right)\dot{\theta}_1\dot{\theta}_2 \\ &+ \frac{1}{2}\left(m_2\left(a_2/2\right)^2 I_2\right)\dot{\theta}_2^2 \end{aligned} \tag{8.40}$$

The potential energy of the robot can easily be obtained considering the structure of the robot as follows:

$$P = (m_1(a_1/2) + m_2 a_1)\, g\, s_1 + m_2(a_2/2)\, g\, s_{12} \tag{8.41}$$

Given the kinetic and potential energies of the system, the Lagrangian function for the system is expressed as:

$$\mathcal{L} = K - P$$

To obtain the dynamic formulation of the robot, the Euler-Lagrange equation can be applied for each of the generalized coordinates. For the first coordinate, considering $q_1 = \theta_1$ and $Q_1 = \tau_1$, we have:

$$\begin{aligned} &\frac{d}{dt}\left\{m_1(a_1/2) + m_2\left(a_1^2 + (a_2/2)^2\right) + m_2 a_1 a_2 c_2 + I_1 + I_2\right\}\dot{\theta}_1 + \\ &\frac{1}{2}\left(m_2 a_2 + m_2 a_1 a_2 c_2 + 2I_2\right)\dot{\theta}_2 + \left(\frac{1}{2}m_1 a_1 + m_2 a_1\right) g s_1 + \frac{1}{2} m_2 a_2 g s_{12} = \tau_1 \end{aligned}$$

The equation of motion for the first component of the system is obtained through differentiation and simplification.

$$\begin{aligned} &\left(m_1(a_1/2) + m_2\left(a_1^2 + (a_2/2)^2\right) + m_2 a_1 a_2 c_2 + I_1 + I_2\right)\ddot{\theta}_1 + \\ &\left(m_2(a_2/2)^2 + \frac{1}{2}m_2 a_1 a_2 c_2 + I_2\right)\ddot{\theta}_2 - m_2 a_1 a_2 s_2 \dot{\theta}_1\dot{\theta}_2 - \frac{1}{2}m_2 a_1 a_2 s_2 \dot{\theta}_2^2 \\ &\left(\frac{1}{2}m_1 a_1 + m_2 a_1\right) g c_1 + \frac{1}{2}m_2 a_2 g c_{12} = \tau_1 \end{aligned} \tag{8.42}$$

Now, for $q_2 = \theta_2$ and $Q_2 = \tau_2$, applying the Euler-Lagrange equation, we get:

$$\begin{aligned} &\frac{d}{dt}\left\{\left(m_2(a_2/2)^2 + \frac{1}{2}m_2 a_1 a_2 c_2 + I_2\right)\dot{\theta}_1 + \left(m_2(a_2/2)^2 + I_2\right)\dot{\theta}_2\right\} + \\ &\frac{1}{2}m_2 a_2 g c_{12} = \tau_2 \end{aligned}$$

By performing the differentiation and simplification, the equation of motion for the second component of the system is obtained.

$$\begin{aligned} &\left(m_2(a_2/2)^2 + \frac{1}{2}m_2 a_1 a_2 c_2 + I_2\right)\ddot{\theta}_1 + \left(m_2(a_2/2)^2 + I_2\right)\ddot{\theta}_2 \\ &+\frac{1}{2}m_2 a_1 a_2 s_2 \dot{\theta}_1^2 + \frac{1}{2}m_2 a_2 g c_{12} = \tau_2 \end{aligned} \tag{8.43}$$

As observed, equations 8.42 and 8.43 describe the system's motion with two coupled second-order nonlinear differential equations. Using these equations, the robot's dynamic formulation can easily be represented in closed-form.

$$\boldsymbol{M}(\boldsymbol{q})\ddot{q} + \boldsymbol{\mathfrak{v}}(\boldsymbol{q}, \dot{\boldsymbol{q}}) + \boldsymbol{\mathfrak{g}}(\boldsymbol{q}) = \boldsymbol{Q}$$

which in this case, the mass matrix is equal to:

$$\boldsymbol{M}(\boldsymbol{q}) = \begin{bmatrix} M_{11} & M_{12} \\ M_{12} & M_{22} \end{bmatrix} \tag{8.44}$$

where,

$$\begin{aligned} M_{11} &= m_1(a_1/2)^2 + m_2\left(a_1^2 + (a_2/2)^2\right) + m_2a_1a_2c_2 + I_1 + I_2 \\ M_{12} &= m_2(a_2/2)^2 + \frac{1}{2}m_2a_1a_2c_2 + I_2 \\ M_{22} &= m_2(a_2/2)^2 + I_2 \end{aligned}$$

The vector of Coriolis and centrifugal terms, $\boldsymbol{\mathfrak{v}}(\boldsymbol{q}, \dot{\boldsymbol{q}})$ is equal to:

$$\boldsymbol{\mathfrak{v}}(\boldsymbol{q}, \dot{\boldsymbol{q}}) = \begin{bmatrix} -m_2a_1a_2s_2\left(\dot{\theta}_1\dot{\theta}_2 + \frac{1}{2}\dot{\theta}_2^2\right) \\ \frac{1}{2}m_2a_1a_2s_2\dot{\theta}_1^2 \end{bmatrix} \tag{8.45}$$

and the gravitational vector $\boldsymbol{\mathfrak{g}}(\boldsymbol{q})$ is also determined as follows:

$$\boldsymbol{\mathfrak{g}}(\boldsymbol{q}) = \begin{bmatrix} \left(\frac{1}{2}m_1a_1 + m_2a_1\right)c_1 + \frac{1}{2}m_2a_2c_{12} \\ \frac{1}{2}m_2a_2c_{12} \end{bmatrix} g \tag{8.46}$$

As observed in this case study, two parts of the calculations involve significant complexity, and performing them manually, even for the simplest planar robot, can be very time-consuming. The first part involves complex calculations for determining the linear velocities of the link centers of mass. In the next section, the recursive calculation of these velocities and the use of the Jacobian matrix of the links to simplify these calculations will be presented. The second time-consuming part in these calculations is the required differentiation in the Euler-Lagrange equations and their simplification. Fortunately, with the use of symbolic computing toolbox of software like MATLAB® and Python, this part can also be handled by the computer, making these calculations more straightforward.

8.4.3 Recursive Method

As we have seen in the Euler-Lagrange method, calculating the velocity of links' center of mass involves complex computations. In this section, we perform these calculations recursively using the Jacobian matrices of the links. To do this, consider the six-dimensional array of the center of mass linear and angular velocity of the link denoted by $\dot{\boldsymbol{\mathcal{X}}}_{c_i}$. With this definition and considering the Jacobian matrix $\boldsymbol{J}_i(\boldsymbol{q})$ as the mapping between joint space and task space velocities, we can write,

$$\dot{\boldsymbol{\mathcal{X}}}_{c_i} = \boldsymbol{J}_i(\boldsymbol{q})\dot{\boldsymbol{q}} \tag{8.47}$$

where

$$\dot{\boldsymbol{\mathcal{X}}}_{c_i} = \begin{bmatrix} \boldsymbol{v}_{c_i} \\ \boldsymbol{\omega}_i \end{bmatrix} \quad \text{and} \quad \boldsymbol{J}_i(\boldsymbol{q}) = \begin{bmatrix} \boldsymbol{J}_{v_i} \\ \boldsymbol{J}_{\omega_i} \end{bmatrix} \tag{8.48}$$

In this context, $\boldsymbol{J}_i$ denotes the Jacobian matrix for the link, while $\boldsymbol{J}_{v_i}$ and $\boldsymbol{J}_{\omega_i}$ represent submatrices associated with linear and angular velocity, respectively. Consequently, the linear center of mass velocity and the link's angular velocity are derived based on these submatrices within the Jacobian matrices pertaining to the links.

$$\boldsymbol{v}_{c_i} = \boldsymbol{J}_{v_i}(\boldsymbol{q})\dot{\boldsymbol{q}} \qquad \boldsymbol{\omega}_{c_i} = \boldsymbol{J}_{\omega_i}(\boldsymbol{q})\dot{\boldsymbol{q}} \tag{8.49}$$

With this definition, it is possible to obtain the Jacobian matrix of the robot link $\boldsymbol{J}_i(\boldsymbol{q})$, following the approach used for calculating the manipulator Jacobian matrix. It suffices to substitute the center of mass of the relevant link as the robot's end-effector point E and perform the calculations for this point.

Following the general method for determining the Jacobian matrix, the Jacobian matrix of the i-th link in a serial robot, composed of n main revolute and prismatic joints, can be described by its columns as follows.

$$\boldsymbol{J}_i(\boldsymbol{q}) = \left[\ \boldsymbol{J}_i^1 \mid \boldsymbol{J}_i^2 \mid \cdots \mid \boldsymbol{J}_i^j \mid 0 \mid \cdots \mid 0\ \right]_{6\times n} \tag{8.50}$$

Note that only the i-th first column of this matrix is non-zero, as the linear and angular velocities of link i depend solely on the motions of its preceding joints. The columns of the Jacobian matrix of the i-th link for $j \leq i$ in **revolute joints** are given by the equation:

$$\boldsymbol{J}_i^j = \begin{bmatrix} \boldsymbol{z}_{j-1} \times \boldsymbol{p}_{c_i}^{j-1} \\ \boldsymbol{z}_{j-1} \end{bmatrix} \tag{8.51}$$

and in **prismatic joints** by the equation:

$$\boldsymbol{J}_i^j = \begin{bmatrix} \boldsymbol{z}_{j-1} \\ \boldsymbol{0} \end{bmatrix} \tag{8.52}$$

In these equations, as shown in Figure 8.5, $\boldsymbol{z}_{j-1}$ is the axis of rotation (or translation) of the j-th revolute (or prismatic) joint, and the vector $\boldsymbol{p}_{c_i}^{j-1}$ is the *distance vector from the center of mass of link i to the origin of the coordinate system $j-1$*. In these calculations, determining the vectors $\boldsymbol{z}_j$ is usually straightforward. To obtain the vector $\boldsymbol{p}_{c_i}^{j-1}$, according to Figure 8.5, one can use the following vector relationship:

$$\boldsymbol{p}_{c_i}^{j-1} = {}^0\boldsymbol{p}_{c_i} - {}^0\boldsymbol{p}_{j-1} \tag{8.53}$$

Here, ${}^0\boldsymbol{p}_{c_i}$ is the position vector of the center of mass of link i, and ${}^0\boldsymbol{p}_{j-1}$ is the position vector of the origin of the coordinate system $j-1$. Thus, in less complex robotic structures, the vectors forming the Jacobian matrix can be directly obtained using this equation.

If the kinematic structure of the robot is more complex, these vectors can be obtained using Denavit-Hartenberg parameters. For this purpose, for $j = 1, 2, \ldots, n$, the axis vectors $\boldsymbol{z}_{j-1}$ are first obtained recursively using the following relationship.

$$\boldsymbol{z}_{j-1} = {}^0\boldsymbol{R}_{j-1} \begin{bmatrix} 0 \\ 0 \\ 1 \end{bmatrix} \tag{8.54}$$

Here, ${}^0\boldsymbol{R}_{j-1}$ is the rotation matrix from the coordinate system $j-1$ to the reference coordinate system $\{0\}$.

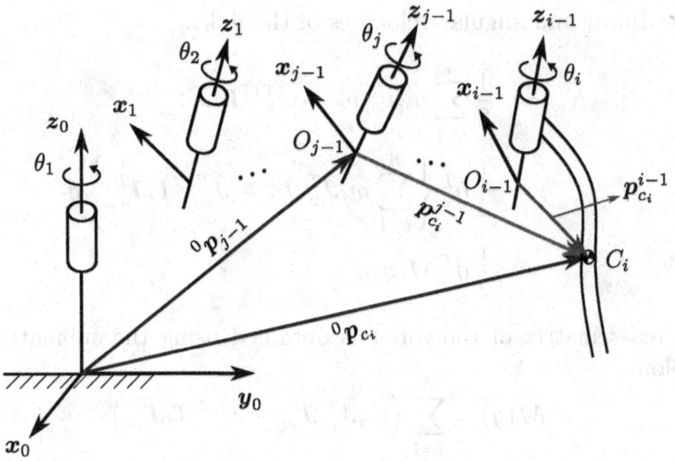

FIGURE 8.5
Representation of vectors used in the general method for determining the Jacobian matrix of the i-th link.

Furthermore, for $j = i-1, \ldots, 2, 1$, the vector $p_{c_i}^{j-1}$ is obtained recursively in the backward direction using the following relationship.

$$ {}^0p_{c_i}^{j-1} = {}^0R_{j-1}\,{}^{j-1}p_j^{j-1} + {}^0p_{c_i}^{j} \tag{8.55}$$

where, by using the Denavit-Hartenberg parameters, we have:

$$ {}^{j-1}p_j^{j-1} = \begin{bmatrix} a_j c_j \\ a_j s_j \\ d_j \end{bmatrix} \tag{8.56}$$

For $j = i-1$, we first determine the value of $p_{c_i}^j$ intuitively, considering the location of the center of mass of link i with respect to the origin of the coordinate system $i-1$. Then, using the rotation matrix ${}^0R_{j-1}$, we proceed with the recursive calculations by computing the vector ${}^0p_{c_i}^j$. Note that in these general equations, all vectors are calculated with respect to the reference coordinate system $\{0\}$, and the link Jacobian matrix with respect to the reference coordinate system ${}^0J_i(q)$ will be obtained. However, all calculations can be performed in another coordinate system, and the Jacobian matrix can be determined with respect to that coordinate system.

By specifying the Jacobian matrix of the links, the submatrices of its linear and angular velocities can be easily separated as follows.

$$ J_i(q) = \begin{bmatrix} J_{v_i} \\ J_{\omega_i} \end{bmatrix} \tag{8.57}$$

where

$$ J_{v_i}(q) = \left[\, J_{v_i}^1 \mid J_{v_i}^2 \mid \cdots \mid J_{v_i}^j \mid 0 \ \cdots \ 0 \,\right]_{3\times n} \tag{8.58}$$

$$ J_{\omega_i}(q) = \left[\, J_{\omega_i}^1 \mid J_{\omega_i}^2 \mid \cdots \mid J_{\omega_i}^j \mid 0 \mid \cdots \mid 0 \,\right]_{3\times n} \tag{8.59}$$

In this way, the kinetic energy of the robot can be rewritten in terms of the submatrices of the Jacobian for linear and angular velocities of the links.

$$\begin{aligned} K &= \frac{1}{2}\sum_{i=1}^{n} m_i \boldsymbol{v}_{c_i}^T \boldsymbol{v}_{c_i} + \boldsymbol{\omega}_i^{T\,C}\boldsymbol{I}_i\boldsymbol{\omega}_i \\ &= \frac{1}{2}\dot{\boldsymbol{q}}^T\left\{\sum_{i=1}^{n} m_i \boldsymbol{J}_{v_i}^T\boldsymbol{J}_{v_i} + \boldsymbol{J}_{\omega_i}^{T\,C}\boldsymbol{I}_i\boldsymbol{J}_{\omega_i}^T\right\}\dot{\boldsymbol{q}} \qquad (8.60) \\ &= \frac{1}{2}\dot{\boldsymbol{q}}^T\boldsymbol{M}(\boldsymbol{q})\dot{\boldsymbol{q}} \end{aligned}$$

Therefore, the mass matrix of the robot is obtained using the submatrices of the links' Jacobian as follows.

$$\boldsymbol{M}(\boldsymbol{q}) = \sum_{i=1}^{n}\left(m_i\boldsymbol{J}_{v_i}^T\boldsymbol{J}_{v_i} + \boldsymbol{J}_{\omega_i}^{T\,C}\boldsymbol{I}_i\boldsymbol{J}_{\omega_i}^T\right) \qquad (8.61)$$

Gravity vector $\mathbf{g}(\mathbf{q})$, in the absence of flexibility in the robot joints, is defined using the potential energy of the system. Assuming gravitational potential energy in the form:

$$P = -\sum_{i=1}^{n} m_i\boldsymbol{g}^T\boldsymbol{p}_{c_i} \qquad (8.62)$$

The gravity vector can be obtained from its partial derivative with respect to the generalized coordinates.

$$\begin{aligned} \mathfrak{g}(\boldsymbol{q}) &= \frac{\partial P(\boldsymbol{q})}{\partial \boldsymbol{q}} = -\sum_{i=1}^{n}\frac{\partial}{\partial \boldsymbol{q}}\left(m_i\boldsymbol{g}^T\boldsymbol{p}_{c_i}\right) \\ &= -\sum_{i=1}^{n} m_i\boldsymbol{J}_{v_i}^T\boldsymbol{g} \qquad (8.63) \end{aligned}$$

It should be noted that in these calculations, the relation $\frac{\partial \boldsymbol{p}_{c_i}}{\partial \boldsymbol{q}} = \boldsymbol{J}_{v_i}(\boldsymbol{q})$ is used. Considering matrix multiplication, the transpose of this matrix is used to ensure the correct dimensions in the matrix multiplication.

As seen, the mass matrix and the gravity vector can be easily obtained by having the Jacobian matrix of the links. By determining these two quantities, the closed-form robot's dynamic formulation, as given in the equation 8.38, can be established, where the Coriolis and centrifugal vector $\mathfrak{v}(\boldsymbol{q}, \dot{\boldsymbol{q}})$ can be calculated based on the relation 8.39. The characteristics and representation of this vector will be given in the subsequent chapter.

8.5 Case Studies

In the subsequent sections, the dynamic formulations of four distinct case studies are presented. Initially, the planar 2<u>R</u> and 3<u>R</u> robots are examined, followed by the dynamic formulation for the SCARA robot and 3 DOF Elbow manipulator.

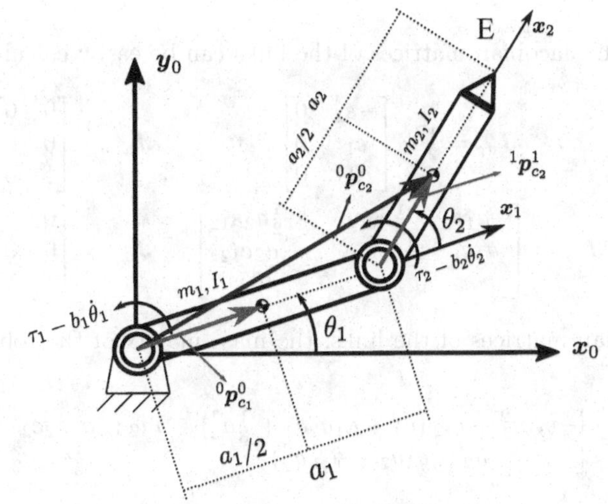

FIGURE 8.6
Schematic of a planar 2R robot and the representation of necessary vectors in determining the Jacobian matrix of its links.

8.5.1 Planar 2R Robot

Case Study 8.2 (Planar 2R Robot) *Consider the planar 2R robot depicted in Figure 8.6. Assume the robot's links are modeled as slender rods and suppose there exist viscous friction torques at the joints. Utilize the recursive Euler-Lagrange method to formulate the dynamic motion equations of the robot and determine its mass matrix, gravity vector, and Coriolis and centrifugal vector.*

Solution: As seen in Figure 8.6, the length of the robot's links is denoted by a_i, their mass by m_i, and the torque applied at the joints by τ_i. The center of mass of the links is positioned at mid-length, $a_i/2$. Assuming slender rods in the links, their moment of inertia about their center of mass is given by $I_i = \frac{1}{12}m_i a_i^2$.

In this robot, generalized coordinates can be represented by the joint angle vector $\boldsymbol{q} = [\theta_1, \theta_2]^T$. Now, considering viscous friction torques at the joints, the generalized forces become $\boldsymbol{Q} = [\tau_1 - b_1\dot{\theta}_1, \tau_2 - b_2\dot{\theta}_2]^T$, where b_i is the linear friction coefficient at the joints.

Next, to determine the kinetic energy, the Jacobian matrices of the links are calculated. For this purpose, the vectors representing the distance from the center of mass of the links to their corresponding joints relative to the reference coordinate system $\{0\}$ are first obtained.

$$
{}^0\boldsymbol{p}^0_{c_1} = \frac{1}{2}a_1\begin{bmatrix} c_1 \\ s_1 \\ 0 \end{bmatrix}, \quad {}^1\boldsymbol{p}^1_{c_2} = \frac{1}{2}a_2\begin{bmatrix} c_{12} \\ s_{12} \\ 0 \end{bmatrix} \rightarrow {}^0\boldsymbol{p}^0_{c_2} = \begin{bmatrix} a_1c_1 + \frac{1}{2}a_2c_{12} \\ a_1s_1 + \frac{1}{2}a_2s_{12} \\ 0 \end{bmatrix}
$$

By this means, the Jacobian matrices of the links can be easily calculated.

$$J_{v_1} = \frac{1}{2}a_1\begin{bmatrix} -s_1 & 0 \\ c_1 & 0 \\ 0 & 0 \end{bmatrix}, \qquad J_{\omega_1} = \begin{bmatrix} 0 & 0 \\ 0 & 0 \\ 1 & 0 \end{bmatrix}$$

$$J_{v_2} = \begin{bmatrix} -a_1s_1 - \frac{1}{2}a_2s_{12} & -\frac{1}{2}a_2s_{12} \\ a_1c_1 + \frac{1}{2}a_2c_{12} & \frac{1}{2}a_2c_{12} \\ 0 & 0 \end{bmatrix}, \quad J_{\omega_2} = \begin{bmatrix} 0 & 0 \\ 0 & 0 \\ 1 & 1 \end{bmatrix}$$

Given the Jacobian matrices of the links, the mass matrix of the robot can be determined.

$$M(q) = \begin{bmatrix} \frac{1}{3}m_1a_1^2 + m_2\left(a_1^2 + a_1a_2c_2 + \frac{1}{3}a_2^2\right) & m_2\left(\frac{1}{2}a_1a_2c_2 + \frac{1}{3}a_2^2\right) \\ m_2\left(\frac{1}{2}a_1a_2c_2 + \frac{1}{3}a_2^2\right) & \frac{1}{3}m_2a_2^2 \end{bmatrix}$$

The gravity vector $\mathfrak{g}(q)$ can also be easily computed based on the Jacobian matrices of the links as follows.

$$\mathfrak{g}(q) = -\sum_{i=1}^{n} m_i J_{v_i}^T g = \begin{bmatrix} \left(\frac{1}{2}m_1 + m_2\right)a_1c_1 + \frac{1}{2}m_2a_2c_{12} \\ \frac{1}{2}m_2a_2c_{12} \end{bmatrix} g$$

The Coriolis and centrifugal vector $\mathfrak{v}(q, \dot{q})$ is also determined as:

$$\mathfrak{v}(q, \dot{q}) = \begin{bmatrix} -m_2a_1a_2s_2\left(\dot{\theta}_1\dot{\theta}_2 + \frac{1}{2}\dot{\theta}_2^2\right) \\ \frac{1}{2}m_2a_1a_2s_2\dot{\theta}_1^2 \end{bmatrix}$$

By determining the required matrices and vectors, the dynamic formulation for this robot can be rewritten in closed-form as follows.

$$M(q)\ddot{q} + \mathfrak{v}(q, \dot{q}) + \mathfrak{g}(q) = Q$$

8.5.2 Planar 3$\underline{\text{R}}$ Robot

Case Study 8.3 (Planar 3$\underline{\text{R}}$ Robot) *Consider the planar 3$\underline{R}$ robot shown in Figure 8.7. Similar to the previous example, assume the robot's links are modeled as slender rods, and suppose there exists linear friction torques at the joints. Utilize the recursive Euler-Lagrange method to formulate the dynamic formulation of the robot and determine its mass matrix, gravity, and Coriolis and centrifugal vectors.*

Solution: Similar to the previous example, assuming slender rods in the links, their inertia about their center of mass is given by $I_i = \frac{1}{12}m_ia_i^2$. In this robot, generalized coordinates can be represented by the joint angle vector $q = [\theta_1, \theta_2, \theta_3]^T$. Considering viscous friction torques at the joints, the generalized forces become $Q = [\tau_1 - b_1\dot{\theta}_1, \tau_2 - b_2\dot{\theta}_2, \tau_3 - b_3\dot{\theta}_3]^T$, where b_i is the linear friction coefficient at the joints. In order to determine the kinetic energy, the Jacobian matrices of the links are calculated. For this purpose, the vectors representing the distance from the center of mass of the links to their corresponding joints relative to the reference coordinate system $\{0\}$ are first obtained.

$$ {}^0\boldsymbol{p}^0_{c_1} = \frac{1}{2}a_1 \begin{bmatrix} c_1 \\ s_1 \\ 0 \end{bmatrix}, {}^1\boldsymbol{p}^1_{c_2} = \frac{1}{2}a_2 \begin{bmatrix} c_{12} \\ s_{12} \\ 0 \end{bmatrix} \rightarrow {}^0\boldsymbol{p}^0_{c_2} = \begin{bmatrix} a_1c_1 + \frac{1}{2}a_2c_{12} \\ a_1s_1 + \frac{1}{2}a_2s_{12} \\ 0 \end{bmatrix} $$

In a similar manner, employing recursive computations, we will obtain:

$$ {}^2\boldsymbol{p}^2_{c_3} = \frac{1}{2}a_3 \begin{bmatrix} c_{123} \\ s_{123} \\ 0 \end{bmatrix}, {}^1\boldsymbol{p}^1_{c_3} = \begin{bmatrix} a_2c_{12} + \frac{1}{2}a_3c_{123} \\ a_2s_{12} + \frac{1}{2}a_3s_{123} \\ 0 \end{bmatrix}, {}^0\boldsymbol{p}^0_{c_3} = \begin{bmatrix} a_1c_1 + a_2c_{12} + \frac{1}{2}a_3c_{123} \\ a_1s_1 + a_2s_{12} + \frac{1}{2}a_3s_{123} \\ 0 \end{bmatrix} $$

Thus, the Jacobian matrices of the links can be readily calculated.

$$ \boldsymbol{J}_{v_1} = \frac{1}{2}a_1 \begin{bmatrix} -s_1 & 0 & 0 \\ c_1 & 0 & 0 \\ 0 & 0 & 0 \end{bmatrix}, \quad \boldsymbol{J}_{v_2} = \begin{bmatrix} -a_1s_1 - \frac{1}{2}a_2s_{12} & -\frac{1}{2}a_2s_{12} & 0 \\ a_1c_1 + \frac{1}{2}a_2c_{12} & \frac{1}{2}a_2c_{12} & 0 \\ 0 & 0 & 0 \end{bmatrix} $$

$$ \boldsymbol{J}_{v_3} = \begin{bmatrix} -a_1s_1 - a_2s_{12} - \frac{1}{2}a_3s_{123} & -a_2s_{12} - \frac{1}{2}a_3s_{123} & -\frac{1}{2}a_3s_{123} \\ a_1c_1 + a_2c_{12} + \frac{1}{2}a_3c_{123} & a_2c_{12} + \frac{1}{2}a_3c_{123} & \frac{1}{2}a_3c_{123} \\ 0 & 0 & 0 \end{bmatrix} $$

$$ \boldsymbol{J}_{\omega_1} = \begin{bmatrix} 0 & 0 & 0 \\ 0 & 0 & 0 \\ 1 & 0 & 0 \end{bmatrix}, \quad \boldsymbol{J}_{\omega_2} = \begin{bmatrix} 0 & 0 & 0 \\ 0 & 0 & 0 \\ 1 & 1 & 0 \end{bmatrix}, \quad \boldsymbol{J}_{\omega_3} = \begin{bmatrix} 0 & 0 & 0 \\ 0 & 0 & 0 \\ 1 & 1 & 1 \end{bmatrix} $$

By using the Jacobian matrices of the links, the mass matrix of the robot, which is a symmetric matrix, can be computed.

$$ \boldsymbol{M}(\boldsymbol{q}) = \sum_{i=1}^{3} m_i \left(\boldsymbol{J}^T_{v_i}\boldsymbol{J}_{v_i} + \boldsymbol{J}^T_{\omega_i}\,{}^C\boldsymbol{I}_i\boldsymbol{J}^T_{\omega_i} \right) = \begin{bmatrix} M_{11} & M_{12} & M_{13} \\ & M_{22} & M_{23} \\ & & M_{33} \end{bmatrix} $$

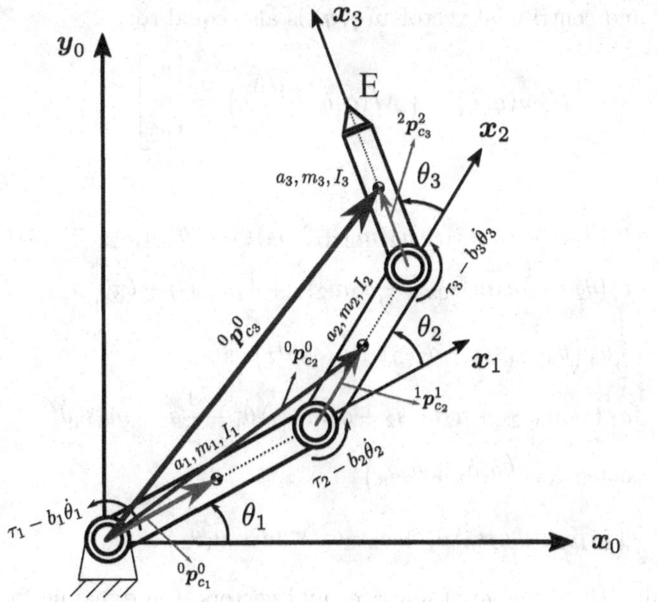

FIGURE 8.7
Schematic of a planar *3R* robot and the representation of necessary vectors in determining the Jacobian matrix of its links.

where

$$\begin{aligned}
M_{11} &= \left(\frac{1}{3}m_1 + m_2 + m_3\right)a_1^2 + \left(\frac{1}{3}m_2 + m_3\right)a_2^2 + m_3a_3^2 \\
&\quad (m_2 + 2m_3)\,a_1a_2c_2 + a_3m_3(a_1c_{23} + a_2c_3) \\
M_{12} &= \left(\frac{1}{3}m_2 + m_3\right)a_2^2 + m_3a_3\left(\frac{1}{3}a_3 + \frac{1}{2}a_2c_{23}\right) + \\
&\quad \frac{1}{2}m_2a_1a_2c_2 + m_3a_2(a_1c_2 + a_3c_3) \\
M_{13} &= \left(\frac{1}{3}a_3 + \frac{1}{2}a_2c_{23} + \frac{1}{2}a_2c_3\right)m_3a_3 \\
M_{22} &= \left(\frac{1}{3}m_2 + m_3\right)a_2^2 + \frac{1}{3}m_3a_3^2 + m_3a_2a_3c_3 \\
M_{23} &= \left(\frac{1}{3}a_3 + \frac{1}{2}a_2c_3\right)m_3a_3 \\
M_{33} &= \frac{1}{3}m_3a_3^2
\end{aligned}$$

The gravity vector $\mathfrak{g}(\boldsymbol{q})$ can also be easily calculated based on the Jacobian matrices of the links as follows.

$$\begin{aligned}
\mathfrak{g}(\boldsymbol{q}) &= -\sum_{i=1}^{3} m_i \boldsymbol{J}_{v_i}^T \boldsymbol{g} \\
&= \begin{bmatrix} \frac{1}{2}m_1a_1c_1 + m_2(a_1c_1 + \frac{1}{2}a_2c_{12}) + m_3\left(a_2c_{12} + \frac{1}{2}a_3c_{123}\right) \\ \frac{1}{2}m_2a_2c_{12} + m_3\left(a_2c_{12} + \frac{1}{2}a_3c_{123}\right) \\ \frac{1}{2}m_3a_3c_{123} \end{bmatrix} g
\end{aligned}$$

The Coriolis and centrifugal vector $\mathfrak{v}(\boldsymbol{q}, \dot{\boldsymbol{q}})$ is also equal to:

$$\mathfrak{v}(\boldsymbol{q}, \dot{\boldsymbol{q}}) = \left(\dot{\boldsymbol{M}}(\boldsymbol{q})\dot{\boldsymbol{q}} - \frac{\partial K}{\partial \boldsymbol{q}}\right) = \begin{bmatrix} \mathfrak{v}_1 \\ \mathfrak{v}_2 \\ \mathfrak{v}_2 \end{bmatrix}$$

where

$$\begin{aligned}
\mathfrak{v}_1 &= -\dot{\theta}_1(\dot{\theta}_2a_1(a_3m_3s_{23} + a_2(m_2 + 2m_3)s_2) + \dot{\theta}_3(a_1s_{23} + a_2s_3)a_3m_3) \\
&\quad -\dot{\theta}_2(\dot{\theta}_2a_1(\frac{1}{2}a_3m_3s_{23} + \frac{1}{2}a_2m_2s_2 + a_2m_3s_2) + \dot{\theta}_3(\frac{1}{2}a_1s_{23} + a_2s_3)a_3m_3) \\
&\quad -\frac{1}{2}\dot{\theta}_3\left(\dot{\theta}_3\left(a_1s_{23} + a_2s_3\right) + a_1s_{23}\dot{\theta}_2\right)a_3m_3 \\
\mathfrak{v}_2 &= \frac{1}{2}a_1\left(a_3m_3s_{23} + a_2m_2s_2 + a_2m_3s_2\right)\dot{\theta}_1^2 - \frac{1}{2}a_2a_3m_3s_3\dot{\theta}_3^2 \\
&\quad -a_2a_3m_3s_3\left(\dot{\theta}_1\dot{\theta}_3 + \dot{\theta}_2\dot{\theta}_3\right) \\
\mathfrak{v}_3 &= \frac{1}{2}\left(\left(a_1s_{23} + a_2s_3\right)\dot{\theta}_1^2 + a_2s_3\dot{\theta}_2^2 + 2a_2s_3\dot{\theta}_1\dot{\theta}_2\right)a_3m_3
\end{aligned}$$

By determining the designated matrices and vectors, the dynamic formulation of this robot can be rewritten in a closed-form as follows.

$$\boldsymbol{M}(\boldsymbol{q})\ddot{\boldsymbol{q}} + \mathfrak{v}(\boldsymbol{q}, \dot{\boldsymbol{q}}) + \mathfrak{g}(\boldsymbol{q}) = \boldsymbol{Q}$$

where,

$$Q = \begin{bmatrix} \tau_1 - b_1\dot{\theta}_1 \\ \tau_2 - b_2\dot{\theta}_2 \\ \tau_3 - b_3\dot{\theta}_3 \end{bmatrix}$$

8.5.3 SCARA Robot

Case Study 8.4 (SCARA Robot) *Consider the SCARA robot shown in Figure 8.8. Assume that the rotational arm in the fourth link, and the load carried by the robot is integrated into the third joint. Consequently, consider the length of the third link as a_3, and assume its center of mass is at the middle of this link. Furthermore, model the robot's links as slender rods, consider their center of mass at their midpoint, and neglect friction in the robot joints. Utilize the recursive Euler-Lagrange method to formulate the dynamic formulation for the robot and determine its mass matrix, gravity, and Coriolis and centrifugal vectors.*

Solution: In this robot, generalized coordinates can be represented as $\boldsymbol{q} = [\theta_1, \theta_2, d_3]^T$, while the generalized forces along the generalized coordinates are $\boldsymbol{Q} = [\tau_1, \tau_2, f_3]^T$. Assuming the links as slender rods, the inertia matrix of the links relative to the center

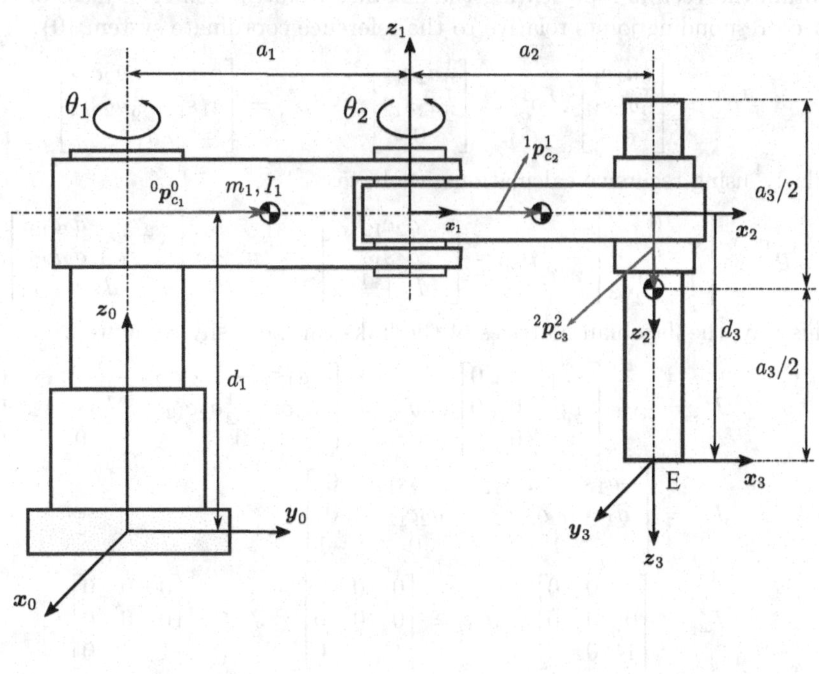

FIGURE 8.8
Schematic of a SCARA robot and the representation of necessary vectors in determining the Jacobian matrix of its links.

of mass coordinate system is given by:

$$ {}^C\boldsymbol{I}_1 = \frac{1}{12}m_1a_1^2\begin{bmatrix}0&0&0\\0&1&0\\0&0&1\end{bmatrix},\quad {}^C\boldsymbol{I}_2 = \frac{1}{12}m_2a_2^2\begin{bmatrix}0&0&0\\0&1&0\\0&0&1\end{bmatrix} $$

$$ {}^C\boldsymbol{I}_3 = \frac{1}{12}m_3a_3^2\begin{bmatrix}1&0&0\\0&1&0\\0&0&0\end{bmatrix} $$

By using the rotation matrix, the inertia matrix of the links in the reference coordinate system $\{0\}$ can be obtained as follows:

$$ {}^0\boldsymbol{I}_1 = {}^0\boldsymbol{R}_1\,{}^C\boldsymbol{I}_1\,{}^0\boldsymbol{R}_1^T = \frac{1}{12}m_1a_1^2\begin{bmatrix}s_1^2&-s_1c_1&0\\-s_1c_2&c_1^2&0\\0&0&1\end{bmatrix} $$

$$ {}^0\boldsymbol{I}_2 = {}^0\boldsymbol{R}_2\,{}^C\boldsymbol{I}_2\,{}^0\boldsymbol{R}_2^T = \frac{1}{12}m_2a_2^2\begin{bmatrix}s_{12}^2&-s_{12}c_{12}&0\\-s_{12}c_{12}&c_{12}^2&0\\0&0&1\end{bmatrix} $$

$$ {}^0\boldsymbol{I}_3 = {}^0\boldsymbol{R}_3\,{}^C\boldsymbol{I}_3\,{}^0\boldsymbol{R}_3^T = \frac{1}{12}m_3a_3^2\begin{bmatrix}1&0&0\\0&1&0\\0&0&0\end{bmatrix} $$

To determine the dynamic formulation using recursive Euler-Lagrange method, it is required to calculate the Jacobian matrices of the links. For this purpose, we first need to obtain the vectors representing the distance from the center of mass of the links to their corresponding joints relative to the reference coordinate system $\{0\}$.

$$ {}^0\boldsymbol{p}^0_{c_1} = \begin{bmatrix}\frac{1}{2}a_1c_1\\\frac{1}{2}a_1s_1\\d_1\end{bmatrix},\ {}^1\boldsymbol{p}^1_{c_2} = \begin{bmatrix}\frac{1}{2}a_2c_{12}\\\frac{1}{2}a_2s_{12}\\0\end{bmatrix} \rightarrow {}^0\boldsymbol{p}^0_{c_2} = \begin{bmatrix}a_1c_1+\frac{1}{2}a_2c_{12}\\a_1s_1+\frac{1}{2}a_2s_{12}\\d_1\end{bmatrix} $$

Similarly, using recursive calculations, we have:

$$ {}^2\boldsymbol{p}^2_{c_3} = \begin{bmatrix}0\\0\\d_3-\frac{1}{2}a_3\end{bmatrix} \rightarrow {}^1\boldsymbol{p}^1_{c_3} = \begin{bmatrix}a_2c_{12}\\a_2s_{12}\\-d_3+\frac{1}{2}a_3\end{bmatrix} \rightarrow {}^0\boldsymbol{p}^0_{c_3} = \begin{bmatrix}a_1c_1+a_2c_{12}\\a_1s_1+a_2s_{12}\\d_1-d_3+\frac{1}{2}a_3\end{bmatrix} $$

In this way, the Jacobian matrices of the links can be easily computed.

$$ \boldsymbol{J}_{v_1} = \frac{1}{2}a_1\begin{bmatrix}-s_1&0&0\\c_1&0&0\\0&0&0\end{bmatrix},\quad \boldsymbol{J}_{v_2} = \begin{bmatrix}-a_1s_1-\frac{1}{2}a_2s_{12}&-\frac{1}{2}a_2s_{12}&0\\a_1c_1+\frac{1}{2}a_2c_{12}&\frac{1}{2}a_2c_{12}&0\\0&0&0\end{bmatrix} $$

$$ \boldsymbol{J}_{v_3} = \begin{bmatrix}-a_1s_1-a_2s_{12}&-a_2s_{12}&0\\a_1c_1+a_2c_{12}&a_2c_{12}&0\\0&0&-1\end{bmatrix} $$

$$ \boldsymbol{J}_{\omega_1} = \begin{bmatrix}0&0&0\\0&0&0\\1&0&0\end{bmatrix},\quad \boldsymbol{J}_{\omega_2} = \begin{bmatrix}0&0&0\\0&0&0\\1&1&0\end{bmatrix},\quad \boldsymbol{J}_{\omega_3} = \begin{bmatrix}0&0&0\\0&0&0\\1&1&0\end{bmatrix} $$

By using the obtained Jacobian matrices of the links, the mass matrix of the robot, which is a symmetric matrix, is determined.

$$ \boldsymbol{M}(\boldsymbol{q}) = \sum_{i=1}^{3} m_i\left(\boldsymbol{J}^T_{v_i}\boldsymbol{J}_{v_i} + \boldsymbol{J}^T_{\omega_i}\,{}^C\boldsymbol{I}_i\boldsymbol{J}^T_{\omega_i}\right) = \begin{bmatrix}M_{11}&M_{12}&0\\&M_{22}&0\\\cdot&&1\end{bmatrix} $$

where

$$\begin{aligned} M_{11} &= \frac{1}{3}m_1a_1^2 + m_2\left(a_1^2 + a_1a_2c_2 + \frac{1}{3}a_2^2\right) + m_3\left(a_1^2 + 2a_1a_2c_2 + a_2^2\right) \\ M_{12} &= m_2\left(\frac{1}{2}a_1a_2c_2 + \frac{1}{3}a_2^2\right) + m_3\left(a_1a_2c_2 + a_2^2\right) \\ M_{22} &= \frac{1}{3}m_2a_2^2 + m_3a_2^2 \end{aligned}$$

The gravity vector $\mathfrak{g}(\boldsymbol{q})$ can also be easily calculated based on the Jacobian matrices of the links as follows.

$$\mathfrak{g}(\boldsymbol{q}) = -\sum_{i=1}^{3} m_i \boldsymbol{J}_{v_i}^T \boldsymbol{g} = \begin{bmatrix} 0 \\ 0 \\ -m_3 g \end{bmatrix}$$

The vector of Coriolis and centrifugal forces $\mathfrak{v}(\boldsymbol{q}, \dot{\boldsymbol{q}})$ is also equal to:

$$\mathfrak{v}(\boldsymbol{q}, \dot{\boldsymbol{q}}) = \left(\dot{\boldsymbol{M}}(\boldsymbol{q})\dot{\boldsymbol{q}} - \frac{\partial K}{\partial \boldsymbol{q}}\right) = \left(\frac{1}{2}m_2 + m_3\right) a_1a_2s_2 \begin{bmatrix} \dot{\theta}_2^2 + 2\dot{\theta}_1\dot{\theta}_2 \\ \dot{\theta}_1^2 \\ 0 \end{bmatrix}$$

By determining the necessary matrices and vectors, the dynamic motion equations of the robot can be represented in the following closed-form.

$$\boldsymbol{M}(\boldsymbol{q})\ddot{\boldsymbol{q}} + \mathfrak{v}(\boldsymbol{q}, \dot{\boldsymbol{q}}) + \mathfrak{g}(\boldsymbol{q}) = \boldsymbol{Q}$$

8.5.4 3DOF Elbow Manipulator

Case Study 8.5 (3DOF Elbow Manipulator Robot) *Consider the 3DOF Elbow manipulator shown in Figure 8.9. Suppose the first link of the robot has a mass of* m_1 *and an inertia moment* $I_{zz} = I_1$*. Consider the second and third links of the robot as slender rods. Assume the center of mass for all links is at their midpoint and neglect friction in the robot joints. Utilize the recursive Euler-Lagrange method to determine the dynamic formulation of the robot and calculate its mass matrix, gravity, and Coriolis and centrifugal vectors.*

Solution: Considering the information provided, the generalized coordinates for this robot can be represented as $\boldsymbol{q} = [\theta_1,\ \theta_2,\ \theta_3]^T$. The generalized force vector, based on the generalized coordinates, is denoted as $\boldsymbol{Q} = [\tau_1,\ \tau_2,\ \tau_3]^T$. Assuming slender rods for the links, the inertia matrix for these links with respect to their center of mass can be obtained.

$${}^C\boldsymbol{I}_1 = I_1 \begin{bmatrix} 0 & 0 & 0 \\ 0 & 0 & 0 \\ 0 & 0 & 1 \end{bmatrix}, \quad {}^C\boldsymbol{I}_2 = \frac{1}{12}m_2a_2^2 \begin{bmatrix} 0 & 0 & 0 \\ 0 & 1 & 0 \\ 0 & 0 & 1 \end{bmatrix}$$

$${}^C\boldsymbol{I}_3 = \frac{1}{12}m_3a_3^2 \begin{bmatrix} 0 & 0 & 0 \\ 0 & 1 & 0 \\ 0 & 0 & 1 \end{bmatrix}$$

Now, using the rotation matrix, the inertia moments of the links in the reference coordinate system 0 are obtained as follows:

$$
\begin{aligned}
{}^0\boldsymbol{I}_1 = {}^0\boldsymbol{R}_1\,{}^C\boldsymbol{I}_1\,{}^0\boldsymbol{R}_1^T &= I_1 \begin{bmatrix} s_1^2 & -s_1c_1 & 0 \\ -s_1c_2 & c_1^2 & 0 \\ 0 & 0 & 0 \end{bmatrix} \\
{}^0\boldsymbol{I}_2 = {}^0\boldsymbol{R}_2\,{}^C\boldsymbol{I}_2\,{}^0\boldsymbol{R}_2^T &= \frac{1}{12}m_2a_2^2 \begin{bmatrix} c_1^2s_2^2 + s_1^2 & -c_1s_1c_2^2 & -c_1c_2s_2 \\ -c_1s_1c_2^2 & s_1^2s_2^2 + c_1^2 & -\,s_1c_2s_2 \\ -c_1c_2s_2 & -s_1c_2s_2 & c_2^2 \end{bmatrix} \\
{}^0\boldsymbol{I}_3 = {}^0\boldsymbol{R}_3\,{}^C\boldsymbol{I}_3\,{}^0\boldsymbol{R}_3^T &= \frac{1}{12}m_3a_3^2 \begin{bmatrix} c_1^2s_{23}^2 + s_1^2 & -c_1s_1c_{23}^2 & -c_1c_{23}s_{23} \\ -c_1s_1c_{23}^2 & s_1^2s_{23}^2 + c_1^2 & -\,s_1c_{23}s_{23} \\ -c_1c_{23}s_{23} & -s_1c_{23}s_{23} & c_{23}^2 \end{bmatrix}
\end{aligned}
$$

Next, it is necessary to calculate the Jacobian matrices of the links. For this purpose, the vectors representing the distance from the center of mass of the links to their corresponding joints with respect to the reference coordinate system $\{0\}$ are determined using recursive calculations.

$$
{}^0\boldsymbol{p}_{c_1}^0 = \begin{bmatrix} 0 \\ 0 \\ -d_1/2 \end{bmatrix}, {}^1\boldsymbol{p}_{c_2}^1 = {}^0\boldsymbol{p}_{c_2}^0 = \frac{a_2}{2}\begin{bmatrix} c_1c_2 \\ s_1c_2 \\ s_2 \end{bmatrix}
$$

Similarly,

$$
{}^2\boldsymbol{p}_{c_3}^2 = \frac{a_3}{2}\begin{bmatrix} c_1c_{23} \\ s_1c_{23} \\ s_{23} \end{bmatrix} \to {}^1\boldsymbol{p}_{c_3}^1 = {}^0\boldsymbol{p}_{c_3}^0 = \begin{bmatrix} c_1\left(a_2c_2 + \frac{a_3}{2}c_{23}\right) \\ s_1\left(a_2c_2 + \frac{a_3}{2}c_{23}\right) \\ a_2s_2 + \frac{a_3}{2}s_{23} \end{bmatrix}
$$

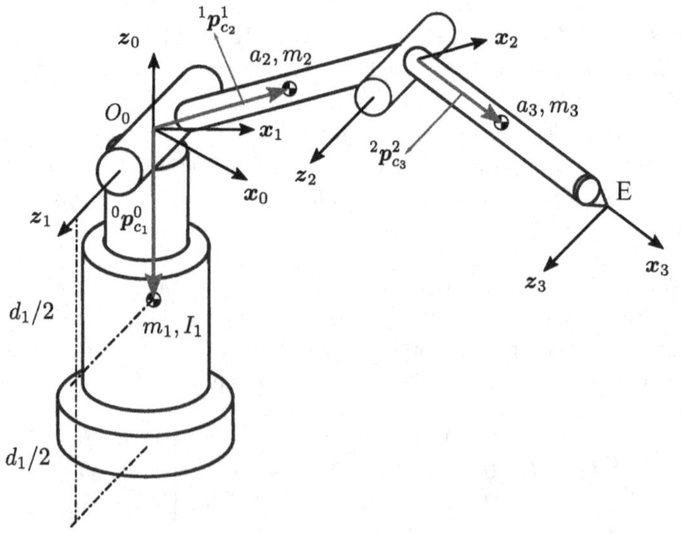

FIGURE 8.9
Schematic of a 3DOF Elbow manipulator and the representation of necessary vectors in determining the Jacobian matrix of its links.

By this means, the Jacobian matrices of the links can be easily obtained.

$$\boldsymbol{J}_{v_1} = \begin{bmatrix} 0 & 0 & 0 \\ 0 & 0 & 0 \\ 0 & 0 & 0 \end{bmatrix}, \quad \boldsymbol{J}_{v_2} = \begin{bmatrix} -\frac{1}{2}a_2 s_1 c_2 & -\frac{1}{2}a_2 c_1 s_2 & 0 \\ \frac{1}{2}a_2 c_1 c_2 & -\frac{1}{2}a_2 s_1 s_2 & 0 \\ 0 & \frac{1}{2}a_2 c_2 & 0 \end{bmatrix}$$

$$\boldsymbol{J}_{v_3} = \begin{bmatrix} -s_1\left(a_2 c_2 + \frac{1}{2}a_3 c_{23}\right) & -c_1\left(a_2 s_2 + \frac{1}{2}a_3 s_{23}\right) & -\frac{1}{2}a_3 c_1 s_{23} \\ c_1\left(a_2 c_2 + \frac{1}{2}a_3 c_{23}\right) & -s_1\left(a_2 s_2 + \frac{1}{2}a_3 s_{23}\right) & -\frac{1}{2}a_3 s_1 s_{23} \\ 0 & a_2 c_2 + \frac{1}{2}a_3 c_{23} & \frac{1}{2}a_3 c_{23} \end{bmatrix}$$

$$\boldsymbol{J}_{\omega_1} = \begin{bmatrix} 0 & 0 & 0 \\ 0 & 0 & 0 \\ 1 & 0 & 0 \end{bmatrix}, \quad \boldsymbol{J}_{\omega_2} = \begin{bmatrix} 0 & s_1 & 0 \\ 0 & -c_1 & 0 \\ 1 & 0 & 0 \end{bmatrix}, \quad \boldsymbol{J}_{\omega_3} = \begin{bmatrix} 0 & s_1 & s_1 \\ 0 & -c_1 & -c_1 \\ 1 & 0 & 0 \end{bmatrix}$$

Now, by using the Jacobian matrices of the links, the symmetric mass matrix of the robot can be determined.

$$\boldsymbol{M}(\boldsymbol{q}) = \sum_{i=1}^{3} m_i \left(\boldsymbol{J}_{v_i}^T \boldsymbol{J}_{v_i} + \boldsymbol{J}_{\omega_i}^T {}^C\boldsymbol{I}_i \boldsymbol{J}_{\omega_i}^T\right) = \begin{bmatrix} M_{11} & 0 & 0 \\ & M_{22} & M_{23} \\ \cdot & & M_{33} \end{bmatrix}$$

where

$$\begin{aligned} M_{11} &= \left(\frac{1}{3}m_2 + m_3\right) a_2^2 c_2^2 + \frac{1}{3}m_3 a_3^2 c_{23}^2 + m_3 a_2 a_3 c_2 c_{23} \\ M_{22} &= \left(\frac{1}{3}m_2 + m_3\right) a_2^2 + \frac{1}{3}m_3 a_3^2 + m_3 a_2 a_3 c_3 \\ M_{23} &= \frac{1}{3}m_3 a_3^2 + \frac{1}{2}m_3 a_2 a_3 c_3 \\ M_{33} &= \frac{1}{3}m_3 a_3^2 \end{aligned}$$

The gravity vector $\mathfrak{g}(\boldsymbol{q})$ can also be calculated based on the Jacobian matrices of the links as follows.

$$\mathfrak{g}(\boldsymbol{q}) = -\sum_{i=1}^{3} m_i \boldsymbol{J}_{v_i}^T \boldsymbol{g} = \begin{bmatrix} \frac{1}{2}m_2 a_2 c_1 c_2 + m_3 c_1\left(a_2 c_2 + \frac{1}{2}a_3 c_{23}\right) \\ -\frac{1}{2}m_2 a_2 s_1 s_2 - m_3 s_1\left(a_2 s_2 + \frac{1}{2}a_3 s_{23}\right) \\ -\frac{1}{2}m_3 a_3 s_1 s_{23} \end{bmatrix} g.$$

The vector of Coriolis and centrifugal forces $\mathfrak{v}(\boldsymbol{q}, \dot{\boldsymbol{q}})$ is also determined through symbolic calculations.

$$\mathfrak{v}(\boldsymbol{q}, \dot{\boldsymbol{q}}) = \left(\dot{\boldsymbol{M}}(\boldsymbol{q})\dot{\boldsymbol{q}} - \frac{\partial K}{\partial \boldsymbol{q}}\right) = \begin{bmatrix} \mathfrak{v}_1 \\ \mathfrak{v}_2 \\ \mathfrak{v}_2 \end{bmatrix}$$

where

$$\begin{aligned} \mathfrak{v}_1 = & -\frac{1}{3}m_2 a_2^2 \sin(2\theta_2)\dot{\theta}_1\dot{\theta}_2 \\ & -m_3\left(\frac{1}{3}a_3^2 \sin 2(\theta_2 + \theta_3) + a_2 a_3 \sin(2\theta_2 + \theta_3) + a_2^2 \sin 2\theta_2\right)\dot{\theta}_1\dot{\theta}_2 \\ & -m_3\left(\frac{1}{3}a_3^2 \sin 2(\theta_2 + \theta_3) + \frac{1}{2}a_2 a_3 \sin(2\theta_2 + \theta_3) + \frac{1}{2}a_2 a_3 s_3\right)\dot{\theta}_1\dot{\theta}_3 \end{aligned}$$

$$\begin{aligned}
\mathfrak{v}_2 &= m_3\left(\frac{1}{3}a_3^2 c_{23}s_{23} + \frac{1}{2}a_2a_3c_{23}s_2 + \frac{1}{2}a_2a_3s_{23}c_2\right)\dot{\theta}_1^2 \\
&\quad + \left(\frac{1}{3}m_2 + m_3\right)a_2^2 s_2 c_2\dot{\theta}_1^2 - m_3a_2a_3s_3\dot{\theta}_2\dot{\theta}_3 - \frac{1}{2}m_3a_2a_3s_3\dot{\theta}_3^2 \\
\mathfrak{v}_3 &= \frac{1}{2}m_3a_2a_3s_{23}c_2\dot{\theta}_1^2 + \frac{1}{2}m_3a_2a_3s_3\dot{\theta}_2^2 + \frac{1}{6}m_3a_3^2\sin 2(\theta_2+\theta_3)\,\dot{\theta}_1^2
\end{aligned}$$

By determining the necessary matrices and vectors, the dynamic formulation of this robot can be written in the following closed-form.

$$\boldsymbol{M}(\boldsymbol{q})\ddot{\boldsymbol{q}} + \boldsymbol{\mathfrak{v}}(\boldsymbol{q},\dot{\boldsymbol{q}}) + \boldsymbol{\mathfrak{g}}(\boldsymbol{q}) = \boldsymbol{Q}$$

Problems

8.1 Consider the two-degrees-of-freedom <u>RP</u> robot shown in Figure 8.10. Assume the robot links as slender rods with their center of mass located at their midpoint lengths, and consider viscous friction in the joints. Using the Euler-Lagrange method, derive the dynamic equations of the robot in closed-form and obtain the robot's mass matrix $\boldsymbol{M}(\boldsymbol{q})$, gravitational vector $\boldsymbol{\mathfrak{g}}(\boldsymbol{q})$, and the vector for Coriolis and centrifugal torques $\boldsymbol{\mathfrak{v}}(\boldsymbol{q},\dot{\boldsymbol{q}})$.

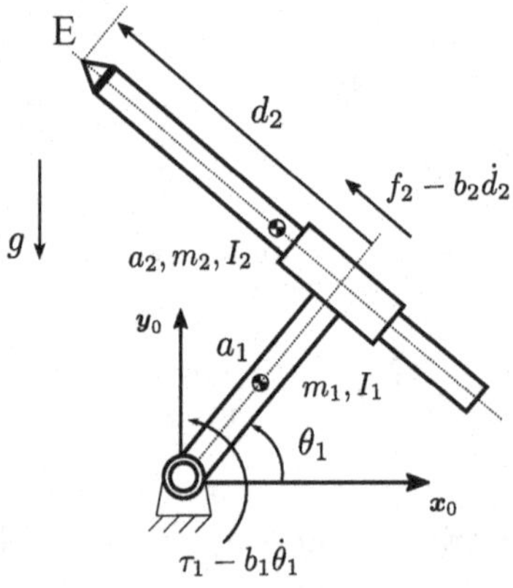

FIGURE 8.10
2-DOF robot of Problem 8.1

8.2 Consider the three-degrees-of-freedom <u>PRP</u> robot illustrated in Figure 8.11. Assume the robot links are slender rods with their center of mass located at their midpoints, and consider viscous friction in the joints. Utilizing the Euler-Lagrange method, deduce the dynamic equations of the robot in closed-form and determine the robot's mass matrix $\boldsymbol{M}(\boldsymbol{q})$, gravitational vector $\boldsymbol{\mathfrak{g}}(\boldsymbol{q})$, and the vector representing Coriolis and centrifugal torques $\boldsymbol{\mathfrak{v}}(\boldsymbol{q},\dot{\boldsymbol{q}})$.

8.3 Consider the 3-DOF <u>PRR</u> robot depicted in Figure 8.12. Assume the robot links are slender rods with their center of mass located at the midpoint of their lengths, and consider viscous friction in the joints. Using the Euler-Lagrange method, derive the

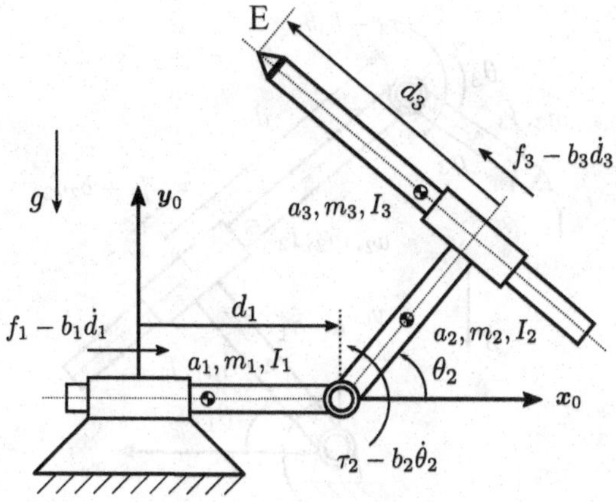

FIGURE 8.11
3-DOF robot of Problem 8.2

dynamic formulation of the robot in closed-form and obtain the robot's mass matrix $\boldsymbol{M}(\boldsymbol{q})$, gravitational vector $\mathfrak{g}(\boldsymbol{q})$, and the vector for Coriolis and centrifugal torques $\mathfrak{v}(\boldsymbol{q}, \dot{\boldsymbol{q}})$.

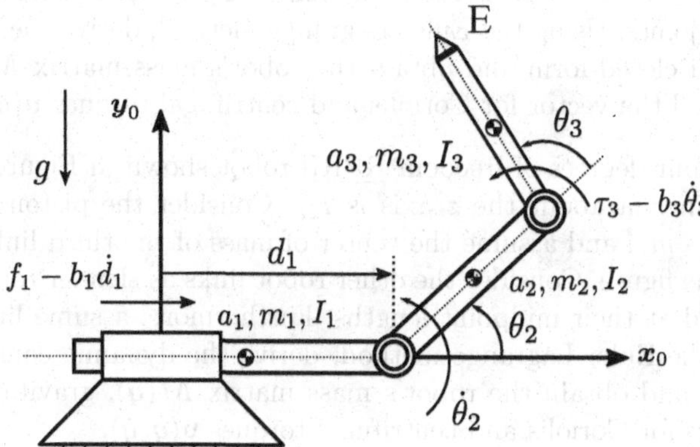

FIGURE 8.12
3-DOF robot of Problem 8.3

8.4 Consider the three-degrees-of-freedom <u>RPR</u> robot shown in Figure 8.13. Assume that the robot links are slender rods with their center of mass located at the midpoint of their lengths, and consider viscous friction in the joints. Employing the Euler-Lagrange method, derive the dynamic equations of the robot in closed-form and obtain the robot's mass matrix $\boldsymbol{M}(\boldsymbol{q})$, gravitational vector $\mathfrak{g}(\boldsymbol{q})$, and the vector for Coriolis and centrifugal torques $\mathfrak{v}(\boldsymbol{q}, \dot{\boldsymbol{q}})$.

8.5 Consider the two-degrees-of-freedom <u>RR</u> robot shown in Figure 8.14. Assume the rotary base's inertia about the z axis is I_{zz}. Consider the robot links as slender rods with their center of mass located at their midpoint lengths. Furthermore, assume viscous friction in the joints. Using the Euler-Lagrange method, derive the dynamic equations of the robot in closed-form and obtain the robot's mass matrix $\boldsymbol{M}(\boldsymbol{q})$, gravitational vector $\mathfrak{g}(\boldsymbol{q})$, and the vectors for Coriolis and centrifugal torques $\mathfrak{v}(\boldsymbol{q}, \dot{\boldsymbol{q}})$.

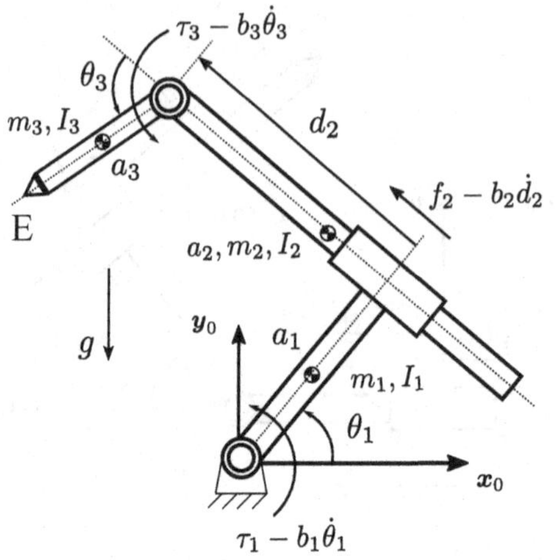

FIGURE 8.13
3-DOF robot of Problem 8.4

8.6 Consider the four-degrees-of-freedom <u>RRRP</u> robot shown in Figure 8.15. Assume the rotary base's inertia about the z axis is I_{zz}, and the robot links as slender rods with their center of mass located at their midpoint lengths. Furthermore, assume viscous friction in the joints. Using the Euler-Lagrange method, derive the dynamic equations of the robot in closed-form and obtain the robot's mass matrix $\boldsymbol{M}(\boldsymbol{q})$, gravitational vector $\mathfrak{g}(\boldsymbol{q})$, and the vector for Coriolis and centrifugal torques $\mathfrak{v}(\boldsymbol{q}, \dot{\boldsymbol{q}})$.

8.7 Consider the four-degrees-of-freedom <u>RPRR</u> robot shown in Figure 8.16. Assume the rotary base's inertia about the z axis is I_{zz}. Consider the piston link of the second joint with mass m_p, and assume the center of mass of the third link is at the location indicated in the figure. Consider the other robot links as slender rods with their center of mass located at their midpoint lengths. Furthermore, assume linear friction in the joints. Using the Euler-Lagrange method, derive the dynamic equations of the robot in closed-form and obtain the robot's mass matrix $\boldsymbol{M}(\boldsymbol{q})$, gravitational vector $\mathfrak{g}(\boldsymbol{q})$, and the vectors for Coriolis and centrifugal torques $\mathfrak{v}(\boldsymbol{q}, \dot{\boldsymbol{q}})$.

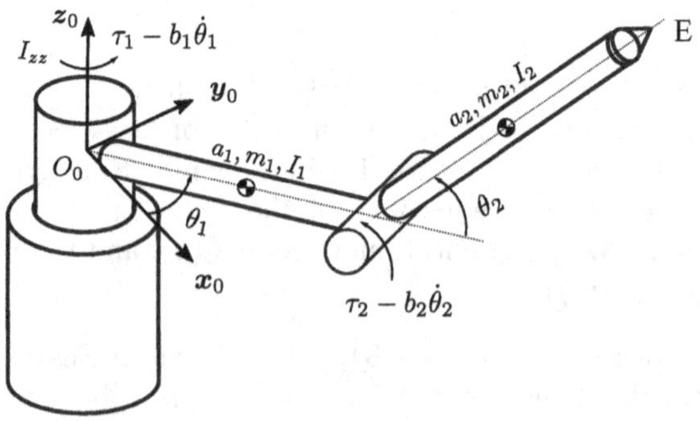

FIGURE 8.14
Two degrees-of-freedom robot of Problem 8.5

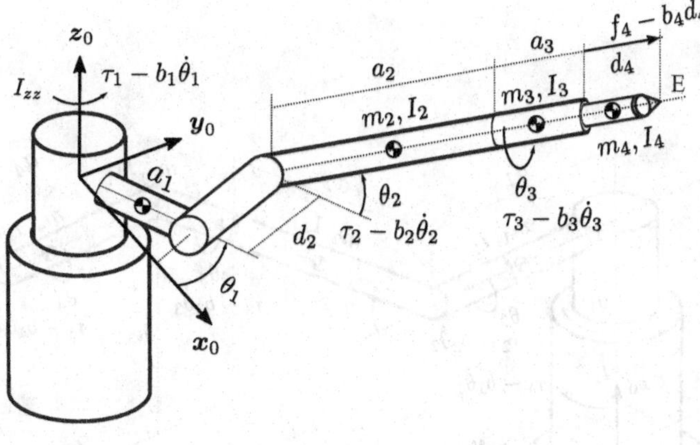

FIGURE 8.15
4-DOF robot of Problem 8.6

8.8 Consider the four-degrees-of-freedom 4R robot shown in Figure 8.17. Assume the rotary base's inertia about the z axis is I_{zz}. Consider the center of mass of the third link at the location indicated in the figure and assume the other robot links are slender rods with their center of mass located at their midpoint lengths. Also, assume viscous friction in the joints. Using the Euler-Lagrange method, derive the dynamic equations of the robot in closed-form and obtain the robot's mass matrix $\boldsymbol{M}(\boldsymbol{q})$, gravitational vector $\mathfrak{g}(\boldsymbol{q})$, and the vectors for Coriolis and centrifugal torques $\mathfrak{v}(\boldsymbol{q}, \dot{\boldsymbol{q}})$.

8.9 Consider the four-degrees-of-freedom RRPR robot shown in Figure 8.18. Assume the rotary base's inertia about the z-axis is I_{zz}. Consider the center of mass of the second link at the location indicated in the figure and assume the other robot links are slender rods with their center of mass located at their midpoint lengths. Furthermore, assume viscous friction in the joints. Using the Euler-Lagrange method, derive the dynamic equations of the robot in closed-form and obtain the robot's mass matrix $\boldsymbol{M}(\boldsymbol{q})$, gravitational vector $\mathfrak{g}(\boldsymbol{q})$, and the vectors for Coriolis and centrifugal torques $\mathfrak{v}(\boldsymbol{q}, \dot{\boldsymbol{q}})$.

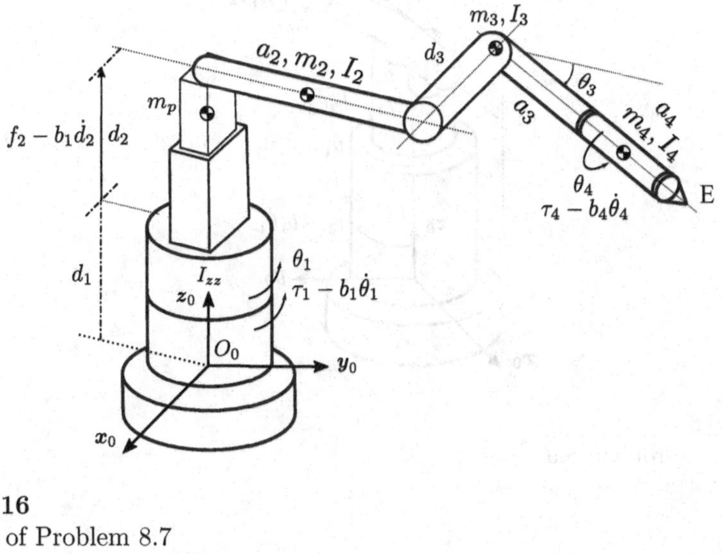

FIGURE 8.16
4-DOF robot of Problem 8.7

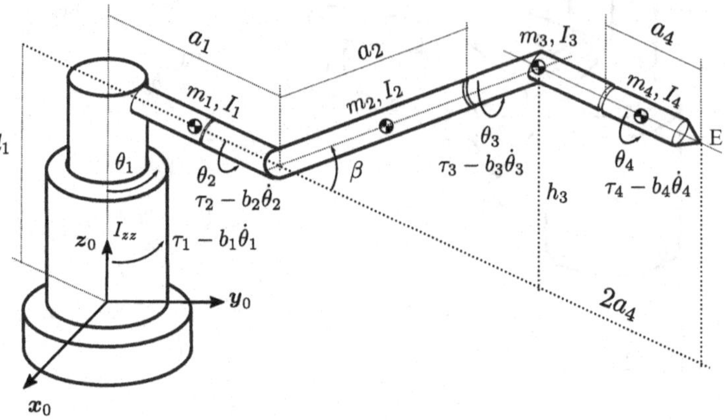

FIGURE 8.17
4-DOF robot of Problem 8.8

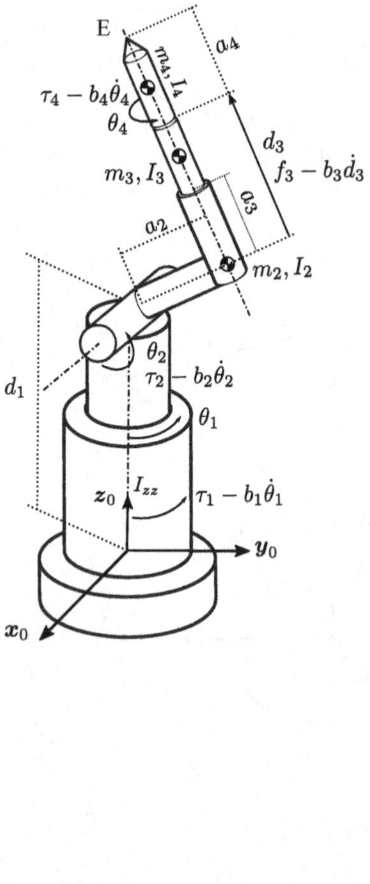

FIGURE 8.18
4-DOF robot of Problem 8.9

9

Dynamics Properties and Representations

This chapter elaborates on the properties of dynamics formulation and their advanced representations, by providing an insightful exploration of various key concepts. The chapter commences with an introduction, setting the stage for a comprehensive discussion on the dynamic aspects of the subject matter. Attention then shifts to the significance of the mass matrix in dynamics formulations. The chapter proceeds to examine the Christoffel matrix, detailing its derivation and exploring the Kronecker product. The chapter further extends its purview to the linear regression representation of dynamics formulations, emphasizing its role in calibration and adaptation. This part elucidates nuanced aspects, providing a deeper understanding of linear regression representation.

A substantial portion of the chapter is dedicated to actuator dynamics, where the discussion encompasses a range of topics including the intricacies of permanent magnet DC motors, the role of the motor drivers, the significant impact of gearboxes on the overall robot dynamics with actuators. This section not only provides a detailed examination of individual components but also integrates them into a holistic understanding of the dynamic behavior of robots when coupled with actuators. The chapter concludes with a number of suggested problems, inviting readers to engage with the material on a practical level, reinforcing their comprehension through problem-solving exercises. In essence, this Chapter serves as a comprehensive exploration of the properties and formulations in dynamics, offering a detailed and interconnected view of the key components and their roles in shaping the dynamics of robotic systems.

9.1 Introduction

As elaborated in the preceding chapter, the Euler-Lagrange method enables the derivation of the motion equations for serial robots in a closed form. The universality of this method in establishing motion equations, along with the matrices and vectors shaping its closed-form, gives rise to conditions that govern these equations, irrespective of the robot's degrees of freedom and its kinematic structure. These inherent characteristics play a pivotal role in the design and analysis of diverse model-based control approaches. In this section, we elaborate on the process of determining dynamic formulation for a generic robot, examining the noteworthy features associated with the dynamic equations.

As demonstrated in Section 8.4.2, the closed-form dynamics for a robot can be written as follows:

$$\boldsymbol{M}(\boldsymbol{q})\ddot{\boldsymbol{q}} + \boldsymbol{\mathfrak{v}}(\boldsymbol{q}, \dot{\boldsymbol{q}}) + \boldsymbol{\mathfrak{g}}(\boldsymbol{q}) = \boldsymbol{Q} \tag{9.1}$$

In this equation, the robot's mass matrix is defined based on the Jacobian matrices of the

DOI: 10.1201/9781003491415-9

links as:

$$M(q) = \sum_{i=1}^{n} \left(m_i J_{v_i}^T J_{v_i} + J_{\omega_i}^T {}^C I_i J_{\omega_i}^T \right) \tag{9.2}$$

Furthermore, the gravity vector $\mathfrak{g}$, is defined based on the potential energy of the system as:

$$\mathfrak{g}(q) = \frac{\partial P(q)}{\partial q} \tag{9.3}$$

where it can be derived in terms of the Jacobian matrices of the links as follows:

$$\mathfrak{g}(q) = -\sum_{i=1}^{n} m_i J_{v_i}^T g \tag{9.4}$$

Moreover, the vector of Coriolis and centrifugal terms $\mathfrak{v}(q, \dot{q})$ can be obtained from the following relationship:

$$\begin{aligned} \mathfrak{v}(q, \dot{q}) &= \dot{M}(q)\dot{q} - \frac{\partial K}{\partial q} \\ &= \dot{M}(q)\dot{q} - \frac{1}{2}\frac{\partial}{\partial q}\left(\dot{q}^T M(q)\dot{q}\right) \end{aligned} \tag{9.5}$$

In the upcoming sections, we will explore the attributes of the mass matrix and the Coriolis and centrifugal vector, by introducing the Christoffel matrix. Additionally, various alternative representations for dynamics formulation will be introduced, serving practical purposes in the stability analysis of closed-loop robot motion, as well as in the calibration and adaptive control of the robot.

9.2 Properties of Mass Matrix

The mass matrix of the robot, denoted as $M(q)$, is derived from the system's kinetic energy using the relationship given in 8.36. Although this $n \times n$ matrix depends on the robot's configuration q, given that the kinetic energy of the system is always non-negative, the mass matrix will be symmetric and positive definite. Therefore, this matrix is invertible for all values of q and is bounded from above and below:

$$\underline{\lambda}(q)\, I_{n\times n} \le M(q) \le \overline{\lambda}(q)\, I_{n\times n} \tag{9.6}$$

In this equation, $I_{n\times n}$ is the identity matrix with dimensions $n \times n$. $\overline{\lambda}$ and $\underline{\lambda}$ are positive values representing the maximum and minimum eigenvalues of the mass matrix across the entire workspace of the robot, respectively. Similarly, it can be shown that the inverse of the mass matrix is also bounded.

$$\frac{1}{\underline{\lambda}(q)} I_{n\times n} \ge M^{-1}(q) \ge \frac{1}{\overline{\lambda}(q)} I_{n\times n} \tag{9.7}$$

The boundedness property of the robot's mass matrix can also be interpreted by its bounded norm.

$$\underline{M} \le \|M(q)\| \le \overline{M} \tag{9.8}$$

In serial robots with revolute joints, the bounds on the mass matrix $\overline{M}$ and $\underline{M}$ are constant positive values; that is, they are not functions of joint variables. These properties make them suitable for stability analysis in closed-loop control and simulation of robot motion using the inverse of the mass matrix.

9.3 Christoffel Matrix

Deriving the Coriolis and centrifugal vector 9.5 in the dynamics formulation of serial robots constitutes the most intricate part of the calculations. In this section, by employing the Kronecker product in matrices, we replace this vector with a matrix containing Christoffel symbols, which we refer to as the "Christoffel matrix" in this book, and we will elucidate the unique features of this matrix. It can be demonstrated that the Coriolis and centrifugal vector can be substituted by the product of the Christoffel matrix $\boldsymbol{C}(\boldsymbol{q}, \dot{\boldsymbol{q}})$ and the vector of angular joint velocities as follows:

$$\mathfrak{v}(\boldsymbol{q}, \dot{\boldsymbol{q}}) = \boldsymbol{C}(\boldsymbol{q}, \dot{\boldsymbol{q}})\dot{\boldsymbol{q}} \tag{9.9}$$

In this representation, the Christoffel matrix is not uniquely determined, and there are various representations for it. To determine the Christoffel matrix, the audience needs to be familiar with the definition and properties of the Kronecker product [109]. A brief overview of the necessary concepts is provided in the following section, with more detailed information given in Appendix A.

9.3.1 Kronecker Product

The Kronecker product, denoted by $\otimes$, defines a multiplication operation on two matrices of arbitrary dimensions, resulting in a block matrix. If matrix $\boldsymbol{A}$ has dimensions $m \times n$ and matrix $\boldsymbol{B}$ has dimensions $p \times q$, then the Kronecker product of these two matrices, denoted as $\boldsymbol{A} \otimes \boldsymbol{B}$, is a matrix with dimensions $mp \times nq$. It is formed as a block matrix by taking the product of the elements of the first matrix with the second matrix.

$$\boldsymbol{A} \otimes \boldsymbol{B} = \begin{bmatrix} a_{11}\boldsymbol{B} & a_{12}\boldsymbol{B} & \cdots & a_{1n}\boldsymbol{B} \\ a_{21}\boldsymbol{B} & a_{22}\boldsymbol{B} & \cdots & a_{2n}\boldsymbol{B} \\ \vdots & \vdots & \ddots & \vdots \\ a_{m1}\boldsymbol{B} & a_{m2}\boldsymbol{B} & \cdots & a_{mn}\boldsymbol{B} \end{bmatrix} \tag{9.10}$$

The Kronecker product exhibits bilinearity and associativity properties:

$$\begin{gathered} \alpha(\boldsymbol{A} \otimes \boldsymbol{B}) = (\alpha\boldsymbol{A}) \otimes \boldsymbol{B} = \boldsymbol{A} \otimes (\alpha\boldsymbol{B}) \;\; ; \;\; \alpha \in R \\ \boldsymbol{A} \otimes (\boldsymbol{B} + \boldsymbol{C}) = \boldsymbol{A} \otimes \boldsymbol{B} + \boldsymbol{A} \otimes \boldsymbol{C} \\ (\boldsymbol{A} + \boldsymbol{B}) \otimes \boldsymbol{C} = \boldsymbol{A} \otimes \boldsymbol{C} + \boldsymbol{B} \otimes \boldsymbol{C} \\ (\boldsymbol{A} \otimes \boldsymbol{B}) \otimes \boldsymbol{C} = \boldsymbol{A} \otimes (\boldsymbol{B} \otimes \boldsymbol{C}) \end{gathered}$$

Furthermore, for matrices with appropriate dimensions, their compound product possesses the following property.

$$(\boldsymbol{A} \otimes \boldsymbol{B})(\boldsymbol{C} \otimes \boldsymbol{D}) = \boldsymbol{AC} \otimes \boldsymbol{BD} \tag{9.11}$$

The matrix $(\boldsymbol{A} \otimes \boldsymbol{B})$ is invertible, if and only if, matrices $\boldsymbol{A}$ and $\boldsymbol{B}$ are invertible. In this case, the inverse of this matrix is calculated as follows.

$$(\boldsymbol{A} \otimes \boldsymbol{B})^{-1} = \boldsymbol{A}^{-1} \otimes \boldsymbol{B}^{-1} \tag{9.12}$$

Moreover, the transpose in this product is distributive.

$$(\boldsymbol{A} \otimes \boldsymbol{B})^{T} = \boldsymbol{A}^{T} \otimes \boldsymbol{B}^{T} \tag{9.13}$$

Another feature in the Kronecker product is utilized in simplifying dynamic relations. Considering the definition of the vector $\boldsymbol{x} \in \mathbb{R}^n$, and since the Kronecker product is not commutative in general, we have $\boldsymbol{I}_{n\times n} \otimes \boldsymbol{x} \neq \boldsymbol{x} \otimes \boldsymbol{I}_{n\times n}$, where $\boldsymbol{I}_{n\times n}$ represents the identity matrix of dimensions $n \times n$. However, it can be shown that the following relation holds for any arbitrary vector $\boldsymbol{x} \in \mathbb{R}^n$.

$$(\boldsymbol{I}_{n\times n} \otimes \boldsymbol{x})\,\boldsymbol{x} = (\boldsymbol{x} \otimes \boldsymbol{I}_{n\times n})\,\boldsymbol{x} \tag{9.14}$$

Now, let us define the time derivative and partial derivatives of matrices using the Kronecker product and investigate the application of the Kronecker product in examining the gradients of composite matrices. Consider the matrix $\boldsymbol{A}(\boldsymbol{q})$, according to the convention, the partial derivative of this matrix with respect to the variable $\boldsymbol{q}$ can be expressed in the form of a block-column matrix as follows.

$$\frac{\partial \boldsymbol{A}(\boldsymbol{q})}{\partial \boldsymbol{q}} = \begin{bmatrix} \frac{\partial \boldsymbol{A}}{\partial q_1} \\ \vdots \\ \frac{\partial \boldsymbol{A}}{\partial q_n} \end{bmatrix} \tag{9.15}$$

With this definition, the time derivative of a matrix can be obtained using the Kronecker product, as elaborated in [109].

$$\frac{d\boldsymbol{A}}{dt} = \frac{\partial \boldsymbol{A}}{\partial \boldsymbol{q}}^T (\boldsymbol{I}_{n\times n} \otimes \dot{\boldsymbol{q}}) \tag{9.16}$$

Now, if we seek to find the relative derivative of the product of two matrices, we can utilize the following relation.

$$\frac{\partial}{\partial \boldsymbol{q}}(\boldsymbol{A}(\boldsymbol{q})\boldsymbol{B}(\boldsymbol{q})) = (\boldsymbol{I}_{n\times n} \otimes \boldsymbol{A})\frac{\partial \boldsymbol{B}}{\partial \boldsymbol{q}} + \frac{\partial \boldsymbol{A}}{\partial \boldsymbol{q}}\boldsymbol{B} \tag{9.17}$$

9.3.2 Christoffel Matrix Derivation

We initiate the calculations to determine the Christoffel matrix using Kronecker product. Equation 9.5 can be rewritten using the Kronecker product as follows.

$$\mathfrak{v}(\boldsymbol{q}, \dot{\boldsymbol{q}}) = \left[\frac{\partial \boldsymbol{M}}{\partial \boldsymbol{q}}^T (\boldsymbol{I}_{n\times n} \otimes \dot{\boldsymbol{q}}) - \frac{1}{2}\left(\boldsymbol{I}_{n\times n} \otimes \dot{\boldsymbol{q}}^T\right)\frac{\partial \boldsymbol{M}}{\partial \boldsymbol{q}}\right]\dot{\boldsymbol{q}}$$

In this calculation, we first utilize the relations 9.15 and 9.16, and then apply the equality $\frac{\partial \dot{\boldsymbol{q}}}{\partial \boldsymbol{q}} = 0$. Using this relation, the first representation of the Christoffel matrix $\boldsymbol{C}_1(\boldsymbol{q}, \dot{\boldsymbol{q}})$ is computed.

$$\mathfrak{v}(\boldsymbol{q}, \dot{\boldsymbol{q}}) = \boldsymbol{C}_1(\boldsymbol{q}, \dot{\boldsymbol{q}})\dot{\boldsymbol{q}} \tag{9.18}$$

where

$$\boldsymbol{C}_1(\boldsymbol{q}, \dot{\boldsymbol{q}}) = \boldsymbol{V}(\boldsymbol{q}, \dot{\boldsymbol{q}}) - \frac{1}{2}\boldsymbol{U}(\boldsymbol{q}, \dot{\boldsymbol{q}}) \tag{9.19}$$

$$\boldsymbol{V}(\boldsymbol{q}, \dot{\boldsymbol{q}}) = \dot{\boldsymbol{M}}(\boldsymbol{q}) = \frac{\partial \boldsymbol{M}}{\partial \boldsymbol{q}}^T (\boldsymbol{I}_{n\times n} \otimes \dot{\boldsymbol{q}}) \tag{9.20}$$

$$\boldsymbol{U}(\boldsymbol{q}, \dot{\boldsymbol{q}}) = \left(\boldsymbol{I}_{n\times n} \otimes \dot{\boldsymbol{q}}^T\right)\frac{\partial \boldsymbol{M}}{\partial \boldsymbol{q}} \tag{9.21}$$

As mentioned, the decomposition of the vector $\mathfrak{v}(\boldsymbol{q}, \dot{\boldsymbol{q}})$ does not uniquely lead to a Christoffel matrix, and alternative representation of this matrix can be obtained. The most commonly used representation of the Christoffel matrix is calculated in a way that, by using it the matrix $\dot{M} - 2C$ becomes skew-symmetric. This version of the Christoffel matrix can be obtained as follows.

$$\begin{aligned} C(q,\dot{q})\dot{q} &= \dot{M}(q)\dot{q} - \frac{1}{2}\left(I_{n\times n}\otimes\dot{q}^T\right)\frac{\partial M}{\partial q}\dot{q} \\ &= \frac{1}{2}\frac{\partial M^T}{\partial q}\left(I_{n\times n}\otimes\dot{q}\right)\dot{q} + \frac{1}{2}\frac{\partial M^T}{\partial q}\left(I_{n\times n}\otimes\dot{q}\right)\dot{q} - \frac{1}{2}\left(I_{n\times n}\otimes\dot{q}^T\right)\frac{\partial M}{\partial q}\dot{q} \\ &= \frac{1}{2}V(q,\dot{q})\dot{q} + \frac{1}{2}\frac{\partial M^T}{\partial q}\left(\dot{q}\otimes I_{n\times n}\right)\dot{q} - \frac{1}{2}\left(I_{n\times n}\otimes\dot{q}^T\right)\frac{\partial M}{\partial q}\dot{q} \\ &= \frac{1}{2}\left[V(q,\dot{q}) + U^T(q,\dot{q}) - U(q,\dot{q})\right]\dot{q} \end{aligned}$$

In these calculations, initially, the matrix $\dot{M} = V$ is divided into two equal parts. Then, the relation in 9.14 is utilized for simplification. Consequently, the Christoffel matrix is given by:

$$C(q,\dot{q}) = \frac{1}{2}\left[V(q,\dot{q}) + U^T(q,\dot{q}) - U(q,\dot{q})\right] \tag{9.22}$$

In this representation the matrices V and U are defined in the equations 9.20 and 9.21. With this definition, the computational complexity for determining the Christoffel matrix does not increase significantly, as we still need to obtain the matrices V and U as before. However, this version of the Christoffel matrix has an advantage over other versions, as it results in the skew-symmetry of the matrix $\dot{M} - 2C$.

$$\begin{aligned} \dot{M} - 2C &= V - \left[V + U^T - U\right] \\ &= U - U^T \end{aligned}$$

This implies that for any vector x, the scalar value $x^T(\dot{M} - 2C)x$ is equal to zero. This is in contrast to using other representations of the Christoffel matrix, where the matrix $\dot{M} - 2C$ will not necessarily be skew-symmetric. Instead, the relationship $\dot{q}^T(\dot{M} - 2C)\dot{q}$ will only be zero for the vector $\dot{q}$. By using the Christoffel matrix, the closed-form dynamics formulation for the robot can be rewritten as follows.

$$M(q)\ddot{q} + C(q,\dot{q})\dot{q} + \mathfrak{g}(q) = Q \tag{9.23}$$

This representation and the discussed skew-symmetric property will be directly applied in the subsequent chapters of the book, particularly in the stability analysis of the closed-loop robot motion with model-based controllers.

Case Study 9.1 (Planar 2<u>R</u> Robot) *Consider the planar 2<u>R</u> robot shown in Figure 8.7. The Dynamic formulations for this system are obtained in Example 8.3. Calculate the Christoffel matrix for this robot and verify the skew-symmetry of the matrix $\dot{M} - 2C$ in this representation.*

Solution: Initially, based on the convention in 9.15, obtain the derivative of the mass matrix in a matrix of appropriate dimensions.

$$\frac{\partial M}{\partial q} = \begin{bmatrix} 0 & 0 \\ 0 & 0 \\ -a_1a_2s_2m_2 & -\frac{1}{2}a_1a_2s_2m_2 \\ -\frac{1}{2}a_1a_2s_2m_2 & 0 \end{bmatrix}_{4\times 2}$$

Next, the matrices $\boldsymbol{V}$ and $\boldsymbol{U}$ are computed using the Kronecker product and the relationships given in 9.20 and 9.21, respectively.

$$\begin{aligned}\boldsymbol{V}(\boldsymbol{q},\dot{\boldsymbol{q}}) &= \dot{\boldsymbol{M}}(\boldsymbol{q}) = \frac{\partial \boldsymbol{M}^T}{\partial \boldsymbol{q}}\left(\boldsymbol{I}_{n\times n}\otimes\dot{\boldsymbol{q}}\right) = \begin{bmatrix}0 & -a_1a_2s_2m_2\left(\dot{\theta}_1+\frac{1}{2}\dot{\theta}_2\right)\\ 0 & -\frac{1}{2}a_1a_2s_2m_2\dot{\theta}_1\end{bmatrix}\\ \boldsymbol{U}(\boldsymbol{q},\dot{\boldsymbol{q}}) &= \left(\boldsymbol{I}_{n\times n}\otimes\dot{\boldsymbol{q}}^T\right)\frac{\partial \boldsymbol{M}}{\partial \boldsymbol{q}} = \begin{bmatrix}0 & 0\\ -a_1a_2s_2m_2\left(\dot{\theta}_1+\frac{1}{2}\dot{\theta}_2\right) & -\frac{1}{2}a_1a_2s_2m_2\dot{\theta}_1\end{bmatrix}\end{aligned}$$

By this means, the Christoffel matrix is easily calculated.

$$\begin{aligned}\boldsymbol{C}(\boldsymbol{q},\dot{\boldsymbol{q}}) &= \frac{1}{2}\left[\boldsymbol{V}(\boldsymbol{q},\dot{\boldsymbol{q}})+\boldsymbol{U}^T(\boldsymbol{q},\dot{\boldsymbol{q}})-\boldsymbol{U}(\boldsymbol{q},\dot{\boldsymbol{q}})\right]\\ &= \begin{bmatrix}0 & -a_1a_2s_2m_2\left(\dot{\theta}_1+\frac{1}{2}\dot{\theta}_2\right)\\ \frac{1}{2}a_1a_2s_2m_2\left(\dot{\theta}_1+\frac{1}{2}\dot{\theta}_2\right) & -\frac{1}{4}a_1a_2s_2m_2\dot{\theta}_1\end{bmatrix}\end{aligned}$$

To verify the calculations, we first compute $\boldsymbol{C}(\boldsymbol{q},\dot{\boldsymbol{q}})\dot{\boldsymbol{q}}$ and compare it with the Coriolis and centrifugal vector $\mathfrak{v}(\boldsymbol{q},\dot{\boldsymbol{q}})$ calculated in Example 8.2. They yield the same result.

$$\boldsymbol{C}(\boldsymbol{q},\dot{\boldsymbol{q}})\dot{\boldsymbol{q}} = \mathfrak{v}(\boldsymbol{q},\dot{\boldsymbol{q}}) = \begin{bmatrix}-a_1a_2s_2m_2\left(\dot{\theta}_1\dot{\theta}_2+\frac{1}{2}\dot{\theta}_2^2\right)\\ \frac{1}{2}a_1a_2s_2m_2\dot{\theta}_1^2\end{bmatrix}$$

Next, we calculate the matrix $\dot{\boldsymbol{M}}-2\boldsymbol{C}$.

$$\dot{\boldsymbol{M}}(\boldsymbol{q})-2\boldsymbol{C}(\boldsymbol{q},\dot{\boldsymbol{q}}) = \begin{bmatrix}0 & a_1a_2s_2m_2\left(\dot{\theta}_1+\frac{1}{2}\dot{\theta}_2\right)\\ -a_1a_2s_2m_2\left(\dot{\theta}_1+\frac{1}{2}\dot{\theta}_2\right) & 0\end{bmatrix}$$

It can be observed that, as expected, this matrix is skew-symmetric.

Case Study 9.2 (Planar 3R Robot) *Consider the planar 3R robot shown in Figure 8.7. The dynamics formulations for this system are obtained in Case Study 8.3. Calculate the Christoffel matrix for this robot and verify the skew-symmetry of the matrix $\dot{\boldsymbol{M}}-2\boldsymbol{C}$ in this representation.*

Solution: Similar to the previous example, we start by taking the derivative of the mass matrix and then determine the matrices $\boldsymbol{V}$ and $\boldsymbol{U}$ using symbolic computations. With these matrices, the Christoffel matrix is calculated as follows.

$$\begin{aligned}\boldsymbol{C}(\boldsymbol{q},\dot{\boldsymbol{q}}) &= \frac{1}{2}\left[\boldsymbol{V}(\boldsymbol{q},\dot{\boldsymbol{q}})+\boldsymbol{U}^T(\boldsymbol{q},\dot{\boldsymbol{q}})-\boldsymbol{U}(\boldsymbol{q},\dot{\boldsymbol{q}})\right]\\ &= \begin{bmatrix}C_{11} & C_{12} & C_{13}\\ C_{21} & C_{22} & C_{23}\\ C_{31} & C_{32} & C_{33}\end{bmatrix}\end{aligned}$$

where,

$$
\begin{aligned}
C_{11} &= 0 \\
C_{12} &= -a_1\left(a_2 s_2 m_2 + (2a_2 s_2 + a_3 s_{23}) m_3\right)\left(\dot\theta_1 + \frac{1}{2}\dot\theta_2\right) - \frac{1}{2}a_1 a_3 s_{23} m_3 \dot\theta_3 \\
C_{13} &= -a_3(a_1 s_{23} + a_2 s_3) m_3 \left(\dot\theta_1 + \frac{1}{2}\dot\theta_3\right) - \frac{1}{2}a_3(a_1 s_{23} + 2a_2 s_3) m_3 \dot\theta_2 \\
C_{21} &= \frac{1}{2}a_1\left(a_2 s_2(m_2 + 2m_3) + a_3 s_{23} m_3\right)\left(\dot\theta_1 + \frac{1}{2}\dot\theta_2\right) + \frac{1}{4}a_1 a_3 s_{23} m_3 \dot\theta_3 \\
C_{22} &= -\frac{1}{4}a_1\left(a_2 m_2 s_2 + 2a_2 m_3 s_2 + a_3 m_3 s_{23}\right)\dot\theta_1 \\
C_{23} &= -a_3 m_3 \left(\left(a_2 s_3 + \frac{1}{4}a_1 s_{23}\right)\dot\theta_1 + a_2 s_3 \dot\theta_2 + \frac{1}{2}a_2 s_3 \dot\theta_3\right) \\
C_{31} &= a_3 m_3 \left(\frac{1}{2}(a_1 s_{23} + a_2 s_3)\dot\theta_1 + \frac{1}{4}(a_1 s_{23} + 2a_2 s_3)\dot\theta_2 + \frac{1}{4}(a_1 s_{23} + a_3 s_3)\dot\theta_3\right) \\
C_{32} &= \frac{1}{2}a_3 m_3 \left(\left(a_2 s_3 - \frac{1}{2}a_1 s_{23}\right)\dot\theta_1 + a_2 s_3 \dot\theta_2 + \frac{1}{2}a_2 s_3 \dot\theta_3\right) \\
C_{33} &= -\frac{1}{4}a_3 m_3 \left((a_1 s_{23} + a_2 s_3)\dot\theta_1 + a_2 s_3 \dot\theta_2\right)
\end{aligned}
$$

To verify the calculations, we first compute $\boldsymbol{C}(\boldsymbol{q}, \dot{\boldsymbol{q}})\dot{\boldsymbol{q}}$ and compare it with the Coriolis and centrifugal vector $\boldsymbol{\mathfrak{v}}(\boldsymbol{q}, \dot{\boldsymbol{q}})$ calculated in Example 8.2. As expected the results of the computations are identical. Next, we calculate the matrix $\dot{\boldsymbol{M}} - 2\boldsymbol{C}$.

$$
\dot{\boldsymbol{M}}(\boldsymbol{q}) - 2\boldsymbol{C}(\boldsymbol{q}, \dot{\boldsymbol{q}}) = \begin{bmatrix} 0 & D_{12} & D_{13} \\ -D_{12} & 0 & D_{23} \\ -D_{13} & -D_{23} & 0 \end{bmatrix}
$$

where

$$
\begin{aligned}
D_{12} &= a_1\left(a_2 m_2 s_2 + m_3(2a_2 s_2 + a_3 s_{23})\right)\left(\dot\theta_1 + \frac{1}{2}\dot\theta_2\right) + \frac{1}{2}a_1 a_3 m_3 s_{23}\dot\theta_3 \\
D_{13} &= a_3 m_3 \left((a_1 s_{23} + a_2 s_3)\left(\dot\theta_1 + \frac{1}{2}\dot\theta_3\right) + \frac{1}{2}(a_1 s_{23} + 2a_2 s_3)\dot\theta_2\right) \\
D_{23} &= a_2 a_3 m_3 s_3(\dot\theta_1 + \dot\theta_2 + \frac{1}{2}\dot\theta_3)
\end{aligned}
$$

As expected, the obtained matrix is skew-symmetric.

Case Study 9.3 (SCARA Robot) *Consider the SCARA robot shown in Figure 8.8. The dynamics formulation for this robot is obtained in Case Study 8.4. Calculate the Christoffel matrix for this robot and verify the skew-symmetry of the matrix* $\dot{\boldsymbol{M}} - 2\boldsymbol{C}$ *in this representation.*

Solution: Initially, based on the convention in 9.15, we obtain the derivative of the mass matrix in a matrix of appropriate dimensions. Then, using the Kronecker product and the relationships in 9.20 and 9.21, we determine the matrices $\boldsymbol{V}$ and $\boldsymbol{U}$, respectively.

$$\begin{aligned}
\boldsymbol{V}(\boldsymbol{q},\dot{\boldsymbol{q}}) &= \dot{\boldsymbol{M}}(\boldsymbol{q}) = \frac{\partial \boldsymbol{M}^T}{\partial \boldsymbol{q}}\left(\boldsymbol{I}_{n\times n}\otimes \dot{\boldsymbol{q}}\right) \\
&= \begin{bmatrix} 0 & -\frac{1}{2}a_1a_2s_2(m_2+2m_3)(2\dot{\theta}_1-\dot{\theta}_2) & 0 \\ 0 & \frac{1}{2}a_1a_2s_2(m_2+2m_3)\dot{\theta}_1 & 0 \\ 0 & 0 & 0 \end{bmatrix} \\
\boldsymbol{U}(\boldsymbol{q},\dot{\boldsymbol{q}}) &= \left(\boldsymbol{I}_{n\times n}\otimes \dot{\boldsymbol{q}}^T\right)\frac{\partial \boldsymbol{M}}{\partial \boldsymbol{q}} \\
&= \begin{bmatrix} 0 & 0 & 0 \\ -\frac{1}{2}a_1a_2s_2(m_2+2m_3)(2\dot{\theta}_1-\dot{\theta}_2) & -\frac{1}{2}a_1a_2s_2(m_2+2m_3)\dot{\theta}_1 & 0 \\ 0 & 0 & 0 \end{bmatrix}
\end{aligned}$$

By this means, the Christoffel matrix is easily calculated.

$$\boldsymbol{C}(\boldsymbol{q},\dot{\boldsymbol{q}}) = \frac{1}{2}\left[\boldsymbol{V}(\boldsymbol{q},\dot{\boldsymbol{q}}) + \boldsymbol{U}^T(\boldsymbol{q},\dot{\boldsymbol{q}}) - \boldsymbol{U}(\boldsymbol{q},\dot{\boldsymbol{q}})\right] =$$

$$\begin{bmatrix} 0 & -\frac{1}{2}a_1a_2s_2(m_2+2m_3)(2\dot{\theta}_1-\dot{\theta}_2) & 0 \\ -\frac{1}{4}a_1a_2s_2(m_2+2m_3)(2\dot{\theta}_1-\dot{\theta}_2) & \frac{1}{4}a_1a_2s_2(m_2+2m_3)\dot{\theta}_1 & 0 \\ 0 & 0 & 0 \end{bmatrix}$$

To verify the calculations, we first compute $\boldsymbol{C}(\boldsymbol{q},\dot{\boldsymbol{q}})\dot{\boldsymbol{q}}$ and compare it with the Coriolis and centrifugal vector $\mathfrak{v}(\boldsymbol{q},\dot{\boldsymbol{q}})$ calculated in Example 8.4. The results are identical. Next, we calculate the matrix $\dot{\boldsymbol{M}} - 2\boldsymbol{C}$.

$$\dot{\boldsymbol{M}}(\boldsymbol{q}) - 2\boldsymbol{C}(\boldsymbol{q},\dot{\boldsymbol{q}}) =$$

$$\begin{bmatrix} 0 & \frac{1}{2}a_1a_2s_2(m_2+2m_3)(2\dot{\theta}_1-\dot{\theta}_2) & 0 \\ -\frac{1}{2}a_1a_2s_2(m_2+2m_3)(2\dot{\theta}_1-\dot{\theta}_2) & 0 & 0 \\ 0 & 0 & 0 \end{bmatrix}.$$

As anticipated, the results of the calculations show that the derived matrix is skew-symmetric.

Case Study 9.4 (3DOF Elbow Manipulator) *Consider the three-degrees-of-freedom Elbow manipulator shown in Figure 8.9. The dynamics formulation for this robot is obtained in Case Study 8.5. Calculate the Christoffel matrix for this robot and verify the skew-symmetry of the matrix* $\dot{\boldsymbol{M}} - 2\boldsymbol{C}$ *in this representation.*

Solution: Similar to the previous case study, we start by taking the derivative of the mass matrix and then determine the matrices $\boldsymbol{V}$ and $\boldsymbol{U}$ using symbolic computations. With these calculations, the Christoffel matrix is calculated as follows.

$$\boldsymbol{C}(\boldsymbol{q},\dot{\boldsymbol{q}}) = \begin{bmatrix} C_{11} & C_{12} & C_{13} \\ C_{21} & C_{22} & C_{23} \\ C_{31} & C_{32} & C_{33} \end{bmatrix}$$

where

$$
\begin{aligned}
C_{11} &= 0 \\
C_{12} &= -\frac{1}{3}\left(a_2^2 m_2 \sin 2\theta_2 + a_3^2 m_3 \sin 2(\theta_2+\theta_3)\right)\dot{\theta}_1 - \\
&\qquad\qquad m_3\left(a_2^2 \sin 2\theta_2 + a_2 a_3 \sin(2\theta_2+\theta_3)\right)\dot{\theta}_1 \\
C_{13} &= -\frac{1}{3} a_3 m_3 s_{23}(2a_3 c_{23} + 3a_2 c_2)\dot{\theta}_1 \\
C_{21} &= \frac{1}{6}\left(a_2^2 m_2 \sin 2\theta_2 + a_3^2 m_3 \sin 2(\theta_2+\theta_3)\right)\dot{\theta}_1 + \\
&\qquad\qquad \frac{1}{2} m_3\left(a_2^2 \sin 2\theta_2 + a_2 a_3 \sin(2\theta_2+\theta_3)\right)\dot{\theta}_1 \\
C_{22} &= 0 \\
C_{23} &= -\frac{1}{2} a_2 a_3 m_3 s_3(2\dot{\theta}_2 + \dot{\theta}_3) \\
C_{31} &= \frac{1}{6} a_3 m_3 s_{23}\left(2a_3 c_{23} + 3a_2 c_2\right)\dot{\theta}_1 \\
C_{32} &= \frac{1}{4} a_2 a_3 m_3 s_3\left(2\dot{\theta}_2 + \dot{\theta}_3\right) \\
C_{33} &= -\frac{1}{4} a_2 a_3 m_3 s_3 \dot{\theta}_2
\end{aligned}
$$

To verify the calculations, we first compute $\boldsymbol{C}(\boldsymbol{q},\dot{\boldsymbol{q}})\dot{\boldsymbol{q}}$ and compare it with the Coriolis and centrifugal vector $\mathfrak{v}(\boldsymbol{q},\dot{\boldsymbol{q}})$ calculated in Example 8.5. The results are identical. Next, we will proceed with calculating the matrix $\dot{\boldsymbol{M}} - 2\boldsymbol{C}$.

$$
\dot{\boldsymbol{M}}(\boldsymbol{q}) - 2\boldsymbol{C}(\boldsymbol{q},\dot{\boldsymbol{q}}) = \begin{bmatrix} 0 & D_{12} & D_{13} \\ -D_{12} & 0 & D_{23} \\ -D_{13} & -D_{23} & 0 \end{bmatrix}
$$

where

$$
\begin{aligned}
D_{12} &= \frac{1}{3}\left(a_2^2 m_2 \sin 2\theta_2 + a_3^2 m_3 \sin 2(\theta_2+\theta_3)\right)\dot{\theta}_1 + \\
&\qquad\qquad m_3\left(a_2^2 \sin 2\theta_2 + a_2 a_3 \sin(2\theta_2+\theta_3)\right)\dot{\theta}_1 \\
D_{13} &= \frac{1}{3} a_3 m_3 s_{23}(2a_3 c_{23} + 3a_2 c_2)\dot{\theta}_1 \\
D_{23} &= \frac{1}{2} a_2 a_3 m_3 s_3\left(2\dot{\theta}_2 + \dot{\theta}_3\right)
\end{aligned}
$$

As anticipated, the calculation results confirm that the obtained matrix is skew-symmetric.

9.4 Linear Regression Representation

The dynamics formulation of serial robots can be generally represented by the relation 9.23. In order to use these equations in dynamic simulation or model-based control design, it is essential to identify the kinematic and dynamic parameters of the robot. Various calibration methods have been developed in the robotics literature to accurately determine these parameters [110, 111, 112]. However, due to the complexity of dynamic equations, the robot calibration process is highly challenging in practice.

Fortunately, it can be demonstrated that the nonlinear and coupled dynamic equations of serial robots, in terms of $\boldsymbol{q}$ and its derivatives, can be rewritten in the form of linear regression with respect to the kinematic and dynamic parameters of the robot. If the parameter vector of the robot can be represented in the ℓ-dimensional space by the symbol $\boldsymbol{\Phi} \in \mathbb{R}^{\ell}$, it should be noted that the dimension of this space is not unique. In general, in addition to Denavit-Hartenberg parameters, the dynamic parameters of each link consist of ten components, including the mass of the link, its three center of mass components, and six components of the link's inertia matrix. However, depending on the type of robot motion, some of these parameters may not be excited and may not appear in the robot's dynamic equations. Determining the minimum dimension of the parameter space is an interesting research topic that has been addressed in the robotics literature [113].

By determining the dimension of the robot's parameter space, the dynamic equations can be rewritten as a linear regression with respect to the parameters.

$$\boldsymbol{M}(\boldsymbol{q})\ddot{q} + \boldsymbol{C}(\boldsymbol{q},\dot{\boldsymbol{q}})\dot{\boldsymbol{q}} + \boldsymbol{\mathfrak{g}}(\boldsymbol{q}) = \boldsymbol{\mathcal{Y}}(\boldsymbol{q},\dot{\boldsymbol{q}},\ddot{\boldsymbol{q}})\boldsymbol{\Phi} \tag{9.24}$$

In this relation, the matrix $\boldsymbol{\mathcal{Y}}(\boldsymbol{q},\dot{\boldsymbol{q}},\ddot{\boldsymbol{q}})$ is considered as the regression matrix, and $\boldsymbol{\Phi}$ is the vector of robot parameters determined through experimentation. Having the regression form of the robot's dynamic equations, the robot parameters can be obtained through experiments using least squares methods. This form is also employed for adaptive control in robots [100, 101].

Case Study 9.5 (Planar 2$\underline{\text{R}}$ Robot) *Consider the planar 2$\underline{R}$ robot depicted in Figure 8.6. The dynamics formulation of the robot, assuming the robot links are treated as slender rod, and viscous friction exists at the robot joints, have been analyzed in Case Study 8.2. In this robot, determine the necessary kinematic and dynamic parameters and obtain the linear regression form of the robot's dynamic formulation.*

Solution: Consider the inertia matrix of the robot reported in Case Study 8.2. If we categorize the elements composing this matrix in the following order,

$$\begin{aligned}
\Phi_1 &= \frac{1}{3}m_1a_1^2 + m_2\left(a_1^2 + \frac{1}{3}a_2^2\right)\\
\Phi_2 &= \frac{1}{2}m_2a_1a_2\\
\Phi_3 &= \frac{1}{3}m_2a_2^2
\end{aligned}$$

It is possible to rewrite the components of the inertia matrix using these parameters.

$$\begin{aligned}
M_{11} &= \Phi_1 + 2\Phi_2c_2\\
M_{12} = M_{21} &= \Phi_3 + \Phi_2c_2\\
M_{22} &= \Phi_3
\end{aligned}$$

Now consider the gravity vector, and intuitively obtain the following independent parameters for expressing this vector in a linear regression form.

$$\begin{aligned}
\Phi_4 &= \frac{1}{2}m_1a_1 + m_2a_1\\
\Phi_5 &= \frac{1}{2}m_2a_2
\end{aligned}$$

Then, the components of the gravitational vector can be rewritten using these parameters.

$$\begin{aligned} \mathfrak{g}_1 &= (\Phi_4 c_1 + \Phi_5 c_{12})g \\ \mathfrak{g}_2 &= \Phi_5 c_{12} g \end{aligned}$$

To represent the Coriolis and centrifugal vectors in a linear regression form, there is no need to introduce new parameters. They can be obtained using the parameters defined earlier as follows.

$$\begin{aligned} \mathfrak{v}_1 &= -\Phi_2 s_2 \left(2\dot{\theta}_1\dot{\theta}_2 + \dot{\theta}_2^2\right) \\ \mathfrak{v}_2 &= 2\Phi_2 s_2 \dot{\theta}_1^2 \end{aligned}$$

Considering the presence of viscous friction at the joints, it is possible to add friction coefficients as additional model parameters in a linear regression form.

$$\begin{aligned} \Phi_6 &= b_1 \\ \Phi_7 &= b_2 \end{aligned}$$

Considering these parameters, the linear regression matrix in this robot is obtained as follows.

$$\mathcal{Y}(q,\dot{q},\ddot{q}) = \begin{bmatrix} \ddot{\theta}_1 & c_2(2\ddot{\theta}_1 + \ddot{\theta}_2) - s_2(\dot{\theta}_1^2 + 2\dot{\theta}_1\dot{\theta}_2) & \ddot{\theta}_2 & c_1 g & c_{12} g & \dot{\theta}_1 & 0 \\ 0 & c_2\ddot{\theta}_1 + s_2\dot{\theta}_1^2 & \ddot{\theta}_1 + \ddot{\theta}_2 & 0 & c_{12} g & 0 & \dot{\theta}_2 \end{bmatrix}$$

where,

$$\boldsymbol{\Phi} = \begin{bmatrix} \frac{1}{3}m_1 a_1^2 + m_2\left(a_1^2 + \frac{1}{3}a_2^2\right) \\ \frac{1}{2}m_2 a_1 a_2 \\ \frac{1}{3}m_2 a_2^2 \\ \frac{1}{2}m_1 a_1 + m_2 a_1 \\ \frac{1}{2}m_2 a_2 \\ b_1 \\ b_2 \end{bmatrix}$$

By determining the regression matrix and parameter vector, the robot's dynamics formulation can be rewritten in the form of linear regression.

$$\begin{bmatrix} \tau_1 \\ \tau_2 \end{bmatrix} = \boldsymbol{M}(q)\ddot{q} + \boldsymbol{C}(q,\dot{q})\dot{q} + \mathfrak{g}(q) = \mathcal{Y}(q,\dot{q},\ddot{q})\boldsymbol{\Phi}$$

Case Study 9.6 (Planar 3R Robot) *Examine the planar 3R robot depicted in Figure 8.7. The dynamic equations of the robot, considering the robot links as slender rods, have been scrutinized in Case Study 8.3. Disregarding the joint friction torques, identify the required kinematic and dynamic parameters, and derive the linear regression form of the robot's dynamic equations.*

Solution: Examine the dynamic equations of the robot presented in Case Study 8.3. By utilizing the components comprising the inertia matrix, gravity and Coriolis and centrifugal vectors, the subsequent kinematic and dynamic parameters can be defined.

$$\begin{aligned}
\Phi_1 &= \left(\frac{1}{3}m_1 + m_2 + m_3\right)a_1^2 + \left(\frac{1}{3}m_2 + m_3\right)a_2^2 + m_3 a_3^2 \\
\Phi_2 &= (m_2 + 2m_3)\,a_1a_2, \ \ \Phi_3 = m_3a_1a_2, \ \ \Phi_4 = m_3a_2a_3, \ \ \Phi_5 = m_3a_1a_3, \ \ \Phi_6 = m_2a_1a_2 \\
\Phi_7 &= \left(\frac{1}{3}m_2 + m_3\right)a_2^2, \ \ \Phi_8 = \frac{1}{3}m_3a_3^2 \\
\Phi_9 &= \frac{1}{2}(m_1 + m_2)\,a_1, \ \ \Phi_{10} = \frac{1}{2}(m_2 + 2m_3)\,a_2, \ \ \Phi_{11} = \frac{1}{3}m_3a_3
\end{aligned}$$

By this means, the components of the inertia matrix can be rewritten using these parameters.

$$\begin{aligned}
M_{11} &= \Phi_1 + \Phi_2c_2 + \Phi_3c_3 + \Phi_4c_{23} \\
M_{12} &= \frac{1}{2}\Phi_2c_2 + \Phi_4\left(\frac{1}{2}c_{23} + c_3\right) + \Phi_7 + \Phi_8 \\
M_{13} &= \frac{1}{2}\Phi_4\,(c_3 + c_{23}) + \Phi_8 \\
M_{22} &= \Phi_4c_3 + \Phi_7 + \Phi_8, \ \ M_{23} = \frac{1}{2}\Phi_4c_3 + \Phi_8, \ \ M_{33} = \Phi_8
\end{aligned}$$

Similarly, the components of the gravity vector can be intuitively calculated using these parameters.

$$\begin{aligned}
\mathfrak{g}_1 &= (\Phi_9c_1 + \Phi_{10}c_{12} + \Phi_{11}c_{123})g \\
\mathfrak{g}_2 &= (\Phi_{10}c_{12} + \Phi_{11}c_{123})g \\
\mathfrak{g}_3 &= \Phi_{11}c_{123}g
\end{aligned}$$

Furthermore, the Coriolis and centrifugal vector can be expressed in linear regression form using the defined parameters.

$$\begin{aligned}
\mathfrak{v}_1 &= -\Phi_2s_2\left(\dot\theta_1\dot\theta_2 + \frac{1}{2}\dot\theta_2^2\right) - \Phi_4s_3\left(\dot\theta_1\dot\theta_3 + \dot\theta_2\dot\theta_3 + \frac{1}{2}\dot\theta_3^2\right) - \\
&\quad \frac{1}{2}\Phi_5s_{23}\left(3\left(\dot\theta_1\dot\theta_2 + \dot\theta_2\dot\theta_3\right) + \frac{1}{2}\dot\theta_3^2\right) \\
\mathfrak{v}_2 &= \frac{1}{2}\phi_3s_2\dot\theta_1^2 - \Phi_4s_3\left(\dot\theta_1\dot\theta_3 + \dot\theta_2\dot\theta_3 + \frac{1}{2}\dot\theta_3^2\right) + \frac{1}{2}\Phi_5s_{23}\dot\theta_1^2 + \frac{1}{2}\Phi_6s_2\dot\theta_1^2 \\
\mathfrak{v}_3 &= \frac{1}{2}\phi_4s_3\left(\dot\theta_1 + \dot\theta_2\right)^2 + \frac{1}{2}\Phi_5s_{23}\dot\theta_1^2
\end{aligned}$$

Disregarding the joint friction, the components of the first row of the linear regression matrix in this robot can be obtained as follows.

$$\begin{aligned}
\mathcal{Y}(1,1) &= \ddot\theta_1, \ \ \mathcal{Y}(1,2) = c_2\ddot\theta_1 + \frac{1}{2}c_2\ddot\theta_2s_2\left(\dot\theta_1\dot\theta_2 + \frac{1}{2}\dot\theta_2^2\right), \ \ \mathcal{Y}(1,3) = c_3 \\
\mathcal{Y}(1,4) &= c_{23}\ddot\theta_1 + \left(\frac{1}{2}c_{23} + c_3\right)\ddot\theta_2 + \frac{1}{2}(c_3 + c_{23})\ddot\theta_3 - s_3\left(\dot\theta_1\dot\theta_3 + \dot\theta_2\dot\theta_3 + \frac{1}{2}\dot\theta_3^2\right) \\
\mathcal{Y}(1,5) &= -\frac{1}{2}s_{23}\left(3\left(\dot\theta_1\dot\theta_2 + \dot\theta_2\dot\theta_3\right) + \frac{1}{2}\dot\theta_3^2\right) \\
\mathcal{Y}(1,6) &= 0, \ \ \mathcal{Y}(1,7) = \ddot\theta_2, \ \ \mathcal{Y}(1,8) = \ddot\theta_2 + \ddot\theta_3 \\
\mathcal{Y}(1,9) &= c_1g, \ \ \mathcal{Y}(1,10) = c_{12}g, \ \ \mathcal{Y}(1,11) = c_{123}g
\end{aligned}$$

Similarly, the components of the second row of the linear regression matrix in this robot can be obtained as follows.

$$\begin{aligned}
\mathcal{Y}(2,1) &= 0, \;\; \mathcal{Y}(2,2) = \frac{1}{2}c_2\ddot{\theta}_1 \;\; \mathcal{Y}(2,3) = \frac{1}{2}s_2\ddot{\theta}_1^2 \\
\mathcal{Y}(2,4) &= \left(\frac{1}{2}c_{23} + c_3\right)\ddot{\theta}_1 + c_3\ddot{\theta}_2 + \frac{1}{2}c_3\ddot{\theta}_3 - s_3\left(\dot{\theta}_1\dot{\theta}_3 + \dot{\theta}_2\dot{\theta}_3 + \frac{1}{2}\dot{\theta}_3^2\right) \\
\mathcal{Y}(2,5) &= \frac{1}{2}s_{23}\ddot{\theta}_1^2, \;\; \mathcal{Y}(2,6) = \frac{1}{2}s_2\ddot{\theta}_1^2, \;\; \mathcal{Y}(2,7) = \ddot{\theta}_1 + \ddot{\theta}_2 \\
\mathcal{Y}(2,8) &= \ddot{\theta}_1 + \ddot{\theta}_2 + \ddot{\theta}_3, \;\; \mathcal{Y}(2,9) = 0, \;\; \mathcal{Y}(2,10) = c_{12}g, \;\; \mathcal{Y}(2,11) = c_{123}g
\end{aligned}$$

And finally, the non-zero components of the third row of the linear regression matrix in this robot are equal to:

$$\begin{aligned}
\mathcal{Y}(3,4) &= \frac{1}{2}(c_3 + c_{23})\ddot{\theta}_1 + \frac{1}{2}c_3\ddot{\theta}_2 + \frac{1}{2}s_3\left(\dot{\theta}_1 + \dot{\theta}_2\right)^2 \\
\mathcal{Y}(3,5) &= -\frac{1}{2}s_{23}\ddot{\theta}_1^2, \;\; \mathcal{Y}(3,8) = \ddot{\theta}_1 + \ddot{\theta}_2 + \ddot{\theta}_3, \;\; \mathcal{Y}(3,11) = c_{123}g
\end{aligned}$$

By determining the regression matrix and parameter vector, the robot's dynamics formulation can be rewritten in the following linear regression form.

$$\begin{bmatrix} \tau_1 \\ \tau_2 \\ \tau_3 \end{bmatrix} = \boldsymbol{M}(q)\ddot{q} + \boldsymbol{C}(q,\dot{q})\dot{q} + \mathfrak{g}(q) = \boldsymbol{\mathcal{Y}}(q,\dot{q},\ddot{q})\boldsymbol{\Phi}$$

9.5 Dynamic Formulation Represented in Task Space

So far, the dynamic equations of serial robots and their characteristics have been derived in the joint space of the robot. The formulation of dynamic equations in serial robots in this form is quite common and is directly used in joint space control methods. However, in some control methods, it is necessary to represent the dynamic equations of the system in the task space and based on the variable $\boldsymbol{\mathcal{X}}$. In such cases, it is possible to achieve these equations using the concept of Jacobian matrix mapping.

The general form of the dynamic equations of a robot with n fully actuated joints in the task space can be represented as follows.

$$\boldsymbol{\mathcal{F}} - \boldsymbol{\mathcal{F}}_d = \boldsymbol{M}_x(q)\ddot{\boldsymbol{\mathcal{X}}} + \mathfrak{v}_x(q,\dot{q}) + \mathfrak{g}_x(q) \tag{9.25}$$

In this equation, $\boldsymbol{\mathcal{F}} - \boldsymbol{\mathcal{F}}_d$ represents the wrench applied to the robot's end-effector, where $\boldsymbol{\mathcal{F}}$ is obtained using the transpose Jacobian matrix mapping from the actuator torques. If there exist external forces and friction torques in the links and end-effector of the robot, by using the appropriate mapping of the corresponding Jacobian matrices, one may introduce all these factors as disturbances in the dynamic equation in the form of the disturbance wrench $\boldsymbol{\mathcal{F}}_d$. In this case, the generalized force entering the system can be expressed as follows:

$$\boldsymbol{Q} = \boldsymbol{J}^T(q)(\boldsymbol{\mathcal{F}} - \boldsymbol{\mathcal{F}}_d) \tag{9.26}$$

The matrices and vectors in the dynamic equations in the task space, represented with the index x in Equation 9.25, can be easily calculated using the robot Jacobian matrix and based on the relation given in Equation 9.1. To do so, multiply both sides of Equation 9.1 from the left side by the inverse transpose Jacobian matrix $\boldsymbol{J}^{-T}$.

$$\boldsymbol{J}^{-T}\boldsymbol{Q} = \boldsymbol{J}^{-T}\boldsymbol{M}\ddot{q} + \boldsymbol{J}^{-T}\mathfrak{v} + \boldsymbol{J}^{-T}\mathfrak{g} \tag{9.27}$$

Now, utilize the mapping of the Jacobian matrix and its derivative in the transformation from joint variables to task space variables.

$$\dot{\mathcal{X}} = \boldsymbol{J}(q)\dot{q} \quad \rightarrow \quad \ddot{\mathcal{X}} = \dot{\boldsymbol{J}}(q)\dot{q} + \boldsymbol{J}(q)\ddot{q} \tag{9.28}$$

By solving this equation for $\ddot{q}$, we will have:

$$\ddot{q} = \boldsymbol{J}^{-1}\ddot{\mathcal{X}} - \boldsymbol{J}^{-1}\dot{\boldsymbol{J}}\dot{q} \tag{9.29}$$

By substituting the equations 9.26 and 9.29 into 9.27, we can write:

$$\mathcal{F} - \mathcal{F}_d = \boldsymbol{J}^{-T}\boldsymbol{M}\left(\boldsymbol{J}^{-1}\ddot{\mathcal{X}} - \boldsymbol{J}^{-1}\dot{\boldsymbol{J}}\dot{q}\right) + \boldsymbol{J}^{-T}\mathfrak{v} + \boldsymbol{J}^{-T}\mathfrak{g} \tag{9.30}$$

Thus, the general form of the dynamic equations of a robot with n fully actuated joints in the task space can be represented using the equation 9.25, where:

$$\begin{aligned} \boldsymbol{M}_x(q) &= \boldsymbol{J}^{-T}(q)\boldsymbol{M}(q)\,\boldsymbol{J}^{-1}(q), \\ \mathfrak{v}_x(q,\dot{q}) &= \boldsymbol{J}^{-T}(q)\left\{\mathfrak{v}(q,\dot{q}) - \boldsymbol{M}(q)\boldsymbol{J}^{-1}(q)\dot{\boldsymbol{J}}(q)\dot{q}\right\} \\ \mathfrak{g}_x(q) &= \boldsymbol{J}^{-T}(q)\mathfrak{g}(q) \end{aligned} \tag{9.31}$$

It is noted that the dynamic equations for a robot in the task space can be formulated using the inverse Jacobian matrix. Consequently, in scenarios where the robot lacks full actuation or encounters singular configurations, these equations are not efficiently applicable for determining the dynamic equations. Furthermore, it is important to highlight that the involvement of the Jacobian matrix and its derivatives in these computations adds complexity to the nonlinear nature of the robot's dynamic equations in the task space. Therefore, dynamic equations in joint space are commonly favored in serial robots.

9.6 Actuator Dynamics

The desired trajectory for motion in robots is achieved through the execution of control commands in the joints of the robot. Serial robots utilize various types of actuators, including pneumatic, hydraulic, and electric actuators. Nowadays, in the majority of industrial robots, except for specific cases requiring exceptionally high power, electric actuators are commonly employed. Therefore, this section focuses on these types of actuators.

Electric actuators used in robots come in various forms, including permanent magnet and brushless DC motors, and different types of AC motors. The torque production structure in permanent magnet DC motors is simpler to be dynamically analyzed, while it represents the general characteristics of typical electric motors. Hence, the dynamics of these types of actuators, called as electric motors in this section, will be examined in the following section.

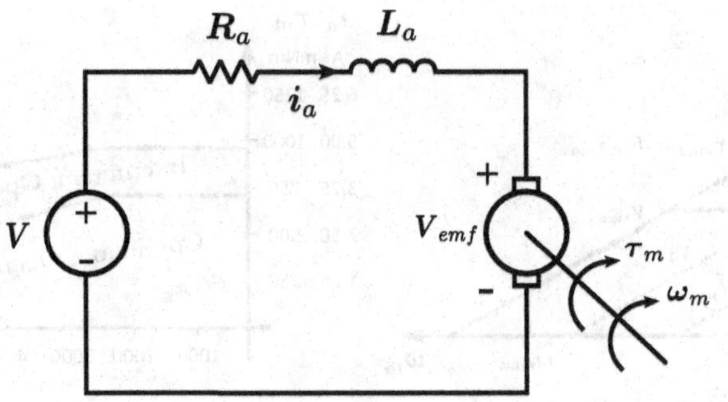

FIGURE 9.1
Schematic of the electromechanical model of a permanent magnet DC motor.

9.6.1 Permanent Magnet DC Motors

This electric motor typically consists of a rotor formed by a large number of current-carrying wires wound cylindrically. The stator of the motor, using a permanent magnet, generates the necessary magnetic flux. Two key quantities in the modeling of permanent magnet electric motors are the produced torque in the motor, denoted as τ_m, and the back electromotive force (EMF) voltage, denoted as V_{emf}. It can be demonstrated that the motor's torque is proportional to the armature current i_a, and its back EMF voltage is proportional to the motor's rotational speed ω_m, determined by the following relationships.

$$\tau_m = K_i\, i_a \qquad V_{emf} = K_v\, \omega_m \tag{9.32}$$

where, K_i is the current constant, and K_v is the voltage constant of the electric motor. If standard SI units are used, these two constants are equal to each other, and the symbol K_m is commonly used as an alternative to represent the motor's electrical constant. Thus, the model of a permanent magnet electric motor can be derived as an electromechanical system by combining electrical and mechanical dynamics.

Consider the electrical model of the motor shown in Figure 9.1. In this model, the motor armature is powered by a DC voltage, and based on the wound wire structure, the rotor is modeled as a resistor with a coefficient of R_a and a coil with inductance L_a. The back electromotive force (EMF) of the motor, proportional to the motor's rotational speed, is also considered in this model. In this way, the electrical model of a PMDC motor can easily be obtained as follows.

$$V = L_a \frac{di_a}{dt} + R_a\, i_a + V_{emf}$$

By rewriting this model in the Laplace domain, we will have:

$$V(s) - V_{emf}(s) = (L_a s + R_a) I_a(s) \quad \rightarrow \quad I_a(s) = \frac{V(s) - V_{emf}(s)}{L_a s + R_a}$$

Now, let's add the mechanical motion dynamics to the model to obtain the electromechanical model of the motor.

$$\tau_m(s) = K_m I_a = K_m \frac{V - K_m \omega_m}{L_a s + R_a} \tag{9.33}$$

In this equation, the motor constant K_m is used to determine both the motor torque τ_m and the back electromotive force V_{emf}, utilizing the relation in Equation 9.32. As observed in

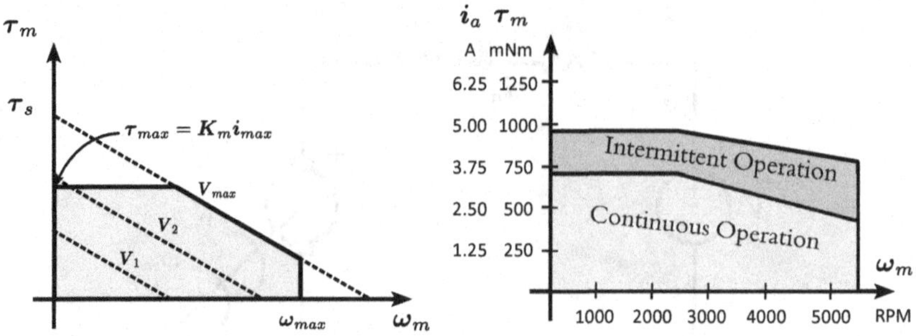

FIGURE 9.2
Technical characteristics of the electric motor in the form of a torque-speed graph.

this equation, an increase in the motor supply voltage leads to an additional motor torque, resulting in rotor acceleration and an increase in its speed. The law of Lenz is accurately modeled here, showing that with an increase in the motor speed, the motor torque decreases, and thus, the motor speed will reach its equilibrium value. The inductive property of the motor also acts as a first-order low-pass filter, smoothing out rapid changes in torque.

In the design of electric motors, the inductance of the motor coil, L_a, is typically much smaller than the motor resistance, R_a, i.e., $L_a/R_a \ll 1; (\approx 10^{-3})$. By neglecting the motor inductance, Equation 9.33 can be suitably simplified to the following algebraic equation:

$$\tau_m = \frac{K_m}{R_a}(V - K_m\omega_m) \tag{9.34}$$

This relationship, illustrated in Figure 9.2, elucidates the fundamental principles of electric motor operation. As observed in the left side of the figure, an increase in the motor supply voltage from V_1 to the tolerable maximum voltage of the motor, V_{max}, allows for higher speed and torque output from the motor. Now, considering a supply voltage less than its maximum limit, denoted by V_2, an increase in motor speed leads to a decrease in the obtained torque. This occurs while the maximum motor torque is generated when the motor speed is zero. Thus, it is possible to interpret the motion of the motor with a constant supply voltage as receiving constant power from it. In this interpretation, as the speed of motion increases, the output torque decreases proportionally.

In Figure 9.2, while illustrating this physical interpretation, the maximum speed limit ω_{max}, maximum torque τ_{max}, and stall torque τ_s of the electric motor are also shown. It should be noted that the maximum current in the motor is determined based on the tolerance of the rotor coil currents, while exceeding it will damage the motor. The maximum torque that the motor can produce is determined by $\tau_{max} = K_m i_{max}$. The stall torque, denoted as τ_s, represents the torque achievable by the motor under ideal conditions, when the motor speed approaches zero and the maximum allowable voltage is applied.

On the other hand, the allowable voltage in the motor coil wires is limited, and the maximum applicable voltage to the motor also determines the maximum speed limit of the motor ω_{max}. Thus, in addition to the linear relationship between torque and motor speed, the limits of these two quantities should also be considered when using an electric motor. Ultimately, the operating range of the electric motor is shown on the hashed curve in the left side of Figure 9.2, which is commonly used for selecting a suitable motor in robotics applications.

Typically, the speed-torque diagram of electric motors is presented in their technical specifications. An example of such a diagram for an electric motor is shown on the right side of Figure 9.2. In this diagram, the working range of the motor is displayed in two

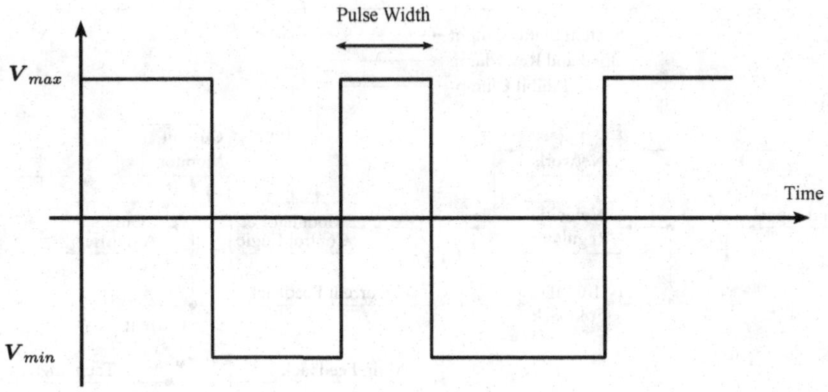

FIGURE 9.3
Voltage output of pulse width modulation converter in electric motors.

regimes: continuous and intermittent operations. If the motor is consistently in operation, it is necessary to use the continuous operating limits for its selection. However, if the motor is used intermittently for producing motion and starts and stops periodically, larger limits of intermittent operation can be used.

9.6.2 Motor Drivers

While applying DC voltage to an electric motor can set it in motion, a motor driver circuit is used to control the motor's movement and protect it against excessive currents. Given the significance of employing driver circuits in implementing motor control methods, a brief description of these drive circuits is provided.

The driver circuit for a permanent magnet DC motor typically utilizes pulse width modulation (PWM) converters. In these converters, as shown in Figure 9.3, the motor supply voltage is generated by pulses with a controllable duty cycle. If the pulse width is equal to half of the period, the effective voltage applied to the motor is zero. By increasing the pulse width to more than 50% of the cycle time, the applied voltage becomes positive, and by decreasing it to less than 50%, the motor's applied voltage becomes negative. Usually, frequencies exceeding 25 kHz are used to produce this modulated signal to ensure that its noise is not audible.

The block diagram illustrating the operation of a motor drive circuit for a specific type of electric motor is shown in Figure 9.4. As seen in this figure, to control the motion of an electric motor, feedback on often taken from the motor's speed and current to regulate its movement at least in two adjustable modes. The primary mode used in robot control is the current or torque mode in the motor drive circuit. In this mode, the feedback is taken from the armature current, and it is compared and controlled with the desired current commanded to the motor. Thus, the command applied to the motor drive circuit is the desired amount of motor current, which will be proportional to the torque produced by the electric motor. This section of the drive circuit realizes internal current feedback in the motor and is typically employed with a protective circuit to prevent the motor from producing current beyond the permissible limit. Implementing the motor supply current is achieved by controlling PWM converters and using a lead-lag controller. Given that the required torque in robot joints is determined by the robot control system and applied to the actuators, the use of the current mode as the primary method of motor drive is considered in robot control methods.

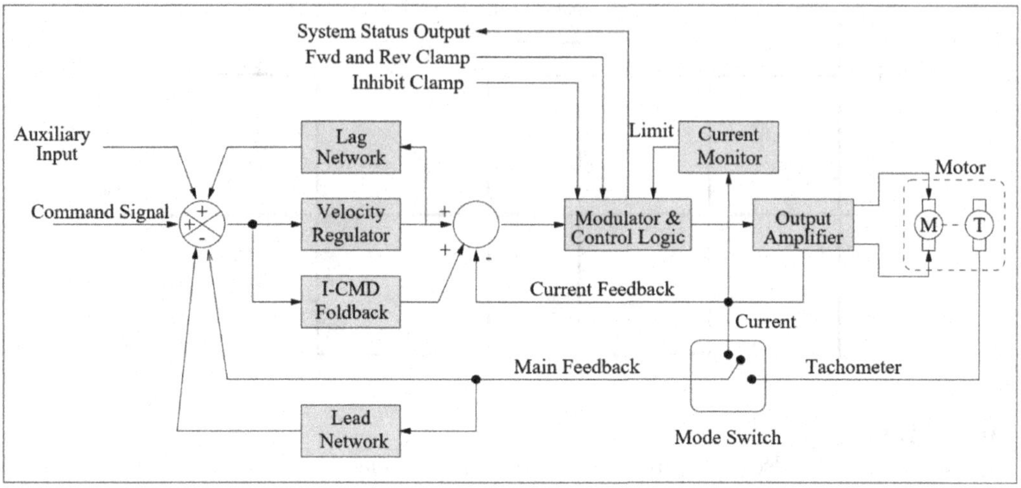

FIGURE 9.4
Block diagram illustrating the operation of a driver circuit for an electric motor.

Another operating mode in electric motors is the voltage or speed mode. In this mode, the motor speed is controlled by regulating the supply voltage. The control action involves feedback of the motor speed, comparing it with the desired speed setpoint. As depicted in this figure, the motor speed feedback is compared with the desired speed set by the user, and a phase-locked loop (PLL) or a PID controller adjusts the effective voltage applied to the motor. This external feedback loop is integrated into the motor drive circuit and is utilized when speed control of the motor is the primary objective. Alternatively, similar to speed measurement, position measurement using an encoder can be employed to control the final motion of the motor.

In the control of torque (or current) and speed (or voltage) in electric motors, the electrodynamic relationship of the motor must be considered. It is important to note that due to the significant load imposed on the motor by the robot linkage connected to it, expecting accurate execution of speed or position commands to the robot without controlling the motor current is not feasible. This is because the parameters controlling the position in the drive circuit are set without applying the required motor torque, and only in the presence of a disturbance torque that allocates a small percentage of the motor power to itself can the control system handle and mitigate it. This scenario is exacerbated when the motor load is increased, and the disturbance torque becomes predominant compared to the required torque for the dominant motion, leading to unsatisfactory motion tracking in the robot.

This issue is also seen in current (or torque) control, but its intensity is much lower. By commanding current to the electric motor, the motor current increases, and the motor accelerates. The increase in speed in the motor, according to Lenz's law, creates a back electromotive force, reducing the effective voltage in the robot. Like the previous case, this disturbance acts as an interference in the control circuit of the motor drive. However, the percentage of this disturbance is much lower than in the previous case since the robot's movements are not as fast, and the level of interference is significantly reduced. Thus, in a well-tuned drive circuit with properly adjusted control parameters, this disturbance is effectively mitigated.

In current and speed control systems, the bandwidth of the drive circuit can be analyzed by examining the frequency response. As an example, the frequency response of a permanent

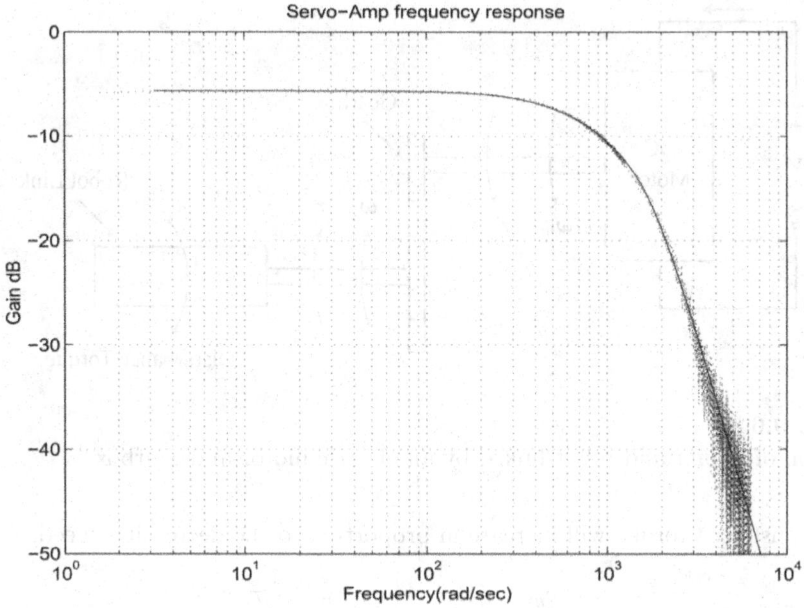

FIGURE 9.5
Frequency response of the electric motor and its drive circuit in torque mode.

magnet DC motor with its drive circuit in torque mode is shown in Figure 9.5. This figure is obtained through experimentation on a specific electric motor [114] and demonstrates that the electric motor and its drive circuit in torque mode exhibit ideal torque behavior with a bandwidth of 300 radians per second (approximately 100 Hz). This bandwidth is more than ten times the bandwidth required for common movements in industrial robots. With this description, when utilizing the motor drive in the current mode, the model of the motor can easily be considered as a linear system generating the required torque based on the motor torque constant.

$$\tau_m = K_m i_c, \tag{9.35}$$

In which i_c represents the commanded torque applied to the motor drive circuit.

9.6.3 Gearbox

Electric motors often generate maximum torques close to one Newton meter at speeds exceeding 2000 revolutions per minute (RPM). However, in many industrial robots, we require torques close to 200 Newton meters at speeds lower than 10 RPM. This adaptation is well achieved by using suitable gearboxes. Describing the types of gearboxes available and how they function is beyond the scope of this book. Still, due to their significant impact on robot dynamics, we focus on the dynamics formulation of the robot in the presence of actuators with gearbox in this section.

For this purpose, consider the motion of a single joint along with the robot link, created by an electric motor and gearbox, as depicted in Figure 9.6. As observed in this figure, by controlling the motor current, the required torque for the robot joint's motion is generated. Through a gearbox, symbolically represented using two gears in this figure, the motion is transfered from the motor to the load while reducing the motor speed with a ratio of η. If we idealize the gearbox and neglect power losses in it, according to the power conservation

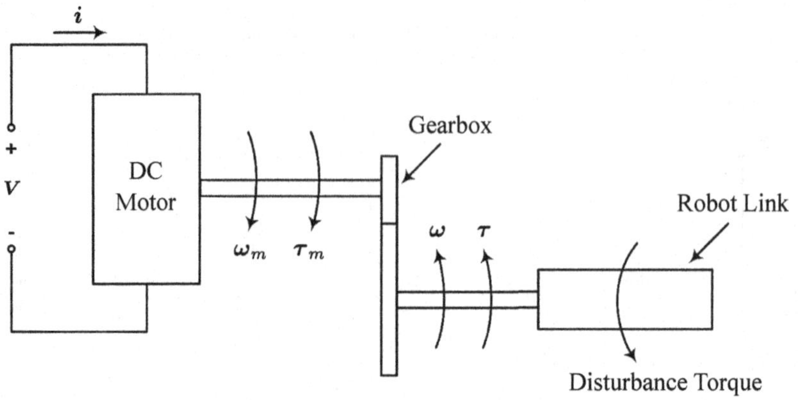

FIGURE 9.6
The motion of a joint and robot linked by an electric motor and gearbox.

law, the transferred torque will increase in proportion to the decreasing speed.

$$\omega_m = \eta\,\omega \quad \rightarrow \quad \tau_m = \frac{1}{\eta}\tau \tag{9.36}$$

In this context, the index m is used to represent quantities on the motor side or the high-speed part. τ_m represents the torque transferred to the robot joint on the motor side. The quantities on the joint side, which constitute the low-speed section of the system, are presented without using an index.

If we represent the motor's inertia with I_m and its viscous friction coefficient with b_m, and similarly represent the inertia of the robot link with I and its viscous friction coefficient with b, the dynamic equations of motion can be written in two parts, high-speed and low-speed, as follows.

$$\begin{aligned} \tau_m &= I_m\dot{\omega}_m + b_m\omega_m + \frac{1}{\eta}\tau \\ \tau &= I\dot{\omega} + b\,\omega \end{aligned}$$

Note that in this equation, τ is the torque transferred to the robot joint by the gearbox, causing its motion. Thus, in the first equation, a part of the torque from the electric motor induces the motion of the motor armature, and a part of it, after being transferred by the gearbox, leads to the movement of the robot link. Considering that the ratio of the speed and acceleration of the motor and the robot joint is given by the relationship 9.36, these two equations can be combined as follows.

$$\tau_m = I_m\dot{\omega}_m + b_m\omega_m + \frac{1}{\eta}\left(I\dot{\omega} + b\omega\right)$$

In this way, we will have,

$$\text{Motor Side} \qquad \tau_m = \left(I_m + \frac{1}{\eta^2}I\right)\dot{\omega}_m + \left(b_m + \frac{1}{\eta^2}b\right)\omega_m$$

or

$$\text{Robot Side} \qquad \tau = \left(\eta^2 I_m + I\right)\dot{\omega} + \left(\eta^2 b_m + b\right)\omega$$

If the effective inertia coefficient is denoted by I_e and the effective friction coefficient is denoted by b_e as follows:

$$I_e = (\eta^2 I_m + I), \qquad b_e = (\eta^2 b_m + b) \tag{9.37}$$

then, the torque relationship on the robot side can be compactly written as follows.

$$\tau = I_e\dot{\omega} + b_e\omega \tag{9.38}$$

This relationship highlights the important feature of using a gearbox in the dynamic analysis of the robot. By using a gearbox, the small inertia of the motor armature and its viscous friction coefficient will affect the robot's dynamics with a very large factor of η^2. If we consider the gearbox transmission ratio as $\eta = 200$, then the dynamic effect of the actuator with a factor of $\eta^2 = 40000$ will dominate the robot nonlinear dynamics. This implies a reduction in the influence of the nonlinear and coupled dynamics of the robot's compared to the linear dynamics of the motor. Therefore, the use of linear controllers such as PID in the robot becomes reasonable. It is noteworthy that over 99% of controllers employed in industrial robots are PID.

However, the gearbox transmission ratio varies for different robots and ranges from 30 to 10,000. In robots that require high speeds in their workspace, a smaller transmission ratio close to 30 is used. This is while in space robots that experience very low speeds in their workspace, a conversion ratio of over 1000 is used. The typical transmission ratio in industrial robotic arms is between 50 and 250. A greater gearbox transmission ratio results in a greater attenuation of the nonlinear dynamics of the robot, allowing for the use of simpler linear models. On the other hand, if a gearbox is not used in the actuator structure (i.e., direct-drive motors), or if a conversion ratio less than 10 is used due to the need for higher speed, the nonlinear dynamics of the robot will dominate, and the use of nonlinear control methods becomes more crucial. We will elaborate on a more detailed analysis of the overall dynamics of the robot with a typical actuator and when employing a gearbox with a transmission ratio ranging from 50 to 250 in the subsequent section.

9.7 Robot Dynamics with Actuators

Consider an n-degrees-of-freedom robot that employs electric motors and gearboxes in its joints. In general, the robot's dynamics in the presence of viscous friction in the joints can be expressed based on Equation 9.23.

$$\boldsymbol{\tau} - \boldsymbol{b}\dot{\boldsymbol{q}} = \boldsymbol{M}(\boldsymbol{q})\ddot{\boldsymbol{q}} + \boldsymbol{C}(\boldsymbol{q}, \dot{\boldsymbol{q}})\dot{\boldsymbol{q}} + \boldsymbol{g}(\boldsymbol{q}) \tag{9.39}$$

By incorporating the dynamics of the robot's joint actuators, this equation is transformed to

$$\boldsymbol{\tau} - \boldsymbol{b}_e\dot{\boldsymbol{q}} = \boldsymbol{M}_e(\boldsymbol{q})\ddot{\boldsymbol{q}} + \boldsymbol{C}(\boldsymbol{q}, \dot{\boldsymbol{q}})\dot{\boldsymbol{q}} + \boldsymbol{g}(\boldsymbol{q}) \tag{9.40}$$

where

$$\boldsymbol{M}_e(\boldsymbol{q}) = \boldsymbol{M}(\boldsymbol{q}) + \begin{bmatrix} \eta_1^2 I_{m_1} & 0 & \cdots & 0 \\ 0 & \eta_2^2 I_{m_2} & \cdots & 0 \\ \vdots & 0 & \ddots & \vdots \\ 0 & 0 & \cdots & \eta_n^2 I_{m_n} \end{bmatrix}, \quad \boldsymbol{b}_e = \begin{bmatrix} b_1 + \eta_1^2 b_{m_1} \\ b_2 + \eta_2^2 b_{m_2} \\ \vdots \\ b_n + \eta_n^2 b_{m_n} \end{bmatrix} \tag{9.41}$$

in which, η_i denote the gear ratio of the i'th joint. As evident in these equations, in the presence of a gearbox, the dynamics of the actuator are amplified with the factor η_i^2 and contribute significantly to the robot's dynamics.

Now, if the gear ratio is large, $\eta_i \gg 1$ (typically between 30 to 100), then the dynamics of the actuator dominate. In this case, the dynamics of the robot and the actuator together

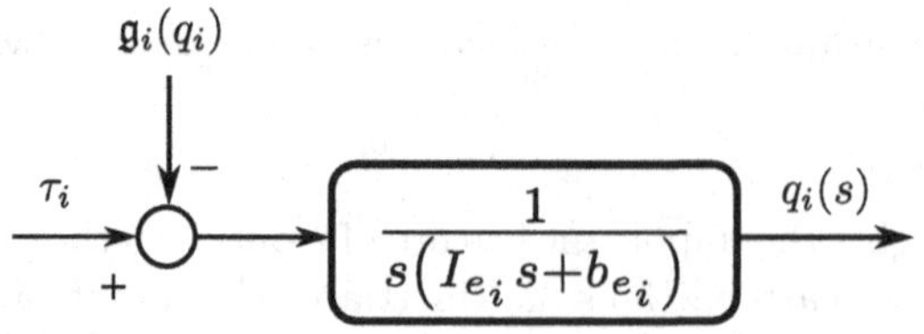

FIGURE 9.7
Block diagram of the linear model for each joint of the robot in the presence of gearbox dynamics with a very large transmission ratio.

form a set of decoupled differential equations. The overall dynamics consist of n independent differential equations and can be written in the following form.

$$\tau_i - b_{e_i}\dot{q}_i \approx \left(M_{ii}(q) + \eta_i^2 I_{m_i}\right)\ddot{q}_i + \mathfrak{v}_i(\boldsymbol{q}, \dot{\boldsymbol{q}}) + \mathfrak{g}_i(\boldsymbol{q}) \tag{9.42}$$

If the gear ratio is significantly large, $\eta_i \ggg 1$ (a number much larger than 100), then the dynamics of the actuator completely dominate, and the nonlinear dynamics of the robot can be neglected. In this case, the effective inertia and viscous friction coefficient can be approximated as follows.

$$\begin{aligned} I_{e_i} &= M_{ii} + \eta_i^2 I_{m_i} \approx \eta_i^2 I_{m_i} \\ b_{e_i} &= b_i + \eta_i^2 b_{m_i} \approx \eta_i^2 b_{m_i} \end{aligned}$$

The effect of the robot's dynamics in these equations has been greatly attenuated and is practically neglected because the contributions of the vector terms such as Coriolis, centrifugal, and gravitational vector compared to $I_{e_i}\ddot{q}_i + b_{e_i}\dot{q}_i$ are negligible. Therefore, these terms can be disregarded. However, considering the direct effect of gravity on the tracking error of the system, it is better to account for its influence in the relevant differential equation. Thus, the dynamics of the robot and the actuator transform into a system of linear and non-coupled differential equations. This system consists of n independent linear differential equations, and for each joint, it will be in the following form.

$$\tau_i = I_{e_i}\ddot{q}_i + b_{e_i}\dot{q}_i + \mathfrak{g}_i(q_i) \tag{9.43}$$

With this description, each joint of the robot can be modeled with a linear transformation function in the presence of disturbance torque due to gravity, as illustrated in the block diagram in Figure 9.7. By simplifying the robot's dynamic equation due to the actuator's dynamics with a gearbox to this form, linear control methods can be employed to design an appropriate linear controller for the robot. This will be discussed in detail in Chapter 11.

Problems

9.1 Consider the two-degrees-of-freedom <u>RP</u> robot shown in Figure 9.8. The robot dynamics has been investigated in Problem 8.1. Using the equation 9.22, determine the Christoffel matrix $\boldsymbol{C}(\boldsymbol{q}, \dot{\boldsymbol{q}})$ for this robot and verify the skew-symmetry of the matrix $\dot{\boldsymbol{M}} - 2\boldsymbol{C}$.

9.2 Consider the three-degrees-of-freedom <u>PRP</u> robot illustrated in Figure 9.9. The dynamics of the robot have been explored in Problem 8.2. Utilizing equation 9.22, ascertain

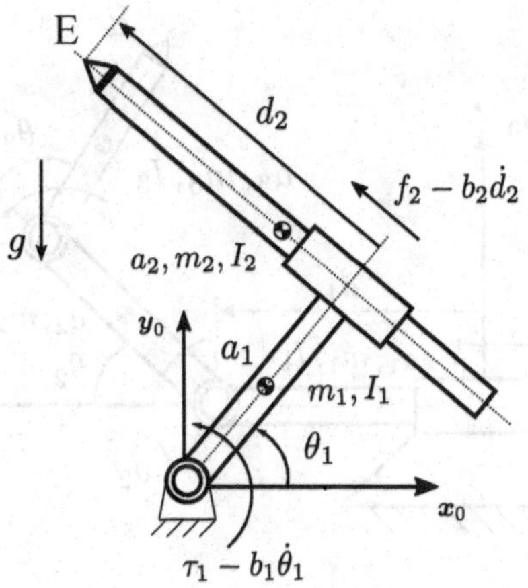

FIGURE 9.8
2-DOF Robot of Problem 9.1

the Christoffel matrix $\boldsymbol{C}(\boldsymbol{q}, \dot{\boldsymbol{q}})$ for this robot and confirm the skew-symmetric nature of the matrix $\dot{\boldsymbol{M}} - 2\boldsymbol{C}$.

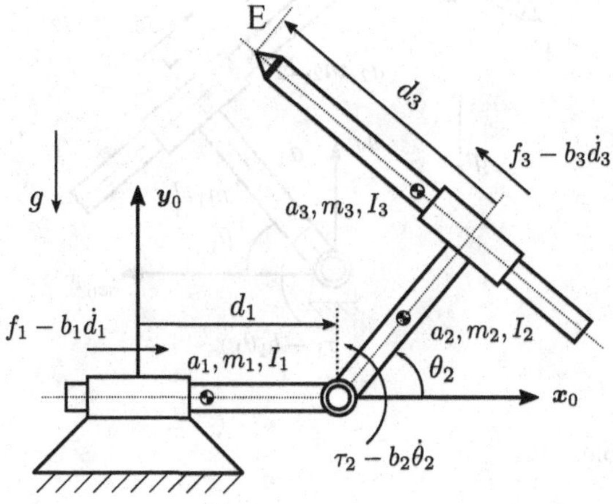

FIGURE 9.9
3-DOF Robot of Problem 9.2

9.3 Consider the 3-DOF <u>P</u>R<u>R</u> robot depicted in Figure 9.10. Investigate the dynamics of the robot in Problem 8.3. Use equation 9.22 to determine the Christoffel matrix $\boldsymbol{C}(\boldsymbol{q}, \dot{\boldsymbol{q}})$ for this robot and validate the skew-symmetry of the matrix $\dot{\boldsymbol{M}} - 2\boldsymbol{C}$.

9.4 Consider the three-degrees-of-freedom <u>R</u>P<u>R</u> robot shown in Figure 9.11.The robot dynamics has been investigated in Problem 8.4. Using the equation 9.22, determine the Christoffel matrix $\boldsymbol{C}(\boldsymbol{q}, \dot{\boldsymbol{q}})$ for this robot and verify the skew-symmetry of the matrix $\dot{\boldsymbol{M}} - 2\boldsymbol{C}$.

9.5 Consider the two-degrees-of-freedom <u>R</u>R robot shown in Figure 9.12. The dynamics of the robot have been explored in Problem 8.5. Utilizing equation 9.22, ascertain the Christoffel matrix $\boldsymbol{C}(\boldsymbol{q}, \dot{\boldsymbol{q}})$ for this robot and confirm the skew-symmetric nature of the matrix $\dot{\boldsymbol{M}} - 2\boldsymbol{C}$.

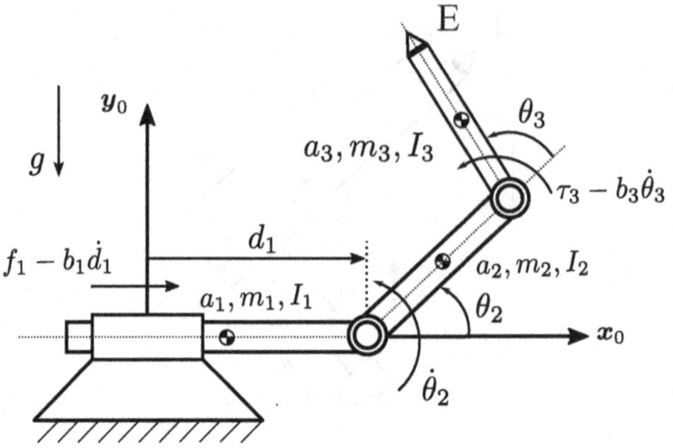

FIGURE 9.10
3-DOF Robot of Problem 9.3

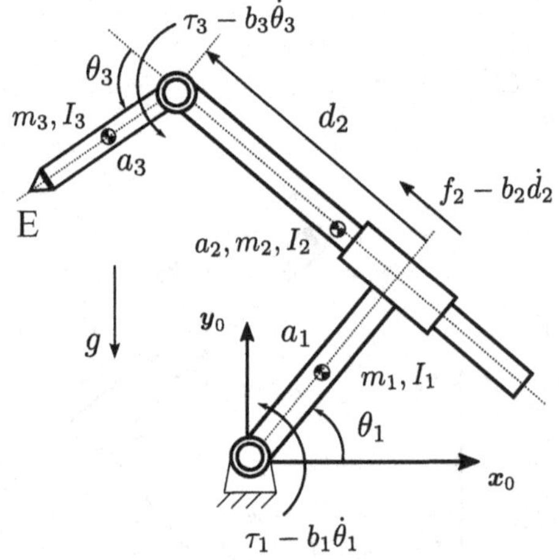

FIGURE 9.11
3-DOF Robot of Problem 9.4

9.6 Consider the four-degrees-of-freedom RRRP robot shown in Figure 9.13. Investigate the dynamics of the robot in Problem 8.6. Use equation 9.22 to determine the Christoffel matrix $\boldsymbol{C}(\boldsymbol{q}, \dot{\boldsymbol{q}})$ for this robot and validate the skew-symmetry of the matrix $\dot{\boldsymbol{M}} - 2\boldsymbol{C}$.

9.7 Consider the four-degrees-of-freedom robot RPRR shown in Figure 9.14. The robot dynamics has been investigated in Problem 8.7. Using the equation 9.22, determine the Christoffel matrix $\boldsymbol{C}(\boldsymbol{q}, \dot{\boldsymbol{q}})$ for this robot and verify the skew-symmetry of the matrix $\dot{\boldsymbol{M}} - 2\boldsymbol{C}$.

9.8 Consider the four-degrees-of-freedom 4R robot shown in Figure 9.15. The dynamics of the robot have been explored in Problem 8.8. Utilizing equation 9.22, ascertain the Christoffel matrix $\boldsymbol{C}(\boldsymbol{q}, \dot{\boldsymbol{q}})$ for this robot and confirm the skew-symmetric nature of the matrix $\dot{\boldsymbol{M}} - 2\boldsymbol{C}$.

9.9 Consider the four-degrees-of-freedom RRPR robot shown in Figure 9.16. Investigate the dynamics of the robot in Problem 8.9. Use equation 9.22 to determine the Christof-

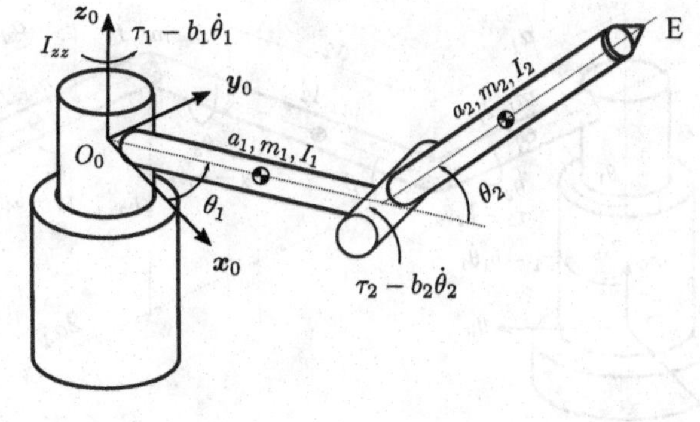

FIGURE 9.12
Two degrees-of-freedom Robot of Problem 9.5

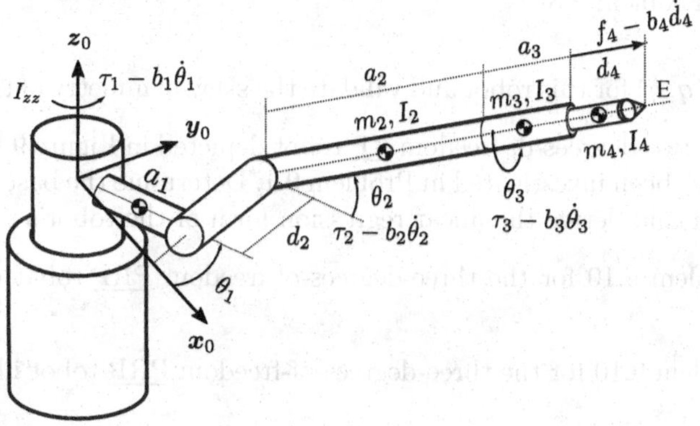

FIGURE 9.13
4-DOF Robot of Problem 9.6

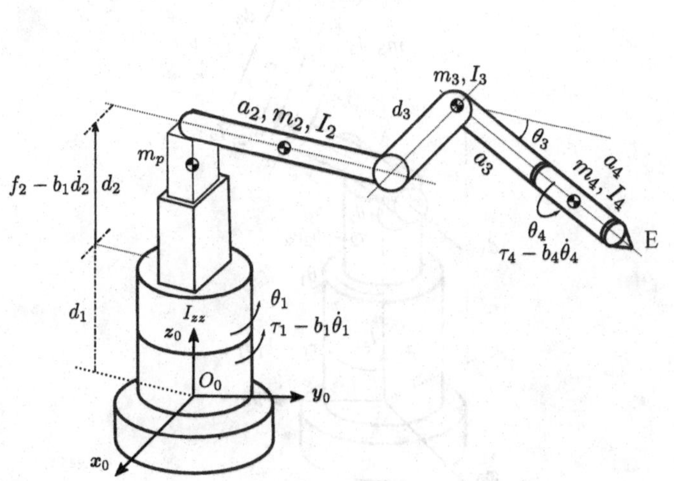

FIGURE 9.14
4-DOF Robot of Problem 9.7

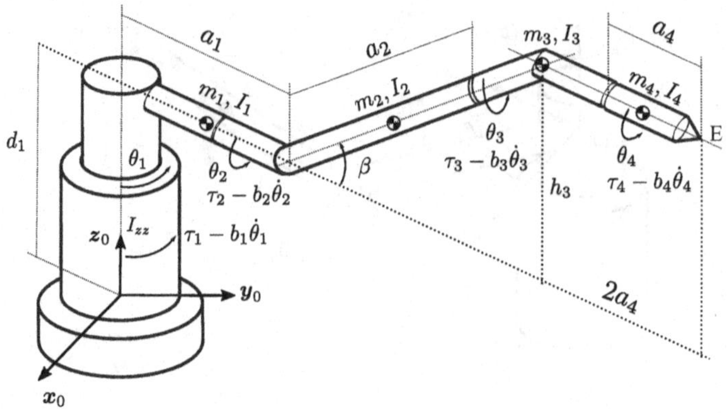

FIGURE 9.15
4-DOF Robot of Problem 9.8

fel matrix $\boldsymbol{C}(\boldsymbol{q}, \dot{\boldsymbol{q}})$ for this robot and validate the skew-symmetry of the matrix $\dot{\boldsymbol{M}} - 2\boldsymbol{C}$.

9.10 Consider the two-degrees-of-freedom RP robot depicted in Figure 9.8. The dynamics of the robot have been investigated in Problem 9.1. Determine the base inertia parameters for this robot and derive the linear regression form of the robot's dynamics.

9.11 Repeat Problem 9.10 for the three-degrees-of-freedom PRP robot depicted in Figure 9.9.

9.12 Revisit Problem 9.10 for the three-degrees-of-freedom PRR robot illustrated in Figure 9.10.

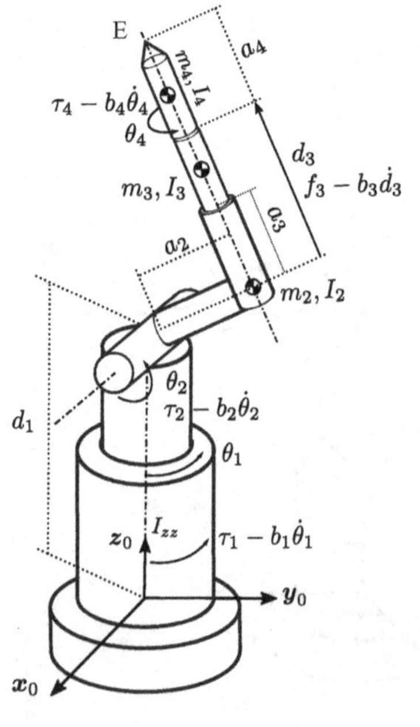

FIGURE 9.16
4-DOF Robot of Problem 9.9

9.13 Reperform the task specified in Problem 9.10 with the 3DOF <u>RP</u>R robot depicted in Figure 9.11.

9.14 Repeat Problem 9.10 for the two-degrees-of-freedom <u>RR</u> robot illustrated in Figure 9.12.

9.15 Revisit Problem 9.10 for the four-degrees-of-freedom <u>RRRP</u> robot illustrated in Figure 9.13.

9.16 Repeat Problem 9.10 for the four-degrees-of-freedom <u>RPRR</u> robot depicted in Figure 9.14.

9.17 Reperform the task specified in Problem 9.10 with the four-degrees-of-freedom 4<u>R</u> robot depicted in Figure 9.15.

9.18 Repeat Problem 9.10 for the four-degrees-of-freedom <u>RRPR</u> robot illustrated in Figure 9.16.

10

Simulation and Calibration

This chapter introduces fundamental concepts essential for the understanding and implementation of robotic systems. It begins by elaborating the intricate process of motion planning, a crucial aspect of robot simulation. The chapter explores various trajectory planning methods, such as cubic polynomials, cubic polynomials with via points, quintic polynomials, linear trajectories with parabolic blends, and trajectory planning in task space. This comprehensive overview equips readers with diverse tools for shaping the desired motion profiles of robotic systems.

Furthermore, the chapter explores the realm of motion simulation, focusing on both forward and inverse dynamics. The intricate dynamics of the planar 3$\underline{R}$ robot are simulated to provide readers with practical insights into these simulation techniques. Furthermore, the discussion includes an exploration of inverse dynamics simulation for this robot, shedding light on the complexities involved in predicting robot motion accurately.

A significant aspect of robot design is dynamic calibration, ensuring that the predicted motion aligns precisely with the actual robot behavior. The chapter introduces readers to calibration methods, emphasizing linear regression models and the least squares method. This calibration process is essential for refining dynamic models, ensuring that the model parameters are accurately identified. The discussion extends to the evaluation of the calibrated model and its application to the planar 2$\underline{R}$ robot, providing a practical perspective on the calibration process.

Throughout the chapter, the integration of theoretical concepts with practical applications enhances the reader's understanding of simulation, motion planning, and calibration, setting the stage for a deeper exploration of these topics in the field of robotics.

10.1 Introduction

Simulating the motion of a robot is a widely employed technique by designers prior to the actual implementation and construction of the robot. Through the use of engineering software, designers can closely simulate the motion behavior of robots, allowing for a comprehensive investigation and evaluation. In the realm of robot motion simulation, two prevalent methods are commonly employed.

In the initial approach, the simulation relies on the dynamic formulation of the robot, with the numerical solution of the governing differential equations implemented through software. This method directly incorporates the dynamic equations specific to the robot under consideration. Due to the computational intricacies associated with determining the dynamic equations governing robot motion, several highly efficient engineering software have been developed. In the second approach, these programs are utilized to simulate robot motion without the explicit need for deriving dynamic formulation, by using only the kinematic structure of the robot and the dynamic characteristics of its components.

DOI: 10.1201/9781003491415-10

While the second category of simulators is advantageous for visualizing and analyzing robot motion, unfortunately, it does not provide explicit dynamic formulations to the designer. As a result, incorporating robot dynamics into model-based control methods becomes impractical. This is the reason why, despite the usefulness of these simulation methods, this chapter centers on simulation approaches grounded in the dynamic model of the robot, enabling a thorough examination and analysis.

In the simulation of robot motion, the first step involves defining the trajectory of the robot's motion. This process, referred to as motion trajectory planning in robotics literature, can take place either in joint space or task space. Given that the dynamic equations of serial robots are typically established in joint space, it is logical for trajectory generation to occur in this space, as well. The upcoming sections of this chapter will explore various methods of trajectory generation in robots.

Similar to our examination of both forward and inverse kinematics processes in robot kinematic analysis, in dynamic simulation, these two processes take the form of forward and inverse dynamic simulations in robots. The subsequent sections will elaborate on the characteristics of each of these methods, provide details related to dynamic simulation in robots, and explore various applications of simulating these processes.

Another crucial aspect of practical robot design is kinematic and dynamic calibration. The accuracy of a robot's dynamic model in predicting its motion relies on the precise identification of model parameters. In the context of robotics, the process of identifying these model parameters is termed robot calibration. There exists a diverse array of calibration methods for robots, with extensive research depth, making it impractical to cover all cases in this brief overview. Given the innovative approach presented in this book, where linear models in robots are developed, special attention has been given to the calibration of dynamic models for these linear structures. The implementation of robot calibration using linear regression models and the determination of model parameters through least squares methods will be thoroughly elaborated. Furthermore, by introducing consistency measures the determined model will be evaluated, and the obtained parameters will be examined.

10.2 Motion Planning

In this section, we will explore methods for designing and generating motion trajectories in serial robots. To start, it's crucial to establish a precise definition of the motion path and its distinction to the trajectory. Consider the scenario where we aim to move the robot's end-effector through three-dimensional space, navigating from the initial point $\boldsymbol{x}_0$ to the final point $\boldsymbol{x}_f$. The path from the starting point to the endpoint can be defined by the position function connecting these two points. Various functions such as linear, circular, or helical, among others, can be employed to traverse this path. Regardless of the chosen function, the path is expressed in terms of spatial variables x, y, z and does not explicitly depend on time. This is in contrast to the motion trajectory, which incorporates time, and considers the speed and acceleration of movement along the desired path. In the design of robot motion trajectories, the objective is to determine the task space variables $\boldsymbol{\mathcal{X}}(t)$ or the joint space variables $\boldsymbol{q}(t)$ over time.

With this differentiation, it becomes evident that, whether in the joint or task space of the robot, the initial step in designing a trajectory is to create a preferred path that ensures the robot avoids collisions with its surroundings. Subsequently, the temporal aspects of the motion path need to be taken into account, and the duration of the journey from the starting point to the endpoint must be determined. Numerous methods have been developed for robot path planning, particularly in mobile robots, to circumvent obstacles [115, 116, 117]. Due to the breadth of these methods, we avoid detailing them here.

In the context of motion planning for industrial robots, it is evident that utilizing a teaching pendant to establish the robot's motion points serves to simplify the path generation process. The teaching pendant allows for the manual movement of the robot to a specified point while recording the position and orientation of the robot's end-effector. Consequently, based on the desired path for the robot's end-effector, the key motion points are identified, and the corresponding joint information is documented. These recorded points then facilitate trajectory planning, wherein suitable time functions can be determined to smoothly guide the robot through these specified points [118].

For this purpose, let's consider the scenario where we aim to transition from the initial point $\boldsymbol{q}(t_0) = \boldsymbol{q}_0$ to the final point $\boldsymbol{q}(t_f) = \boldsymbol{q}_f$ within the time interval $[t_0, t_f]$. Additionally, let's assume that the joint velocities at the initial and final points are given as $\dot{\boldsymbol{q}}(t_0) = \boldsymbol{\omega}_0$ and $\dot{\boldsymbol{q}}(t_f) = \boldsymbol{\omega}_f$, respectively. This implies that designing the path in the robot involves identifying a suitable time function $\boldsymbol{q}(t)$ that satisfies these specified boundary conditions. Through the application of interpolation methods, it is possible to determine various functions that meet these criteria. In the subsequent section, we will concentrate on and explore this issue using three distinct methods.

10.2.1 Cubic Polynomials

Assume the initial and final conditions of the trajectory of the robot's joint vector are given.

$$\begin{aligned} \boldsymbol{q}(t_0) &= \boldsymbol{q}_0 \qquad & \boldsymbol{q}(t_f) &= \boldsymbol{q}_f \\ \dot{\boldsymbol{q}}(t_0) &= \boldsymbol{\omega}_0 \qquad & \dot{\boldsymbol{q}}(t_f) &= \boldsymbol{\omega}_f \end{aligned}$$

In the following equation, the index i is used to introduce the motion variable for each of the joints.

$$\begin{aligned} \boldsymbol{q}_0 &= \begin{bmatrix} q_{1_0} & q_{2_0} & \cdots & q_{n_0} \end{bmatrix} \qquad & \boldsymbol{q}_f &= \begin{bmatrix} q_{1_f} & q_{2_f} & \cdots & q_{n_f} \end{bmatrix} \\ \boldsymbol{\omega}_0 &= \begin{bmatrix} \omega_{1_0} & \omega_{2_0} & \cdots & \omega_{n_0} \end{bmatrix} \qquad & \boldsymbol{\omega}_f &= \begin{bmatrix} \omega_{1_f} & \omega_{2_f} & \cdots & \omega_{n_f} \end{bmatrix} \end{aligned}$$

By using cubic polynomials as a function of time, it is possible to design a continuous function for traversing this trajectory for each of the robot's joints, denoted as $q_i(t)$.

$$q_i(t) = p_0 + p_1 t + p_2 t^2 + p_3 t^3 \tag{10.1}$$

$$\dot{q}_i(t) = p_1 + 2p_2 t + 3p_3 t^2 \tag{10.2}$$

In order to satisfy the motion boundary conditions at the initial and final times, four algebraic relations are obtained in determining the four associated parameters p_i for the desired cubic polynomial. These relations can be rewritten in the form of the following matrix equation.

$$\begin{bmatrix} 1 & t_0 & t_0^2 & t_0^3 \\ 1 & t_f & t_f^2 & t_f^3 \\ 0 & 1 & 2t_0 & 3t_0^2 \\ 0 & 1 & 2t_f & 3t_f^2 \end{bmatrix} \begin{bmatrix} p_0 \\ p_1 \\ p_2 \\ p_3 \end{bmatrix} = \begin{bmatrix} q_{i_0} \\ q_{i_f} \\ \omega_{i_0} \\ \omega_{i_f} \end{bmatrix} \tag{10.3}$$

If $t_0 \neq t_f$, the matrix in 10.3 is invertible, and by determining its inverse, the coefficients of the cubic polynomial, in general, can be obtained.

$$\begin{aligned} p_0 &= q_{i_0} \quad , \quad p_1 = \omega_{i_0} \\ p_2 &= \frac{3}{t_f^2}\left(q_{i_f} - q_{i_0}\right) - \frac{1}{t_f}\left(2\omega_{i_0} + \omega_{i_f}\right) \\ p_3 &= -\frac{2}{t_f^3}\left(q_{i_f} - q_{i_0}\right) + \frac{1}{t_f^2}\left(\omega_{i_0} + \omega_{i_f}\right) \end{aligned} \tag{10.4}$$

In many applications, it is common to initiate the robot's motion at the starting point with zero velocity and come to a gradual stop at the endpoint. In this case, $\omega_{i_0} = \omega_{i_f} = 0$, and the cubic polynomial trajectory simplifies to the following equation.

$$q_i(t) = q_{i_0} + \frac{3}{t_f^2}\left(q_{i_f} - q_{i_0}\right) t^2 - \frac{2}{t_f^3}\left(q_{i_f} - q_{i_0}\right) t^3 \tag{10.5}$$

If this motion trajectory is utilized in the robot's joints, a continuous and smooth motion will be achieved in each joint as it moves from the initial point to the endpoint. Consequently, the robot's end-effector movement will also be smooth and continuous.

Case Study 10.1 (Cubic Trajectory Planning) *Consider the planar 2R robot, as shown in Figure 8.4. The goal is for the robot joints to reach the final state* $\mathbf{q}_f = [90,\ -45]^T$ *degrees within four seconds, starting from the initial state* $\mathbf{q}_0 = [10,\ -10]^T$ *degrees. Design the desired trajectory of joint motion using a cubic polynomial and plot the joint movements as well as the final position of the robot's end-effector in the* $x - y$ *plane.*

Solution: Using the initial and final states of the joints, the path of their motion is obtained using Equation 10.5.

$$\begin{aligned} q_1(t) &= 10 + 15t^2 - 2.5t^3 \\ q_2(t) &= -10 - 6.56t^2 + 1.09t^3 \end{aligned}$$

To determine the relationships the velocity and acceleration of the joints, differentiate these equations with respect to time:

$$\begin{aligned} \dot{q}_1(t) = 30t - 7.5t^2 \ \rightarrow \ \ddot{q}_1(t) &= 30 - 15t \\ \dot{q}_2(t) = -13.13t + 3.28t^2 \ \rightarrow \ \ddot{q}_2(t) &= -13.13 + 6.56t \end{aligned}$$

The motion trajectories of the joints and their first and second derivatives are illustrated in Figure 10.1. As observed in this figure, the motion of the first joint, represented by the solid line (in blue), and the motion of the second joint, shown by the dashed line (in black), smoothly transition from their initial desired angles to their final desired angles over the course of four seconds. The angle of the first joint is positive, ranging from 10 degrees to 90 degrees, while the angle of the second joint is consistently negative, varying from -10 degrees to -45 degrees smoothly over time-based on the cubic polynomial function.

The angular velocity of the first joint, depicted by the solid line (in blue) in the middle graph, is also smooth and consistently positive, forming a parabolic function with respect to time. Its maximum value occurs at 2 seconds, reaching 30 degrees

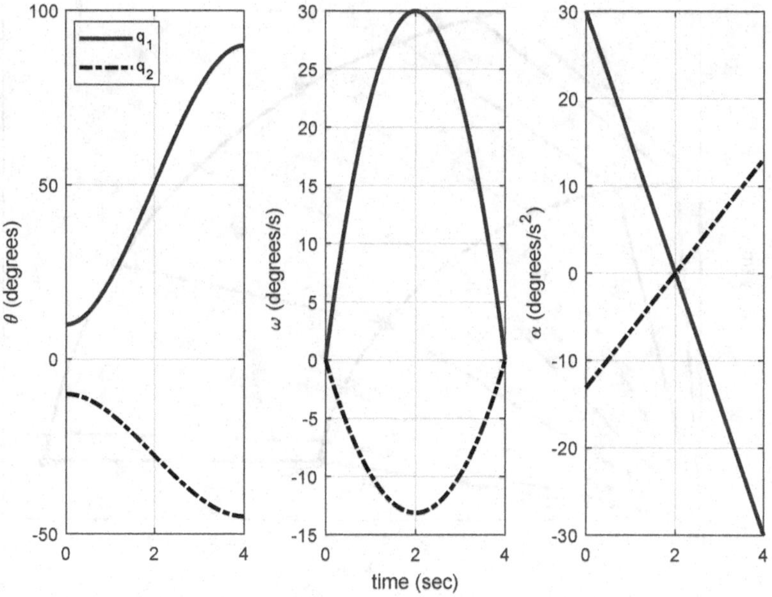

FIGURE 10.1
The designed trajectory of joint motion, along with their angular velocity and acceleration.

per second. Meanwhile, the angular velocity of the second joint, illustrated by the dashed line (in black), is consistently negative, with its maximum value occurring at 2 seconds and reaching -13.13 degrees per second. Both joint accelerations, shown in the rightmost graph, are linear with the first joint reaching a maximum value of 30 degrees per second squared and the second joint achieving a maximum value of 13.13 degrees per second squared.

Using the robot's forward kinematics and the obtained angular motions of the first and second joints, one can easily determine the motion of the robot's end-effector in the $x - y$ plane. The results of these calculations are highlighted in Figure 10.2, where the positions of the links are shown at 9 states with a time separation of 0.5 seconds. As evident in this figure, the end-effector's motion is also smooth, reaching its initial and final positions seamlessly, with lower motion speed at the beginning and end and maximum speed in the intermediate time.

10.2.2 Cubic Polynomials with Via Points

So far, the trajectory design from the initial point to the final point has been studied using cubic polynomials. However, in the trajectory design for robot motion, passing through multiple intermediate points is often desired. Consider, for example, an industrial robot like SCARA, used to move components from a conveyor belt to their packaging location. In this case, for the proper functioning of the robot, the robot's motion trajectory must pass through several points, including picking from the conveyor belt, lifting the piece sufficiently, rotating the robot arm to place the part in its appropriate location, and finally placing the part in the packaging area.

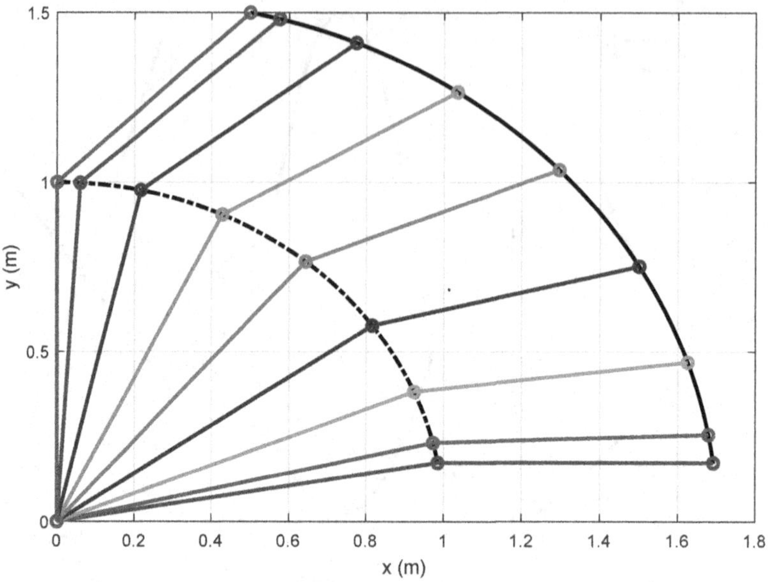

FIGURE 10.2
The result of the motion of the robot's links and end-effector by applying the designed trajectory.

In the trajectory design using intermediate points, it is usually not necessary to consider zero velocity at via points. If non-zero velocity is desired at intermediate points, the general trajectory equation with non-zero velocity, given in 10.4, can be used, requiring the specification of the velocity at intermediate points. Although the user can have complete freedom in choosing the velocity at intermediate points and determine them at their discretion, sometimes, especially in trajectories with multiple via points, this freedom can be cumbersome for the user, and it is convenient to automatically make these selections.

For automatic selection of the trajectory velocity at via points, the average speed between two consecutive points on the motion path, as illustrated in Figure 10.3, can be used. To do this, connect the path points with straight lines and choose the average speed between two via points as the appropriate velocity for intermediate points. This simple method facilitates the user's selection of velocity at intermediate points and allows for its automatic execution.

For a better understanding of the subject, let's assume, according to Figure 10.4, we want to design a cubic motion trajectory for a robot's motion with an intermediate point. If the initial and final conditions are given as before, and the motion state at the intermediate point, denoted by index v, the velocity at the intermediate point is automatically determined based on the following equation.

$$\omega_v = \dot{q}_v = \frac{1}{2}\left(\frac{q_v - q_0}{t_v - t_0} + \frac{q_f - q_v}{t_f - t_v}\right) \tag{10.6}$$

In this way, the initial, intermediate, and final conditions of the motion trajectory are given as follows.

$$\begin{array}{lll} q(t_0) = q_0 & q(t_v) = q_v & q(t_f) = q_f \\ \dot{q}(t_0) = \omega_0 & \dot{q}(t_v) = \omega_v & \dot{q}(t_f) = \omega_f \end{array}$$

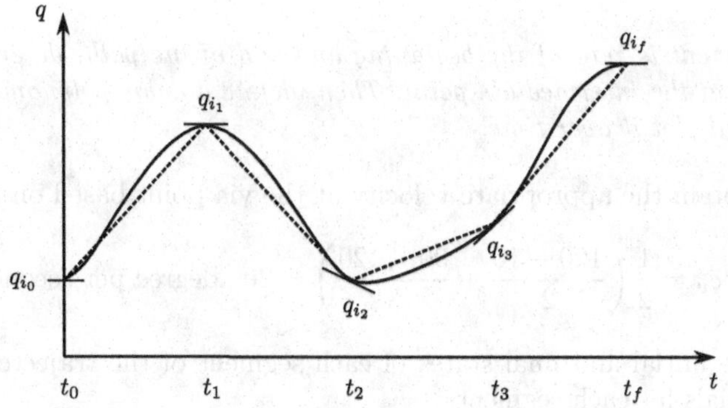

FIGURE 10.3
Determination of the velocity at via points in robot trajectory planning.

With this description, the trajectory consists of two segments described by a piecewise function composed of two cubic polynomials.

$$q(t) = \begin{cases} p_{1_0} + p_{1_1}t + p_{1_2}t^2 + p_{1_3}t^3 & t_0 < t \leq t_v \\ p_{2_0} + p_{2_1}t + p_{2_2}t^2 + p_{2_3}t^3 & t_v < t \leq t_f \end{cases}$$

The parameters of each polynomial in this equation can be obtained based on the general relation given in 10.4. In these calculations, it is sufficient to substitute the initial and final conditions as well as the time of traversing the trajectory for each segment into this equation. Similarly, if we use multiple via points in trajectory design, we just need to first calculate the velocities at these via points and then determine the parameters of the polynomial for each segment of the trajectory.

Case Study 10.2 (Cubic Trajectory Planning with Via Point) *Assume in the initial link of the planar 2R robot, as in Case Study 10.1, we want to design a desired trajectory that reaches the final state of* $q_f = 90$ *degrees in four seconds from the initial condition* $q_0 = 10$ *degrees. This should be done while passing through the point* $q_v = 120$ *degrees in the middle of the trajectory and at the time of two seconds. If the*

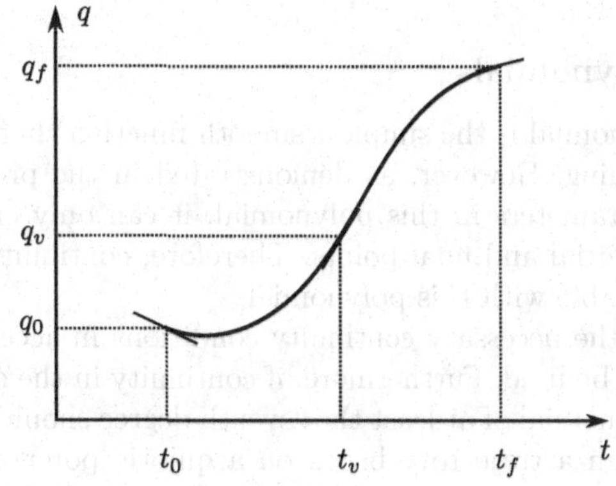

FIGURE 10.4
A segment of the trajectory with one via point.

speed of movement is zero at the beginning and end of the path, determine the appropriate velocity at the intermediate point. Then, define a cubic polynomial trajectory for this motion and plot it over time.

Initially, we obtain the appropriate velocity at the via point based on Equation 10.6.

$$\dot{q}_v = \frac{1}{2}\left(\frac{120-10}{2} + \frac{90-120}{2}\right) = 20 \quad \text{degree per second}$$

Now, using the initial and final states of each segment of the trajectory, we calculate cubic polynomials for each segment.

$$q(t) = \begin{cases} 10 + 72.5t^2 - 22.5t^3 & 0 < t \leq 2 \\ 120 + 20t - 42.5t^2 + 12.5t^3 & 2 < t \leq 4 \end{cases}$$

By differentiating this equation to determine the relationships for velocity and acceleration, we have:

$$\dot{q}(t) = \begin{cases} 145 - 67.5t^2 & 0 < t \leq 2 \\ 20 - 85t + 375t^2 & 2 < t \leq 4 \end{cases}$$

$$\ddot{q}(t) = \begin{cases} 145t - 135t & 0 < t \leq 2 \\ -85 + 750t & 2 < t \leq 4 \end{cases}$$

The path of motion and its first and second derivatives are illustrated in Figure 10.5. In this figure, the top graph depicts the continuous trajectory curve, smoothly covering the motion path from an angle of 0 to 90 degrees, passing through the intermediate point of 120 degrees. The middle graph represents the velocity profile, which is a continuous function. The motion starts and ends with zero velocity, reaching a velocity of 20 degrees per second at 2 seconds.

What is observed in the acceleration profile at the bottom graph, is the presence of linear acceleration in both segments of the path. However, it is noticeable that the acceleration is not continuous, and there are discontinuities at the initial, intermediate, and final points of motion. To address this issue, higher-degree polynomials, discussed in the next section, can be utilized.

10.2.3 Quintic Polynomials

The use of a cubic polynomial is the simplest smooth function that can be chosen for robot motion trajectory planning. However, as demonstrated in the previous Case Study, with only four adjustable parameters in this polynomial, it can only ensure interpolation with four conditions at the initial and final points. Therefore, continuity in acceleration and its derivatives is not achievable with this polynomial.

If we aim to satisfy the necessary continuity conditions in acceleration, at least a fifth-degree polynomial must be used. Furthermore, if continuity in the derivative of acceleration (jerk) is required, a polynomial of at least the seventh degree should be used. In this section, we explore how to design a trajectory based on a quintic polynomial, and similarly, this concept can be extended to higher-degree polynomials.

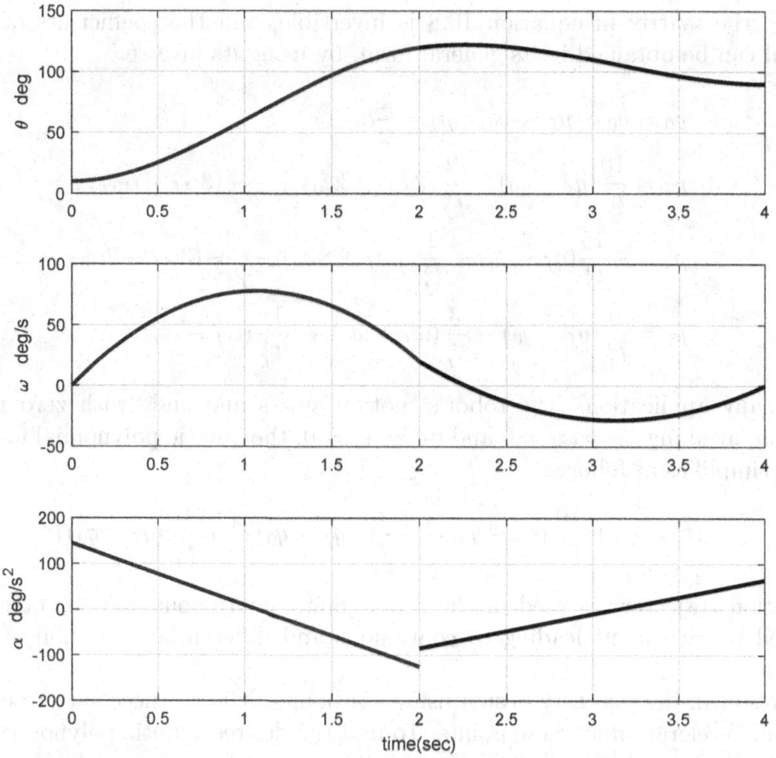

FIGURE 10.5
Cubic trajectory planning with one via point.

Suppose the initial and final conditions for the motion trajectory, velocity, and acceleration of a robot joint are given with the following values.

$$\begin{aligned} q(t_0) &= q_0 & q(t_f) &= q_f \\ \dot{q}(t_0) &= \omega_0 & \dot{q}(t_f) &= \omega_f \\ \ddot{q}(t_0) &= \alpha_0 & \ddot{q}(t_f) &= \alpha_f \end{aligned}$$

And we want to design a continuous function over time using fifth-degree polynomials for traversing this motion trajectory.

$$\begin{aligned} q_i(t) &= p_0 + p_1 t + p_2 t^2 + p_3 t^3 + p_4 t^4 + p_5 t^5 \\ \dot{q}_i(t) &= p_1 + 2p_2 t + 3p_3 t^2 + 4p_4 t^3 + 5p_5 t^4 \\ \ddot{q}_i(t) &= 2p_2 t + 6p_3 t + 12p_4 t^2 + 20p_5 t^3 \end{aligned} \tag{10.7}$$

In order to satisfy the motion, velocity, and acceleration conditions at the initial and final times, six algebraic equations are derived to determine the six parameters p_i of the quintic polynomial. These equations can be rewritten in matrix form as follows:

$$\begin{bmatrix} 1 & t_0 & t_0^2 & t_0^3 & t_0^4 & t_0^5 \\ 1 & t_f & t_f^2 & t_f^3 & t_f^4 & t_f^5 \\ 0 & 1 & 2t_0 & 3t_0^2 & 4t_0^3 & 5t_0^4 \\ 0 & 1 & 2t_f & 3t_f^2 & 4t_f^3 & 5t_f^4 \\ 0 & 0 & 2 & 6t_0 & 12t_0^2 & 20t_0^3 \\ 0 & 0 & 2 & 6t_f & 12t_f^2 & 20t_f^3 \end{bmatrix} \begin{bmatrix} p_0 \\ p_1 \\ p_2 \\ p_3 \\ p_4 \\ p_5 \end{bmatrix} = \begin{bmatrix} q_0 \\ q_f \\ \omega_0 \\ \omega_f \\ \alpha_0 \\ \alpha_f \end{bmatrix} \tag{10.8}$$

If $t_0 \neq t_f$, the matrix in equation 10.8 is invertible, and the coefficients of the quintic polynomial can be obtained in its general form, by using its inverse.

$$\begin{aligned}
p_0 &= q_0, \quad p_1 = \omega_0, \quad p_2 = \frac{1}{2}\alpha_0 \\
p_3 &= \frac{10}{t_f^3}(q_f - q_0) - \frac{2}{t_f^2}(2\omega_f + 3\omega_0) + \frac{1}{2t_f}(3\alpha_f - \alpha_0) \\
p_4 &= -\frac{15}{t_f^4}(q_f - q_0) + \frac{1}{t_f^3}(7\omega_f + 8\omega_0) - \frac{1}{2t_f^2}(3\alpha_f - 2\alpha_0) \\
p_5 &= \frac{6}{t_f^5}(q_f - q_0) - \frac{3}{t_f^4}(\omega_f + \omega_0) + \frac{1}{2t_f^3}(\alpha_f - \alpha_0)
\end{aligned} \tag{10.9}$$

Since in many applications, the robot's motion starts and ends with zero velocity and acceleration, meaning $\omega_0 = \omega_f = 0$ and $\alpha_0 = \alpha_f = 0$, the quintic polynomial for the motion trajectory simplifies as follows:

$$q(t) = q_0 + \frac{10}{t_f^3}(q_f - q_0)\,t^3 - \frac{15}{t_f^4}(q_f - q_0)\,t^4 + \frac{6}{t_f^5}(q_f - q_0)\,t^5 \tag{10.10}$$

If this motion trajectory is used in the robot joints, continuous and smooth motion will be achieved in each joint, leading to continuous and differentiable motion of the robot's end-effector.

As mentioned, in trajectory design using via points, it is not necessary to consider zero velocity and acceleration at these points. To find the desired quintic polynomial trajectory, it is sufficient to use the general relationship for trajectory design with non-zero velocity and acceleration, given in equation 10.9. To use this equation, the velocity and acceleration at the intermediate points need to be specified. Similar to the method of automatically determining the velocity of the motion trajectory at via points, the average velocity and acceleration between two consecutive points on the motion trajectory can be used.

Case Study 10.3 (Quintic Trajectory Planning) *Consider the planar 2R robot shown in Figure 8.4. It is desired that the robot joints, starting from the initial configuration* $\boldsymbol{q}_0 = [10, -10]^T$ *degrees, reach the final configuration* $\boldsymbol{q}_f = [90, -45]^T$ *degrees within four seconds. Design and plot the desired motion trajectories of the joints using a quintic polynomial and compare it with the motion trajectory based on cubic polynomials.*

With the initial and final joint given, the motion trajectory of the joints can be obtained using the equation 10.5.

$$\begin{aligned}
q_1(t) &= 10 + 12.5t^3 - 4.69t^4 + 0.47t^5 \\
q_2(t) &= -10 - 5.47t^3 + 2.05t^4 - 0.20t^5
\end{aligned}$$

By taking derivatives of these equations, we can determine the relationships for the angular velocity and acceleration of the joints:

$$\begin{aligned}
\dot{q}_1(t) &= 37.5t^2 - 18.75t^3 + 2.34t^4 \\
\ddot{q}_1(t) &= 75t - 56.25t^2 + 9.38t^3
\end{aligned}$$

$$\begin{aligned}
\dot{q}_2(t) &= -16.41t^2 + 8.20t^3 - 1.03t^4 \\
\ddot{q}_2(t) &= -32.81t + 24.61t^2 - 4.10t^3
\end{aligned}$$

The motion trajectory of the joints and their first and second derivatives are shown in Figure 10.6. As observed in this figure, the motion of the first joint, represented by the solid line (in blue), and the motion of the second joint, represented by the dashed line (in black), smoothly follows their initial angles and reaches their final angles within four seconds. The angle of the first joint is positive, changing from 10 to 90 degrees smoothly over time. Meanwhile, the angle of the second joint is always negative, changing from −10 to −45 degrees smoothly over time based on the quintic polynomial function.

The angular velocity of the first joint, plotted in the middle graph with a solid line (in blue), is continuous, always positive, and follows a fourth-degree polynomial function over time. Its maximum speed will be reached at 2 seconds, reaching 37.5 degrees per second. On the other hand, the angular velocity of the second joint, shown with a dashed line (in black), is always negative. Its maximum speed will be reached at 2 seconds, reaching 16.4 degrees per second.

Both angular accelerations are third-degree polynomial functions over time, shown in the rightmost graph, start from zero while regularly oscillate, return to zero. The acceleration of the first joint changes sign in the first half of the trajectory, being positive, and then becomes negative in the second half. The absolute maximum and minimum values of acceleration are both 28.9 degrees per second squared. The angular acceleration of the second joint changes sign, becoming negative in the first half and then positive. The absolute maximum and minimum values of acceleration are both 12.63 degrees per second squared.

Comparing this designed motion with the one in Case Study 10.1, it is noticeable that in this Case Study, the motion starts slowly, and at the end of the motion, it smoothly comes to rest with zero acceleration. The slope of the motion is steeper, resulting in a higher maximum speed.

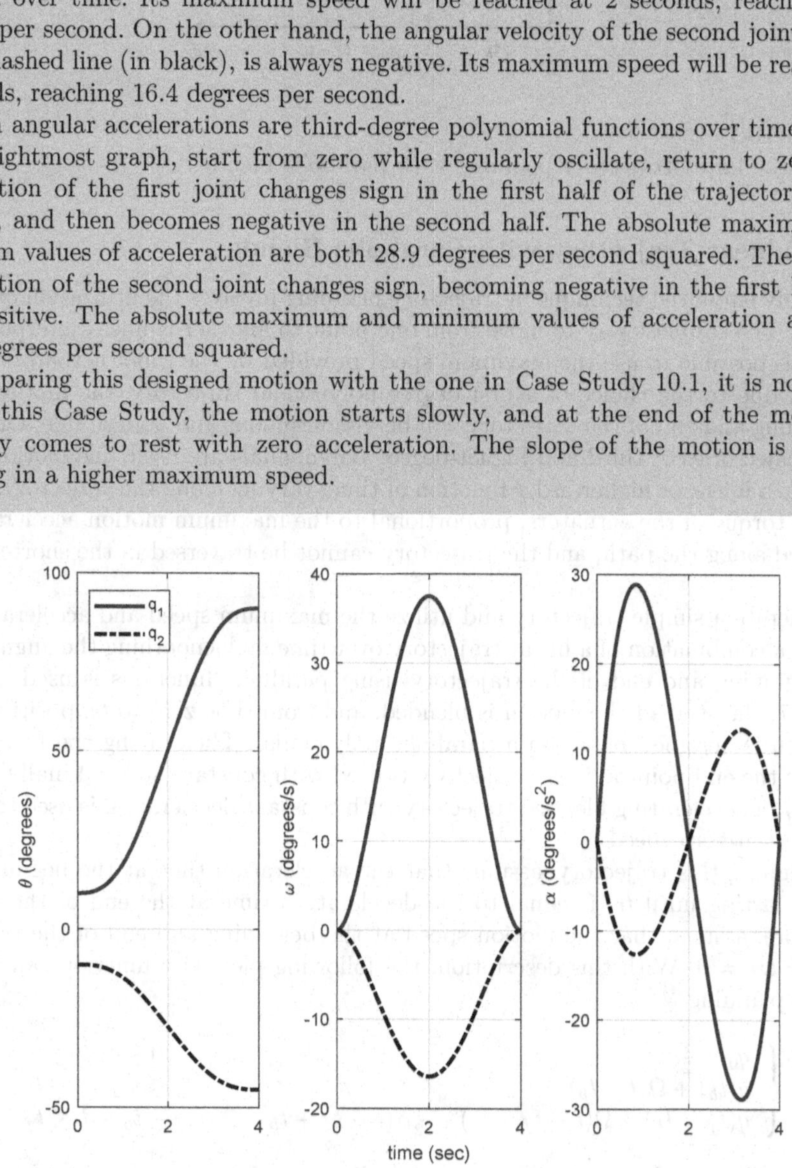

FIGURE 10.6
The designed motion trajectory of the joints using quintic polynomials and their angular velocity and acceleration.

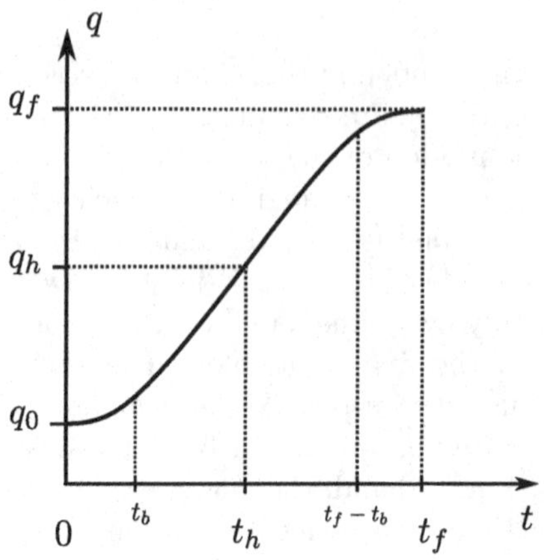

FIGURE 10.7
Details of the linear trajectory planning with parabolic blends.

10.2.4 Linear Trajectory with Parabolic Blends

A commonly employed technique in trajectory planning involves the utilization of a blended trajectory. The simplest way to move from one point to another is linear interpolation. Although it is possible to use the maximum speed provided by the robot actuators along this trajectory, due to the choice of a first-degree polynomial trajectory, the motion speed at the beginning and end of the trajectory will be discontinuous and undesirable. On the other hand, if trajectories of third and higher-degree polynomials are used, the motion acceleration will be a linear or higher-order function of time, varying along the trajectory. Thus, the maximum torque of the actuators, proportional to the maximum motion acceleration, cannot be used along the path, and the trajectory cannot be traversed in the shortest possible time.

To generate a simple trajectory and utilize the maximum speed and acceleration of the actuators, a combination of a linear trajectory over time and smoothing the angular motion at the beginning and end of the trajectory using parabolic functions is used. As seen in Figure 10.7, the start of the motion is blended, and from time zero to t_b, positive constant acceleration is designed based on a parabolic polynomial. Then, along the trajectory and just before the end point a linear trajectory is used with constant speed. Finally, from time $t_f - t_b$ to t_f, a decelerating blended trajectory with constant deceleration is used to smoothly come to rest at zero speed.

In designing this trajectory, assume that the acceleration time at the beginning of the trajectory lasting until t_b, is equal to the deceleration time at the end of the trajectory. Furthermore, assume that the motion speed at the beginning and end of the trajectory is zero, $\omega_0 = \omega_f = 0$. With this description, the following piecewise function can be used in trajectory planning.

$$q(t) = \begin{cases} q_0 + \frac{1}{2}\alpha t^2 & 0 \le t \le t_b \\ q(t_b) + \Omega(t - t_b) & t_b \le t \le t_f - t_b \\ q(t_f - t_b) + \Omega(t - t_f + t_b) - \frac{1}{2}\alpha(t - t_f + t_b)^2 & t_f - t_b \le t \le t_f \end{cases} \tag{10.11}$$

Where Ω is the speed of movement along the linear trajectory and α is the acceleration along the blended trajectory. In order to create a continuous and smooth motion at time t_b, the speeds of the two blended and linear trajectory segments at the blending time t_b must be equal.

$$\Omega = \alpha t_b \;\rightarrow\; t_b = \frac{\Omega}{\alpha} \tag{10.12}$$

The trajectory planning in this method is possible in two ways. In the first case, suppose that in addition to q_0 and q_f, the time to traverse the trajectory t_f is also given in advance. In this case, one can consider the speed and acceleration limits in the joint actuators, and if the maximum speed and acceleration of the joint actuators are higher than the required limits, the trajectory can be traversed within the given time. To determine the speed limit of the joint actuator, it is sufficient to assume that the motion is entirely linear. In this case, the necessary speed limit for traversing the trajectory is obtained as follows

$$\Omega_{max} > \frac{q_f - q_0}{t_f} \tag{10.13}$$

In this equation, the minimum speed required to traverse the trajectory at a constant speed is determined, and the necessary speed limit for the joint actuator is determined based on that. To determine the necessary acceleration limit of the actuator, assume that the motion is entirely blended over half of the trajectory. In this case:

$$\alpha_{max} > 2\frac{(q_f - q_0)/2}{t_f^2/4} = 4\frac{q_f - q_0}{t_f^2} \tag{10.14}$$

In this equation, the minimum acceleration required to traverse half of the trajectory at a constant acceleration is determined, and the necessary acceleration limit for the joint actuator is calculated based on that. If these conditions are satisfied, the piecewise trajectory function 10.11 can be simplified as follows. By determining $q(t_b)$ initially, the blending time t_b can be obtained from the equation 10.12.

$$q(t_b) = q_0 + \frac{1}{2}\Omega t_b$$

Similarly, $q(t_f - t_b)$ can be obtained in a similar manner.

$$\begin{aligned} q(t_f - t_b) &= q(t_b) + \Omega(t_f - t_b) \\ &= q_0 + \left(t_f - \frac{3}{2}t_b\right)\Omega \end{aligned}$$

Substitute these relations into Equation 10.11.

$$q(t) = \begin{cases} q_0 + \frac{1}{2}\alpha t^2 & 0 \le t \le t_b \\ q_0 + \Omega\left(t - \frac{1}{2}t_b\right) & t_b \le t \le t_f - t_b \\ q_0 + \Omega\left(t - \frac{1}{2}t_b\right) - \frac{1}{2}\alpha\left(t - t_f + t_b\right)^2 & t_f - t_b \le t \le t_f \end{cases} \tag{10.15}$$

Thus, the velocity along the trajectory can also be determined from the following piecewise function.

$$\dot{q}(t) = \begin{cases} \alpha t & 0 \le t \le t_b \\ \Omega & t_b \le t \le t_f - t_b \\ \Omega - \alpha(t - t_f + t_b) & t_f - t_b \le t \le t_f \end{cases} \tag{10.16}$$

In these equations, t_b, Ω, α must be chosen in such a way that, while satisfying the maximum speed and acceleration conditions, the trajectory is traversed within the time t_f. To determine these quantities appropriately, first express the velocity based on length of trajectory and determine the blending time t_b based on the movement to half of the trajectory.

$$q_h = \frac{1}{2}(q_f + q_0) = q_0 + \Omega(t_h - \frac{1}{2}t_b) \;\rightarrow\; \Omega = \frac{q_f - q_0}{t_f - t_b} \tag{10.17}$$

If t_b is given, Ω can be easily determined based on this relationship. If the motor's speed and acceleration limits are provided, an appropriate option is to set the ratio of speed to maximum acceleration.

$$t_b = \frac{\Omega_{max}}{\alpha_{max}} \tag{10.18}$$

However, keep in mind that this is not the only option and various values can be selected for t_b. Nevertheless, it is crucial to ensure the condition $t_b \leq \frac{1}{2}t_f$ is satisfied when choosing an alternative. By selecting t_b, the motion speed Ω is obtained from the relationship in 10.17, and the motion acceleration is chosen as follows.

$$\alpha = \frac{\Omega}{t_b} \tag{10.19}$$

By determining these values, the piecewise function in 10.15 can easily be used in trajectory planning, and the function in 10.16 can be utilized to determine the speed profile.

The second method of trajectory planning using the linear trajectory with parabolic blends is when the time to traverse the trajectory is not specified, and we want to determine the minimum time to traverse the trajectory based on the maximum speed and acceleration constraints of the actuator. In this case, the motion speed and acceleration are set equal to their possible maximum values in the actuator.

$$\Omega = \Omega_{max} \quad , \quad \alpha = \alpha_{max}$$

The blending time can be obtained using the equation 10.12.

$$t_b = \frac{\Omega}{\alpha}$$

Using these values, the optimum time to cover half of the trajectory, $t_h = t_f/2$, is calculated with the following equation.

$$\frac{1}{2}(q_f + q_0) = q_0 + \Omega\left(t_h - \frac{\Omega}{2\alpha}\right)$$

By this means, the optimal time to cover the entire trajectory, t_f, is determined.

$$t_f = \frac{1}{\Omega}(q_f - q_0) + \frac{\Omega}{\alpha} \tag{10.20}$$

By determining these values, one can easily use the piecewise function 10.15 for trajectory planning and the function 10.16 to determine the trajectory velocity.

Case Study 10.4 (Linear Trajectory with Parabolic Blends) *Consider the planar 2$\underline{R}$ robot shown in Figure 8.4. The goal is for the robot joints to reach the final state $\boldsymbol{q}_f = [90, -45]^T$ in degrees within four seconds, starting from the initial state $\boldsymbol{q}_0 = [10, -10]^T$. If the maximum speed of joint motion is 30 degrees per second and the maximum acceleration is 40 degrees per square second, design and plot the desired joint motion trajectory using linear trajectories with parabolic blends by selecting an appropriate time t_b.*

Solution: First, considering the trajectory traversal time and the speed and acceleration limits required for the joints, we examine the feasibility of designing the trajectory based on conditions 10.13 and 10.14. Given that the range of motion for joint one is greater, it is sufficient to check these conditions for this joint.

$$\begin{aligned}\Omega_{max} &> \frac{90-10}{4} = 20 \\ \alpha_{max} &> 4 \cdot \frac{90-10}{16} = 20\end{aligned}$$

Since the speed and acceleration limits of the joint actuators are greater than the required values, designing the trajectory within four seconds is feasible. Now, using Equation 10.12, obtain the blending time.

$$t_b = \frac{30}{40} = 0.75$$

Using these values along with the initial and final positions of the joints, the trajectory of their motion is obtained using Equation 10.15.

$$\begin{aligned}q_1(t) &= \begin{cases} 10 + 16.41t^2 & 0 \le t \le 0.75 \\ 10 + 24.62(t - 0.375) & 0.75 \le t \le 3.25 \\ 10 + 24.62(t - 0.375) - 16.41(t - 3.25)^2 & 3.25 \le t \le 4 \end{cases} \\ q_2(t) &= \begin{cases} -10 - 7.18t^2 & 0 \le t \le 0.5 \\ -10 - 10.77\,(t - 0.375) & 0.75 \le t \le 3.25 \\ -10 - 10.77\,(t - 0.375) + 7.18(t - 3.25)^2 & 3.25 \le t \le 4 \end{cases}\end{aligned}$$

The velocity of motion is also calculated based on Equation 10.16.

$$\begin{aligned}\dot{q}_1(t) &= \begin{cases} 32.82t & 0 \le t \le 0.75 \\ 24.62 & 0.75 \le t \le 3.25 \\ 24.62 - 32.82(t - 3.25) & 3.25 \le t \le 4 \end{cases} \\ \dot{q}_2(t) &= \begin{cases} -14.36t & 0 \le t \le 0.75 \\ -10.77 & 0.75 \le t \le 3.25 \\ -10.77 + 14.36(t - 3.25) & 3.25 \le t \le 4 \end{cases}\end{aligned}$$

The trajectory of joint movements and their angular velocity and acceleration are illustrated in Figure 10.8. As observed in this figure, the movement of the first joint is represented by the solid line (in blue), while the second joint's movement is depicted by the dashed line (in black). Similar to previous examples, both joints smoothly reach their final positions in four seconds, starting from their initial angles. The angular velocity of the first joint, shown by the solid line (in blue) in the middle plot, indicates that it starts with an increasing linear speed, maintains a constant speed in the middle of the trajectory, and then decelerates with a linear speed toward the end of the trajectory. This trend is more apparent in the angular acceleration, displayed in the rightmost plot of this figure. The acceleration of the first joint is constant and positive at the beginning of the trajectory, becomes zero in the middle of the trajectory, and then remains constant and negative toward the end of the trajectory.

Case Study 10.5 (Linear-Parabolic Trajectory with Optimal Time)
Consider the planar $2\underline{R}$ *robot from the previous Case Study. The goal is for the*

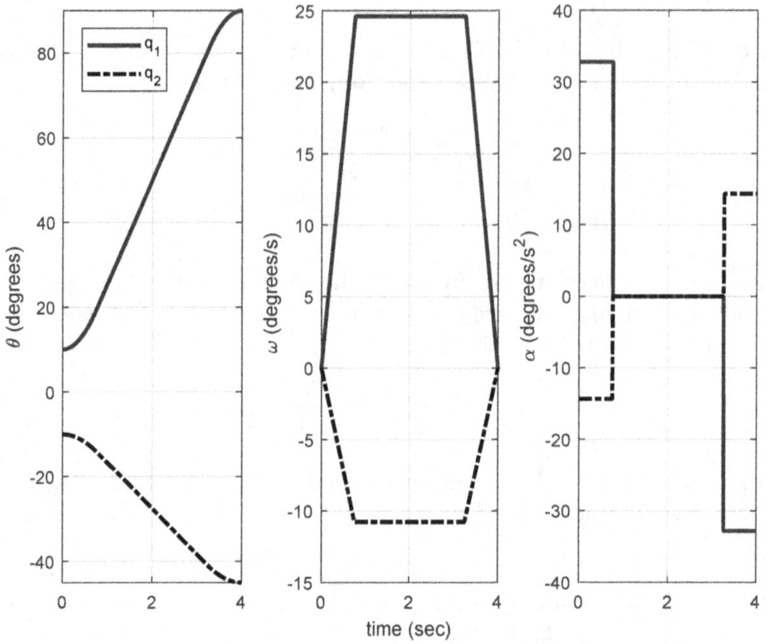

FIGURE 10.8
Linear trajectory with parabolic blends: joint motion, and angular velocity and acceleration.

robot joints to reach the final configuration $\boldsymbol{q}_f = [90, , -45]^T$ *from the initial state* $\boldsymbol{q}_0 = [10, -10]^T$ *in the shortest possible time. Assuming the maximum angular velocity of the joints is* 30 *degrees per second and the maximum angular acceleration is* 40 *degrees per square second, determine the optimal time of motion* t_f*, design the optimal linear trajectory with parabolic blend, and plot it.*

Solution: First, considering the maximum velocity and acceleration constraints in the joints, the minimum time of motion for each joint is determined based on the equation 10.20.

$$
\begin{aligned}
t_{f_1} &= \frac{1}{30}(90-10) + \frac{30}{40} = 3.42 \\
t_{f_2} &= \frac{1}{-30}(-45+10) + \frac{30}{40} = 1.92
\end{aligned}
$$

As observed, it is necessary to select a minimum time of 3.42 seconds for both joints to complete their motion, ensuring that both joints can complete their motion successfully. Thus, the velocity and acceleration of the first joint are set to the maximum actuator constraints, $\Omega_1 = 30$ and $\alpha_1 = 40$. Then, considering the required motion ratio γ for the second joint relative to the first one, these values are adjusted accordingly for the second joint.

$$
\begin{aligned}
\gamma &= \frac{-45+10}{90-10} = -0.4375 \\
\Omega_2 &= \gamma\,\Omega_1 = -13.125 \\
\alpha_2 &= \gamma\,\alpha_1 = -17.5
\end{aligned}
$$

Next, the blend time is determined based on the joint motions using Equation 10.18.

$$t_b = \frac{30}{40} = 0.75$$

With these values and the initial and final joint states, the motion trajectory is obtained using Equation 10.15.

$$q_1(t) = \begin{cases} 10 + 20t^2 & 0 \le t \le 0.75 \\ 10 + 30(t - 0.375) & 0.75 \le t \le 2.67 \\ 10 + 30(t - 0.375) - 20(t - 3.25)^2 & 2.67 \le t \le 3.42 \end{cases}$$

$$q_2(t) = \begin{cases} -10 - 8.75t^2 & 0 \le t \le 0.75 \\ -10 - 13.125\,(t - 0.375) & 0.75 \le t \le 2.67 \\ -10 - 13.125\,(t - 0.375) + 8.75(t - 3.25)^2 & 2.67 \le t \le 3.42 \end{cases}$$

The motion velocity can also be calculated using Equation 10.16.

$$\dot{q}_1(t) = \begin{cases} 40t, & 0 \le t \le 0.75 \\ 30, & 0.75 \le t \le 2.67 \\ 30 - 40(t - 3.25) & 2.67 \le t \le 3.42 \end{cases}$$

$$\dot{q}_2(t) = \begin{cases} -17.5t, & 0 \le t \le 0.75 \\ -13.125, & 0.75 \le t \le 2.67 \\ -13.125 + 17.5(t - 3.25) & 2.67 \le t \le 3.42 \end{cases}$$

You can observe the motion trajectory of the joints, as well as their angular velocity and acceleration, in Figure 10.9. As depicted in this figure, the motion of the first joint

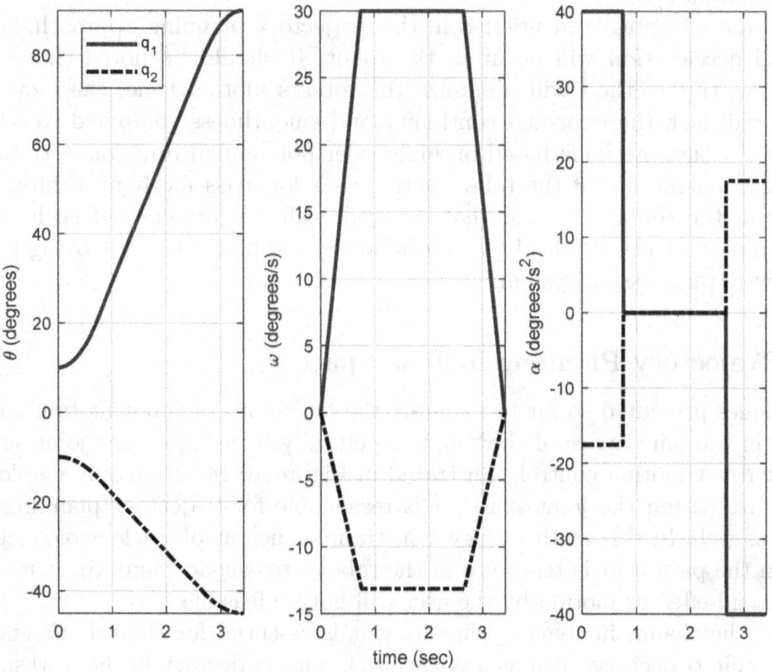

FIGURE 10.9
The linear-parabolic optimal time trajectory for joint movements and their angular velocity and acceleration.

is represented by the solid line (in blue), and the motion of the second joint is denoted by the dashed line (in black). As expected and similar to previous examples, both joints smoothly reach their final positions in the minimum time of 3.42 seconds. The angular velocity of the first joint, illustrated by the solid line (in blue) in the middle graph, indicates a constant maximum velocity throughout the mid-trajectory of the motion. Additionally, the acceleration of the first joint starts at the maximum allowable value at the beginning of the trajectory. In this way, the optimal time trajectory has been successfully designed.

In the manual movement of robots using the teaching pendant, this trajectory generation method is usually adopted. In this type of movement, by actuating the lever or motion key of each joint, the command to initiate the movement of that joint is prompted. In practice, upon receiving the motion command from the operator, the motor of that joint is briefly stimulated with constant acceleration until the joint's movement speed reaches the desired speed (and less than the maximum speed). While the stimulation button remains engaged, the joint maintains its motion at a consistent speed. Once the operator disengages the motion command, the robot undergoes maximum deceleration to swiftly come to a stop.

On the other hand, this method can also be used in optimal-time motion trajectory planning. In this application, the movement of the joint begins with the maximum possible torque in the initial motion using the maximum acceleration command. This command will continue until either the robot's speed reaches its maximum, or the movement time reaches half of t_f. In the first case, the robot will continue its movement at its maximum speed, and in the second case, the deceleration with its maximum value will occur. In both cases, the motor along the trajectory will exhibit either the maximum acceleration or speed, resulting in optimal motion time.

Despite the advantages mentioned in this trajectory planning approach, sudden acceleration and deceleration will occur in the robot. It should be noted that although the application of this method will minimize the robot's motion time, the movement of the robot arm will lack the expected continuity and smoothness compared to what happens in trajectory design methods based on high-order polynomial functions. Although, due to the mechanical structure of the robot, acting as a low-pass mechanical filter, sudden accelerations in the robot are somewhat reduced, still the presence of such discontinuous acceleration movements will lead to a reduction in the useful life of moving parts and the robot's power transmission system.

10.2.5 Trajectory Planning in Task Space

The techniques presented so far in planning the trajectory of robot motion have been implemented in the joint space of the robot, specifically focusing on the joint variables $\boldsymbol{q}(t)$. Given that robot motion control, illustrated in Figure 10.14, commonly employs a closed-loop structure within the joint space, it is reasonable for trajectory planning to occur in this space as well. In this methodology, constraints, such as obstacle avoidance, cannot be imposed on the path that is traversed in the robot's task space, and direct modification of the robot's end-effector motion by the user will not be feasible.

On the other hand, in some applications, it is essential for the robot's end-effector to follow a specific trajectory, such as a linear or circular trajectory in the workspace. In such cases, all trajectory planning methods can be implemented on the task space variables $\boldsymbol{\mathcal{X}}(t)$. It should be noted that in this case, if a control structure based on joint space is used in the robot, the designed trajectories in the task space must be mapped to the joint space using the robot's inverse kinematics. Naturally, the computational workload for this mapping process should be considered, and trajectory design in the task space should only be utilized when necessary. If a straight line is the desired focus in the robot's end-effector trajectory design in the task space, it can easily be performed without the need for via points. However, if circular, sinusoidal, or similar trajectorys are desired, the motion of the robot's end-effector can be designed on the desired trajectory by selecting a sufficient number of via points. The design of a straight line in a planar $2\underline{R}$ robot will be examined in the following Case Study of this section.

Case Study 10.6 (Trajectory Planning in Task Space) *Consider the planar $2\underline{R}$ robot depicted in Figure 8.4. The objective is for the robot's joints to reach the final state $\boldsymbol{q}_f = [90, -45]^T$ degrees from the initial state $\boldsymbol{q}_0 = [10, -10]^T$ degrees within four seconds. Design the desired motion trajectory in the task space using a cubic polynomial in such a way that the end-effector's movement follows a straight line in the $x-y$ plane. Determine and plot the velocity and acceleration of the end-effector in the two motion axes, as well as the velocity and acceleration of each joint. Illustrate the robot's end-effector motion in the $x-y$ plane to confirm the straight line of the motion trajectory.*

Solution: First, by employing the forward kinematics of the robot and the initial and final positions of the joints, we obtain the initial and final points of the motion trajectory in the task space.

$$
\begin{aligned}
x_{e_0} &= a_1 \cos q_{1_0} + a_2 \cos(q_{1_0} + q_{2_0}) = 1.6918 \\
y_{e_0} &= a_1 \sin q_{1_0} + a_2 \sin(q_{1_0} + q_{2_0}) = 0.1736 \\
x_{e_f} &= a_1 \cos q_{1_f} + a_2 \cos(q_{1_f} + q_{2_f}) = 0.5 \\
y_{e_f} &= a_1 \sin q_{1_f} + a_2 \sin(q_{1_f} + q_{2_f}) = 1.5
\end{aligned}
$$

Next, the motion trajectory of the robot's end-effector can be obtained using the cubic polynomial relation 10.5 as follows.

$$
\begin{aligned}
x_e(t) &= 1.6918 - 0.2235t^2 + 0.0372t^3 \\
y_e(t) &= 0.1736 + 0.2487t^2 - 0.0414t^3
\end{aligned}
$$

The relationships for the velocity and acceleration of the robot's end-effector can be easily obtained by appropriately differentiating these equations.

$$
\begin{aligned}
\dot{x}_e(t) = -0.447t + 0.117t^2 \;\rightarrow\; \ddot{x}_e(t) &= -0.447 + 0.2235t \\
\dot{y}_e(t) = 0.4974t - 0.1243t^2 \;\rightarrow\; \ddot{y}_e(t) &= 0.4974 - 0.2487t
\end{aligned}
$$

You can observe the components of the robot's end-effector motion, as well as its velocity and acceleration, in Figure 10.10. As evident in this figure, the end-effector's motion along the x direction is represented by the solid line (in blue), and along the y direction by the dashed line (in black). As expected, the end-effector smoothly travels from its initial values to its final values within four seconds. The velocity profile in the x direction, plotted in the middle graph with a solid line (in blue), is continuously negative and follows a quadratic function of time. Meanwhile, the velocity of the end-effector along the y direction, represented by the dashed line (in black), is consistently positive. The acceleration of the end-effector in both x and y directions, depicted in the rightmost graph, is also linear.

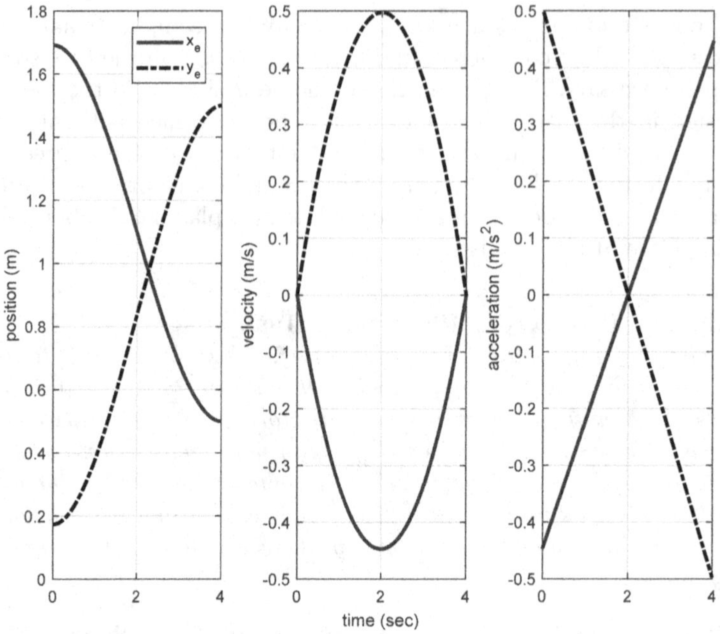

FIGURE 10.10
The trajectory planning in the robot's task space.

By utilizing the inverse kinematics of the robot and having the end-effector's motion, we can easily obtain the motion of the first and second joints and plot them. Performing these calculations demonstrates that the joint angles will smoothly transition from their initial to final values, but their motion will no longer be a cubic polynomial function as it is shown in Figure 10.11.

Figure 10.12 illustrates how the robot's end-effector moves and the configuration of the links in 9 instances with a time interval of 0.5 seconds. As observed in this figure, the robot's end-effector follows a completely linear trajectory in the $x - y$ plane. Notably, the velocity is especially slower at the start and at the end, while it reaches its maximum in the intermediate interval.

10.3 Motion Simulation

By deriving the dynamic formulation of the robot and designing appropriate joint trajectories, the robot's motion can be simulated by applying torques to the joint actuators. Simulating the dynamic motion of the robot is crucial for analyzing its movement and designing a suitable controller. Through the simulation of dynamic robot motion prior to construction, it is possible to analyze the dynamic behavior and motion characteristics of robot components. The method of simulating dynamic robot motion using the robot's forward dynamics will be explained further in section 10.3.1. On the other hand, simulating the dynamic motion of the robot can also be advantageous in the selection of sensors and

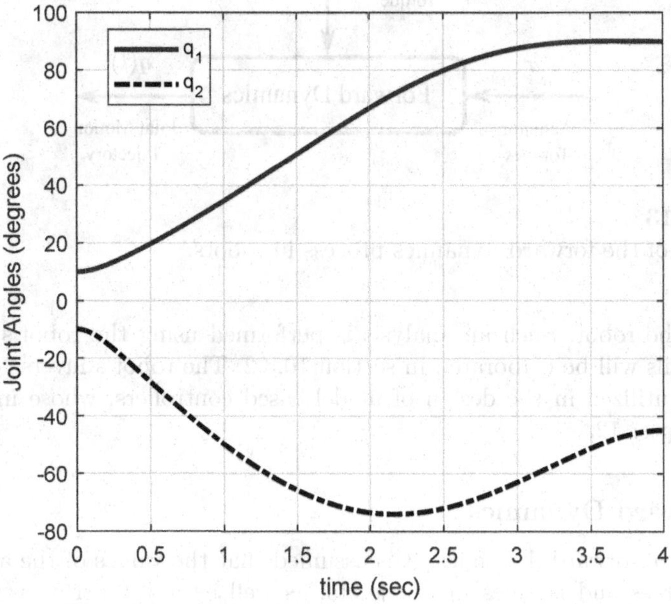

FIGURE 10.11
Joint movements based on the designed trajectory in the robot's task space by using linear trajectory.

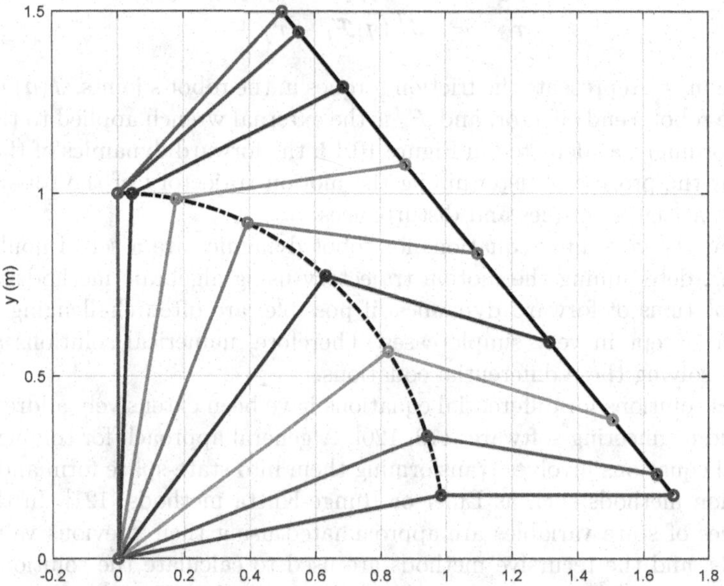

FIGURE 10.12
The result of the link and end-effector motion during the straight line maneuver of the robot's end-effector.

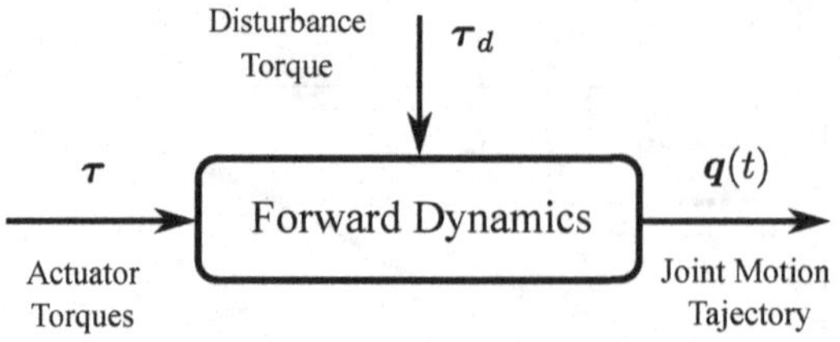

FIGURE 10.13
Block diagram of the forward dynamics process in robots.

actuators for the robot. Such an analysis is performed using the robot's inverse dynamics, whose details will be elaborated in section 10.3.2. The robot's inverse dynamics is also fundamentally utilized in the design of model-based controllers, whose intricacies will be detailed in chapter 12.

10.3.1 Forward Dynamics

In the analysis of forward dynamics, it is assumed that the values of the actuator torques, and friction forces and torques in the joints, as well as any external wrench applied to the robot and its end-effector, are known, and it is desirable to determine the dynamic motion of the robot and the performed trajectory of its motion. If we denote the robot's actuator torques as $\boldsymbol{\tau}$ and encapsulate all the other friction torques and the mapping of all wrenches applied to the robot and its end-effector in joint space by $\boldsymbol{\tau}_d$, the governing dynamic equation for the robot's motion can be rewritten as follows.

$$\begin{aligned} \boldsymbol{\tau} + \boldsymbol{\tau}_d &= \boldsymbol{M}(\boldsymbol{q})\ddot{\boldsymbol{q}} + \boldsymbol{C}(\boldsymbol{q}, \dot{\boldsymbol{q}})\dot{\boldsymbol{q}} + \mathfrak{g}(\boldsymbol{q}) \\ \boldsymbol{\tau}_d &= \boldsymbol{J}^T(\boldsymbol{q})\boldsymbol{\mathcal{F}}_e - \boldsymbol{\tau}_f + \cdots \end{aligned} \tag{10.21}$$

In this equation, $\boldsymbol{\tau}_f$ represents the friction torques in the robot's joints, $\boldsymbol{J}(\boldsymbol{q})$ is the Jacobian matrix of the robot's end effector, and $\boldsymbol{\mathcal{F}}_e$ is the external wrench applied to the robot's end-effector. Accordingly, as depicted in Figure 10.13, the forward dynamics of the robot can be introduced as the process of determining the motion trajectory of the robot based on the applied input actuator torques and disturbances.

Given that the governing equations for robot dynamics are a set of nonlinear differential equations, determining the motion trajectory using algebraic methods is not feasible. Analytical solutions of forward dynamics, if possible, are often challenging and not readily achievable except in very simple cases. Therefore, numerical solutions are commonly employed for solving these differential equations.

Numerical solutions for differential equations have been extensively addressed in various references and engineering software [119, 120]. A general approach for solving such systems of differential equations involves transforming them into state-space form and using numerical integration methods such as Euler or Runge-Kutta methods [121]. In these methods, the derivatives of state variables are approximated using their previous values over small time intervals, and the recursive methods are used to calculate the solution to the differential equation with sufficient accuracy. In software like MATLAB®, this approach can be implemented using the `ode` family of commands.

Transforming the robot's dynamic equations into state-space form is also straightforward for serial robots. Considering the invertibility property of the mass matrix discussed in section 9.2, the dynamic equations of serial robots can be easily transformed into state-space form in two steps. Firstly, using the inverse of the mass matrix, the equation 10.21

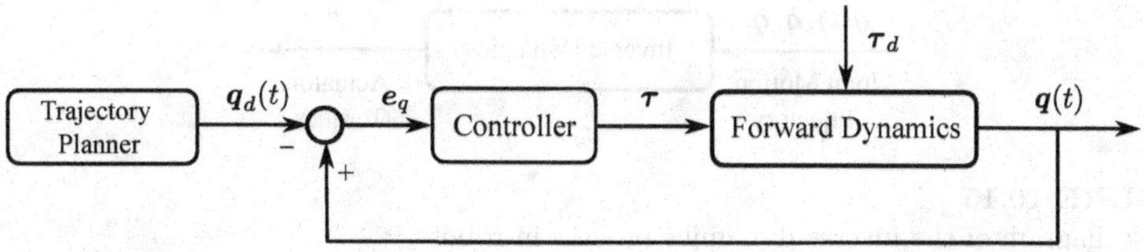

FIGURE 10.14
Block diagram of the closed-loop control process of a robot in tracking the desired trajectory.

can be explicitly written in terms of $\ddot{q}$ as follows.

$$\ddot{q} = M^{-1}(q)\,(\tau + \tau_d - C(q,\dot{q})\dot{q} - g(q)) \tag{10.22}$$

This equation represents a second-order system of differential equations. This system can be transformed by introducing a state variable:

$$x = \begin{bmatrix} x_1 \\ x_2 \end{bmatrix} = \begin{bmatrix} q \\ \dot{q} \end{bmatrix} \tag{10.23}$$

into a state-space representation.

$$\begin{aligned} \dot{x}_1 &= \dot{q} = x_2 \\ \dot{x}_2 &= \ddot{q} = M^{-1}(q)\,(\tau + \tau_d - C(q,\dot{q})\dot{q} - g(q)) \end{aligned} \tag{10.24}$$

In engineering software, numerically solving these state-space equations with specified initial conditions $x(0) = [x_1(0), x_2(0)]^T$ is easily accomplished using numerical differential equation solvers such as the Runge-Kutta method. An example of such numerical solution for a robot will be performed in the continuation of this section, and the results will be analyzed.

Forward dynamics can also be utilized in the simulation of closed-loop robot motion. If motion tracking is the goal, feedback loops are commonly used in robot motion control, as illustrated in the block diagram in Figure 10.14. As seen in this diagram, by employing feedback loops and designing an appropriate controller, the desired motion tracking of joints can be achieved. In this diagram, the desired robot motion trajectory is generated by the trajectory planner block, for which some trajectory generation methods are presented in section 10.2.

The primary characteristic of this diagram lies in its utilization of a feedback loop. This loop dynamically compares the real trajectory $q(t)$ of the robot with the desired trajectory $q_d(t)$ in real-time, and applies the tracking error e_q to the robot's control block. The robot controller, utilizing various methods as a function of the tracking error, issues torque commands to the robot's actuators to minimize the tracking error. Consequently, this approach enables the continuous online monitoring and control of the tracking error.

Irrespective of the specific type and design of control law implemented in the robot, details of which will be elaborated on in upcoming chapters, the closed-loop structure stands out as the fundamental element of motion control. It ensures the achievement of the desired tracking even in the presence of disturbances and model uncertainties. Adjusting the controller parameters within this structure typically involve simulating the closed-loop dynamics of the robot, as illustrated in the block diagram presented in Figure 10.14. The numerical simulation of this structure will be conducted on a robot in the subsequent sections of this chapter, and the results will be evaluated.

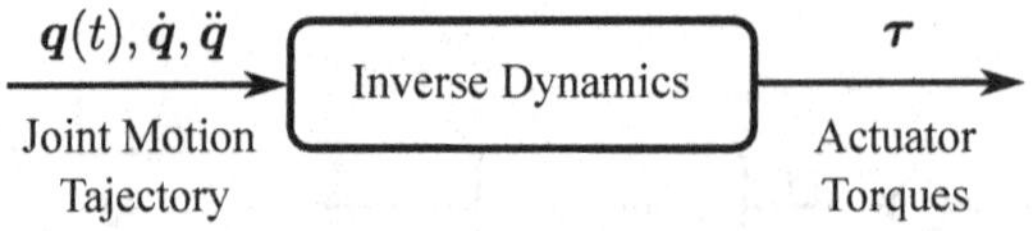

FIGURE 10.15
Block diagram of the inverse dynamics process in robots.

10.3.2 Inverse Dynamics

In the analysis of inverse dynamics for a robot, it is assumed that the trajectory of the robot's motion, denoted as $q(t)$, and its derivatives are given, and the objective is to determine the joint torques that produces such a predefined motion. Thus, the inverse dynamics of the robot can be described as the process of determining the necessary joint torques to produce a motion along a predefined trajectory.

Despite the governing equations of robot dynamics being a set of nonlinear differential equations, assuming knowledge of the motion trajectory and its derivatives eliminates the need for integration in determining the required actuator torques. In this case, it is sufficient to calculate the following algebraic relationship.

$$\tau = M(q)\ddot{q} + C(q, \dot{q})\dot{q} + g(q)$$

Through inverse dynamics simulation, one can leverage the simulation results to select appropriate sensors and actuators for the robot. To achieve this, the required motions of the robot in the workspace are pre-planned, and the inverse dynamics calculate the necessary joint torques to produce such a desired motion. Considering that the calculation of these torques assumes precise robot motion without tracking errors, in the absence of noise and disturbances, it is necessary to adjust the calculated torques and joint velocities with a suitable and greater-than-one scaling factor. This ensures that the maximum power required by the actuators and their maximum speed and torque during motion are determined. Based on the maximum power requirement, the suitable actuator type is selected, and based on the maximum torque and speed requirements, the appropriate combination of actuator and gearbox is determined.

Inverse dynamics of the robot are widely used in model-based control methods. Since the robot's dynamic formulation is nonlinear and coupled, using the inverse dynamics of the system in robot control allows these equations to be linearized through feedback. This control method, known as *inverse dynamics control*, has served as the fundamental basis for the development of model-based control methods in robots. The application of inverse dynamics control and enhanced methods based on it will be explored in Chapter 12.

10.4 Motion Simulation Case Study: The Planar 3R Robot

In this section, we will focus on the direct and inverse dynamics simulations of the planar 3R robot as a Case Study, and we will examine the simulation results. For this means, consider the planar 3R robot represented in Figure 8.7 with the kinematic and mass parameters given in Table 10.1. Assume that all links of the robot are slender rods, and their center of mass is located at the link mid-length. The motion trajectory of the robot arms is designed based on cubic polynomials, while the initial and final values of each joint's motion are reported in this table. The duration of this maneuver is also set to one second.

TABLE 10.1
Kinematic and dynamic parameters of the planar 3R robot

Quantity	Unit	Notation	Link 1	Link 2	Link 3
Length	m	a_i	1.0	0.7	0.3
Mass	Kg	m_i	10	7	3
Center of Mass	m	m_{c_i}	0.5	0.35	0.15
Inertia Moment	$Kg \cdot m^2$	${}^{C}I_i$	0.833	0.286	0.023
Initial Angle	deg	θ_{i_0}	0	0	0
Final Angle	deg	θ_{i_f}	60	45	30
Maneuver Time	sec	t_f	1	1	1

10.4.1 Inverse Dynamic Simulation

Initially, consider the dynamic equations of the robot based on the dynamic formulation obtained in Case Study 8.3. Through simulation, determine the power required by the actuators, as well as their maximum speed and torque. To achieve this, first design the trajectories of the robot's joints using cubic polynomials, and then ascertain the velocity and acceleration of the joint movements. Furthermore, according to the block diagram in Figure 10.15, obtain the necessary torques for executing this motion by inputting these values into the inverse dynamics equation of the robot.

The simulation results of the robot's inverse dynamics, considering the parameters provided in Table 11.1, are depicted in Figure 10.16. The simulation result corresponding to the first joint is depicted by a solid line (in blue), the second joint by a dashed line (in black), and the third joint by a dash-dotted line (in red). The motion of the joints is shown on the left side of the graph, while the determined torques from the inverse dynamics of the robot are displayed on the right side.

As seen in the left-side plot of this figure, the joint movements smoothly transition from the initial to the final angles. The required torque to generate this motion in the first joint is higher than the others and ranges within approximately −150 to 350 Newton-meters. Given that the acceleration of joint motions is linear, the changes in the required torque are also close to a linear function. Similarly, the required torque in the second and third joints, owing to the significantly shorter lengths of these links compared to the first link, is noticeably less than the required torque in the first joint. To select a suitable motor for this robot, we determine the required power for performing this maneuver by multiplying the torque by the angular velocity of each joint and plot it over time, as shown in Figure 10.17. In this figure, the required power for each joint is plotted on the left side, and the required torque versus the angular velocity of the joints is displayed on the right side. As evident in this figure, the required power for the first joint is approximately 275 watts, while the power for the second joint is about 75 watts, and the power needed for the third joint is estimated at 5 watts. It is natural for the required power to be directly dependent on the speed of motion, and the power required for other expected motion maneuvers of the robot should be similarly simulated. By conducting these simulations for a set of desired trajectories of the robot, the maximum required power is obtained, taking into account an appropriate safety margin.

Determining the maximum power helps specify the type and class of motor needed. Motor selection is typically based on its technical specifications, often represented by the torque-speed curve. These specifications for the three joints of the robot are depicted in the right-side plot of Figure 10.17. In this figure, the curve with a solid line (in blue) indicates that the maximum torque required for the motor of the first joint is approximately 350

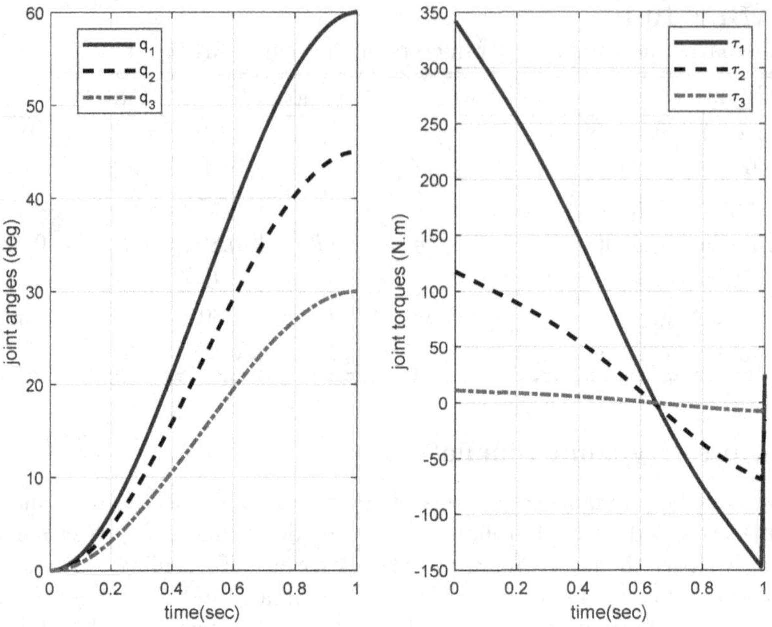

FIGURE 10.16
Simulation of joint motions and the required torques for executing this maneuver using the inverse dynamics of a 3<u>R</u> robot.

Newton-meters at a speed close to zero radians per second. Furthermore, the maximum speed of the actuator is around 1.2 radians per second which corresponds to about 12 revolutions per minute. Considering that the typical generated torque of 300 watts electric motors is around two Newton-meter and their speed is about 3000 revolutions per minute, it is natural to use a suitable gearbox with a conversion ratio of about 200 to provide such torque/speed requirements in the robot. Harmonic drives are a suitable choice for these types of gearboxes, widely used in industrial robots.

10.4.2 Closed-Loop Dynamic Simulation

Now, we will study the closed-loop simulation of the robot according to the block diagram shown in Figure 10.14. For this purpose, it is necessary to design the desired trajectory $\boldsymbol{q}_d$ using the trajectory planner block, and the tracking error $\boldsymbol{e}_q$ should be applied to the robot control system. Various controllers can be used in this structure. Since the main focus in this section is on simulating the closed-loop system, and the details of designing a specific control are not the main concern, we will use a PD controller as the simplest option commonly employed in industrial robots.

In this controller, the control torque $\boldsymbol{\tau}$ can be determined using the following relation:

$$\boldsymbol{\tau} = \boldsymbol{K}_p(\boldsymbol{q}_d - \boldsymbol{q}) + \boldsymbol{K}_d(\dot{\boldsymbol{q}}_d - \dot{\boldsymbol{q}}) \tag{10.25}$$

where $\boldsymbol{K}_p$ is the proportional gain matrix, and $\boldsymbol{K}_d$ is the derivative gain matrix in the controller. By this means, the control torque $\boldsymbol{\tau}$ is generated from the weighted sum of the tracking error and its derivative. Typically, the gain matrices in the controller are chosen to be diagonal with large gains to achieve proper tracking performance.

In the closed-loop simulation of the robot, consider the kinematic and dynamic parameters of the robot and the desired motion trajectory as given in Table 10.1. Furthermore,

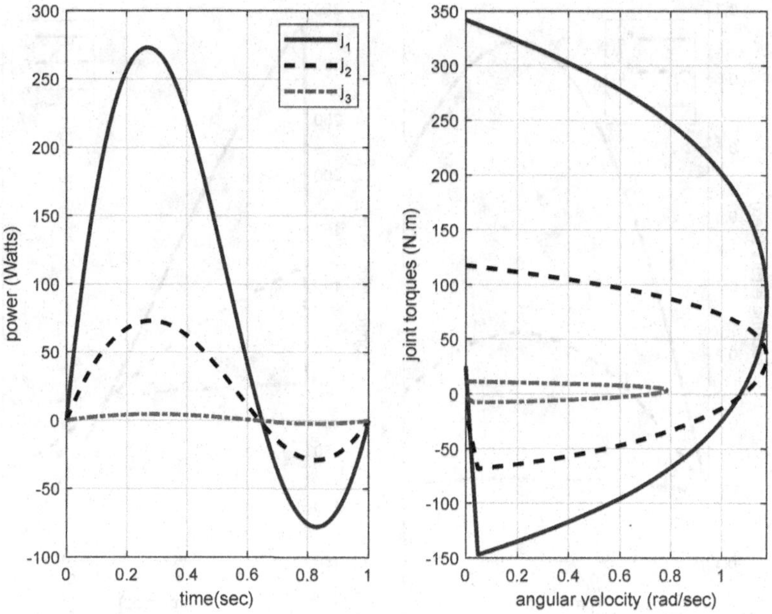

FIGURE 10.17
Power required for the joint actuators to perform the desired motion maneuver of the 3R robot.

set the gain matrices of the controller as follows:

$$\boldsymbol{K}_p = 10^3 \boldsymbol{I}_{3\times 3}, \quad \boldsymbol{K}_d = 10^4 \boldsymbol{I}_{3\times 3}$$

The simulation of the motion in this case is performed using the equation 10.24, where the torque $\boldsymbol{\tau}$ is calculated from the equation 10.25, and the disturbance torque $\boldsymbol{\tau}_d$ is assumed to be zero. The numerical solution of these state equations, with initial conditions $\boldsymbol{q}(0) = \boldsymbol{q}_0, \dot{\boldsymbol{q}}(0) = 0$, is carried out using numerical differential equation solvers in MATLAB software, employing the `ode23s` function among the available methods in the software. This choice is made to take advantage of the large control gains in the control structure, ensuring the differential equations are solved with more ease and high accuracy.

Using this control structure, the joints accurately follow the desired trajectory shown in Figure 10.16. To demonstrate the tracking accuracy, the simulation result of this control structure is shown in Figure 10.18. In the left plot of this figure, the tracking error is shown, and in the right plot, the applied torque to the robot is displayed. The tracking error in the first joint is represented by the solid (blue) line, and in the second joint by the dashed (black) line, and in the third joint by the dash-dotted (red) line. As observed, the maximum tracking error in the first joint is negligible and about 0.7 degrees, while that in the other two joints is significantly lower. Furthermore, it can be observed from these graphs that the steady-state error in this control structure has not reached zero. If necessary to eliminate the residual error, a PID controller can be used.

The control effort or the applied torques by the actuators in the robot joints are shown in the right chart, respectively with the solid (blue) line, dashed (black) line, and dash-dotted (red) line for the first to third joints. If you compare the applied torques in this case with the torques obtained from the inverse dynamics method shown in Figure 10.16, you will notice that the overall trend and the magnitude of torques closely match the two cases. The main reason for this consistency is the use of feedback control and the proper design of

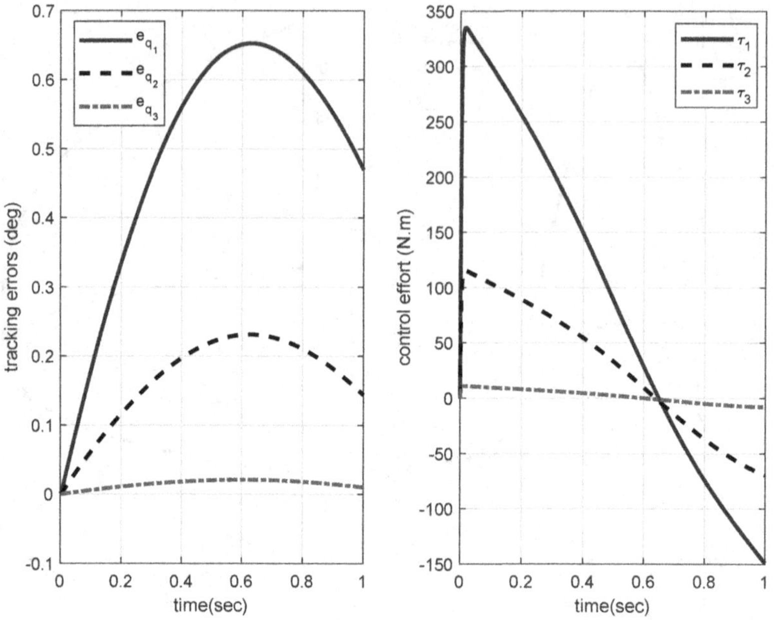

FIGURE 10.18
Simulation of the tracking error of the joints and the applied torque to the robot in the closed-loop structure.

the control system, which results in minimal tracking error. Thus, in this case as well, the maximum torque required is in the first joint and is less than 350 Newton-meters applied at the beginning of the maneuver. Additionally, considering that the joint accelerations are linear, the required torques exhibit behavior close to a linear function.

10.5 Dynamic Calibration

So far, we have examined methods for modeling the dynamics of robots and their actuators. The dynamic model of robots, determined by any method, includes both kinematic and dynamic parameters, and the accuracy of predicting the dynamic behavior of a robot relies on the precise knowledge of these parameters. In the robotics literature, the precise identification of these parameters is studied under the title of "robot calibration."

Robot calibration methods have been developed for two main aspects: kinematic and dynamic parameters. While the fundamental principles for both are similar, the implementation details differ. In kinematic calibration, the focus is on the robot's Denavit-Hartenberg parameters. In contrast, dynamic calibration methods aim to determine mass properties, which generally consist of ten parameters for each link, including the link mass, three parameters defining the center of mass, and six parameters describing the inertia matrix. Moreover, the dynamic calibration process can also involve examining and estimating frictional forces and disturbances affecting the robot dynamics.

It should be noted that depending on how the robot moves, many of the dynamic parameters may remain unexcited and challenging to determine in practice. For example, if we consider a planar robot, the dynamic parameters of each link reduce to four parameters, namely, the link mass, two parameters defining the center of mass in the plane, and only one parameter describing the inertia matrix about out-of-plane motion. By reducing the general dynamic parameters to a limited number, we will be able to identify the base inertial parameters (BIP) [122]. These base inertial parameters are the fundamental inertial parameters that can be experimentally identified in dynamic calibration methods.

The first and simplest method for estimating the kinematic and dynamic parameters of a robot is to use computer-aided design (CAD) software to approximate nominal values for these parameters. Although these software tools are capable of accurately determining these parameters in the three-dimensional model of robot components, they usually lack precision in determining the accurate parameters of complex robot structures composed of multiple moving parts [123]. These nominal values can often be used as initial values in the robot model and in dynamic simulation.

The primary methods for robot calibration are based on linear regression models. In this approach, the robot's kinematic or dynamic model is expressed as a linear regression in terms of base parameters, and least squares linear regression methods are employed to determine these parameters. To implement these methods in dynamic calibration, motion and torque data are recorded during a maneuver, and then these data are utilized in least squares methods to determine the parameters of the linear regression model of the robot.

Calibration methods can be readily applied to complex robot models obtained in the form of linear regression models, as derived in Equation 9.24. Given the breadth of the topic, this section will explore an example of this calibration applied to the linear model derived in the previous chapter. This example aims to provide a better understanding of the general principles of robot dynamic calibration. The implementation of linear regression methods is applicable to various parametric models of robots that are represented in linear regression form.

In the upcoming section dynamic calibration of the specified model for only one joint of the robot, assuming linear dynamics derived in Section 9.7, will be detailed. Evaluating the accuracy of the obtained linear model is a subject that must be empirically addressed. For this purpose, in Section 10.5.3, methods for evaluating the obtained linear model are studied to determine the conditions under which the use of a complete multivariable and nonlinear model is necessary for the robot. In such cases, the developed method in this section can be applied to the dynamic calibration of more complex models.

10.5.1 Linear Regression Models

Consider the linear model of one joint of the robot, given in Equation 9.43. For simplicity, let's rewrite this equation by omitting the index i as follows.

$$\tau = I_e\ddot{q} + b_e\dot{q} + \mathfrak{g}(q) \tag{10.26}$$

Although this model is a differential equation with respect to the motion variables of the robot q, it has a linear algebraic relationship with respect to the dynamic parameters of the robot. Transforming this model into a linear regression form with calibration parameters allows for leveraging methods rooted in least squares error to effectively determine these parameters. For this purpose, in several cases, model 10.26 is formulated and identified.

In the first case, assume that the gravitational torque is a constant value $\mathfrak{g}(q) = \mathfrak{g}$. Furthermore, suppose that during a typical maneuver in the robot's workspace, motion information $q, \dot{q}, \ddot{q}$ and the robot's actuator torque τ are recorded at a number of time instances $i = 1, 2, \ldots, N$ throughout the maneuver.

$$\boldsymbol{\tau} = \begin{bmatrix} \tau_1 \\ \tau_2 \\ \vdots \\ \tau_N \end{bmatrix}, \quad \boldsymbol{q} = \begin{bmatrix} q_1 \\ q_2 \\ \vdots \\ q_N \end{bmatrix}, \quad \dot{\boldsymbol{q}} = \begin{bmatrix} \dot{q}_1 \\ \dot{q}_2 \\ \vdots \\ \dot{q}_N \end{bmatrix}, \quad \ddot{\boldsymbol{q}} = \begin{bmatrix} \ddot{q}_1 \\ \ddot{q}_2 \\ \vdots \\ \ddot{q}_N \end{bmatrix}, \quad i = 1, 2, \ldots, N \tag{10.27}$$

In this scenario, if we represent the calibration parameters with the vector $\boldsymbol{\Phi}$ and the linear regression matrix with $\boldsymbol{\mathcal{Y}}$, similar to what has been explained in Section 9.4, the robot joint model can be expressed in a general linear regression form.

$$\boldsymbol{\tau} = \boldsymbol{\mathcal{Y}}(\boldsymbol{q}, \dot{\boldsymbol{q}}, \ddot{\boldsymbol{q}})\boldsymbol{\Phi} \tag{10.28}$$

where,

$$\boldsymbol{\tau} = \begin{bmatrix} \tau_1 \\ \tau_2 \\ \vdots \\ \tau_N \end{bmatrix}, \quad \boldsymbol{\mathcal{Y}} = \begin{bmatrix} \ddot{q}_1 & \dot{q}_1 & 1 \\ \ddot{q}_2 & \dot{q}_2 & 1 \\ \vdots & \vdots & \vdots \\ \ddot{q}_N & \dot{q}_N & 1 \end{bmatrix}, \quad \Phi = \begin{bmatrix} I_e \\ b_e \\ \mathfrak{g} \end{bmatrix} \tag{10.29}$$

Given this relationship, we observe that the vector of calibration parameters for a robot joint consists of only three components. This is despite the fact that, for the identification of these three parameters, one can utilize information from the motion data of the robot at N time instances.

In the second case, let's assume that the gravitational torque is a function of the robot's motion q. For instance, considering the first component of the gravity vector for a planar 2$\underline{\text{R}}$ robot, as reported in Equation 8.46, the gravitational torque of its first joint would be equivalent to

$$\tau_{g_1} = \frac{1}{2} a_1 m_1 g \cos q_1 = \mathfrak{g} \cos q_1$$

Therefore, the gravity vector can be considered as an unknown parameter $\mathfrak{g}$ multiplied to the cosine of the joint angle. Thus, by assuming logging the motion data, such as the previous case 10.27, and the linear regression model based on the relationship 10.28, the linear regression matrix $\boldsymbol{\mathcal{Y}}$ is written as follows.

$$\boldsymbol{\tau} = \begin{bmatrix} \tau_1 \\ \tau_2 \\ \vdots \\ \tau_N \end{bmatrix}, \quad \boldsymbol{\mathcal{Y}} = \begin{bmatrix} \ddot{q}_1 & \dot{q}_1 & \cos q_1 \\ \ddot{q}_2 & \dot{q}_2 & \cos q_2 \\ \vdots & \vdots & \vdots \\ \ddot{q}_N & \dot{q}_N & \cos q_N \end{bmatrix}, \quad \Phi = \begin{bmatrix} I_e \\ b_e \\ \mathfrak{g} \end{bmatrix} \tag{10.30}$$

The difference in this relationship from the previous case lies solely in the gravitational torque model, which is not considered constant and is incorporated as a function of the cosine q in the regression matrix. If the gravitational torque relationship is formed from a more complex function, this part of the regression matrix can be adjusted according to the corresponding model.

In the third case, suppose that in logging the robot's motion data, only the angular motion of joint q is measured, and the joint's velocity and acceleration are not directly available. This is common in industrial robots where only encoders are typically used to measure the joint angle. In this case, various methods can be used to obtain the joint's velocity and acceleration using motion data, and this information can be used in the regression matrix.

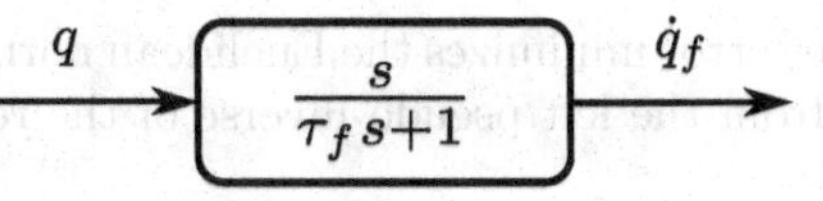

FIGURE 10.19
Block diagram of filtered derivative.

One effective approach in this regard is the use of derivative filters. Consider the block diagram of the simplest derivative filter in Figure 10.19. This filter can be easily implemented in software like MATLAB, Simulink or Python. The transfer function of this filter is

$$G(s) = \frac{s}{\tau_f s + 1} \tag{10.31}$$

As observed, the Laplace transform variable s in this transfer function is the derivative operator in the Laplace space. If this operator is used alone, signal noise will be amplified during differentiation, leading to undesirable results. The denominator of the transfer function shown in Figure 10.19 represents a very simple first-order low-pass filter with a time constant t_f. This part of the transfer function is used to reduce noise in the differentiated signal. Choosing an appropriate time constant for the filter is dependent on the information stored in the signal and the characteristics of its noise. If we consider the cutoff frequency of the filter as $\omega_f = 1/\tau_f$, it is advisable to choose this frequency to be larger than ten times the joint motion bandwidth ω_{bw} and smaller than one-tenth of the main noise frequency ω_n in the measured signal.

$$10\,\omega_{bw} < \omega_f < 0.1\,\omega_n \tag{10.32}$$

In robotics applications, by choosing $20 < \omega_f < 10^3$ or equivalently, $10^{-3} < \tau_f < 0.05$, a suitable filter can be obtained. The implementation of this filter can be easily done in the discrete-time domain using the `filtfilt` command in MATLAB. The advantage of using this command is that it completely eliminates the inherent time delay in applying low-pass filters. However, the execution of this command may not be real-time and is only feasible on stored and recorded data.

Using this filter, the filtered joint angular velocity $\dot{q}_f$ can be obtained based on the stored joint motion data. By applying this filter again to the resulting data, the filtered joint angular acceleration $\ddot{q}_f$ can also be determined. With the determination of these quantities, the linear regression matrix $\mathcal{Y}$ is obtained as follows.

$$\tau = \begin{bmatrix} \tau_1 \\ \tau_2 \\ \vdots \\ \tau_N \end{bmatrix}, \quad \mathcal{Y} = \begin{bmatrix} \ddot{q}_{f1} & \dot{q}_{f1} & \cos q_1 \\ \ddot{q}_{f2} & \dot{q}_{f2} & \cos q_2 \\ \vdots & \vdots & \vdots \\ \ddot{q}_{fN} & \dot{q}_{fN} & \cos q_N \end{bmatrix}, \quad \Phi = \begin{bmatrix} I_e \\ b_e \\ \mathfrak{g} \end{bmatrix} \tag{10.33}$$

10.5.2 Least Squares Method

By modeling the robot dynamics in the form of linear regression, the values of robot's dynamic parameters can be obtained using least squares error methods. For this purpose, consider the main linear regression relationship 10.28. Given that the number of calibration parameters p is significantly less than the number of points where the input and output system data is recorded and stored, the regression matrix will be a rectangular matrix with the number of rows N, much greater than the number of its columns. This means that the number of equations is much greater than the number of unknowns, and there is no solution that satisfies all equations.

Solving the least squares error minimizes the Euclidean norm of the error in all equations. This solution is obtained from the left pseudo-inverse of the regression matrix.

$$\hat{\boldsymbol{\Phi}} = \mathcal{Y}^{\dagger}\,\boldsymbol{\tau} \tag{10.34}$$

$$\mathcal{Y}^{\dagger} = \left(\mathcal{Y}^{T}\mathcal{Y}\right)^{-1}\mathcal{Y}^{T} \tag{10.35}$$

Given the large number of rows in the regression matrix and the possibility of having more parameters than the basic inertia parameters in the general robot calibration problem, the direct calculation of the inverse matrix $\mathcal{Y}^{T}\mathcal{Y}$ is not recommended. Usually, appropriate computational methods are used when the condition number of the matrix is large. One efficient method used in the `pinv` command in MATLAB is the Householder reflection method. In this method, the matrix $\mathcal{Y}$ is decomposed as follows.

$$\mathcal{Y}_{N\times p}\boldsymbol{P}_{p\times p} = \boldsymbol{Q}_{N\times p}\boldsymbol{R}_{p\times p}$$

In this expression, $\boldsymbol{P}$ is a permutation matrix, $\boldsymbol{Q}$ is an orthogonal matrix, and $\boldsymbol{R}$ is an upper triangular matrix. Using this decomposition, the pseudo-inverse of the regression matrix is easily calculated as follows.

$$\mathcal{Y}^{\dagger} = \boldsymbol{P}\boldsymbol{R}^{-1}\boldsymbol{Q}^{T} \tag{10.36}$$

In this expression, the inverse of the matrix $\boldsymbol{R}$ can be easily determined through recursive calculations.

10.5.3 Evaluation of Calibrated Model

Evaluation of the accuracy of the obtained model is crucial in any modeling procedure. In dynamic calibration, this becomes even more important for the linear model, as the multivariable and nonlinear robot model is transformed into several linear and decoupled models. Depending on the gear ratio used in the robot joint, this linear model may not be accurate enough, and it may not be possible to use it effectively in the design of a linear controller.

To evaluate the accuracy of the obtained model, the calibration results can be used in multiple experiments. If the assumed model and the robot parameters are perfectly matched, the robot's trajectory of motion will not affect the accuracy of the calculations. However, both the linear model and the calibrated dynamic parameters have limited accuracy; thus, the calibration process must be repeated in several different experiments.

For this purpose, by considering various motion maneuvers in the robot's workspace, data recorded from the input and output of each of these motion maneuvers is stored as an independent experiment. Then, by modeling the robot in the form of linear regression for each of these experiments, the calibrated dynamic parameters of the robot are determined. If the calculation results in different experiments are relatively close to each other, it can be concluded that the assumed model is close enough to reality, and the calibrated dynamic parameters of the robot are reliable.

However, due to the presence of measurement noise and structural uncertainty in the system model, accurate match is never achieved, and the results obtained from different experiments are always different. In this case, the average of the results obtained from the calibration of each of the parameters in different experiments can be considered as the best estimate of the nominal value of that parameter. In addition, the standard deviation of the results obtained for each of the parameters in different experiments can also be used as a suitable statistical quantity to assess the confidence level of the obtained result.

In robotics literature, the "Consistency Measure (CM)" is introduced for qualitative evaluation of parametric calibration results. The consistency measure is the ratio of the standard deviation to the mean of the estimated parameter in different experiments [124].

$$\%\text{CM} = \frac{\text{Standard Deviation}}{\text{Mean}} \cdot 100 \tag{10.37}$$

Through multiple calibration experiments, it has been empirically demonstrated that if the value of the consistency measure in practical calibration experiments is less than 30%, the used model and the obtained parameters are entirely reliable and can be used in the design of a linear controller. With an increase in the value of the consistency measure to around 80%, there is less confidence in the estimated parameters. This means that the model used in the dynamic calibration process is not accurate enough, and it needs refinement through engineering judgment. If the consistency measure is greater than %100, it indicates that either the employed model or the experimental data used does not have an acceptable level of accuracy, and the obtained results are not reliable.

Therefore, the process of calibrating the dynamic parameters of the robot is usually done in several stages, with the results being evaluated at each stage. It should be noted that determining the model parameters in the dynamic calibration section is solely for use in linear controllers or developed model-based controllers, and the obtained model is by no means applicable to predict the exact dynamic behavior of the system. Given this goal, the selection of appropriate motion maneuvers in the robot is also entirely different from a scenario where dynamic identification of the robot is used to predict its motion behavior.

In the literature of dynamic identification of general system motions, the use of persistently exciting (PE) inputs is recommended. These types of inputs are chosen in a way that continuously excites all dynamic modes of the system. However, in the calibration of linear robot model parameters, this type of input is not suitable, because it excites the nonlinear dynamics of the robot, and it will not lead to a suitable linear model. The consequence of employing this type of excitation will be seen in the large consistency measure obtained during robot calibration.

Considering that the linear model is only used in the robot's workspace for common motion control, it is essential that the motion maneuvers of the robot are also designed in this workspace with similar characteristics to the main robot motion. Thus, for robots performing repetitive motions, the design of motion maneuvers that are consistent with them is suggested, and it is expected that the result of robot calibration will have a suitable consistency measure.

10.5.4 Calibration Case Study: Planar 2R Robot

In this section, we will investigate the calibration of the planar 2R robot as a Case Study, and report the identified parameters of the linear model. To do this, consider the 2R robot shown in Figure 8.6, with its kinematic and mass parameters presented in Table 10.2. Assume that all robot links are slender rods and their center of mass is located at the link length midpoints. Furthermore, assume that motors with suitable gearboxes and a conversion ratio of 200 are used in the robot joints. The linear friction coefficient in the robot joints and in its motor actuators is also provided in Table 10.2.

To calibrate the linear model of the robot, we first simulate the motion maneuvers of the robot in the closed-loop structure, utilizing the nonlinear robot model in the presence of the actuator dynamics according to the equation 9.39. This simulation is performed in accordance with the block diagram shown in Figure 10.14. For this purpose, various motion maneuvers are considered, and a *PD* controller with the following large gain matrices is employed, to accurately track the desired motion trajectories.

$$\boldsymbol{K}_p = 10^6 \boldsymbol{I}_{2\times 2}, \quad \boldsymbol{K}_d = 10^8 \boldsymbol{I}_{2\times 2}$$

TABLE 10.2
Kinematic and dynamic parameters of the planar 2<u>R</u> robot.

Quantity	Unit	Notation	Link 1	Link 2
Length	m	a_i	1.0	0.7
Mass	Kg	m_i	10	7
Center of Mass	m	m_{c_i}	0.5	0.35
Inertia Moment	$Kg \cdot m^2$	${}^{C}I_i$	0.833	0.286
Gearbox Ratio	–	η_i	200	200
Motor Inertia Moment	Kgm^2	${}^{C}I_{m_i}$	0.1	0.07
Joint Viscous Friction	$\frac{Nm.s}{rad}$	b_i	0.1	0.1
Motor Viscous Friction	$\frac{Nm.s}{rad}$	b_{m_i}	0.001	0.001

As an example, among the twelve tested motion maneuvers, one is shown in Figure 10.20. In the left plot of this figure, the motion maneuver is depicted, and in the right plot, the applied torques on the robot joints are shown. The motion maneuver of the first joint of the robot is represented by the solid line (in blue), and the second joint, it is represented by the dashed line (in black). The control efforts or applied torques by the actuators on the robot joints are also illustrated in the corresponding type (and colors) for the first and second joints, respectively. As observed in this figure, the first joint moves from zero to 30 degrees in one second, then smoothly reaches −60 degrees over two seconds, and finally returns to zero degrees in another two seconds. A similar motion maneuver has been considered for the second joint, but with different reaching angles.

To conduct various calibration experiments, similar to this motion maneuver but with shorter duration and different motion angles, twelve different experiments have been simulated on the robot. The motion and torque data for each joint, considering actuator dynamics, have been recorded and saved. Since all the information about the system's state variables and torque inputs is available in the simulation, only the $\ddot{q}_f$ needs to be calculated. By using the introduced derivative filter in Equation 10.31 and setting the filter parameter $\tau_f = 0.05$, the required information for this quantity has also been obtained and recorded for each experiment.

By having the motion maneuver data for each joint $q_i, \dot{q}_i, \ddot{q}_{f_i}$ and the torque generating the motion τ_i, we can rewrite the linear model for each joint in the form of regression. In this regard, consider the nonlinear gravitational vector relationship as a function of the joint variables. By simplifying the model intentionally, assume that the gravitational torque of joint one is only a function of the cosine of the joint angle of the first joint $\tau_{g_1} = \mathfrak{g}_1 \cos q_1$ and this quantity in the second joint is a function of the cosine of the sum of the angles of the first and second joints $\tau_{g_2} = \mathfrak{g}_2 \cos(q_1 + q_2)$. With this assumption, the regression model of the robot joints can be obtained with the following matrices and vectors.

$$\tau_1 = \begin{bmatrix} \tau_{1_1} \\ \tau_{1_2} \\ \vdots \\ \tau_{1_N} \end{bmatrix}, \quad \mathcal{Y}_1 = \begin{bmatrix} \ddot{q}_{f_{1_1}} & \dot{q}_{1_1} & \cos q_{1_1} \\ \ddot{q}_{f_{1_2}} & \dot{q}_{1_2} & \cos q_{1_2} \\ \vdots & \vdots & \vdots \\ \ddot{q}_{f_{1_N}} & \dot{q}_{1_N} & \cos q_{1_N} \end{bmatrix}, \quad \Phi_1 = \begin{bmatrix} I_{e_1} \\ b_{e_1} \\ \mathfrak{g}_1 \end{bmatrix}$$

$$\tau_2 = \begin{bmatrix} \tau_{2_1} \\ \tau_{2_2} \\ \vdots \\ \tau_{2_N} \end{bmatrix}, \quad \mathcal{Y}_2 = \begin{bmatrix} \ddot{q}_{f_{2_1}} & \dot{q}_{2_1} & (\cos q_{1_1} + \cos q_{2_1}) \\ \ddot{q}_{f_{2_2}} & \dot{q}_{2_2} & (\cos q_{1_2} + \cos q_{2_2}) \\ \vdots & \vdots & \vdots \\ \ddot{q}_{f_{2_N}} & \dot{q}_{2_N} & (\cos q_{1_N} + \cos q_{2_N}) \end{bmatrix}, \quad \Phi_2 = \begin{bmatrix} I_{e_2} \\ b_{e_2} \\ \mathfrak{g}_2 \end{bmatrix}$$

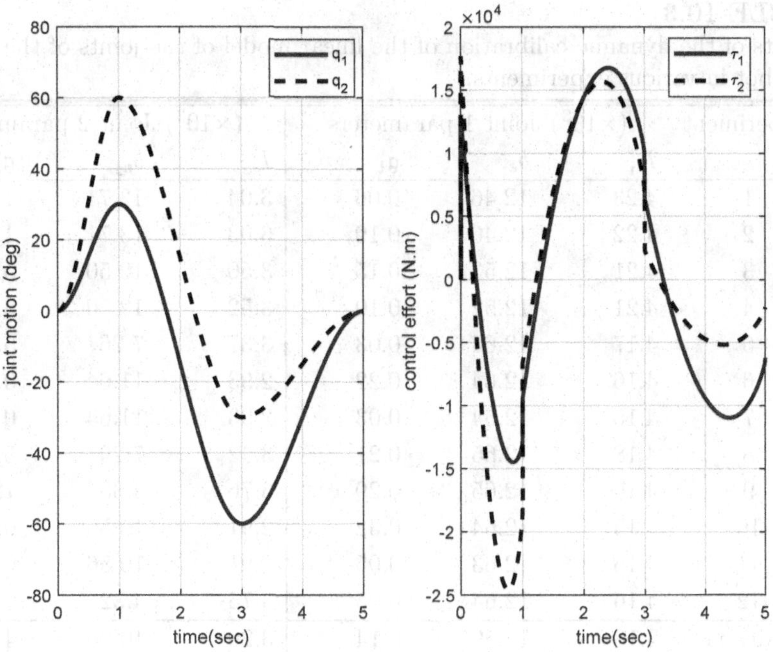

FIGURE 10.20
Simulation of a sample motion maneuver and the applied torques on the joints of the planar 2R robot in a closed-loop structure.

The vector of joint model parameters has been obtained using data from various experiments and the least squares error method, as reported in Table 10.3. In this table, for the sake of simplicity in writing, the joint parameters are presented times 10^3. For example, the moment of inertia in the first row is written as 4.23, while its actual value is 4.23×10^{-3}.

As observed in this table, despite the nonlinear dynamics of the robot, the calibration results show good consistency measure except for a few cases. The effective inertia moment of the first joint is obtained as 4.18×10^{-3} with a negligible standard deviation of 0.03 and a consistency measure of 0.7%. Similar results can be observed for the moment of inertia of the second joint and the linear friction coefficient of the first joint, albeit with slightly less consistency.

This is while the consistency measure for the friction coefficient of the second joint is around 34%, but the consistency measure for the gravitational torque of the robot is in the range of 71% to 101%. These results precisely align with the intentional simplification in the gravitational model in the linear regression of the joints. Thus, it can be concluded that the identified linear model has good consistency, especially in the parameters related to the effective inertia, but the selected simple model for determining the gravitational torque in the robot is not sufficiently reliable.

From a control perspective, this implies that the identified linear model and the estimated parameters can be confidently used for linear controller design. However, if it is necessary to minimize the tracking error close to zero, it would be necessary to use the nonlinear robot model and the gravity vector. The application of gravity compensation by nonlinear controllers will be studied in detail in Chapter 12.

TABLE 10.3
Results of the dynamic calibration of the linear model of the joints of the planar 2R robot in various experiments.

Experiment	($\times 10^3$) Joint 1 parameters			($\times 10^3$) Joint 2 parameters		
	I_{e_1}	b_{e_1}	$\mathfrak{g}_1$	I_{e_2}	b_{e_2}	$\mathfrak{g}_2$
1	4.23	12.46	0.06	3.04	13.71	1.31
2	4.22	12.46	0.19	3.03	13.71	1.34
3	4.21	12.53	0.15	3.50	12.50	5.53
4	4.21	12.52	0.10	3.52	12.50	5.68
5	4.17	12.64	0.03	3.37	7.26	5.76
6	4.16	12.64	0.22	2.93	11.64	0.21
7	4.16	12.64	0.03	2.93	11.64	0.16
8	4.18	12.65	0.21	3.42	7.29	5.88
9	4.16	12.65	0.29	3.76	4.35	12.20
10	4.15	12.64	0.32	2.91	10.56	0.63
11	4.15	12.63	0.05	2.91	10.56	0.70
12	4.16	12.64	0.05	3.75	4.32	12.11
Mean	4.18	12.59	0.14	3.26	10.00	4.29
STD	0.03	0.07	0.10	0.33	3.37	4.36
%CM	0.70	0.59	71.31	10.21	33.70	101.6

Problems

10.1 Consider the two-degrees-of-freedom RP robot shown in Figure 10.21. The dynamic equations of the robot are examined in Example 8.1. The kinematic and dynamic parameters of the robot are given according to the following values, and it is assumed that the robot joints smoothly traverse a trajectory from the initial state $\boldsymbol{q}_0$ passing

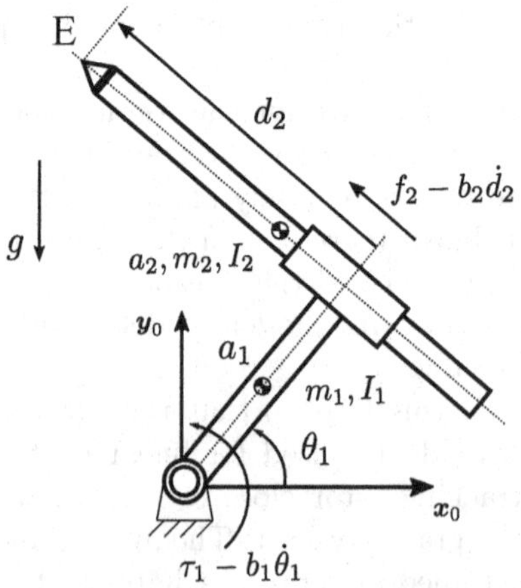

FIGURE 10.21
2-DOF robot of Problem 10.1

through the intermediate state $\boldsymbol{q}_v$ to the final state $\boldsymbol{q}_f$.

$$a_1 = 0.7\,m,\;\; a_2 = 1\,m,\;\; m_1 = 7\,Kg,\;\; m_2 = 10\,Kg,\;\; g = 9.81\,m/s^2,$$
$$b_1 = 0.5\,N.m.s/rad,\;\; b_2 = 0.2\,N.s/m,\;\; t_0 = 0\,s,\;\; t_v = 1\,s,\;\; t_f = 2\,s,$$
$$\boldsymbol{q}_0 = [0,\ 0]^T,\;\; \boldsymbol{q}_v = [90(deg),\ 1\,(m)]^T,\;\; \boldsymbol{q}_f = [60(deg),\ \ 0.7\,(m)]^T.$$

Initially, obtain an appropriate trajectory for the joint motion using a cubic polynomial. Then, through the inverse dynamics simulation of the robot, determine the required torque (or force) in the joint actuators. Calculate the power required in the robot's actuators and, by plotting the speed/torque (or force) characteristic of the actuators, identify the maximum speed and torque (or force) of the robot's actuators.

10.2 Repeat Problem 10.1 using a quintic polynomial trajectory design, and compare and analyze the obtained maximum power and torque values.

10.3 Revisit Problem 10.1 by employing a linear trajectory with parabolic blends. Subsequently, compare and analyze the maximum power and torque values acquired.

10.4 Consider the PRP robot with three degrees of freedom, illustrated in Figure 10.22. The dynamic analysis of the robot is explored in Problem 8.2. The robot's kinematic and dynamic parameters are provided as follows, and it is presumed that the robot joints smoothly follow a trajectory from the initial state $\boldsymbol{q}_0$ through the intermediate state $\boldsymbol{q}_v$ to the final state $\boldsymbol{q}_f$.

$$a_1 = 1\,m,\; a_2 = 0.7\,m,\; a_3 = 0.5\,m,\; m_1 = 10\,Kg,\; m_2 = 7\,Kg,\; m_3 = 5\,Kg,$$
$$g = 9.81\,m/s^2,\;\; b_1 = b_3 = 0.5\,N.m.s/rad,\;\; b_2 = 0.2\,N.s/m,$$
$$t_0 = 0\,s,\;\; t_v = 1\,s,\;\; t_f = 2\,s,\;\; \boldsymbol{q}_0 = [0,\ 0,\ 0]^T,$$
$$\boldsymbol{q}_v = [1\,(m)\ 90(deg),\ 0.5\,(m)]^T,\;\; \boldsymbol{q}_f = [0.7\,(m)\ 60(deg),\ 0.25\,(m)]^T.$$

Firstly, derive a suitable trajectory for the joint motion using a cubic polynomial.

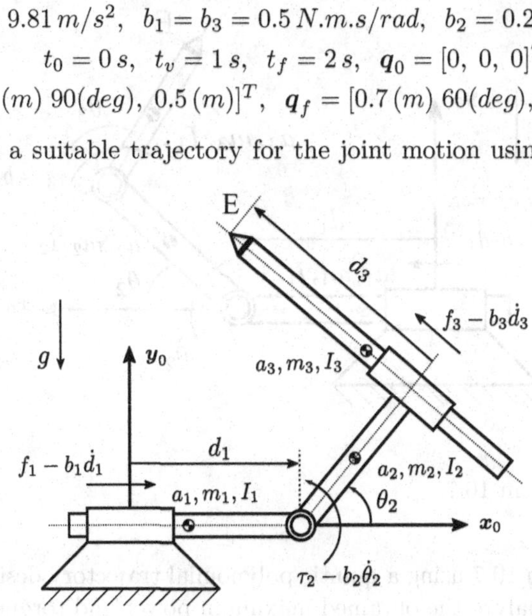

FIGURE 10.22
3-DOF robot of Problem 10.4

Subsequently, employ inverse dynamics simulation for the robot to ascertain the necessary torque (or force) in the joint actuators. Compute the power demanded by the robot's actuators and, through plotting the speed/torque (or force) profile of the actuators, determine the maximum speed and torque (or force) exerted by the robot's actuators.

10.5 Revisit Problem 10.4 employing a quintic polynomial trajectory design, then compare and analyze the resulting maximum power and torque values.

10.6 Repeat Exercise 10.4 using a linear trajectory with parabolic blends, then compare and analyze the obtained maximum power and torque values.

10.7 Examine the PRR robot, which features three degrees of freedom, depicted in Figure 10.23. Problem 8.3 elaborates on the dynamic analysis of the robot. The kinematic and dynamic parameters of the robot are given below, assuming that the robot joints seamlessly trace a trajectory from the initial state $\boldsymbol{q}_0$ through the intermediate state $\boldsymbol{q}_v$ to the final state $\boldsymbol{q}_f$.

$$a_1 = 1\,m,\ a_2 = 0.7\,m,\ a_3 = 0.5\,m,\ m_1 = 10\,Kg,\ m_2 = 7\,Kg,\ m_3 = 5\,Kg,$$
$$g = 9.81\,m/s^2,\ b_1 = 0.5\,N.m.s/rad,\ b_2 = b_3 = 0.2\,N.s/m,$$
$$t_0 = 0\,s,\ t_v = 1\,s,\ t_f = 2\,s,\ \boldsymbol{q}_0 = [0,\ 0,\ 0]^T,$$
$$\boldsymbol{q}_v = [1\,(m)\ 90(deg),\ -30\,(deg]^T,\ \boldsymbol{q}_f = [0.7\,(m)\ 60(deg),\ 30\,(deg)]^T.$$

Initially, determine an appropriate trajectory for the joint motion using a cubic polynomial. Then, utilize inverse dynamics simulation for the robot to determine the required torque (or force) in the joint actuators. Calculate the power required by the robot's actuators and, by plotting the speed/torque (or force) characteristics of the actuators, identify the maximum speed and torque (or force) exerted by the robot's actuators.

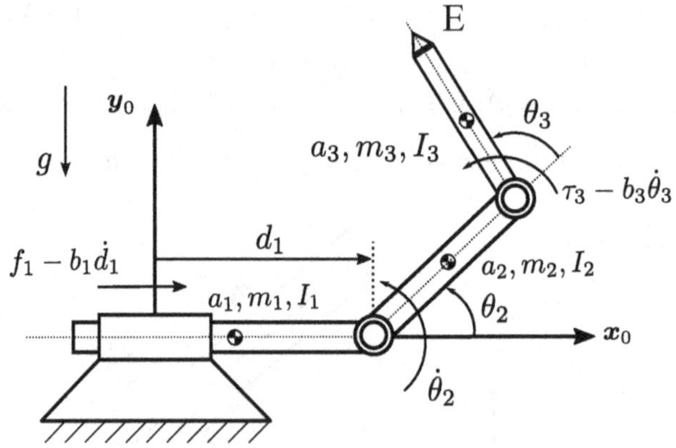

FIGURE 10.23
3-DOF robot of Problem 10.7

10.8 Repeat Problem 10.7 using a quintic polynomial trajectory design, and subsequently, compare and analyze the obtained maximum power and torque values.

10.9 Revisit Problem 10.7 employing a linear trajectory with parabolic blends, and subsequently, compare and analyze the resulting maximum power and torque values.

10.10 Take a look at the RPR robot, characterized by three degrees of freedom, as illustrated in Figure 10.24. Problem 8.4 explores the dynamic analysis of the robot. The kinematic and dynamic parameters of the robot are provided below, assuming that the robot joints smoothly follow a trajectory from the initial state $\boldsymbol{q}_0$ through the intermediate state $\boldsymbol{q}_v$ to the final state $\boldsymbol{q}_f$.

$$a_1 = 0.7\,m,\ a_2 = 1\,m,\ a_3 = 0.3\,m,\ m_1 = 7\,Kg,\ m_2 = 10\,Kg,\ m_3 = 3\,Kg,$$
$$g = 9.81\,m/s^2,\ b_1 = b_3 = 0.5\,N.m.s/rad,\ b_2 = 0.2\,N.s/m,$$
$$t_0 = 0\,s,\ t_v = 1\,s,\ t_f = 2\,s,\ \boldsymbol{q}_0 = [0,\ 0,\ 0]^T,$$
$$\boldsymbol{q}_v = [90\,(deg),\ 1(m),\ -30\,(deg)]^T,\ \boldsymbol{q}_f = [60\,(deg),\ 0.7(m),\ 30\,(deg)]^T.$$

Establish a suitable trajectory for the joint motion by employing a cubic polynomial. Subsequently, employ inverse dynamics simulation for the robot to ascertain the necessary torque (or force) in the joint actuators. Calculate the power demanded by the robot's actuators and, through plotting the speed/torque (or force) characteristics of the actuators, determine the maximum speed and torque (or force) exerted by the robot's actuators.

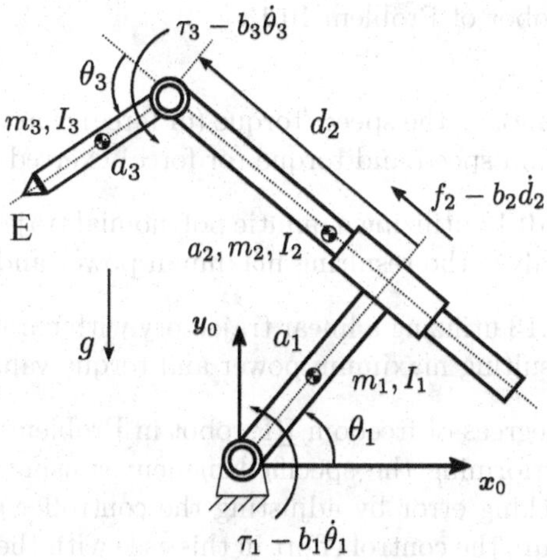

FIGURE 10.24
3-DOF robot of Problem 10.10

10.11 Revisit Problem 10.10 with a quintic polynomial trajectory design, and then compare and analyze the resulting maximum power and torque values.

10.12 Repeat Problem 10.10 utilizing a linear trajectory with parabolic blends, and afterward, compare and analyze the resulting maximum power and torque values.

10.13 Examine the <u>R</u>R robot, featuring two degrees of freedom, depicted in Figure 10.25. Problem 8.5 explores the dynamic analysis of the robot. The kinematic and dynamic parameters of the robot are provided below, assuming that the robot joints smoothly traverse a trajectory from the initial state $\boldsymbol{q}_0$ through the intermediate state $\boldsymbol{q}_v$ to the final state $\boldsymbol{q}_f$.

$$a_1 = 1\,m,\ a_2 = 0.7\,m,\ m_1 = 10\,Kg,\ m_2 = 7\,Kg,\ g = 9.81\,m/s^2,$$
$$I_{zz} = 0.02\,Kgm^2,\ b_1 = b_2 = 0.5\,N.m.s/rad,\ t_0 = 0\,s,\ t_v = 1\,s,\ t_f = 2\,s,$$
$$\boldsymbol{q}_0 = [0,\ 0]^T,\ \boldsymbol{q}_v = [90(deg),\ 30\,(deg)]^T,\ \boldsymbol{q}_f = [60(deg),\ -30\,(deg)]^T.$$

Determine an appropriate trajectory for the joint motion using a cubic polynomial. Then, utilize inverse dynamics simulation for the robot to determine the required torque (or force) in the joint actuators. Calculate the power required by the robot's

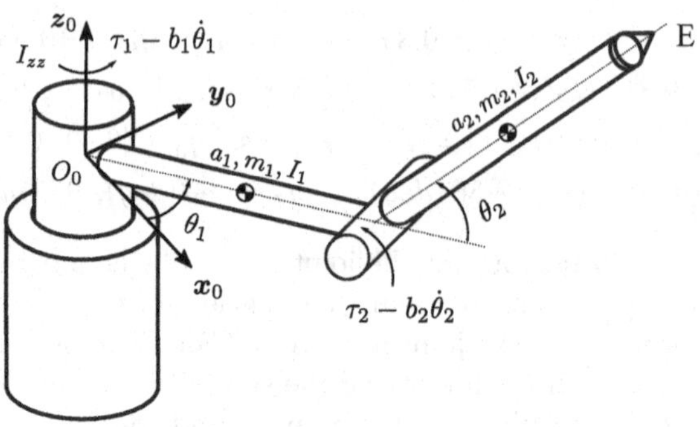

FIGURE 10.25
Two degrees-of-freedom robot of Problem 10.13

actuators, and by plotting the speed/torque (or force) characteristics of the actuators, identify the maximum speed and torque (or force) exerted by the robot's actuators.

10.14 Reiterate Problem 10.13 utilizing a quintic polynomial trajectory design, then proceed to compare and analyze the resulting maximum power and torque values.

10.15 Revisit Problem 10.13 utilizing a linear trajectory with parabolic blends, and compare and analyze the resulting maximum power and torque values.

10.16 Consider the two-degrees-of-freedom RP robot in Problem 10.1. Simulate the closed-loop system for performing the specified maneuver using a PD controller. Achieve an appropriate tracking error by adjusting the controller gain matrices and plot it. Additionally, compare the control effort in this case with the inverse dynamics torques (or forces) and analyze the results.

10.17 Repeat Problem 10.16 on the three-degrees-of-freedom PRP robot shown in Figure 10.22.

10.18 Revisit Problem 10.16 for the three-degrees-of-freedom PRP robot depicted in Figure 10.23.

10.19 Repeat Problem 10.16 for the three-degrees-of-freedom RPR robot illustrated in Figure 10.24.

10.20 Reiterate Problem 10.16 for the two-degrees-of-freedom RR robot depicted in Figure 10.25.

10.21 Consider the two-degrees-of-freedom RP robot in Problem 10.1. Assume that gearboxes with a ratio of 200 are used in the robot joints. Determine the robot's dynamics in the presence of the actuator and simulate its inverse dynamics. Compare and analyze the results with the inverse dynamics simulation of the robot without the actuator.

Furthermore, Simulate the closed-loop system of the robot with the actuator using a PD controller to perform the specified maneuver. Achieve an appropriate tracking error by adjusting the controller gain matrices and plot it. Additionally, compare the control effort in this case with the inverse dynamics torques (or forces) and analyze the results.

10.22 Repeat Problem 10.21 on the three-degrees-of-freedom PRP robot shown in Figure 10.22.

10.23 Revisit Problem 10.21 for the PRR robot with three degrees of freedom, as illustrated in Figure 10.23.

10.24 Reiterate Problem 10.21 for the threes-degree-of-freedom RPR robot depicted in Figure 10.24.

10.25 Repeat Problem 10.21 for the two-degrees-of-freedom RR robot shown in Figure 10.25.

10.26 Consider the two-degrees-of-freedom RP robot in Problem 10.1. In the presence of the actuator dynamics, perform corresponding motion maneuvers on the system and record input-output data in various experiments. Derive the joint model in the form of linear regression in the presence of appropriate gravity force and identify the model parameters. Set the inertia values and viscous friction coefficient of the actuator in the model to one-tenth of those considered in the robot joint. Determine the model accuracy and parameter compatibility by calculating consistency measure for different experiments. Iteratively refine the model and achieve satisfactory consistency measure in the calibrated parameters, and report the results of this process.

10.27 Repeat Problem 10.26 on the three-degrees-of-freedom PRP robot shown in Figure 10.22.

10.28 Revisit Problem 10.26 for the three-degrees-of-freedom PRR robot depicted in Figure 10.23.

10.29 Reiterate Problem 10.26 for the three-degrees-of-freedom RPR robot illustrated in Figure 10.24.

10.30 Reiterate Problem 10.26 for the two-degrees-of-freedom RR robot illustrated in Figure 10.25.

Part IV

Control

11

Linear Motion Control

This chapter explores the context of linear motion control for robotics applications, beginning with an introduction that sets the stage for a comprehensive understanding of the subject. The chapter elaborates on various robot motion control topologies, including motion control in joint space, motion control in task space, and the integration of task space control with joint variable measurement. Each topology is carefully examined, providing insights into their applications and implementation intricacies.

A significant portion of the chapter focuses on individual joint control in robots, highlighting the magic of feedback and its transformative impact. It will be demonstrated that leveraging feedback loops, regardless of the controller type, can facilitate the attainment of multiple objectives within a control system. A key reason behind the success of control systems in practical applications is the ability to achieve desirable tracking performance in the presence of disturbances and uncertainties in the system model through online observation of system states via feedback. In the context of linear controller design, this chapter covers PD control, PID control, and provides an insightful discussion on performance limitations, taking into account factors such as sensor noise and actuator saturation. The inclusion of feed-forward control and multivariable PID control further enriches the discussion, presenting a comprehensive perspective on control strategies for individual joints.

The practical application of the concepts is illustrated through the motion simulation of a planar 3R robot as a case study. The chapter investigates the performance of PID and PD control in this context, shedding light on their effectiveness. Furthermore, the chapter explores the implications of measurement noise and actuator saturation on motion simulation, providing valuable insights into the real-world challenges and considerations in linear motion control. The chapter closes with a collection of problems, providing readers with an opportunity to put their knowledge into practice and enhance their comprehension of the discussed concepts.

11.1 Introduction

A seemingly lifeless robot body springs to life when control systems generate desired motions into it. Serving as the software brain of the robot, the robot controller assesses the position of its links and end-effector using sensor information and issues control commands to the robot's actuators to attain the intended motion trajectory. Consequently, the robot seems to exhibit autonomous intent in its actions, even though it lacks true autonomy, as the robot control system has adeptly orchestrated the desired motion tracking.

In the present era, as robotic applications continue to expand, robots find diverse applications in various aspects of daily life. The broad range of these applications has spurred the development of different control structures for robots. Broadly classified, robotic applications fall into two main categories. In the first category, robots exhibit minimal interaction

DOI: 10.1201/9781003491415-11

with their surrounding environment and function as autonomous agents for the movement of components and equipment. This category includes many industrial robots employed in diverse industries such as packaging, assembly, and even space exploration. In these applications, the primary emphasis is the motion control of the robot's links and end-effector to accurately follow desired trajectories. While this fundamental control structure is prevalent in most robots, additional extensions are incorporated into the robot control system in certain applications.

The second category pertains to applications where the robot engages with its environment and involves force interactions. If for any reason, it becomes essential for the robot's end-effector to apply a desired wrench to the environment, relying solely on motion control is inadequate. The robot's control system must incorporate various force control methods. To illustrate the contrast, let's consider an industrial robot utilized in two closely related applications. In the first scenario, the robot traverses the entire surface of a blackboard, successfully reaching any designated point using the motion control system. If in the same application, the goal is to erase or write on the blackboard, simple motion control is no longer effective, and applying a desired force, particularly in the direction perpendicular to the blackboard, becomes necessary. Now, envision the robot interacting with a human; in applications like rehabilitation or exoskeleton robots, force interaction between the robot and the human needs to be finely tuned to ensure human safety. In such cases, the design of the required controllers becomes more intricate.

This chapter elaborates on the examination and design of fundamental linear "motion control" methods in serial robots, which serve as essential control techniques in the context of robotics. In the subsequent chapters, multivariable and nonlinear control schemes, along with force control in robots, will be examined in detail. Further insights into advanced motion and force control techniques in robots can be obtained from additional references in this field [1, 125].

11.2 Motion Control Topologies

To study robot motion control, it is essential to first introduce the arrangement of control structures in robots, known as motion control topologies, before elaborating on design methodologies. A common feature among all contemporary robot control topologies is the integration of feedback loops within their structure. The magic of feedback in precisely tracking the desired motion trajectory, while also mitigating disturbances impacting the robot in the presence of uncertain models, renders its distinctive features undeniable. Consequently, the selection of motion variables measured in the feedback loop becomes imperative, and the control design space across diverse control topologies should be elucidated.

11.2.1 Motion Control in Joint Space

The predominant motion control topology in serial robots is depicted in Figure 11.1. In this configuration, the control of the robot's motion occurs within the joint space. To manage the robot's motion and follow the desired trajectory $\boldsymbol{q}_d$, all joint motions encompassed in the vector $\boldsymbol{q}$ must be continuously measured and integrated into the feedback loop. Upon receiving the tracking error vector $\boldsymbol{e}_q = \boldsymbol{q}_d - \boldsymbol{q}$ and leveraging its understanding of the robot's dynamic structure, the robot controller computes the necessary torques in the joints $\boldsymbol{\tau}$ and applies them to the robot through the actuators.

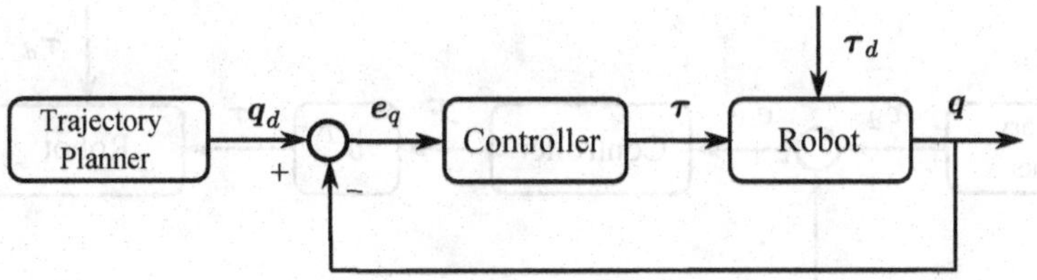

FIGURE 11.1
Motion control topology of a robot in joint space.

This topology offers distinctive features that make it well-suited for deployment in various industrial robots. As illustrated in Figure 11.1, the measurement of joint variables and the corresponding feedback form the backbone of the motion control system in this topology, seamlessly fitting with the use of traditional measurement systems in industrial robots. Typically, these robots employ electric motors for joint actuation, and furthermore, utilize optical encoders for accurate measurement of joint motion. Additionally, in this topology, the controller's output generates actuator torques, easily applied to them through motor drive circuits. Furthermore, if it becomes necessary to incorporate the dynamic model of the robot and its actuators in controller design, the model obtained in the joint space will naturally be employed in the robot modeling and calibration process.

The only contemplative point in this topology is designing the desired trajectory in the joint space, which may not align with the direct needs of users in typical robotics applications. This is because, in many industrial applications, tracking the desired motion of the robot's end-effector in the task space is required. For example, consider a robot used in the automotive body production line for welding specific sections of the body. It is natural that the path of the welding tool attached to the robot's end-effector is defined in the robot's task space.

In practice, there are two distinct approaches to address this challenge. In the first approach, the use of a teaching pendant allows the robot's end-effector path in the task space to be traversed only once manually. Throughout this path, both the movement points of the robot's end-effector and the motion information of the robot's joints are meticulously recorded. Subsequently, if the topology depicted in Figure 11.1 is employed for robot control, designing the trajectory based on the initially and finally recorded points, while considering an ample number of intermediate points in the joint space, is sufficient. This designed trajectory is then incorporated into the control structure. With a satisfactory number of intermediate points selected, the robot's end-effector can traverse the desired path without concerns about colliding with surrounding obstacles. The second approach shifts the focus to motion control in the task space, a concept that will be explored in the subsequent section.

11.2.2 Motion Control in Task Space

If, for any reason, the user does not approve the desired trajectory in the robot's joint space, it becomes necessary to design the trajectory in the task space. In this scenario, the application of motion control topology in the joint space is not applicable, and instead, the motion control topology in the task space, illustrated in Figure 11.2, should be employed. In this topology, the control of robot motion is executed in the task space.

Initially, the desired motion of the robot's end-effector, represented by the vector $\mathcal{X}_d$ in this figure, is determined by planning the trajectory in the task space. Furthermore, the continuous measurement of the motion information of the robot's end-effector $\mathcal{X}$, is

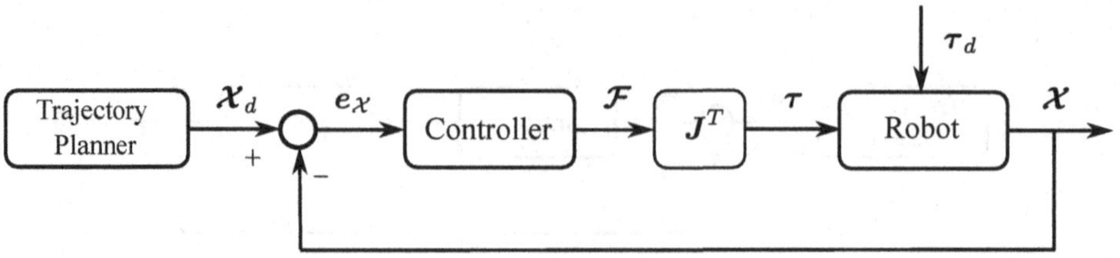

FIGURE 11.2
Motion control topology of a robot in the task space.

utilized within the feedback loop. Consequently, the robot controller receives the tracking error vector in the task space, denoted as $e_\mathcal{X} = \mathcal{X}_d - \mathcal{X}$. Based on its knowledge of the robot's dynamic structure, the controller computes the necessary wrench in the task space $\mathcal{F}$. Subsequently, through the use of the transpose Jacobian matrix J^T, the required torques τ are applied to the robot via the actuators to generate such motion.

In this topology, the measurement of motion variables and the controller design are conducted in the task space. As seen in Figure 11.2, the measurement of task variables and their feedback forms the essential foundation of the motion control system in this topology. Consequently, this approach unlike the previous topology, requires more complex technologies for position measurement and orientation of the robot's end-effector, which is economically much more expensive than measuring joint variables. Furthermore, in this topology, the controller is designed in the task space. If the design of the controller necessitates the dynamic model of the robot and its actuators, the mapping of the dynamic model in the robot's task space must be utilized, which is significantly more complex than modeling and calibrating the robot in the joint space.

Despite the challenges in designing and implementing this topology, if tracking the robot's motion trajectory in the task space is the primary concern, it is natural that direct measurement and the design of a controller that directly measures and minimizes the tracking error in this space are more advantageous. The use of the transpose Jacobian matrix in this structure is a crucial approach in the controller structure in the task space, replacing the initial ideas of using the inverse Jacobian matrix in a similar structure with this topology [126]. In the robotics literature, among the family of controllers designed in the task space, this type of controller is referred to as an "operational controller" or a "Jacobian transpose controller" [127].

11.2.3 Motion Control in Task Space with Joint Variable Measurement

The necessary technology and the considerably high cost associated with the accurate measurement of the robot's end-effector position and orientation in real-time, challenge the proposed control topology in the task space. This challenge can be mitigated by leveraging the robot's forward kinematics. As depicted in Figure 11.3, rather than directly measuring the task space variables of the robot $\mathcal{X}$, it is possible to obtain the motion variables in the task space by measuring the joint space variables and utilizing the robot's forward kinematics. In this topology, the expense of measuring the robot's position in the task space is significantly reduced by utilizing the motion data from the joint actuators and relying on precise knowledge of the robot's kinematics and its calibrated parameters. Through this approach, accurate tracking of the desired motion of the robot in the task space can be achieved.

While this structure eliminates the need for direct and costly measurement of task space variables in the robot, its reliance on precise knowledge of the robot's kinematics

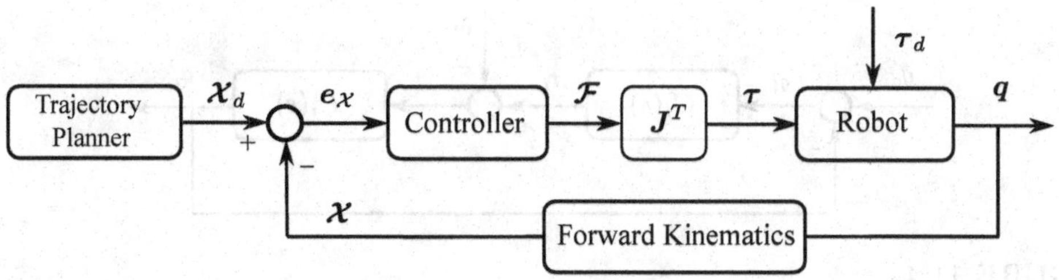

FIGURE 11.3
Robot motion control topology in the task space while using joint variable measurement.

and the necessity for kinematic calibration introduces new challenges in the design and implementation of motion controllers for the robot. Therefore, in practical applications, the first topology and the design and implementation of robot controllers in the joint space are typically preferred. The topology introduced in this subsection is only employed when precision tracking the desired motion trajectory in the task space are crucial requirement. With this description, the subsequent sections will discuss methods for designing robot controllers in the joint space, and those interested in analysis and design techniques in the task space controllers can refer to references [4, Section 9.3.2] and [128].

11.3 Individual Joint Control

As detailed in Section 9.6.3, industrial robots often use electric motors equipped with gearboxes. This results in a significant attenuation of the multivariable and nonlinear dynamics of the robot, making the linear dynamics of the robot's actuators predominant. Consequently, in more than 99% of industrial robots, simple linear controllers such as PID are employed with remarkable precision and reliability. In this chapter, by utilizing the dominant model of actuator dynamics and employing the motion control topology in the joint space, we elaborate on the analysis and design of various linear controllers for serial robots.

To achieve this, suppose that the joints of the target serial robot are actuated by electric motors along with a gearbox with a very large transmission ratio. As shown in Figure 9.7, the block diagram of each robot joint can be separately modeled as a linear system along with the disturbance torque due to gravity. In this representation, each robot joint is modeled separately with the following second-order transfer function:

$$\boldsymbol{P}_i(s) = \frac{q_i(s)}{\tau_i(s)} = \frac{1}{s(I_{e_i} s + b_{e_i})} \tag{11.1}$$

where I_{e_i} is the effective inertia of the robot and actuator at joint i, and b_{e_i} is the effective viscous friction coefficient of the robot and actuator at this joint. Additionally, the gravity torque $\mathfrak{g}_i(q_i)$ is applied as a disturbance input to the system.

If the joint space control topology is applied to the system by using a linear controller $\boldsymbol{C}_i(s)$ for each joint of the robot individually, the closed-loop block diagram for each joint of the robot can be depicted as shown in Figure 11.4. In this diagram, the transfer function of each robot joint, $\boldsymbol{P}_i(s)$, is modeled using the relationship given in Eq. 11.1 through experimental calibration. For the robot controller $\boldsymbol{C}_i(s)$, common industrial structures such as PD, PID, or state feedback can be employed, or initiatives can be taken to completely eliminate disturbance in the input using feed-forward control law or weight compensation.

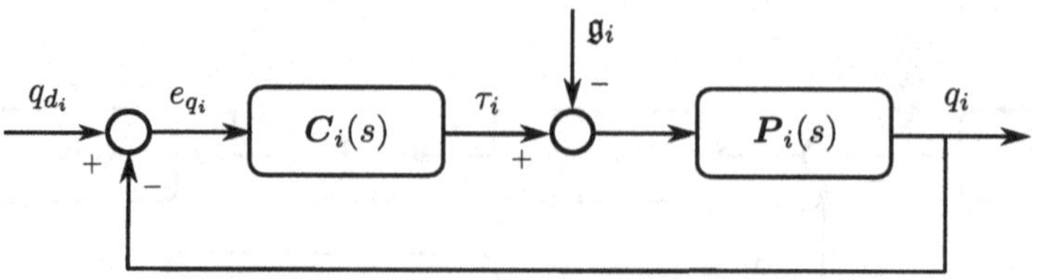

FIGURE 11.4
Block diagram of individual control of each joint of the robot using its linear model in the presence of actuator dynamics with gearbox.

The main goal within a robot's motion control system is the accurate tracking of the intended motion trajectory. However, accomplishing this objective inherently requires achieving additional crucial goals. Internal stability, especially when feedback loops are present, stands as an essential condition that must be taken into account when analyzing the dynamic behavior of the closed-loop system. Instability represents a potential threat to both the system's well-being and the users engaging with it.

On the other hand, the presence of disturbances in the system, as observed in the diagram in Figure 11.4, undoubtedly has an impact on the tracking accuracy of the closed-loop system. Eliminating or reducing the effect of disturbances on the system's performance, which is manifested in the robot's accuracy in tracking the desired trajectory, is another crucial goal that should never be overlooked. Another goal that should be addressed in the successful implementation of control systems is the insensitivity of the closed-loop system's performance to the model used in controller design. As detailed in the previous chapter, the dynamic behavior of the robot can only be modeled accurately with complex nonlinear and multivariable models. However, in the presence of actuator dynamics in serial robots, we have intentionally replaced such a complex model with a number of simple and single input single output linear transfer functions. It is evident that if such a model is used in the controller design process which is implemented on the real system, the closed-loop system's behavior will not necessarily be consistent with what is predicted in the model.

11.3.1 Magic of Feedback

From the very beginning, control system designers have faced a significant challenge in striving to achieve multiple objectives within linear control systems. Remarkably, it is noteworthy to acknowledge that the use of feedback loops, irrespective of the type of controller, can enable the achievement of the aforementioned multiple objectives in a control system. It could be argued that this is perhaps the most pivotal reason behind the success of control systems in practical applications. Through online observation of the system states, control systems can effectively achieve desirable tracking, mitigating the impact of disturbances and uncertainties in the system model.

Given the importance of this issue in the structure of linear control systems, let us explore and verify this property of feedback, which we refer to as the "magic of feedback" in this book, through simple mathematical analysis. For the sake of simplicity in writing equations and making them more general, we deliberately refrain from writing the index i. let us mathematically examine the dynamic behavior of the tracking error of the closed-loop system, denoted as e_q, considering the desired trajectory input q_d, disturbance input $\mathfrak{g}$, and uncertainty in the system modeling.

To achieve this, we first derive the transfer function of the tracking error of the closed-loop system, e_q, with respect to the desired input q_d. Easily and by simplifying block diagrams in linear systems, it can be demonstrated that this transfer function is equal to:

$$\boldsymbol{S}(s) = \frac{e_q(s)}{q_d(s)} = \frac{1}{1 + \boldsymbol{C}(s)\boldsymbol{P}(s)} \tag{11.2}$$

In control literature, $\boldsymbol{S}(s)$ is called the sensitivity transfer function. It is evident that achieving a small tracking error $e_q \approx 0$, occurs only when the value of this transfer function is close to zero within the desired bandwidth of the system. Regardless of the control structure used, if the control is high-gain, the denominator of the transfer function $\boldsymbol{S}(s)$ becomes significantly large, resulting in a very small value for $\boldsymbol{S}(s)$ within this bandwidth.

Now, to examine the effect of disturbance on the tracking error, based on the tracking error transfer function of the closed-loop system e_q with respect to the input $\mathfrak{g}$, we will have:

$$\boldsymbol{D}(s) = \frac{e_q(s)}{\mathfrak{g}(s)} = \frac{\boldsymbol{P}(s)}{1 + \boldsymbol{C}(s)\boldsymbol{P}(s)} = \boldsymbol{P}(s)\,\boldsymbol{S}(s) \tag{11.3}$$

This transfer function is denoted by $\boldsymbol{D}(s)$, representing the transfer function of disturbance effect on the tracking performance. Due to the presence of a feedback loop in the control structure, this transfer function is proportional to the sensitivity transfer function. It is evident that achieving a small tracking error $e_q \approx 0$, in the presence of disturbances still occurs when the controller is high-gain within the required bandwidth. In this case, as the denominator of the disturbance transfer function grows, the magnitude of this transfer function becomes very small.

The simultaneous realization of two crucial and distinct goals in control systems, facilitated only by the existence of a feedback loop, is an extraordinary accomplishment in the control system. Although it experiences a change in intensity compared to the sensitivity transfer function alone, owing to the presence of the plant transfer function $\boldsymbol{P}(s)$, achieving effective tracking of the desired path in the presence of input disturbances is still feasible through the use of a high-gain controller.

If achieving the third objective is also feasible using a high-gain controller, the magic of feedback will be even more pronounced. This objective can be examined by mathematically analyzing the impact of model uncertainties on the system output. For this purpose, we use the analysis of the sensitivity of the input-output transfer function with respect to the model variations. The input-output transfer function of the system, denoted as $\boldsymbol{T}(s)$, can be expressed as follows:

$$\boldsymbol{T}(s) = \frac{q(s)}{q_d(s)} = \frac{\boldsymbol{C}(s)\boldsymbol{P}(s)}{1 + \boldsymbol{C}(s)\boldsymbol{P}(s)} \tag{11.4}$$

Calculate the normalized variations of this transfer function with respect to the changes in the system model $\boldsymbol{P}(s)$. This is calculated by partially differentiating this transfer function with respect to $\boldsymbol{P}(s)$, and then normalizing it. The expression is as follows:

$$\text{Normalized Sensitivity Function} = \boldsymbol{S}_T^P = \frac{\partial \boldsymbol{T}(s)/\boldsymbol{T}(s)}{\partial \boldsymbol{P}(s)/\boldsymbol{P}(s)} \tag{11.5}$$

Note that in this expression, by dividing the partial derivative of the numerator and denominator with respect to their respective arguments, the normalized sensitivity function is obtained. With some calculation and simplification, this transfer function can be written as follows.

$$
\begin{aligned}
\boldsymbol{S}_T^P &= \frac{\partial \boldsymbol{T}(s)}{\partial \boldsymbol{P}(s)} \cdot \frac{\boldsymbol{P}(s)}{\boldsymbol{T}(s)} \\
&= \frac{\boldsymbol{C}(s)(1+\boldsymbol{P}(s)\boldsymbol{C}(s)) - \boldsymbol{C}(s)\boldsymbol{P}(s)\boldsymbol{C}(s)}{(1+\boldsymbol{P}(s)\boldsymbol{C}(s))^2} \cdot \frac{\boldsymbol{P}(s)(1+\boldsymbol{P}(s)\boldsymbol{C}(s))}{\boldsymbol{P}(s)\boldsymbol{C}(s)} \\
&= \frac{1+\boldsymbol{P}(s)\boldsymbol{C}(s)}{(1+\boldsymbol{P}(s)\boldsymbol{C}(s))^2} \\
&= \frac{1}{1+\boldsymbol{P}(s)\boldsymbol{C}(s)} \\
&= \boldsymbol{S}(s)
\end{aligned}
$$

It is astonishing to observe that the normalized sensitivity function $\boldsymbol{S}_T^P$ is exactly equivalent to the sensitivity transfer function of the system $\boldsymbol{S}(s)$ defined in Equation 11.2 and introduced in the tracking error analysis. This magic is only witnessed in closed-loop control systems. Thus, by utilizing a high-gain controller, all the desired objectives, including tracking the desired trajectory in the presence of disturbances and uncertainty in the modeling, can be achieved. The magic of feedback has enabled achieving satisfactory performance even with simple controllers. Among the various linear controllers that can be employed in this structure, a few examples will be studied in the upcoming sections.

11.3.2 PD Control

The PD controller is the simplest and most effective linear controller that aims to ensure the tracking of the desired input trajectory. This controller applies the weighted sum on the input error and its derivative as the control effort required to actuate the robot's motion.

$$\tau_i(t) = k_{p_i} e_{q_i}(t) + k_{d_i} \dot{e}_{q_i}(t) \tag{11.6}$$

In this relation, k_{p_i} is the proportional gain of the controller, and k_{d_i} is its derivative gain. This controller directly utilizes the product of the tracking error and the proportional gain to generate the necessary control effort, aiming for a rapid reduction in the tracking error. The use of the derivative of the tracking error serves as a form of prediction of the future state of the tracking error, contributing to the establishment of appropriate transient behavior. This controller, in Laplace space, is expressed as follows with two control gains, k_{p_i}, k_{d_i}:

$$\boldsymbol{C}_i(s) = k_{p_i} + k_{d_i} s \tag{11.7}$$

Thus, using the PD controller, the block diagram of the closed-loop system for each joint in the presence of external disturbance is depicted in Figure 11.5. With this block diagram, the transfer function of the motion trajectory for each joint, $q_i(s)$, with respect to the desired trajectory input $q_{d_i}(s)$ and the disturbance input $\mathfrak{g}_i(s)$, can be obtained.

$$
\begin{aligned}
q_i(s) &= \frac{k_{p_i} + k_{d_i} s}{\Delta_i(s)} q_{d_i}(s) - \frac{1}{\Delta_i(s)} \mathfrak{g}_i(s) \\
\Delta_i(s) &= I_{e_i} s^2 + (b_{e_i} + k_{d_i}) s + k_{p_i}
\end{aligned} \tag{11.8}
$$

The tracking error can also be expressed as follows:

$$
\begin{aligned}
e_{q_i}(s) &= q_{d_i}(s) - q_i(s) \\
&= \frac{I_{e_i} s^2 + b_{e_i} s}{\Delta_i(s)} q_{d_i}(s) + \frac{1}{\Delta_i(s)} \mathfrak{g}_i(s)
\end{aligned} \tag{11.9}
$$

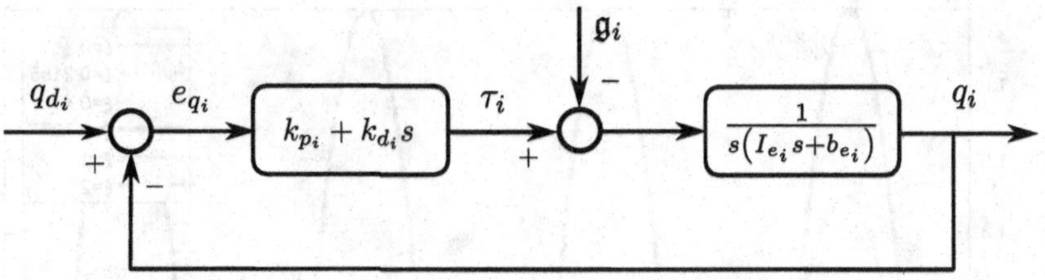

FIGURE 11.5
Block diagram of individual control for each joint of the robot using a linear model and PD controller.

As evident in these two equations, the dynamics of the tracking output and tracking error are both second-order systems whose characteristics are determined by the characteristic equation of the closed-loop system $\Delta_i(s)$. The gains of the PD controller directly appear in this characteristic equation and consequently affect the system's output behavior. To ensure system stability, it is sufficient to choose both controller gains to be positive. Furthermore, if the controller gains are chosen to be sufficiently large, the roots of the characteristic equation will be located farther away from the imaginary axis in the complex plane, resulting in a faster transient response. Now, if the controller gains are chosen in a way that the characteristic equation has complex conjugate roots, the system's output and, consequently, the tracking error of the system will exhibit oscillatory behavior. In this way, by examining the behavior of second-order systems, the dynamic behavior of the tracking error in the system can be adjusted as desired.

For a standard analysis, rewrite the characteristic equation of the closed-loop system in the following standard form.

$$\Delta_i(s) = I_{e_i}(s^2 + 2\xi_i\omega_{n_i}s + \omega_{n_i}^2) \tag{11.10}$$

where ω_{n_i} is the natural frequency of the closed-loop system in joint i, and ξ_i is the damping ratio. These primary parameters, which shape the dynamics of the second-order system, are adjustable using the control gains. The relationship between the control gains and these primary parameters is obtained as follows.

$$\omega_{n_i}^2 = \frac{k_{p_i}}{I_{e_i}} \quad \longrightarrow \quad \omega_{n_i} = \sqrt{k_{p_i}/I_{e_i}} \tag{11.11}$$

$$2\xi_i\omega_{n_i} = \frac{b_{e_i} + k_{d_i}}{I_{e_i}} \quad \longrightarrow \quad \xi_i = \frac{1}{2}\frac{b_{e_i} + k_{d_i}}{\sqrt{k_{p_i}I_{e_i}}} \tag{11.12}$$

In these equations, it is observed that by increasing the proportional gain k_{p_i}, the natural frequency of the closed-loop dynamic system increases. The natural frequency is proportional to the system's response speed, allowing the system to reach its steady-state more quickly. On the other hand, the derivative gain in the controller k_{d_i} is proportional to the damping ratio of the system. Assuming the controller gains are positive and the stability of the closed-loop system is maintained, if the ratio k_{d_i} to k_{p_i} is chosen such that the damping ratio falls within the interval $0 \leq \xi_i \leq 1$, the characteristic roots of the equation become complex conjugates, leading to an oscillatory behavior in the closed-loop system. In this interval, reducing the derivative gain k_{d_i} results in a smaller damping ratio, leading to more pronounced oscillations in the system. However, if the ratio k_{d_i} to k_{p_i} is selected in a way that the damping ratio exceeds one, the characteristic roots become real, resulting in a slower but oscillation-free behavior of the system. Therefore, by designing or adjusting the

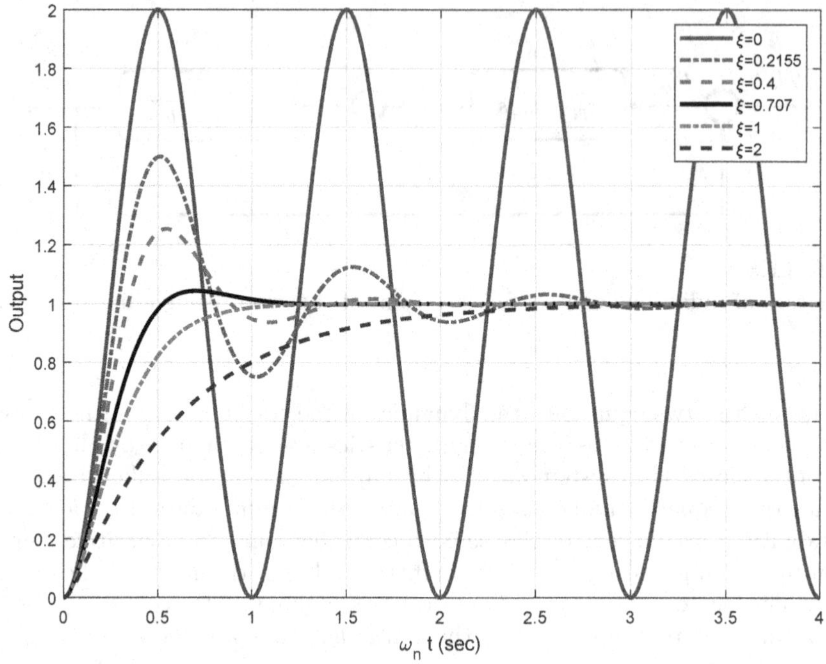

FIGURE 11.6
Response of a second-order system with different natural frequencies and damping ratios to a unit step input.

proportional and derivative gains in the controller, one can achieve the desired transient behavior.

Figure 11.6 illustrates the closed-loop system's response to a unit step input for various damping ratios in the range of zero to two. As observed from this figure, decreasing the damping ratio increases the level of oscillations, to the extent that with a damping ratio of zero, the system exhibits persistent oscillations and approaches instability. In robotic systems, selecting a damping ratio of $\xi_i = 0.707$ will result in complex roots of the characteristic equation positioned along the 45 degrees angle on the complex plane, and the system's time response will converge to the desired value with minimal oscillations and maximum speed. Therefore, tuning control gains to achieve such a response is highly desirable. The desired natural frequency in robotic systems is also adjusted based on the required bandwidth and the saturation limits of the actuator, typically ranging between one to five radians per second.

On the other hand, using the "final value Theorem," the steady-state behavior of the tracking error can also be determined [129].

$$e_{i_{ss}} = \lim_{t\to\infty} e_{q_i}(t) = \lim_{s\to 0} s\, e_{q_i}(s) \tag{11.13}$$

If the input disturbance to the system, denoted as $\mathfrak{g}_i(t)$, is considered as a constant value $\mathfrak{g}_i$, in the Laplace domain, this disturbance can be treated as a step input with an amplitude of $\mathfrak{g}_i$, having the Laplace transform $\mathfrak{g}_i(s) = \mathfrak{g}_i/s$. In this case, using the tracking error transfer function 11.9 and applying the final value Theorem 11.13, the steady-state error due to the input disturbance can be written as follows:

$$\begin{aligned} e_{i_{ss}} &= \lim_{s\to 0} s\,\frac{1}{\Delta_i(s)}\frac{\mathfrak{g}_i}{s} \\ &= \lim_{s\to 0} \frac{\mathfrak{g}_i}{I_{e_i}s^2 + (b_{e_i} + k_{d_i})s + k_{p_i}} \\ &= \frac{\mathfrak{g}_i}{k_{p_i}} \end{aligned} \tag{11.14}$$

This relationship clearly demonstrates that if the disturbance caused by gravitational forces on the robot tends toward zero over time ($\mathfrak{g}_i \to 0$), the steady-state error caused by it also becomes zero. This is in contrast to the scenario where this disturbance does not diminish over time and converges to a constant value; in this case, the steady-state error does not vanish, but using a large proportional gain $k_{p_i} \gg 1$ significantly reduces its magnitude. This conclusion aligns with the analysis conducted regarding the magic of feedback and the recommendation to use large gains in the controller.

In robotic systems, the disturbance acting on the robot links due to the gravitational force usually does not converge to zero as time evolves, and for precise tracking at steady-state, special attention must be paid to this issue. One approach to mitigate steady-state error is by incorporating the integral of error into the feedback loop, or by employing a PID controller, which will be discussed in Section 11.3.3. Although using the integral of error will cause the steady-state error to converge to zero, it will decrease the response speed of the closed-loop system. For this reason, in industrial robots, other compensation methods such as mechanical or hydro-pneumatic weight compensation are often employed in robot design to make more efficient use of actuators with lower power and cost. Furthermore, force/torque compensation in the robot due to gravitational forces can be achieved using a feed-forward loop or predominant inverse dynamics control, which will be discussed in sections 11.3.5 and 12.3, respectively.

11.3.3 PID Control

The PID controller is the most predominant linear controller used in industrial processes and robotics for the purposes of desirable trajectory tracking. In this controller, in addition to the proportional and derivative terms, the integral of error is also utilized in calculating the control effort.

$$\tau_i(t) = k_{p_i} e_{q_i}(t) + k_{d_i}\dot{e}_{q_i}(t) + k_{i_i}\int_0^t e_{q_i}(\sigma)d\sigma \tag{11.15}$$

In this equation, k_{p_i} represents the proportional gain of the controller, k_{d_i} is the derivative gain, and k_{i_i} is the integral gain. This controller utilizes the integral of tracking error to reduce the steady-state error of the system toward zero. In Laplace space, this controller is characterized by three control gains as follows:

$$\boldsymbol{C}_i(s) = k_{p_i} + k_{d_i}s + \frac{k_{i_i}}{s} \tag{11.16}$$

In this way, the block diagram of the closed-loop system for each joint in the presence of external disturbances can be illustrated using the PID controller, as shown in Figure 11.7. Utilizing this block diagram, the dynamics of the motion for each joint $q_i(s)$ can be determined with respect to the desired trajectory input $q_{d_i}(s)$, and the disturbance input $\mathfrak{g}_i(s)$, as follows.

$$\begin{aligned} q_i(s) &= \frac{k_{d_i}s^2 + k_{p_i}s + k_{i_i}}{\Omega_i(s)} q_{d_i}(s) - \frac{s}{\Omega_i(s)}\mathfrak{g}_i(s) \\ \Omega_i(s) &= I_{e_i}s^3 + (b_{e_i} + k_{d_i})s^2 + k_{p_i}s + k_{i_i} \end{aligned} \tag{11.17}$$

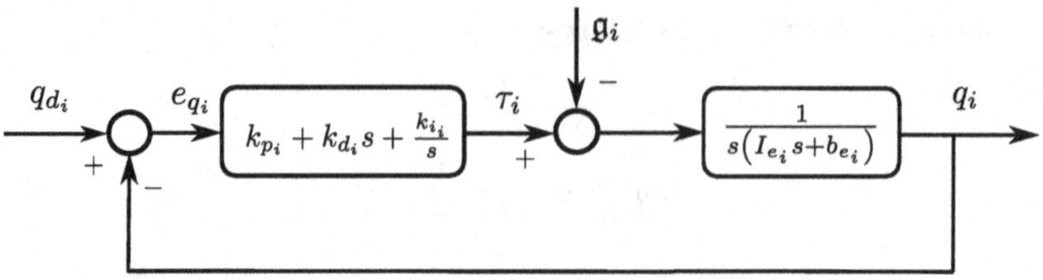

FIGURE 11.7
Block diagram of individual control for each joint of the robot using a linear model and PID controller.

The tracking error can also be calculated using the following relationship:

$$\begin{aligned} e_{q_i}(s) &= q_{d_i}(s) - q_i(s) \\ &= \frac{I_{e_i}s^3 + b_{e_i}s^2}{\Omega_i(s)} q_{d_i}(s) + \frac{s}{\Omega_i(s)} \mathfrak{g}_i(s) \end{aligned} \tag{11.18}$$

As observed, utilizing the PID controller results in the dynamics of the tracking output and tracking error being of the third order, characterized by the characteristic equation $\Omega_i(s)$. PID controller gains directly affect this characteristic equation coefficients and, consequently, the behavior of the system's output. By using the Routh-Hurwitz stability criterion, it can be demonstrated that, for the closed-loop system to be stable, the controller gains must all be positive and satisfy the following condition.

$$k_{i_i} < \frac{b_{e_i} + k_{d_i}}{I_{e_i}} k_{p_i} \tag{11.19}$$

Moreover, if the controller gains are appropriately selected to satisfy stability conditions, and are chosen sufficiently large, the roots of the characteristic equation will be located farther from the imaginary axis, resulting in a faster transient response in the system. If the controller gains are chosen in a way that the characteristic equation has complex roots, the system output and, consequently, the tracking error will exhibit oscillations.

One of the most important features of the PID controller is the elimination of steady-state error. To investigate this, consider Laplace transforming the tracking error for a unit step reference input $q_{d_i}(s) = 1/s$ and a step disturbance with a constant gravitational force $\mathfrak{g}_i(s) = \mathfrak{g}_i/s$.

$$e_{q_i}(s) = \frac{I_{e_i}s^2 + b_{e_i}s}{\Omega_i(s)} + \frac{\mathfrak{g}_i}{\Omega_i(s)} \tag{11.20}$$

It can be easily understood that by using the final value Theorem, in the steady-state, the tracking error will be equal to zero.

$$e_{i_{ss}} = \lim_{s\to 0} s\, e_{q_i}(s) = 0 \tag{11.21}$$

It should be noted that this feature comes with a slower response in trajectory tracking in the closed-loop system. Various methods for tuning the parameters of PID controllers have been proposed in the control system literature. However, we refrain from repeating them here and invite interested readers to explore references [130, 131] for further details.

11.3.4 Performance Limitations: Sensor Noise and Actuator Saturation

As mentioned in Section 11.3, by using the definition of the sensitivity transfer function $\boldsymbol{S}(s)$ in closed-loop systems, it can be demonstrated that effective trajectory tracking in the

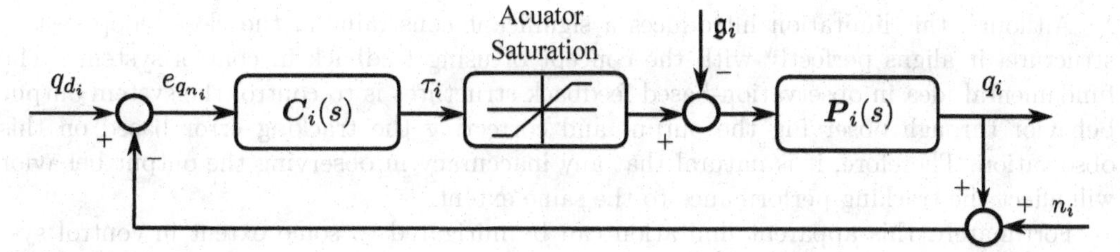

FIGURE 11.8
Block diagram of individual control for each joint of the robot using in the presence of measurement noise and actuator saturation.

presence of simultaneous disturbance and uncertainty in the system modeling is achievable with proper high-gain controller. However, it should be noted that in the presence of limitations in the robot's sensors and actuators, the performance of the closed-loop system will be affected. Therefore, in the presence of such limitations, measures should be taken in the implementation of control systems, which will be discussed in this section.

Let's begin by theoretically exploring the influence of measurement noise in sensors and actuator saturation on the performance of the closed-loop system. To achieve this, examine the block diagram of the closed-loop system presented in Figure 11.4. Redraw this diagram in the format shown in Figure 11.8, taking into account the presence of measurement noise and actuator saturation. As depicted in this figure, the observation and control of joint motion are affected by the measurement noise n_i. Furthermore, the control effort commanded by the controller is constrained by the actuator saturation block, as there are limitations on applying large currents in electric motors.

With the introduction of measurement noise n_i in the output transfer function and the tracking error of the system, the output transfer function of the system obtained from the relation $q_i = \boldsymbol{C}_i(s)\boldsymbol{P}_i(s)(q_{d_i} - q_i - n_i)$ can be expressed in terms of three inputs q_{d_i}, $\mathfrak{g}_i$, and n_i as follows.

$$q_i(s) = \boldsymbol{T}_i(s)q_{d_i}(s) - \boldsymbol{P}(s)\boldsymbol{S}_i(s)\mathfrak{g}_i(s) - \boldsymbol{T}_i(s)n_i(s) \tag{11.22}$$

In this relation, the sensitivity transfer function $\boldsymbol{S}_i(s)$ and its complement $\boldsymbol{T}_i(s)$ in each joint of the robot are defined by the following equations.

$$\boldsymbol{S}_i(s) = \frac{1}{1 + \boldsymbol{C}_i(s)\boldsymbol{P}_i(s)}, \quad \boldsymbol{T}_i(s) = \frac{\boldsymbol{C}_i(s)\boldsymbol{P}_i(s)}{1 + \boldsymbol{C}_i(s)\boldsymbol{P}_i(s)} \tag{11.23}$$

$$\boldsymbol{S}_i(s) + \boldsymbol{T}_i(s) = 1 \tag{11.24}$$

Considering that the sum of the sensitivity transfer function $\boldsymbol{S}_i(s)$ and its complement $\boldsymbol{T}_i(s)$ is equal to one, $\boldsymbol{T}_i(s)$ is referred to as the complementary sensitivity transfer function in control literature. Now, by calculating the actual tracking error, which is calculated from the relation $e_{q_i} = q_{d_i} - q_i$, we have,

$$e_{q_i}(s) = \boldsymbol{S}_i(s)q_{d_i}(s) + \boldsymbol{P}(s)\boldsymbol{S}_i(s)\mathfrak{g}_i(s) + \boldsymbol{T}_i(s)n_i(s) \tag{11.25}$$

The third term in Equations 11.22 and 11.25 indicates that measurement noise appears in the tracking error through the complementary sensitivity transfer function $\boldsymbol{T}_i(s)$. This means that if a high-gain controller is used in the closed-loop structure, according to the definition, its value approaches one. Thus, the effect of measurement noise will be introduced into the output and tracking error without attenuation.

Although this limitation introduces a significant constraint in the closed-loop system structure, it aligns perfectly with the concept of using feedback in control systems. The fundamental idea in observation-based feedback structures is to control the system output behavior through observing the output and correcting the tracking error based on this observation. Therefore, it is natural that any inaccuracy in observing the output behavior will affect the tracking performance to the same extent.

Fortunately, this apparent limitation can be mitigated to some extent in control systems. This is because measurement noise in sensors often has high-frequency content. If the frequency content of sensor noise falls outside the desirable bandwidth for system tracking, it is possible to keep the sensitivity transfer function very small in this bandwidth by applying large control gains only at the system bandwidth frequencies. Meanwhile, by reducing these gains outside the system bandwidth, the complementary sensitivity transfer function can be minimized in this range, reducing the impact of sensor noise on tracking error. Designing a controller with these features is feasible in the frequency domain using loop shaping methods [132], while simple PD and PID controllers typically struggle to meet this objective.

Now, let's examine the theoretical effect of actuator saturation on tracking performance. For this purpose, consider the required control effort τ_i as the output of the closed-loop control system, and obtain its transfer function with respect to the three inputs of the system using the block diagram shown in Figure 11.8. To do this, you can use the relation $\tau_i = \boldsymbol{C}_i(s)(q_{d_i} - q_i - n_i)$ and simplify it to the following expression.

$$\tau_i = \boldsymbol{P}_i^{-1}(s)\boldsymbol{T}_i(s)\,(q_{d_i} - n_i) + \boldsymbol{T}_i(s)\mathfrak{g}_i \tag{11.26}$$

If we examine this relation carefully, it becomes apparent that the inverse of the main system transfer function, $\boldsymbol{P}_i^{-1}(s)$, enters into the transfer function of the control effort. Considering that in most dynamic systems and linear models of robot arms, the main system transfer function becomes very small at frequencies higher than its bandwidth, the product of $\boldsymbol{P}^{-1}(s)\boldsymbol{T}_i(s)$ in this frequency range will become very large. Therefore, the control effort required to track input signals with frequency content outside the system bandwidth becomes notably large. In the controller design, this should be considered, and the bandwidth of the closed-loop system should not exceed the bandwidth of the open-loop system.

Equation 11.26 also reveals another important fact. The second term in this equation shows that the product of disturbance torque and the complementary sensitivity transfer function directly affects the required control effort. Since the sensitivity function of the system is designed to be very small in the system bandwidth to achieve desirable tracking error; therefore, the control effort required to compensate for disturbances caused by gravitational forces in the system will be considerable. This issue has been previously studied in the context of compensating for disturbance torques in systems, emphasizing the preference for using mechanical or hydro-pneumatic weight compensation systems to avoid the need for high-power actuators in the robot structure.

The result of this theoretical analysis of control system design can be summarized as follows. Although increasing the control gain to reduce tracking error in the presence of disturbances and uncertainties in the robot model is important, this control gain should only be increased within the desired bandwidth of the system to reduce the impact of measurement noise and the required control effort. Measurement noise directly affects the tracking error of the system; therefore, the use of appropriate filters to reduce the effect of measurement noise is recommended.

In tuning proportional, derivative, and integral gains of industrial controllers, attention to the required control effort and avoiding it from being in the saturation range is crucial. Thus, it is recommended to determine the proportional and derivative control gains using

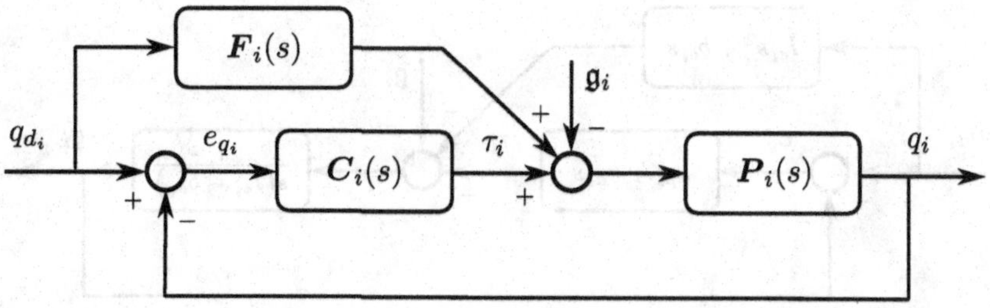

FIGURE 11.9
Block diagram of a linear control system using feedback loop and feed-forward controller.

the desired damping ratio $\xi_i = 0.707$ and keep this ratio in tuning the bandwidth of the closed loop system. Subsequently, gradually increase the natural frequency of the closed-loop system, and consequently the bandwidth of the system, in such a way that it does not significantly exceed the open-loop system bandwidth and keeps the required control effort in the appropriate range. Furthermore, adjust the integral gain in a way that, while reducing the steady-state error and maintaining the stability of the closed-loop system, it does not have a significant effect on the system's time response speed.

11.3.5 Feed-Forward Control

In linear control systems, to enhance performance and reduce tracking error in the presence of disturbances, one can leverage a feed-forward controller in addition to feedback control loops. This control structure, in the presence of disturbances, is illustrated in Figure 11.9. As depicted in this figure, alongside the feedback controller $\boldsymbol{C}_i(s)$ in the feedback loop, a second feedforward controller $\boldsymbol{F}_i(s)$ is employed in the feed-forward path. By utilizing the additional degree of freedom introduced by this feed-forward controller in the closed-loop system structure, tracking errors can be made smoother, and the impact of disturbances on steady-state error can be completely eliminated.

To analyze the system and design the feed-forward controller, the output transfer function of the system $q_i(s)$ in the presence of disturbances and the feed-forward controller $\boldsymbol{F}_i(s)$ can be expressed as follows.

$$q_i(s) = \frac{(\boldsymbol{C}_i(s) + \boldsymbol{F}_i(s))\boldsymbol{P}_i(s)}{1 + \boldsymbol{C}_i(s)\boldsymbol{P}_i(s)} q_{d_i}(s) - \frac{\boldsymbol{P}_i(s)}{1 + \boldsymbol{C}_i(s)\boldsymbol{P}_i(s)} \mathfrak{g}_i(s) \tag{11.27}$$

In this way, the tracking error transfer function is obtained as follows.

$$e_{q_i}(s) = \frac{1 - \boldsymbol{F}_i(s)\boldsymbol{P}_i(s)}{1 + \boldsymbol{C}_i(s)\boldsymbol{P}_i(s)} q_{d_i}(s) + \frac{\boldsymbol{P}_i(s)}{1 + \boldsymbol{C}_i(s)\boldsymbol{P}_i(s)} \mathfrak{g}_i(s) \tag{11.28}$$

Compare this equation with the case in which no feed-forward controller is utilized,

$$e_{q_i}(s) = \frac{1}{1 + \boldsymbol{C}_i(s)\boldsymbol{P}_i(s)} q_{d_i}(s) + \frac{\boldsymbol{P}_i(s)}{1 + \boldsymbol{C}_i(s)\boldsymbol{P}_i(s)} \mathfrak{g}_i(s) \tag{11.29}$$

As observed in Equation 11.28, in the presence of a feed-forward controller, the numerator of the tracking error transfer function becomes $1 - \boldsymbol{F}_i(s)\boldsymbol{P}_i(s)$. By appropriately selecting the feed-forward controller $\boldsymbol{F}_i(s)$ such that this expression is set to zero, and considering the stability of the closed-loop system, the tracking error will be reduced to zero for any

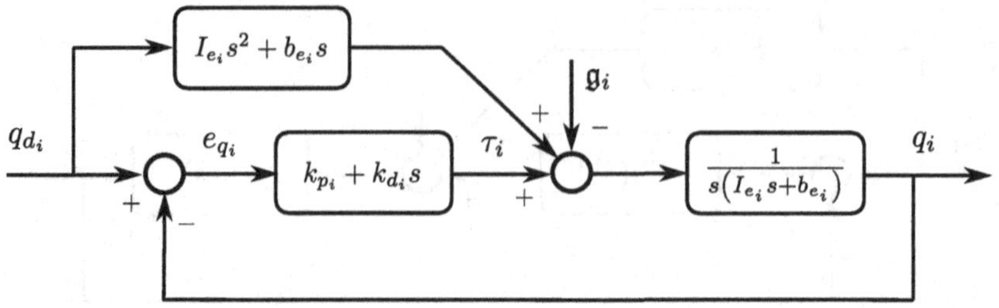

FIGURE 11.10
Block diagram of individual control for each robot joint using PD and feed-forward controllers.

arbitrary input. This scenario only occurs when the feed-forward controller can be set equal to the inverse of the system transfer function.

$$\boldsymbol{F}_i(s) = \boldsymbol{P}_i^{-1}(s) \tag{11.30}$$

In many systems, including the linear model of individual robot joints, the system transfer function is stable and minimum-phase, and its inverse can be used as a feed-forward controller.

By reducing the tracking error to zero for any desired motion trajectory in the robot, the effect of disturbances is also reduced. Thus, by considering a feed-forward controller with a large gain in the frequency range where the disturbance due to gravitational force is applied to the robot, the steady-state error caused by the disturbance can be reduced. If the disturbance torque becomes zero over time, the tracking error will also asymptotically converge to zero. Integral gain controllers can also be used to reduce steady-state error and improve tracking performance.

Now, implement this theory to the control of the robot by adding the following feed-forward controller to the control structure of each robot joint with a PD controller. According to the calibrated linear model of the robot, the feed-forward controller can be considered as a transfer function $\boldsymbol{F}_i(s) = I_{e_i}s^2 + b_{e_1}s$ based on Equation 11.30. Considering Figure 11.10, it can be shown that using this controller, the dynamics of the tracking error follows the second-order differential equation as follows.

$$I_{e_i}\ddot{e}_{q_i} + (b_{e_i} + k_{d_i})\,\dot{e}_{q_i} + k_{p_i} = \mathfrak{g}_i(t) \tag{11.31}$$

In this case, the dynamics of the tracking error assume a highly desirable form, and in the absence of disturbances or in the presence of decaying disturbances, the steady-state error will tend toward zero for any arbitrary input. If the disturbance torque entering the robot joints can be accurately modeled and calibrated, or if it can be accurately estimated using estimation methods, its effect on the dynamics of the tracking error can be completely compensated. This compensation method will be thoroughly studied in the general case in the context of inverse dynamics control in Section 12.4.

11.3.6 Multivariable PID Control

Due to the popularity of the PID controller in industry, in this section, we introduce this controller in a multivariable structure. Considering the excellent performance of this type of controller on each joint of the robot, it can also be applied in a multivariable form. In this case, as observed in the block diagram in Figure 11.11, the n-dimensional vector $\boldsymbol{q}$ is used in the feedback loop, and the multivariable PID controller is constructed using

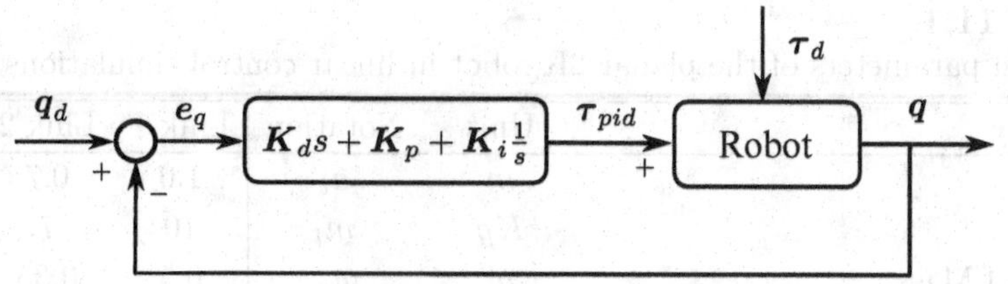

FIGURE 11.11
Block diagram of multivariable PID control for a robot.

the proportional matrix $\boldsymbol{K}_p$, the derivative matrix $\boldsymbol{K}_d$, and the integral matrix $\boldsymbol{K}_i$. If the gain matrices are assumed to be diagonal, this control structure perfectly aligns with the individual PID control of the joints. However, if these matrices are not diagonal, information from the motion of other joints is also used in controlling each joint, making the controller structure multivariable.

The structure of this controller is simple and can be easily implemented. The adjustment of its control gains requires minimal robot modeling and calibration. Achieving satisfactory performance in tracking the desired trajectory is possible by using high gains and conducting a tuning process. However, it has been observed by experience that tuning the controller gains can be challenging and may involve significant trial and error, especially when non-diagonal matrices are used in this type of controller.

Given that parameter tuning is conducted for one or a few specific desired motion trajectories, the tracking error for traversing other trajectories in the workspace may not be as favorable. This challenge becomes more pronounced, particularly in the case of steady-state tracking errors. The complete elimination of disturbance torques caused by the gravity vector in different scenarios poses a challenging task, and achieving zero steady-state error for all trajectories poses a significant hurdle.

To overcome these challenges, we will enhance this simple structure by incorporating multivariable and nonlinear control methods. The specifics of this expansion will be discussed in the next chapter, presented in a step-by-step way. Nevertheless, it's essential to highlight that for robots equipped with electric actuators and gearboxes, the utilization of linear controllers remains highly effective. Therefore, implementing PID controller stands as a foundational and widely embraced design approach to attain the necessary tracking performance.

11.4 Motion Control Simulation of Planar 3<u>R</u> Robot

In this section, we present the simulation outcomes of motion control for the planar 3<u>R</u> robot, serving as an illustrative case study among the robots examined in this book. We apply different linear control techniques introduced in this chapter to this robot, followed by a comprehensive comparison and analysis.

For this purpose, consider the 3<u>R</u> robot depicted in Figure 8.7 with kinematic and mass parameters as listed in Table 11.1. Assume that the robot links are all considered as slender rods, and their center of mass is located at the link's mid lengths. Additionally, the motion trajectory of the robot arms is designed based on quintic polynomial functions, and the initial, intermediate, and final values of each joint's motion are presented in Table 11.1. As shown in Figure 11.12, the duration of this maneuver is set to four seconds, with the joints

TABLE 11.1
Simulation parameters of the planar 3<u>R</u> robot in linear control simulations

Quantity	Unit	Notation	Link 1	Link 2	Link 3
Length	m	a_i	1.0	0.7	0.3
Mass	Kg	m_i	10	7	3
Center of Mass	m	m_{c_i}	0.5	0.35	0.15
Inertia Moment	$Kg \cdot m^2$	${}^{C}I_i$	0.833	0.286	0.023
Initial Angle	deg	θ_{i_0}	0	0	0
Intermediate Angle	deg	θ_{i_v}	90	−60	30
Final Angle	deg	θ_{i_f}	0	0	0
Maneuver Time	sec	t_f	4	4	4
Step Disturbance Amplitude	$N.m$	τ_{d_i}	50	20	2
Disturbance Injection Time	sec	t_d	1	2	3

reaching their maximum angles in the first two seconds. In the next two seconds, the joints reach their final angle values, and in this state, the motion comes to a smooth stop with zero velocity and acceleration. The motion of all three joints is considered simultaneous to observe the effect of multivariable dynamics on the behavior of the robot on its controller performance.

Consider the dynamic formulation of the robot based on the obtained relationships in Case Study 8.3 and the Christoffel matrix determined in Case Study 9.2, expressed in the matrix form 9.40. To investigate the nonlinear behavior of the system, we neglect the effects of the actuator dynamics and the gearbox. Thus, the dynamic equation of the system is formulated as:

$$\boldsymbol{\tau} + \boldsymbol{\tau}_d = \boldsymbol{M}(\boldsymbol{q})\ddot{\boldsymbol{q}} + \boldsymbol{C}(\boldsymbol{q}, \dot{\boldsymbol{q}})\dot{\boldsymbol{q}} + \boldsymbol{\mathfrak{g}}(\boldsymbol{q}) \tag{11.32}$$

This equation is used in the simulation of the robot's motion.

To perform motion control simulations of the robot, we use the most common motion control topology in serial robots, as illustrated in Figure 11.1. In this topology, the design of the desired trajectory and motion control of the robot is carried out in the joint space. Accordingly, the desired joint trajectories are determined with the vector $\boldsymbol{q}_d$, and the robot controller, based on the received tracking error vector $\boldsymbol{e}_q = \boldsymbol{q}_d - \boldsymbol{q}$, calculates the required joint torques $\boldsymbol{\tau}$ and applies them to the robot. In the motion control simulation, the effect of disturbance torques in the joints $\boldsymbol{\tau}_d$, measurement noise of joint motions, and actuator saturation will also be studied.

11.4.1 PID Control

The PID controller is the most predominant controller used in trajectory tracking in industrial robots. In this section, we study the tracking performance of the desired trajectory in the planar 3<u>R</u> robot using this controller, which is implemented in a multivariable form. In this case, as seen in the block diagram in Figure 11.11, the $\boldsymbol{q}$ vector is used in the feedback loop, and the multivariable PID controller is constructed using proportional matrix $\boldsymbol{K}_p$, derivative matrix $\boldsymbol{K}_d$, and integral matrix $\boldsymbol{K}_i$. In this simulation, diagonal matrices are used in the design of the PID controller.

In the simulation of the closed-loop motion of the robot, the kinematic and dynamic parameters of the robot, along with the desired motion trajectory, are chosen according to Table 11.1. Additionally, the gravitational vector is set to $\boldsymbol{g} = [0,\ -9.81]^T$. Furthermore, the

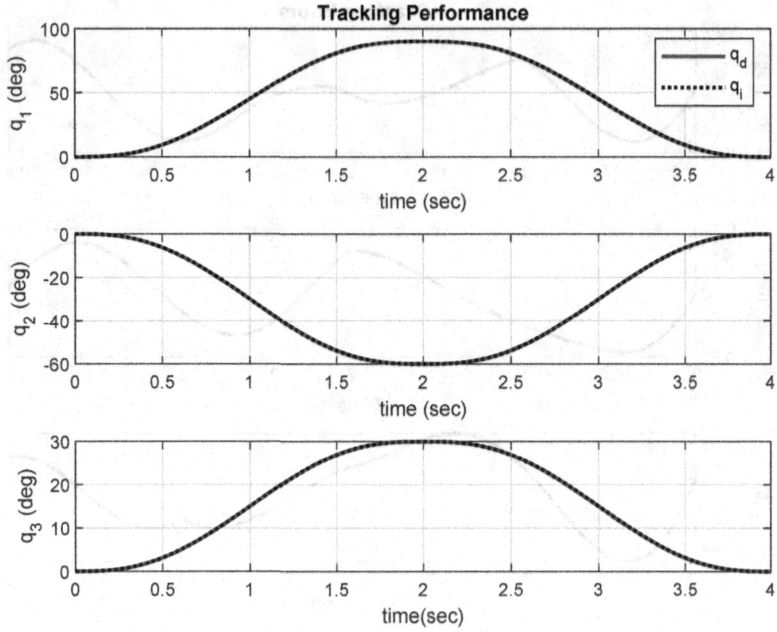

FIGURE 11.12
Simulation of the desired and the actual trajectory of the joints using the PID controller.

control gain matrices have been tuned through trial and error and are set to the following values, in order to ensure desired tracking in all three joints, while the required torques of the joints remain within the power capabilities of the electric actuators commonly used in this types of robots.

$$\boldsymbol{K}_p = \begin{bmatrix} 8 & 0 & 0 \\ 0 & 3 & 0 \\ 0 & 0 & 0.2 \end{bmatrix} \times 10^4, \;\; \boldsymbol{K}_d = \boldsymbol{K}_p, \;\; \boldsymbol{K}_i = 10\boldsymbol{K}_p \tag{11.33}$$

The motion simulation in this scenario is performed using the relation 10.24, where the control effort, denoted as $\boldsymbol{\tau}_{pid}$, is calculated using the PID controller and based on the block diagram depicted in Figure 11.11. Disturbance torques $\boldsymbol{\tau}_d$ are applied as step torques in the three joints. According to Table 11.1, in the first second, a step torque disturbance with an amplitude of 50 $N.m$ is applied to the first joint, while in the 2nd second, a step torque disturbance with an amplitude of 20 $N.m$ is applied to the second joint. Finally, in the third second, a step torque disturbance with an amplitude of 2 $N.m$ is applied to the third joint.

Numerical solution of the state equations with the initial conditions $\boldsymbol{q}(0) = \boldsymbol{q}_0, \dot{\boldsymbol{q}}(0) = 0$ may be computed using the `ode23s` command in MATLAB® software or similar commands in Python. This approach facilitates the solution of differential equations with greater ease and accuracy, considering the large control gains used in the control structure.

Utilizing this control structure, the results demonstrate a precise tracking of the desired trajectory of the quintic polynomial for joint motion, as depicted in Figure 11.12. As observed in this figure, the trajectory of the three robot joints, illustrated in three separate plots, closely matches the desired motion trajectory, with negligible errors.

For a more detailed representation of tracking accuracy, the tracking errors for each of the three joints in this simulation are displayed in Figure 11.13. As evident, by appropriately

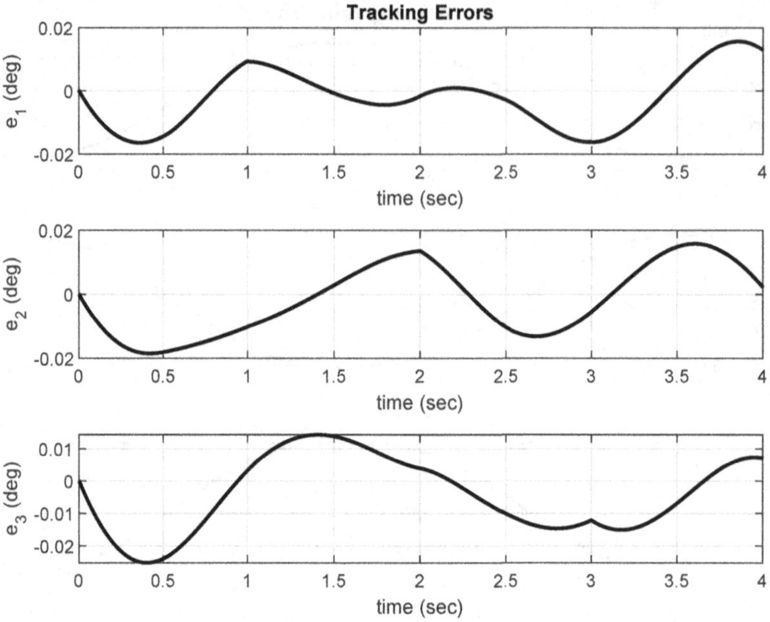

FIGURE 11.13
Tracking error of the PID controller in the presence of disturbance torques.

tuning the control gains of the PID controller, the tracking errors for all three joints are minimized to very small values, reaching around 0.02 degrees. Notably, this achievement is realized in the presence of considerable and non-vanishing disturbance torques applied to each joint. The change in tracking error behavior over time during the application of disturbance torques in the first second for the first joint, 2nd second for the second joint, and third second for the third joint is well illustrated in the tracking error plots of these joints. Achieving such satisfactory performance is solely possible through the magic of feedback and the proper tuning of the PID controller gains.

The control efforts or torques applied by the actuators in the joints of the robot are shown, respectively, for the first to third joints in three plots in Figure 11.14. In these plots, the disturbance torque entering the joints is represented by the dotted line, and the control effort resulting from the PID controller is represented by the solid line. As observed in this figure, the required control effort for the first joint is less than 300 $N.m$, for the second joint is less than 100 $N.m$, and for the third joint is less than 10 $N.m$. Considering that the maximum torque due to gravity force is applied to the robot arm at the beginning of the motion, it is evident that approximately half of the maximum required torque in this robot is consumed to compensate for the effect of gravity. Furthermore, the requirement for an instant increase in control effort at the moment of applying disturbance torque, without direct awareness of its occurrence, is clearly evident in the plots. Based on the points discussed in Chapter 10, it is possible to determine the required control effort and the speed of joint motion and proceed with selecting suitable motors and gearboxes for each joint.

11.4.2 PD Control

In order to investigate the impact of the integral term in the PID control, let us repeat the closed-loop system simulation in the presence of disturbance with a PD controller. For

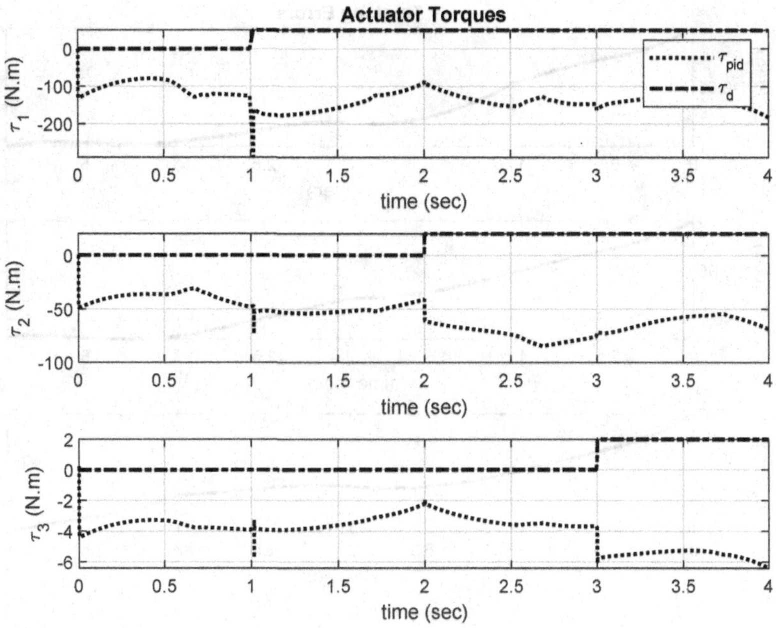

FIGURE 11.14
Control effort of the PID controller in the presence of disturbance torques.

this purpose, it is sufficient to set the integral gain matrix to zero, i.e., $\boldsymbol{K}_i = 0$. In this case, the tracking error of the closed-loop system, as observed in Figure 11.15, increases by approximately twenty times, reaching around 0.1 degrees in all three joints. As shown in this figure, the steady-state error in all three joints' movements noticeably increases, and at the end of the maneuver, it will not converge to zero.

The control effort required in the PD control is also depicted in Figure 11.16. By comparing the control effort level in this figure with what was reported for the PID control in Figure 11.14, it can be observed that, except for the times when the disturbance torque is introduced to the system, the control efforts in both methods are quite similar. This comparative analysis effectively highlights the preference for utilizing the PID controller over PD in the closed-loop system structure.

11.4.3 The Effect of Measurement Noise

As previously discussed, achieving appropriate tracking in the control structure relies on the magic of feedback and precise tuning of controller gains. To quantitatively examine this issue, we first introduce measurement noise into the feedback loop by perturbing the joint angles and assess its impact on proper trajectory tracking. For this purpose, the joint angle measurement signal is contaminated with Gaussian noise at a signal-to-noise ratio of one percent, and the PID control simulation is repeated.

Although the control gains are precisely tuned, and the measurement noise is minimal, the trajectory tracking error in this situation, illustrated in Figure 11.17, is notably impacted, escalating from 0.02 degrees in the previous case to 2 degrees. This substantial effect is primarily attributed to the utilization of a high-gain PID controller, which significantly amplifies even the slightest observation errors, directly influencing the trajectory tracking performance.

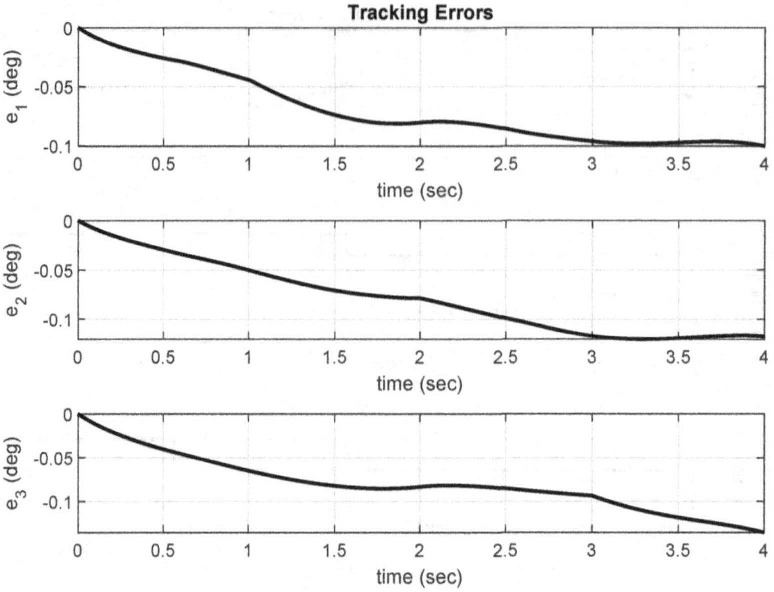

FIGURE 11.15
Tracking error of the PD controller in the presence of disturbance torques.

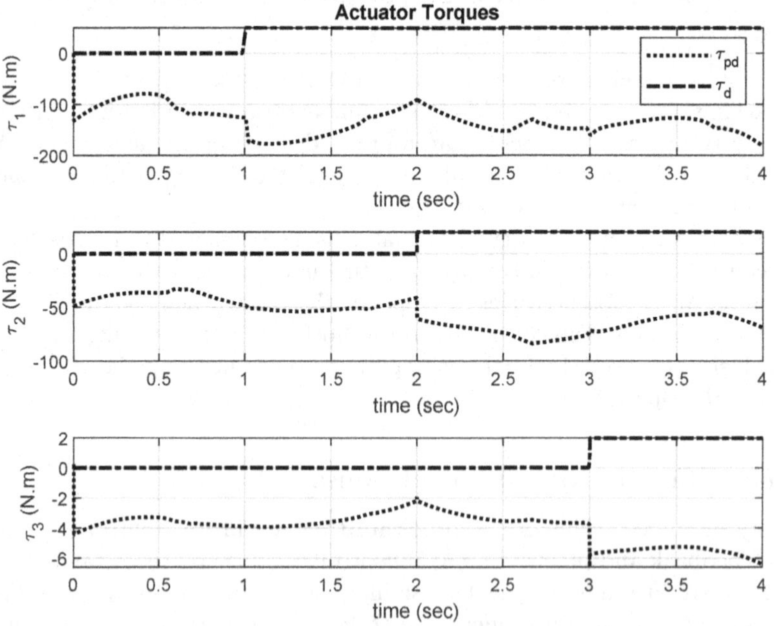

FIGURE 11.16
Control effort of the PD controller in the presence of disturbance torques.

The corresponding control effort is also depicted in Figure 11.18. This visualization clearly shows the intensified impact of measurement noise due to the substantial gains of the controller and its direct influence on the required actuator torques. Examining the frequency characteristics of the signals presented in this figure indicates that reaching the simulated

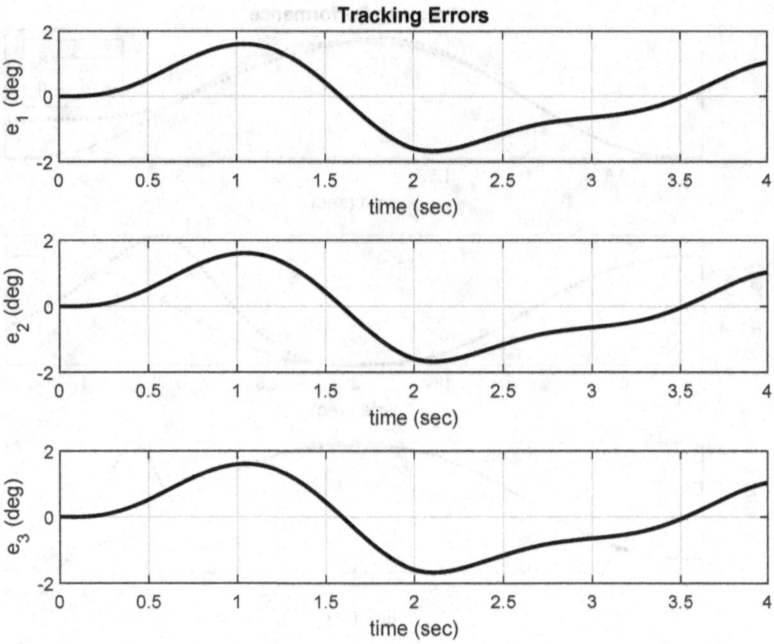

FIGURE 11.17
Tracking error of the PID controller in the presence of disturbance torques and measurement noise.

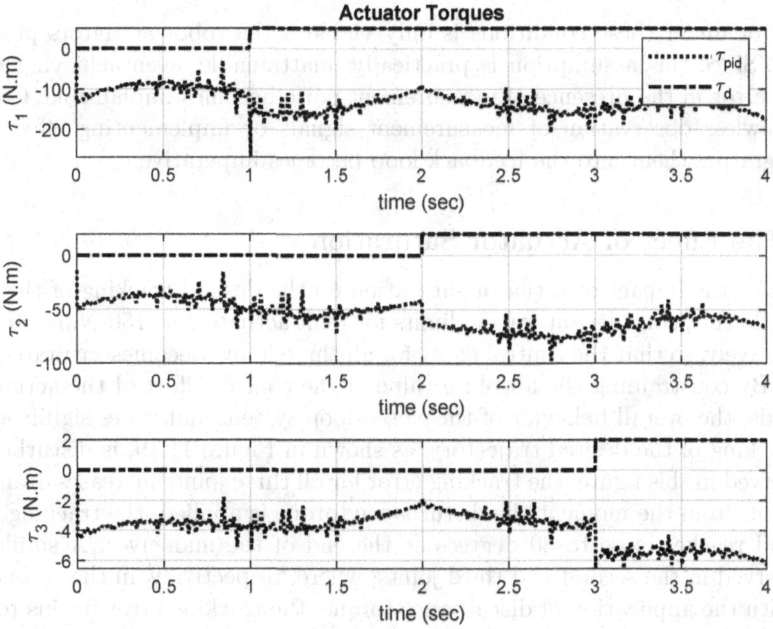

FIGURE 11.18
Control effort of the PID controller in the presence of disturbance torques and measurement noise.

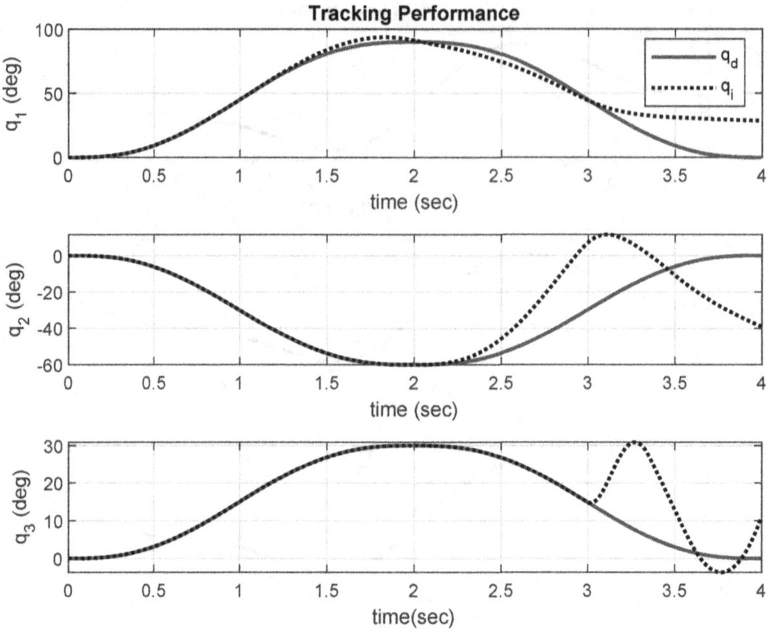

FIGURE 11.19
Tracking performance of the PID controller in the presence of disturbance torques and actuator saturation.

tracking error under these conditions is only viable if the robot actuators possess a wide bandwidth. Since this assumption is practically unattainable, even achieving two degrees of tracking error in the presence of measurement noise becomes implausible. Consequently, ensuring flawless observation of measurement signals or implementing effective filtering before integrating them into the feedback loop becomes imperative.

11.4.4 The Effect of Actuator Saturation

Now, examine the impact of actuator saturation on the desired tracking of the closed-loop system. To do this, consider saturation limits for joint actuators as 150 $N.m$, 60 $N.m$, and 5 $N.m$ respectively, so that the control effort for all three joints becomes saturated during the maneuver. By constraining the absolute value of the control effort of the actuators within these bounds, the overall behavior of the closed-loop system undergoes significant changes, and the tracking of the desired trajectory, as shown in Figure 11.19, is disturbed.

As observed in this figure, the tracking error for all three joints increases dramatically. In the first joint, from the moment the disturbance torque is applied, the tracking error starts growing and reaches close to 30 degrees at the end of the maneuver. A similar behavior can be observed in the second and third joints, where, respectively, in the second and third seconds, with the application of disturbance torque, the tracking error begins to grow. The steady-state tracking error in the second joint reaches nearly 40 degrees, and in the third joint, the tracking error becomes oscillatory from the third second with a range exceeding 30 degrees.

The primary reason for the degradation of the closed-loop system's performance in this situation can be observed in the saturated control effort plots in Figure 11.20. As depicted in this figure, the actuator of the first joint is saturated during the time intervals of one to two

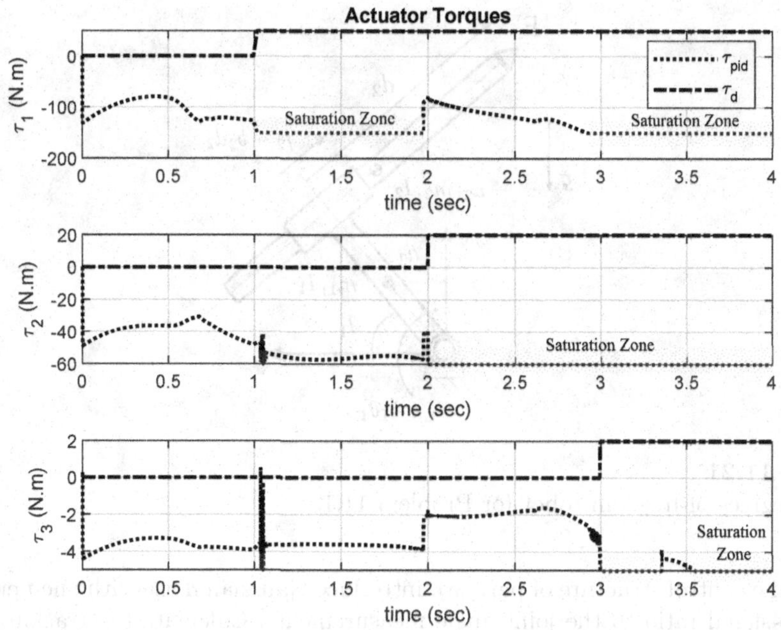

FIGURE 11.20
Control effort of the PID controller in the presence of disturbance torques and actuator saturation.

seconds and three to four seconds. Similarly, the actuator for the second joint experiences saturation in the time interval of two to four seconds, and the actuator for the third joint is saturated in the time interval of three to four seconds. It is precisely during these time intervals that the tracking performance of the closed-loop system undergoes significant degradation.

Problems

11.1 Consider the two-degrees-of-freedom <u>RP</u> robot depicted in Figure 11.21. The dynamic model of the robot has been determined in Problems 8.1 and 9.1, and the numerical values of the robot parameters are used for simulating its motion in Problem 10.1.

a) In this robot, adjust the control gains matrices of the PID controller to track the desired trajectory with negligible tracking error. Simulate the closed-loop system's motion, obtain the required control effort, and plot it.

b) Apply a step disturbance torque with a maximum amplitude of 20% of the required control effort, to each joint for the case described in part (a). Repeat the closed-loop system simulation, and calculate and plot the system's tracking error.

c) To investigate the impact of the integral term in the PID control, set the integral gain matrix to zero and simulate the tracking error in the closed-loop system. Compare it with the situation described in part (a). Compare the required control effort in these two scenarios.

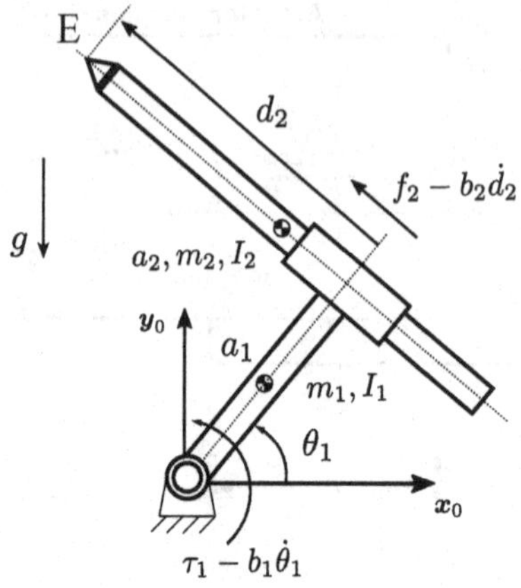

FIGURE 11.21
The two-degrees-of-freedom robot for Problem 11.1.

d) In the control structure of part (a), introduce Gaussian noise with one percent noise-to-signal ratio to the joint angle measurements. Calculate the tracking error and required control effort in this noisy condition and compare it with the noise-free situation.

e) Limit the control effort of the joints to 80% of the required amount of control effort in the situation described in part (a). Simulate the saturated actuator results in the closed-loop system tracking and compare it with the unsaturated actuator scenario.

11.2 Consider the three-degrees-of-freedom PRP robot illustrated in Figure 11.22. The dynamic model of the robot has been determined in Problems 8.2 and 9.2, and the numerical values of the robot parameters are used for simulating its motion in Problem 10.4. Repeat Problem 11.1 on this robot.

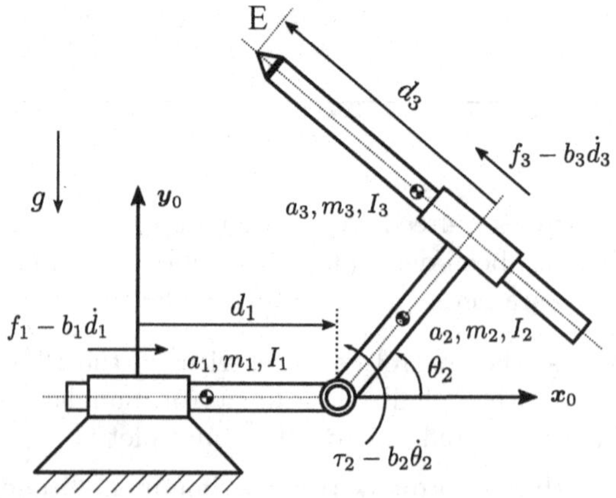

FIGURE 11.22
The three-degrees-of-freedom robot for Problem 11.2.

11.3 Examine the three-degrees-of-freedom PRR robot depicted in Figure 11.23. The robot's dynamic model has been established in Problems 8.3 and 9.3, and the numer-

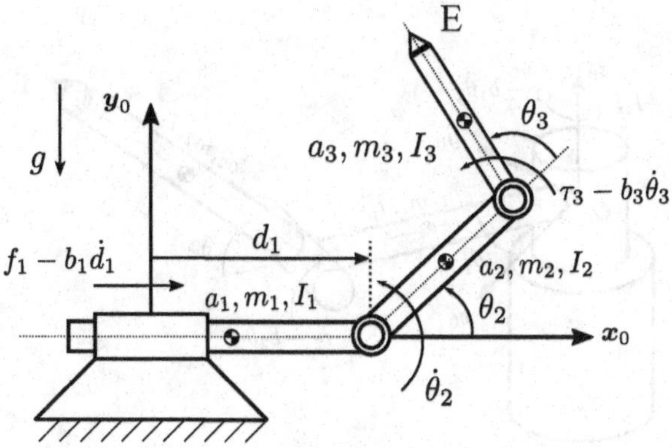

FIGURE 11.23
The three-degrees-of-freedom robot for Problem 11.3.

ical values of its parameters have been employed to simulate its motion in Problem 10.7. Repeat Problem 11.1 for this robot.

11.4 Investigate the three-degrees-of-freedom <u>RPR</u> robot shown in Figure 11.24. The robot's dynamic model has been defined in Problems 8.4 and 9.4, and the numerical values of its parameters have been used to simulate its motion in Problem 10.10. Repeat Problem 11.1 for this robot.

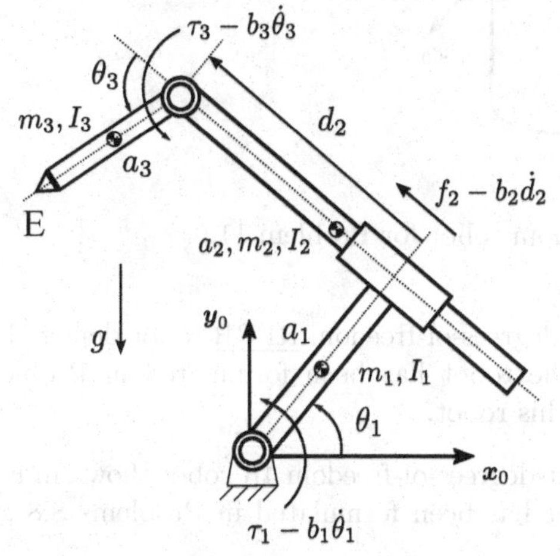

FIGURE 11.24
The three-degrees-of-freedom robot for Problem 11.4.

11.5 Examine the two-degrees-of-freedom <u>RR</u> robot depicted in Figure 11.25. The dynamic model of the robot has been established in Problems 8.5 and 9.5, and the numerical values of its parameters have been utilized to simulate its motion in Problem 10.13. Repeat Problem 11.1 for this robot.

11.6 Investigate the four-degrees-of-freedom <u>RRRP</u> robot shown in Figure 11.26. The dynamic model of the robot has been formulated in Problems 8.6 and 9.6. Repeat Problem 11.1 for this robot.

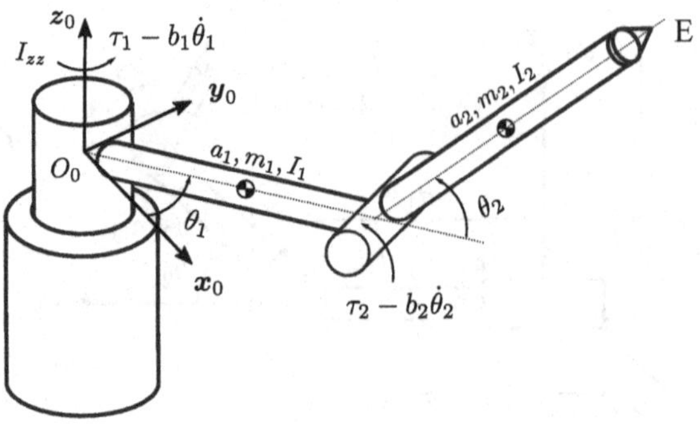

FIGURE 11.25
The two-degrees-of-freedom robot for Problem 11.5.

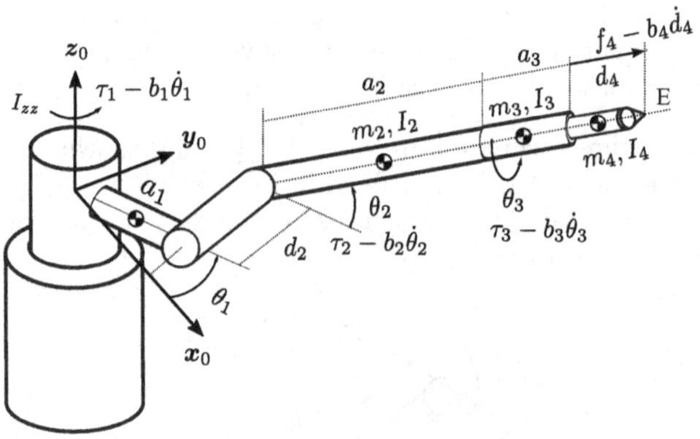

FIGURE 11.26
The four-degrees-of-freedom robot for Problem 11.6.

11.7 Examine the four-degrees-of-freedom RPRR robot depicted in Figure 11.27. The dynamic model of the robot has been formulated in Problems 8.7 and 9.7. Repeat Problem 11.1 for this robot.

11.8 Investigate the four-degrees-of-freedom 4R robot shown in Figure 11.28. The dynamic model of the robot has been formulated in Problems 8.8 and 9.8. Repeat Problem 11.1 for this robot.

11.9 Examine the four-degrees-of-freedom RRPR robot depicted in Figure 11.29. The dynamic model of the robot has been formulated in Problems 8.9 and 9.9. Repeat Problem 11.1 for this robot.

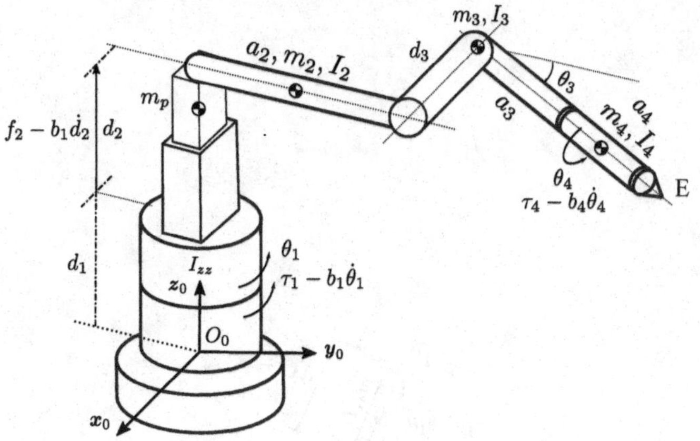

FIGURE 11.27
The four-degrees-of-freedom robot for Problem 11.7.

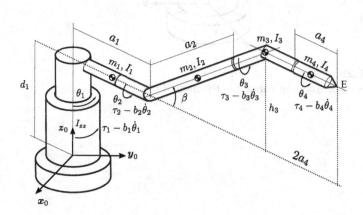

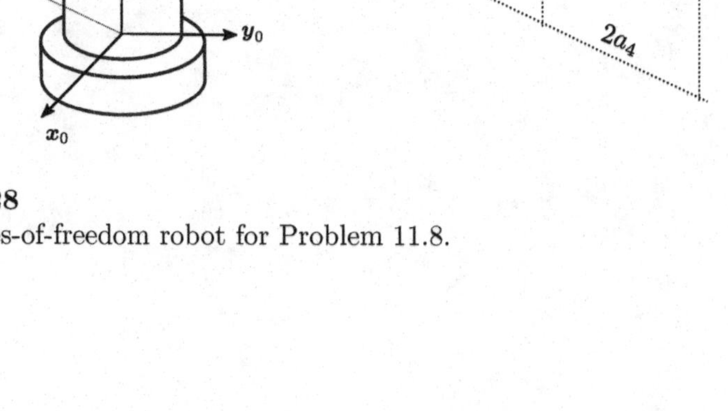

FIGURE 11.28
The four-degrees-of-freedom robot for Problem 11.8.

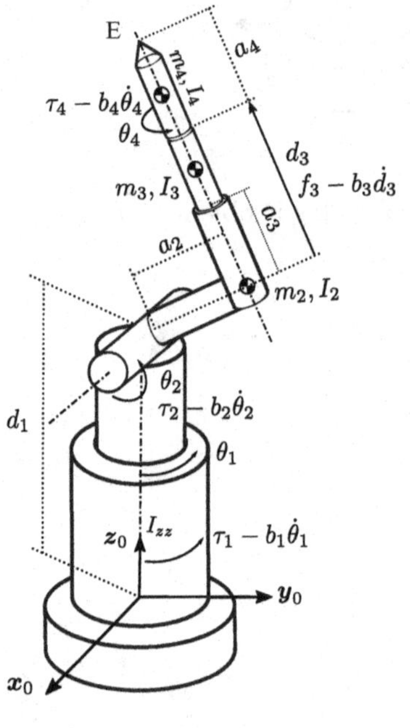

FIGURE 11.29
The four-degrees-of-freedom robot for Problem 11.9.

12

Nonlinear Motion Control

This chapter explores a range of advanced control methods for achieving precise motion control in robotic systems. It begins with an introduction, laying the foundation for exploring control techniques. The chapter covers multivariable PD control with feed-forward, focusing on strategies to enhance control for superior motion tracking. The fundamental controller discussed is the inverse dynamics controller (IDC), recognized for establishing a basis for system linearization through feedback and considered a fundamental nonlinear controller in robotics. The discussion progresses to partial feedback linearization and IDC in task space, offering a comprehensive overview of these techniques.

Inverse dynamics control calculates and implements a corrective torque on robot actuators based on the system's inverse dynamics, significantly improving linear controllers' performance in the closed-loop system. This approach is fundamental in robot control, forming the basis for many other methods. A significant portion of the chapter explores robust and adaptive control, addressing challenges such as model uncertainty in tracking performance. Concepts such as robust inverse dynamics control, chattering mitigation, and adaptive inverse dynamics control are examined in detail.

The chapter concludes with a summary of the various motion control methodologies discussed. To illustrate practical application, a case study on nonlinear control of a planar 3R robot is examined, comparing PID control with feed-forward loop, inverse dynamics control, partial feedback linearization, and robust inverse dynamics control. Problems at the end of the chapter encourage readers to apply acquired knowledge, enhancing their understanding and proficiency in implementing these advanced control techniques. Overall, the chapter provides a comprehensive overview of advanced control techniques, serving as a valuable resource for readers interested in nonlinear motion control in robotics.

12.1 Introduction

As mentioned earlier, incorporating a gearbox into the robot's actuators is anticipated to mitigate the nonlinear and multivariable aspects of the robot's dynamics; consequently, justifying the use of linear controllers. However, if the gear ratio applied to the robot's actuators is not sufficiently large, or if direct-drive actuators are employed in the robot, the complete elimination of nonlinear dynamics is not guaranteed, and the robot's dynamics will persist as a multivariable and nonlinear system. In such a scenario, by utilizing the relation 9.40, the dynamic equation of the robot, encompassing its actuator dynamics, in the presence of disturbance torque $\boldsymbol{\tau}_d$ can be rephrased as follows.

$$\boldsymbol{\tau} + \boldsymbol{\tau}_d = \boldsymbol{M}_e(\boldsymbol{q})\ddot{\boldsymbol{q}} + \boldsymbol{C}(\boldsymbol{q}, \dot{\boldsymbol{q}})\dot{\boldsymbol{q}} + \boldsymbol{b}_e\dot{\boldsymbol{q}} + \boldsymbol{g}(\boldsymbol{q}) \tag{12.1}$$

DOI: 10.1201/9781003491415-12

where

$$\boldsymbol{M}_e(\boldsymbol{q}) = \boldsymbol{M}(q) + \begin{bmatrix} \eta_1^2 I_{m_1} & 0 & \cdots & 0 \\ 0 & \eta_2^2 I_{m_2} & \cdots & 0 \\ \vdots & 0 & \ddots & \vdots \\ 0 & 0 & \cdots & \eta_n^2 I_{m_n} \end{bmatrix}, \quad \boldsymbol{b}_e = \begin{bmatrix} b_1 + \eta_1^2 b_{m_1} \\ b_2 + \eta_2^2 b_{m_2} \\ \vdots \\ b_n + \eta_n^2 b_{m_n} \end{bmatrix} \tag{12.2}$$

In this dynamic model, joint variables are represented by the vector $\boldsymbol{q}(t)$ with dimension n, input torques of the actuators by the vector $\boldsymbol{\tau}$, the equivalent mass matrix of the robot by $\boldsymbol{M}_e(\boldsymbol{q})$, the Christoffel matrix by $\boldsymbol{C}(\boldsymbol{q}, \dot{\boldsymbol{q}})$, equivalent viscous friction vector by $\boldsymbol{b}_e$, gravitational vector by $\mathfrak{g}(\boldsymbol{q})$, and disturbance torques at the joints by the vector $\boldsymbol{\tau}_d$. The determination of the matrices and vectors constituting this dynamic model is extensively discussed in Chapters 8 and 9. Consequently, the robot's dynamics are modeled by a set of coupled and nonlinear multivariable differential equations. Employing this nonlinear model in robot control provides an opportunity to apply methods of designing nonlinear controllers for this system, aiming to achieve favorable tracking performance characteristics.

In practice, if individually implemented linear controllers on the joints fail to provide satisfactory performance in tracking the desired trajectory, nonlinear controllers are employed. In the domain of nonlinear controller design, the utilization of Laplace transforms and linear stability analysis methods, such as the Routh-Hurwitz method and Nyquist criterion, is not applicable. Instead, control design in the time domain and stability analysis using Lyapunov-based methods become essential. In the upcoming sections of this chapter, we progress from the design of the multivariable PD controller introduced in the previous chapter. Initially, we enhance this structure by incorporating a nonlinear feedforward term. The fundamental controller discussed later in this chapter is the inverse dynamics controller, known for establishing a basis for system linearization through feedback, and recognized as a fundamental nonlinear controller in robotics [133]. The detailed discussion of the design of robust and adaptive controllers, along with the simulation results of these developed methods on a typical robot, will be presented in Section 12.8.

12.2 Multivariable PD control with Feed-Forward

The primary method to extend the multivariable PD controller involves the utilization of the feed-forward term to specifically reduce steady-state tracking errors. This method requires minimal knowledge of the dynamic model of the system. Suppose the dynamic model of the robot, obtained through the methods detailed in Chapters 8 and 9, is available along with the nominal values of the robot's parameters. Under this assumption, the matrices and vectors constituting the robot's dynamic model in the nominal state, described by notations $\hat{\boldsymbol{M}}_e(\boldsymbol{q})$, $\hat{\boldsymbol{b}}_e$, $\hat{\boldsymbol{C}}(\boldsymbol{q}, \dot{\boldsymbol{q}})$, and $\hat{\mathfrak{g}}(\boldsymbol{q})$, are accessible. In this representation, the notation $(\hat{\cdot})$ denotes the nominal values of these quantities. By precise dynamic calibration of the robot, these nominal values closely approximate the actual values, and they can be used with sufficient reliability in the robot's control structure.

In this approach, to compensate for the impact of nonlinear terms in the robot's dynamics, a torque calculated based on the robot's dynamics is utilized in the feed-forward loop of the control system. For this purpose, real-time information about joint variables is not employed. Instead, as illustrated in Figure 12.1, the robot's dynamics are computed using the desired motion trajectory and its derivatives $\boldsymbol{q}_d, \dot{\boldsymbol{q}}_d, \ddot{\boldsymbol{q}}_d$, and are utilized in the feed-forward loop. It's worth noting that if the controller's performance is satisfactory, the

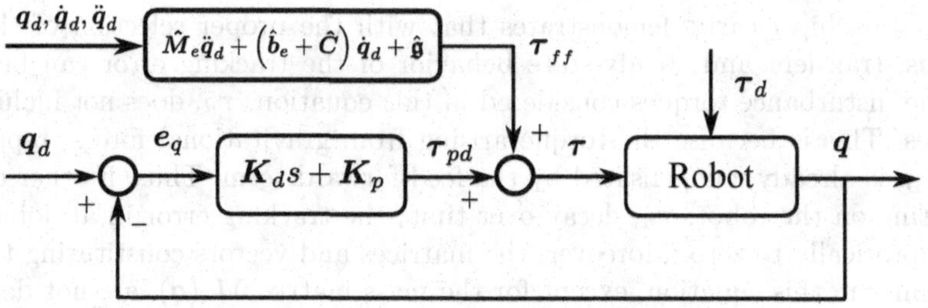

FIGURE 12.1
Block diagram of the multivariable PD control for the robot using a feed-forward controller.

actual motion variables will also track this desired motion trajectory with minimal tracking error, and we can expect a suitable dynamic compensation in the control system.

Accordingly, the feed-forward torque τ_{ff} is computed using the following relation.

$$\tau_{ff} = \hat{M}_e(q_d)\ddot{q}_d + \left(\hat{b}_e + \hat{C}(q_d, \dot{q}_d)\right)\dot{q}_d + \hat{\mathfrak{g}}(q_d) \tag{12.3}$$

This term is added to the control effort resulting from the PD controller, denoted by τ_{pd} in this figure. In this method, since the calculation of the feed-forward torque τ_{ff} is not performed in an online manner, it can be pre-calculated in advance and based on the designed motion trajectory of the robot. This is in contrast to the PD controller, which is implemented in the system using the online information of the robot's motion in the feedback loop.

This control structure, given its utilization of the nonlinear dynamic model of the system, falls into the category of model-based controllers. To analyze the impact of employing this model-based control method, derive the dynamic equation of the closed-loop system.

$$M_e(q)\ddot{q} + C(q, \dot{q})\dot{q} + b_e\dot{q} + \mathfrak{g}(q) = \tau_{ff} + \tau_{pd} + \tau_d \tag{12.4}$$

Substitute the values of τ_{pd} and τ_{ff} and simplify to arrive at the following equation:

$$\begin{aligned} &M_e(q)\ddot{q} + C(q, \dot{q})\dot{q} + b_e\dot{q} + \mathfrak{g}(q) = \\ &= \hat{M}_e(q_d)\ddot{q}_d + \left(\hat{b}_e + \hat{C}(q_d, \dot{q}_d)\right)\dot{q}_d + \hat{\mathfrak{g}}(q_d) + K_d\dot{e}_q + K_p e_q + \tau_d \end{aligned} \tag{12.5}$$

If the knowledge of the system model is accurate enough, and if the values of the matrices and vectors constituting the dynamic model are properly identified through dynamic calibration of the robot, we may assume:

$$\hat{M}_e = M_e \quad , \quad \hat{C} = C \quad , \quad \hat{b}_e = b_e \quad , \quad \hat{\mathfrak{g}} = \mathfrak{g} \tag{12.6}$$

Furthermore, if the performance of the robot controller is satisfactory and the actual motion variables also track the desired motion trajectory with minimal tracking error, it is reasonable to assume that the following approximations are true.

$$q(t) \approx q_d(t) \quad , \quad \dot{q}(t) \approx \dot{q}_d(t) \tag{12.7}$$

In this case, the differential equation governing the closed-loop system in the presence of the feed-forward loop will be simplified to the following equation:

$$M_e\ddot{e}_q + K_d\dot{e}_q + K_p e_q + \tau_d = 0 \tag{12.8}$$

This relationship clearly demonstrates that with the proper selection of the PD controller gains, transient and steady-state behavior of the tracking error can be effectively shaped. The disturbance torques considered in this equation, $\boldsymbol{\tau}_d$, does not include gravitational forces. This is because the torque arising from gravitational forces, represented by the vector $\mathbf{g}$, is already compensated by the feed-forward term. Thus, if other disturbance torques acting on the robot, $\boldsymbol{\tau}_d$, decay over time, the tracking error in all joints will converge asymptotically to zero. Moreover, the matrices and vectors constituting the tracking error dynamics in this equation, except for the mass matrix $\boldsymbol{M}_e(\boldsymbol{q})$, are not dependent on the kinematic configuration of the robot in its workspace. This makes adjusting the PD controller gains much easier compared to the case without feed-forward.

However, notice that the desirable features of this control structure can be achieved only when the assumptions based on the complete knowledge about the system dynamics (Equation 12.6) and the desired input tracking (Equation 12.7) are satisfied. As seen in Chapter 8, deriving the dynamic formulations of industrial robots is a very complex and time-consuming task. Calibration of these equations and determining the model parameters are also a prohibited and time-consuming task. Under these circumstances, Equation 12.5 practically demonstrates the dynamic behavior of the tracking error in a more appropriate manner, where nonlinear and multivariable behavior is still present. In this case, considering the error of accurately realizing the dynamic behavior of the system as a significant disturbance vector applied to the system, reaching to a zero steady-state tracking error by using this structure becomes infeasible.

Even if the knowledge of the dynamic model of the system is accurate, achieving precise trajectory tracking through controller design is not easily realized. This introduces another significant error to the dynamics of the tracking error system. Another notable shortcome of this structure is its reliance on the robot's mass matrix, $\boldsymbol{M}_e(\boldsymbol{q})$, resulting in a dependence of the tracking performance on the robot's configuration in the workspace. Consequently, tuning the gains of the PD controller becomes more complicated in practice. To overcome these challenges, the fundamental method of inverse dynamics control has been developed, which will be explored in the following section.

12.3 Inverse Dynamics Control (IDC)

As mentioned before, the multivariable PD control structure, even with the use of a feed-forward term, does not enable achieving desirable tracking error in the presence of disturbances and uncertainty in the system model. In this section, we introduce inverse dynamics control, a method devised by researchers to tackle these challenges.

In the robotics literature, this control method is also referred to as the "Computed Torque (CT) method" [133]. On the other hand in control literature, a comprehensive category of approaches, similar to this method, that utilize inverse dynamics to linearize the system dynamics and achieve desirable tracking error in the feedback loop, is termed "Feedback Linearization Method" [134, 135]. In this book, acknowledging that both titles are entirely suitable for introducing this method, we will use the term Inverse Dynamics Control (IDC), which has been repeatedly used in the robotics literature [136, 137, 138].

In this approach, by utilizing the inverse dynamics of the system, the generated torque to compensate the system dynamics is calculated and added as a corrective term to the multivariable PD or PID controller designed before. By using this computed torque, in the ideal scenario, the multivariable and nonlinear dynamics of the robot are linearized, and the performance of linear controllers in the closed-loop system is significantly enhanced. It

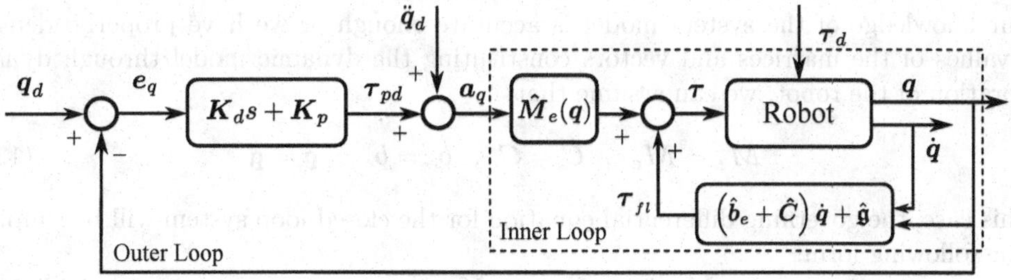

FIGURE 12.2
Block diagram of inverse dynamics control of the robot in joint space.

is important to note that this approach is considered as a fundamental method in robot control, and many other methods have been grounded and developed based on its extension.

For a detailed discussion, let's consider the overall structure of the inverse dynamics control in the block diagram shown in Figure 12.2. This structure employs two feedback loops, referred to as the inner and outer loops. Focusing on the inner control loop, we observe that based on the robot's inverse dynamics and utilizing feedback from joint variables at each instant of time, the torque τ_{fl} is computed and applied to the robot as a corrective term. In this computation, the Christoffel matrix, the gravitational vector of the robot, and the effective friction torques in the joints are utilized.

The outer loop consists of the usual PD or PID controller designed before for the motion control of the robot. Additionally, by imposing the nominal mass matrix of the robot $\hat{M}(q)$ on the desired motion path, the acceleration vector of the robot's desired motion $a(q)$ is utilized as another input to the inner loop. It's important to note that in practice, the actual values of the matrices and vectors of the dynamic model of the system are not available, and their nominal values are used in the calculation of the feedback linearization torque τ_{fl}. The application of this calculated torque on the system diminishes the influence of nonlinear dynamic terms of the system, hence this method is renowned as feedback linearization in control literature.

By this means, and using the block diagram shown in Figure 12.2, the dynamic equation of the closed-loop system can be obtained step by step, in the following order:

$$\tau = \hat{M}_e(q) a_q + \tau_{fl} \tag{12.9}$$

By substituting τ_{fl} into it, we will have:

$$\tau = \hat{M}_e(q) a_q + \left(\hat{b}_e + \hat{C}(q,\dot{q})\right)\dot{q} + \hat{g}(q) \tag{12.10}$$

which, in general, the vector a_q is in the form of an acceleration term with a dimension of n, calculated as follows.

$$a_q = \ddot{q}_d + K_d \dot{e}_q + K_p e_q \tag{12.11}$$

Due to utilizing the nonlinear dynamic model of the system, this control structure falls into the category of model-based controllers. To analyze the performance of this control method, obtain the closed-loop dynamic equation of the system in the presence of the PD controller and the inner feedback linearizing loop as follows.

$$M_e(q)\ddot{q} + \left(C(q,\dot{q}) + b_e\right)\dot{q} + g(q) = \tau_{fl} + \tau_{pd} + \tau_d \tag{12.12}$$

By substituting the values of τ_{pd} and τ_{fl} and simplifying, we arrive at the following equation.

$$\begin{aligned} &M_e(q)\ddot{q} + \left(C(q,\dot{q}) + b_e\right)\dot{q} + g(q) = \\ &= \hat{M}_e(q)\left(\ddot{q}_d + K_d \dot{e}_q + K_p e_q\right) + \left(\hat{b}_e + \hat{C}(q,\dot{q})\right)\dot{q} + \hat{g}(q) + \tau_d \end{aligned} \tag{12.13}$$

If our knowledge of the system model is accurate enough or we have properly identified the values of the matrices and vectors constituting the dynamic model through dynamic calibration of the robot, we can assume that

$$\hat{\boldsymbol{M}}_e = \boldsymbol{M}_e \ , \ \ \hat{\boldsymbol{C}} = \boldsymbol{C} \ , \ \ \hat{\boldsymbol{b}}_e = \boldsymbol{b}_e \ , \ \ \hat{\mathfrak{g}} = \mathfrak{g} \tag{12.14}$$

In this case, the governing differential equation for the closed-loop system will be simplified to the following form.

$$\boldsymbol{M}_e \left(\ddot{\boldsymbol{e}}_q + \boldsymbol{K}_d \dot{\boldsymbol{e}}_q + \boldsymbol{K}_p \boldsymbol{e}_q \right) + \boldsymbol{\tau}_d = 0 \tag{12.15}$$

Given the positive-definiteness of the robot's mass matrix, the dynamics of the tracking error can be characterized using the following second-order differential equation system:

$$\ddot{\boldsymbol{e}}_q + \boldsymbol{K}_d \dot{\boldsymbol{e}}_q + \boldsymbol{K}_p \boldsymbol{e}_q = -\boldsymbol{M}_e^{-1} \boldsymbol{\tau}_d \tag{12.16}$$

This equation clearly illustrates that by appropriately selecting gains for the PD controller matrices, one can achieve desirable transient and steady-state behavior of tracking error across the entire workspace. The disturbance torques considered in this equation, denoted as τ_d, does not encompass gravitational forces. This is because the torque resulting from gravitational forces, represented by $\mathfrak{g}$, has already been compensated within the inner feedback loop of the system.

Consequently, if other disturbance torques acting on the robot τ_d diminish over time, the tracking error will asymptotically converge to zero for all joints. Additionally, as evident from this equation, the matrices and vectors constituting the dynamics of the tracking error are independent of the kinematic configuration and the end-effector's pose, simplifying the tuning of the PD controller gains across the entire workspace of the robot. With this perspective, inverse dynamics control emerges as a highly effective method in the control of serial robots, serving as a foundational approach in the development of more complex control strategies.

The desirable features of this control structure are only attainable when assumptions based on complete knowledge of the system's dynamics (Equation 12.14) remain valid. As discussed in Chapters 8 and 9, deriving the dynamics formulation of industrial robots is a highly complex and time-consuming task, with the calibration of these equations and determining model parameters requiring another time-intensive effort. Under these practical conditions, equation 12.13 reflects the dynamic behavior of tracking error more realistically, where all the nonlinearities are not eliminated from the dynamics formulation. With a slight simplification, this dynamic behavior can be expressed in the following form.

$$\begin{aligned} &\ddot{\boldsymbol{e}}_q + \boldsymbol{K}_d \dot{\boldsymbol{e}}_q + \boldsymbol{K}_p \boldsymbol{e}_q = \\ &\boldsymbol{M}_e^{-1} \left\{ \left(\boldsymbol{M}_e - \hat{\boldsymbol{M}}_e \right) \ddot{q} + \left(\boldsymbol{b}_e - \hat{\boldsymbol{b}}_e \right) \dot{q} + \left(\boldsymbol{C} - \hat{\boldsymbol{C}} \right) \dot{q} + \left(\mathfrak{g} - \hat{\mathfrak{g}} \right) - \boldsymbol{\tau}_d \right\} \end{aligned} \tag{12.17}$$

In this equation, the nonlinear and multivariable behavior in the final error dynamics persists. However, considering the close proximity of the actual values of the matrices and vectors of the robot's dynamic model to their nominal values, the outcome of inaccurate identification of the dynamic model can be regarded as a disturbance vector with a small amplitude. Essentially, this disturbance vector is applied to the system in addition to the external disturbance vector τ_d. Consequently, the significance of the linear controller in the outer loop becomes more evident. This controller is tasked with mitigating the effects of disturbances and uncertainties in the modeling from the system's tracking error. In general, the outer loop controller does not necessarily need to remain as simple as the basic linear PD controller, and the utilization of PID controllers and even linear $\mathcal{H}_\infty$ controllers are seen in many robotics applications [139].

Inverse dynamics control is a highly regarded and extensively utilized method in the robotics literature. Its significance stems from the linearization of the robot's nonlinear dynamics, achieved by incorporating the mass matrix along the desired trajectory. With these considerations, the dynamics of the tracking error become independent of the robot's configuration and the pose of its end-effector. Consequently, tuning simple linear controllers or designing linear robust control methods becomes advantageous. However, since nonlinear dynamics are not completely eliminated in practice, the use of nonlinear robust control methods designed based on the Lyapunov theorem can be considered in the development of this controller. This approach, extensively discussed in Section 12.6.2, introduces another corrective term to the control structure, enhancing the robot's performance in tracking the desired trajectories.

Moreover, given the model-reference nature of inverse dynamics control and the ability to express the robot's model in a generalized form of linear regression, adaptive control approaches have also garnered attention in enhancing this method. In adaptive control methods, the robot's control begins with the nominal model, and over time by using tracking error data, this model expressed in the form of linear regression is adapted to reduce tracking error in the closed-loop system. The design of the adaptation law and stability analysis of the closed-loop system in this approach are also based on the Lyapunov theorem, details of which will be explored in Section 12.6.4.

12.4 Partial Feedback Linearization

Inverse dynamics control has unique theoretical features that have made it a successful method. However, it should be noted that for the complete implementation of this method, it is necessary to determine the dynamic equations of the robot and calibrate its parameter values through kinematic calibration and dynamic identification. This introduces two significant challenges in implementing this approach. First, determining the robot's dynamic formulation in closed form, especially determining the Christoffel matrix, is very complex and time-consuming. This issue becomes evident, particularly in robots with more than three degrees of freedom.

On the other hand, robot calibration requires obtaining the dynamic equations in the form of linear regression and designing appropriate experiments to determine these parameters. If the nonlinear dynamics of the robot are well identified, employing feedback linearization can provide a very efficient performance in the system. However, implementing inverse dynamics control, given the aforementioned challenges, is computationally expensive and time-consuming. In this section, we present a streamlined version of inverse dynamics control that substantially decreases computational expenses, all while preserving fundamental aspects of the closed-loop system, including stability and the convergence of the steady-state tracking error to zero.

In this method, known as "Partial Feedback Linearization" in the control community, there is no need to determine and calibrate the mass and Christoffel matrices. Only the gravity vector is employed to linearize a portion of the nonlinear dynamics that directly affects the steady-state tracking error. By this means, this approach may be categorized into weight compensation methods using feedback. Since the gravity vector in the system's dynamic equations is solely determined by the potential energy of the system, calculating this vector is significantly simpler compared to determining the mass and Christoffel matrices. Furthermore, for gravity vector calibration, static experiments and utilizing the equilibrium states of the robot arms in various configurations can also be employed, reducing the cost and time required for calibrating parameters related to the gravity vector.

Thus, in the simplest version of this method, the input torque vector is generated solely using the system's gravity vector in the inner loop and a PD controller in the outer loop.

$$\boldsymbol{\tau} = \boldsymbol{\tau}_{pd} + \boldsymbol{\tau}_{fl} \tag{12.18}$$

where

$$\boldsymbol{\tau}_{pd} = \boldsymbol{K}_d \dot{\boldsymbol{e}}_q + \boldsymbol{K}_p \boldsymbol{e}_q \ , \ \ \boldsymbol{\tau}_{fl} = \hat{\mathfrak{g}}(\boldsymbol{q}) \tag{12.19}$$

To eliminate the steady-state tracking error, complete knowledge of the gravity vector, as the only term used in the inner loop linearization with feedback, must be available. In this case, by assuming that:

$$\hat{\mathfrak{g}}(\boldsymbol{q}) = \mathfrak{g}(\boldsymbol{q}) \tag{12.20}$$

the dynamic behavior of a closed-loop system in the absence of disturbance torques $\boldsymbol{\tau}_d$ is expressed as follows.

$$\boldsymbol{M}_e(\boldsymbol{q})\ddot{\boldsymbol{q}} + (\boldsymbol{C}(\boldsymbol{q}, \dot{\boldsymbol{q}}) + \boldsymbol{b}_e)\,\dot{\boldsymbol{q}} = \boldsymbol{K}_d \dot{\boldsymbol{e}}_q + \boldsymbol{K}_p \boldsymbol{e}_q \tag{12.21}$$

Observe that the dynamic behavior of the system remains multivariable and nonlinear, with only the effect of the gravity vector removed. Nevertheless, despite the simplicity of this control method, its performance, especially in steady-states, is highly desirable. To validate this claim and analyze the closed-loop stability, we utilize the Lyapunov stability theorem. A brief review of stability analysis in nonlinear systems is presented in Appendix B.

To examine the tracking error behavior in a steady-state, it is necessary for the desired trajectory to converge to a constant value over time; therefore, such analysis can be performed by assuming $\dot{\boldsymbol{q}}_d = 0$. For the analysis of the dynamic behavior of the closed-loop system, the total energy of the closed-loop system is considered as the Lyapunov function candidate.

$$V(\boldsymbol{q}) = \frac{1}{2}\dot{\boldsymbol{q}}^T \boldsymbol{M}(\boldsymbol{q})\dot{\boldsymbol{q}} + \frac{1}{2}\boldsymbol{e}_q^T \boldsymbol{K}_p \boldsymbol{e}_q \tag{12.22}$$

What is noticeable in this relation is that the proposed Lyapunov function is composed of two terms: the first term represents the kinetic energy of the system, and the second term represents the potential energy induced by the proportional controller gain, analogous to a mechanical spring storing the system's energy. The velocity vector of the system's motion appears in the kinetic energy term, while in the potential energy term, the tracking error vector is present.

The proposed Lyapunov function is positive in all points of the workspace, except in points where the traversed trajectory has reached its desired value and has come to a rest, i.e., $\boldsymbol{q} = \boldsymbol{q}_d \rightarrow \dot{\boldsymbol{q}} = 0$. In such case the Lyapunov function is zero, and thus, the necessary positive-definite condition for the Lyapunov function candidate in nonlinear systems is satisfied.

In the Lyapunov theorem, stability analysis and the steady-state behavior of tracking error are carried out by calculating and analyzing the derivative of the Lyapunov function with respect to time. If the derivative of this positive-definite function, is negative-definite, the stability of the closed-loop system is guaranteed, and the tracking error asymptotically converges to zero.

Considering the desired trajectory where $\dot{\boldsymbol{q}}_d = 0$, the derivative of the Lyapunov function with respect to time is expressed as follows:

$$\dot{V}(\boldsymbol{q}) = \dot{\boldsymbol{q}}^T \boldsymbol{M}(\boldsymbol{q})\ddot{\boldsymbol{q}} + \frac{1}{2}\dot{\boldsymbol{q}}^T \dot{\boldsymbol{M}}(\boldsymbol{q})\dot{\boldsymbol{q}} + \dot{\boldsymbol{e}}_q^T \boldsymbol{K}_p \boldsymbol{e}_q \tag{12.23}$$

At this stage of stability analysis, the dynamic equations of the closed-loop system come into the picture. Substitute the expression $\boldsymbol{M}(\boldsymbol{q})\ddot{\boldsymbol{q}}$ using the relation in 12.21 and simplify:

$$\dot{V}(\boldsymbol{q}) = \dot{\boldsymbol{q}}^T\left(\boldsymbol{K}_d\dot{\boldsymbol{e}}_q + \boldsymbol{K}_p\boldsymbol{e}_q - (\boldsymbol{C}(\boldsymbol{q},\dot{\boldsymbol{q}}) + \boldsymbol{b}_e)\,\dot{\boldsymbol{q}}\right) + \frac{1}{2}\dot{\boldsymbol{q}}^T\dot{\boldsymbol{M}}(\boldsymbol{q})\dot{\boldsymbol{q}} + \dot{\boldsymbol{e}}_q^T\boldsymbol{K}_p\boldsymbol{e}_q$$

Given the assumption of a constant desired trajectory in steady-state, the derivative of the tracking error is equal to $\dot{\boldsymbol{e}}_q = \dot{\boldsymbol{q}}_d - \dot{\boldsymbol{q}} = -\dot{\boldsymbol{q}}$. Therefore,

$$\begin{aligned}
\dot{V}(\boldsymbol{q}) &= -\dot{\boldsymbol{q}}^T\boldsymbol{K}_d\dot{\boldsymbol{q}} + \dot{\boldsymbol{q}}^T\boldsymbol{K}_p\boldsymbol{e}_q + \frac{1}{2}\dot{\boldsymbol{q}}^T\left(\dot{\boldsymbol{M}} - 2\boldsymbol{C}\right)\dot{\boldsymbol{q}} - \dot{\boldsymbol{q}}^T\boldsymbol{b}_e\dot{\boldsymbol{q}} - \dot{\boldsymbol{q}}^T\boldsymbol{K}_p\boldsymbol{e}_q \\
&= -\dot{\boldsymbol{q}}^T\boldsymbol{K}_d\dot{\boldsymbol{q}} + \frac{1}{2}\dot{\boldsymbol{q}}^T\left(\dot{\boldsymbol{M}} - 2\boldsymbol{C}\right)\dot{\boldsymbol{q}} - \dot{\boldsymbol{q}}^T\boldsymbol{b}_e\dot{\boldsymbol{q}} \\
&= -\dot{\boldsymbol{q}}^T\left(\boldsymbol{K}_d + \boldsymbol{b}_e\right)\dot{\boldsymbol{q}} \leq 0
\end{aligned} \tag{12.24}$$

Equation 12.24 is derived based on the skew-symmetry property of the matrix $\dot{\boldsymbol{M}} - 2\boldsymbol{C}$. Here, the importance of rewriting robot dynamic equations in closed form and utilizing the Christoffel matrix becomes more apparent. With this property, it is possible to derive the negative semi-definite derivative of the Lyapunov function without directly using the mass matrix $\boldsymbol{M}(q)$ and the Christoffel matrix $\boldsymbol{C}(\boldsymbol{q},\dot{\boldsymbol{q}})$. Utilizing the Krasovski-Lasalle's Theorem, the Lyapunov function's negative semi-definite derivative can be used to prove the asymptotic stability of the system and the convergence of the tracking error toward zero.

To demonstrate that, we will show that the case where $\dot{\boldsymbol{q}} = 0$ and $\boldsymbol{q} \neq \boldsymbol{q}_d$ can be never reached. For this means, consider the Lyapunov function in the state where $\dot{V} \equiv 0$. In this case, according to Equation 12.24, $\dot{\boldsymbol{q}} \equiv 0$ must be true. In this scenario, the dynamic equation describing the behavior of the closed-loop system, given by Equation 12.21, simplifies to a more straightforward form of

$$\boldsymbol{K}_p\boldsymbol{e}_q = 0 \tag{12.25}$$

This implies that the tracking error will inevitably be reached to zero, $\boldsymbol{e}_q = 0$. Consequently, according to the Krasovski-Lasalle's Theorem, in a steady-state where the robot has reached its desired trajectory, $\boldsymbol{q} = \boldsymbol{q}_d$, and the tracking error of the closed-loop system will asymptotically converge toward zero.

This analysis clearly demonstrates that even if the mass matrix and Christoffel matrix are not directly used in the linearization of the robot through feedback, and although the dynamics of the closed-loop system will not become completely linear, the PD controller, along with compensating for the gravity vector $\mathbf{g}(\boldsymbol{q})$, will force the tracking error for constant trajectories to asymptotically converge toward zero.

It is important to note that this analysis solely addresses the steady-state behavior of the system, and the transient behavior of tracking error dynamics may not be as optimal as in the case of full inverse dynamics control. Furthermore, it should be emphasized that this control algorithm represents the simplest form of partial linearization through feedback in the absence of complete knowledge of the robot's dynamics. To shape the transient behavior of tracking error dynamics toward desired values, the utilization of available mass or Christoffel matrices, if accessible, or stronger controllers such as PID and $\mathcal{H}_\infty$ controller can provide significant assistance. However, if, for any reason, the cost of determining complete dynamic formulation or conducting calibration experiments is prohibitive, this simple control structure can ensure the main properties of inverse dynamics control, namely the stability of the robot's motion and convergence of the tracking error toward zero.

12.5 Inverse Dynamics Control in Task Space

As discussed in the section 9.5, using the Jacobian matrix mapping, one can obtain the dynamic equations in task space according to the relationship 9.25. However, we observed that considering the presence of the Jacobian matrix and its derivatives in the mass matrix, Christoffel matrix, and gravity vector in the task space, in accordance with the relationship 9.31, makes the nonlinear dynamic equations of the robot in the workspace more complex compared to the dynamic equations in joint space. Therefore, it is recommended that when using inverse dynamics control in task space, a simpler control structure shown in Figure 12.3 be used.

As seen in this proposed structure, the inner loop of inverse dynamics control has not changed, and the dynamic equations determined in joint space are still in use for feedback linearization. With this precaution, there is no need to use very complex matrices of dynamic equations obtained in the task space, and only the components of the outer control loop are carefully designed in task space.

Examining this control structure through the lens of control topologies, reveals its foundation in the third topology outlined in Section 11.2.3 and its alignment with the configuration depicted in Figure 11.3. In this topology, the variables measured in joint space are utilized, proving significantly more economical in terms of measurement costs compared to measurements in task space variables. Moreover, employing direct kinematics facilitates the transformation of joint variables into task space, enabling the feedback loop within the task space. Consequently, the direct observation and control of the desired trajectory in task space become feasible, enabling effective implementation of objectives such as collision avoidance in the workspace or adherence to a desired-shaped path and similar goals.

The primary distinction of this configuration from inverse dynamics control in joint space lies in the implementation of a PD controller in task space, guided by the tracking error $\boldsymbol{e}_{\mathcal{X}}$ within this space. In this manner, the resultant torque output from this controller, denoted as $\boldsymbol{\tau}_{pd}$, is implemented within the task space. Simultaneously, the generated acceleration variable, serving as the input for the inner loop denoted as $\boldsymbol{a}_{\mathcal{X}}$, is also interpreted within the task space. Leveraging the Jacobian matrix and its derivatives, this acceleration variable, by its very nature, can be translated into a joint space variable denoted by $\boldsymbol{a}_{\boldsymbol{q}}$. For this purpose, the mapping of the Jacobian matrix and its derivatives can be utilized as follows:

$$\begin{aligned} \dot{\boldsymbol{\mathcal{X}}} &= \boldsymbol{J}(\boldsymbol{q})\dot{\boldsymbol{q}} \\ \ddot{\boldsymbol{\mathcal{X}}} &= \boldsymbol{J}(\boldsymbol{q})\ddot{\boldsymbol{q}} + \dot{\boldsymbol{J}}(\boldsymbol{q})\dot{\boldsymbol{q}} \end{aligned} \tag{12.26}$$

In this case, the variable $\boldsymbol{a}_{\boldsymbol{q}}$ is obtained using this mapping as follows:

$$\boldsymbol{a}_{\boldsymbol{q}} = \boldsymbol{J}^{-1}(\boldsymbol{q})\left(\boldsymbol{a}_{\mathcal{X}} - \dot{\boldsymbol{J}}(\boldsymbol{q})\dot{\boldsymbol{q}}\right) \tag{12.27}$$

Furthermore, with the proposed structure of this method illustrated in Figure 12.3, $\boldsymbol{a}_{\mathcal{X}}$ is calculated as follows:

$$\boldsymbol{a}_{\mathcal{X}} = \ddot{\boldsymbol{\mathcal{X}}}_d + \boldsymbol{K}_d\dot{\boldsymbol{e}}_{\mathcal{X}} + \boldsymbol{K}_p\boldsymbol{e}_{\mathcal{X}} \tag{12.28}$$

Consider the case where the knowledge of the system model is sufficiently accurate and the dynamic calibration of the robot identifies the values of matrices and vectors forming the dynamic model correctly. In this case, the governing differential equation for the closed-loop system, using this control structure, in the absence of disturbance torques, will be simplified to the following form.

$$\ddot{\boldsymbol{e}}_{\mathcal{X}} + \boldsymbol{K}_d\dot{\boldsymbol{e}}_{\mathcal{X}} + \boldsymbol{K}_p\boldsymbol{e}_{\mathcal{X}} = 0 \tag{12.29}$$

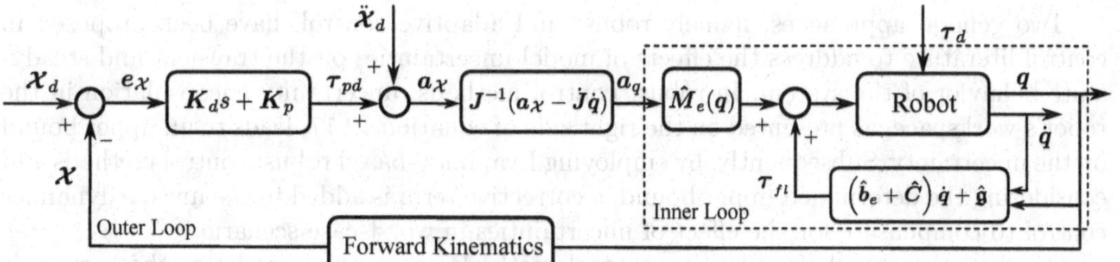

FIGURE 12.3
Block diagram of inverse dynamics control for the robot in task space.

This equation clearly illustrates that by choosing appropriate gains for the PD controller, one can effectively shape the transient behavior and steady-state tracking error throughout the entire workspace. Just like in the case of inverse dynamics control in joint space, the matrices and vectors governing the tracking error dynamics in this equation remain unaffected by the kinematic arrangement and the robot's configuration in the workspace. Consequently, the adjustment of PD controller gains is made more accessible.

As depicted in Figure 12.3, the design of inverse dynamics control in task space has been revamped with minimal alterations from its original form. This redesign removes the necessity to derive robot dynamic equations in task space, with only the outer loop incorporating tracking errors in task space. Moreover, the integration of forward kinematics in the feedback loop eliminates the explicit measurement of motion variables in task space. Consequently, the PD controller is effectively designed in task space, and its coefficients are fine-tuned according to the tracking error in this space.

Despite these considerations, the adoption of inverse dynamics control in task space is not as widespread as its counterpart in joint space. The primary reason for this limitation stems from the method's reliance on forward kinematics and the Jacobian matrix, along with its derivatives, in the feedback loop. Consequently, uncertainties in the kinematic calibration of the robot can substantially impact the performance of this method. Furthermore, the use of the Jacobian matrix inverse in equation 12.27 renders this method unsuitable for underactuated robots, redundantly actuated robots or those exhibiting singular configurations. In such cases, the tracking performance experiences a significant decline even in regions of the workspace where the robot approaches a singular state.

12.6 Robust and Adaptive Control

Despite its widespread use in various applications, inverse dynamics control encounters two primary drawbacks. Firstly, its performance relies on complete knowledge of the dynamic model of the system and the calibration of its parameters. This is challenging in practice, as despite efforts to determine an accurate system model, the existence of model uncertainties is inevitable. Secondly, implementing this approach can be constrained by the need to integrate a precise system model featuring diverse nonlinear and multivariable expressions in the closed-loop form. Applying such a complex model to linearize the system through feedback significantly amplifies computational complexity. Consequently, this necessitates more powerful and expensive hardware for successful implementation.

Two general approaches, namely robust and adaptive control, have been proposed in control literature to address the effects of model uncertainties on the transient and steady-state behavior of the system. In robust control methods, uncertainty encapsulation in the robot's workspace, as presented on the right side of equation 12.17, leads to an upper bound on the uncertainty. Subsequently, by employing Lyapunov-based robust control methods and considering the determined upper bound, a corrective term is added to the inverse dynamics control to compensate for the effect of uncertainties in worst-case scenario.

On the other hand, in adaptive control methods, a nominal model of the system is initially used in the inverse dynamics control loop. By observing the dynamics tracking error and updating the model parameters based on the regression form of the model, the model parameters are gradually adjusted for desirable tracking. Typically, the adaptation law for model parameters in this method is also designed based on the Lyapunov stability analysis of the closed-loop system.

It is noteworthy that in both robust and adaptive control methods, the nonlinear dynamic structure of the robot is considered, and the need for complete linearization in dynamic equations is eliminated. Although the stability of the closed-loop system in the presence of uncertainties is analyzed using Lyapunov-based methods in both approaches, the compensation of uncertainties in these two methods is entirely different. In the robust control approach, a fixed controller is used to compensate for uncertainty in its worst-case effect, requiring significant control effort. However, in adaptive control methods, a controller with time-varying coefficients is used to gradually adjust the model parameters, while ensuring convergence in parameter adaptation.

The choice between these two approaches depends on the interpretation of uncertainties. If the complete structure of the model is assumed to be known, and uncertainties are limited to parameter variations, adaptive methods will be more effective. On the other hand, if intentional neglect or inadvertent ignorance of a part of the model dynamics is considered as uncertainty, robust control methods will be more suitable. A combination of these two general approaches is also proposed under the titles of Adaptive Robust Control and Robust Adaptive Control, while the details of these approaches are beyond the scope of this book [140, 141].

12.6.1 Model Uncertainty

The term "model uncertainty" refers to the disparity between the behavior of the nominal model and the actual behavior of the system. As evident in this interpretation, the manner in which the mathematical modeling of a system is performed is crucial for providing a suitable mathematical encapsulation of model uncertainty. If a system is modeled with a linear model using a transfer function, it is natural for the interpretation of uncertainty in the Laplace space, expressible in frequency domain. On the other hand, if the dynamic model of a robot is expressed based on the state space and in the form of a nonlinear system, the interpretation of uncertainty in the system can manifest as disturbance torques in the governing differential equation of the system.

In general, uncertainties existing in a system can be classified into two main categories: structured and unstructured uncertainties. In structured uncertainty, it is assumed that the nominal model of the system has the capability to reproduce the dynamic behavior of the system, even though the model parameters may not be accurately calibrated or may change over time and under environmental conditions. Since the structure of the system is known in this type of uncertainty, it is referred to as structured uncertainty. Meanwhile, as this category of uncertainties is based on the lack of precise knowledge about model parameters, it is also known as parametric uncertainty.

Conversely, numerous uncertainties encountered in actual systems stem from an incomplete understanding of the dynamic system behavior, deliberate simplifications, or inaccuracies introduced while modeling certain dynamic behaviors during the modeling procedure. Take, for instance, an electrical resistor: the linearity of the voltage-current relationship holds true if it works under specific conditions, excluding high frequencies, where the resistor shows inductive or capacitive characteristics. It is inherent that the inductive behavior of a resistor cannot be precisely elucidated by meticulously calibrating the linear relation between voltage-current and resistance value.

Similar to this example, many components and devices working in real systems exhibit such characteristics. We often model these devices with linear models, even though their linear behavior is only valid in a small range of their inputs. This type of modeling error is called unstructured model uncertainty because in this type of uncertainty, the structure of the model is ambiguous, not its parameters.

Another example of such uncertainty in model structure is observed in the partial feedback linearization control method. As mentioned earlier, assuming that the dynamic model of the system accurately reconstructs the system's behavior, complexities in determining and calibrating the mass and Christoffel matrices have deliberately been avoided by partially linearizing the dynamic behavior of the system. Consequently, in the closed-loop system model, nonlinear expressions persist, and linear control methods alone have to compensate for the lingering effects of this nonlinear dynamic behavior in the system.

In adaptive control methods, considering the need for a reliable nominal model in the system and attempting to adjust model parameters based on tracking errors, structured uncertainties are taken into account. This is while, in robust control methods, the bound on the modeling uncertainty with respect to its main dynamic behavior is encapsulated, and its effect is compensated for as a structured model uncertainty in the worst-case. Consequently, in these methods, unstructured uncertainties are employed in the system modeling.

In the realm of robotic systems, when a significant amount of time and energy is invested in precise system modeling and calibration, one can presume that the robot is afflicted solely with parametric uncertainties. Under such circumstances, adaptive control, leveraging linear regression models, can effectively offset the influence of uncertain system parameters. This stands in contrast to scrutinizing the unmodeled nonlinear behavior of the closed-loop system and the perturbing effects it introduces to the system. These terms are delineated on the right side of Equation 12.17, and in the robust control approach, the bounds of these terms are ascertained in the worst-case scenario. Consequently, robust control methods take precedence in mitigating the impact of uncertainties on the system's tracking error. The ensuing sections will explore the robust inverse dynamics control method first, followed by an exploration of the adaptive inverse dynamics control method.

12.6.2 Robust Inverse Dynamics Control

As elaborated, the efficiency of the widely-used inverse dynamics control method is contingent when complete information about the robot's dynamic model, including matrices, vectors, and constituent parameters, is available. However, in practice, this assumption is rarely met, and attaining satisfactory tracking is achievable only through the refinement of this method, taking into account uncertainties in the model. In the robust inverse dynamics control method, by leveraging the structure of inverse dynamics control and the inclusion of a corrective term will offset the impacts of model uncertainties, ensuring both stability and desirable tracking.

In order to better understand this approach, as depicted in Figure 12.4, the corrective term $\boldsymbol{\delta}_a$ is added to the typical inverse dynamic structure. In this scenario, the applied

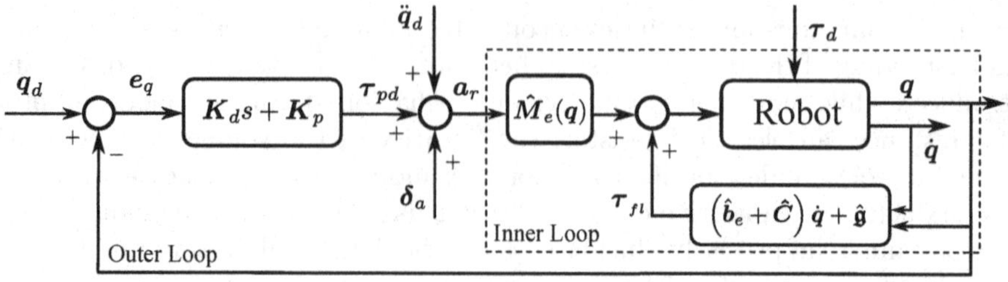

FIGURE 12.4
Block diagram of robust inverse dynamics control for the robot.

torques to the robot can be expressed as follows.

$$\tau = \hat{M}_e(q)a_r + \left(\hat{b}_e + \hat{C}(q, \dot{q})\right)\dot{q} + \hat{\mathfrak{g}}(q) \tag{12.30}$$

In which the input a_r, generally having a dimension of n and representing acceleration, is made robust by the addition of the corrective term δ_a.

$$a_r = \ddot{q}_d + K_d\dot{e}_q + K_p e_q + \delta_a \tag{12.31}$$

Designing a corrective term in a way that preserves the stability of the closed-loop system in the presence of uncertainties while ensuring satisfactory performance is a subject addressed in robust control methods for nonlinear systems [134].

If the notation $(\hat{\cdot})$ is used to represent the nominal value of the quantity $(\cdot)$, and the error of adaptation, representing the difference between the nominal value and the actual value, is denoted by $(\tilde{\cdot}) = (\hat{\cdot}) - (\cdot)$, the error of parameter uncertainty for the robot can be expressed as follows.

$$\tilde{M}_e = \hat{M}_e - M_e \ ; \ \tilde{C} = \hat{C} - C \ ; \ \tilde{b}_e = \hat{b}_e - b_e \ ; \ \tilde{\mathfrak{g}} = \hat{\mathfrak{g}} - \mathfrak{g} \tag{12.32}$$

Similarly, the notion $(\tilde{\cdot})$ is used to denote the error in the motion variables.

$$\tilde{q} = q - q_d = -e_q \ ; \ \dot{\tilde{q}} = \dot{q} - \dot{q}_d = -\dot{e}_q \tag{12.33}$$

Note that this notation is very common in nonlinear control systems, even though with this definition, the tracking error enters subsequent calculations with a negative sign. With this description, the dynamics of the closed-loop system using this control structure can be written as follows.

$$\ddot{q} = a_r + \eta(q, \dot{q}, a_r) \tag{12.34}$$

In this equation, η is a function that represents the level of uncertainty in the modeling, and can be determined by:

$$\eta = M_e^{-1}\left(\tilde{M}_e a_r + \left(\tilde{b}_e + \tilde{C}\right)\dot{q} + \tilde{\mathfrak{g}}\right) \tag{12.35}$$

The corrective term in robust control, δ_a, should be designed in a way to reduce the impact of uncertainty η on the closed-loop system behavior. For this purpose, by rewriting the relationship in Equation 12.31 using the notation $(\tilde{\cdot})$,

$$a_r = \ddot{q}_d - K_d\dot{\tilde{q}} - K_p\tilde{q} + \delta_a \tag{12.36}$$

and substituting it into the robot dynamics, we have:

$$\ddot{\tilde{q}} + \boldsymbol{K}_d\dot{\tilde{q}} + \boldsymbol{K}_p\tilde{q} = \boldsymbol{\delta}_a + \boldsymbol{\eta} \tag{12.37}$$

In this context, the tracking error dynamics exhibit second-order characteristics, with its transient response being adaptable through the adjustable gains of the PD controller. Examining the other facet of this relation, where the impacts of uncertainties in the robot's model, represented by $\boldsymbol{\eta}$, and the corrective term of the robust controller, $\boldsymbol{\delta}_a$, are incorporated, the tracking error dynamics persists in its nonlinear and multivariable nature, despite the resemblance to multivariable linear systems.

If $\boldsymbol{\eta}$ were to be known, compensating for the effects of model uncertainties could be easily accomplished by selecting the corrective term $\boldsymbol{\delta}_a = -\boldsymbol{\eta}$. However, it's crucial to acknowledge that $\boldsymbol{\eta}$ represents all structural and parametric uncertainties in the system and, as a result, cannot be precisely determined. The use of the term "uncertainty" underscores this aspect, since this representation encompasses all unmodeled dynamics, intentional or unintentional simplifications in system modeling, and imprecise knowledge of model parameters during calibration experiments.

Furthermore, as inferred from Equation 12.35, this uncertainty is dependent on the system's state variables $\boldsymbol{q}, \dot{\boldsymbol{q}}$, and $\boldsymbol{a}_r$. Hence, the upper bound of this term can only be determined through experiments under various operational conditions within the workspace. The corrective term, employing high gains, counteracts the impact of this uncertainty in the worst-case scenario. This process is executed using the Lyapunov redesign method [134]. By defining the state vector as follows,

$$\boldsymbol{\varepsilon} = \begin{bmatrix} \tilde{q} \\ \dot{\tilde{q}} \end{bmatrix} = \begin{bmatrix} q - q_d \\ \dot{q} - \dot{q}_d \end{bmatrix} \tag{12.38}$$

The representation of the closed-loop dynamic of the robot in state space can be written as:

$$\dot{\boldsymbol{\varepsilon}} = \boldsymbol{A}\boldsymbol{\varepsilon} + \boldsymbol{B}\left(\boldsymbol{\delta}_a + \boldsymbol{\eta}\right) \tag{12.39}$$

where

$$\boldsymbol{A} = \begin{bmatrix} \boldsymbol{0} & \boldsymbol{I} \\ -\boldsymbol{K}_p & -\boldsymbol{K}_d \end{bmatrix}, \quad \boldsymbol{B} = \begin{bmatrix} \boldsymbol{0} \\ \boldsymbol{I} \end{bmatrix} \tag{12.40}$$

The depiction of the system's state space serves to elucidate the modification term $\boldsymbol{\delta}_a$ within the robustness framework. Essential for the stability of the tracking error dynamics is the stability of its linear component, which is characterized by the matrix $\boldsymbol{A}$. To ensure stability, it is imperative to tailor the gains of the PD controller such that the matrix $\boldsymbol{A}$ is Hurwitz. Consequently, all eigenvalues of the matrix $\boldsymbol{A}$ will be strictly positioned to the left of the imaginary axis.

This condition can be ensured using the Lyapunov equation. By appropriately choosing the gains of the PD controller, a symmetric positive-definite matrix $\boldsymbol{P}$ will exist such that, for any symmetric positive-definite matrix $\boldsymbol{Q}$, the Lyapunov equation holds:

$$\boldsymbol{A}^T\boldsymbol{P} + \boldsymbol{P}\boldsymbol{A} = -\boldsymbol{Q} \tag{12.41}$$

Notice that, the primary part of the error equation, incorporating nonlinear and uncertain expressions, still remains. In order to stabilize this component, the corrective term $\boldsymbol{\delta}_a$ must be designed in such a way to satisfy the Lyapunov theorem conditions for the closed-loop system. To achieve this, we define the Lyapunov function candidate for the system using the weighted Euclidean norm of the state error vector of the system.

$$V = \boldsymbol{\varepsilon}^T\boldsymbol{P}\boldsymbol{\varepsilon} \tag{12.42}$$

In which, the matrix $\boldsymbol{P}$ is determined by solving the Lyapunov equation represented in Equation 12.41. The Lyapunov function in Eq. 12.42 is positive-definite, and if its derivative is negative-definite, the stability of the equilibrium point of the closed-loop system is guaranteed. By taking the derivative of this function with respect to time, we can write:

$$\begin{aligned} \dot{V} &= \dot{\boldsymbol{\varepsilon}}^T \boldsymbol{P} \boldsymbol{\varepsilon} + \boldsymbol{\varepsilon}^T \boldsymbol{P} \dot{\boldsymbol{\varepsilon}} \\ &= \boldsymbol{\varepsilon}^T \left(\boldsymbol{A}^T \boldsymbol{P} + \boldsymbol{P} \boldsymbol{A} \right) \boldsymbol{\varepsilon} + 2 \boldsymbol{\varepsilon}^T \boldsymbol{P} \boldsymbol{B} (\boldsymbol{\delta}_a + \boldsymbol{\eta}) \\ &= -\boldsymbol{\varepsilon}^T \boldsymbol{Q} \boldsymbol{\varepsilon} + 2 \boldsymbol{\varepsilon}^T \boldsymbol{P} \boldsymbol{B} (\boldsymbol{\delta}_a + \boldsymbol{\eta}) \end{aligned} \tag{12.43}$$

Note that in the first equation of the tracking error, dynamic equation 12.39 is substituted, and the second equation is simplified using the Lyapunov equation 12.41.

The first term in the equation 12.43 is negative-definite. To ensure the stability of the closed-loop system, $\boldsymbol{\delta}_a$ must be designed in a way that, for the most stringent conditions of uncertainty $\boldsymbol{\eta}$, the second term remains negative-definite. To achieve this, assume that the upper bound of uncertainty $\boldsymbol{\eta}$ has been determined through various experiments on the system and can be represented as a scalar and positive function of state variables and time, denoted as $\rho(\boldsymbol{\varepsilon}, t)$.

$$\|\boldsymbol{\eta}\| \leq \rho(\boldsymbol{\varepsilon}, t) \tag{12.44}$$

By introducing the notion

$$\boldsymbol{v} = \boldsymbol{B}^T \boldsymbol{P} \boldsymbol{\varepsilon} \tag{12.45}$$

The second term in the equation 12.43 is rewritten as $2\boldsymbol{v}^T(\boldsymbol{\delta}_a + \boldsymbol{\eta})$. This expression is zero only when $\boldsymbol{v} = \boldsymbol{0}$, and it is negative-definite when the corrective term is defined as follows:

$$\boldsymbol{\delta}_a = -\rho \frac{\boldsymbol{v}}{\|\boldsymbol{v}\|} \tag{12.46}$$

Certainly, this choice has been made among various options proposed in the control literature [134]. To prove this claim, the Schwartz inequality is employed, and subsequently, the negative-definiteness of this expression is demonstrated as follows.

$$\begin{aligned} \boldsymbol{v}^T(\boldsymbol{\delta}_a + \boldsymbol{\eta}) &= \boldsymbol{v}^T(-\rho \frac{\boldsymbol{v}}{\|\boldsymbol{v}\|} + \boldsymbol{\eta}) \\ &\leq -\rho \|\boldsymbol{v}\| + \|\boldsymbol{v}\| \|\boldsymbol{\eta}\| \\ &\leq \|\boldsymbol{v}\| (-\rho + \|\boldsymbol{\eta}\|) \\ &\leq 0 \end{aligned} \tag{12.47}$$

The last inequality is derived given the relation in Equation 12.44, where we have $\|\boldsymbol{\eta}\| \leq \rho$. Thus the derivative of the Lyapunov function given in Equation 12.43 simplifies to

$$\dot{V} \leq -\boldsymbol{\varepsilon}^T \boldsymbol{Q} \boldsymbol{\varepsilon} < 0 \tag{12.48}$$

These calculations clearly demonstrate that by designing the corrective term based on the relationship in Equation 12.46, stability of the closed-loop system is guaranteed under the worst-case uncertainty conditions.

Let us rewrite and analyze the final corrective term designed in this method.

$$\boldsymbol{\delta}_a = \begin{cases} -\rho \frac{\boldsymbol{v}}{\|\boldsymbol{v}\|} & \text{if} \ \ \|\boldsymbol{v}\| \neq 0 \\ 0 & \text{if} \ \ \|\boldsymbol{v}\| = 0 \end{cases} \tag{12.49}$$

The value of the vector $\boldsymbol{v}$ is obtained by calculating $\boldsymbol{v} = \boldsymbol{B}^T \boldsymbol{P}\boldsymbol{\varepsilon}$, where the symmetric positive-definite matrix $\boldsymbol{P}$ is determined by solving the Lyapunov equation given in Equation 12.41. Furthermore, the uncertainty bound of the model is represented by the scalar variable ρ, where $|\eta| \leq \rho$.

As observed in this design approach, the appropriate choice of the corrective term is made based on stability analysis and the application of the Lyapunov theorem. This method is widely used and effective, having been validated in numerous other applications [142]. However, considering the nature of determining the corrective term in this approach, there are various choices available, and using the Euclidean norm of the vector $\boldsymbol{v}$ in the form presented in Eq. 12.49 is just one option among many.

Another noteworthy consideration in the design path of this robust controller is the assurance of stability. However, in practice, achieving the desired performance in robust trajectory tracking is essential. Does stability assurance also guarantee the attainment of the desired tracking performance? This is not true in general, however, to describe how we may achieve robust performance, it is necessary to revisit the discussion on the magic of feedback and the utilization of high gains, as addressed in Chapter 11. As illustrated there, in linear systems, employing feedback in conjunction with high-gain controllers enables the attainment of three key objectives: desired tracking, disturbance rejection, and resilience against model inaccuracies. While these fundamental principles remain applicable in nonlinear control systems, establishing their validity is not as straightforward as in linear systems. The crucial factor in achieving the desired performance lies in the utilization of high-gain controllers.

Does the corrective term have a high gain? In general, ρ should cover all uncertainties arising from the uncertainty in system modeling. If the amount of uncertainty is small, there is generally no need to implement robust methods, and inverse dynamics control itself is considered a suitable and optimal choice for robot control. The choice of large gains often stems from the practical difficulty of precisely determining the upper bound of uncertainty through experiments. Despite sufficient theoretical groundwork for defining this bound, a large value is considered and then adjusted to achieve the desired tracking. In this way, while ensuring the stability of the closed-loop system, because of using such high gains in the controller structure, the desired tracking is achieved in the presence of disturbances and uncertainties in the model.

This situation turns the selection of the ρ coefficient, which practically measures the level of uncertainty in the system, into one of the control challenges in addition to tuning the gains of the linear PD controller. In practice, to escape from any complexity and ensure the stability of the closed-loop system, a large value is chosen for this coefficient. It is natural that the limitations of sensors and actuators, mentioned in linear control systems and discussed in Section 11.3.4, are present here as well, and should be taken into account.

Another aspect observed in the nature of the corrective term is its piecewise and non-smooth character in computation and implementation. To illustrate this point, let's consider the simplified form of the equation 12.49 in the single variable case. In this case, the expression $\boldsymbol{v}/\|\boldsymbol{v}\| = v/|v|$ implies using only the sign of the signal, not its magnitude, denoted by (v), $v/|v| = \text{sgn(v)}$. Considering the large gain used in this controller ρ, with a negative sign, if the value of the signal v derived from the tracking error is positive, a large negative term is applied, if the tracking error is negative, a large positive term is applied, and only if the tracking error is exactly zero, a zero term is applied as the corrective term to the system. The multivariable form of this expression used in Eq. 12.49 employs the weighted signs of its elements instead of the signal value, ensuring the angle of the control vector is preserved.

Controllers with similar characteristics are widely used in industrial control systems. In the control of industrial processes, especially in applications where programmable logic controllers (PLCs) are used, on-off control structures similar to the type of control structure used here, receive attention [143]. Furthermore, in optimal control of systems, the *bang-bang* controller, which precisely has the same switching structure between two large positive and negative values, is employed [144]. In the context of nonlinear control systems, this type of control structure is referred to as "Variable Structure Control (VSC)" or "Sliding Mode Control (SMC)," and it has been used in various and diverse applications [145, 146]. The common feature of these controllers is the high-gain switching nature, based on the tracking error in the system.

This feature is highly effective and popular in reducing the impact of modeling uncertainty on the tracking error, and for this reason, it is suggested as the adopted choice in robust inverse dynamics control in this book. Nevertheless, research development in this area to provide robust control algorithms is significant, to the extent that various books have been written on this topic [147, 148]. One limitation of the proposed method in practical applications is the phenomenon of chattering, which causes unintended oscillations. Further explanations on this issue and how to mitigate it will be discussed in this following section.

12.6.3 Chattering Mitigation

As observed, the corrective term in robust inverse dynamics control, represented as $\boldsymbol{\delta}_a$ in Equation 12.49, is a piecewise and discontinuous function with respect to time. It solely reaches zero when the tracking error achieves zero; otherwise, given the large gain of ρ, its magnitude becomes significantly positive or negative.

In theory, functions such as this one, serving as switching functions to generate control effort, demonstrate effectiveness in achieving asymptotic tracking stability. However, as depicted in Figure 12.5, practical implementation encounters challenges due to noise affecting the measurement of joint variables and calculation of tracking error. As it is shown in this figure, the measured value contaminated with noise never precisely reaches zero. This signifies that in practical scenarios, asymptotic tracking stability is lost, and a persistent error will remain in the system's closed-loop output.

On the other hand, high-gain control in the input of the system implies exciting fast dynamics in the system and creating small but persistent oscillations on the system's output. In control literature, this phenomenon is referred to as "chattering." As depicted in Figure 12.5, chattering not only impedes the convergence of tracking error toward zero but also introduces noticeable and unwanted oscillations in the system's motion through the application of high-frequency input. This will not only reduce the lifespan of the robot's components and equipment but also give rise to undesirable tracking performance. While the use of high gain in the corrective term makes the system's output behavior more robust, it also has a significant impact on the occurrence of the chattering phenomenon.

To mitigate the effect of chattering, various methods have been proposed in the literature. Among these methods, continuous approximations of the switching function such as saturation, hyperbolic tangent, or fuzzy functions have been suggested [149, 150]. Additionally, the use of higher-order sliding mode control methods [151, 152] can mitigate the effect of chattering under specific conditions.

To elaborate on this issue, notice that since the discontinuity in the corrective term that causes chattering occur in the vicinity of zero error, smoothing this term can significantly help to combat this phenomenon. The simplest type of smoothing is to replace the signum function (sgn), to a saturation function (sat). In this approach, the infinite slope of the signum function is replaced with a large but bounded slope to transform the discontinuous function into a continuous one. If a differentiable alternative is desired, the hyperbolic tangent function can be used. By using the saturation function, when the tracking error is

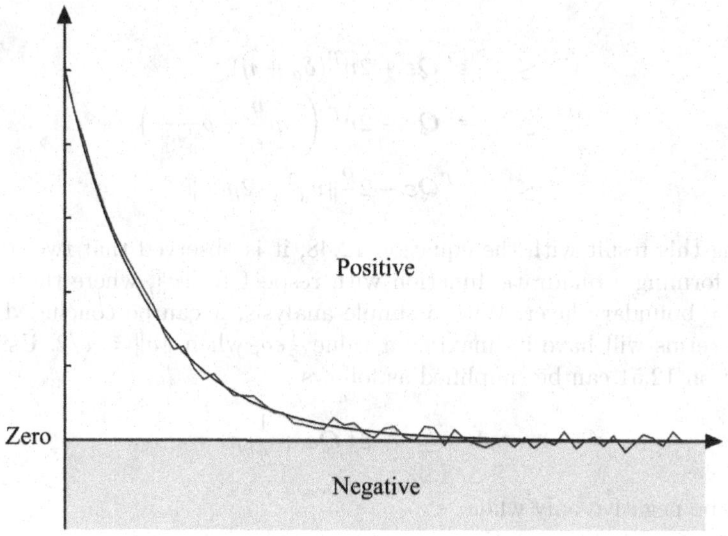

FIGURE 12.5
Creation of the chattering phenomenon in high-gain switching control on a variable contaminated with measurement noise.

within a range close to zero (referred to as the boundary layer), the control effort smoothly transitions from its positive limit (ρ) to its negative limit ($-\rho$). Assuming the use of the saturation function, the corrective term is modified as follows:

$$\delta_a = \begin{cases} -\rho\frac{\boldsymbol{v}}{\|\boldsymbol{v}\|} & \text{if} \quad \|\boldsymbol{v}\| > \epsilon \\ -\rho\frac{\boldsymbol{v}}{\epsilon} & \text{if} \quad \|\boldsymbol{v}\| \le \epsilon \end{cases} \tag{12.50}$$

In this relation, ϵ represents the boundary layer width, as if the norm of the vector $\boldsymbol{v}$ falls within it ($|\boldsymbol{v}| \le \epsilon$), the discontinuous switching function is replaced by the linear function $-\rho\frac{\boldsymbol{v}}{\epsilon}$. It is evident that choosing a smaller ϵ will result in a larger slope for the smoothed linear function, making this approximation closer to the signum function. However, it should be noted that to eliminate the chattering phenomena, the boundary layer width ϵ must be chosen large enough to cover the range of measurement noise. In practice, by initially selecting a very small value for ϵ, it is gradually increased to observe avoiding undesirable oscillations caused by chattering in the system's output.

Although this minor correction of the corrective term is very effective in mitigating the chattering phenomenon, the stability analysis performed on the closed-loop system is no longer valid. It is necessary to revisit and rectify the stability analysis of the closed-loop system using the modified corrective term 12.50. For this purpose, consider the previous Lyapunov function candidate 12.42 and differentiate it with respect to time. For $|\boldsymbol{v}| \ge \epsilon$, no change occurs in the control effort compared to before. Therefore, it can be concluded that in this interval, $\dot{V} < 0$.

However, upon entering the boundary layer $|\boldsymbol{v}| \le \epsilon$, the control effort of the system and, consequently, the derivative of the Lyapunov function with respect to time changes as follows:

$$\begin{aligned}\dot{V} &\leq -\boldsymbol{\varepsilon}^T\boldsymbol{Q}\boldsymbol{\varepsilon} + 2\boldsymbol{v}^T(\boldsymbol{\delta}_a + \boldsymbol{\eta}) \\ &\leq -\boldsymbol{\varepsilon}^T\boldsymbol{Q}\boldsymbol{\varepsilon} + 2\boldsymbol{v}^T\left(-\rho\frac{\boldsymbol{v}}{\epsilon} + \rho\frac{\boldsymbol{v}}{\|\boldsymbol{v}\|}\right) \\ &\leq -\boldsymbol{\varepsilon}^T\boldsymbol{Q}\boldsymbol{\varepsilon} - 2\frac{\rho}{\epsilon}\|\boldsymbol{v}\|^2 + 2\rho\|\boldsymbol{v}\| \end{aligned} \tag{12.51}$$

By comparing this result with the equation 12.48, it is observed that two new terms have been added, forming a quadratic function with respect to $\|\boldsymbol{v}\|$, where these terms can be positive in the boundary layer. With a simple analysis, it can be concluded that the sum of these two terms will have its maximum value $\frac{1}{2}\epsilon\rho$, when $\|\boldsymbol{v}\| = \epsilon/2$. Using this upper bound, equation 12.51 can be simplified as follows.

$$\dot{V} \leq -\boldsymbol{\varepsilon}^T\boldsymbol{Q}\boldsymbol{\varepsilon} + \frac{1}{2}\epsilon\rho \tag{12.52}$$

Thus, $\dot{V}$ will be negative only when:

$$\boldsymbol{\varepsilon}^T\boldsymbol{Q}\boldsymbol{\varepsilon} > \frac{1}{2}\epsilon\rho \tag{12.53}$$

On the other hand, we know that $\boldsymbol{\varepsilon}^T\boldsymbol{Q}\boldsymbol{\varepsilon}$ has the following upper and lower bounds:

$$\underline{\lambda}(\boldsymbol{Q})\|\boldsymbol{\varepsilon}\| \leq \boldsymbol{\varepsilon}^T\boldsymbol{Q}\boldsymbol{\varepsilon} \leq \bar{\lambda}(\boldsymbol{Q})\|\boldsymbol{\varepsilon}\| \tag{12.54}$$

where $\underline{\lambda}(\boldsymbol{Q})$ and $\bar{\lambda}(\boldsymbol{Q})$ are the minimum and maximum eigenvalues of the matrix $\boldsymbol{Q}$, respectively. Therefore, with some conservatism, the following condition must be satisfied for $\dot{V}$ to remain negative-definite.

$$\underline{\lambda}(\boldsymbol{Q})\|\boldsymbol{\varepsilon}\| > \frac{1}{2}\epsilon\rho \tag{12.55}$$

This condition can be simplified to the following final form:

$$\|\boldsymbol{\varepsilon}\| > \sqrt{\frac{\epsilon\rho}{2\underline{\lambda}(\boldsymbol{Q})}} \doteq \delta_e \tag{12.56}$$

This analysis demonstrates that if the tracking error norm lies outside the ball B_δ with radius δ_e, the derivative of the Lyapunov function will remain negative definite. This will cause all state error trajectories in this space to eventually enter the ball B_δ. However, it should be noted that asymptotic convergence of the error toward zero will not occur because the derivative of the Lyapunov function is not negative definite globally, and its negative-definite behavior near the origin is not certain. This type of stability is referred to in control literature as "Uniformly Ultimately Bounded (UUB) stability," which in this book, we briefly refer to as UUB stability. For further study on the definition of this stability and its properties, refer to Appendix B. Furthermore, for a deeper analysis of the final steady-state error, we need to study UUB stability using upper and lower bounds of the Lyapunov function. Interested readers are recommended to refer to the reference [134, Section 4.8] for more details.

Nevertheless, what is directly inferred from the equation 12.56 is that the final steady-state error in the system state variables is proportional to the product of the uncertainty level ρ and the width of the boundary layer ϵ. This implies that to reduce the steady-state error, a better understanding of the system is required, and as much as possible, the boundary layer width should be minimized. This reveals the trade-off between mitigating the chattering phenomena and reducing the ultimate steady-state error in the system.

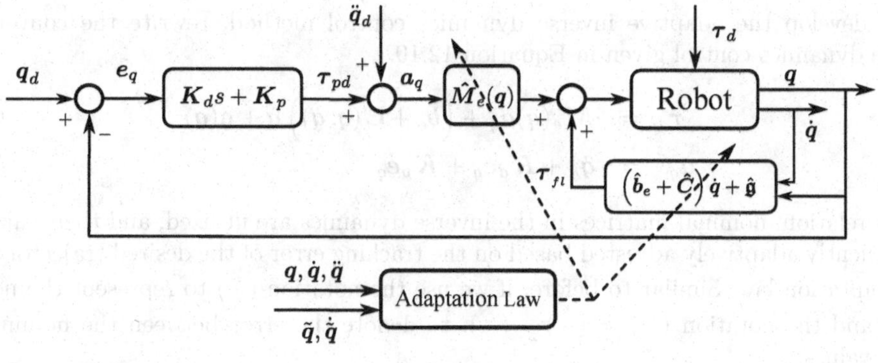

FIGURE 12.6
Block diagram of Adaptive Inverse Dynamics Control (AIDC) for the robot.

12.6.4 Adaptive Inverse Dynamics Control

One of the enhanced methods based on Inverse Dynamics Control (IDC) is its adaptive version. In Adaptive Inverse Dynamics Control (AIDC), the complete dynamic formulation of the system are used in the inverse dynamics. Initially, nominal values of matrices and vectors are utilized in their computation. Then, the numerical values of the model's dynamic parameters are adaptively adjusted to minimize the tracking error in the closed-loop system. Ideally, this adaptation aims to converge the system errors asymptotically toward zero.

Note that the objective in both robust and adaptive control methods is the same, but the approach to achieving it differs. The goal of both methods is to eliminate the effect of model uncertainty on the final behavior of the closed-loop system. In robust control, a fixed controller with large gains is used to mitigate the effect of model uncertainty in the worst-case scenario. This is in contrast to adaptive control, where the identified model parameters are traced using tracking error, and the controller adapts its structure over time. Thus, in addition to the stability of the closed-loop system in adaptive inverse dynamics control, the convergence of parameter estimation becomes crucial.

Hence, in robust control, there is no need for precise knowledge of the model structure, and only the structural uncertainty bounds in the model need to be determined. This is in contrast to adaptive control, which requires a proper understanding of the model structure in the form of linear regression, and only the model parameters are considered to be structurally uncertain. Considering this possibility of parameter estimation divergence, robust control methods are preferred in real-world applications.

In the adaptive inverse dynamics control as illustrated in Figure 12.6, we rely on the structure of inverse dynamics control and assuming that the robot's inverse dynamic equations are known in the nominal state. Then we use the adaptation law to identify the real-time parameters of the model in linear regression form and employ it in inverse dynamics to compensate for the effects of uncertain model structure in desired tracking. Fortunately, as demonstrated in Section 9.4, by determining the dimension of base inertial parameter space, it is possible to rewrite the dynamic equations in the form of linear regression based on the parameters.

$$M_e(q)\ddot{q} + (b_e + C(q,\dot{q}))\,\dot{q} + \mathfrak{g}(q) = \mathcal{Y}(q,\dot{q},\ddot{q})\Phi \tag{12.57}$$

where the matrix $\mathcal{Y}(q,\dot{q},\ddot{q})$ serves as the regressor, and Φ is the vector of robot base inertial parameters. Having the regression form of equations, it is possible to experimentally determine the robot parameters using least squares error methods.

To develop the adaptive inverse dynamics control method, rewrite the conventional inverse dynamics control given in Equation 12.10.

$$\tau = \hat{M}_e(q)a_q + \left(\hat{b}_e + \hat{C}(q,\dot{q})\right)\dot{q} + \hat{\mathfrak{g}}(q) \tag{12.58}$$

$$a_q = \ddot{q}_d + K_d\dot{e}_q + K_p e_q \tag{12.59}$$

In this relation, nominal matrices in the inverse dynamics are utilized, and their values are subsequently adaptively adjusted based on the tracking error of the desired trajectory using the adaptation law. Similar to before, if we use the notation $(\hat{\cdot})$ to represent the nominal value, and the notation $(\tilde{\cdot}) = (\hat{\cdot}) - (\cdot)$, to denote the error between the nominal and actual values:

$$\tilde{M}_e = \hat{M}_e - M_e \ ; \ \tilde{C} = \hat{C} - C \ ; \ \tilde{b}_e = \hat{b}_e - b_e \ ; \ \tilde{\mathfrak{g}} = \hat{\mathfrak{g}} - \mathfrak{g}$$

Furthermore, we will use the notation $(\tilde{\cdot})$ to denote the error in the joint variables.

$$\tilde{q} = q - q_d = -e_q \ ; \ \dot{\tilde{q}} = \dot{q} - \dot{q}_d = -\dot{e}_q$$

With this description, the dynamics of the closed-loop system in the absence of disturbance torques can be written as follows:

$$\begin{aligned} M_e\ddot{q} + (C + b_e)\dot{q} + \mathfrak{g} &= \tau \\ &= \hat{M}_e\left(\ddot{q}_d - K_d\dot{\tilde{q}} - K_p\tilde{q}\right) + \left(\hat{b}_e + \hat{C}\right)\dot{q} + \hat{\mathfrak{g}} \end{aligned} \tag{12.60}$$

By adding and subtracting $\hat{M}_e\ddot{q}$ on the right side of this equation and simplifying, we can write:

$$\tilde{M}_e\ddot{q} + \left(\tilde{C} + \tilde{b}_e\right)\dot{q} + \tilde{\mathfrak{g}} = \hat{M}_e\left(\ddot{\tilde{q}} + K_d\dot{\tilde{q}} + K_p\tilde{q}\right)$$

Accordingly, the dynamic tracking error equation can be written based on the linear regression form of system dynamics.

$$\begin{aligned} \ddot{\tilde{q}} + K_d\dot{\tilde{q}} + K_p\tilde{q} &= \hat{M}_e^{-1}\left(\tilde{M}_e\ddot{q} + \left(\tilde{C} + \tilde{b}_e\right)\dot{q} + \tilde{\mathfrak{g}}\right) \\ &= \hat{M}_e^{-1}\mathcal{Y}(q,\dot{q},\ddot{q})\tilde{\Phi} \end{aligned} \tag{12.61}$$

To facilitate the appropriate design of the adaptation law, as before rewrite the error dynamic formulation in state space. By defining the state vector ε, in the form:

$$\varepsilon = \begin{bmatrix} \tilde{q} \\ \dot{\tilde{q}} \end{bmatrix} = \begin{bmatrix} q - q_d \\ \dot{q} - \dot{q}_d \end{bmatrix} \tag{12.62}$$

the dynamic state-space representation of the closed-loop system can be written as follows:

$$\dot{\varepsilon} = A\varepsilon + BZ\tilde{\Phi} \tag{12.63}$$

where

$$A = \begin{bmatrix} 0 & I \\ -K_p & -K_d \end{bmatrix} , \quad B = \begin{bmatrix} 0 \\ I \end{bmatrix} , \quad Z = \hat{M}_e^{-1}\mathcal{Y}(q,\dot{q},\ddot{q}) \tag{12.64}$$

The representation of the system's state space demonstrates the adaptation problem-solving approach effectively. Similar to what was expressed in the analysis of robust inverse dynamics control, the necessary condition for the stability of the tracking error dynamics is that its linear part be stable. The linear part of this dynamics is represented by the matrix $\boldsymbol{A}$, and for stability, it is necessary to choose the control gains of the PD controller in such a way that all the eigenvalues of the matrix $\boldsymbol{A}$ will lie strictly to the left of the imaginary axis.

This condition can be ensured using the Lyapunov equation. With the appropriate selection of the gains of the PD controller, there will exist a symmetric positive-definite matrix $\boldsymbol{P}$ such that, for any arbitrary symmetric positive-definite matrix $\boldsymbol{Q}$, the following Lyapunov equation holds true:

$$\boldsymbol{A}^T\boldsymbol{P} + \boldsymbol{P}\boldsymbol{A} = -\boldsymbol{Q} \tag{12.65}$$

However, the main part of the error equation, which incorporates nonlinear and uncertain terms, $\boldsymbol{B}\boldsymbol{\mathcal{Z}}\tilde{\boldsymbol{\Phi}}$, is still present. To ensure the convergence of the model parameter estimation and the stability of the closed-loop system, the adaptive law must be designed in a way that the effect of this nonlinear term does not lead the system to instability. Here, we also employ Lyapunov stability theory. For this purpose, we propose a Lyapunov function candidate for the system using the Euclidean norm of the weighted state vector in addition to the weighted estimation error vector of the parameters:

$$V = \boldsymbol{\varepsilon}^T\boldsymbol{P}\boldsymbol{\varepsilon} + \tilde{\boldsymbol{\Phi}}^T\boldsymbol{\Gamma}\tilde{\boldsymbol{\Phi}} \tag{12.66}$$

In this equation, the matrix $\boldsymbol{P}$ is determined by solving the Lyapunov equation given in 12.65. Additionally, the weighting matrix $\boldsymbol{\Gamma}$ is another symmetric positive-definite matrix used in the adaptive law. The inclusion of the system state variables $\boldsymbol{\varepsilon}$ along with the parameter estimation variables $\boldsymbol{\Phi}$ in the Lyapunov function ensures the comprehensive examination of the closed-loop system's stability, incorporating both the desired controller and the adaptation law simultaneously.

Considering the positive-definiteness of the Lyapunov candidate function, if its derivative is negative-definite, the stability of the closed-loop system is guaranteed. Therefore, to achieve this, we will differentiate this function with respect to time.

$$\begin{aligned}\dot{V} &= \boldsymbol{\varepsilon}^T\left(\boldsymbol{A}^T\boldsymbol{P} + \boldsymbol{P}\boldsymbol{A}\right)\boldsymbol{\varepsilon} + 2\tilde{\boldsymbol{\Phi}}^T\boldsymbol{\mathcal{Z}}^T\boldsymbol{B}^T\boldsymbol{P}\boldsymbol{\varepsilon} + 2\tilde{\boldsymbol{\Phi}}^T\boldsymbol{\Gamma}\dot{\tilde{\boldsymbol{\Phi}}} \\ &= -\boldsymbol{\varepsilon}^T\boldsymbol{Q}\boldsymbol{\varepsilon} + 2\tilde{\boldsymbol{\Phi}}^T\left(\boldsymbol{\mathcal{Z}}^T\boldsymbol{B}^T\boldsymbol{P}\boldsymbol{\varepsilon} + \boldsymbol{\Gamma}\dot{\tilde{\boldsymbol{\Phi}}}\right)\end{aligned} \tag{12.67}$$

Note that in the first equation of the tracking error dynamics 12.61 is substituted, and the second equation is simplified using the Lyapunov equation 12.65.

The first term in equation 12.67 is negative-definite, and to ensure the stability of the closed-loop system, the adaptive law can be designed in such a way that the second term equals zero. In this case, the adaptive law can be formulated as follows:

$$\dot{\hat{\boldsymbol{\Phi}}} = \boldsymbol{\Gamma}^{-1}\boldsymbol{\mathcal{Z}}^T\boldsymbol{B}^T\boldsymbol{P}\boldsymbol{\varepsilon} \tag{12.68}$$

By utilizing this adaptive law, the second term in the equation 12.67 is eliminated, and considering the derivative of the Lyapunov function, it will become negative-definite with respect to the state variables.

$$\dot{V} = -\boldsymbol{\varepsilon}^T\boldsymbol{Q}\boldsymbol{\varepsilon} < 0 \tag{12.69}$$

Given that the derivative of the Lyapunov function is negative-definite, the stability of the closed-loop system is guaranteed, while the parameter estimation error remains bounded.

Figure 12.6 illustrates the structure of the adaptive inverse control. As observed in this figure, the structure of this control method is entirely similar to inverse dynamics control, with the difference that, based on the adaptive law presented in equation 12.68, the robot's inverse dynamics equations will be adapted. In this control structure, in addition to the PD controller matrices, the positive-definite matrix $\boldsymbol{\Gamma}$ must also be tuned. The larger the norm of this matrix, the less the intensity of model parameter adaptation, resulting in more stability, fewer oscillations in state variables, albeit at the cost of reduced response speed and increased time to attenuate the effects of parameter uncertainties in the transient behavior of the system.

To implement the adaptive law, we need the second derivative of the robot joint motions $\ddot{q}$ and the inverse of the estimated effective mass matrix $\hat{\boldsymbol{M}}$ in the calculation of $\boldsymbol{\mathcal{Z}}$. The requirement for joint accelerations is a significant limitation that complicates the implementation of this approach. If the gains of the linear controllers are properly tuned, it is possible to use the desired joint accelerations $\ddot{q}_d$ instead of the actual motions, effectively mitigating this constraint. Although this correction practically alleviates this limitation, it complicates the stability analysis and the guarantee of the convergence of the estimated parameters. Other methods have been developed to address these limitations in the robotics control literature, such as passivity-based approaches that overcome both of the aforementioned constraints [4, Section 9.4].

12.7 Summary of Motion Control Methods

In this chapter, various methods for nonlinear and multivariable control of serial robots have been introduced. In this section, we will provide a summary of these methods for a quick reference, obtaining a general understanding, and comparing them.

- Dynamic Model

The dynamic model governing the motion of a robot in joint space can generally be represented as follows:

$$\boldsymbol{M}_e(\boldsymbol{q})\ddot{\boldsymbol{q}} + \boldsymbol{C}(\boldsymbol{q},\dot{\boldsymbol{q}})\dot{\boldsymbol{q}} + \boldsymbol{b}_e\dot{\boldsymbol{q}} + \boldsymbol{\mathfrak{g}}(\boldsymbol{q}) = \boldsymbol{\tau} + \boldsymbol{\tau}_d$$

These equations can be written in the robot's task space in terms of the variable $\boldsymbol{\mathcal{X}}$ as follows:

$$\boldsymbol{M}_x(\boldsymbol{q})\ddot{\boldsymbol{\mathcal{X}}} + \boldsymbol{\mathfrak{v}}_x(\boldsymbol{q},\dot{\boldsymbol{q}}) + \boldsymbol{\mathfrak{g}}_x(\boldsymbol{q}) = \boldsymbol{\mathcal{F}} + \boldsymbol{\mathcal{F}}_d$$

These two models are related through the Jacobian matrix and its derivative.

$$\begin{aligned}
\boldsymbol{M}_x(\boldsymbol{q}) &= \boldsymbol{J}^{-T}(\boldsymbol{q})\boldsymbol{M}_e(\boldsymbol{q})\,\boldsymbol{J}^{-1}(\boldsymbol{q}) \\
\boldsymbol{\mathfrak{v}}_x(\boldsymbol{q},\dot{\boldsymbol{q}}) &= \boldsymbol{J}^{-T}(\boldsymbol{q})\left\{\boldsymbol{C}(\boldsymbol{q},\dot{\boldsymbol{q}}) + \boldsymbol{b}_e - \boldsymbol{M}(\boldsymbol{q})\boldsymbol{J}^{-1}(\boldsymbol{q})\dot{\boldsymbol{J}}(\boldsymbol{q})\right\}\dot{\boldsymbol{q}} \\
\boldsymbol{\mathfrak{g}}_x(\boldsymbol{q}) &= \boldsymbol{J}^{-T}(\boldsymbol{q})\boldsymbol{\mathfrak{g}}(\boldsymbol{q})
\end{aligned}$$

In addition, the dynamics formulation in joint space can be rewritten as a linear regression form with respect to a vector composed of the base model parameters $\boldsymbol{\Phi}$.

$$\boldsymbol{M}_e(\boldsymbol{q})\ddot{\boldsymbol{q}} + (\boldsymbol{C}(\boldsymbol{q},\dot{\boldsymbol{q}}) + \boldsymbol{b}_e)\,\dot{\boldsymbol{q}} + \boldsymbol{\mathfrak{g}}(\boldsymbol{q}) = \boldsymbol{\mathcal{Y}}(\boldsymbol{q},\dot{\boldsymbol{q}},\ddot{\boldsymbol{q}})\boldsymbol{\Phi}$$

- Multivariable PD Control

The simplest method for multivariable control in serial robots. If this controller is implemented in the joint space of the robot, the control effort is calculated as follows:

$$\boldsymbol{\tau} = \boldsymbol{\tau}_{pd} = \boldsymbol{K}_d \dot{\boldsymbol{e}}_q + \boldsymbol{K}_p \boldsymbol{e}_q$$

In this controller, the gain matrices $\boldsymbol{K}_d$ and $\boldsymbol{K}_p$ are adjusted through trial and error to achieve desired transient behavior and steady-state tracking error.

- Multivariable PD Control with Feed-Forward

To alleviate the limitations of multivariable PD control, the feed-forward torque $\boldsymbol{\tau}_{ff}$ can be added to the control effort of the closed-loop system as follows:

$$\boldsymbol{\tau} = \boldsymbol{\tau}_{pd} + \boldsymbol{\tau}_{ff}$$

where

$$\boldsymbol{\tau}_{ff} = \hat{\boldsymbol{M}}_e(\boldsymbol{q}_d)\ddot{\boldsymbol{q}}_d + \left(\hat{\boldsymbol{b}}_e + \hat{\boldsymbol{C}}(\boldsymbol{q}_d, \dot{\boldsymbol{q}}_d)\right)\dot{\boldsymbol{q}}_d + \hat{\mathfrak{g}}(\boldsymbol{q}_d)$$

- Inverse Dynamics Control

In inverse dynamics control, the dynamic equations of the robot are used to compute a feedback linearizing torque. If we implement this control method in the joint space, the required control effort is computed by:

$$\boldsymbol{\tau} = \hat{\boldsymbol{M}}_e(\boldsymbol{q})\boldsymbol{a}_q + \boldsymbol{\tau}_{fl}$$

where

$$\boldsymbol{a}_q = \ddot{\boldsymbol{q}}_d + \boldsymbol{K}_d \dot{\boldsymbol{e}}_q + \boldsymbol{K}_p \boldsymbol{e}_q$$

$$\boldsymbol{\tau}_{fl} = \left(\hat{\boldsymbol{b}}_e + \hat{\boldsymbol{C}}(\boldsymbol{q}, \dot{\boldsymbol{q}})\right)\dot{\boldsymbol{q}} + \hat{\mathfrak{g}}(\boldsymbol{q})$$

In this case, the dynamics of the tracking error in the presence of uncertainty in the model may be determined by:

$$\ddot{\boldsymbol{e}}_q + \boldsymbol{K}_d \dot{\boldsymbol{e}}_q + \boldsymbol{K}_p \boldsymbol{e}_q = \boldsymbol{M}_e^{-1}\left(\tilde{\boldsymbol{M}}_e \ddot{\boldsymbol{q}} + \tilde{\boldsymbol{b}}_e \dot{\boldsymbol{q}} + \tilde{\boldsymbol{C}}\dot{\boldsymbol{q}} + \tilde{\mathfrak{g}} - \boldsymbol{\tau}_d\right)$$

- Partial Feedback Linearization

In order to reduce the computational cost of inverse dynamics control, one can place only a portion of the robot's dynamic equations in the feedback linearization loop. If we compensate only the torque due to gravity forces, the required control effort is equal to:

$$\boldsymbol{\tau} = \boldsymbol{K}_d \dot{\boldsymbol{e}}_q + \boldsymbol{K}_p \boldsymbol{e}_q + \hat{\mathfrak{g}}(\boldsymbol{q})$$

To analyze the stability of the closed-loop system, one can use this Lyapunov function.

$$V(\boldsymbol{q}) = \frac{1}{2}\dot{\boldsymbol{q}}^T \boldsymbol{M}(\boldsymbol{q})\dot{\boldsymbol{q}} + \frac{1}{2}\boldsymbol{e}_q^T \boldsymbol{K}_p \boldsymbol{e}_q$$

The stability analysis of the closed-loop system using the Lyapunov theorem demonstrates that this control method is capable of ensuring the stability of the closed-loop system while driving the steady-state tracking error asymptotically to zero.

- Inverse Dynamics Control in Task Space

By utilizing the mapping of the Jacobian matrix and its derivative, inverse dynamics control can be implemented in the task space. In this case, the required control effort is determined by:

$$\boldsymbol{\tau} = \hat{\boldsymbol{M}}_e(\boldsymbol{q})\boldsymbol{a}_q + \boldsymbol{\tau}_{fl}$$

where, as before,

$$\boldsymbol{\tau}_{fl} = \left(\hat{\boldsymbol{b}}_e + \hat{\boldsymbol{C}}(\boldsymbol{q}, \dot{\boldsymbol{q}})\right)\dot{\boldsymbol{q}} + \hat{\boldsymbol{g}}(\boldsymbol{q})$$

However, $\boldsymbol{a}_q$ is calculated based on the tracking error in the task space and the mapping of the Jacobian matrix and its derivative.

$$\boldsymbol{a}_q = \boldsymbol{J}^{-1}(\boldsymbol{q})\left(\boldsymbol{a}_{\mathcal{X}} - \dot{\boldsymbol{J}}(\boldsymbol{q})\dot{\boldsymbol{q}}\right)$$
$$\boldsymbol{a}_{\mathcal{X}} = \ddot{\boldsymbol{\mathcal{X}}}_d + \boldsymbol{K}_d\dot{\boldsymbol{e}}_{\mathcal{X}} + \boldsymbol{K}_p\boldsymbol{e}_{\mathcal{X}}$$

- Robust Inverse Dynamics Control

In order to mitigate the impact of uncertainties in the robot's dynamic model, a corrective term is added to the control effort in robust inverse dynamics control. In this case,

$$\boldsymbol{\tau} = \hat{\boldsymbol{M}}_e(\boldsymbol{q})\boldsymbol{a}_r + \left(\hat{\boldsymbol{b}}_e + \hat{\boldsymbol{C}}(\boldsymbol{q}, \dot{\boldsymbol{q}})\right)\dot{\boldsymbol{q}} + \hat{\boldsymbol{g}}(\boldsymbol{q})$$
$$\boldsymbol{a}_r = \ddot{\boldsymbol{q}}_d + \boldsymbol{K}_d\dot{\boldsymbol{e}}_q + \boldsymbol{K}_p\boldsymbol{e}_q + \boldsymbol{\delta}_a$$

where the corrective and robust term $\boldsymbol{\delta}_a$ is obtained using the Lyapunov redesign method. In this case, the dynamics of the tracking error can be represented in the state space.

$$\dot{\boldsymbol{\varepsilon}} = \boldsymbol{A}\boldsymbol{\varepsilon} + \boldsymbol{B}(\boldsymbol{\delta}_a + \boldsymbol{\eta})$$

where $\boldsymbol{\eta}$ represents the bounded modeling uncertainty, and the matrices of this model are given by:

$$\boldsymbol{A} = \begin{bmatrix} \boldsymbol{0} & \boldsymbol{I} \\ -\boldsymbol{K}_p & -\boldsymbol{K}_d \end{bmatrix}, \quad \boldsymbol{B} = \begin{bmatrix} \boldsymbol{0} \\ \boldsymbol{I} \end{bmatrix}$$

The corrective and robust term $\boldsymbol{\delta}_a$ is calculated as follows:

$$\boldsymbol{\delta}_a = \begin{cases} -\rho\frac{\boldsymbol{v}}{\|\boldsymbol{v}\|} & \text{if} \quad \|\boldsymbol{v}\| > \epsilon \\ -\rho\frac{\boldsymbol{v}}{\epsilon} & \text{if} \quad \|\boldsymbol{v}\| \le \epsilon \end{cases}$$

where $\boldsymbol{v}$ is defined by the relation $\boldsymbol{v} = \boldsymbol{B}^T\boldsymbol{P}\boldsymbol{\varepsilon}$, and $\boldsymbol{P}$ is obtained from solving the Lyapunov equation introduced in the equation 12.41. In this case, by considering a boundary layer of width ϵ the chattering phenomenon is mitigated, while the system achieves UUB stability in the presence of bounded uncertainties.

- Adaptive Inverse Dynamics Control

In the adaptive version of inverse dynamics control, a nominal model of the system is used in feedback linearization, and the model parameters are adaptively updated using the tracking error. The dynamics of the error model in this case can be written in state-space form as follows:

$$\dot{\boldsymbol{\varepsilon}} = \boldsymbol{A}\boldsymbol{\varepsilon} + \boldsymbol{B}\boldsymbol{\mathcal{Z}}\tilde{\boldsymbol{\Phi}}$$

where

$$\boldsymbol{A} = \begin{bmatrix} \boldsymbol{0} & \boldsymbol{I} \\ -\boldsymbol{K}_p & -\boldsymbol{K}_d \end{bmatrix}, \quad \boldsymbol{B} = \begin{bmatrix} \boldsymbol{0} \\ \boldsymbol{I} \end{bmatrix}, \quad \boldsymbol{\mathcal{Z}} = \hat{\boldsymbol{M}}_e^{-1} \boldsymbol{\mathcal{Y}}(\boldsymbol{q}, \dot{\boldsymbol{q}}, \ddot{\boldsymbol{q}})$$

Using Lyapunov analysis and considering the Lyapunov function candidate:

$$V = \boldsymbol{\varepsilon}^T \boldsymbol{P} \boldsymbol{\varepsilon} + \tilde{\boldsymbol{\Phi}}^T \boldsymbol{\Gamma} \tilde{\boldsymbol{\Phi}}$$

The adaptation law is obtained in the following form:

$$\dot{\hat{\boldsymbol{\Phi}}} = \boldsymbol{\Gamma}^{-1} \boldsymbol{\mathcal{Z}}^T \boldsymbol{B}^T \boldsymbol{P} \boldsymbol{\varepsilon}$$

Using this control structure ensures the stability of the closed-loop system, and the estimation error of the parameters will remain bounded.

12.8 Nonlinear Control of Planar 3R Robot

In this section, we report on the simulation of nonlinear motion control of the planar 3R robot as a case study deliberated in this book. We implement various control methods introduced in this chapter on this case study and we conduct an extensive comparison and analysis of the outcomes.

For this means, let's examine the three-link revolute 3R robot illustrated in Figure 8.7, characterized by kinematic and mass parameters outlined in Table 12.1. For the purpose of this analysis, we assume that the robot links are slender rods, and their center of mass is positioned at the mid-lengths of the respective links. The motion trajectory for the robot arms is designed using quintic polynomial functions, where the initial, intermediate, and final values for each joint's motion are given in Table 12.1. As indicated in the table, the maneuver spans four seconds, with the joints reaching their maximum angles within the initial two seconds. Subsequently, over the next two seconds, the joints attain their final angle values, leading to a smooth cessation of motion with zero velocity and acceleration. To assess the impact of multivariable dynamics on the controller's performance, the motion of all three joints is considered simultaneous.

TABLE 12.1
Simulation parameters of the planar 3R robot in nonlinear control simulations

Quantity	Unit	Notation	Link 1	Link 2	Link 3
Length	m	a_i	1.0	0.7	0.3
Mass	Kg	m_i	10	7	3
Center of Mass	m	m_{c_i}	0.5	0.35	0.15
Moment of Inertia	$Kg \cdot m^2$	${}^C I_i$	0.833	0.286	0.023
Initial Angle	deg	θ_{i_0}	0	0	0
Intermediate Angle	deg	θ_{i_v}	90	−60	30
Final Angle	deg	θ_{i_f}	0	0	0
Maneuver Time	sec	t_f	4	4	4
Step Disturbance Amplitude	$N.m$	τ_{d_i}	50	20	2
Disturbance Injection Time	sec	t_d	1	2	3

Examine the dynamic formulation of the robot represented in matrix form, using the established relationships in Case Study 8.3 and the Christoffel matrix derived in Case Study 9.2. To explore the nonlinear characteristics of the system, we intentionally disregard the influences of the actuator and gearbox dynamics. Consequently, the dynamic equation of the system can be expressed as follows, which is used in the simulation of the robot motion.

$$\boldsymbol{\tau} + \boldsymbol{\tau}_d = \boldsymbol{M}(\boldsymbol{q})\ddot{q} + \boldsymbol{C}(\boldsymbol{q}, \dot{\boldsymbol{q}})\dot{\boldsymbol{q}} + \mathfrak{g}(\boldsymbol{q}) \tag{12.70}$$

To conduct motion control simulations for the robot, we adopt the prevalent motion control topology commonly used in serial robots, depicted in Figure 11.1. In this configuration, the design of the desired trajectory and motion control of the robot are executed in the joint space. Consequently, the desired joint trajectories are defined using the vector $\boldsymbol{q}_d$, and the robot controller uses the tracking error $\boldsymbol{e}_q = \boldsymbol{q}_d - \boldsymbol{q}$ and the understanding of the robot's dynamic structure to compute and apply the necessary actuator torques $\boldsymbol{\tau}$ to the robot. In the motion control simulation, the impact of disturbance torques in the joints $\boldsymbol{\tau}_d$, will also be investigated.

12.8.1 PID Control with Feed-Forward

In the simulations of the previous chapter, it was observed that achieving satisfactory performance in trajectory tracking could be realized by fine-tuning the gains of the PID controller. However, adjusting the controller gains in a way that maintains appropriate tracking error throughout the entire workspace requires a considerable amount of trial and error. Moreover, in the presence of measurement noise and actuator saturation, trajectory tracking will be severely affected.

In this section, we will investigate the impact of adding a feed-forward term based on the dynamic model of the system to the previously tuned PID controller. As discussed in this chapter, this method requires complete knowledge of the dynamic model of the system and calibrated robot parameters. However, since precise identification of kinematic and dynamic parameters involves various calibration experiments and comes with a significant computational cost, in this study, we assume these parameters are not accurately known. In the conducted simulations, we consider parameter uncertainties at a level of 50% to examine the response of the closed-loop system with the feed-forward controller under these severe conditions.

With this introduction, in this simulation, we use the tuned gains of the PID controller as the linear control simulation:

$$\boldsymbol{K}_p = \begin{bmatrix} 8 & 0 & 0 \\ 0 & 3 & 0 \\ 0 & 0 & 0.2 \end{bmatrix} \times 10^4, \quad \boldsymbol{K}_d = \boldsymbol{K}_p, \quad \boldsymbol{K}_i = 10\boldsymbol{K}_p \tag{12.71}$$

and apply the feed-forward term to the system using the following computed torque.

$$\boldsymbol{\tau}_{ff} = \hat{\boldsymbol{M}}(\boldsymbol{q}_d)\ddot{\boldsymbol{q}}_d + \hat{\boldsymbol{C}}(\boldsymbol{q}_d, \dot{\boldsymbol{q}}_d)\dot{\boldsymbol{q}}_d + \hat{\mathfrak{g}}(\boldsymbol{q}_d) \tag{12.72}$$

In this equation, we assume that the matrices and vectors constituting the dynamic model of the robot are calculated when their link lengths and masses are set to 50% of their actual values. The desired motion trajectory of the joints is considered as before what was presented in Figure 11.12. Additionally, disturbance torques $\boldsymbol{\tau}_d$ are applied as step torques in the three joints. According to Table 12.1, in the first second, a step torque disturbance with an amplitude of 50 $N.m$ is applied to the first joint, while in the 2nd second, a step torque disturbance with an amplitude of 20 $N.m$ is applied to the second joint. Finally, in

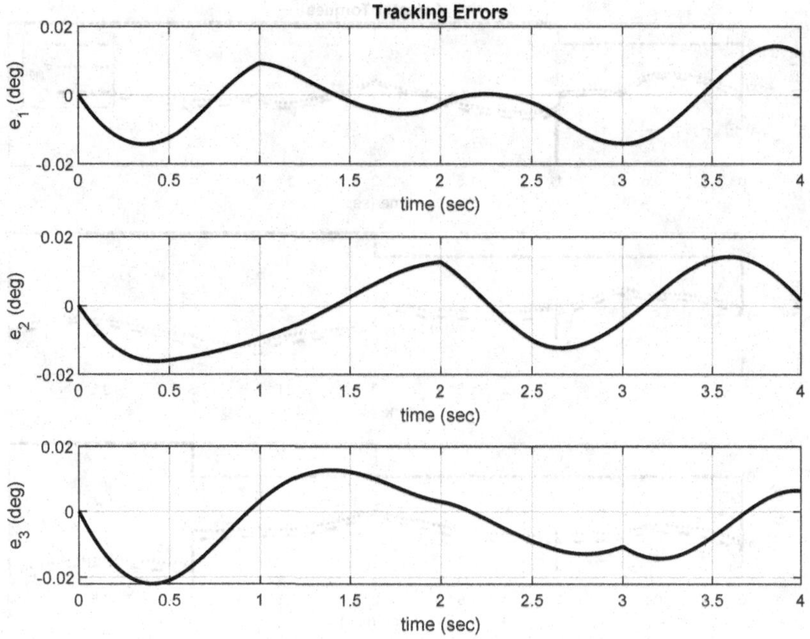

FIGURE 12.7
Trajectory tracking error of the PID controller along with feed-forward term in the presence of 50% uncertainty in model parameters.

the third second, a step torque disturbance with an amplitude of $2\,N.m$ is applied to the third joint. The effects of measurement noise and actuator saturation are neglected in this simulation.

The trajectory tracking error of the closed-loop system in this scenario, as depicted in Figure 12.7, has reached less than 0.02 degrees in all three joints, which is set to the same level achieved in the pure PID control. This observation is also evident in the torque exerted on the robot actuators, illustrated in Figure 12.8. In this figure, the disturbance torque $\boldsymbol{\tau}_d$ is represented by the dash-dotted line, the torque generated by the PID controller is depicted with the dotted line, and the final control effort, composed of the sum of the PID torque and the linear feedback term $\boldsymbol{\tau}_{pid} + \boldsymbol{\tau}_{fl}$, is shown with the dashed line.

According to this figure, it can be understood that the torque generated by the PID control constitutes the main portion of the applied torque to the actuators, and surpasses the influence of the feed-forward term throughout the entire maneuver. This indicates the negligible impact of the feed-forward term on the performance of the closed-loop system in this scenario. In other words, it may be concluded that the input disturbance in the joints and the considerable parameter uncertainty considered in the system parameter estimation are entirely compensated by the power of the PID controller.

12.8.2 Inverse Dynamics Control

The overall configuration of the inverse dynamics control is outlined in the block diagram depicted in Figure 12.2. This configuration employs both internal and external feedback loops. Within the internal loop, the torque $\boldsymbol{\tau}_{fl}$ is computed at each instant based on feedback from joint variables and is then employed as a compensatory term on the robot. Consequently,

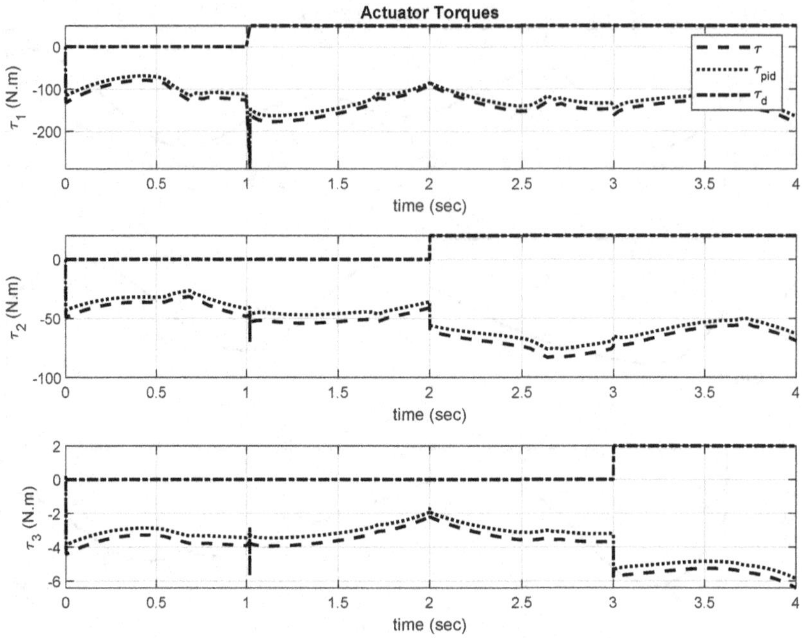

FIGURE 12.8
Control effort of the PID controller along with feed-forward term in the presence of 50% uncertainty in model parameters.

the exerted control effort on the system is equivalent to,

$$\boldsymbol{\tau} = \hat{\boldsymbol{M}}(\boldsymbol{q})\boldsymbol{a}_q + \boldsymbol{\tau}_{fl} \tag{12.73}$$

where

$$\begin{aligned} \boldsymbol{\tau}_{fl} &= \hat{\boldsymbol{C}}(\boldsymbol{q}, \dot{\boldsymbol{q}})\dot{\boldsymbol{q}} + \hat{\mathfrak{g}}(\boldsymbol{q}) \qquad (12.74) \\ \boldsymbol{a}_q &= \ddot{\boldsymbol{q}}_d + \boldsymbol{K}_d\dot{\boldsymbol{e}}_q + \boldsymbol{K}_p\boldsymbol{e}_q + \boldsymbol{K}_i\int_0^t \boldsymbol{e}_q(\tau)d\tau \qquad (12.75) \end{aligned}$$

Considering that, in this control method, the PID controller gain matrices are multiplied by the robot's inertia matrix $\hat{\boldsymbol{M}}(\boldsymbol{q})$, the values of these gains should be adjusted differently compared to the previous control methods to enable a fair comparison. In simulating this control approach, by employing a less trial and error process compared to the previous scenario, the PID controller gain matrices can be tuned more effectively as follows.

$$\boldsymbol{K}_p = \begin{bmatrix} 1 & 0 & 0 \\ 0 & 2 & 0 \\ 0 & 0 & 0.2 \end{bmatrix} \times 10^4 \ , \ \boldsymbol{K}_d = \boldsymbol{K}_p \ , \ \boldsymbol{K}_i = 10\boldsymbol{K}_p \tag{12.76}$$

By utilizing these gain matrices, the necessary matching for tracking error and control effort, such as the PID controller with feed-forward, can be achieved. By this means, comparing the results of these methods will be more meaningful.

In this simulation, the desired joint motion trajectory and a step disturbance are considered for all three robot joints, similar to the previous scenario, while neglecting the effects of measurement noise and actuator saturation. Additionally, we assume that the model parameters are not precisely known, and in the simulations, a 50% uncertainty is considered to examine the system's response with the inverse dynamics controller under these conditions.

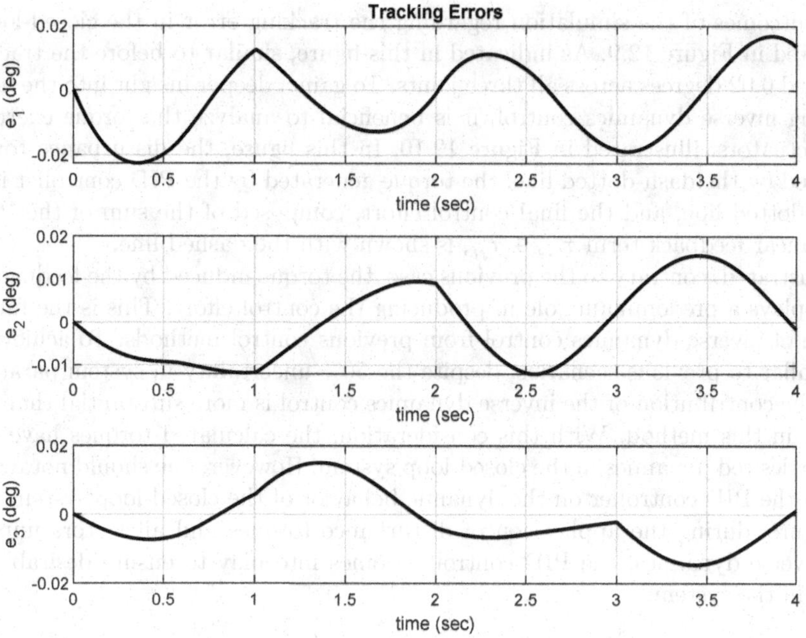

FIGURE 12.9
Tracking error of inverse dynamics control using PID controller in the presence of 50% uncertainty in model parameters.

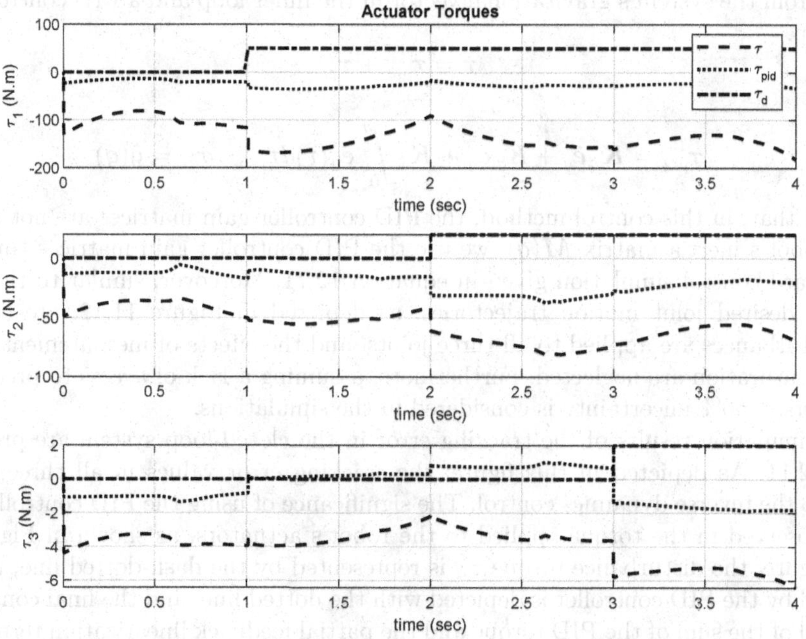

FIGURE 12.10
Control effort of inverse dynamics control using PID controller in the presence of 50% uncertainty in model parameters.

The outcomes of the simulation regarding the tracking error in the closed-loop system are depicted in Figure 12.9. As indicated in this figure, similar to before the tracking error has reached 0.02 degrees across all three joints. To gain a deeper insight into the importance of utilizing inverse dynamics control, it is beneficial to analyze the torque exerted on the robot's actuators, illustrated in Figure 12.10. In this figure, the disturbance torque $\boldsymbol{\tau}_d$ is represented by the dash-dotted line, the torque generated by the PID controller is depicted with the dotted line, and the final control effort, composed of the sum of the PID torque and the linear feedback term $\boldsymbol{\tau}_{pid} + \boldsymbol{\tau}_{fl}$, is shown with the dashed line.

As illustrated, contrary to the previous case, the torque induced by the feedback linearizing term plays a predominant role in producing the control effort. This is the fundamental difference of inverse dynamics control from previous control methods. To achieve tracking errors similar to previous scenarios, despite the 50% uncertainty in system parameter estimation, the contribution of the inverse dynamics control is more substantial than the linear controller in this method. With this consideration, the calculated torques have effectively tuned the desired dynamics in the closed-loop system. However, one should not overlook the impact of the PID controller on the dynamic behavior of the closed-loop system. As shown in the figure, during the application of disturbance torques and all factors unpredictable by the inverse dynamics, the PID controller comes into play to ensure desirable dynamic behavior in the system.

12.8.3 Partial Feedback Linearization

In this section, we examine the simulation results by implementing the partial feedback linearization method on the robot. In this approach, the input torque vector to the robot benefits from the system's gravitational vector in the inner loop and a PID controller in the outer loop.

$$\boldsymbol{\tau} = \boldsymbol{\tau}_{pid} + \boldsymbol{\tau}_{fl} \tag{12.77}$$

where

$$\boldsymbol{\tau}_{pid} = \boldsymbol{K}_d \dot{\boldsymbol{e}}_q + \boldsymbol{K}_p \boldsymbol{e}_q + \boldsymbol{K}_i \int_0^t \boldsymbol{e}_q(\tau) d\tau \quad , \quad \boldsymbol{\tau}_{fl} = \hat{\mathfrak{g}}(\boldsymbol{q}) \tag{12.78}$$

Given that, in this control method, the PID controller gain matrices are not multiplied by the robot's inertia matrix $\hat{\boldsymbol{M}}(\boldsymbol{q})$, we use the PID controller gain matrices tuned in the feed-forward control simulation given in equation 12.71. Moreover, similar to the previous case, the desired joint motion trajectories, as depicted in Figure 11.12, are considered, while disturbances are applied to all three joints, and the effects of measurement noise and actuator saturation are neglected. Furthermore, assuming a lack of access to precise model parameters, a 50% uncertainty is considered in the simulations.

The simulation results of the tracking error in the closed-loop system are presented in Figure 12.11. As depicted in this figure, the tracking error values in all three joints are similar to the inverse dynamics control. The significance of using the PID controller is more clearly observed in the torque applied to the robot's actuators, as shown in Figure 12.12. In this figure, the disturbance torque $\boldsymbol{\tau}_d$ is represented by the dash-dotted line, the torque generated by the PID controller is depicted with the dotted line, and the final control effort, composed of the sum of the PID torque and the partial feedback linearization term is shown with the dashed line.

According to this figure, the torque induced by the PID control is more dominant than the gravity compensation term compared to the previous case. Therefore, the input disturbance in the joints and the significant uncertainty considered in estimating the gravity-induced torque are mostly compensated by the PID controller. This observation echoes

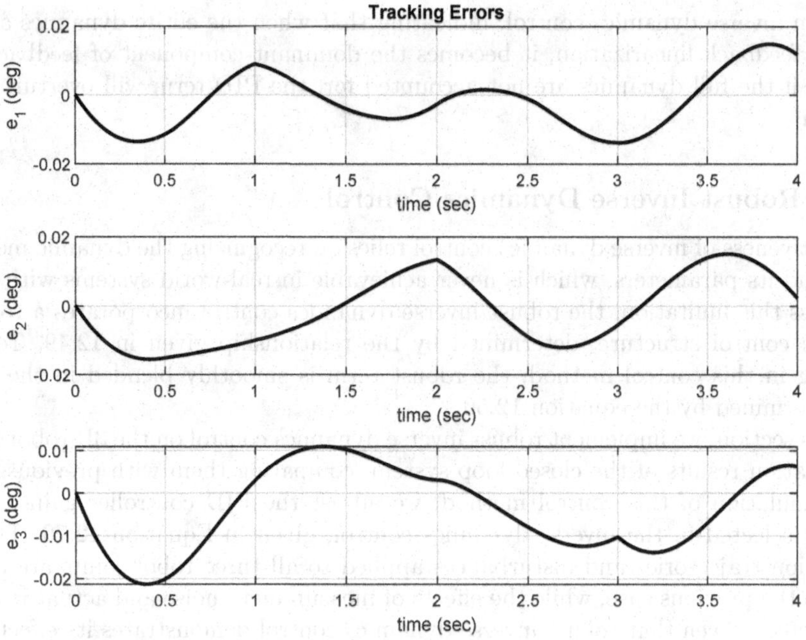

FIGURE 12.11
Tracking error of partial feedback linearization method using PID controller in the presence of 50% uncertainty in model parameters.

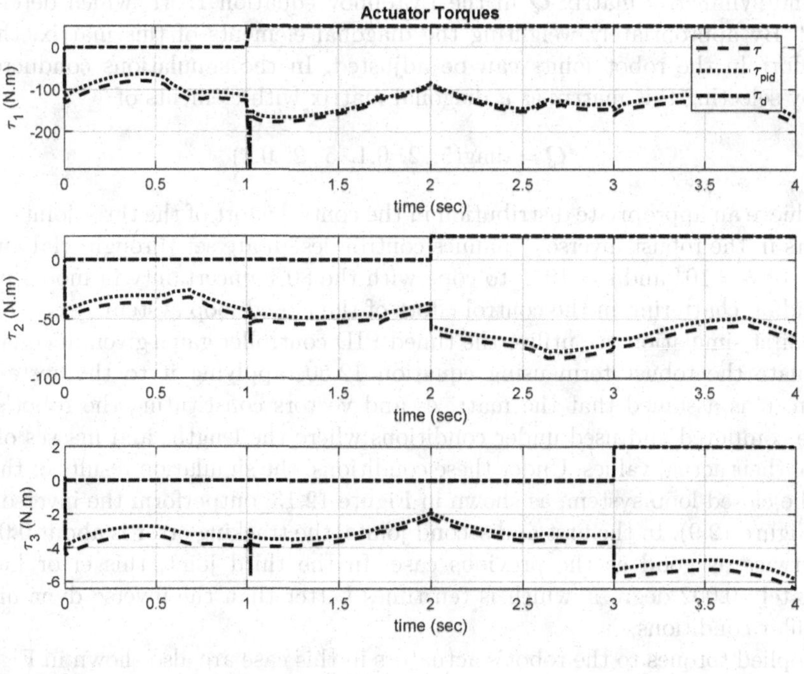

FIGURE 12.12
Control Effort of partial feedback linearization method using PID controller in the presence of 50% uncertainty in model parameters.

findings in inverse dynamics control, indicating that when the entire dynamics are considered into feedback linearization, it becomes the dominant component of feedback control. However, if the full dynamics are not accounted for, the PID term will overturn the compensation.

12.8.4 Robust Inverse Dynamics Control

The effectiveness of inverse dynamics control relies on recognizing the dynamic model of the system and its parameters, which is never achievable in real-world systems with precision. To address this limitation, the robust inverse dynamics control incorporates a robust term $\boldsymbol{\delta}_a$ in the control structure, determined by the relationship given in 12.49. To mitigate chattering in this control method, the robust term is smoothly blended in the boundary layer, determined by the equation 12.50.

In this section, we implement robust inverse dynamics control on the 3$\underline{\text{R}}$ robot and study the simulation results of the closed-loop system, comparing them with previous scenarios. In the simulation of this control method, we adjust the PID controller gains, similar to what was selected in the inverse dynamics control, given in Equation 12.76. The desired joint motion trajectories and disturbances applied to all three robot joints are considered similar to the previous case, while the effects of measurement noise and actuator saturation are neglected. Given that robust inverse dynamics control demonstrates its effectiveness in situations where the precise dynamic model of the robot is not available, we assume that the model parameters are not precisely known. In two different simulations, we consider uncertainties of 50% and 80% to compare the system's response in these two cases.

The tunable parameters in the robust inverse dynamics control include the positive-definite and symmetric matrix $\boldsymbol{Q}$ in the Lyapunov equation 12.41, which determines the matrix $\boldsymbol{P}$. By appropriately weighting the diagonal elements of this matrix, the ratio of control effort in the robot joints can be adjusted. In the simulations conducted in this section, by selecting this matrix as a diagonal matrix with elements of

$$\boldsymbol{Q} = \text{diag}(5,\ 2,\ 0.1,\ 5,\ 2,\ 0.1) \tag{12.79}$$

we can achieve an appropriate distribution in the control effort of the three joints. The other parameters in the robust inverse dynamics control design are set through trial and error to the values of $\rho = 10^4$ and $\epsilon = 10^{-3}$ to cope with the 80% uncertainty in model parameters while avoiding chattering in the control effort of the closed-loop system.

In the first simulation, we utilize the tuned PID controller gains given in equation 12.76 and calculate the robust term using equation 12.50, applying it to the system. In this simulation, it is assumed that the matrices and vectors constituting the robot's dynamic model are computed and used under conditions where the lengths and masses of the links are 50% of their actual values. Under these conditions, the simulation results of the tracking error of the closed-loop system, as shown in Figure 12.13, outperform the inverse dynamics control (Figure 12.9). In the first and second joints, the tracking error is about 0.01 degrees, which is twice as good as the previous case. In the third joint, this error has reached approximately 0.002 degrees, which is ten times better than the inverse dynamics control under similar conditions.

The applied torques to the robot's actuators in this case are also shown in Figure 12.14. In this figure, the disturbance torque $\boldsymbol{\tau}_d$ is represented by the dash-dotted line, the torque induced by the PID controller is denoted by the dotted line, the torque resulting from the robust term is illustrated by the solid line (in blue), and the final control effort, composed of the sum of the PID torque, the feedback linearization term, and the robust term $\boldsymbol{\tau}_{pid} + \boldsymbol{\tau}_{fl} + \boldsymbol{\delta}_a$, is depicted by the dashed line. As observed in this figure and in the first and third joint

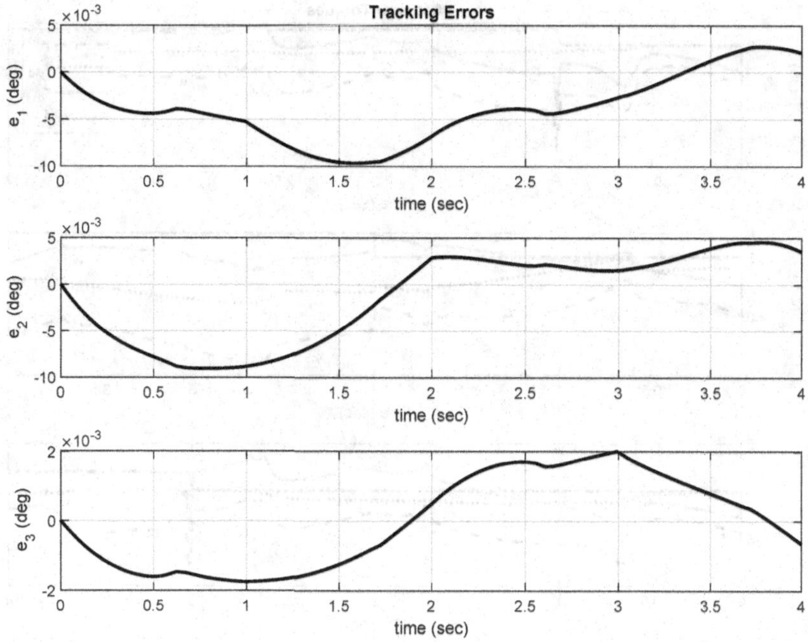

FIGURE 12.13
Tracking error of robust inverse dynamic method using PID controller in the presence of 50% uncertainty in model parameters.

plots, in the first second and before application of the disturbance torque to the system, the control effort is equally distributed by the feedback linearization term and the robust term. This is in contrast to the limited impact of the torque induced by the PID controller in mitigating the steady-state tracking error. The torque distribution in the second joint is different, and before the application of the disturbance torque in two seconds, the sum of the robust term and the PID controller torque constitutes half of the control effort, and the feedback linearization term accounts for the other half. However, it is evident that the main contributors to the achieved performance in the closed-loop system are the feedback linearization control along with the robust term.

To study the effect of uncertainty on the performance of this control method, we repeat the closed-loop system simulation under very challenging conditions of 80% uncertainty in the system parameters. In this simulation, all conditions are considered similar to the previous case, and only in the calculation of $\boldsymbol{\tau}_{fl}$, the matrices and vectors constituting the robot's dynamic model are used under conditions where the lengths and masses of the links are 20% of their actual values. The simulation results of the tracking error of the closed-loop system under this uncertainty condition are shown in Figure 12.15. The tracking error is two times worse compared to the 50% uncertainty scenario. However, the significance of using the robust term is clearly evident in the torque profiles applied to the robot's actuators, as depicted in Figure 12.16. In this figure too, the disturbance torque $\boldsymbol{\tau}_d$ is represented by the dash-dotted line, the torque induced by the PID controller is denoted by the dotted line, the torque resulting from the robust term is illustrated by the solid line (in blue), and the final control effort, composed of the sum of the PID torque, the feedback linearization term, and the robust term $\boldsymbol{\tau}_{pid} + \boldsymbol{\tau}_{fl} + \boldsymbol{\delta}_a$, is depicted by the dashed line.

As shown in this figure, given the 80% uncertainty in estimating the system parameters, the contribution of the feedback linearization torque has significantly decreased, and the

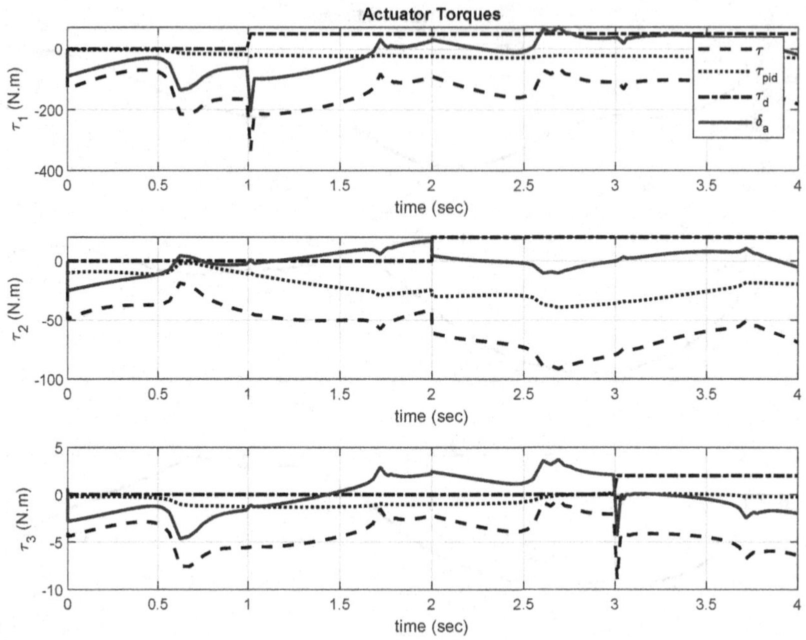

FIGURE 12.14
Control effort of robust inverse dynamic method using PID controller in the presence of 50% uncertainty in model parameters.

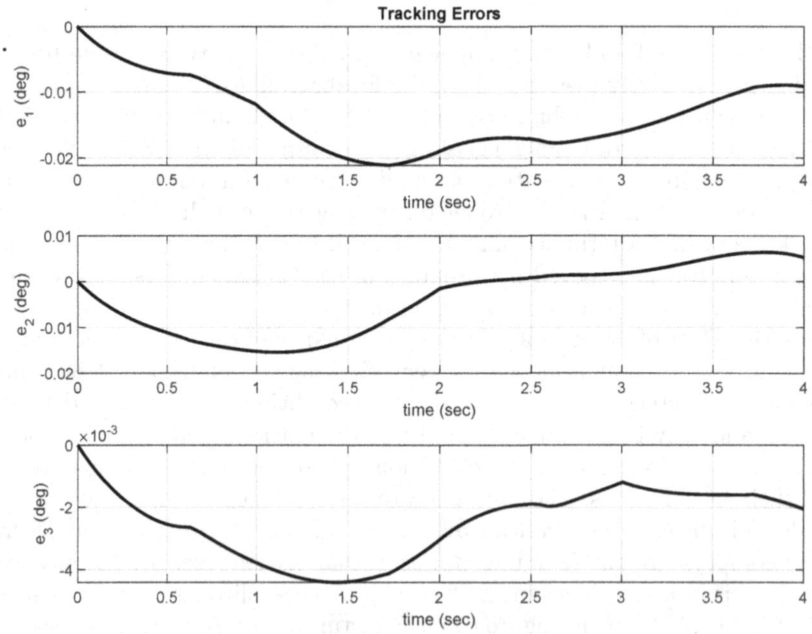

FIGURE 12.15
Tracking error of robust inverse dynamic method using PID controller in the presence of 80% uncertainty in model parameters.

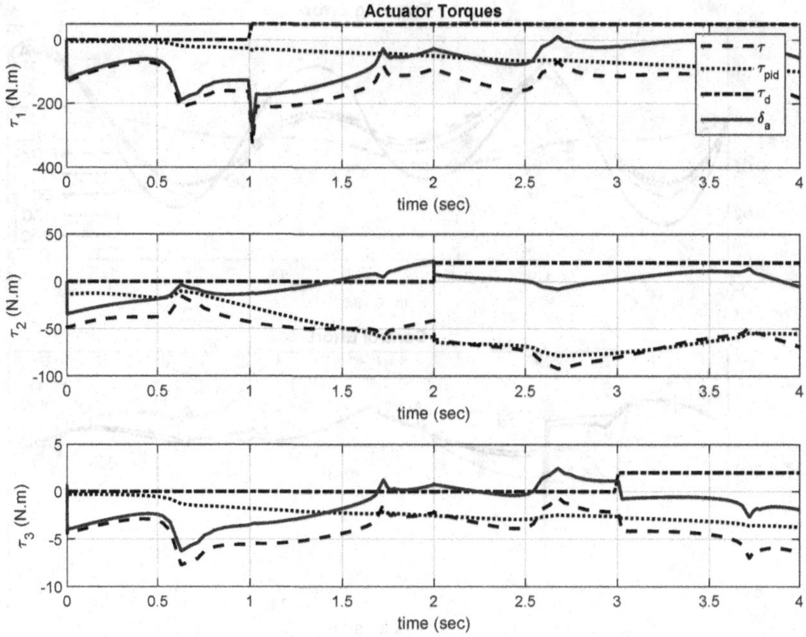

FIGURE 12.16
Control effort of robust inverse dynamic method using PID controller in the presence of 80% uncertainty in model parameters.

primary role in generating control effort is played first by the robust term and then by the torque induced by the PID controller. Interestingly, unlike all previous cases, at the moment of applying the disturbance torque, it is the robust term that quickly adapts and copes with the disturbance effect, while the PID control torque, considering the large gain of the integral gain matrix, directly addresses the steady-state error. This observation serves as a strong confirmation of the robustness of the PID controller and the efficiency of the robust term in the robust inverse dynamics control structure. Even in the presence of an 80% uncertainty in the robot's parameters, the system exhibits satisfactory tracking performance.

Certainly, attention should be paid to the control effort in this structure as well. Considering the large gains used in the robust term, one would expect larger torque values in the actuators. This is well illustrated in the torque required for the first joint, as seen in this figure. At the moment of applying the disturbance torque on the first joint and in the first second, the required control effort reaches nearly 300 $N.m$, which is almost twice the control effort in the inverse dynamics control case. It is essential to note that the tuning of control gains has been considered the same in different scenarios to allow for a meaningful comparison between various cases. Adjusting the gains will inevitably alter the system's response characteristics, such as tracking error magnitude and required control effort.

12.8.5 Comparative Study

In order to compare and evaluate the performance of the proposed control methods, the simulation results of the robot's motion using various control methods are presented in this section. For this purpose, assume that in model-based methods, accurate information about the robot's parameters is not available, and in the conducted simulations, there is a 50% uncertainty in the parameters. Additionally, consider that the desired trajectory of motion

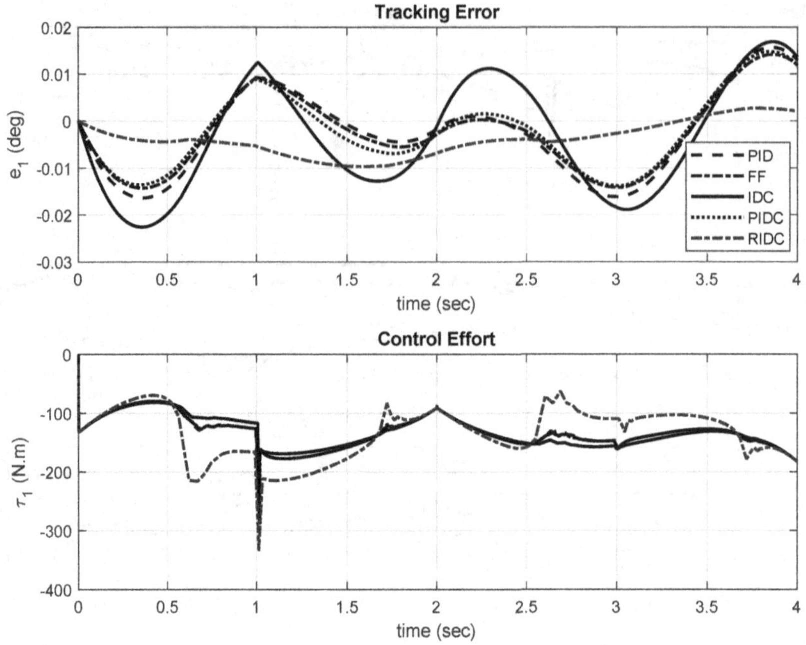

FIGURE 12.17
Comparison of tracking error and control effort in the first joint of the robot using various control methods in the presence of disturbance and 50% uncertainty in model parameters.

and disturbance entering each of the three joints of the robot is similar to before, and the effects of measurement noise and actuator saturation are neglected.

A comparison of tracking error and control effort in the three joints of the robot is presented in Figures 12.17, 12.18, and 12.19. In each of these figures, the graphs related to tracking error are shown in the upper part, and the required control effort is illustrated in the lower part, allowing for simultaneous examination of tracking error and control effort to achieve the desired performance.

In all the plots, the simulation results of the pure PID control method are represented by the dashed line, the feed-forward method by the dash-dotted line, inverse dynamics control by the solid line, partial feedback linearization method by the dotted line, and robust inverse dynamics control method by the dash-dotted line (in blue). The tracking error in terms of degrees and the required control effort in terms of $N.m$ are plotted in the graphs. A general overview reveals that the pure PID control, the feed-forward method, and partial feedback linearization yield similar results, and by using these methods, tracking error and required control effort are very close to each other.

Two alternative control approaches that stand out from these methodologies are the inverse dynamics control method and its robust counterpart. Across all three joints, the robust inverse dynamics control method demonstrates notably superior tracking error compared to other methods, albeit at the cost of necessitating increased control effort. Conversely, the inverse dynamics control method exhibits enhanced performance in comparison to other methods, particularly in the initial moments of motion and when disturbance torques are absent from the robot joints. Moreover, the associated control effort in this method does not significantly surpass that in other methodologies.

When comparing control methodologies, it is essential to consider additional factors, where, within the detailed discussion provided as follows, only a few will be highlighted.

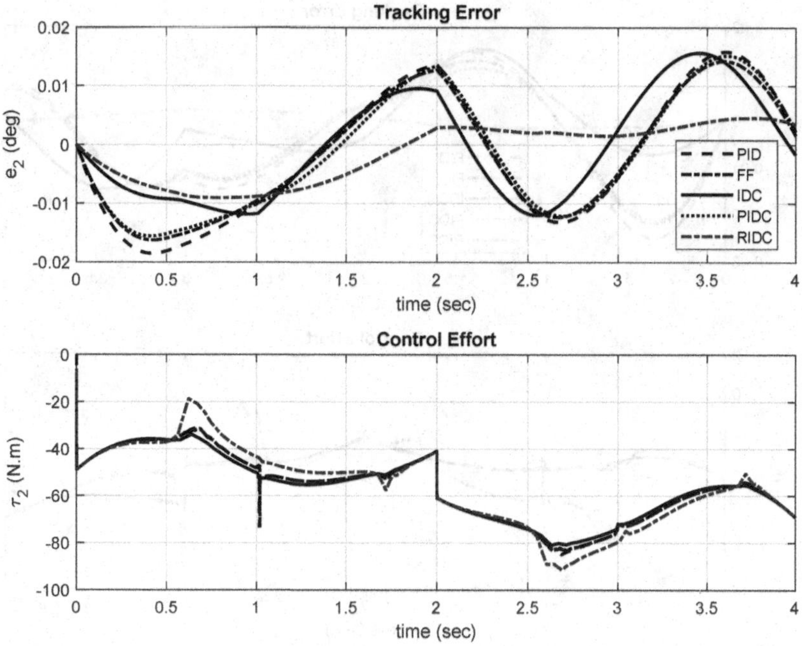

FIGURE 12.18
Comparison of tracking error and control effort in the second joint of the robot using various control methods in the presence of disturbance and 50% uncertainty in model parameters.

One critical aspect is the necessary balance between control effort and tracking error, a consideration applicable to all methods. In the analyzed case study within this section, it was observed that, with increased control effort, the performance of robust inverse dynamics control surpasses that of other methods significantly. This is noteworthy, given that in other control methodologies, the PID controller primarily manages the control contribution, and meticulous tuning of its coefficients is performed to maintain the required control effort within the feasible range for the actuators. If there were no such constraints on tuning the PID controller's gains and the feasibility of applying more substantial control efforts, akin to what is incorporated in robust inverse dynamics control, one could naturally anticipate a more favorable tracking error.

On the other hand, all the analyses and comparisons conducted in these simulations have focused on a particular predefined motion trajectory. This trajectory, intended for the specific robot, is relatively stringent, and characterized by high speed and acceleration. Naturally, if the desired motion trajectory exhibits lower speed and acceleration, the control effort needed will be reduced. Conversely, with a decrease in the time taken to traverse the trajectory and an increase in speed and acceleration, the control system will require more effort. Additionally, the spatial arrangement of the motion trajectory within the robot's workspace is of paramount importance. If the robot's geometry remains predominantly horizontal during the desired motion, it will encounter greater gravitational forces compared to a vertical orientation. Moreover, if the motion trajectory is situated near the workspace boundaries where singular configurations occur, achieving the desired motion will necessitate more substantial torques. In practical terms, thoroughly examining the motion type and the spatial coverage of the robot's workspace should be a focal point in further simulations during the design of an actual robot.

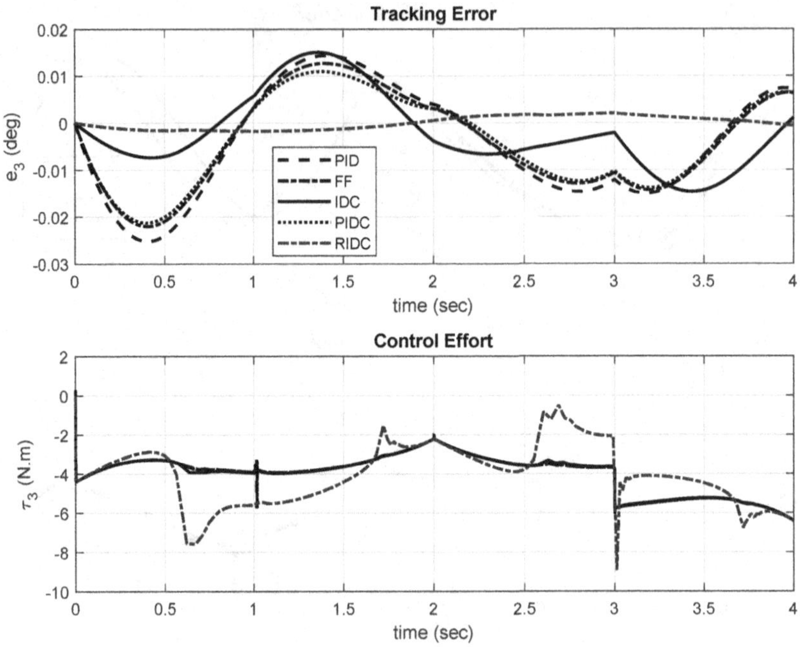

FIGURE 12.19
Comparison of tracking error and control effort in the third joint of the robot using various control methods in the presence of disturbance and 50% uncertainty in model parameters.

Another aspect to consider is the influence of disturbances, constraints due to actuator saturation, measurement noise, and uncertainties within the model, consequently affecting the structure of the control system. In the simulations presented in this chapter, disturbances were modeled as step functions. While this type of disturbance activates all the system's primary frequencies, its magnitude directly influences the necessary control effort. These simulations aimed to ensure that the disturbance magnitude represents a percentage of the torque applicable by the robot's actuators, enabling a comparison with the impact of gravitational forces on the joints. In the real robot's design, the nature and magnitude of disturbances should be determined based on the operational conditions, and these disturbances should be integrated into the simulations. Moreover, the saturation levels of the actuators must be defined according to the maximum torque they can apply and utilized in the simulations. Considering actuator limitations, control gains, and other parameters should be fine-tuned to mitigate actuator saturation as much as possible.

The impact of measurement noise is a crucial factor directly influencing the performance of the closed-loop system. In the corresponding simulation, the signal-to-noise ratio was deliberately chosen to allow for a clear observation of noise effects on trajectory tracking. Notably, no filters were applied to the measurement mechanism structure. In practical implementation, careful consideration must be given to sensor selection for the robot's control structure, ensuring minimal noise accompanies the measurements. Additionally, the recommendation is to incorporate analog or digital filters in the measured quantities. Optical encoders, combined with suitable interface circuits and equipped with digital filters, are commonly employed for measuring joint motions to reduce measurement noise. It's important to highlight that the direct measurement of joint velocities is generally avoided, favoring the use of digital filters in tandem with appropriate derivatives. It's noteworthy that the computation of derivatives expands the scope of measurement noise, explaining why joint motion acceleration is seldom utilized in the control structure.

Concerning uncertainties in system modeling, it's important to highlight that when considering parameter uncertainties, the system's simplest form of uncertainty is modeled without incorporating structural details. In actual systems, unmodeled dynamics manifest as uncertainty lacking a defined structure, making simulation challenging. Therefore, in nonlinear system modeling, the impact of unstructured uncertainty is addressed by introducing a constrained disturbance torque. The extent of this disturbance directly correlates with the level of uncertainty in the system model, posing a challenge for accurate determination. In essence, simulating the closed-loop system allows for the analysis and exploration of the system's overall behavior and initial adjustments to control gains. In practical terms, conducting performance tests on the constructed robot prototype enables precise tuning of gains, control parameters, and tracking errors.

Problems

12.1 Consider the two-degrees-of-freedom robot RP illustrated in Figure 12.20. The PID control for this robot has been designed in Problem 11.1.

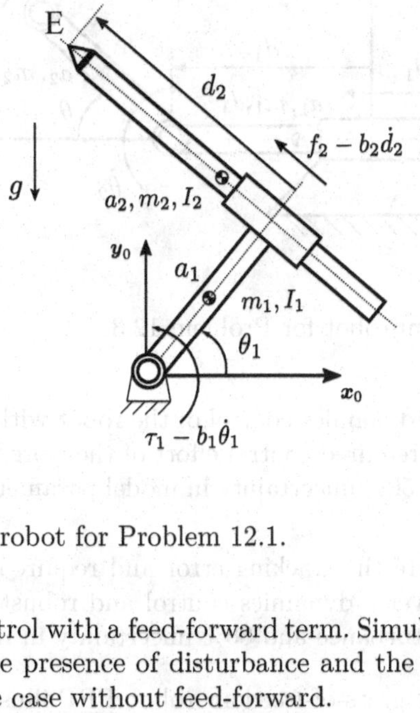

FIGURE 12.20
The two-degrees-of-freedom robot for Problem 12.1.

a) Enhance the PID control with a feed-forward term. Simulate the closed-loop system's tracking error in the presence of disturbance and the required control effort, and compare it with the case without feed-forward.

b) Implement inverse dynamics control on this robot and adjust the control gains matrices to ensure that the desired motion trajectory is tracked with an error similar to case (a). Simulate the tracking error and required control effort of the closed-loop system in the presence of disturbance and 50% uncertainty in model parameters and compare it with case (a).

c) Simulate partial feedback linearization control on this robot. Analyze the tracking error and required control effort of the closed-loop system in the presence of disturbance and 50% uncertainty in model parameters and compare it with case (b).

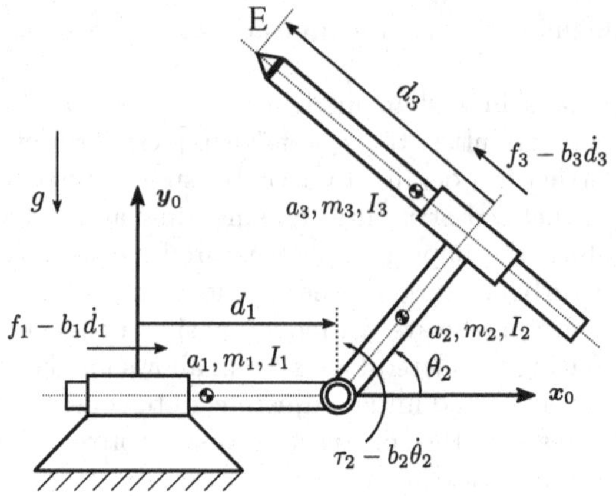

FIGURE 12.21
The three-degrees-of-freedom robot for Problem 12.2.

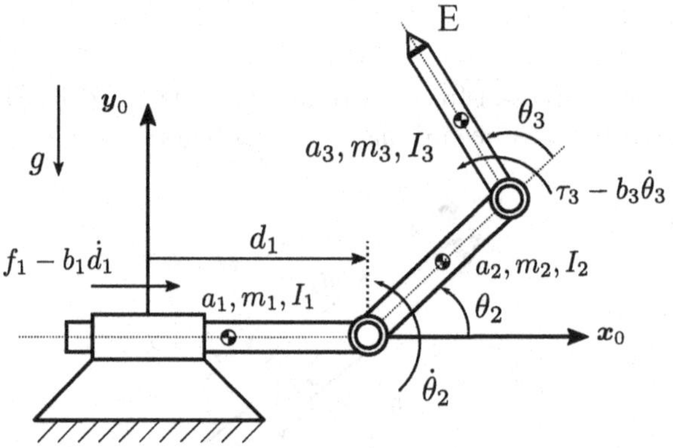

FIGURE 12.22
The three-degrees-of-freedom robot for Problem 12.3.

d) Enhance the inverse dynamics control of the robot with a robust term. Simulate the tracking error and required control effort of the closed-loop system in the presence of disturbance and 50% uncertainty in model parameters and compare it with case (b).

e) Simulate and compare the tracking error and required control effort of the closed-loop system for inverse dynamics control and robust inverse dynamics control in the presence of disturbance and 80% uncertainty in model parameters.

12.2 Consider the three-degrees-of-freedom robot PRP illustrated in Figure 12.21. The PID control for this robot has been designed in Problem 11.2. Repeat Problem 12.1 for this robot.

12.3 Examine the three-degrees-of-freedom robot PRR depicted in Figure 12.22. The PID control designed for this robot in Problem 11.3. Reiterate Problem 12.1 for this robot.

12.4 Investigate the three-degrees-of-freedom RPR robot shown in Figure 12.23. The PID control designed for this robot in Problem 11.4. Repeat Problem 12.1 for this robot.

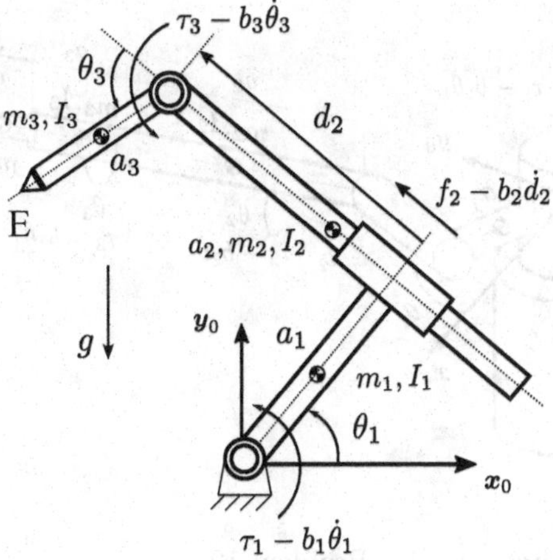

FIGURE 12.23
The three-degrees-of-freedom robot for Problem 12.4.

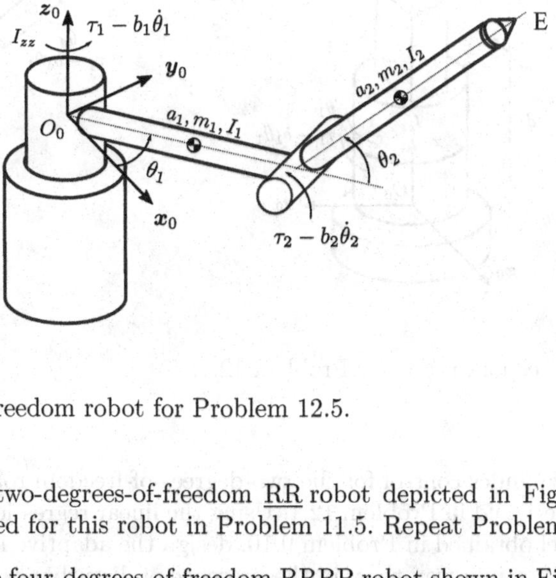

FIGURE 12.24
The two-degrees-of-freedom robot for Problem 12.5.

12.5 Examine the two-degrees-of-freedom <u>RR</u> robot depicted in Figure 12.24. The PID control designed for this robot in Problem 11.5. Repeat Problem 12.1 for this robot.

12.6 Investigate the four-degrees-of-freedom <u>RRRP</u> robot shown in Figure 12.25. The PID control designed for this robot in Problem 11.6. Reiterate Problem 12.1 for this robot.

12.7 Examine the four-degrees-of-freedom <u>RPRR</u> robot depicted in Figure 12.26. The PID control designed for this robot in Problem 11.7. Repeat Problem 12.1 for this robot.

12.8 Investigate the four-degrees-of-freedom 4<u>R</u> robot shown in Figure 12.27. The PID control designed for this robot in Problem 11.8. Reiterate Problem 12.1 for this robot.

12.9 Examine the four-degrees-of-freedom <u>RRPR</u> robot depicted in Figure 12.28. The PID control designed for this robot in Problem 11.9. Repeat Problem 12.1 for this robot.

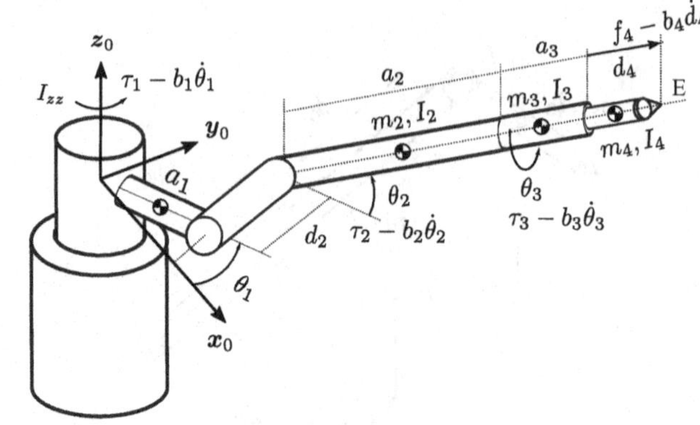

FIGURE 12.25
The four-degrees-of-freedom robot for Problem 12.6.

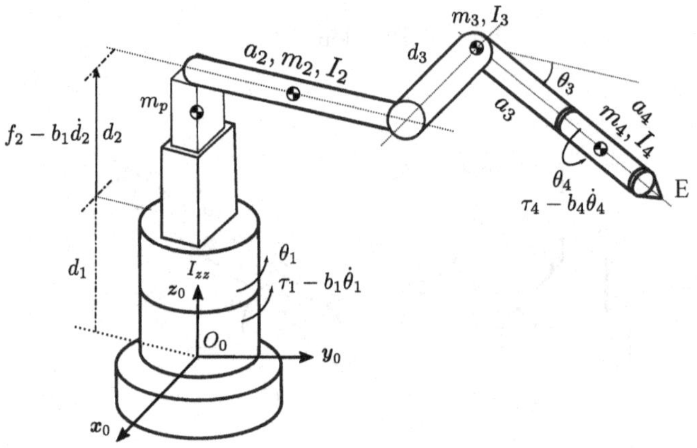

FIGURE 12.26
The four-degrees-of-freedom robot for Problem 12.7.

12.10 The inverse dynamics control for the two-degrees-of-freedom robot <u>R</u>P shown in Figure 12.20, is designed in Problem 12.1. Using the linear regression form of the robot's dynamic model obtained in Problem 9.10, design the adaptive inverse dynamics control for this robot in a way that, in the presence of disturbance and 20% uncertainty in model parameters, the adaptation of parameters converges, and the tracking error of the robot is proportionate to the result of tracking with inverse dynamics control. Compare the tracking error and control effort in these two scenarios.

If the uncertainty in model parameters increases to 50%, is it still possible to achieve convergence in parameter adaptation and appropriate tracking of motion using adaptive inverse dynamics control? Investigate this scenario by adjusting control gains appropriately and report the simulation results.

12.11 Repeat Problem 12.10 for the three-degrees-of-freedom <u>PRP</u> robot, which has its dynamic model's linear regression form obtained in Problem 9.11.

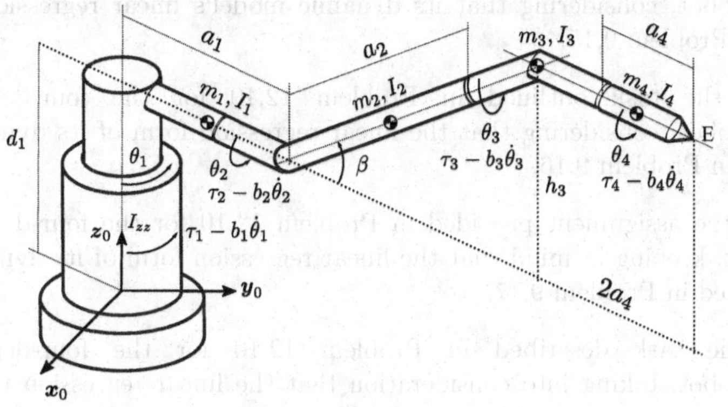

FIGURE 12.27
The four-degrees-of-freedom robot for Problem 12.8.

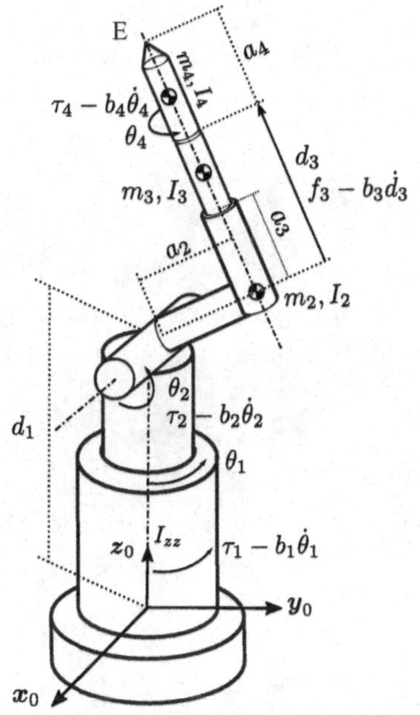

FIGURE 12.28
The four-degrees-of-freedom robot for Problem 12.9.

12.12 Reiterate the task outlined in Problem 12.10 for the three-degrees-of-freedom <u>PRR</u> robot, considering that its dynamic model's linear regression form was derived in Problem 9.12.

12.13 Repeat the Problem 12.10 for the three-degrees-of-freedom <u>RPR</u> robot, taking into account that the linear regression form of its dynamic model was obtained in Problem 9.13.

12.14 Repeat the task outlined in Problem 12.10 for the two-degrees-of-freedom <u>RR</u> robot, considering that the linear regression form of its dynamic model was derived in Problem 9.14.

12.15 Reiterate the task outlined in Problem 12.10 for the four-degrees-of-freedom RRRP robot, considering that its dynamic model's linear regression form was derived in Problem 9.15.

12.16 Repeat the task outlined in Problem 12.10 for the four-degrees-of-freedom RPRR robot, considering that the linear regression form of its dynamic model was derived in Problem 9.16.

12.17 Repeat the assignment provided in Problem 12.10 for the four-degrees-of-freedom 4R robot, keeping in mind that the linear regression form of its dynamic model was established in Problem 9.17.

12.18 Redo the task described in Problem 12.10 for the four-degrees-of-freedom RRPR robot, taking into consideration that the linear regression representation of its dynamic model was deduced in Problem 9.18.

13

Force Control

This chapter explores various methodologies and techniques employed in the force control of robotic systems. It begins with an introduction, laying the groundwork for understanding force control principles. The chapter then explores different force control topologies, including cascade control and force feedback in both outer and inner loops. Each topology is carefully examined, providing insights into their applications and implementation intricacies. The concept of compliance control is then explored, elucidating the similarities and differences between compliance control and position control of robotic systems. Following this, the methodology for implementing compliance control is scrutinized. Initially, the focus is on single-degree-of-freedom compliance control, which is subsequently expanded into a more intricate multivariable control strategy.

The main course of study progresses with the direct force control method, which entails elaborating on the design of cascade control and conducting a comprehensive performance analysis of closed-loop systems. Additionally, it discusses hybrid position-force control techniques, showcasing their effectiveness in handling force-related tasks. Furthermore, to exemplify the practical application of these methods, the chapter examines a case study on the force control methodologies elaborated in this chapter on a planar 3$\underline{R}$ robot. It examines both compliance control and direct force control methods within this context, demonstrating their applications and effectiveness. Furthermore, the incorporation of problems at the end of the chapter encourages readers to apply the knowledge they have acquired, thereby enhancing their understanding and proficiency in implementing these advanced control techniques in practical scenarios. Overall, the chapter offers a comprehensive overview of force control principles, methodologies, and their practical implementations in robotic systems.

13.1 Introduction

In the preceding two chapters, the control of robot motion has been examined through a range of linear and nonlinear methodologies. This control framework proves to be effective when the robot operates freely within the workspace, aiming to trace a predefined trajectory. In such scenarios, the robot's end-effector remains untouched with its surroundings, and there is no wrench interaction between them. Nevertheless, in numerous industrial settings, it becomes imperative for the robot's end-effector to allow a specified motion trajectory, while it is engaged with its environment. The efficiency of robots in applications like milling, grinding, assembly, and even surgery is accessible only when, along with executing accurate movements, the management of wrench exchanged between the robot's end-effector and its environment is precisely regulated.

To understand how a robot operates in such applications, let's consider a robot designed for tasks like window cleaning or writing on a blackboard. In both scenarios, relying solely on positional control is inadequate. This is due to minor tracking errors in the robot's

DOI: 10.1201/9781003491415-13

end-effector movement, leading to potential issues such as drifting away from the surface or applying excessive pressure. Consequently, in the window cleaning scenario, there's a risk of glass breakage, and in the writing task, having the pen slightly lifted from the surface would compromise the accurate depiction of the desired text. In these applications, exclusive dependence on positional control is deemed inappropriate. Instead, it becomes crucial to ensure that the force exerted by the robot's end-effector on the environment is adequately controlled, particularly in the direction perpendicular to the plane of motion.

In force control methods, the required torques in the robot actuators are provided in a way that, while the robot follows the desired trajectory, a desired interactive wrench is also generated between the robot's end-effector and the environment. In such applications, the robot's end-effector is in contact with its surrounding environment, and as a result of this contact, a considerable interactive wrench is exchanged between the robot and the environment. These issues and challenges are studied in the context of appropriate force control schemes. In this control method, the desired robot motion trajectory and the degree of end-effector compliance in the environment are adjusted in such a way that the required interactive wrench reaches its desired level.

When the robot's end-effector makes contact with its surroundings, the geometry of the contact surface introduces specific motion constraints on the robot's trajectory. In these scenarios, relying solely on position control proves inefficient, and the utilization of high gains may entail certain risks. This concern becomes particularly pronounced in surgical robots and robots engaged in human interaction. The accuracy of position controllers in industrial robots hinges on employing high gains and maintaining a relatively rigid robot structure. While these attributes ensure satisfactory performance when the robot moves freely in its environment, the situation changes as soon as the robot's end-effector encounters the environment, leading to motion constraints and an increase in tracking error.

In response to this unintentional tracking error, the controller tirelessly endeavors to diminish this error. Consequently, the mechanical components of the robot may incur damage, or an obstacle in the robot's surrounding environment may suffer harm due to the application of such formidable forces. In scenarios involving human interaction, where the robot's structure is typically more rigid than that of humans, this could result in human injury. In such instances, it becomes imperative to equip the robot with force sensors in the end-effector and incorporate force control methods into the robot's control structure.

To measure the wrench applied to the robot's end-effector, state-of-the-art technologies have been employed to develop wrench measurements in six axes. Additionally, torque applied to the robot's joints can be measured using torque sensors, and by employing the transpose of its Jacobian matrix, it can be mapped into the task space. Different force control topologies for robots have been introduced based on how force is measured in joint or task space. The use of these sensors does not imply the elimination of the need for position sensors since, in most applications, the desired precision of the robot's motion must also be controlled. However, relying solely on position sensors is insufficient, and it is necessary to incorporate motion information along with interactive wrenches in the feedback loop.

Considering the type and application of position and force sensors, various topologies have been developed for force control in advanced serial robots, which will be explored further in Section 13.2. In most of these topologies, both position and force sensors are utilized. However, the hierarchy of their utilization depends on the primary and secondary objectives of the control system. The connection of these hierarchies with the intended goals in robotics applications will also be examined in Section 13.2.1.

13.2 Force Control Topologies

In the interaction of the robot with its surrounding environment, force control involves determining the necessary torques in the robot's actuators in a way that, while traversing the desired trajectory, the interaction wrench between the robot's end-effector and the environment is also controlled. This is achieved by measuring the interactive wrench between the robot and the environment and simultaneously adjusting the robot's motion trajectory. Suppose the robot's joint motion variables are represented by the vector $\boldsymbol{q}$, and the vector of torques in the robot's actuators is denoted by $\boldsymbol{\tau}$. Additionally, represents the wrench introduced to the robot's end-effector in interaction with the environment by the six-dimensional array $\boldsymbol{\mathcal{F}}_e = [\boldsymbol{f}_e, \boldsymbol{n}_e]^T$. Considering the mapping of the Jacobian transpose matrix one may write:

$$\boldsymbol{\tau}_e = \boldsymbol{J}^T(\boldsymbol{q})\boldsymbol{\mathcal{F}}_e \tag{13.1}$$

This interaction wrench can be added to the dynamic formulation of the robot in free motion.

$$\boldsymbol{\tau} + \boldsymbol{\tau}_d = \boldsymbol{M}_e(\boldsymbol{q})\ddot{\boldsymbol{q}} + \boldsymbol{C}(\boldsymbol{q}, \dot{\boldsymbol{q}})\dot{\boldsymbol{q}} + \boldsymbol{b}_e\dot{\boldsymbol{q}} + \mathfrak{g}(\boldsymbol{q}) + \boldsymbol{\tau}_e \tag{13.2}$$

In this relation, the effective mass matrix of the robot $\boldsymbol{M}_e(\boldsymbol{q})$ and the effective damping torque vector $\boldsymbol{b}_e$ are defined in equation 12.2. Furthermore, the Christoffel matrix is denoted by $\boldsymbol{C}(\boldsymbol{q}, \dot{\boldsymbol{q}})$, the gravitational force vector by $\mathfrak{g}(\boldsymbol{q})$, and the disturbance torques applied to the robot joints are represented by the vector $\boldsymbol{\tau}_d$. Although in this equation, only the term $\boldsymbol{\tau}_e$ has been added compared to the previous cases, considering the significant magnitude of interaction torques in comparison to inertia and gravitational forces, the required torque by the actuators increases noticeably.

Before designing force control in this robot, it is necessary to examine the implementation topologies. By force control topology, we mean formulating a suitable structure that allows the calculation of the necessary torques for the robot's motion while controlling the interaction wrench to the environment. In various force control methods for serial robots, it is essential to compute the required torques for the actuators to follow the desired trajectory while the robot's end-effector is in wrench interaction with the environment. In this structure, based on the measurement of interaction forces, the desired trajectory of the robot's motion slightly changes, adapting to variations in the indentation of the robot's end-effector in the environment, and consequently controlling the desired interactive wrench in the robot's end-effector.

In general, the desired motion trajectory of the robot in the joint space, represented by the vector $\boldsymbol{q}_d$, is of the same dimension as the joint variable vector in the robot, denoted as $\boldsymbol{q}$. As discussed in the motion control chapters, it is common for the vector of joint variables $\boldsymbol{q}$ to be directly measured for practical and economical considerations.

The use of force sensor information is essential in force control topologies. If six-axis wrench sensors are used at the end-effector or wrist of the robot for this purpose, the applied wrench to the robot's end-effector by the environment, denoted as $\boldsymbol{\mathcal{F}}_e$, can be measured. Using the transpose of the Jacobian matrix and the relationship in Equation 13.1, this force can be mapped to the joint space. Furthermore, if the desired interaction force in the task space is represented as $\boldsymbol{\mathcal{F}}_d$, the same relationship can be used to map it to the joint space and the vector $\boldsymbol{\tau}_d$. Although commercially produced six-axis wrench sensors are available, their significant cost may lead to the use of joint torque sensors as an alternative. Moreover, state observers and measurements of electric motors current can provide a suitable estimation of the applied torques at the joints.

Given the concurrent access to both motion and force feedback from the robot, the effective incorporation of this feedback into a control framework will be explored and analyzed in

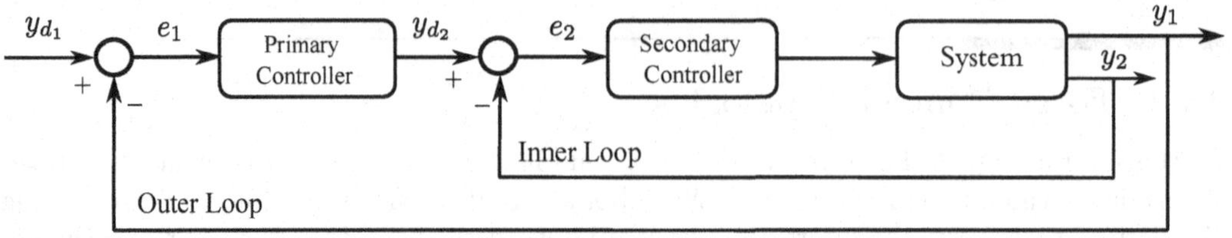

FIGURE 13.1
Block diagram of a cascade control with two nested loops.

the upcoming section. The discussion will focus on cascade control structures. Furthermore, major force control topologies in serial robots, derived from this control framework, will be subject to further investigation.

13.2.1 Cascade Control

As mentioned earlier, in the design of force control in robots, two simultaneous objectives with different priorities are considered. In situations where the robot interacts with its environment, adjusting the level of interactive wrench becomes the primary control objective. This is while keeping the robot's motion along the desired trajectory is consistently a secondary goal in the design. When the contact between the robot's end-effector and the environment is lost, the robot undergoes free motion, and tracking the desired trajectory becomes the primary design objective.

To satisfy these dual objectives concurrently, the cascade control structure is commonly employed in control systems [130]. This control structure is utilized when two sets of system state variables are measured, while simultaneously a primary and a secondary goal is pursuing. In this control structure, the primary goal is achieved effectively, but fulfilling the secondary goal is contingent upon certain assumptions and precise controller design.

As depicted in Figure 13.1, cascade control is established through the incorporation of two or more nested loops. Within this framework, the principal control objective centers around the measurement and utilization of the primary system variable within the outer loop. The main controller is situated within the outer loop, and the tracking error of the primary variable is addressed by this feedback loop. The inner loop within this configuration is employed to attain the second design objective, under some considerations outlined. Within this inner loop, another state variable in the system is measured, compared with the output of the initial loop, while it is subjected to control. A second controller is also crafted within this loop to optimize the fulfillment of the designer's second objective. Consequently, the first state variable, denoted in this figure by the notation y_1, is fed back into the outer loop, while the second state variable y_2 is incorporated in the inner loop. State variables can be employed in implementing multivariable cascade control in both loops. Furthermore, it is possible to augment more number of nested cascade loops; however, the usage of more than three nested loops is not prevalent in industrial applications.

In order to design controllers for this structure, controllers for the inner loop are usually designed initially with a wide bandwidth. This design ensures that the inner closed-loop system operates very quickly, allowing for effective tracking with the design of the outer loop controller. Considering the benefits of feedback, the use of the inner loop reduces model uncertainty, coupling in the system's state variables, and disturbance effects simultaneously. Consequently, designing the outer loop controller becomes more straightforward, aiming to achieve optimal tracking of the main state variable(s) of the system. It is crucial to establish a meaningful relationship between the first and second state variables and avoid dynamic behavioral conflicts to fulfill both the primary and secondary objectives in the system through the inner loop.

For instance, in the control of industrial processes, a two-loop control of pressure and temperature is employed in the boiler drums of thermal power plants. If controlling the temperature in the drum is crucial, the feedback of this variable is placed in the outer loop [153]. This is done while pressure control can be utilized as the secondary objective in the system by substituting it into the feedback of the inner loop. Similarly, in the multi-rotor drone flight control, a cascade control with three nested loops of acceleration, speed, and robot position is used. Considering the special significance of controlling the robot's position, this variable is measured and fed back in the outer feedback loop [154].

Within the field of robotics, variables incorporated in cascade loops encompass the robotic joint space motion vector $\boldsymbol{q}$ and the wrench exerted on the robot's end-effector $\boldsymbol{\mathcal{F}}_e$, or its mapping $\boldsymbol{\tau}_e$ within the joint space. Considering that the central objective of control in the closed-loop system revolves around force or motion tracking, the pertinent variable is assigned to the outer loop, while the complementary variable finds its place in the inner loop. The subsequent sections will elaborate on the discussion of the control topology in these two scenarios, accompanied by a detailed exploration of their characteristics.

13.2.2 Force Feedback in Outer Loop

If the primary objective within the force control cascade structure is to control the interactive wrench of the robot with the environment, it is essential to incorporate the measured wrench in the outer loop. In such a situation, should position feedback be integrated into the inner loop, the secondary control objective extends beyond merely tracking the interactive wrench of the robot with the environment to encompass trajectory tracking of the robot. In this context, let's consider a scenario where the robot's joint space motion vector $\boldsymbol{q}$ and the six-dimensional array representing the interactive wrench of the robot with the environment in the task space $\boldsymbol{\mathcal{F}}_e$, are measured. Recognizing that the dynamic equations of the robot are more conveniently derived in the joint space, our focus shifts toward designing force and position control within this space.

For this purpose, consider the force control topology in serial robots as presented in the block diagram in Figure 13.2. As observed in this figure, the robot's joint space motion vector $\boldsymbol{q}$ and the vector representing the interactive wrench of the robot with the environment in the task space $\boldsymbol{\mathcal{F}}_e$, are measured and fed back in two nested loops. Considering that the main objective, in this case, is force control, the interactive wrench of the robot is mapped to the joint space using the transpose Jacobian matrix and placed in the outer feedback loop.

Furthermore, to achieve the trajectory tracking goal, the position feedback of the joint motions is placed in the inner loop. Using the Jacobian transpose mapping, the design of both position and force controllers is performed in the joint space to enhance simplicity. Moreover, if for any reason, it is not possible to measure the interactive wrench of the robot in the task space, the measurement of the interactive torque in the joint space $\boldsymbol{\tau}_e$ can easily replace $\boldsymbol{\mathcal{F}}_e$ by removing the Jacobian transpose block in the outer loop.

The primary input in this cascade control structure is the desired interactive torque in the joint space, denoted as $\boldsymbol{\tau}_d$. Therefore, it is evident that ensuring force tracking is pursued as the main goal in this structure. On the other hand, it is observed that the desired trajectory of the robot joints $\boldsymbol{q}_d$ is not independently introduced in this structure. Instead, it is influenced by the force controller in the outer loop. Thus, if the objectives of force control and robot motion control are in harmony, these two objectives will be fulfilled within this control topology. However, if for some reason, these two objectives conflict with each other, this topology will prioritize force tracking as the primary objective, and evidently, precise trajectory tracking may not be achieved.

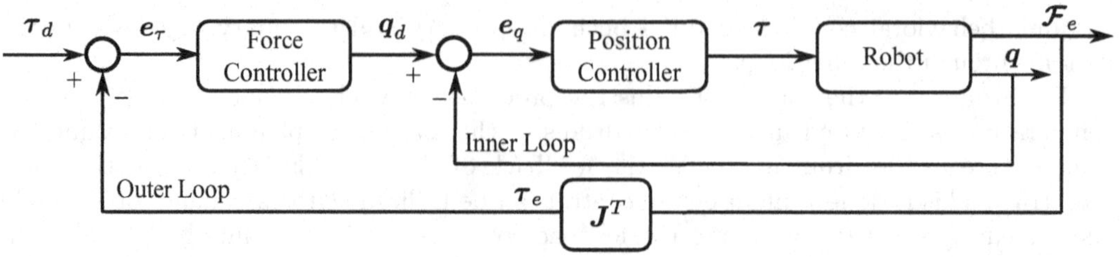

FIGURE 13.2
Block diagram of cascade control: Force feedback in the outer loop and position feedback in the inner loop.

Examining the inner loop of this control topology reveals that this part is exactly analogous to robot motion control, while various methods have been developed to design suitable controllers for it. Inverse dynamics control, as the primary model-based method among these approaches, introduces how to linearize the closed-loop system with feedback. Regardless of the specific control method used in the inner loop, the magic of feedback in the inner loop reduces model uncertainty, coupling in the system's state variables, and disturbance effects simultaneously. With this advantage, achieving wrench tracking of the robot's interaction with the environment in the outer loop becomes more straightforward and more efficient.

The introduced topology in this section is mainly used in the structure of direct force control, the details of which will be discussed in Section 13.4. The term "direct force control" implies that the primary objective in this control method is to track the interactive wrench between the robot and the environment. Simultaneously, the stability of the robot's motion is guaranteed, and whenever contact between the robot and the environment is lost, the robot's motion control becomes the focal point. For this purpose, various controllers are employed in the inner and outer loops, some of which will be discussed in Section 13.4.

13.2.3 Force Feedback in Inner Loop

If the primary goal in the robot's cascade control structure is to track its motion trajectory, it is necessary to place this quantity in the outer loop. In this case, if force feedback is included in the inner loop, in addition to tracking the desired motion trajectory of the robot, achieving suitable dynamics between force and the robot's motion will also be pursued as the second control objective. In this scenario, suppose as before, the vector representing the robot's joint space motion in $\boldsymbol{q}$ along with the six-dimensional array representing the interactive wrench of the robot with the environment in the task space, $\mathcal{F}_e$, is measured. Thus, using the Jacobian transpose matrix, the mapping of interactive force in the joint space can be obtained.

In this case, consider the robot control topology as presented in the block diagram in Figure 13.3. As observed in this figure, the robot's joint space motion vector $\boldsymbol{q}$ and the vector representing the interactive wrench of the robot with the environment in the workspace are measured and fed back in two nested loops. Given that the main objective in this case is to track the desired motion trajectory of the robot, the position feedback of the joint motions is placed in the outer feedback loop. Furthermore, the interactive wrench of the robot $\mathcal{F}_e$, mapped to the joint space $\boldsymbol{\tau}_e$, using the Jacobian transpose is utilized in the inner loop. The design of both position and force control is performed in the joint space using the transpose Jacobian matrix mapping to enhance ease of implementation. Moreover, if for any reason, it is not possible to measure the interactive compliance of the robot in the workspace, the measurement or estimation of the interactive torque in the joint space $\boldsymbol{\tau}_e$ can easily replace $\mathcal{F}_e$ by removing the Jacobian transpose block in the inner loop.

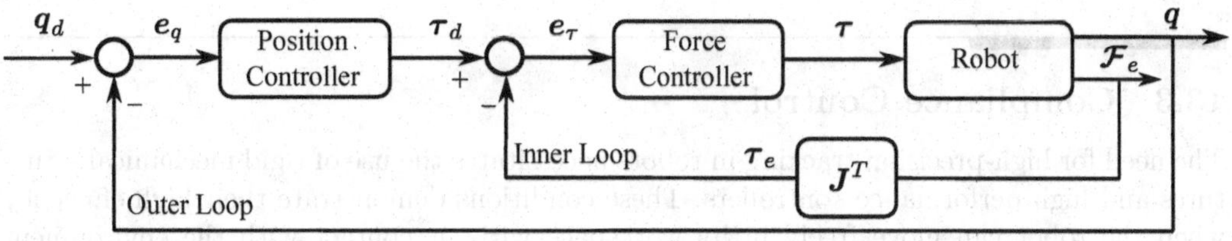

FIGURE 13.3
Block diagram of cascade control: Position feedback in the outer loop and force feedback in the inner loop.

In this control topology, when the robot is not in contact with the environment, the structure of the robot's outer loop ensures the tracking of the robot's motion trajectory. However, whenever the robot came in contact with the environment, it is natural that due to the high stiffness of the environment, unrestricted motion of the robot's end-effector would not be feasible. In this case, the interactive forces between the robot and the environment become significantly larger than the inertial forces and even dominate the gravitational forces acting on the robot. Thus, considering the presence of nested loops in the control structure, the dynamic relationship between force and position will be adjusted based on the characteristics of the position controller.

The main input to the control in this nested structure is the desired motion trajectory of the robot joints, denoted by $\boldsymbol{q}_d$. It should be noted that when the robot is in contact with its surroundings, the desired trajectory of motion in the robot must be carefully selected, and unconditional tracking the desired trajectory without constraints is not possible. On the other hand, it is observed that the desired interaction torque in the robot is not introduced independently into this control structure. Instead, it is derived from the result obtained from the position controller in the outer loop and applied as the input to the inner loop. This implies that the desired interactive force between the robot and its surrounding environment is automatically calculated based on the desired motion trajectory and the stiffness of the environment and is applied to the inner loop. Thus, an appropriate dynamic relationship between motion and force variables of the robot is established.

The introduced topology in this section is utilized in the impedance control structure, whose details will be discussed in the upcoming chapter. While various structures are employed in designing force and position controllers in this topology, in the following chapter, a fundamental method based on inverse dynamic control will be introduced.

The topologies presented in this chapter represent only two types of control topologies envisioned for force control in robots. These two topologies, used more than others, are formulated based on how the position and force quantities are measured, allowing for the use of simpler dynamic and kinematic models in their implementation. It is obvious that in special applications and in the presence of constraints or preferences for other types of measurement systems, these basic topologies can easily be extended. However, for the sake of readability and ease of learning basic concepts, we refrain to give their detailed information in this section.

To develop force control methods for robots, we will start with the simplest design structure called compliance control. In this structure, force sensors are not utilized, and the concept of robot compliance and its relation with precise motion control will be explained. With this structure, the static relationship between force and motion in the controlled robot is established, and its extension to a dynamic relationship is facilitated. Then, direct force control will be studied using the topology described in Section 13.2.2.

13.3 Compliance Control

The need for high-precision tracking in robots necessitates the use of rigid mechanical structures and high-performance controllers. These conditions demonstrate their high efficiency when the robot can move freely in its workspace without contact with the environment or potential obstacles. To achieve this, the range of robot movement in industrial environments is usually equipped with a suitable fence, preventing the entry of humans or any other obstacles into the robot's workspace. However, in cases where the robot interacts with its surrounding environment, these conditions are not feasible, and the use of motion control in such situations may lead to damage to the robot's mechanical structure or the obstacle.

Imagine that while the robot is freely moving in its workspace, an obstacle is appeared in its predetermined trajectory. In this scenario, by using a rigid mechanical structure and high gain controllers in the robot, the smallest deviation from the pre-defined trajectory will be compensated for by the significant torques in the actuators. Considering the imposed constraints on the robot's movement trajectory, the robot strives with its maximum capability to remove this obstacle from its path. If the rigidity of the robot exceeds that of the obstacle, serious damage will be inflicted on the obstacle; otherwise, the robot itself or its mechanical components will be damaged. In the case of interaction between the robot and a human, the risk of serious injury to the human is more critical.

In such cases, incorporating the necessary compliance in the robot's structure will prevent the occurrence of such complications. Reducing the stiffness of the robot and adding compliance to it can be achieved in the mechanical structure of the robot or through its controller. Adding compliance to robots is a new trend that is being considered in the structure design of modern robots. Structural compliance in a robot may arise due to the presence of lightweight and flexible components in the mechanical structure of the robot, similar to what is used in space robots [155, 156], or as a result of the existence of flexible and compliant tools in minimally invasive surgeries [157, 158], or the use of novel power transmission technologies [124, 159]. Although the presence of compliance in the robot's structure will improve the robot's interaction with its environment, designing appropriate control for it will be more complex.

Another method of adding compliance to the robot is through its control system. By designing an appropriate controller for the robot, the robot's compliance can be adjusted in a programmed manner. In this case, flexible mechanical components are not used in the robot's structure, instead, a suitable control structure and appropriate gains are utilized to create the necessary compliance in the robot. In this way, the mechanical structure of the robot can be considered completely rigid, and the level of compliance needed is adjusted by tuning the control gains of the robot. Such control structures in robotics literature are known as compliance control [160] or stiffness control [161], which we use the first term in this book.

The topology of compliance control is entirely similar to the control topologies discussed in Section 11.2, for position control in serial robots. In compliance control, only the motion variables in the joint or task space of the robot are measured and utilized in the feedback loop, eliminating the need for force sensors and additional nested loops. However, it is assumed that the robot is in contact with the environment and experiences interactive wrenches, which are estimated using soft body models. By doing so, interactive deformation between the robot and the environment enters the dynamic model of the system according to Equation 13.2. The stability of motion and transient and steady-state performance shall be analyzed using the governing dynamic formulation.

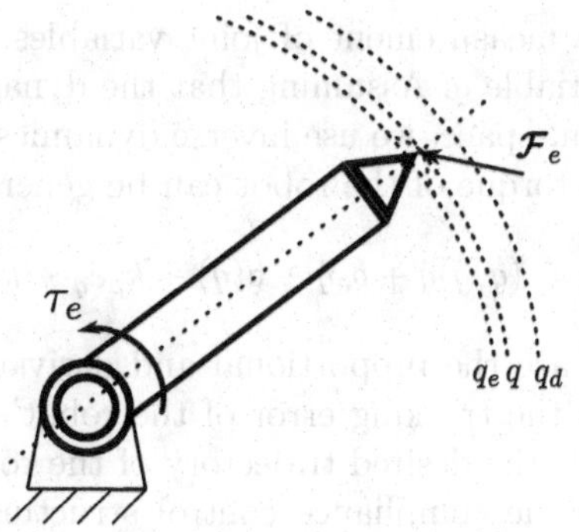

FIGURE 13.4
Force interaction between the robot end-effector and a soft environment in the compliance control of a robot joint.

13.3.1 Compliance Control in one Joint

To illustrate the concept of compliance control, consider a robotic arm with a single degree of freedom interacting with its surrounding environment. As shown in Figure 13.4, the robot link rotates about its revolute joint, represented by the scalar variable q. Meanwhile, an obstacle is placed in the path of the robot's movement, and the robot's end-effector interacts with it, experiencing a wrench due to this collision. Assume the the stiffness of the robot is much larger than the environmental stiffness, and as a result upon contact between the robot end-effector and the environment, a small deformation occurs in the environment, and interactive wrench $\mathcal{F}_e$ is applied to the robot's end-effector based on the extent of this deformation in different directions.

For this simple case where the robot has only one rotational degree of freedom, consider the resulting wrench is limited to the applied torque on the robot joint in one-dimension being calculated by the Kelvin-Voigt model.

$$\tau_e = \kappa_e\, \Delta q = \kappa_e(q - q_e) \tag{13.3}$$

In this context, the environmental stiffness coefficient in the joint space, denoted as κ_e, serves to represent the environment stiffness model in one degree of freedom. Note that the actual trajectory of the robot end-effector is denoted with the variable q and the position of the obstacle is denoted with the variable q_e in this illustration. It is natural that, considering the large magnitude of the environmental stiffness coefficient, these two variables should be close, while there difference is represented by the variable Δq.

Rewrite the complete dynamic model of the robot in the presence of this interactive torque and in the absence of any disturbance. In this case, Equation 13.2 can be simplified into the following scalar equation.

$$\tau = m_e\ddot{q} + c(q,\dot{q})\dot{q} + b_e\dot{q} + \mathfrak{g}(q) + \tau_e \tag{13.4}$$

In this equation, τ represents the actuator torque of the robot, m_e is the effective mass of the robot, $c(q,\dot{q})$ is the simplified Christoffel coefficient in the direction of motion q, b_e is the equivalent viscous friction coefficient, $\mathfrak{g}(q)$ is the gravity-induced term, and τ_e is the mapping of the external wrench applied to the robot in the joint space.

To introduce the compliance control method in this robot, we use the fundamental inverse dynamics control approach. Advanced robust and adaptive control methods can then be built upon this fundamental approach. The implementation of the inverse dynamics control method requires the measurement of joint variables, which in this case is simply represented by the scalar variable q. Assuming that the dynamic model information of the system is available in the joint space, we use inverse dynamics control in the feedback loop. By this means, the actuator torque of the robot can be generated as follows.

$$\tau = c(q,\dot{q})\dot{q} + b_e\dot{q} + \mathfrak{g}(q) + k_p e_q + k_d \dot{e}_q \tag{13.5}$$

In this equation, k_p and k_d are the proportional and derivative control gains of a single-variable PD controller, e_q is the tracking error of the robot's motion calculated using the relation $e_q = q_d - q$, and q_d is the desired trajectory of the robot.

As mentioned earlier, in the compliance control structure of the robot, force sensors are not used, and only the joint motions are measured. Therefore, the interactive torque in this case is calculated from Equation 13.3, and to ensure the robot's contact with the environment, the robot's trajectory should be considered slightly inside the environment, as shown in Figure 13.4. If the desired trajectory is not designed with this consideration, the robot may lose contact with the environment, and the possibility of applying the necessary torque to the environment and adjusting the robot's compliance in this interactive motion will be lost.

To examine the characteristics of this control method and compare it with motion control, substitute Equation 13.5 into the robot's motion dynamics 13.4, while using the interaction torque given in 13.3. Assuming complete knowledge of the robot's dynamics, the nonlinear dynamics of the robot is linearized by feedback, and with some simplifications, the following equation is obtained.

$$m_e\ddot{q} + \kappa_e(q - q_e) = k_p(q_d - q) + k_d(\dot{q}_d - \dot{q}) \tag{13.6}$$

Collect the motion variables terms all in one side and rewrite the relationship 13.6 as:

$$m_e\ddot{q} + k_d\dot{q} + (k_p + \kappa_e)q = \kappa_e q_e + k_p q_d + k_d \dot{q}_d \tag{13.7}$$

Note that, with complete knowledge of the robot's dynamics, the obtained relationship is fully linearized, and linear systems analysis can be applied to determine transient response and steady-state behavior. To this end, derive the Laplace transform of the relationship 13.7 in the s domain.

$$q(s) = \frac{(k_p + k_d s)q_d(s) + \kappa_e q_e(s)}{m_e s^2 + k_d s + (k_p + \kappa_e)} \tag{13.8}$$

In order to analyze the steady-state behavior of the system, consider the two inputs to the system as step functions. In this case, the environment input is represented with the step function $q_e(s) = \bar{q}_e/s$, where $\bar{q}_e$ is the steady-state value of this input. Similarly, assume the desired trajectory input is replaced by the step function $q_d(s) = \bar{q}_d/s$, with its steady-state value denoted by $\bar{q}_d$. By this means, the steady-state value of the robot's motion variable can be obtained using the final value Theorem [162]:

$$\bar{q} = \lim_{t\to\infty} q(t) = \lim_{s\to 0} s\, q(s) \tag{13.9}$$

Since the inputs of the relationship 13.8 are considered as step functions, by using the final value Theorem, the motion steady-state value $\bar{q}$ is determined as follows.

$$\begin{aligned} \bar{q} &= \lim_{s\to 0} \frac{(k_p + k_d s)\bar{q}_d + \kappa_e \bar{q}_e}{m_e s^2 + k_d s + (k_p + \kappa_e)} \\ &= \frac{k_p \bar{q}_d + \kappa_e \bar{q}_e}{k_p + \kappa_e} \end{aligned} \tag{13.10}$$

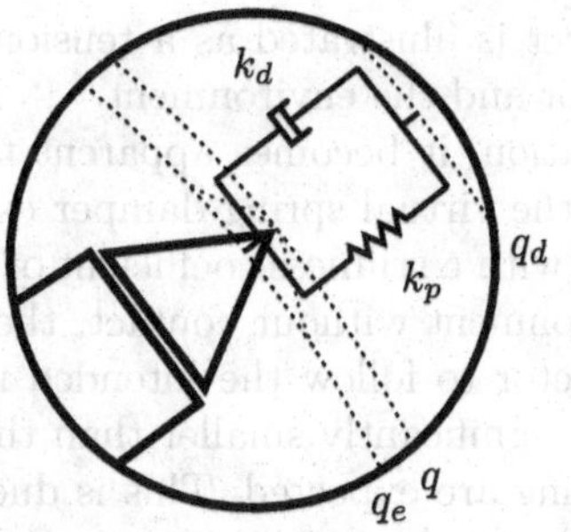

FIGURE 13.5
Interplay of forces between the robot end-effector and a soft environment in compliance control: a virtual spring-damper model.

Let us calculate the steady-state value of the interaction torque $\bar{\tau}_e$ using the relationship 13.3.

$$\begin{aligned}\bar{\tau}_e &= \kappa_e(\bar{q} - \bar{q}_e) \\ &= \frac{\kappa_e(k_p\bar{q}_d + \kappa_e\bar{q}_e)}{k_p + \kappa_e} - \frac{\kappa_e\bar{q}_e(k_p + \kappa_e)}{k_p + \kappa_e} \\ &= \frac{\kappa_e k_p(\bar{q}_d - \bar{q}_e)}{k_p + \kappa_e} \end{aligned} \tag{13.11}$$

Note that the environment stiffness κ_e is significantly larger than the proportional gain of the controller k_p, i.e., $\kappa_e \gg k_p$. In this way, the steady-state value of the interaction torque can be approximated using the following relationship.

$$\begin{aligned}\bar{\tau}_e &\simeq \frac{\kappa_e k_p}{\kappa_e}(\bar{q}_d - \bar{q}_e) \\ &\simeq k_p(\bar{q}_d - \bar{q}_e)\end{aligned} \tag{13.12}$$

This relationship effectively highlights a fundamental characteristic of compliance control in robots. Despite the substantial stiffness of the environment denoted by κ_e, the incorporation of compliance control in robots results in the steady-state relationship between the interaction torque $\bar{\tau}_e$ and the slight deformation introduced in the environment $\bar{q}_d - \bar{q}_e$ being proportional to the proportional gain of the controller k_p, rather than the stiffness of the environment, κ_e. Consequently, by suitable selection of this gain, the interaction torque between the robot and the elastic deformation of the environment can be adjusted as soft as required in the steady-state. This implies that, through the utilization of feedback loops and the adjustment of the proportional gain in a PD controller within the robot, one can render the interaction with the environment as gentle as desired, even when the robot possesses an entirely rigid mechanical structure. For this reason, this control structure is acknowledged in the robotics literature as compliance control.

We underscore the significance of crafting a suitable desired motion trajectory in the context of this control method. As delineated in the relationship 13.12, for the proper adjustment of the robot's compliance during its interactive motion with the environment, it becomes imperative to contemplate a desired motion trajectory slightly within the environment, ensuring that the robot establishes, and keeps contact with the environment. To enhance comprehension of this concept, the relationship is illustrated as a virtual spring-damper model in Figure 13.5. This visual representation portrays the interactive force between the robot joint and the environment functioning as a linear spring and damper in task space. By this means, this effect is illustrated as a tension/compression force interchange between the robot's end-effector and the environment.

With this visual representation, it becomes apparent that when the robot comes into contact with its environment, the virtual spring-damper exerts a force on the robot's end-effector, sustaining the contact with a stiffness coefficient of k_p. In parallel, when the robot's joint moves freely in the environment without contact, the virtual spring-damper imparts a force on the robot's end-effector to follow the intended motion trajectory. However, because of utilizing control gains significantly smaller than those employed in motion control schemes, minor errors in tracking are expected. This is due to the deliberate choice of the control gain k_p not to be excessively large in this case. Consequently, this control method facilitates the adjustment of the interaction force between the robot and the environment or the desired compliance in the steady-state interaction.

Despite the apparent similarity in the implementation topology of this control method to motion control employing an inverse dynamics structure, these two approaches exhibit significant distinctions in terms of design and efficiency. In the context of inverse dynamics control, the underlying assumption is that the robot operates freely within its workspace without any contact with the surrounding environment. Consequently, linear control gains in this method are intentionally selected to be high, aiming to achieve suitable tracking performance. Conversely, compliance control diverges from this goal in controller design, deliberately avoiding the use of excessively large control gains. This is done to ensure an appropriate interaction force between the robot and the environment, thereby preventing any potential damage to the environment or mechanical components of the robot.

Additionally, it's crucial that in scenarios where prior knowledge suggests the robot will make contact with an object, the planned motion trajectory must be carefully crafted considering the object's shape. This ensures that the robot's contact point is situated slightly within the object. As a result, in this control method, trajectory tracking is achieved with a lower level of precision while still ensuring adherence to the desired compliance specified by the designer for the robot's interaction with the environment.

13.3.2 Multivariable Compliance Control

The fundamental concepts of the compliance control method in a single-degree-of-freedom robot have been examined, and the steady-state behavior of the closed-loop system has been analyzed using Laplace transform and the final value Theorem. This is while in the general case, where the robotic dynamical model is multivariable and nonlinear, the analysis based on linear theory is not applicable and nonlinear system theories needs to be considered.

For this purpose, consider a serial robot in the general case, where its motion in the joint space is represented by the vector $\boldsymbol{q}$ and the twist of its end-effector in the task space is represented by a six-dimensional array denoted by $\boldsymbol{\mathcal{X}}$. Furthermore, assume that the robot's end-effector is in contact with its environment during its motion, and the wrench $\boldsymbol{\mathcal{F}}_e$ is applied to it. This interactive wrench can generally be described by a six-dimensional array formed by the combination of interactive forces and torques of the environment: $\boldsymbol{\mathcal{F}}_e = \begin{bmatrix}\boldsymbol{F}_e & \boldsymbol{n}_e\end{bmatrix}^T$.

Suppose the mechanical stiffness of the robot is much greater than that of the hitting object, and as a result of the contact, it is the object that undergoes twist deformation $\boldsymbol{\mathcal{X}} - \boldsymbol{\mathcal{X}}_e$. This deformation, considering the general Kelvin-Voigt linear model, generates the following interactive wrench on the robot's end-effector.

$$\boldsymbol{\mathcal{F}}_e = \boldsymbol{K}_e \Delta \boldsymbol{\mathcal{X}} = \boldsymbol{K}_e (\boldsymbol{\mathcal{X}} - \boldsymbol{\mathcal{X}}_e) \tag{13.13}$$

In this relation, $\Delta\boldsymbol{\mathcal{X}}$ represents the twist deformation of the object in six dimensions, and $\boldsymbol{K}_e$ is the stiffness matrix of the object. This matrix is defined based on three elements for translational deformation and three elements for rotational deformation. By this means, the matrix $\boldsymbol{K}_e$ incorporates the stiffness coefficients of the environment with respect to the six deformation components $\Delta\boldsymbol{\mathcal{X}}$, and generates the interactive wrench $\boldsymbol{\mathcal{F}}_e$ applied to the robot's end-effector by the environment.

Considering that compliance control in robots utilizes the control topology in the joint space, we will use the mapping of the interactive force in the joint space. To achieve this, note that by using the transpose of the Jacobian matrix, the interactive wrench in the task space $\boldsymbol{\mathcal{F}}_e$ can be mapped to the interactive torque in the joint space denoted by $\boldsymbol{\tau}_e$.

$$\boldsymbol{\tau}_e = \boldsymbol{J}^T(\boldsymbol{q})\boldsymbol{\mathcal{F}}_e = \boldsymbol{J}^T(\boldsymbol{q})\boldsymbol{K}_e\Delta\boldsymbol{\mathcal{X}} \tag{13.14}$$

Furthermore, map the task space twist deformation $\Delta\boldsymbol{\mathcal{X}}$ to the joint space using the Jacobian matrix.

$$\Delta\boldsymbol{\mathcal{X}} = \boldsymbol{J}(\boldsymbol{q})\Delta\boldsymbol{q} \tag{13.15}$$

By substituting the relationship 13.15 into 13.14 and defining the stiffness matrix of the environment in the joint space with the matrix $\boldsymbol{\mathcal{K}}_e$, obtain the interactive torque resulting from the contact of the robot's end-effector with the environment as following.

$$\boldsymbol{\tau}_e = \boldsymbol{\mathcal{K}}_e\Delta\boldsymbol{q} \tag{13.16}$$

In this equation, $\boldsymbol{\mathcal{K}}_e$ represents the mapping of the stiffness matrix of the environment in the joint space, and $\Delta\boldsymbol{q}$ is the deformation of the environment mapped into the joint space:

$$\boldsymbol{\mathcal{K}}_e = \boldsymbol{J}^T(\boldsymbol{q})\boldsymbol{K}_e\boldsymbol{J}(\boldsymbol{q}), \quad \Delta\boldsymbol{q} = \boldsymbol{q} - \boldsymbol{q}_e \tag{13.17}$$

Now, rewrite the general equation of robot motion dynamics, as stated in Equation 13.2 in the absence of disturbance torques, using 13.16.

$$\boldsymbol{\tau} = \boldsymbol{M}_e(\boldsymbol{q})\ddot{\boldsymbol{q}} + \boldsymbol{C}(\boldsymbol{q},\dot{\boldsymbol{q}})\dot{\boldsymbol{q}} + \boldsymbol{b}_e\dot{\boldsymbol{q}} + \boldsymbol{\mathfrak{g}}(\boldsymbol{q}) + \boldsymbol{\mathcal{K}}_e\Delta\boldsymbol{q} \tag{13.18}$$

In this equation, $\boldsymbol{M}_e(\boldsymbol{q})$ is the effective mass matrix of the robot, $\boldsymbol{C}(\boldsymbol{q},\dot{\boldsymbol{q}})$ is the Christoffel matrix, $\boldsymbol{b}_e$ is the vector of effective viscous friction torques, and $\boldsymbol{\mathfrak{g}}(\boldsymbol{q})$ is the gravity-induced torque vector of the robot.

As mentioned in the previous chapter, various motion control methods can be incorporated into the design of compliance control in robots. To examine the characteristics of compliance control and compare it with motion control methods, here we employ the method of partial feedback linearization. In the simplest alternative of this method, only the gravity-induced torque is compensated in the feedback loop, and a linear PD controller is used in the outer loop.

$$\boldsymbol{\tau} = \boldsymbol{\mathfrak{g}}(\boldsymbol{q}) + \boldsymbol{K}_p\boldsymbol{e}_q + \boldsymbol{K}_d\dot{\boldsymbol{e}}_q \tag{13.19}$$

In this relation, $\boldsymbol{K}_p$ and $\boldsymbol{K}_d$ denote the matrices of the proportional and derivative gains in the multivariable PD controller, and $\boldsymbol{e}_q$ denote the vector of motion tracking error calculated from the relationship $\boldsymbol{e}_q = \boldsymbol{q}_d - \boldsymbol{q}$, while the desired motion trajectory of the robot is given in joint space.

As previously stated, compliance control in robots does not involve the use of force sensors; instead, only joint motions are monitored. Consequently, the interaction force is computed using the relationship 13.13. To guarantee contact between the robot and the environment, it is essential to design the robot's motion trajectory slightly within the object, as discussed earlier. Failure to consider this while planning the desired trajectory may result in the robot losing contact with the environment, making it impossible to exert the required force on the object and regulate the robot's compliance during this interactive motion.

To examine the characteristics of this control method and compare it with robot motion control, obtain the closed-loop formulation of the robot by substituting the relationship 13.19 into the robot dynamics 13.18. Assuming complete knowledge of the gravitational torques in the robot's dynamics, derive the dynamics of the closed-loop system as follows.

$$\boldsymbol{M}_e(\boldsymbol{q})\ddot{\boldsymbol{q}} + \boldsymbol{C}(\boldsymbol{q},\dot{\boldsymbol{q}})\dot{\boldsymbol{q}} + \boldsymbol{b}_e\dot{\boldsymbol{q}} + \boldsymbol{\mathcal{K}}_e(\boldsymbol{q}-\boldsymbol{q}_e) = \boldsymbol{K}_p\boldsymbol{e}_q + \boldsymbol{K}_d\dot{\boldsymbol{e}}_q \tag{13.20}$$

Note that the dynamics of the closed-loop system of the robot remain multivariable and nonlinear. To analyze asymptotic stability and the steady-state behavior of the system, tools for analyzing nonlinear systems need to be employed.

For stability analysis of the steady-state behavior, it is essential to consider the two system inputs to be bounded and converge to constant values in the steady-state. To achieve this, consider the desired input trajectory $\boldsymbol{q}_d$ as a constant function and assume that the mapping of the environmental geometry in the joint space of the robot $\boldsymbol{q}_e$ be formulated as a constant function as well. This assumption allows for the analysis of the steady-state behavior of the closed-loop system using the direct Lyapunov method [134].

Consider the following Lyapunov function candidate for the asymptotic behavior analysis of the system's state variables.

$$V(\boldsymbol{q}) = \frac{1}{2}\dot{\boldsymbol{q}}^T\boldsymbol{M}\dot{\boldsymbol{q}} + \frac{1}{2}\boldsymbol{e}_q^T\boldsymbol{K}_p\boldsymbol{e}_q + \frac{1}{2}(\boldsymbol{q}-\boldsymbol{q}_e)^T\boldsymbol{\mathcal{K}}_e(\boldsymbol{q}-\boldsymbol{q}_e) \tag{13.21}$$

This Lyapunov function candidate consists of three terms. The first term represents the kinetic energy of the robot's motion, the second term represents the potential energy due to proportional control, and the third term represents the stored energy due to the elastic deformation of the environment. Therefore, this function is positive definite, and to analyze asymptotic stability, it is necessary to calculate the time derivative of this function. Considering the constant inputs in this scenario, calculate the derivative of the Lyapunov function as follows.

$$\dot{V}(\boldsymbol{q}) = \dot{\boldsymbol{q}}^T\boldsymbol{M}_e\ddot{\boldsymbol{q}} + \frac{1}{2}\dot{\boldsymbol{q}}^T\dot{\boldsymbol{M}}\dot{\boldsymbol{q}} + \dot{\boldsymbol{e}}_q^T\boldsymbol{K}_p\boldsymbol{e}_q + \dot{\boldsymbol{q}}^T\boldsymbol{\mathcal{K}}_e(\boldsymbol{q}-\boldsymbol{q}_e) \tag{13.22}$$

Use the closed-loop system's dynamic equation 13.20 to substitute $\boldsymbol{M}_e\ddot{\boldsymbol{q}}$, and simplify to reach to:

$$\begin{aligned}\dot{V} &= \dot{\boldsymbol{q}}^T\left(-(\boldsymbol{C}+\boldsymbol{b}_e)\dot{\boldsymbol{q}} - \boldsymbol{\mathcal{K}}_e(\boldsymbol{q}-\boldsymbol{q}_e) + \boldsymbol{K}_p\boldsymbol{e}_q + \boldsymbol{K}_d\dot{\boldsymbol{e}}_q\right) \\ &\quad + \frac{1}{2}\dot{\boldsymbol{q}}^T\dot{\boldsymbol{M}}_e\dot{\boldsymbol{q}} + \dot{\boldsymbol{e}}_q^T\boldsymbol{K}_p\boldsymbol{e}_q + \dot{\boldsymbol{q}}^T\boldsymbol{\mathcal{K}}_e(\boldsymbol{q}-\boldsymbol{q}_e)\end{aligned}$$

To further simplify this equation, consider that, the inputs are constant, $\dot{\boldsymbol{q}}_d = \dot{\boldsymbol{q}}_e = \boldsymbol{0}$; therefore, the relationship $\dot{\boldsymbol{e}}_q = (\dot{\boldsymbol{q}}_d - \dot{\boldsymbol{q}}) = -\dot{\boldsymbol{q}}$ is true.

$$\begin{aligned}\dot{V} &= \frac{1}{2}\dot{\boldsymbol{q}}^T(\dot{\boldsymbol{M}}_e - 2\boldsymbol{C})\dot{\boldsymbol{q}} - \dot{\boldsymbol{q}}^T\boldsymbol{\mathcal{K}}_e(\boldsymbol{q}-\boldsymbol{q}_e) + \dot{\boldsymbol{q}}^T\boldsymbol{K}_p\boldsymbol{e}_q \\ &\quad -\dot{\boldsymbol{q}}^T\boldsymbol{b}_e\dot{\boldsymbol{q}} - \dot{\boldsymbol{q}}^T\boldsymbol{K}_d\dot{\boldsymbol{q}} + \dot{\boldsymbol{q}}^T\boldsymbol{\mathcal{K}}_e(\boldsymbol{q}-\boldsymbol{q}_e) - \dot{\boldsymbol{q}}^T\boldsymbol{K}_p\boldsymbol{e}_q \\ &= \frac{1}{2}\dot{\boldsymbol{q}}^T(\dot{\boldsymbol{M}}_e - 2\boldsymbol{C})\dot{\boldsymbol{q}} - \dot{\boldsymbol{q}}^T(\boldsymbol{K}_d + \boldsymbol{b}_e)\dot{\boldsymbol{q}} \\ &= -\dot{\boldsymbol{q}}^T(\boldsymbol{K}_d + \boldsymbol{b}_e)\dot{\boldsymbol{q}} \leq 0\end{aligned} \tag{13.23}$$

In this simplification, the property of skew-symmetry in the matrix $(\dot{\boldsymbol{M}}_e - 2\boldsymbol{C})$ has been utilized. This analysis demonstrates that the derivative of the Lyapunov function is negative semi-definite, and to prove asymptotic convergence of the tracking error toward zero, the Krasovski-Lasalle's Theorem needs to be applied.

Given the negative semi-definite property of the derivative of the Lyapunov function, it is evident that when $\dot{q} \neq 0$, the Lyapunov function is strictly decreasing. Therefore, the convergence analysis of the tracking error when $\dot{q} = \ddot{q} = 0$, should be considered. In this case, from the equation 13.20, it can be inferred that

$$\lim_{t\to\infty} [\mathcal{K}_e(\boldsymbol{q} - \boldsymbol{q}_e) - \boldsymbol{K}_p(\boldsymbol{q}_d - \boldsymbol{q})] = 0 \tag{13.24}$$

Alternatively,

$$\lim_{t\to\infty} [(\mathcal{K}_e + \boldsymbol{K}_p)\boldsymbol{q} - \mathcal{K}_e\boldsymbol{q}_e - \boldsymbol{K}_p\boldsymbol{q}_d] = 0 \tag{13.25}$$

If the steady-state value of the joint motion vector is denoted by $\bar{\boldsymbol{q}}$, and similar notation is used to represent the steady-state value of the object surface $\bar{\mathcal{X}}_e$, the mapping of this contact surface in the joint space can be denoted with the notation $\bar{\boldsymbol{q}}_e$. Interactive wrench in the steady-state is represented by the notation $\bar{\mathcal{F}}_e$, and its mapping in the joint space is denoted by $\bar{\boldsymbol{\tau}}_e$. Thus, utilize the relationship 13.25 in calculating the steady-state value of the joint motion vector.

$$\bar{\boldsymbol{q}} = [\mathcal{K}_e + \boldsymbol{K}_p]^{-1}[\mathcal{K}_e\bar{\boldsymbol{q}}_e + \boldsymbol{K}_p\bar{\boldsymbol{q}}_d] \tag{13.26}$$

Using this relationship, obtain the steady-state value of the interacting torque $\bar{\boldsymbol{\tau}}_e$.

$$\begin{aligned}
\bar{\boldsymbol{\tau}}_e &= \mathcal{K}_e(\bar{\boldsymbol{q}} - \bar{\boldsymbol{q}}_e) \\
&= \mathcal{K}_e[\mathcal{K}_e + \boldsymbol{K}_p]^{-1}[\mathcal{K}_e\bar{\boldsymbol{q}}_e + \boldsymbol{K}_p\bar{\boldsymbol{q}}_d] - \mathcal{K}_e\bar{\boldsymbol{q}}_e \\
&= \mathcal{K}_e[\mathcal{K}_e + \boldsymbol{K}_p]^{-1}\boldsymbol{K}_p(\bar{\boldsymbol{q}}_d - \bar{\boldsymbol{q}}_e)
\end{aligned} \tag{13.27}$$

Note that the environmental stiffness $\mathcal{K}_e$ is significantly larger than the proportional controller gain $\boldsymbol{K}_p$, i.e., $\mathcal{K}_e \gg \boldsymbol{K}_p$. In this way, the steady-state value of the torque resulting from interactive torques can be approximated by the following relationship.

$$\bar{\boldsymbol{\tau}}_e \simeq \boldsymbol{K}_p(\bar{\boldsymbol{q}}_d - \bar{\boldsymbol{q}}_e) \tag{13.28}$$

Similar to what was described in the compliance control of a single-degree-of-freedom robot, this relationship indicates that despite the very large stiffness of the environment $\mathcal{K}_e$, the use of compliance control in the robot causes the relationship between the torque due to interactive forces $\bar{\boldsymbol{\tau}}_e$ and the deformation created in the environment $\bar{\boldsymbol{q}}_d - \bar{\boldsymbol{q}}_e$ in the steady-state to be proportional to the proportional controller gain matrix $\boldsymbol{K}_p$, not the stiffness of the environment. Thus, by adjusting the controller gain, the interaction force between the robot and the elastic displacement of the environment can be controlled as desired in the steady-state. This means that by using the feedback loop and adjusting the proportional gain of the PD controller in the robot, interaction with the environment can be made as compliant as desired, despite using a robot with a completely rigid mechanical structure.

This analysis also shows that by using a partial linear feedback linearization controller and only compensating for the torque due to gravitational forces, the compliance control method will be able to adjust the interactive force-displacement relationship in the steady-state proportionally to the proportional gain matrix $\boldsymbol{K}_p$. In this analysis, the convergence of the asymptotic tracking error toward zero is proven using the skew-symmetric property of the matrix $(\dot{\boldsymbol{M}}_e - 2\boldsymbol{C})$.

While the implementation structure of this control method closely resembles motion control with the partial feedback linearization control architecture of the robot, these two approaches exhibit significant differences in terms of design and performance. In the context of inverse dynamics control design, the assumption is that the robot moves freely within its workspace without contact with its surrounding environment. Consequently, linear controller gains are chosen to be as large as possible to minimize tracking error. However, in the context of robot compliance control, achieving minimal tracking error is not the primary objective in controller design. Instead, controller gains are intentionally kept relatively small to ensure that the interaction force between the robot and the environment remains soft, preventing any potential damage to the environment or mechanical components of the robot.

Moreover, when it is anticipated that the robot will come into contact with its environment, it becomes essential to accurately position the desired trajectory of motion slightly inside the environment, taking into account the shape of the object and the location of contact with the robot. Consequently, in this control methodology, trajectory tracking is executed with less precision. Instead, the focus is on guaranteeing the desired compliance in the interaction between the robot and the environment.

13.4 Direct Force Control

Compliance control is more often employed in robots engaged with the environment compared to motion control methods. Within the compliance control approach, the adjustment of the relationship between the steady-state force-motion relation is achieved through the tuning of control gains, eliminating the necessity for force sensors. Nevertheless, the application of this control method is constrained due to the following limitations, constraining its broad implementation in industrial settings.

As force sensors are not integrated into the compliance control structure, the precise measurement of the interactive force between the robot and the environment is unavailable. Instead, this important variable is estimated by employing the elastic model of the environment. Due to the slight deformations in this case, the accuracy of their estimation relies on the highly precise measurement of the robot's end-effector position, which may not be cost-effective. Moreover, the relationship between the interactive force within the robot and its motion is simplified and adapted in the most straightforward manner, using a linear spring-damper model, and solely in the steady-state. Consequently, the dynamic correlation between force and motion in this approach remains uncontrollable, and its transient behavior cannot be tuned. Furthermore, only the steady-state behavior can be tailored for constant inputs, which is less pertinent in robotics applications where such inputs are infrequent.

In many robotic applications where the robot is in motion along a desired path, the interaction forces of the robot's end-effector with the environment need to be controlled. Consider, for example, a window-cleaning robot, performing calligraphy on a board by a robot, or a robot that accurately grinds surfaces. While the end-effector of the robot must traverse the desired path with precision, it is also necessary to exert an appropriate wrench on the environment during the movement to ensure the intended operation is performed correctly. In such cases, the use of compliance control is not recommended.

If the goal is to precisely monitor and control the interactive wrench between the robot and the environment, it is essential for this wrench to be directly measured and utilized in the feedback loop. Various methods have been developed for designing and implementing control systems in this scenario, and in this section, the direct force control method using inverse dynamics is introduced as a fundamental approach in this field. This way, the key concepts of this control method are introduced, and by understanding them, the reader can explore advanced robust and adaptive control methods in this domain [163, 164].

Note that in addition to the primary goal of tracking the interactive wrench in the direct force control method, when tracking the desired trajectory of the robot is not in conflict with the primary objective, it can be considered as the secondary objective. For this reason, the topology of cascad control is employed in this method. Since tracking the interacting wrench is the main objective, as depicted in Figure 13.2, force feedback is utilized in the outer loop, and position feedback is employed in the inner loop of the cascad control. The design of position and force controllers in this structure is considered in the joint space for easier implementation. Therefore, the Jacobian transpose matrix is used in the outer feedback loop to map the interactive wrench of the robot and the environment into joint space. In general, six-dimensional wrench sensors capable of measuring the forces and torques exerted on the robot, are typically used at the robot's wrist or end-effector. Additionally, the motion of the robot's joints is commonly measured by position sensors and utilized in the inner feedback loop.

13.4.1 Design of Cascade Control

Consider a scenario where a robot with multiple degrees of freedom is in wrench interaction with its surrounding environment. The motion of the robot's arms denoted by the vector $\boldsymbol{q}$ and the interactive wrench between the robot and the environment with the six-dimensional array $\boldsymbol{\mathcal{F}}_e = \begin{bmatrix}\boldsymbol{F}_e & \boldsymbol{n}_e\end{bmatrix}^T$, has been measured and utilized in the feedback loops. The dynamic model of the robot, in the absence of disturbance torques, can be rewritten using the equation 13.2 as follows.

$$\boldsymbol{\tau} = \boldsymbol{M}_e(\boldsymbol{q})\ddot{\boldsymbol{q}} + \boldsymbol{C}(\boldsymbol{q},\dot{\boldsymbol{q}})\dot{\boldsymbol{q}} + \boldsymbol{b}_e\dot{\boldsymbol{q}} + \boldsymbol{\mathfrak{g}}(\boldsymbol{q}) + \boldsymbol{\tau}_e \tag{13.29}$$

In this equation, $\boldsymbol{M}_e(\boldsymbol{q})$ represents the effective mass matrix of the robot, $\boldsymbol{C}(\boldsymbol{q},\dot{\boldsymbol{q}})$ is the Christoffel matrix, $\boldsymbol{b}_e$ is the effective viscous friction torque vector, $\boldsymbol{\mathfrak{g}}(\boldsymbol{q})$ is the gravitational vector, and $\boldsymbol{\tau}_e$ is the torque resulting from the interactive wrench, mapped to the joint space using the transpose of the Jacobian matrix, $\boldsymbol{\tau}_e = \boldsymbol{J}^T(\boldsymbol{q})\boldsymbol{\mathcal{F}}_e$.

Develop direct force control in this scenario by employing inverse dynamics control in the inner loop and a force controller in the outer loop. The block diagram for this design is depicted in Figure 13.6. As depicted in this figure, the desired interactive torque $\boldsymbol{\tau}_d$ serves as the primary input for tracking in the outer loop, and the torque tracking error $\boldsymbol{e}_\tau = \boldsymbol{\tau}_d - \boldsymbol{\tau}_m$ is dynamically computed in this loop. In the calculation of this error, the $\boldsymbol{\tau}_m$ vector is employed, representing the mapping of the measured interactive wrench of the robot into the joint space.

It's important to highlight that direct motion tracking is not explicitly integrated into this cascaded control structure because $\boldsymbol{q}_d$ is not introduced as an independent input to the system. Motion tracking, as the secondary objective in this control structure, is computed by utilizing the auxiliary input $\boldsymbol{q}_a$, derived from the output of the force controller in the outer loop, and is fed to the inner loop. Consequently, the robot doesn't strictly adhere to a predefined motion trajectory but moves along a path close to the contact surface, in order to ensure precise tracking of the desired interactive wrench.

In the direct force control's inner loop, inverse dynamics control is utilized, and a simple linear controller is employed in the outer loop. It's worth noting that advanced methods such as linear robust control can also be implemented to enhance the linear controller in this loop [165].

Based on this structure, determine the motor control effort in the inner loop, $\boldsymbol{\tau}$, based on the inverse dynamics control of the robot.

$$\boldsymbol{\tau} = \hat{\boldsymbol{M}}_e(\boldsymbol{q})\boldsymbol{a} + \boldsymbol{\tau}_{fl} \tag{13.30}$$

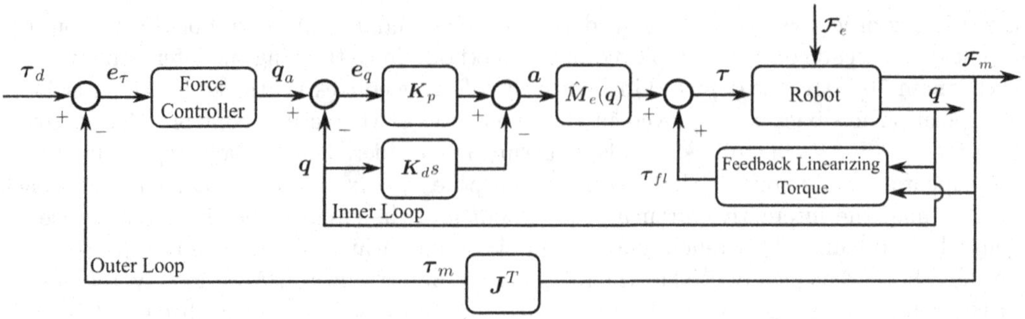

FIGURE 13.6
Block diagram of direct force control: Force feedback in the outer loop and position control in the inner loop of the cascaded control structure.

where

$$\tau_{fl} = \left(\hat{C}(q,\dot{q}) + \hat{b}_e\right)\dot{q} + \hat{\mathfrak{g}}(q) + J^T\mathcal{F}_m \tag{13.31}$$

$$a = K_p(q_a - q) - K_d\dot{q} \tag{13.32}$$

In this context, subtle yet impactful modifications have been introduced to the conventional inverse dynamics control approach. If we consider the gains of the linear controllers in inverse dynamics control as K_p and K_d, the proportional gain matrix is applied to the tracking error matrix, while the derivative gain matrix is multiplied to only the derivative of the robot's motion vector. This seemingly minor adjustment helps mitigate the impact of measurement noise in the force tracking loop.

Another significant modification observed in the inverse dynamics control structure involves compensating for the interactive wrench in the feedback linearization loop. As illustrated in Equation 13.31, the term $J^T\mathcal{F}_m$ is integrated into the feedback linearization torque, deviating from the usual scenario. Here, $\mathcal{F}_m$ represents the measured value of the robot's interactive wrench $\mathcal{F}_e$. Ensuring accurate measurement of the interactive wrench is crucial for effectively compensating the dominant effect of this term in the robot's dynamics. Furthermore, the mapping of this measured quantity, denoted as $\tau_m = J(q)^T\mathcal{F}_m$ in the joint space, is directly employed in the feedback loop of the outer control loop.

Inverse dynamics control comprises two elements: linear feedback linearization control and a linear controller. The input to the linear controller section is the auxiliary motion trajectory vector determined by the force controller. The output of the linear controller is a 6×1 acceleration vector, denoted as a. The joint torques τ are formed according to Equation 13.30, resulting from the product of the estimated robot nominal mass matrix $\hat{M}_e(q)$ by this acceleration vector a, in addition to the calculated feedback linearization torques. Consequently, accurate tracking of the auxiliary motion trajectory generated by the outer loop is achieved while compensating for the nonlinear dynamics of the robot.

The force controller in this configuration is formulated based on the wrench tracking error within the joint space. For this purpose, the tracking error vector, denoted as $e_\tau = \tau_d - \tau_m$, is computed as the disparity between the desired joint torque vector and the measured torque vector of the robot. It's essential to note that, when the measurement of the interactive wrench in the workspace $\mathcal{F}_m$ is available, this quantity is transformed into the joint space using the transpose of the Jacobian matrix, expressed as $\tau_m = J^T(q)\mathcal{F}_m$. Likewise, in the case where the desired interactive wrench for a specific application is provided as $\mathcal{F}_d$, it undergoes mapping to the joint space using the same matrix, denoted as $\tau_d = J^T(q)\mathcal{F}_d$.

If we represent the force controller in joint space with a function $\boldsymbol{C}_\tau$, as depicted in Figure 13.6, the auxiliary motion trajectory vector of the robot, denoted as $\boldsymbol{q}_a$, is determined based on this function in the following manner.

$$\boldsymbol{q}_a = \boldsymbol{C}_\tau(\boldsymbol{\tau}_d - \boldsymbol{\tau}_m) = \boldsymbol{C}_\tau(\boldsymbol{e}_\tau) \tag{13.33}$$

The force controller in this structure can be designed using advanced control theories. For this purpose, the use of robust or adaptive control methods has been suggested in the robotics literature [163, 164]. However, here, we employ a fundamental proportional or proportional-integral (PI) control structure. The incorporation of the derivative of the error in the force controller is not recommended, as force measurements are typically accompanied by noise, and using a derivative term would amplify the effects of the noise. Moreover, when comparing the physical quantity of force with the positional quantity, force is directly proportional to the second derivative of motion. Hence, there is no necessity to introduce an additional derivative term in the force controller structure.

Using the integral term in the force control structure is highly suitable because it ensures that the wrench tracking error converges to zero in steady-state. Moreover, considering that force corresponds to the acceleration of motion, integrating it with respect to time results in a term proportional to the velocity of motion similar to a PD controller used in conventional inverse dynamics method. Utilizing this integral term allows adjusting the transient behavior of wrench tracking to an appropriate form. To leverage the dimensionless control benefits in the cascaded control structure, define the proportional-integral (PI) force controller $\boldsymbol{C}_\tau(\boldsymbol{e}_\tau)$ as follows.

$$\boldsymbol{C}_\tau(\boldsymbol{e}_\tau(t)) = \boldsymbol{K}_P^{-1}\left[\boldsymbol{K}_{\tau_p}\boldsymbol{e}_\tau(t) + \boldsymbol{K}_{\tau_i}\int_0^t \boldsymbol{e}_\tau(\eta)d\eta\right] \tag{13.34}$$

Due to the linearity of this controller, it is possible to rewrite it in the Laplace domain by using the Laplace transform in the s domain.

$$\boldsymbol{C}_\tau(s) = \boldsymbol{K}_P^{-1}\left(\boldsymbol{K}_{\tau_p} + \frac{1}{s}\boldsymbol{K}_{\tau_i}\right)\boldsymbol{e}_\tau(s) \tag{13.35}$$

In this equation, $\boldsymbol{K}_{\tau_p}$ and $\boldsymbol{K}_{\tau_i}$ represent the proportional and integral gain matrices of the force controller, respectively. These matrices are scaled by the inverse of the proportional gain matrix of the position controller $\boldsymbol{K}_P^{-1}$ to counteract its influence on the overall control structure. This ensures that the force controller gains remain independent of the position controller. Moreover, if you prefer to exclusively use a proportional force controller, you can simply set the integral gain matrix to zero, i.e., $\boldsymbol{K}_{\tau_i} = \boldsymbol{0}$.

13.4.2 Analysis of the Closed-Loop System

To analyze the performance of this system, calculate the closed-loop dynamic model of the system using this cascaded control structure. For this purpose, substitute the robot's actuator torques given in Equation 13.31 into the robot's dynamic formulation 13.2, neglecting the disturbance torques.

$$\begin{aligned}\boldsymbol{M}_e(\boldsymbol{q})\ddot{\boldsymbol{q}} + (\boldsymbol{C}(\boldsymbol{q},\dot{\boldsymbol{q}}) + \boldsymbol{b}_e)\dot{\boldsymbol{q}} + \boldsymbol{\mathfrak{g}}(\boldsymbol{q}) + \boldsymbol{J}^T(\boldsymbol{q})\boldsymbol{\mathcal{F}}_e &= \hat{\boldsymbol{M}}_e(\boldsymbol{q})\boldsymbol{a} + \boldsymbol{\mathcal{F}}_{fl} \\ &= \hat{\boldsymbol{M}}(\boldsymbol{q})\left\{\boldsymbol{K}_p(\boldsymbol{q}_a - \boldsymbol{q}) - \boldsymbol{K}_d\dot{\boldsymbol{q}}\right\} + \left(\hat{\boldsymbol{C}}(\boldsymbol{q},\dot{\boldsymbol{q}}) + \hat{\boldsymbol{b}}_e\right)\dot{\boldsymbol{q}} + \hat{\boldsymbol{\mathfrak{g}}}(\boldsymbol{q}) + \hat{\boldsymbol{J}}^T(\boldsymbol{q})\boldsymbol{\mathcal{F}}_m\end{aligned}$$

If we have a complete knowledge of the robot's kinematic and dynamic models, then

$$\hat{\boldsymbol{M}}_e = \boldsymbol{M}_e,\ \hat{\boldsymbol{C}} = \boldsymbol{C},\ \hat{\boldsymbol{b}}_e = \boldsymbol{b}_e,\ \hat{\boldsymbol{\mathfrak{g}}} = \boldsymbol{\mathfrak{g}},\ \hat{\boldsymbol{J}} = \boldsymbol{J}.$$

Furthermore, if the measurement of interactive wrench is accurate and noise-free, then $\mathcal{F}_m = \mathcal{F}_e$. In such instances, by substituting $\boldsymbol{q}_a$ into this relationship and simplifying, the closed-loop system formulation is summarized as follows.

$$\boldsymbol{M}_e(\boldsymbol{q})\ddot{\boldsymbol{q}} = \boldsymbol{M}_e(\boldsymbol{q})\{-\boldsymbol{K}_d\dot{\boldsymbol{q}} - \boldsymbol{K}_p\boldsymbol{q} + \boldsymbol{K}_p\boldsymbol{C}_\tau(\boldsymbol{e}_\tau)\} \tag{13.36}$$

By substituting the relation given in 13.34 into 13.36, we will arrive at the final relation as follows.

$$\ddot{\boldsymbol{q}} + \boldsymbol{K}_d\dot{\boldsymbol{q}} + \boldsymbol{K}_p\boldsymbol{q} = \boldsymbol{K}_{\tau_p}\boldsymbol{e}_\tau(t) + \boldsymbol{K}_{\tau_i}\int_0^t \boldsymbol{e}_\tau(\eta)d\eta \tag{13.37}$$

This relation indicates that when employing direct force control, having complete knowledge of the system dynamics, and precise noise-free measurement of interaction wrench, the closed-loop dynamic equation of the system transforms into a second-order system along with a PI controller. By appropriately choosing the control gains for position and force in the two cascaded loops, not only stability and suitable transient dynamics are regulated in the position controller but also the wrench tracking error converges asymptotically to zero. It's essential to note that when the wrench tracking error is converging to zero, but still there is a error residual, the outer loop adjusts the desired robot trajectory in a manner that steers the wrench tracking error toward zero.

Analyzing the attributes of force control within the cascaded control structure, utilizing the closed-loop dynamic system equation, elucidates the assurance of the primary goal of wrench tracking and the secondary objective of tracking the desired motion. As previously stated, the adjustment of the auxiliary motion vector $\boldsymbol{q}_a$ in the outer control loop governs the regulation of the desired robot motion, contingent on the performance of wrench tracking. The engine of robot motion in the workspace is represented by the right side of the closed-loop dynamic system equation in relation 13.37. This causes the robot's motion to persist until the wrench tracking error in the closed-loop system converges to zero, leading to the deactivation of this motion engine. The inner inverse dynamic control ensures stable and appropriately transient motion for the robot as long as this engine remains active. Consequently, controlled motion is induced in the robot until the wrench tracking error reaches zero. The inclusion of the integral term on the left side of this equation guarantees the gradual convergence of the wrench tracking error toward zero, even when facing disturbances and violating assumptions like precise, noise-free measurement of interactive force.

This control approach finds widespread use in applications where the primary objective is interactive wrench tracking. The success of this method in practical applications can be attributed to the effective linearization of the nonlinear dynamics of the system and the clear distinction between the primary and secondary goals within the cascade control structure. Consequently, the closed-loop characteristics are not contingent on the mechanical structure of the robot or the position of the robot's end-effector, allowing for easy adjustment of control gains for both position and force. Nevertheless, it is crucial to acknowledge that this control method is model-based, necessitating the designer to possess comprehensive knowledge of the robot's kinematic and dynamic models.

Furthermore, in this method, interactive wrench tracking is effectively achieved only when the robot is in contact with the environment. If, for any reason, the contact between the robot and the environment is lost, interactive wrench will suddenly and significantly decrease. To reassure tracking of the desired interactive force, rapid and uncontrollable motions may be generated in the robot. In applications such as human-robot interaction and surgical robots, where the robot's end-effector is not continuously in contact with the environment, it is necessary to control the transient behavior of the robot during contact with the environment. For this purpose, it is recommended to use impedance control methods, which will be detailed in the upcoming chapter.

13.5 Hybrid Position-Force Control

Up to now, methods for position control and direct force control in robots have been studied separately. In applications where the robot freely moves in its workspace and precise motion tracking is desired, the use of position control methods in the robot is recommended. If the robot is constantly in contact with its surrounding environment and the adjustment or tracking of interactive wrench is the focus, direct force control methods would be appropriate.

In fully actuated robots, where the number of robot actuators aligns with its degrees of freedom, it is logical to focus on either motion control or force control objectives at each joint individually. Imagine a scenario where motion tracking is sought in specific degrees of freedom of the robot, while the emphasis is on adjusting or tracking interactive wrench in other degrees of freedom. In such instances, a combination of position control and direct force control methods becomes feasible. The integration of these approaches is viable in a unified structure, provided that the objectives of position and wrench tracking can be discerned in the joint space. This methodology, recognized as hybrid position-force control, was initially introduced in robotics literature in 1981 [166], and has been applied in various applications [167, 168].

To illuminate how this method can be employed, consider the application of robotics in glass cleaning, erasing on a board, or writing on a blackboard. In these applications, the robot's end-effector needs to cover the two-dimensional $x - y$ plane in terms of motion, while simultaneously applying an appropriate force along the perpendicular direction to the z plane. Thus, using either position control or force control alone is not recommended in these applications. Consider a three-degree-of-freedom Elbow manipulator, as depicted in Figure 4.7, is utilized in this application. By this means, the angular motion of the second and third joints provides the necessary motion coverage in the $x - y$ plane, while, the rotational motion of the first joint around the robot's base enables the application of suitable force along the direction perpendicular to the plane. In the hybrid control approach, the force controller in the first joint is combined with position control in the second and third joints. With this description, we refrain from elaborating the details of designing the position and force control, which have been previously discussed in detail, and focus solely on how to integrate these methods into a unified structure in this section.

In order to combine force and position control methods, first define the "Selection Matrix" based on the suitable configuration of force-position hybrid control. The selection matrix, denoted by $\boldsymbol{\mathcal{S}}$, is a diagonal matrix where the diagonal element i is equal to one if the corresponding joint is under force control, and zero otherwise. With this definition, for the three-degree-of-freedom Elbow manipulator in the mentioned application, the selection matrix is given by:

$$\boldsymbol{\mathcal{S}} = \begin{bmatrix} 1 & 0 & 0 \\ 0 & 0 & 0 \\ 0 & 0 & 0 \end{bmatrix}.$$

Thus, force and position control are separately designed for all robot joints, and using the selection matrix, without worrying about their interference, the corresponding control designed for each joint is implemented.

The configuration of the hybrid force-position control is illustrated in the block diagram shown in Figure 13.7. In this figure, inspired by the topology of individual motion and force control in joint space, position and force control are designed as separate and multivariable controllers for all robot joints. Meanwhile, the joint torques resulting from the combination of force and position controllers, selected by the matrices $\boldsymbol{\mathcal{S}}$ and $\boldsymbol{I} - \boldsymbol{\mathcal{S}}$, are combined and applied to each robot joint.

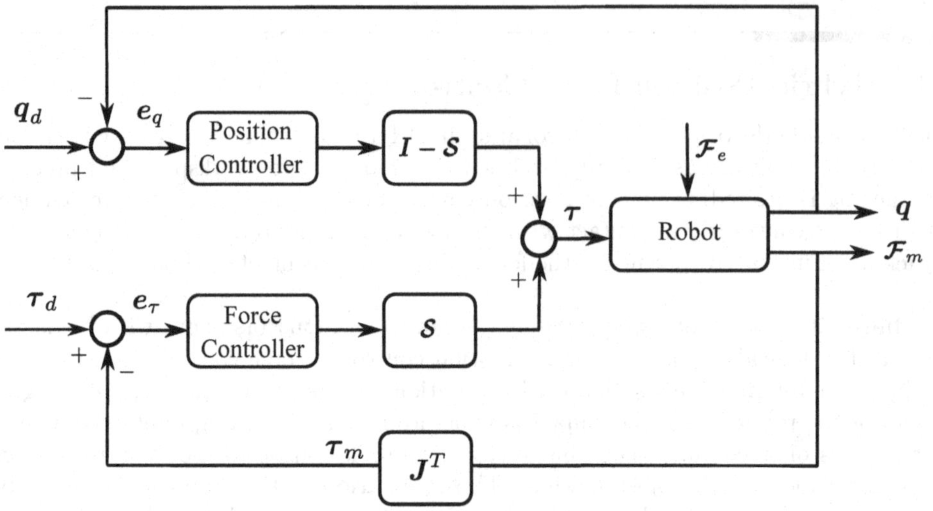

FIGURE 13.7
Block Diagram of Hybrid Force-Position Control: Interplay between two control methods using the selection matrix.

Using this control structure, diverse methods of direct force control and position control can be employed in their respective control loops. If an inverse dynamics control structure is used for designing position control and an inverse dynamics control structure along with a linear controller is used for force control, the hybrid force-position control can be implemented as illustrated in Figure 13.8.

As evident in this figure, the conventional inverse dynamics control structure is considered for designing position control in the upper loop of this block diagram. Meanwhile, for force control design, the use of inverse dynamics control along with compensating interactive wrench is taken into account in the lower feedback loop. By incorporating the product of the interactive torque τ_m with the selection matrix in the linear feedback controller,

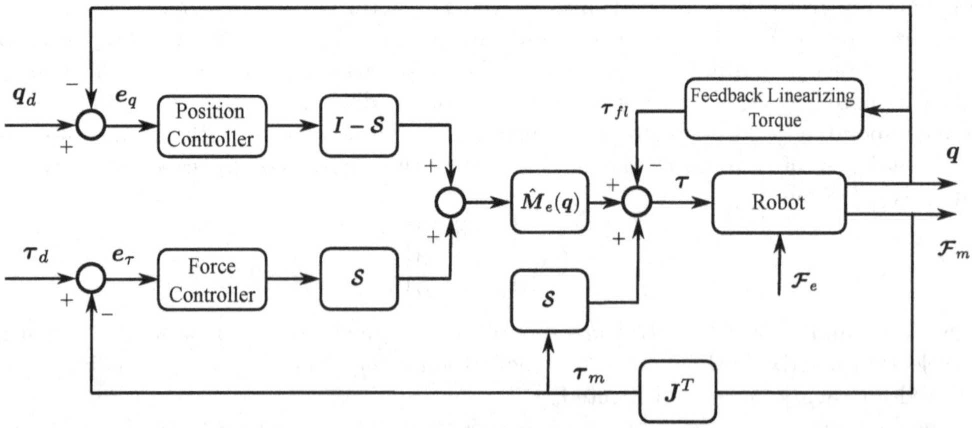

FIGURE 13.8
Block Diagram of Hybrid Force-Position Control: Utilizing inverse dynamics control in designing force and position controllers.

TABLE 13.1
Kinematic and dynamic parameters of the planar 3R robot in force control simulations

Quantity	Unit	Notation	Link 1	Link 2	Link 3
Length	m	a_i	1.0	0.7	0.3
Mass	Kg	m_i	10	7	3
Center of Mass	m	m_{c_i}	0.5	0.35	0.15
Moment of Inertia	$Kg \cdot m^2$	$^C I_i$	0.833	0.286	0.023

compensatory torques will be applied in the joints where force control is designed. With this provision, a hybrid force-position control structure can be designed and implemented by integrating separate designs for position control in the upper loop and direct force control in the lower loop, utilizing various methods developed in each domain.

In this section, the design of hybrid force-position control in the joint space of the robot has been emphasized to facilitate the design and implementation of the controllers. However, this control structure can also be leveraged to combine position and force controllers in the task space. In this approach, depending on the measured quantities in each robot, an appropriate topology needs to be determined, and the combination of controllers needs to be performed using the selection matrix.

13.6 Force Control of the Planar 3R Robot

In this section, we present simulations of force control on the planar 3R robot, serving as an illustrative case study of the robots discussed in this book. We apply different control techniques introduced in this chapter to this robot, subsequently comparing and analyzing the simulations outcomes.

For this purpose, consider the 3R robot depicted in Figure 13.9 and its kinematic and mass parameters as specified in Table 13.1. Assume that all robot links are considered as slender rods, and their center of mass are located at the link's mid length. The dynamics formulation of the robot represented in matrix form 13.2, is derived in the case study 8.3 and the determined Christoffel matrix is reported in case study 9.2. In order to investigate the nonlinear behavior of the system, we neglect the effects of the actuator dynamics. Thus, the following dynamic equation of the system is used in the simulations.

$$\boldsymbol{\tau} + \boldsymbol{\tau}_d = \boldsymbol{M}(q)\ddot{q} + \boldsymbol{C}(q,\dot{q})\dot{q} + \mathfrak{g}(q) + \boldsymbol{J}^T(q)\boldsymbol{\mathcal{F}}_e \tag{13.38}$$

13.6.1 Compliance Control

In force control methods, the interaction between the robot and the environment is a crucial factor that needs to be carefully considered. To address this, in the conducted simulations, we assume a flat obstacle is placed in the path of the robot's end-effector motion. Since the robot's movement is in the plane, we model the cross-sectional surface of this obstacle in the x, y plane. This may be modeled as a line, represented by the following equation.

$$y = y_l + s_l(x - x_l) \tag{13.39}$$

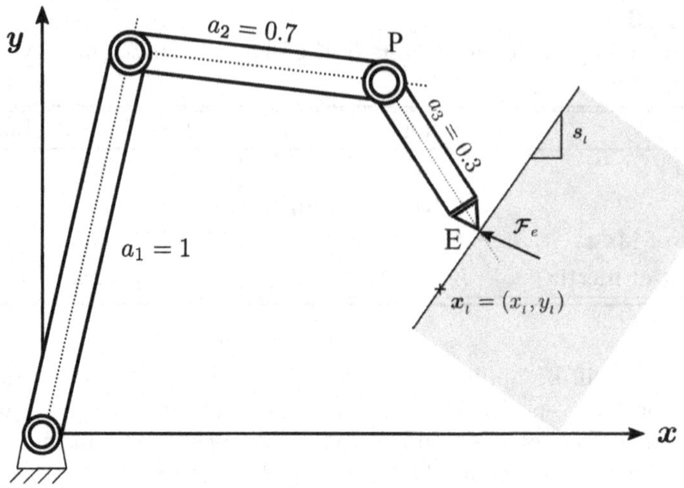

FIGURE 13.9
Schematic of the planar 3R robot interacting with a flat obstacle.

In this equation, a line in the x, y plane is modeled, passing through the point $\boldsymbol{x}_l = (x_l, y_l)$ with a slope of s_l. According to Figure 13.9, by specifying this point and the desired slope, the position and orientation of the flat obstacle in the robot's workspace are determined. In the conducted simulations, the boundary of the object is considered as a line passing through the point $\boldsymbol{x}_l = (1, 1)$ with a slope of 45 degrees ($s_l = 1$). The representation of this line in the robot's workspace is depicted as a dash-dotted line (colored in magenta) with the notation $\boldsymbol{\mathcal{X}}_e$ in Figure 13.10.

The desired trajectory of the robot's motion should also be considered in a way that during the simulation, the robot's end-effector comes into contact with the obstacle and maintains contact over time. Since the positioning of the obstacle in the robot's workspace is more conveniently defined based on the motion variables in the workspace $\boldsymbol{\mathcal{X}}$, in the conducted simulations, the desired trajectory of the robot's motion is initially defined in the task space and then mapped to the joint space, to be used in control methods developed in this space. To achieve this, we define the desired trajectory of the robot's end-effector as another line in the workspace, parallel to and slightly inside the boundary of the obstacle.

In the simulations, it is assumed that the desired path of the robot's end-effector follows a linear trajectory with a distance of $0.1\,m$ away from the boundary along the x and y axes. This line segment is represented as a dashed line (colored blue) annotated with the notation $\boldsymbol{\mathcal{X}}_d$ in Figure 13.10. Furthermore, it is assumed that the desired angle of the third robot link is perpendicular to the plane, ensuring the vertical orientation of the third link throughout the robot's motion. With this definition, the desired starting point in the workspace is considered as $\boldsymbol{\mathcal{X}}_0 = [1.2\,m,\ 1.0\,m,\ -45°]^T$, and the ending point is set to $\boldsymbol{\mathcal{X}}_f = [0.6\,m,\ 0.4\,m,\ -45°]^T$. The time to traverse this path is set to three seconds, and to examine the sustained behavior of the motion, it is assumed that the robot follows the desired path for one more second at the final state. Thus, the simulation time will be equivalent to four seconds, where in the first three seconds of motion, the robot's end-effector traverses the desired path along the boundary of the obstacle, and in the final second, it receives a stop command at its final position.

In the trajectory design, quintic polynomials are used to pass through the initial and final points in the workspace. For this purpose, the desired motion trajectory and its derivatives in the task space are determined:

$$\boldsymbol{\mathcal{X}}_d(t), \dot{\boldsymbol{\mathcal{X}}}_d(t), \ddot{\boldsymbol{\mathcal{X}}}_d(t)$$

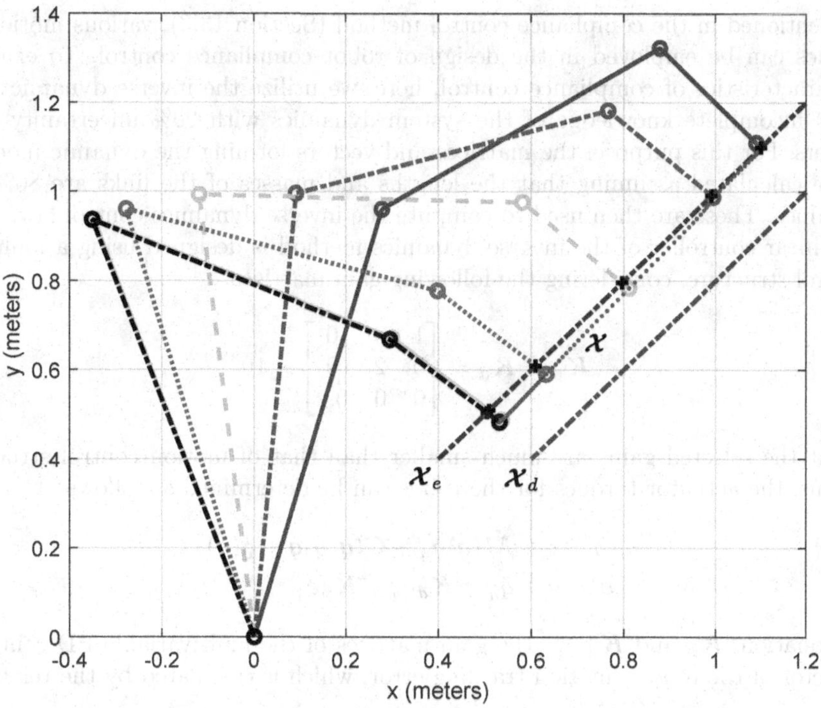

FIGURE 13.10
Illustration of the the placement of the robot and the obstacle in six states in the compliance control method.

Then, the desired motion trajectory in the joint space is obtained using the robot's inverse kinematics:

$$q_d(t) = IK(\mathcal{X}_d(t))$$

and the desired motion velocity in this space is obtained using the inverse Jacobian matrix:

$$\dot{q}_d(t) = J^{-1}(q)\dot{\mathcal{X}}_d(t)$$

while the desired acceleration in the joint space is calculated using the following relation:

$$\ddot{q}_d(t) = J^{-1}(q)\left(\ddot{\mathcal{X}}_d(t) - \dot{J}J^{-1}\dot{\mathcal{X}}\right)$$

To estimate the interactive force between the robot and the environment, first, the geometric location of the robot's end-effector intersection with the environment is determined. Taking into account the elastic model of the environment, the interactive force is then calculated. To achieve this, the intersection point of the boundary line of the obstacle and the center-line of the third link is determined. If a collision occurs, the interactive force is obtained from the following relation.

$$\mathcal{F}_e = K_e(\mathcal{X} - \mathcal{X}_e) \tag{13.40}$$

Given the presence of a flat obstacle, only interactive forces are considered, while potential torques are neglected. Therefore, the assumed stiffness coefficient of the environment in simulations is considered as follows.

$$K_e = \begin{bmatrix} 1 & 0 & 0 \\ 0 & 1 & 0 \\ 0 & 0 & 0 \end{bmatrix} \times 10^4 \tag{13.41}$$

As mentioned in the compliance control method (Section 13.3), various motion control approaches can be employed in the design of robot compliance control. To examine the main characteristics of compliance control, here, we utilize the inverse dynamics method, assuming incomplete knowledge of the system dynamics with 20% uncertainty in model parameters. For this purpose, the matrices and vectors forming the dynamic model of the robot are calculated assuming that the lengths and masses of the links are 80% of their actual values. These are then used to compute the inverse dynamics control law.

The linear controller of the inverse dynamics method is designed using a multivariable PD control structure, considering the following gain matrices.

$$\boldsymbol{K}_p = \boldsymbol{K}_d = \begin{bmatrix} 1 & 0 & 0 \\ 0 & 2 & 0 \\ 0 & 0 & 0.2 \end{bmatrix} \times 10^3 \tag{13.42}$$

Note that the selected gains are much smaller than that of motion control structure. By this means, the actuator torques for the robot can be determined as follows.

$$\begin{aligned} \boldsymbol{\tau} &= \hat{\boldsymbol{M}}(\boldsymbol{q})\boldsymbol{a}_q + \hat{\boldsymbol{C}}(\boldsymbol{q}, \dot{\boldsymbol{q}})\dot{\boldsymbol{q}} + \hat{\boldsymbol{g}}(\boldsymbol{q}) \\ \boldsymbol{a}_q &= \ddot{\boldsymbol{q}}_d + \boldsymbol{K}_d\dot{\boldsymbol{e}}_q + \boldsymbol{K}_p\boldsymbol{e}_q \end{aligned}$$

In this equation, $\boldsymbol{K}_p$ and $\boldsymbol{K}_d$ are the gain matrices of the multivariable PD controller, $\boldsymbol{e}_q$ is the vector of the robot's motion tracking error, which is calculated by the relation:

$$\boldsymbol{e}_q = \boldsymbol{q}_d - \boldsymbol{q}$$

where $\boldsymbol{q}_d$ is the vector of the desired motion trajectory in the joint space, and $\boldsymbol{q}$ is the measured joint position vector.

To examine the effect of disturbance, assume that in the first second, a step disturbance torque is applied to all three joints of the robot with the amplitude of $[-100, -100, -10]^T$ $(N.m)$. The disturbance torque amplitude is chosen to be slightly larger than the interactive forces of the robot. Under these conditions, the simulation results of the closed-loop system are presented in figures 13.10 to 13.14.

In the depiction of the robot's interactive motion with the environment in Figure 13.10, the placement of the robot and the obstacle in the two-dimensional plane (x, y) is illustrated in several stages of movement. As observed in this figure, the robot's configuration at six states from the beginning to the end of its motion are displayed against the boundary of the obstacle in the robot's workspace. In the initial state, the robot is represented by a solid line (colored blue), followed by intermediate states with a dash-dotted line (colored red), then a dashed line (colored green), a dotted line (colored pink), again a solid line (light blue), and finally a dash-dotted line (colored black). To observe how the robot moves and interacts with the environment, the robot's end-effector path is also shown in this figure with a dotted line (colored red).

As evident in this figure, the robot initially starts its movement freely without contact with the obstacle. After colliding with the obstacle, it penetrates slightly inside it and continues its interactive motion with the environment. To clearly find the contact time between the robot's end-effector and the obstacle, Figure 13.11 provides a more accurate depiction. In this figure, the components of the interactive force vector between the robot and the environment, calculated by the relation 13.40, are displayed in the two x and y axes.

As observed in this figure, the robot's end-effector collides with the obstacle at 0.25 seconds and maintains contact with the environment until the end of the simulation time. The interactive force F_{e_x} during the application of disturbance in the first second undergoes

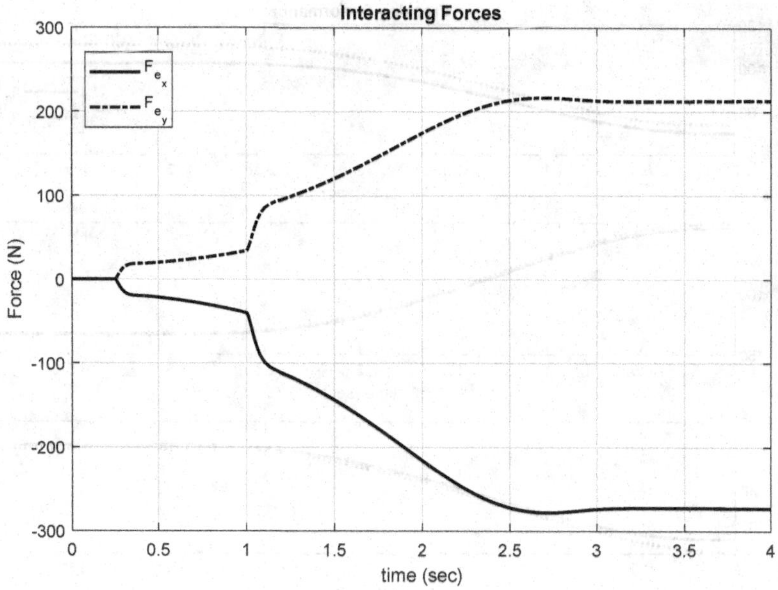

FIGURE 13.11
Interactive force of the robot due to contact with the environment in the compliance control method.

a sharp decrease and reaches its minimum value at 2.68 seconds, to approximately -286 Newtons. Simultaneously, the interactive force F_{e_y} also reaches its maximum value during the same time, to approximately 216 Newtons. The changes in the interactive force in the final second, when the desired motion path of the robot reaches its steady-state, converge to their final values.

The desired motion trajectory of the robot and the actual motion performed by the robot in the joint space are depicted in Figure 13.12. In this figure, the motion of the first joint is shown in the upper graph, the motion of the second joint in the middle graph, and the motion of the third joint in the lower graph. The desired motion trajectory is represented by a solid line, and the actual motion is indicated by a dotted line. The trajectory in the joint space is determined using the inverse kinematics in such a way that the robot's end-effector accomplishes its desired linear path along the boundary of the obstacle in the task space. Consequently, the desired motion trajectory in the first joint changes from an angle of approximately 70 degrees to about 102 degrees during the first three seconds of movement and remains constant in the last second. Additionally, the desired angle for the second joint transitions from around -50 degrees to -135 degrees, while the desired angle for the third joint changes from approximately -65 degrees to about -20 degrees. In this figure, a significant tracking error is evident, which is considerably larger in comparison to motion control methods.

To examine the tracking error more closely, this quantity is illustrated for the three robot joints in Figure 13.13. As observed in this figure, the maximum tracking error in the first joint exceeds 8 degrees, in the second joint exceeds 4 degrees, and in the third joint exceeds 8 degrees. Additionally, it is well-noticeable in this figure that at the first second and during the application of disturbance torque to the three joints, the trend of tracking error changes. This characteristic is entirely expected in the compliance control method, as the primary goal of this approach is to regulate the interactive behavior of the robot with the environment, and the reduction of tracking error is not the main pursuit. It should

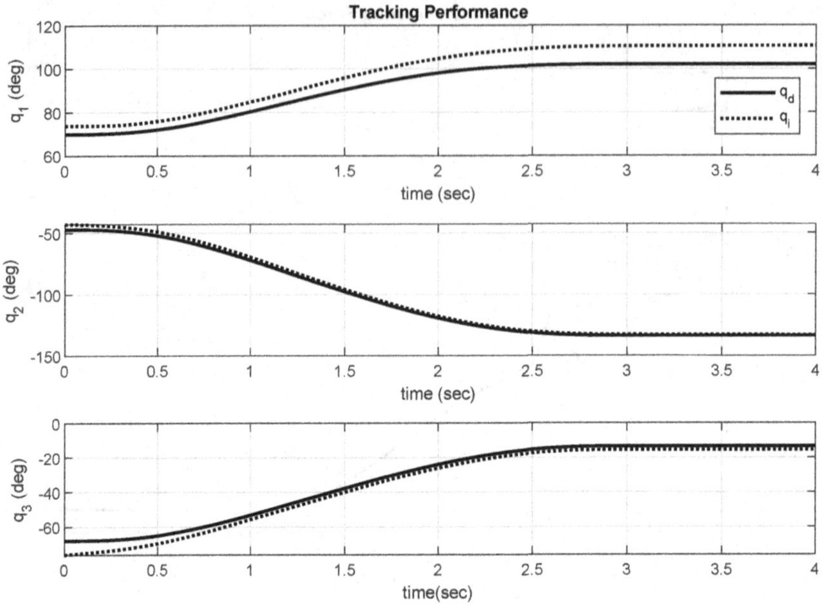

FIGURE 13.12
The desired motion trajectory of the robot and the executed motion of the robot in interaction with the environment in the compliance control method.

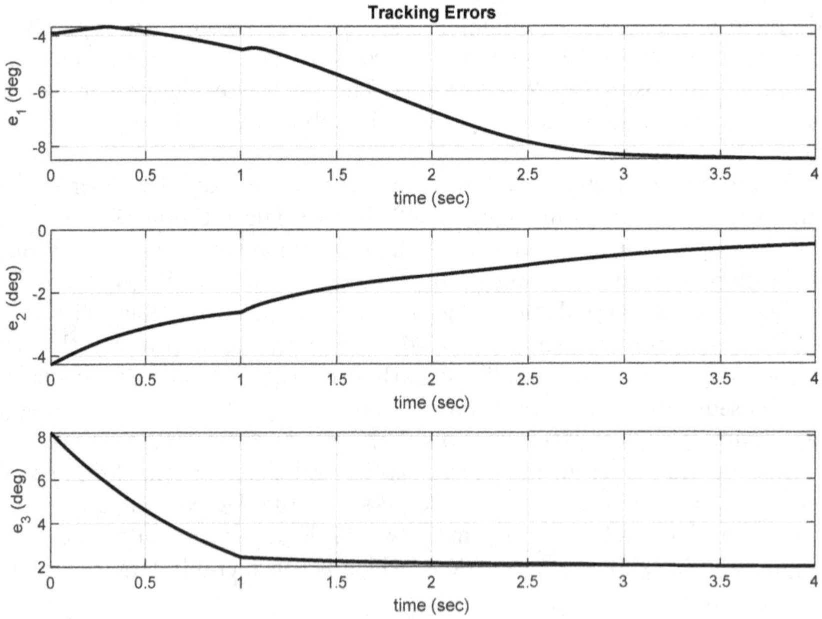

FIGURE 13.13
The tracking error of the robot in the compliance control method.

be noted that the sign of the introduced disturbance torque to the joints is selected in a way that aligns with the interactive force with the environment, ensuring that the contact between the robot's end-effector and the environment is maintained during the application of disturbance.

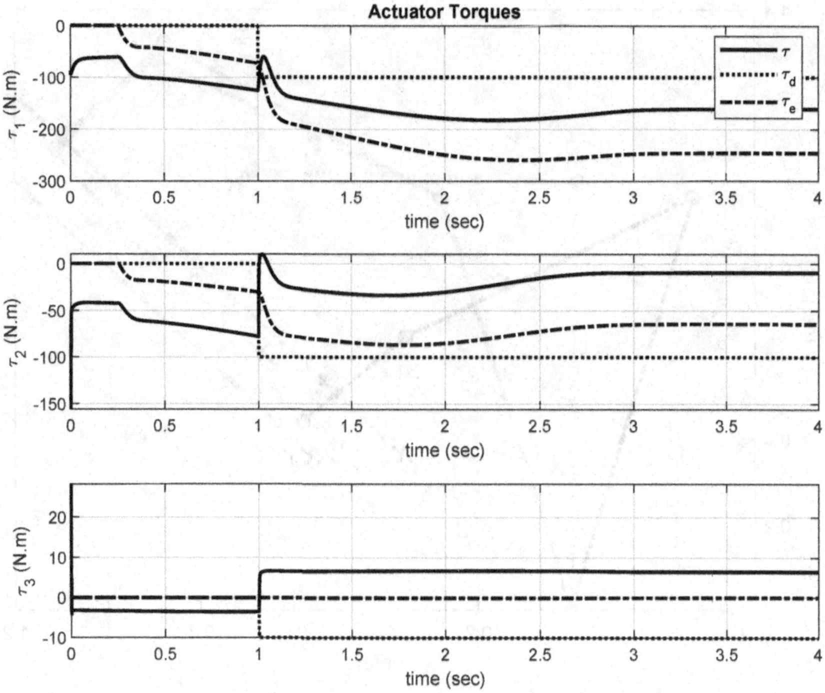

FIGURE 13.14
The control efforts of the robot actuators in the compliance control method.

The torques applied to the robot actuators in this scenario are also shown in Figure 13.14. In this figure, the control effort applied to the joints $\boldsymbol{\tau}$, composed of the sum of the PD torque along with the feedback linearization term $\boldsymbol{\tau}_{pd} + \boldsymbol{\tau}_{fl}$, is represented by a solid line. The disturbance torque $\boldsymbol{\tau}_d$ is indicated with a dotted line, and the torque resulting from the interactive force $\boldsymbol{\tau}_e$ is depicted with a dash-dotted line.

As observed in this figure, the primary contribution to the control effort comes from the interactive force. Despite the fact that this force is not directly measured and fed back in this method, mapping the interactive force in the joint space has been directly incorporated into the control effort to regulate the relationship between force and motion in compliance control. Additionally, it can be noted that during the application of disturbance torque, its effect is directly visible in the control effort. This characteristic arises from the magic of feedback and the linear PD controller in the control structure. It is noteworthy that the required torques in the robot actuators for compliance control are much larger compared to motion control. This is due to the presence of the interactive force between the robot and the environment.

To demonstrate how a designer can appropriately assign the required compliance to the environment by adjusting the controller gains, consider another simulation with all parameters identical to before, but with the controller gains magnified by a factor of 10. Thus, the controller gains are set to:

$$\boldsymbol{K}_p = \boldsymbol{K}_d = \begin{bmatrix} 1 & 0 & 0 \\ 0 & 2 & 0 \\ 0 & 0 & 0.2 \end{bmatrix} \times 10^4 \tag{13.43}$$

Comparing the depiction of the robot's interactive motion with the environment in this case, as shown in Figure 13.15, with the previous case illustrated in Figure 13.10,

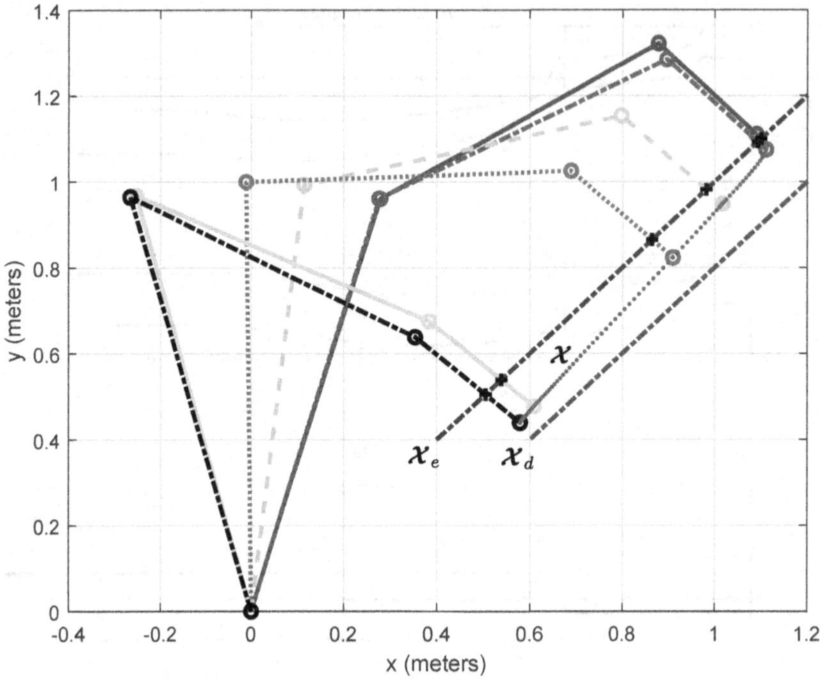

FIGURE 13.15
Robot configuration and its interaction with the environment in the compliance control method with ten times higher controller gains.

it is observed that due to the selection of higher stiffness gains, the robot behaves much stiffer and penetrates deeper into the obstacle. The robot makes contact with the obstacle sooner, resulting in interaction torques increasing by about a factor of three compared to the previous case. Moreover, the trajectory traversed by the robot within the object closely aligns with the desired trajectory, leading to a reduction in tracking error to slightly less than one third of the previous case. Thus, summarily, the stiffness of the robot, calculated by the interactive force divided by the resulting tracking error, is increased by a factor of ten, as commanded.

By examining the simulation results of the compliance control method in this robot and comparing it with the results of the inverse dynamics method in motion control, it becomes evident that although the implementation topology of these two methods is identical, they have fundamental distinctions in terms of design and performance. In the design of the inverse dynamics control, the assumption is that the robot moves freely in its workspace and has no contact with its surrounding environment. Therefore, linear controller gains are selected as large as possible to achieve minimal tracking error for the desired trajectory. However, in the compliance control method, it is assumed that the robot is in contact with the environment or may collide with an obstacle in its workspace. The goal in this method is to regulate the compliance of the robot and to prevent damage to the environment or mechanical components of the robot. Observing the interactive forces in Figure 13.11, despite the relatively high stiffness of the environment ($K_e = 10^4$), the interactive forces experienced by the robot remain within a few hundred Newtons. This effectively prevents any significant damage to the obstacle or the robot's components.

With this elucidation, it's apparent that this control method is applicable when the geometry of the robot's interaction with the environment is predetermined. Assuming the absence of force sensors, the interactive force can be adjusted to the desired stiffness by estimation. In the following section (13.6.2), we simulate a scenario where the position of the obstacle in the environment is not predetermined in the robot's planning trajectory but it is detected using a force sensor. In this case, the emphasis shifts to tracking the interactive wrench between the robot's end-effector and the environment. The behavior of the simulated closed-loop system in wrench tracking is then reported and analyzed, utilizing a direct force control method.

13.6.2 Direct Force Control

In this section, we simulate the direct force control method on the planar 3R̲ robot. For this purpose, and in the calculation of the inverse dynamics terms, we assume that our knowledge of the dynamics of the system is incomplete, with a 20% uncertainty in the model parameters.

With this assumption, the torque applied to the robot's actuators is determined as follows.

$$\begin{aligned} \boldsymbol{\tau} &= \hat{\boldsymbol{M}}_e(\boldsymbol{q})\boldsymbol{a} + \left(\hat{\boldsymbol{C}}(\boldsymbol{q},\dot{\boldsymbol{q}}) + \hat{\boldsymbol{b}}_e\right)\dot{\boldsymbol{q}} + \hat{\mathfrak{g}}(\boldsymbol{q}) + \boldsymbol{J}^T\boldsymbol{\mathcal{F}}_m \\ \boldsymbol{a} &= \boldsymbol{K}_p(\boldsymbol{q}_a - \boldsymbol{q}) - \boldsymbol{K}_d\dot{\boldsymbol{q}} \end{aligned}$$

Furthermore, a linear PI controller has been used in the outer loop of force control scheme.

$$\boldsymbol{q}_a = \boldsymbol{K}_P^{-1}\left(\boldsymbol{K}_{\tau_p} + \frac{1}{s}\boldsymbol{K}_{\tau_i}\right)\boldsymbol{e}_{\tau}$$

The matrices of the efficient linear inverse dynamics controller are determined through trial and error and are set to:

$$\boldsymbol{K}_p = \begin{bmatrix} 1 & 0 & 0 \\ 0 & 1 & 0 \\ 0 & 0 & 1 \end{bmatrix} \times 2000, \quad \boldsymbol{K}_d = \begin{bmatrix} 1 & 0 & 0 \\ 0 & 1 & 0 \\ 0 & 0 & 1 \end{bmatrix} \times 100 \tag{13.44}$$

While the matrices of the force controller have been assigned as following values:

$$\boldsymbol{K}_{\tau_p} = \begin{bmatrix} 1 & 0 & 0 \\ 0 & 10 & 0 \\ 0 & 0 & 0.2 \end{bmatrix} \times 50, \quad \boldsymbol{K}_{\tau_i} = \begin{bmatrix} 1 & 0 & 0 \\ 0 & 10 & 0 \\ 0 & 0 & 0.2 \end{bmatrix} \times 10^3 \tag{13.45}$$

In order to examine the effect of disturbance, it is assumed that in the first second, a step disturbance torque is applied to all three joints of the robot with a range of $[-100, -100, -10]^T N.m$. Considering the interactive forces of the robot, the disturbances has been chosen to be numerically larger than that considered in the motion control scenario and similar to the compliance control simulations.

In this simulation, it is also assumed that there is a flat obstacle in the robot's workspace, modeled using Equation 13.39, while its boundary is considered as a line passing through the point $\boldsymbol{x}_l = (1, 1)$, with a slope of 45 degrees ($s_l = 1$). The boundary of this obstacle is illustrated by the dashed line (colored in magenta) in Figure 13.16.

The desired interactive forces of the robot should be considered in a way that during the simulation, the robot's end-effector hits the obstacle and maintains contact over time. Given that the placement of the obstacle is more conveniently defined in the robot's workspace, in the conducted simulations the interactive forces of the robot are first defined in the task space, and then using the transpose of its Jacobian matrix, it is mapped to the joint space.

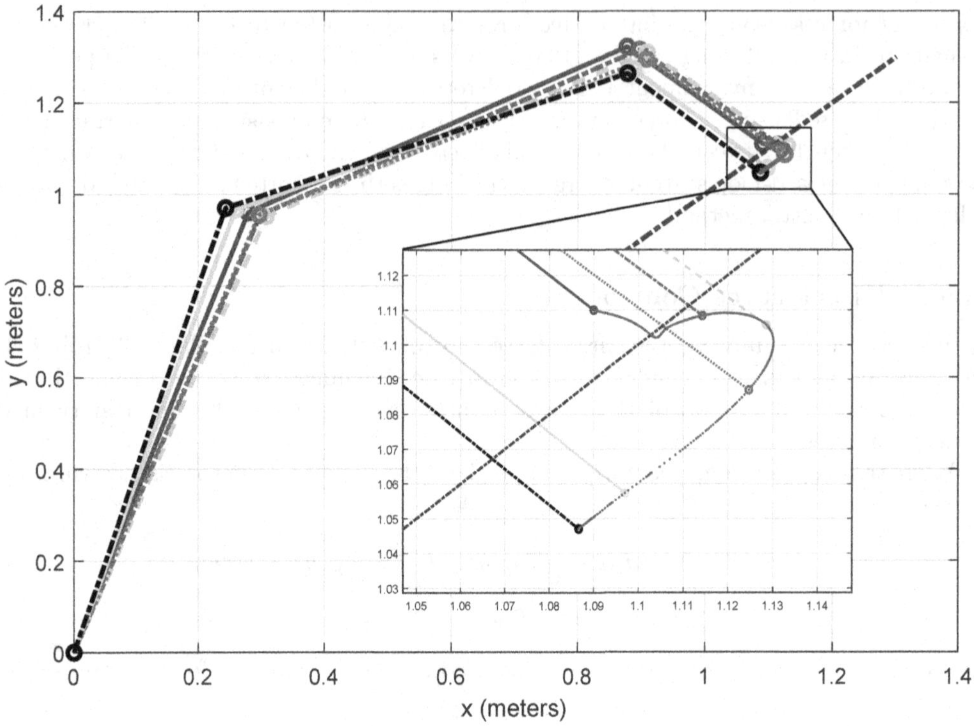

FIGURE 13.16
The configuration of the robot and its interaction with the environment at six stages of motion in the direct force control method.

Considering the flat obstacle in the environment, the desired interactive force of the robot is defined perpendicular to the obstacle surface. For this means a quintic polynomial trajectory is considered for the desired force that starts from $[0, 0, 0]^T$ and ends at $[-200,\ 200,\ 0]^T$, in 3 seconds. To examine the steady-state behavior of the wrench tracking, it is assumed that the robot's interactive force remains constant at this value for one second in the final conditions. Therefore, a simulation time of 4 seconds is considered. Additionally, the initial position of the robot's end-effector is set to $\boldsymbol{\mathcal{X}}_0 = [1.09\,m,\ 1.11\,m,\ -45°]^T$, with a distance of 0.1 meters from the obstacle at the initial time.

To estimate the interactive force between the robot and the environment, similar to compliance control scenario, the intersection point of the robot's end-effector with the environment is geometrically determined. In case of intersection, the interactive force is then obtained from Equation 13.40. Due to the planar nature of the obstacle, only interactive forces are considered, and potential interactive torques are disregarded. The environmental stiffness coefficient is also assumed to be the same as before.

$$\boldsymbol{K}_e = \begin{bmatrix} 1 & 0 & 0 \\ 0 & 1 & 0 \\ 0 & 0 & 0 \end{bmatrix} \times 10^4$$

With the given conditions, the simulation results of the closed-loop system are presented in Figures 13.16 to 13.20. The configuration of the robot and its interaction with the environment in the (x, y) plane at six stages of motion is depicted in Figure 13.16. In this figure, the robot's initial state is shown with a solid line (colored in blue), intermediate states with

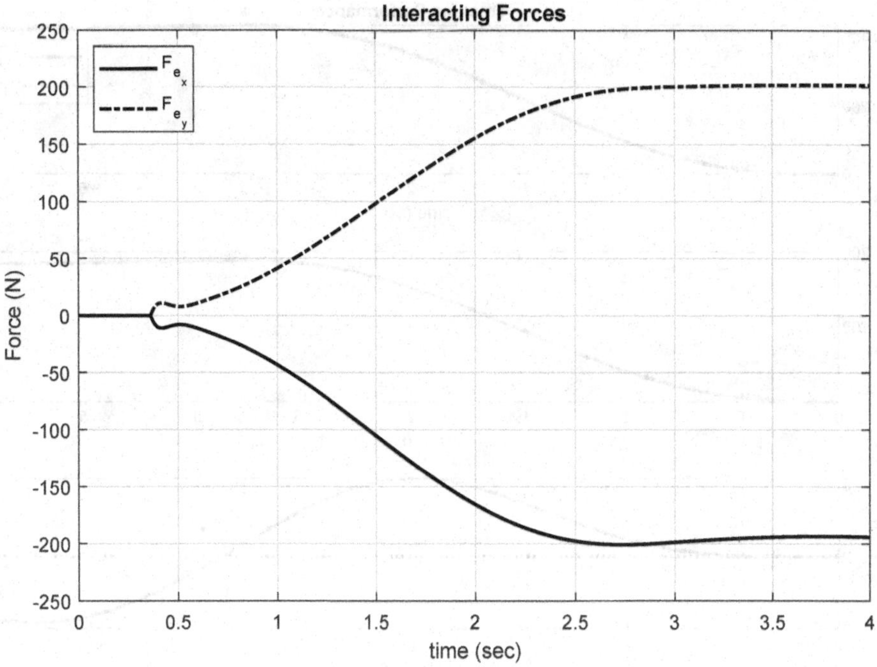

FIGURE 13.17
Interactive force of the robot due to contact with the environment in the direct force control method.

a dash-dotted line (colored in red), then with a dashed line (colored in green), a dotted line (colored in pink), again with a solid line (in light blue), and finally with a dash-dotted line (colored in black). To observe the robot's motion and interaction with the environment, the path of the robot's end-effector with a dotted line (colored in red) is also illustrated in this figure.

Due to the limited movement of the robot's end-effector during interaction with the environment, the position of the robot's end-effector relative to the obstacle is magnified and depicted in this figure. As clearly seen in this magnification, the robot's end-effector is slightly distant from the environment in the initial state. Then, the end-effector moves toward the obstacle, collides with it, and slightly penetrates into it. The motion of the robot's end-effector inside the environment forms a heart shape, while in the final moments of the simulation, the motion of the robot's end-effector runs parallel to the boundary of the obstacle to create equal and oppositely directed components of the desired force in the x and y directions.

The precise time of contact between the robot's end-effector and the obstacle can be observed in Figure 13.17. In this figure, the components of the interactive force vector of the robot with the environment, calculated using Equation 13.40, are shown in the x and y axes. As observed in this figure, the robot's end-effector collides with the obstacle at approximately 0.36 seconds and maintains contact with the environment until the end of the simulation time.

The interactive force $\boldsymbol{F}_{e_y}$ increases slightly after the collision with the environment and then decreases, subsequently, it continues to increase gradually. The decreasing trend of the force after the collision corresponds to the indentation of the robot's end-effector into the environment, as well clearly illustrated in the magnified section of Figure 13.16. Similarly,

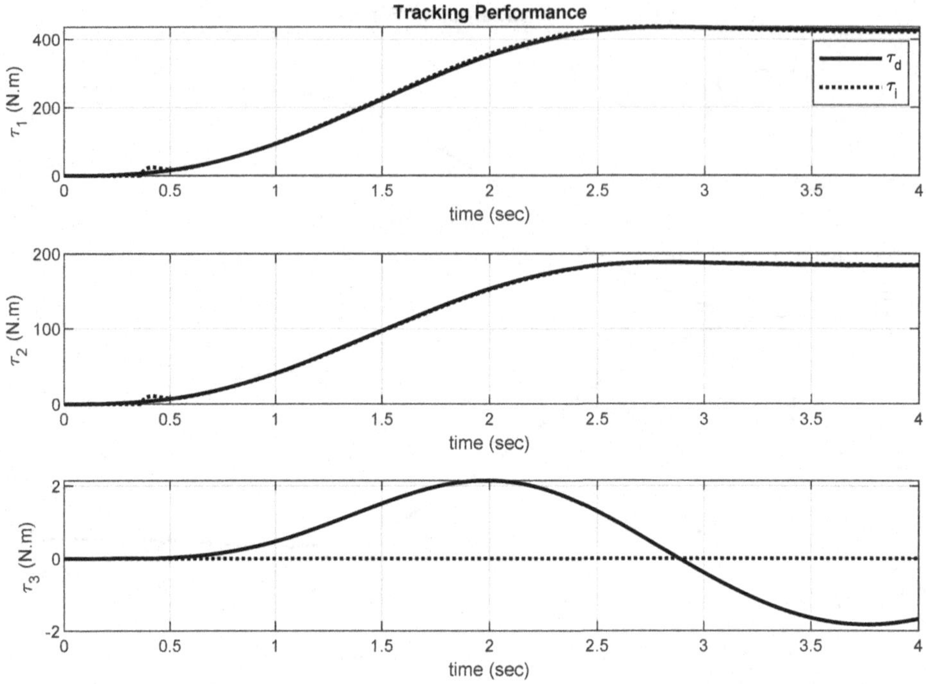

FIGURE 13.18
Desired robot force and interactive force with the environment obtained in the direct force control method.

in Figure 13.17, it is observed that immediately after the collision of the robot's end-effector with the environment, the end-effector penetrates slightly into it, and the interactive force increases. Then, this indentation in the environment decreases, leading to a reduction in the interactive force. In the ongoing simulation, the interactive force gradually increases until it converges to its steady-state in the last second. The maximum value of the absolute force in both the x and y directions is approximately 200 N, while the equal and oppositely signed magnitudes of the interactive force components perpendicular to the surface of the obstacle corresponds to the commanded force.

The desired joint torques along with the actual interactive torques are shown in Figure 13.18. In this figure, the torque of the first joint is displayed in the upper chart, the torque of the second joint in the middle chart, and the torque of the third joint in the lower chart. The desired torque is represented with a solid line, and the actual torque applied to the robot's actuators is shown with a dashed line. As observed in this figure, the tracking of interactive torque is well-performed, and only a minimal tracking error is visible at the moment of collision between the robot's end-effector and the environment.

For a more detailed examination of the torque tracking error, this quantity is illustrated for the three robot joints in Figure 13.19. As seen in this figure, the maximum torque tracking error is approximately 12 N.m for the first joint, around 6 N.m for the second joint, and about 2 N.m for the third joint, which are all negligible compared to the significant amount of the interactive torque. Additionally, it is clearly observed in this figure that at the moment of collision between the robot's end-effector and the obstacle, the applied torques to the joints undergo a significant change. The dynamics of these changes can be adjusted by tuning the control gain matrices. At one second and during the application of disturbance torques to the robot joints, this control method has effectively mitigated

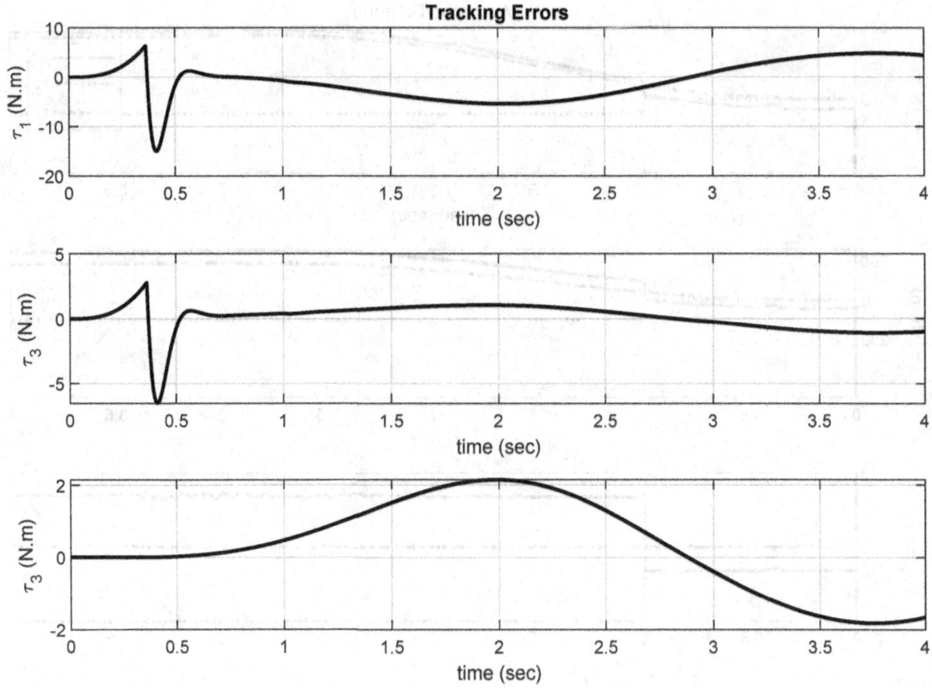

FIGURE 13.19
Tracking error of the robot's interactive force in the direct force control method.

the disturbance without prior knowledge of the disturbance time and nature. This feature is entirely expected in the direct force control method since its main objective is to track interactive force even in the presence of disturbances.

The applied torques to the robot actuators in this simulation are presented in Figure 13.20. In this figure, the control effort entering the joints $\boldsymbol{\tau}$ is represented by a solid line, the disturbance torque $\boldsymbol{\tau}_d$ is denoted by a dotted line, and the torque due to the interactive force $\boldsymbol{\tau}_e$ is illustrated with a dash-dotted line. As observed in this figure, the primary contribution to the control effort comes from the interactive force. In the initial moment of the simulation, due to the free motion of the robot's end-effector, there is a noticeable difference between the interactive force and the desired force, leading to a significant required torque at that moment. In the conducted simulation, the saturation limit for the applied torques to the joints is set to 500 N, which is well-reflected in the upper chart. At the moment of collision between the robot's end-effector and the environment, the interactive torque assumes a large value, requiring increased control effort in the joints. Additionally, during the application of disturbance torque at the first second, the direct force control method has effectively mitigated its effect without the need for its direct measurement. Small oscillations in the applied torques to the actuators are noticeable, and these can be reduced by tuning the control gain matrices.

The effectiveness of wrench tracking in the direct force control method is clearly evident in the simulation results. This is achieved by defining the auxiliary motion vector $\boldsymbol{q}_a$ in the force control loop and adjusting the extent of the end-effector penetration into the obstacle. The application of this control method is widespread, especially in scenarios where the primary objective is interactive wrench tracking. The main reason for the success of this approach lies in the appropriate linearization of the nonlinear dynamics of the system and the separation of the two primary and secondary objectives by the cascaded control

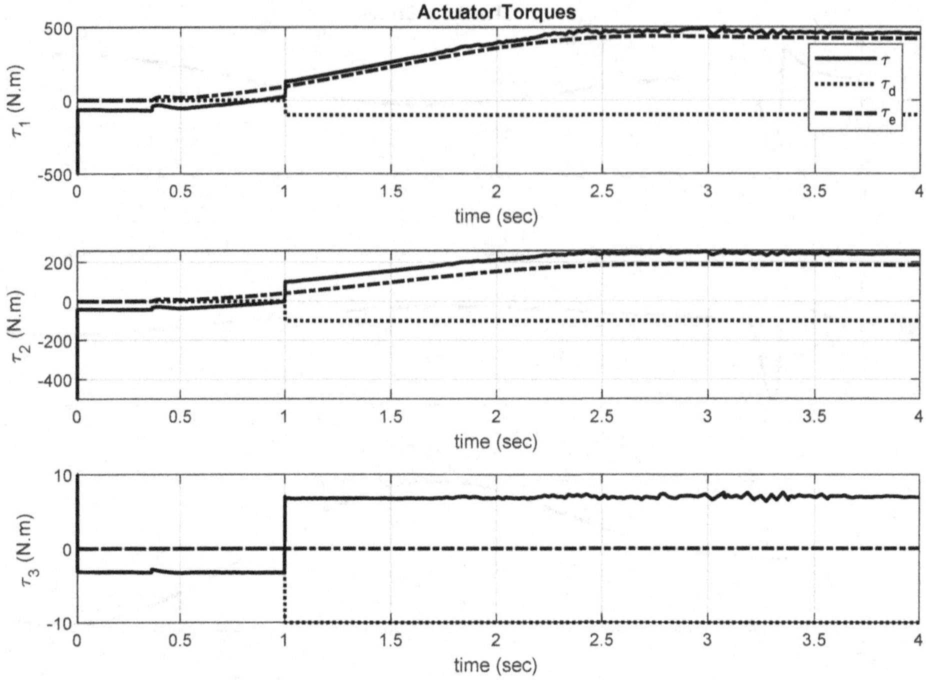

FIGURE 13.20
Control effort of the robot actuators in the direct force control method.

structure. Thus, the closed-loop system characteristics are not heavily dependent on the robot's structure and the location of its end-effector within the robot's workspace, making the tuning of control gains for position and force controllers more straightforward.

However, it should be noted that this control method is model-based and requires a relatively complete understanding of the robot's kinematic and dynamic models. Furthermore, in this method, satisfactory wrench tracking will only be possible as long as the robot maintains continuous contact with the environment. If, for any reason, the contact between the robot and the environment is lost, the interactive forces will suddenly and significantly decrease. To ensure the desired interactive force, rapid and uncontrollable accelerations are generated in the robot joints. In applications such as human-robot interaction and robots used in minimally invasive surgeries, where the robot is not in continuous contact with the environment, it is essential to control the transient behavior of the robot during contact with the environment. For this purpose, the use of impedance control methods detailed in the upcoming chapter, is recommended.

Problems

13.1 Consider the two-degrees-of-freedom <u>RP</u> robot shown in Figure 13.21. The dynamic model of the robot is determined in Problems 8.1 and 9.1. The numerical values of the robot's parameters and simulations of its linear control are explored in Problems 10.1 and 11.1. Furthermore, the motion control of the robot based on inverse dynamics is investigated in Problem 12.1.

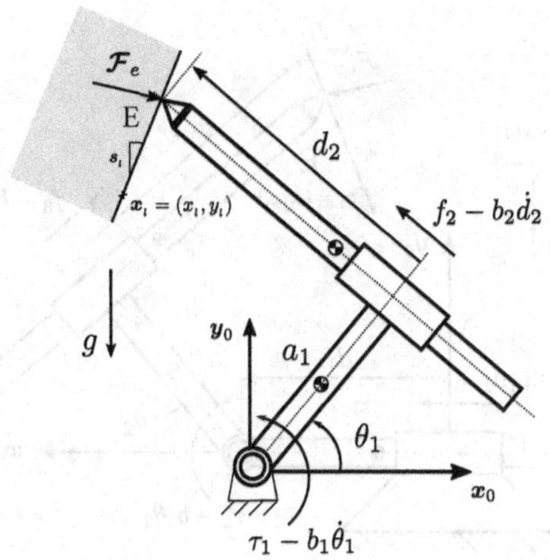

FIGURE 13.21
The two-degrees-of-freedom robot for Problem 13.1.

a) Consider a flat obstacle in the robot's workspace and slightly deviate the desired path of the robot's end-effector inside this obstacle. Assume that the mechanical stiffness of the robot is greater than the obstacle, and its interactive compliance is determined by the relationship 13.13. Calculate the interactive forces with a stiffness coefficient of 10^4 and neglect the interactive torques.

b) Using the compliance control method and considering 20% uncertainty in the kinematic and dynamic parameters of the robot, adjust the control gain matrices in a way that after the robot collides with the obstacle, its motion smoothly continues along the desired path. Simulate the robot's motion in this situation. Illustrate the placement of the robot and the obstacle in a two-dimensional environment (or three-dimensional in next Problems) and show the robot's motion in different states along its path. Display the end-effector's trajectory in interaction with the environment.

c) Obtain the components of interactive forces along the x, y axes (and, if applicable, z) and plot them. Determine the precise time of contact between the robot's end-effector and the obstacle. Furthermore, calculate and plot the tracking error and the required control effort. Analyze and interpret the simulation results.

13.2 Examine the three-degree-of-freedom PRP robot depicted in Figure 13.22. The dynamic model of the robot is derived in Problems 8.2 and 9.2. The numerical values for the robot's parameters and simulations of its motion and linear control are addressed in Problems 10.4 and 11.2. Additionally, the motion control of the robot using inverse dynamics is explored in Problem 12.2. Repeat Problem 13.1 for this robot.

13.3 Consider the PRR robot with three degrees of freedom illustrated in Figure 13.23. The dynamic model of the robot is developed in Problems 8.3 and 9.3. The numerical values of the robot's parameters and simulations of its motion and linear control are addressed in Problems 10.7 and 11.3. Additionally, the motion control of the robot using inverse dynamics is explored in Problem 12.3. Reiterate Problem 13.1 for this specific robot.

13.4 Consider the RPR robot with three degrees of freedom, as depicted in Figure 13.24. The dynamic model of the robot is formulated in Problems 8.4 and 9.4. The numerical

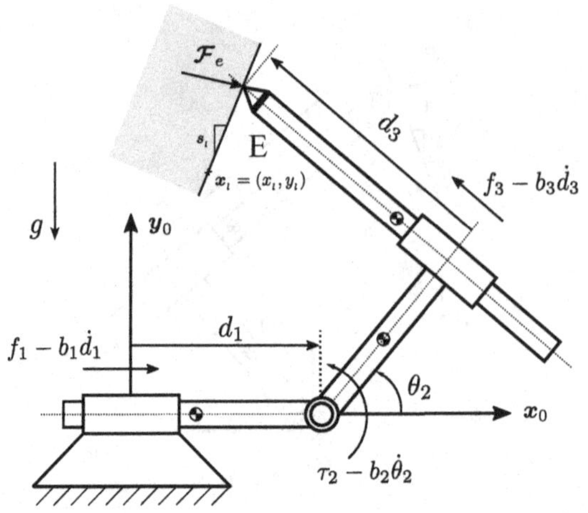

FIGURE 13.22
The three-degrees-of-freedom robot for Problem 13.2.

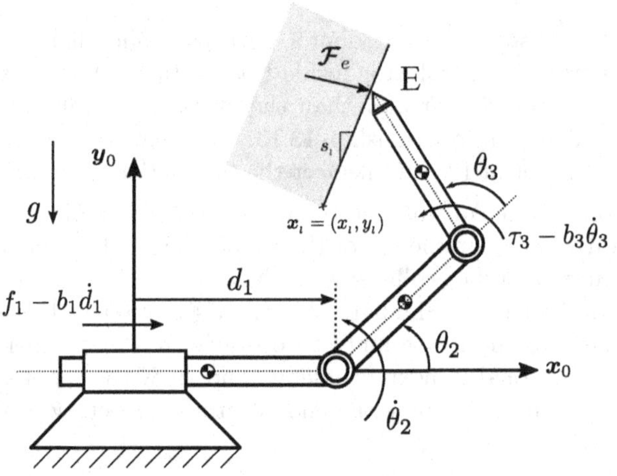

FIGURE 13.23
The three-degrees-of-freedom robot for Problem 13.3.

values for the robot's parameters and simulations of its motion and linear control are discussed in Problems 10.10 and 11.4. Additionally, the motion control of the robot using inverse dynamics is analyzed in Problem 12.4. Repeat Problem 13.1 specifically for this robot.

13.5 Examine the RR robot, featuring two degrees of freedom, as illustrated in Figure 13.25. The dynamic model of the robot is derived in Problems 8.5 and 9.5. The numerical values for the robot's parameters and simulations of its motion and linear control are addressed in Problems 10.13 and 11.5. Additionally, the motion control of the robot using inverse dynamics is investigated in Problem 12.5. Reiterate Problem 13.1 specifically for this robot.

13.6 Examine the four-degrees-of-freedom RRRP robot depicted in Figure 13.26. The dynamic model of the robot is formulated in Problems 8.6 and 9.6. The numerical values

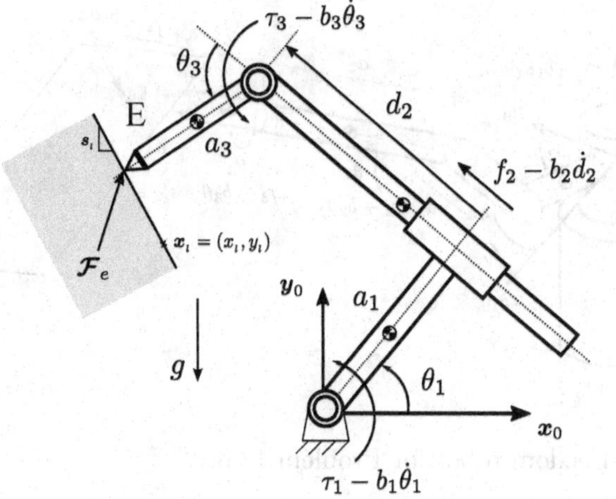

FIGURE 13.24
The three-degrees-of-freedom robot for Problem 13.4.

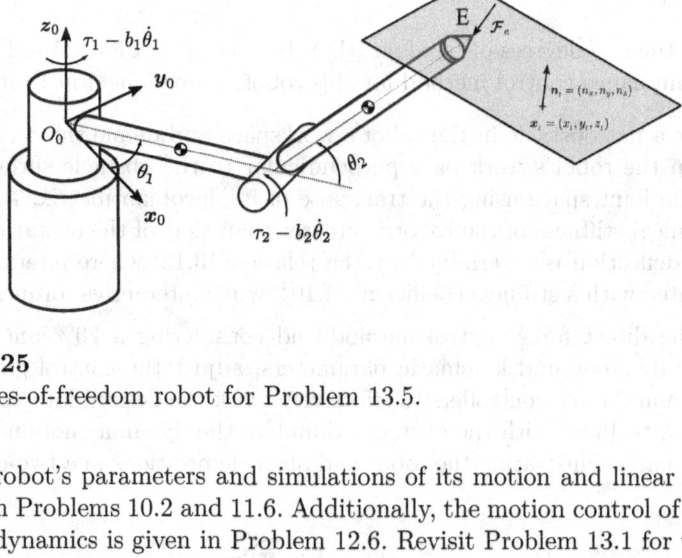

FIGURE 13.25
The two-degrees-of-freedom robot for Problem 13.5.

for the robot's parameters and simulations of its motion and linear control are discussed in Problems 10.2 and 11.6. Additionally, the motion control of the robot using inverse dynamics is given in Problem 12.6. Revisit Problem 13.1 for this robot.

13.7 Examine the <u>RP</u>RR robot with four degrees of freedom, as illustrated in Figure 13.27. The dynamic model of the robot is formulated in Problems 8.7 and 9.7. The numerical values for the robot's parameters and simulations of its motion are discussed in Problems 10.5 and 11.7. Additionally, the motion and linear control control of the robot using inverse dynamics is explored in Problem 12.7. Revisit Problem 13.1 specifically for this robot.

13.8 Consider the four-degrees-of-freedom 4R robot depicted in Figure 13.28. The dynamic model of the robot is formulated in Problems 8.8 and 9.8. The numerical values for the robot's parameters and simulations of its motion and linear control are discussed in Problems 10.8 and 11.8. Additionally, the motion control of the robot using inverse dynamics is analyzed in Problem 12.8. Revisit Problem 13.1 specifically for this robot.

13.9 Consider the four-degrees-of-freedom RRPR robot illustrated in Figure 13.29. The dynamic model of the robot is formulated in Problems 8.9 and 9.9. The numerical

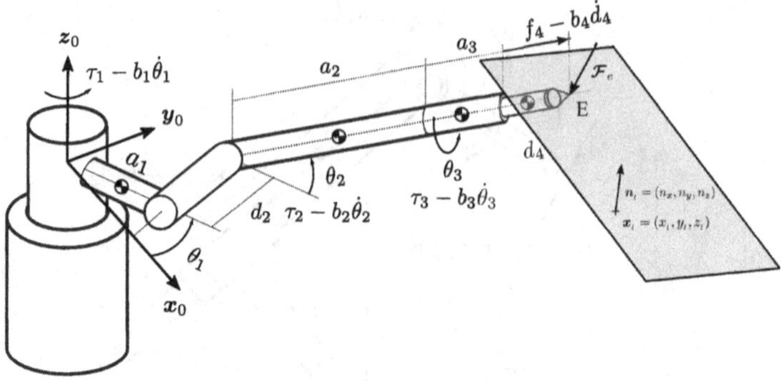

FIGURE 13.26
The four-degrees-of-freedom robot for Problem 13.6.

values for the robot's parameters and simulations of its motion and linear control are discussed in Problems 10.11 and 11.9. Additionally, the motion control of the robot using inverse dynamics is explored in Problem 12.9. Revisit Problem 13.1 specifically for this robot.

13.10 Consider the two-degrees-of-freedom RP robot shown in Figure 13.21 and the details of the compliance control method for this robot, as explained in Problem 13.1.

a) Consider a flat obstacle in the robot's workspace and define the desired interaction force in the robot's workspace perpendicular to the obstacle surface. Then, map it to the joint space using the transpose of its Jacobian matrix. Assume that the mechanical stiffness of the robot is greater than that of the obstacle, and its interactive deflection is determined by the relation 13.13, where interactive forces are calculated with a stiffness coefficient of 10^4, while interactive torques are neglected.

b) Using the direct force control method and considering a 20% uncertainty in the robot's dynamic and kinematic parameters, adjust the control gain matrices for position and force controllers to effectively track the desired interaction force after the robot collides with the obstacle. Simulate the dynamic motion of the robot in this scenario, illustrating the robot and obstacle positions in a two-dimensional en-

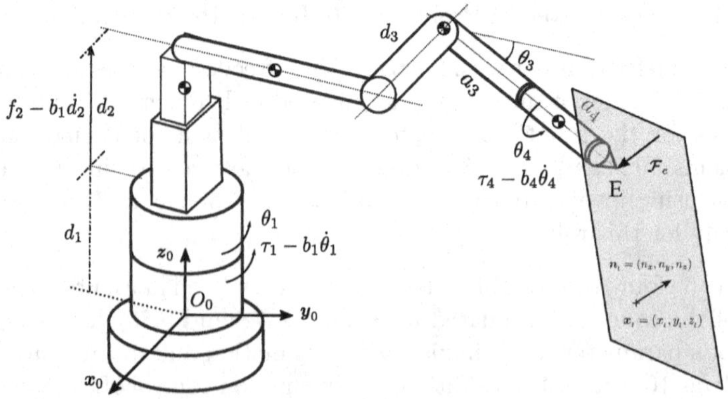

FIGURE 13.27
The four-degrees-of-freedom robot for Problem 13.7.

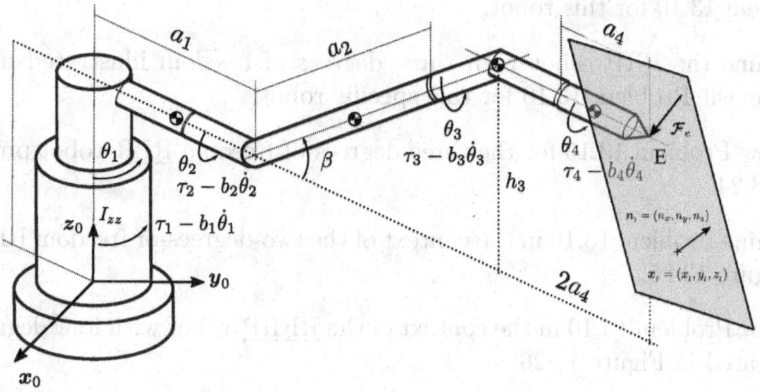

FIGURE 13.28
The four-degrees-of-freedom robot for Problem 13.8.

vironment (or three-dimensional for other Problems) depicting the robot's motion in various states along its interactive path with the environment. Furthermore, plot the path of the robot's end-effector.

c) Calculate the components of the interactive forces along the x, y axes (and z axis if applicable) and plot them. Determine the precise time of contact between the robot's end-effector with the obstacle. Additionally, obtain the tracking error of joint torques and the required control effort. Analyze and visualize the simulation results.

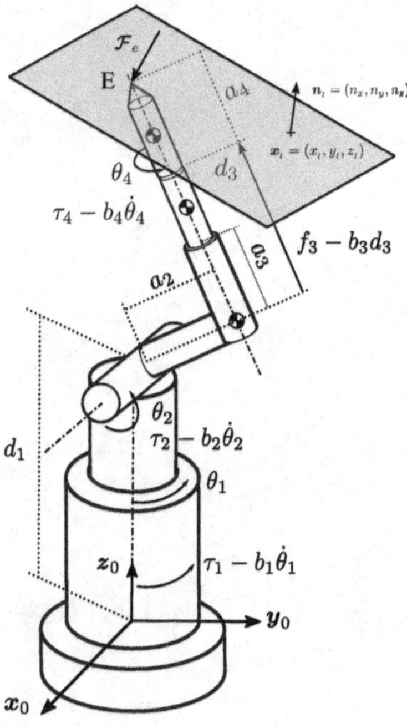

FIGURE 13.29
The four-degrees-of-freedom robot for Problem 13.9.

13.11 Consider the three-degrees-of-freedom PRP robot depicted in Figure 13.22 and repeat Problem 13.10 for this robot.

13.12 Examine the PRP robot with three degrees of freedom illustrated in Figure 13.23 and revisit Problem 13.10 for this specific robot.

13.13 Review Problem 13.10 for the three-degrees-of-freedom RPR robot presented in Figure 13.24.

13.14 Examine Problem 13.10 in the context of the two-degrees-of-freedom RR robot shown in Figure 13.25.

13.15 Review Problem 13.10 in the context of the RRRP robot with four degrees of freedom illustrated in Figure 13.26.

13.16 Examine Problem 13.10 within the framework of the four-degrees-of-freedom RPRR robot presented in Figure 13.27.

13.17 Revisit Problem 13.10 in the context of the four-degrees-of-freedom 4R robot depicted in Figure 13.28.

13.18 Re-examine Problem 13.10 within the framework of the four-degrees-of-freedom RRPR robot as shown in Figure 13.29.

14

Impedance Control

This chapter of the book explores the concept of impedance control, initiating with an introduction that establishes a fundamental grasp of the control scheme concept. The subsequent sections define impedance and scrutinize its complexities. The central tenet of impedance control is thoroughly examined, elucidating its principles and diverse applications. The chapter then continues with a motivating demonstration, and outlines the structural framework of impedance control, providing insights into its organizational facets. The context emphasizes that, through judicious selection of gain matrices in this impedance relation, the interaction dynamics of the robot with its environment can be tuned in both transient and steady-state. Notably, this approach doesn't pursue the individual control of either motion or force; instead, it focuses on managing the dynamic relationship between force and motion.

Furthermore, the chapter addresses a special case within impedance control, leading to the compliance control scheme detailed in the previous chapter. The analysis demonstrates that in stiffness control, the static interaction of force-motion in the robot during the steady-state is examined and adjusted. Conversely, impedance control generalizes this interaction to a dynamic state, allowing for control over both transient and steady-state behavior. A comprehensive summary of force and impedance control methods is presented next in this chapter, providing readers with a synthesized understanding of the discussed concepts in force and impedance control schemes.

The chapter also conducts a specific case study on impedance control, focusing on the planar 3<u>R</u> robot, and offering a practical viewpoint on its implementation. This case study pays special attention to simulating an obstacle in the robot's workspace and tuning controller parameters to achieve the desired impedance dynamics. The chapter concludes with a dedicated section on problems, encouraging readers to apply and solidify their understanding of the presented material through problem-solving exercises.

14.1 Introduction

In the previous chapter, we familiarized ourselves with control techniques tailored for robots interacting with their environment. Among the introduced methods, it was recognized that simultaneously controlling both the robot's configuration vector $\boldsymbol{q}$ and its interactive wrench $\mathcal{F}$ is a complex task and, therefore, is not directly pursued. Instead, the compliant control method was introduced, focusing on adjusting the static relationship between interactive force and resulting deflections. Furthermore, the direct force control method strives to monitor the robot's interactive wrench performance with the environment, indirectly addressing the tracking of the robot's motion. The hybrid force-position control method incorporates direct force control for specific joints while maintaining position control for others.

DOI: 10.1201/9781003491415-14

The main reason why force and position control cannot be performed simultaneously is that the number of actuators in a fully actuated robot with n degrees of freedom is equal to n. Consequently, the number of controllable variables is confined to n. When aiming to independently and simultaneously control both position and force variables, we require $2n$ inputs. Considering this limitation, the impedance control method, which we will describe in this chapter, establishes a dynamic relationship between the motion and force variables under control.

The impedance control method extends the concept of compliant control, wherein the dynamic relationship between the controlled motion and force variables is modified by directly measuring the robot's interactive wrench, as opposed to adjusting a static relationship. Hogan introduced the concept of impedance control in 1985 [169, 170], and it has found extensive application across various domains [171].

Mechanical impedance, in comparison with the well-known concept of electrical impedance, is defined as the relationship between the effort and flow in mechanical structures. Thus, the dynamic relationship between force and motion speed can be referred to as mechanical impedance. In impedance control, the adjustment of dynamic behavior between the motion and force variables is carried out based on the desirable dynamic behavior of these two quantities. It's crucial to highlight that impedance control does not involve the independent or hierarchical control of force and motion; rather, it centers the focus on dynamically adjusting the relationship between these two variables.

In the following sections, mechanical impedance in linear and nonlinear systems is defined. Then, with a motivating demonstration, the importance of using this control method and its conceptual design will be presented. Finally, this control method will be introduced in detail and compared with previous force-centered control methods.

14.2 Definition of Impedance

The notion of impedance in electrical networks is characterized as the resistance of an electrical circuit to the flow of electric current given a specific electric voltage. When extending this concept to other disciplines, the electric voltage is generalized to the broader concept of *effort*, while the electric current is generalized to the general concept of *flow*. To be accurate, electrical impedance can be precisely defined using the voltage-to-current transfer function within a linear electrical network.

Considering that electrical resistance relates voltage to current in a static state, electrical impedance extends this concept to the dynamic behavior of the circuit. As electrical impedance is defined in linear systems with a transfer function, the dynamic characteristics and transient behavior of the system can be explored in its frequency content. Moreover, if a constant electric voltage is applied to an electrical circuit, the electrical impedance will exactly be equal to the electrical resistance of that circuit.

Electrical impedance is represented by the notation $\boldsymbol{Z}(s)$. This quantity, generally a complex variable, can be expressed in polar form using its magnitude and phase angle at each frequency: $|\boldsymbol{Z}|\angle\boldsymbol{Z}$. In this representation, the magnitude of impedance $|\boldsymbol{Z}|$ indicates the ratio of effort amplitude to current amplitude at the desired frequency, and the phase angle $\angle\boldsymbol{Z}$ represents the amount of lead or lag in the phase of input current relative to the phase of output effort. The inverse of electrical impedance is called electrical *admittance*, representing the ratio of current to effort in the Laplace domain.

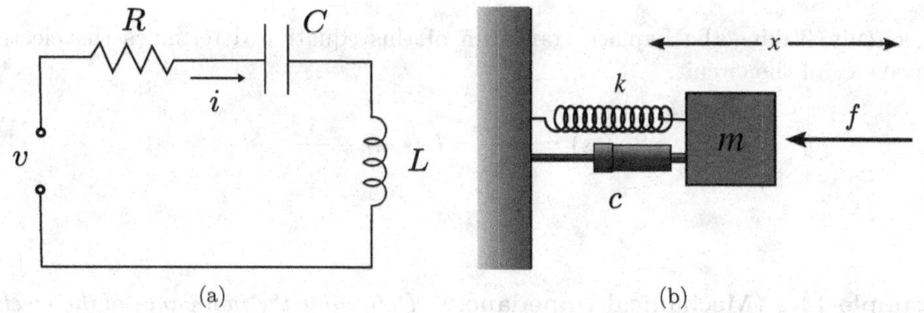

FIGURE 14.1
Comparison of electrical and mechanical impedance. (a) An electrical RLC circuit. (b) A mechanical system with mass, spring, and damper.

In comparison to electrical impedance, mechanical impedance is defined as follows:

Definition: In linear systems, mechanical impedance $\boldsymbol{Z}$ is defined as the ratio of the Laplace transform of mechanical effort to the Laplace transform of mechanical flow:

$$\boldsymbol{Z}(s) = \frac{\boldsymbol{F}(s)}{\boldsymbol{v}(s)} \tag{14.1}$$

where in this context, effort in mechanical systems is interpreted as the applied force $\boldsymbol{F}$, and mechanical flow is interpreted as the velocity of body motion $\boldsymbol{v}$.

This definition has been presented for translational motion of a body in one degree of freedom, and it can be extended to translational motion in other directions and rotational motions. In angular motion, mechanical effort is described by torque, and mechanical flow is described by angular velocity.

In the case of a body motion in six degrees of freedom, the multivariable impedance of the body is defined with the ratio of the six-dimensional array of wrench applied on the body $\boldsymbol{\mathcal{F}}(s)$ to the six-dimensional array representing the motion twist rate denoted by $\dot{\boldsymbol{\mathcal{X}}}(s)$. It should also be noted that in this definition, Laplace transforms of effort and flow have been used, which are applicable only in linear systems. In robots and nonlinear systems, the differential equation relating the effort $\boldsymbol{\mathcal{F}}(t)$ and mechanical motion flow $\dot{\boldsymbol{\mathcal{X}}}(t)$ can be defined as the system impedance.

Example 14.1 (Electrical Impedance) *Obtain the impedance of the RLC electrical circuit shown in Figure 14.1(a).*

Solution: The relationship between voltage and current in an RLC electrical circuit can be determined using the following differential equation.

$$v = L\frac{di}{dt} + Ri + \int_0^t \frac{1}{C} i(\tau) d\tau \tag{14.2}$$

In this equation, L represents the inductance of the inductor, R is the electrical resistance of the circuit, and C denotes the capacitance of the capacitor in the circuit. Furthermore, v represents the supply voltage, and i is the current flowing through the circuit. Taking the Laplace transform of this equation determines the electrical impedance of the circuit.

$$Z(s) = \frac{v(s)}{i(s)} = Ls + R + \frac{1}{Cs} \tag{14.3}$$

Example 14.2 (Mechanical Impedance) *Determine the impedance of the mechanical system composed of a mass-spring-damper, shown in Figure 14.1(b).*

Solution: The relationship between force and motion in the mass-spring-damper mechanical system can be determined using the following differential equation.

$$m\ddot{x} + c\dot{x} + kx = f \tag{14.4}$$

In this equation, m represents the mass of the system, c is the linear viscous damping coefficient of the damper, k is the spring stiffness coefficient, and f is the external excitation force applied to the system. If you take the Laplace transform of this equation, the mechanical impedance of the system can be determined as follows.

$$Z(s) = \frac{f(s)}{v(s)} = ms + c + \frac{k}{s} \tag{14.5}$$

These two examples illustrate that despite structural differences in the physics of electrical and mechanical systems, both exhibit impedance relationships that are similar to each other. This comparison helps electrical engineers comprehend the components of mechanical impedance more easily. By leveraging this analogy, the literature related to the definition of impedance elements in a system is unified. The types of impedance in a system can be classified as follows.

Inductive Impedance: The system has inductive impedance, if and only if, the impedance magnitude at $s = 0$ is equal to zero: $|\boldsymbol{Z}(0)| = 0$.

Resistive Impedance: The system has resistive impedance, if and only if, the impedance magnitude at $s = 0$ is constant and non-zero: $|\boldsymbol{Z}(0)| = R$.

Capacitive Impedance: The system has capacitive impedance, if and only if, the impedance magnitude approaches infinity as $s \to 0$: $\lim_{s\to 0} |\boldsymbol{Z}(s)| = \infty$.

Thus, for the mechanical system in Example 14.2, the mass characterizes the inductive impedance, the viscous damping coefficient characterizes the resistive impedance, and the spring stiffness coefficient characterizes the capacitive impedance. With this definition, the environment with which the robot interacts can be classified into inductive, resistive, and capacitive impedances. An environment with high stiffness can be represented with a large capacitive impedance. On the other hand, an environment with significant structural damping can be represented with a large resistive impedance.

The force-motion relationship in a scenario where a user is holding a surgical haptic device can be modeled by combining inductive, resistive, and capacitive impedances. By defining different impedance elements, the interaction characteristics of users with such a device can be specified. If a user holds the haptic device firmly, its capacitive impedance will dominate over the other two types of impedance. Now, if we want to regulate the dynamic relationship between the robot and the environment, we can adjust the inductive, resistive, and capacitive impedance properties through a control system. In this way, regardless of the user's interaction with the environment or whether the robot is interacting with a stiff or compliant environment, the impedance characteristics of the system can be brought close to the desired values. In the upcoming section, a motivating demonstration is presented to highlight the importance and nature of adjusting impedance.

14.3 Impedance Control Concept

So far, it has been explained that in impedance control methods, the goal is to adjust the dynamic interaction relationship between the robot and its surrounding environment. These methods rely on a suitable control topology that stabilizes the system by measuring force and motion, adjusting the dynamic relationship between these two quantities. To better understand this interactive relationship and desirable impedance values in robotic applications, let's consider the development of this concept by presenting a motivating demonstration.

For this purpose, consider the use of industrial robots in the automated production of fish fillets. In a fish production and packaging factory, there are various types of fish with different sizes, and the production of fish fillets, as one of the premium products of the factory, faces high demand from customers. In the production of fish fillets, the fish is usually sliced along its spinal cord into two parts, and the fish bones are separated from its flesh. Then, it is necessary to carefully remove the fish skin from the flesh. This part of the operation is very precise and delicate, and automating it with robotic systems poses considerable complexity.

To understand the complexities of this operation, let's review how skilled individuals remove the skin from flesh. A skilled person typically places a piece of the separated fish flesh on the cutting table with the skin side down. Then, using a sharp knife, he or she start separating the skin from the fish flesh, beginning from the tail section. The movement of the skilled person's hand is controlled with dexterous angular knife cuts in back-and-forth motions. When the knife moves toward the skin, the skilled person feels more speed and less force, adjusting the knife angle toward the fish flesh. Immediately after the knife penetrates the fish's flesh, he or she feels more force and less motion, tilting the knife toward the skin. This way, by applying precise and rapid back-and-forth movements, the skilled person prevents the knife from tilting excessively toward the skin, avoiding tearing the skin, or conversely, preventing the knife from moving too much toward the flesh and wasting a portion of it.

The method of fish skinning effectively demonstrates the simultaneous sensation of force and motion by the skilled person and the control of the relationship during the motion. If the movement of his or her hand results in feeling more speed with less force, the knife tilts excessively toward the skin, requiring a change in the movement direction. On the other hand, if he or she feels slower motion and increased the force, the knife tends to tilt toward the fish flesh. With extensive practice, the skilled person moves their hand with the necessary agility to experience the desired impedance profile during the motion.

To perform this operation automatically by robotic systems, precise measurement of the robot's motion and force is essential. Measuring the robot's motion alone is not sufficient because fish pieces come in various sizes and shapes. Moreover, the skin of the fish is very thin compared to its flesh dimensions, requiring highly accurate measurement systems to accurately measure both. Therefore, the possibility of defining a predetermined path for the robot's motion is not feasible. Although force control or hybrid force-position control methods work better than motion control methods in this application, practical use of them is not possible due to the need for a desired motion path in these approaches, which is not available beforehand.

If the mechanical impedance of the robot can be adjusted to the desired value during motion, this control method, similar to the skilled hand movements, will be able to separate the skin from the flesh without causing tearing. The desired impedance threshold of the robot in this application can be determined from the impedance of the knife motion on the fish flesh. In a situation where the knife moves more toward the fish's skin, the force decreases, and the speed of movement increases. This implies a proportional decrease in mechanical impedance. The appropriate impedance of the robot's motion can be adjusted by the structure of a multivariable mass-spring-damper model.

$$\boldsymbol{Z}_d(s) = \boldsymbol{M}_d s + \boldsymbol{C}_d + \frac{1}{s}\boldsymbol{K}_d \tag{14.6}$$

In this context, $\boldsymbol{Z}_d$ is the desired impedance of the robot, which has the desired inductive $\boldsymbol{M}_d$, the desired resistive $\boldsymbol{C}_d$, and the desired capacitive impedance $\boldsymbol{K}_d$. Using this simple impedance structure, the fish fillet production process can be automated by industrial robots, while the precise determination of the impedance model components is obtained through experiments.

In some applications, there is no need to change the effective mass matrix of the system $\boldsymbol{M}_e$. In this case, the desired impedance can be determined by specifying the appropriate threshold for the desired resistive impedance $\boldsymbol{C}_d$ and the desired capacitive impedance $\boldsymbol{K}_d$, and using the following relationship.

$$\boldsymbol{Z}_d(s) = \boldsymbol{M}_e s + \boldsymbol{C}_d + \frac{1}{s}\boldsymbol{K}_d \tag{14.7}$$

14.4 Impedance Control Structure

Consider a general scenario where a robot with multiple degrees of freedom is in contact with its surrounding environment and has force interaction. The motion of the robot's joints is represented by the vector $\boldsymbol{q}$, and the interactive force and torque between the robot and the environment is represented by the six-dimensional array of wrench $\boldsymbol{\mathcal{F}}_e = \begin{bmatrix}\boldsymbol{F}_e & \boldsymbol{n}_e\end{bmatrix}^T$, which is measured and used in the feedback loop. The dynamic model of the robot in the absence of disturbance torques can be expressed using the relationship 13.29, which is rewritten here for easier follow up.

$$\boldsymbol{\tau} = \boldsymbol{M}_e(\boldsymbol{q})\ddot{\boldsymbol{q}} + \boldsymbol{C}(\boldsymbol{q},\dot{\boldsymbol{q}})\dot{\boldsymbol{q}} + \boldsymbol{b}_e\dot{\boldsymbol{q}} + \mathfrak{g}(\boldsymbol{q}) + \boldsymbol{\tau}_e. \tag{14.8}$$

In this relation, $\boldsymbol{M}_e(\boldsymbol{q})$ represents the effective mass matrix of the robot, $\boldsymbol{C}(\boldsymbol{q},\dot{\boldsymbol{q}})$ is the Christoffel matrix, $\boldsymbol{b}_e$ is the effective viscous friction torque vector, $\mathfrak{g}(\boldsymbol{q})$ is the gravity vector, and $\boldsymbol{\tau}_e$ is the torque induced by interactive wrench in the task space, mapped using the transpose of the Jacobian matrix in the joint space $\boldsymbol{\tau}_e = \boldsymbol{J}^T(\boldsymbol{q})\boldsymbol{\mathcal{F}}_e$.

Let us design impedance control in this scenario using inverse dynamic control in the inner loop and position controller in the outer loop. A sample block diagram of such a design is illustrated in Figure 14.2. As observed in this figure, the desired joint space trajectory $\boldsymbol{q}_d$ enters as the primary input to the closed-loop system, and the position tracking error $\boldsymbol{e}_q = \boldsymbol{q}_d - \boldsymbol{q}$ is computed in a straightforward manner in the outer loop.

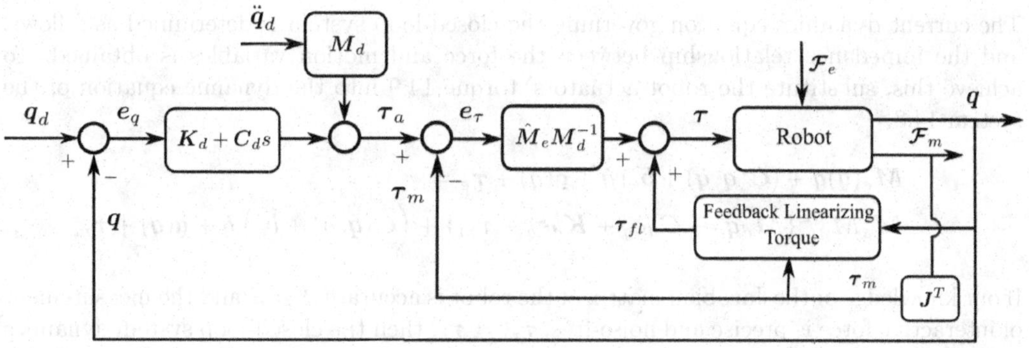

FIGURE 14.2
Block diagram of impedance control: Position feedback in the outer loop and force feedback in the inner loop.

It is also noteworthy that direct force tracking is not incorporated into this control structure, as τ_d does not exist as an independent input. In this control architecture, force tracking, as a secondary objective, is achieved by applying an auxiliary input τ_a, which is the output of the position controller in the outer loop. Thus, the predefined interaction force is not utilized in this control structure; instead, the contact force is designed in a way to regulate the dynamic force-motion dynamics of the robot's end-effector with the environment.

This is achieved in part by direct measurement of interactive robot wrench $\mathcal{F}_m$ in the inner loop, mapping it into the joint space by Jacobian transpose J^T to determine τ_m. In the inner loop, the interaction torque τ_a is applied through the external feedback loop to the inner loop, and the online error in tracking the interactive torque e_τ. Thus, the dynamic force-motion relationship is tuned according to the desired impedance as designed in Equation 14.6.

The motor torque vector in the inner loop τ is determined based on the inverse dynamic control of the robot, which consists of three components. In the inner loop, a feedback linearization torque along with a mass matrix adaptation block $\hat{M}_e(q)$ is incorporated, and in the outer loop, linear position control is considered to adjust the impedance coefficients K_d and C_d. Although various impedance models can be used to define the desired force-motion dynamic relationship, in this structure, tuning of the second-order impedance model 14.6 is considered.

Using the impedance control structure illustrated in Figure 14.2, the actuator torques in the joint space τ is determined by:

$$\tau = \hat{M}_e M_d^{-1} \cdot e_\tau + \tau_{fl} \tag{14.9}$$

where,

$$e_\tau = \tau_a - \tau_m \tag{14.10}$$

$$\tau_a = M_d \ddot{q}_d + C_d \dot{e}_q + K_d e_q \tag{14.11}$$

In Equation 14.11, the term M_d represents the desired mass matrix (inductive impedance matrix) in the dynamic force-motion relationship. Furthermore, C_d and K_d determine the desired resistive and capacitive impedance, respectively.

Similar to the direct force control method, calculate the feedback linearizing torque τ_{fl} while considering the compensatory torque generated by the interactive wrench.

$$\tau_{fl} = \left(\hat{C}(q, \dot{q}) + \hat{b}_e\right) \dot{q} + \hat{\mathfrak{g}}(q) + \tau_m \tag{14.12}$$

The current dynamics equation governing the closed-loop system is determined as follows, and the impedance relationship between the force and motion variables is obtained. To achieve this, substitute the robot actuators' torque 14.9 into the dynamic equation of the system 14.8.

$$\begin{aligned}&\boldsymbol{M}_e(\boldsymbol{q})\ddot{\boldsymbol{q}} + (\boldsymbol{C}(\boldsymbol{q},\dot{\boldsymbol{q}}) + \boldsymbol{b}_e)\,\dot{\boldsymbol{q}} + \boldsymbol{g}(\boldsymbol{q}) + \boldsymbol{\tau}_e = \\ &\hat{\boldsymbol{M}}_e\boldsymbol{M}_d^{-1}\{\boldsymbol{M}_d\ddot{\boldsymbol{q}}_d + \boldsymbol{C}_d\dot{\boldsymbol{e}}_q + \boldsymbol{K}_d\boldsymbol{e}_q - \boldsymbol{\tau}_m\} + \left(\hat{\boldsymbol{C}}(\boldsymbol{q},\dot{\boldsymbol{q}}) + \hat{\boldsymbol{b}}_e\right)\dot{\boldsymbol{q}} + \hat{\boldsymbol{g}}(\boldsymbol{q}) + \boldsymbol{\tau}_m\end{aligned}$$

If our knowledge of the Jacobian matrix of the robot is accurate $\hat{\boldsymbol{J}} = \boldsymbol{J}$ and the measurement of interactive force is precise and noise-free, $\boldsymbol{\tau}_m = \boldsymbol{\tau}_e$, then the closed-loop system dynamics is reduced to:

$$\boldsymbol{M}_e\boldsymbol{M}_d^{-1}(\boldsymbol{M}_d\ddot{\boldsymbol{q}}) = \boldsymbol{M}_e\boldsymbol{M}_d^{-1}\{\boldsymbol{M}_d\ddot{\boldsymbol{q}}_d + \boldsymbol{C}_d\dot{\boldsymbol{e}}_q + \boldsymbol{K}_d\boldsymbol{e}_q - \boldsymbol{\tau}_e\} \tag{14.13}$$

Note that the mass matrix $\boldsymbol{M}_e(\boldsymbol{q})$ is positive definite, and by considering a positive definite matrix $\boldsymbol{M}_d$, the resulting matrix $\boldsymbol{M}_e\boldsymbol{M}_d^{-1}$ will be invertible within the entire robot's workspace. Therefore, the closed-loop dynamic equation of the robot can be summarized as follows.

$$\boldsymbol{M}_d\ddot{\boldsymbol{e}}_q + \boldsymbol{C}_d\dot{\boldsymbol{e}}_q + \boldsymbol{K}_d\boldsymbol{e}_q = \boldsymbol{\tau}_e \tag{14.14}$$

Equation 14.14 confirms that if our knowledge of the dynamic and kinematic equations of the robot is complete, and the interactive wrench measurement is precisely performed without noise, the closed-loop dynamic relation of the system will follow a second-order differential equation, regulating the desired impedance force-motion relation according to 14.6. Thus, by appropriately selecting the matrices in this impedance relation, the interaction dynamics of the robot with the environment can be effectively tailored for both transient and steady-state scenarios. It is natural that, as expected, control of either motion or force alone is not pursued in this method; instead, the dynamic relationship between force and motion is controlled.

If, for any reason, the robot loses contact with the environment during its motion, in this case, the interactive wrench becomes zero ($\boldsymbol{\mathcal{F}}_e = \boldsymbol{0}$), and consequently, the torque resulting from this interaction ($\boldsymbol{\tau}_e = \boldsymbol{0}$) becomes zero as well. In this situation, the closed-loop dynamic equation in 14.14 transforms into a second-order tracking error equation without the presence of any input. Consequently, error tracking is carried out accurately, and the tracking error converges toward zero. However, it should be noted that in this relation, control gain matrices $\boldsymbol{K}_d$ and $\boldsymbol{C}_d$ are designed based on the desired impedance and are usually not tuned with large gains, so perfect trajectory tracking with negligible error is not expected.

In the impedance control method presented in this chapter, the utilization of inverse dynamics and precise measurement of interactive force are considered as two crucial elements in the design. Therefore, the effectiveness of this method can only be expected when the designer has accurate knowledge of the kinematic and dynamic model of the robot and has precisely determined the model parameters using identification methods [172]. However, in practice, the presence of uncertainties, either structured or unstructured, in the system model is inevitable. To address this issue, various robust and adaptive impedance control methods have been developed in the robotics literature [173, 174].

Accurate measurement of interactive wrench is also a challenging and costly subject. The use of six-axis wrench measurements in industrial robots is commonly used in these applications. Although these sensors are usually embedded in the robot's wrist, correction of measurements based on the robot's end-effector dynamics is essential [175]. Additionally, the use of state observers for estimating or mitigating measurement noise proves to be highly effective in practice [176, 177].

Impedance control has garnered widespread acceptance in numerous industrial applications. It proves to be highly effective in industrial applications where the robot's interaction with the desired environment is crucial [178]. Moreover, in teleoperation applications using haptic devices, the use of this control method is frequently suggested in robotics literature [179, 180]. When a robot performs operational tasks in interaction with humans, impedance control is the primary choice [181, 182]. In robotic surgery applications, due to the interaction of the surgeon's hand with the haptic device or the robot's interaction with soft tissue during surgery, impedance control methods are widely used [183, 184]. The use of impedance control in other medical robotics applications, such as robotic rehabilitation [185] and skill transfer from an expert surgeon to a robotic surgery [186, 187], is also frequent.

14.5 Special Case in Impedance Control

A specific type of impedance control can be studied by considering the desired impedance equation 14.7, which provides a suitable physical interpretation. Suppose we do not adjust the inductive impedance of the closed-loop system but instead utilize the inherent mass matrix of the system. Naturally, in adjusting the transient and steady-state behavior of the closed-loop system, as outlined in the equation 14.14, the ratios of the matrices $\boldsymbol{K}_d$ and $\boldsymbol{C}_d$ to the effective mass matrix $\boldsymbol{M}_e$ have a significant impact, and if the system's mass matrix is known, there is no need to adjust it.

This seemingly small modification in the structure of the desired impedance creates a special case in impedance control that has an interesting physical interpretation. To analyze the system's behavior in this case, consider setting the desired mass matrix equal to the intrinsic mass matrix of the system, $\boldsymbol{M}_d = \boldsymbol{M}_e$. In this scenario, the mass matrix adjustment block in the impedance control structure shown in Figure 14.2 simplifies to the identity matrix, $\boldsymbol{M}_e \boldsymbol{M}_d^{-1} = \boldsymbol{I}$.

Therefore, without changing the control structure, this block can be removed from the block diagram of the system. By doing so, it is observed that the interaction torque vector $\boldsymbol{\tau}_m$ is compensated twice in consecutive loops before and after this block's removal. This compensation occurs once with a negative sign in the internal force control feedback and in the equation 14.10 in calculating the interactive torque tracking error $\boldsymbol{e}_\tau = \boldsymbol{\tau}_a - \boldsymbol{\tau}_m$, and once with a positive sign in the force feedback linearization torque in the equation 14.12. Considering the removal of the mass matrix compensation block, the block diagram of the control system can be simplified again, and the interaction torque in the inner loop, along with the compensation term in the force feedback torque, are canceled.

$$\boldsymbol{\tau}_{fl} - \boldsymbol{\tau}_m = \left(\hat{\boldsymbol{C}}(\boldsymbol{q}, \dot{\boldsymbol{q}}) + \hat{\boldsymbol{b}}_e\right) \dot{\boldsymbol{q}} + \hat{\mathfrak{g}}(\boldsymbol{q}) \tag{14.15}$$

The physical interpretation of this special case implies no need for measuring and utilizing force feedback in the inner loop. Thus, the comprehensive structure of impedance control shown in Figure 14.2 simplifies to the block diagram depicted in Figure 14.3, in this specific case. This special case aligns perfectly with the motion control topology in robots, utilizing inverse dynamics control. In this scenario, appropriate interactive gains $\boldsymbol{K}_d$ and $\boldsymbol{C}_d$ are used in the linear part of inverse dynamics control for effective robot-environment

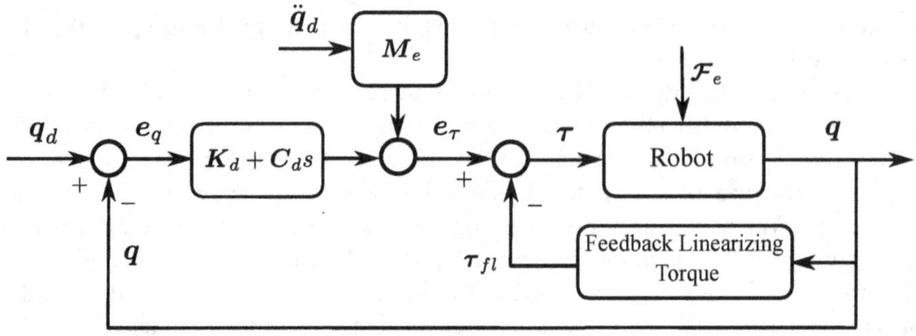

FIGURE 14.3
Block diagram of impedance control in a special case: Removing internal force feedback and transforming impedance control into stiffness control.

interaction. With this explanation, it becomes evident that the impedance control structure converges to compliance control in robots, as detailed in previous chapter.

By providing this physical interpretation of the special case of impedance control, a comparison between compliance control and general impedance control can be made. As previously mentioned, in compliance control, the static interaction of force-motion in the robot in the steady-state is examined and adjusted. This is in contrast to impedance control, where this interaction is generalized to a dynamic state, and both its transient and steady-state behavior is adjusted. It should be noted that this crucial difference in the nature of these two control methods is rooted in the measurement and utilization of interactive wrench between the robot and the environment, leading to the widespread use of impedance control in various robotic applications [171].

14.6 Summary of Force and Impedance Control Methods

In the preceding and current chapters, we have explored diverse force control techniques tailored for serial robots. Given the greater accessibility of the dynamic model of serial robots in joint space, the presented methods are grounded in joint space design. This section aims to concisely summarize these approaches, facilitating a comprehensive overview for effective comparison.

- Dynamics Formulation

The dynamic model governing the motion of a robot in joint space, while it is in force interaction with the environment, is represented as follows.

$$\boldsymbol{\tau} + \boldsymbol{\tau}_d = \boldsymbol{M}_e(\boldsymbol{q})\ddot{\boldsymbol{q}} + \boldsymbol{C}(\boldsymbol{q}, \dot{\boldsymbol{q}})\dot{\boldsymbol{q}} + \boldsymbol{b}_e\dot{\boldsymbol{q}} + \mathfrak{g}(\boldsymbol{q}) + \boldsymbol{\tau}_e.$$

In this equation, $\boldsymbol{\tau}$ represents the actuator torques applied to the robot, $\boldsymbol{\tau}_d$ is the disturbance torque imposed on the robot, $\boldsymbol{M}_e(\boldsymbol{q})$ is the effective mass matrix of the robot, $\boldsymbol{C}(\boldsymbol{q}, \dot{\boldsymbol{q}})$ is the Christoffel matrix, $\boldsymbol{b}_e$ is the effective viscous friction torque vector, and $\mathfrak{g}(\boldsymbol{q})$ is the robot's gravitational vector. Additionally, in this equation, $\boldsymbol{\tau}_e$ is the torque resulting from the interaction wrench of the robot $\boldsymbol{\mathcal{F}}_e$, which is determined by using the Jacobian transpose matrix mapping.

$$\boldsymbol{\tau}_e = \boldsymbol{J}^T(\boldsymbol{q})\boldsymbol{\mathcal{F}}_e,$$

- Compliance Control

Calculate the robot's interaction torque with the environment based on the linear environment model using the following equation.

$$\boldsymbol{\tau}_e = \boldsymbol{\mathcal{K}}_e \Delta \boldsymbol{q}, \quad \Delta \boldsymbol{q} = \boldsymbol{q} - \boldsymbol{q}_e$$

where $\boldsymbol{\mathcal{K}}_e$ is the stiffness matrix of the environment in joint space, and $\Delta \boldsymbol{q}$ is the small deflections of the environment due to the interactive wrench, mapped into joint space. If using partial feedback linearization in the compliance control structure, the actuator torque in the robot is determined by:

$$\boldsymbol{\tau} = \mathfrak{g}(\boldsymbol{q}) + \boldsymbol{K}_p \boldsymbol{e}_q + \boldsymbol{K}_d \dot{\boldsymbol{e}}_q,$$

Assuming complete knowledge of the robot's dynamics, the closed-loop system equation for the robot in the absence of disturbances is summarized as follows.

$$\boldsymbol{M}_e(\boldsymbol{q})\ddot{\boldsymbol{q}} + \boldsymbol{C}(\boldsymbol{q}, \dot{\boldsymbol{q}})\dot{\boldsymbol{q}} + \boldsymbol{b}_e \dot{\boldsymbol{q}} + \boldsymbol{\mathcal{K}}_e(\boldsymbol{q} - \boldsymbol{q}_e) = \boldsymbol{K}_p \boldsymbol{e}_q + \boldsymbol{K}_d \dot{\boldsymbol{e}}_q$$

In this scenario, the asymptotic stability of the tracking error is guaranteed by the direct Lyapunov method, and the steady-state values of the joint motion vector $\bar{\boldsymbol{q}}$ and interaction torque $\bar{\boldsymbol{\tau}}_e$ are determined by the following relationships:

$$\begin{aligned} \bar{\boldsymbol{q}} &= [\boldsymbol{\mathcal{K}}_e + \boldsymbol{K}_p]^{-1}[\boldsymbol{\mathcal{K}}_e \bar{\boldsymbol{q}}_e + \boldsymbol{K}_p \bar{\boldsymbol{q}}_d] \\ \bar{\boldsymbol{\tau}}_e &= \boldsymbol{\mathcal{K}}_e[\boldsymbol{\mathcal{K}}_e + \boldsymbol{K}_p]^{-1}\boldsymbol{K}_p(\bar{\boldsymbol{q}}_d - \bar{\boldsymbol{q}}_e) \end{aligned}$$

Considering that the environmental stiffness $\boldsymbol{\mathcal{K}}_e$ is significantly greater than the proportional gain of the controller $\boldsymbol{K}_p$, the steady-state value of the torque resulting from interactive forces can be approximated by the following equation.

$$\bar{\boldsymbol{\tau}}_e \simeq \boldsymbol{K}_p(\bar{\boldsymbol{q}}_d - \bar{\boldsymbol{q}}_e).$$

- Direct Force Control

The structure of direct force control, based on the block diagram in Figure 13.6, consists of two nested control loops. The actuators' torque vector in the robot is determined as follows.

$$\boldsymbol{\tau} = \hat{\boldsymbol{M}}_e(\boldsymbol{q})\boldsymbol{a} + \boldsymbol{\tau}_{fl}$$

where

$$\begin{aligned} \boldsymbol{\tau}_{fl} &= \left(\hat{\boldsymbol{C}}(\boldsymbol{q}, \dot{\boldsymbol{q}}) + \hat{\boldsymbol{b}}_e\right)\dot{\boldsymbol{q}} + \hat{\mathfrak{g}}(\boldsymbol{q}) + \boldsymbol{J}^T \boldsymbol{\mathcal{F}}_m \\ \boldsymbol{a} &= \boldsymbol{K}_p(\boldsymbol{q}_a - \boldsymbol{q}) - \boldsymbol{K}_d \dot{\boldsymbol{q}} \end{aligned}$$

If we represent the force controller in a general form by the function $\boldsymbol{C}_\tau$, the auxiliary robot motion trajectory vector $\boldsymbol{q}_a$ is determined as follows.

$$\boldsymbol{q}_a = \boldsymbol{C}_\tau(\boldsymbol{e}_\tau), \quad \boldsymbol{e}_\tau = \boldsymbol{\tau}_d - \boldsymbol{\tau}_m$$

If we use a PI controller in the outer force control loop, then:

$$\boldsymbol{C}_\tau(s) = \boldsymbol{K}_P^{-1}\left(\boldsymbol{K}_{\tau_p} + \frac{1}{s}\boldsymbol{K}_{\tau_i}\right)\boldsymbol{e}_\tau(s)$$

Assuming that our knowledge of the system's kinematic and dynamic models is complete, and the measurement of interactive deflection is precise and noise-free, the closed-loop dynamic equation of the system in this scenario can be summarized by the following equation.

$$\ddot{q} + \boldsymbol{K}_d\dot{\boldsymbol{q}} + \boldsymbol{K}_p\boldsymbol{q} = \boldsymbol{K}_{\tau_p}\boldsymbol{e}_\tau(t) + \boldsymbol{K}_{\tau_i}\int_0^t \boldsymbol{e}_\tau(\eta)d\eta.$$

- Impedance Control

If you want the dynamic relationship between interactive force and robot motion to be determined by the following desired impedance:

$$\boldsymbol{Z}_d(s) = \boldsymbol{M}_d s + \boldsymbol{C}_d + \frac{1}{s}\boldsymbol{K}_d$$

Based on the block diagram in Figure 14.2, the impedance control method is implemented using two cascaded control loops. In this method, the vector of robot joint torques is determined as follows.

$$\boldsymbol{\tau} = \hat{\boldsymbol{M}}_e\boldsymbol{M}_d^{-1}\cdot\boldsymbol{e}_\tau + \boldsymbol{\tau}_{fl}$$

where

$$\begin{aligned}
\boldsymbol{e}_\tau &= \boldsymbol{\tau}_a - \boldsymbol{\tau}_m \\
\boldsymbol{\tau}_a &= \boldsymbol{M}_d\ddot{\boldsymbol{q}}_d + \boldsymbol{C}_d\dot{\boldsymbol{e}}_q + \boldsymbol{K}_d\boldsymbol{e}_q \\
\boldsymbol{\tau}_{fl} &= \left(\hat{\boldsymbol{C}}(\boldsymbol{q},\dot{\boldsymbol{q}}) + \hat{\boldsymbol{b}}_e\right)\dot{\boldsymbol{q}} + \hat{\mathfrak{g}}(\boldsymbol{q}) + \boldsymbol{\tau}_m
\end{aligned}$$

Assuming that our knowledge of the system's kinematic and dynamic model is complete, and the measurement of interactive compliance is performed with sufficient precision and without noise, the dynamic equation of the closed-loop system is summarized as follows.

$$\boldsymbol{M}_d\ddot{\boldsymbol{e}}_q + \boldsymbol{C}_d\dot{\boldsymbol{e}}_q + \boldsymbol{K}_d\boldsymbol{e}_q = \boldsymbol{\tau}_e$$

14.7 Impedance Control of the Planar 3R Robot

In this section, we conduct simulations of impedance control on the planar 3R robot, serving as an illustrative case study for the robots discussed in this book. For this purpose, consider the 3R robot shown in Figure 8.7, with kinematic and mass parameters specified in Table 14.1. Assume that all robot links are treated as slender rods, and their center of mass is positioned at the link's mid length. The dynamic equations of the robot are derived in the case study 8.3, and the determined Christoffel matrix is reported in case study 9.2. The dynamic equation of the system is obtained in the matrix form 13.2, and to investigate the nonlinear behavior of the system, we neglect the effects of the actuator dynamics. Thus, the following dynamic formulation of the system is utilized in the simulations.

$$\boldsymbol{\tau} + \boldsymbol{\tau}_d = \boldsymbol{M}(\boldsymbol{q})\ddot{\boldsymbol{q}} + \boldsymbol{C}(\boldsymbol{q},\dot{\boldsymbol{q}})\dot{\boldsymbol{q}} + \mathfrak{g}(\boldsymbol{q}) + \boldsymbol{J}^T(\boldsymbol{q})\boldsymbol{\mathcal{F}}_e, \tag{14.16}$$

If we intend the dynamic relationship between the interactive force and the robot's motion to follow the desired impedance given by:

$$\boldsymbol{Z}_d(s) = \boldsymbol{M}_d s + \boldsymbol{C}_d + \frac{1}{s}\boldsymbol{K}_d$$

TABLE 14.1
Kinematic and dynamic parameters of the planar 3$\underline{R}$ robot in impedance control simulations

Quantity	Unit	Notation	Link 1	Link 2	Link 3
Length	m	a_i	1.0	0.7	0.3
Mass	Kg	m_i	10	7	3
Center of Mass	m	m_{c_i}	0.5	0.35	0.15
Moment of Inertia	$Kg \cdot m^2$	${}^C I_i$	0.833	0.286	0.023

it is necessary to use two cascaded control loops based on the block diagram shown in Figure 14.2. In this way, the vector of motor torques for the robot is determined as follows:

$$\boldsymbol{\tau} \;=\; \hat{\boldsymbol{M}}_e \boldsymbol{M}_d^{-1} \cdot \boldsymbol{e}_\tau + \boldsymbol{\tau}_{fl}$$

where

$$\begin{aligned} \boldsymbol{e}_\tau &= \boldsymbol{\tau}_a - \boldsymbol{\tau}_m \\ \boldsymbol{\tau}_a &= \boldsymbol{M}_d \ddot{\boldsymbol{q}}_d + \boldsymbol{C}_d \dot{\boldsymbol{e}}_q + \boldsymbol{K}_d \boldsymbol{e}_q \\ \boldsymbol{\tau}_{fl} &= \left(\hat{\boldsymbol{C}}(\boldsymbol{q}, \dot{\boldsymbol{q}}) + \hat{\boldsymbol{b}}_e\right) \dot{\boldsymbol{q}} + \hat{\mathfrak{g}}(\boldsymbol{q}) + \boldsymbol{\tau}_m \end{aligned}$$

In the inverse dynamics component, we assume that our knowledge of the system dynamics is incomplete, and there is a 20% uncertainty in the model parameters. The desired impedance matrices for the robot in the simulation are set to:

$$\boldsymbol{M}_d = \begin{bmatrix} 1 & 0 & 0 \\ 0 & 0.1 & 0 \\ 0 & 0 & 0.01 \end{bmatrix}, \; \boldsymbol{C}_d = \begin{bmatrix} 1 & 0 & 0 \\ 0 & 2 & 0 \\ 0 & 0 & 0.2 \end{bmatrix} \times 500, \; \boldsymbol{K}_d = \begin{bmatrix} 1 & 0 & 0 \\ 0 & 2 & 0 \\ 0 & 0 & 0.2 \end{bmatrix} \times 100 \quad (14.17)$$

By this choice, the mass matrix in the desired impedance dynamics of the robot is nearly 20 times larger than the effective mass matrix of the robot itself, and the stiffness coefficient in the impedance dynamics is approximately a hundred times less than the stiffness of the environment. To investigate the effect of disturbance, assume that, in the first second, a disturbance torque of a step function with a range of $[-100, -100, -10]^T$ $(N.m)$ is applied to all three joints of the robot. The disturbance torque amplitude is chosen to be larger than the interactive forces of the robot, as such considered in the force control simulations.

Furthermore, in this simulation, it is assumed that there is a flat obstacle in the robot's workspace, whose boundary is modeled by a line given in equation 13.39, passing through the point $\boldsymbol{x}_l = (1, 1)$ with a slope of 45 degrees ($s_l = 1$), as illustrated in Figure 14.4. In this figure the obstacle boundary is represented by a dash-dotted line (colored in magenta).

The desired path of the robot's movement should also be considered in such a way that during the simulation, the robot's end-effector collide with the obstacle and maintains contact with it over time. Here, similar to the compliance control method, the desired path of the robot's movement in the workspace is defined, and then mapped to the joint space. In the conducted simulations, we assume that the desired path of the robot's end-effector is a line at a distance of $0.1\,m$ along the x-axis and $-0.1\,m$ along the y-axis away from the obstacle boundary. This line is represented as a dash-dotted line (colored in blue) in Figure 14.4. Furthermore, let's assume that the desired angle for the third joint of the robot is perpendicular to the plane to maintain suitable force interaction with respect to the environment during the robot's movement. Thus, as explained in the compliance control method, the desired motion trajectory and its derivatives $\boldsymbol{\mathcal{X}}_d(t), \dot{\boldsymbol{\mathcal{X}}}_d(t), \ddot{\boldsymbol{\mathcal{X}}}_d(t)$ are

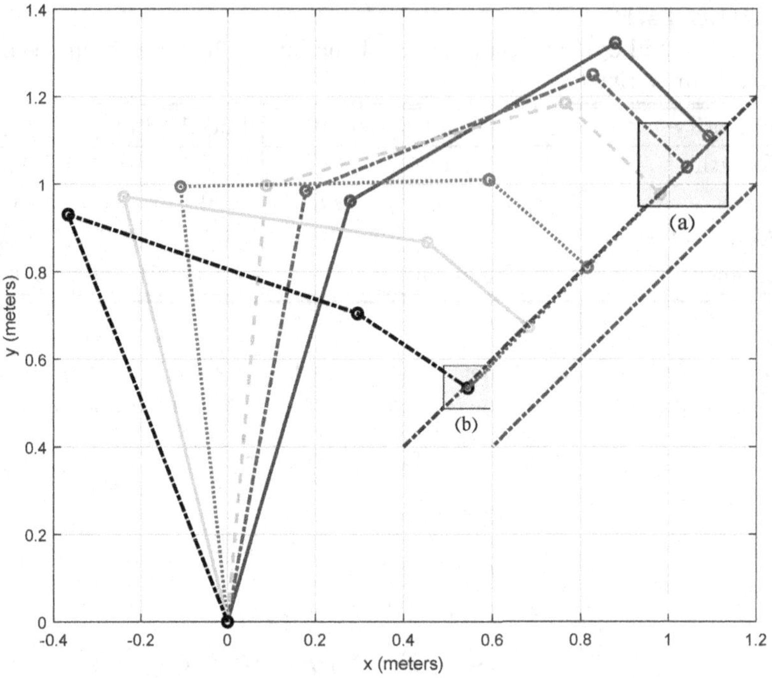

FIGURE 14.4
The configuration of the robot and the obstacle illustrated in multiple stages in the impedance control method.

first determined in the task space. Then, the desired motion trajectory in the joint space is obtained using the inverse kinematics of the robot.

$$\boldsymbol{q}_d(t) = IK(\boldsymbol{\mathcal{X}}_d(t)),$$

The desired velocity of motion in the joint space is obtained using the inverse Jacobian matrix.

$$\dot{\boldsymbol{q}}_d(t) = \boldsymbol{J}^{-1}(\boldsymbol{q})\dot{\boldsymbol{\mathcal{X}}}_d(t),$$

and, the desired acceleration of motion in the joint space is determined using the following relationship.

$$\dot{\boldsymbol{q}}_d(t) = \boldsymbol{J}^{-1}(\boldsymbol{q})\dot{\boldsymbol{\mathcal{X}}}_d(t),$$

To estimate the interactive force between the robot and the environment, determine the geometric location of the robot's end-effector colliding with the environment. In the event of a collision, obtain the interactive force from equation 13.40, where the environment stiffness coefficient is set to

$$\boldsymbol{K}_e = \begin{bmatrix} 1 & 0 & 0 \\ 0 & 1 & 0 \\ 0 & 0 & 0 \end{bmatrix} \times 10^4$$

With these conditions, the simulation results of the closed-loop system are presented in Figures 14.4 to 14.10. The arrangement of the robot and the obstacle in the two-dimensional environment on the (x, y) plane is illustrated in multiple stages of the interactive motion of the robot with the environment in Figure 14.4. As observed in this figure, the robot's state is shown at six stages from the beginning to the end of its motion, interacting with the

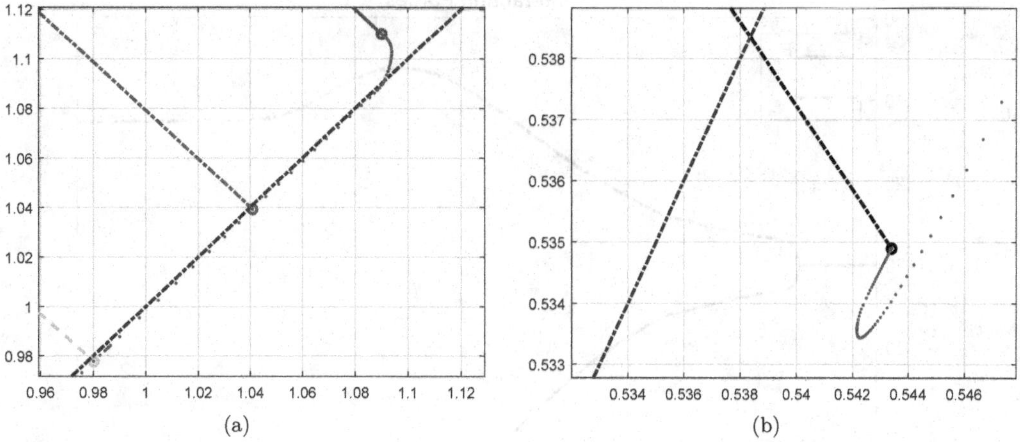

FIGURE 14.5
Magnification of the robot's end-effector interaction with the environment in the impedance control method (a) Collision state with the obstacle (b) Final state.

obstacle. The robot initiates its movement from its initial state represented by the solid line (colored in blue), and after passing through intermediate states, eventually reaches its final state depicted by the dash-dotted line (colored in black). To observe the robot's motion and interaction with the environment, the trajectory of the robot's end-effector with a dotted line (colored in red) is also shown in this figure. Considering the small impedance stiffness coefficient in comparison to the stiffness of the environment, the robot slightly penetrates into the obstacle and then continues its motion parallel to the desired path.

To better observe this impedance interaction, in Figure 14.5, the magnification of the robot's end-effector interaction with the environment is shown in the initial state in Figure 14.5(a) and in the final state of the motion in Figure 14.5(b). As seen in Figure 14.5(a), the robot begins its movement freely and without contact with the obstacle at the start of its motion. After colliding with the obstacle, it slightly penetrates inside it and continues its motion parallel to the desired path. On the other hand, Figure 14.5(b) illustrates the robot's state at the end of its motion. Given the assumption of a fixed desired motion path in the final second of the simulation, this figure effectively depicts the steady-state conditions of the robot's end-effector motion. The final adjustment of the interactive force in this situation is achieved with very small displacements of the robot's end-effector in the steady-state.

The time of contact between the robot's end-effector and the obstacle, along with the interactive forces, can be observed more clearly in Figure 14.6. In this figure, the components of the robot's interactive force vector with the environment, determined using the equation 13.40, are shown in the x, y axes. As observed in this figure, the robot collides with the obstacle at 0.38 seconds and maintains contact with the environment until the end of the simulation time. The symmetry between the interactive forces $\boldsymbol{F}_{e_y}$ and $\boldsymbol{F}_{e_x}$ is disrupted after the disturbance is applied at 1 second, and the minimum interactive force in the x-axis reaches approximately -62 N at 2.68 seconds. At the same time, the interactive force $\boldsymbol{F}_{e_y}$ also reaches its maximum value, approximately 48 N. The variations in interactive forces in the final second, when the robot's desired motion trajectory reaches its steady-state, decrease, and the dynamic behavior of the force converges to a steady-state condition. The interactive forces in this state are significantly lower compared to those in the compliance control method depicted in Figure 13.11. The reason for this is the selection

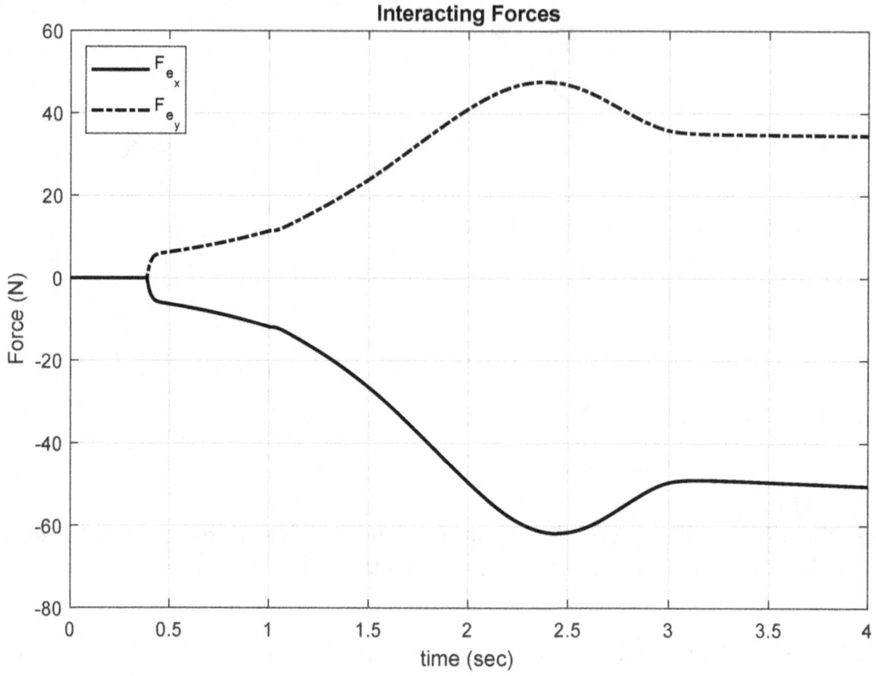

FIGURE 14.6
Interactive force of the robot due to contact with the environment in the impedance control method.

of softer impedance coefficients in this simulation.

The desired motion trajectory of the robot and the actual trajectory in the joint space are shown in Figure 14.7. In this figure, the motion of the first joint is shown in the upper graph, the motion of the second joint in the middle graph, and the motion of the third joint in the lower graph. The desired motion trajectory is represented by solid line, and the actual robot motion is indicated by the dashed line. The desired motion trajectory in the joint space is determined using inverse kinematics in a way that the robot's end-effector follows its desired trajectory along the obstacle boundary. Thus, the desired motion trajectory in the first joint changes from an angle of approximately 70 degrees to about 105 degrees during the first three seconds and remains constant in the last second. Similarly, the desired angle for the second joint changes from approximately -50 degrees to -135 degrees, while the desired angle for the third joint changes from about -65 degrees to approximately -15 degrees. In this figure, a noticeable tracking error is observed, which is significantly larger compared to motion control methods and almost similar to the compliance control method.

The torques applied to the robot's actuators in this condition are also illustrated in Figure 14.9. In this figure, the control efforts applied to the joints $\boldsymbol{\tau}$, represented by the solid line, the disturbance torques $\boldsymbol{\tau}_d$ are displayed with dashed line, and the torques due to the interactive force $\boldsymbol{\tau}_e$ is depicted with dash-dotted line. As observed in this figure, at the beginning of the trajectory, due to a significant difference between the impedance of the robot and the desired impedance, large torques are generated in the joints. In this simulation, we have set the saturation threshold for the actuators to be equal to $[100,\, 100,\, 10]^T$ $(N.m)$, preventing the application of excessively large torques to the actuators.

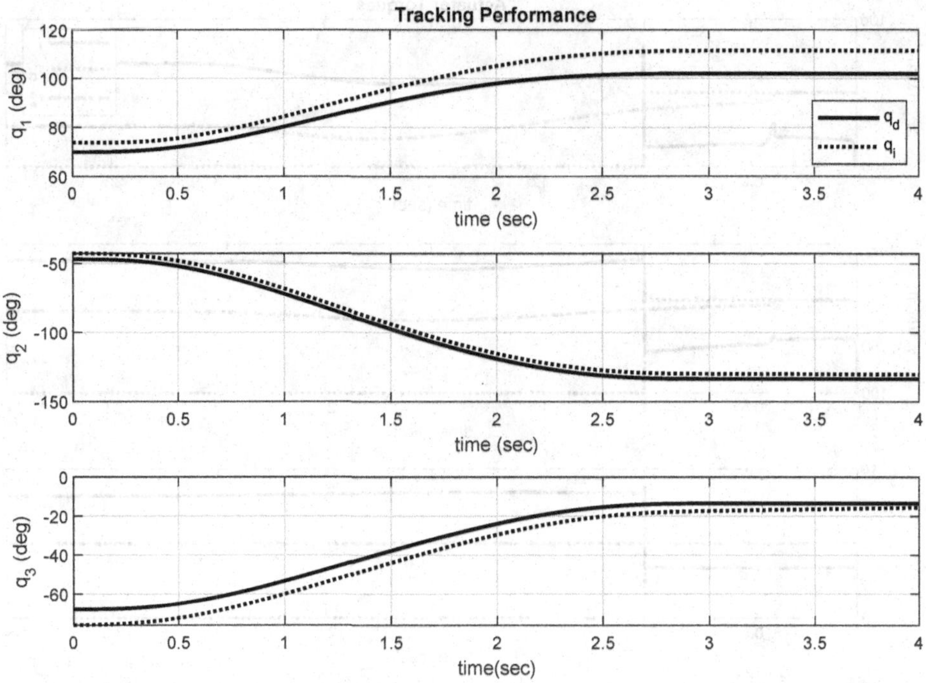

FIGURE 14.7
Desired motion trajectory of the robot and the traversed motion during interaction with the environment in the impedance control method.

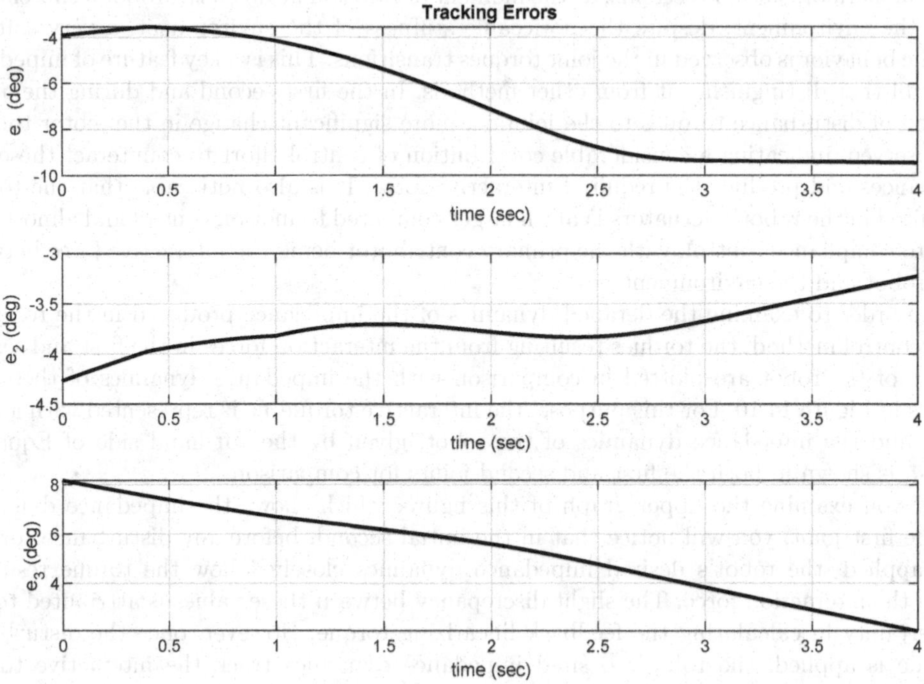

FIGURE 14.8
Tracking error of the robot in the impedance control method.

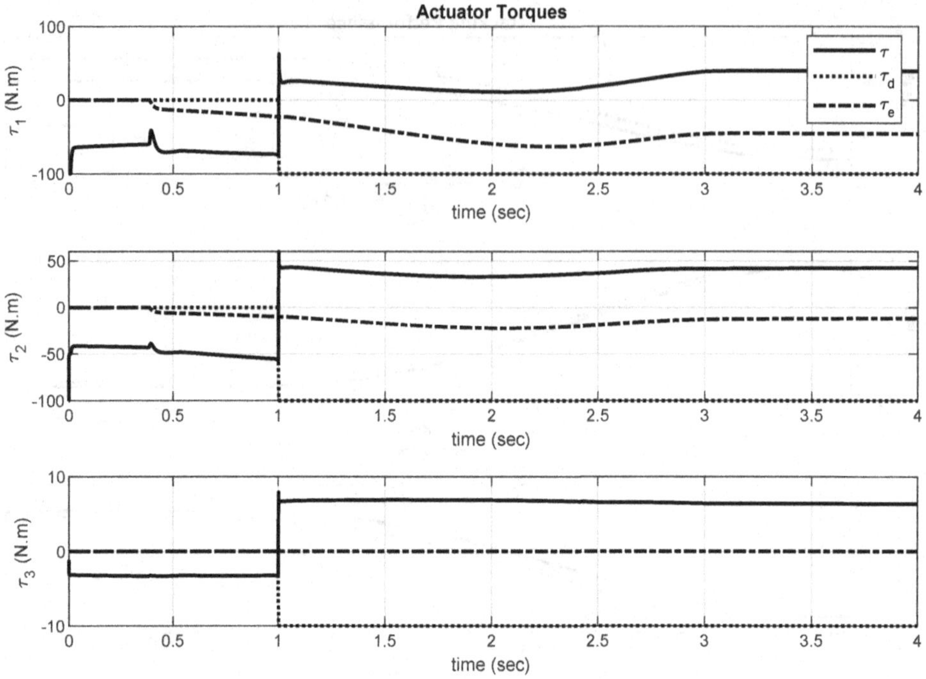

FIGURE 14.9
Control effort of the robot actuators in the impedance control method.

Furthermore, at 0.38 seconds, at the moment of collision between the robot's end-effector and the environment, despite the noticeable stiffness of the environment, a very soft dynamic behavior is observed in the joint torques transitions. This is a key feature of impedance control that distinguishes it from other methods. In the first second and during the application of disturbance torques to the joints, a more significant change in the motor torques is observed, indicating a considerable contribution of control effort to counteract these disturbances and produce the required interactive force. It is also noticeable that the torque required in the robot's actuators is much larger compared to motion control and almost similar to compliance control, with the primary contributor being the interactive force between the robot and the environment.

In order to examine the detailed dynamics of the impedance produced in the robot by this control method, the torques resulting from the interaction forces in the first and second joints of the robot are plotted in comparison with the impedance dynamics of these two joints in Figure 14.10. For this purpose, the interactive torque $\boldsymbol{\tau}_e$ is represented with a solid line, and the impedance dynamics of the robot, given by the left-hand side of Equation 14.14, is shown in both the first and second joints for comparison.

If you examine the upper graph of this figure, which shows the impedance dynamics of the first joint, you will notice that in the initial second, before any disturbance torques are applied, the robot's desired impedance dynamics closely follow the torque resulting from the interaction force. The slight discrepancy between these values is attributed to the uncertainty in calculating the feedback linearizing torque. However, once the disturbance torque is applied, the robot's desired impedance dynamics track the interactive torque with a greater error. A similar trend, but with a larger error, is observed in the lower graph, which pertains to the second joint. The increased error in this graph is due to the more significant influence of the gravity torque, which is not accurately compensated for because of model uncertainty. Nevertheless, it is clear that the requested impedance

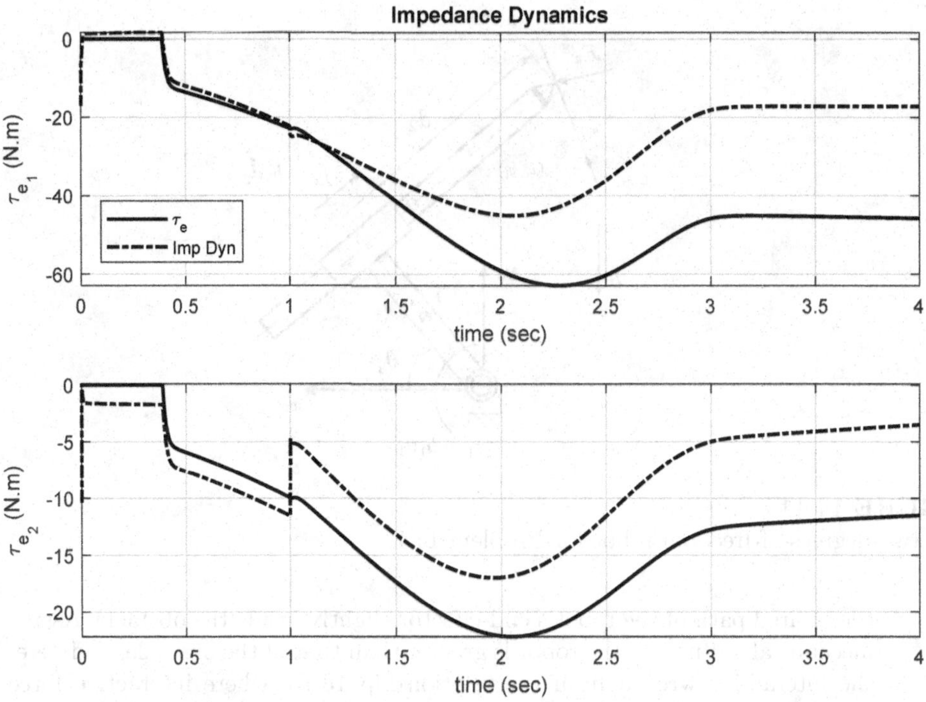

FIGURE 14.10
Comparison of interaction torque and robot impedance in interaction with the environment in the impedance control method.

dynamics are appropriately followed by the robot's motion-force interaction, resulting in the desired impedance dynamics as commanded by the controller.

In impedance control method, it is preferable for the location of the obstacle to be known, and the manner in which the robot's end-effector collide with the environment to be predetermined, based on which a desired motion path is designed. This aspect has been precisely pursued in this simulation. By defining a flat obstacle with its location known in the robot's workspace, the desired motion trajectory of the robot is initially designed in the task space. Subsequently, using inverse kinematics and the Jacobian matrix, it is mapped into the joint space of the robot. Nevertheless, whether in contact or without contact, precise tracking of the pre-determined trajectory is not the main goal of this control method; rather, the desired impedance dynamics are adjusted. As observed in Figure 14.8, in this control method, by sacrificing tracking accuracy of the predetermined motion trajectory up to about 10 degrees, we achieve the desired and a soft impedance interaction between the robot and the environment.

Problems

14.1 Consider the two-degrees-of-freedom <u>RP</u> robot shown in Figure 14.11, and the details of the compliance control method analyzed in Problem 13.1.

a) Considering a flat obstacle in the robot's workspace as shown in the figure, and

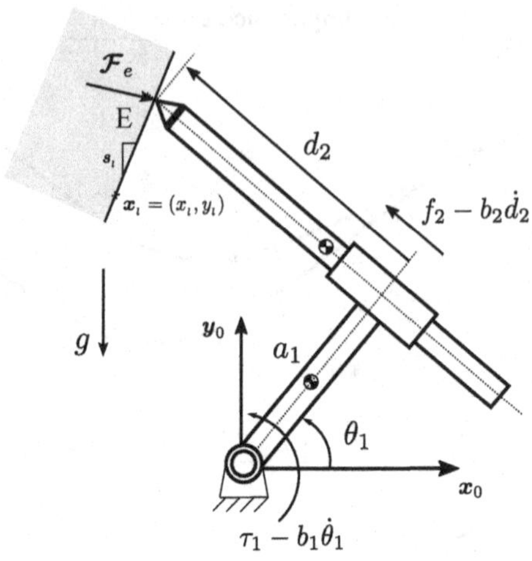

FIGURE 14.11
The two-degrees-of-freedom robot for Problem 14.1.

the desired path of the robot's end-effector slightly inside the obstacle. Assume the mechanical stiffness of the robot is greater than that of the obstacle, and determine the interactive wrench from the relationship 13.13, where interactive forces are calculated with an environment stiffness coefficient of 10^4, while interactive torques are neglected.

b) Utilize the impedance control method, taking into account a 20% uncertainty in the robot's dynamic and kinematic parameters. Adjust the desired impedance matrices such that after the robot collides with the obstacle, the robot's end-effector smoothly follows the desired trajectory. Simulate the robot's motion dynamics in this scenario, depicting the positioning of the robot and the obstacle in a two-dimensional environment (three dimensional in next Problems). Display the robot's movement in various states during its interactive trajectory with the environment. Furthermore, visualize the traversed trajectory of the robot's end-effector.

c) Determine the components of the interactive forces along the x, y axes (and z if applicable), and plot them. Identify the precise time of the end-effector's collision with the obstacle. Furthermore, calculate the tracking error of joint motions and the required control effort. Illustrate and analyze the simulation results.

d) Plot the torques resulting from interactive forces in the joints versus their impedance dynamics. Repeat this process by changing the coefficients of the impedance matrices, and analyze the effect of these changes on the interactive dynamics between the robot and the environment. Visualize and analyze the outcomes of the simulation.

14.2 Examine the <u>PRP</u> robot with three degrees of freedom depicted in Figure 14.12, along with the intricacies of the compliance control approach discussed in Problem 13.2. Iterate through Problem 14.1 for this particular robot.

14.3 Analyze the <u>PRR</u> robot featuring three degrees of freedom as shown in Figure 14.13, along with the details of the compliance control method discussed in Problem 13.3. Repeat Problem 14.1 for this specific robot.

14.4 Examine the <u>RPR</u> robot, which has three degrees of freedom as illustrated in Figure 14.14, and use the details of the compliance control method explored in Problem 13.4. Revisit Problem 14.1 for this particular robot.

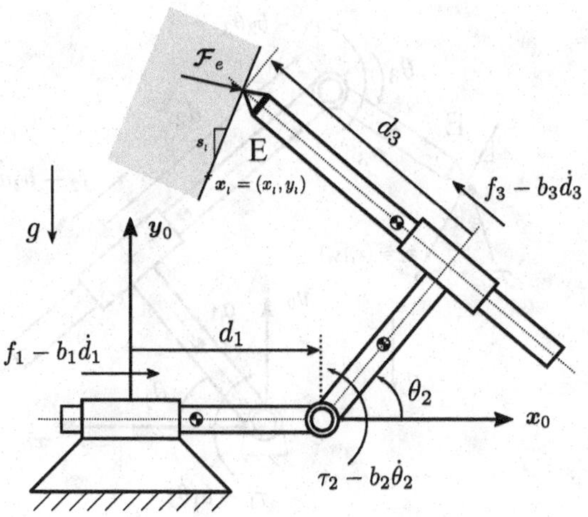

FIGURE 14.12
The three-degrees-of-freedom robot for Problem 14.2.

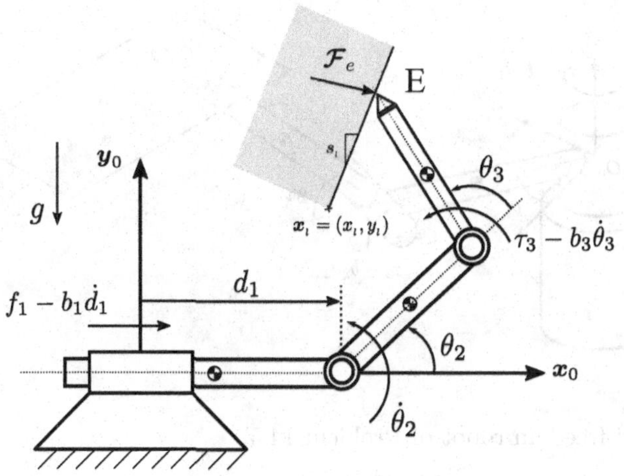

FIGURE 14.13
The three-degrees-of-freedom robot for Problem 14.3.

14.5 Investigate the RR robot, featuring two degrees of freedom as depicted in Figure 14.15, and utilize the specifics of the compliance control method discussed in Problem 13.5. Revisit Problem 14.1 for this specific robotic setup.

14.6 Consider the four-degrees-of-freedom RRRP robot shown in Figure 14.16, and the details of the compliance control method analyzed in Problem 13.6. Revisit Problem 14.1 for this specific robot.

14.7 Examine the RPRR robot with four degrees of freedom depicted in Figure 14.18, along with the intricacies of the compliance control method discussed in Problem 13.8. Revisit Problem 14.1 for this particular robotic configuration.

14.8 Explore the four-degrees-of-freedom 4R robot shown in Figure 14.18, along with the details of the compliance control method scrutinized in Problem 13.8. Revisit Problem 14.1 for this specific robotic setup.

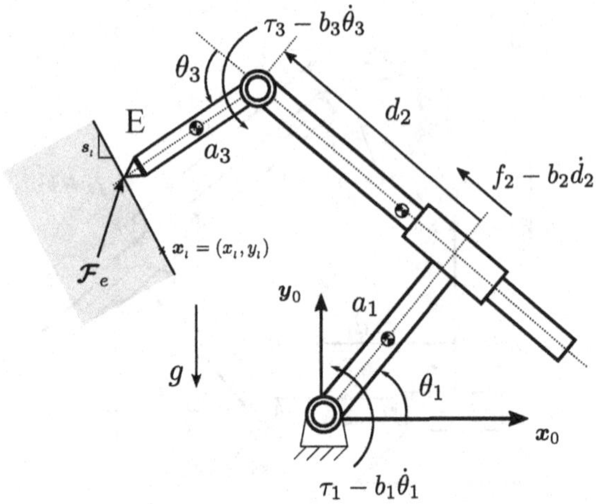

FIGURE 14.14
The three-degrees-of-freedom robot of Problem 14.4.

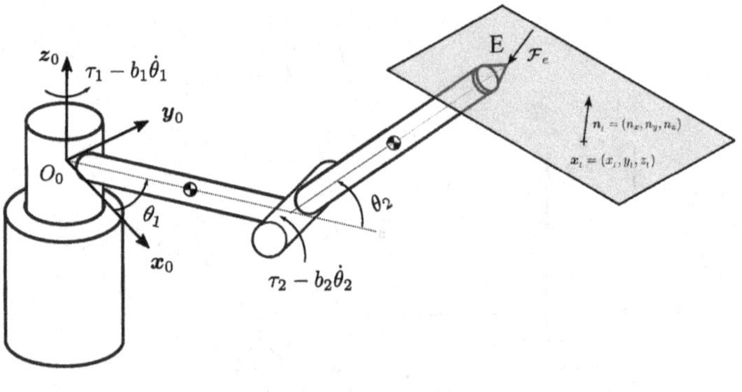

FIGURE 14.15
The two-degrees-of-freedom robot of Problem 14.5.

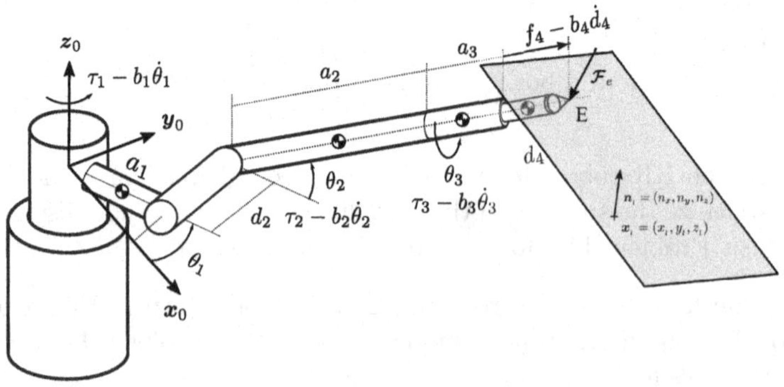

FIGURE 14.16
The four-degrees-of-freedom robot of Problem 14.6.

14.9 Examine the RPRR robot with four degrees of freedom, as depicted in Figure 14.19, and consider the specifics of the compliance control method discussed in Problem

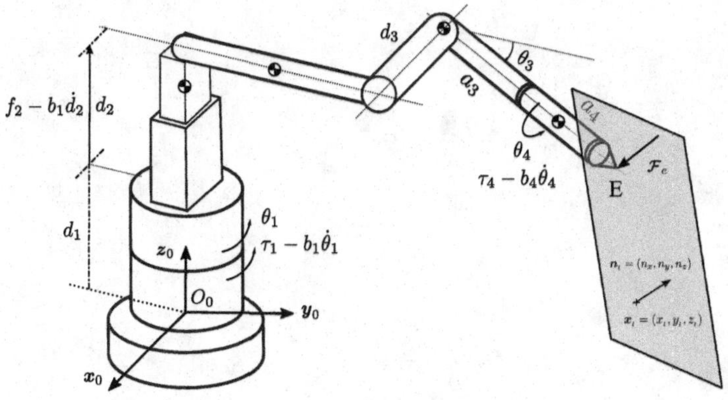

FIGURE 14.17
The four-degrees-of-freedom robot of Problem 14.7.

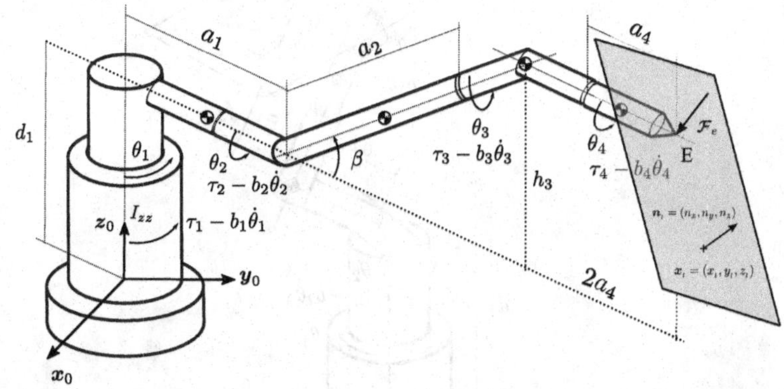

FIGURE 14.18
The four-degrees-of-freedom robot of Problem 14.8.

13.9. Revisit Problem 14.1 for this particular robotic configuration.

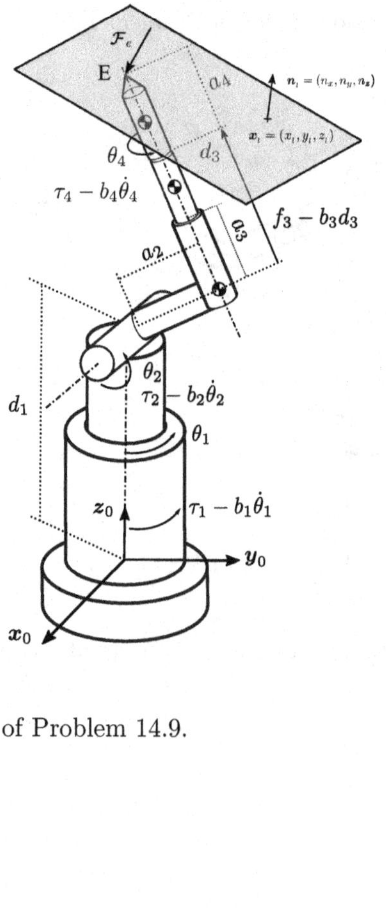

FIGURE 14.19
The four-degrees-of-freedom robot of Problem 14.9.

Part V

Appendices

A

Review on Linear Algebra

In this book, both elementary and advanced concepts in linear algebra are utilized. It is assumed that the reader is familiar with elementary concepts of linear algebra, and for this reason, their explanation is omitted from this appendix. Instead, a brief introduction is provided to introduce advanced concepts. For a more in-depth study of these concepts, interested readers are referred to the original sources in this field [188, 189].

A.1 Vectors and Matrices

A vector is an n-dimensional array of numerical elements represented in column form. In order to represent vectors in this book, lowercase letters are used, and they are written in bold format in the text, such as $\boldsymbol{x}, \boldsymbol{y}, \boldsymbol{z}$.

An n-dimensional vector $\boldsymbol{x}$ belongs to the space $\mathbb{R}^n$ if all its elements belong to the set of real numbers $\mathbb{R}$. Thus, a vector $\boldsymbol{x} \in \mathbb{R}^n$ is defined as follows.

$$\boldsymbol{x} = \begin{bmatrix} x_1 \\ x_2 \\ \vdots \\ x_n \end{bmatrix} \tag{A.1}$$

in which $x_i \in \mathbb{R}, i = 1, 2, \ldots, n$. A row vector is defined as the transpose of a column matrix.

$$\boldsymbol{x}^T = \begin{bmatrix} x_1 & x_2 & \ldots & x_n \end{bmatrix} \tag{A.2}$$

Similarly, a matrix with dimensions $m \times n$ is defined as an array consisting of m rows and n columns. To represent matrices in this book, uppercase letters are used, and they are written in bold format in the text, such as $\boldsymbol{A}, \boldsymbol{B}, \boldsymbol{C}$.

A matrix $\boldsymbol{A}$ with dimensions $m \times n$ belongs to the space $\mathbb{R}^{m\times n}$ if all its elements belong to the set of real numbers $\mathbb{R}$. Thus, a matrix $\boldsymbol{A} \in \mathbb{R}^{m\times n}$ is defined as follows.

$$\boldsymbol{A} = \begin{bmatrix} a_{11} & a_{12} & \cdots & a_{1n} \\ a_{21} & a_{22} & \cdots & a_{2n} \\ \vdots & & & \vdots \\ a_{m1} & a_{m2} & \cdots & a_{mn} \end{bmatrix} \tag{A.3}$$

In which $a_{ij} \in \mathbb{R}, i = 1, 2, \ldots, m; j = 1, 2, \ldots, n$.

A square matrix $\boldsymbol{A}$ with dimensions $n \times n$ is "upper triangular" if for $i > j$, the elements $a_{ij} = 0$; "lower triangular" if for $i < j$, the elements $a_{ij} = 0$; and "diagonal" if for $i \neq j$, the elements $a_{ij} = 0$.

DOI: 10.1201/9781003491415-A

A "diagonal matrix" with all elements equal to one is called an identity matrix and is denoted by $\boldsymbol{I}$. A diagonal matrix with all elements equal to zero is called a zero matrix and is represented by $\mathbf{0}$. The "transpose" of a matrix $m \times n$ is created by swapping its rows and columns and is denoted by $\boldsymbol{A}^T \in \mathbb{R}^{n \times m}$. A square matrix $\boldsymbol{A}$ is "symmetric" if $\boldsymbol{A}^T = \boldsymbol{A}$; "skew-symmetric" if $\boldsymbol{A}^T = -\boldsymbol{A}$; and "orthogonal" if $\boldsymbol{A}^T\boldsymbol{A} = \boldsymbol{A}\boldsymbol{A}^T = \boldsymbol{I}$.

A.2 Vectors and Matrices Arithmetic

The inner product of two vectors, $\boldsymbol{x}$ and $\boldsymbol{y}$, is a scalar value calculated as follows.

$$\boldsymbol{x} \cdot \boldsymbol{y} = \boldsymbol{x}^T\boldsymbol{y} = x_1y_1 + x_2y_2 + \cdots + x_ny_n \tag{A.4}$$

The inner product is commutative if: $\boldsymbol{x} \cdot \boldsymbol{y} = \boldsymbol{y} \cdot \boldsymbol{x}$. Two vectors, $\boldsymbol{x}$ and $\boldsymbol{y}$, are orthogonal if their inner product is equal to zero: $\boldsymbol{x} \cdot \boldsymbol{y} = 0$. The *Euclidean norm* of a vector $\boldsymbol{x} \in \mathbb{R}^n$ is denoted by $\|\boldsymbol{x}\|$, representing the geometric length of the vector, is determined as follows.

$$\|\boldsymbol{x}\| = \sqrt{\boldsymbol{x} \cdot \boldsymbol{x}} = \sqrt{x_1^2 + x_2^2 + \cdots + x_n^2} \tag{A.5}$$

A unit vector is represented by the notation $\hat{\boldsymbol{x}}$ and is determined by dividing the vector by its own Euclidean norm.

$$\hat{\boldsymbol{x}} = \frac{\boldsymbol{x}}{\|\boldsymbol{x}\|} \tag{A.6}$$

The Euclidean norm of vectors has the following properties:

$$|\boldsymbol{x} \cdot \boldsymbol{y}| \leq \|\boldsymbol{x}\| \cdot \|\boldsymbol{y}\| \tag{A.7}$$

$$\|\boldsymbol{x} + \boldsymbol{y}\| \leq \|\boldsymbol{x}\| + \|\boldsymbol{y}\| \tag{A.8}$$

Equation A.7 corresponds to the Cauchy-Schwarz inequality, while Equation A.8 is associated with the well-known triangle inequality.

The cross product of two vectors, $\boldsymbol{x}$ and $\boldsymbol{y}$, results in another vector and is calculated as follows in $\mathbb{R}^3$.

$$\boldsymbol{z} = \boldsymbol{x} \times \boldsymbol{y} = \begin{bmatrix} x_2y_3 - x_3y_2 \\ x_3y_1 - x_1y_3 \\ x_1y_2 - x_2y_1 \end{bmatrix} \tag{A.9}$$

The properties of inner and cross products of vectors can be listed with the following relations.

$$\boldsymbol{x} \times \boldsymbol{x} = \mathbf{0} \tag{A.10}$$

$$\boldsymbol{x} \times \boldsymbol{y} = -\boldsymbol{y} \times \boldsymbol{x} \tag{A.11}$$

$$\boldsymbol{x} \times (\boldsymbol{y} + \boldsymbol{z}) = \boldsymbol{x} \times \boldsymbol{y} + \boldsymbol{x} \times \boldsymbol{z} \tag{A.12}$$

$$\alpha(\boldsymbol{x} \times \boldsymbol{y}) = (\alpha\boldsymbol{x}) \times \boldsymbol{y} = \boldsymbol{x} \times (\alpha\boldsymbol{y}) \tag{A.13}$$

$$\boldsymbol{x} \cdot (\boldsymbol{y} \times \boldsymbol{z}) = \boldsymbol{y} \cdot (\boldsymbol{z} \times \boldsymbol{x}) = \boldsymbol{z} \cdot (\boldsymbol{x} \times \boldsymbol{y}) \tag{A.14}$$

$$(\boldsymbol{x} \cdot \boldsymbol{y})\boldsymbol{z} = (\boldsymbol{x}^T\boldsymbol{y})\boldsymbol{z} = \boldsymbol{z}(\boldsymbol{x}^T\boldsymbol{y}) = (\boldsymbol{z}\boldsymbol{x}^T)\boldsymbol{y} = \boldsymbol{z}\boldsymbol{x}^T\boldsymbol{y} \tag{A.15}$$

Equation A.14 describes the property of the triple product, which is widely used in determining the dynamics equations in robotics. The cross product of two vectors, $\boldsymbol{x} \times \boldsymbol{y}$, can be expressed using the product of a skew-symmetric matrix formed by the elements of the vector $\boldsymbol{x}$, denoted by $\boldsymbol{x}_\times$ or $\boldsymbol{x}^\times$, as given in

$$\boldsymbol{x}_\times \text{ or } \boldsymbol{x}^\times = \begin{bmatrix} 0 & -x_3 & x_2 \\ x_3 & 0 & -x_1 \\ -x_2 & x_1 & 0 \end{bmatrix} \tag{A.16}$$

by vector $\boldsymbol{y}$, as follows.

$$\boldsymbol{x} \times \boldsymbol{y} = \boldsymbol{x}^\times \boldsymbol{y} = -\boldsymbol{y}^\times \boldsymbol{x} \tag{A.17}$$

A set of vectors $\boldsymbol{x}_1, \boldsymbol{x}_2, \ldots, \boldsymbol{x}_n$ is linearly independent, if and only if, the linear combination of these vectors equals zero only when all the coefficients are zero.

$$\sum_{i=1}^{n} \alpha_i \boldsymbol{x}_i = \boldsymbol{0} \Rightarrow \alpha_i = 0 \ \ \forall i \tag{A.18}$$

The rank of a matrix $\boldsymbol{A}$, denoted by rank($\boldsymbol{A}$), is the maximum number of columns or rows of this matrix that are linearly independent. With this definition, the rank of a matrix with dimensions $n \times m$ will never be greater than $\min(m, n)$. The inverse of a square matrix $\boldsymbol{A}$ with dimensions $n \times n$, if it exists, is a matrix of the same dimensions that satisfies the following relationship.

$$\boldsymbol{A}\boldsymbol{A}^{-1} = \boldsymbol{A}^{-1}\boldsymbol{A} = \boldsymbol{I} \tag{A.19}$$

In this equation, $\boldsymbol{I}$ denotes the identity matrix. The inverse of a square matrix, if it exists, could be uniquely computable, if and only if, the matrix is full rank. The inverse of a matrix has the following property.

$$(\boldsymbol{A}^{-1})^{-1} = \boldsymbol{A} \tag{A.20}$$

$$(\boldsymbol{A}\boldsymbol{B})^{-1} = \boldsymbol{B}^{-1}\boldsymbol{A}^{-1} \tag{A.21}$$

The inverse of an orthogonal matrix is equal to its transpose: $\boldsymbol{A}^{-1} = \boldsymbol{A}^T$. If matrices $\boldsymbol{A}$ and $\boldsymbol{C}$ are invertible square matrices, and matrices $\boldsymbol{B}$ and $\boldsymbol{D}$ have appropriate dimensions for the following multiplications, then

$$(\boldsymbol{A} + \boldsymbol{B}\boldsymbol{C}\boldsymbol{D})^{-1} = \boldsymbol{A}^{-1} - \boldsymbol{A}^{-1}\boldsymbol{B}(\boldsymbol{D}\boldsymbol{A}^{-1}\boldsymbol{B} + \boldsymbol{C}^{-1})^{-1}\boldsymbol{D}\boldsymbol{A}^{-1} \tag{A.22}$$

In this equation, the matrix $(\boldsymbol{D}\boldsymbol{A}^{-1}\boldsymbol{B} + \boldsymbol{C}^{-1})$ must be invertible. Furthermore, for a matrix that is partitioned into blocks, the following relationships hold.

$$\begin{bmatrix} \boldsymbol{A} & \boldsymbol{B} \\ \boldsymbol{C} & \boldsymbol{D} \end{bmatrix}^{-1} = \begin{bmatrix} \boldsymbol{A}^{-1} + \boldsymbol{A}^{-1}\boldsymbol{B}\Delta^{-1}\boldsymbol{C}\boldsymbol{A}^{-1} & -\boldsymbol{A}^{-1}\boldsymbol{B}\Delta^{-1} \\ -\Delta^{-1}\boldsymbol{C}\boldsymbol{A}^{-1} & \Delta^{-1} \end{bmatrix} \tag{A.23}$$

$$= \begin{bmatrix} \Omega^{-1} & -\Omega^{-1}\boldsymbol{B}\boldsymbol{D}^{-1} \\ -\boldsymbol{D}^{-1}\boldsymbol{C}\Omega^{-1} & \boldsymbol{D}^{-1} + \boldsymbol{D}^{-1}\boldsymbol{C}\Omega^{-1}\boldsymbol{B}\boldsymbol{D}^{-1} \end{bmatrix} \tag{A.24}$$

In which $\Delta = \boldsymbol{D} - \boldsymbol{C}\boldsymbol{A}^{-1}\boldsymbol{B}$ and $\Omega = \boldsymbol{A} - \boldsymbol{B}\boldsymbol{D}^{-1}\boldsymbol{C}$ must be invertible. In the special case where matrices $\boldsymbol{B}$ or $\boldsymbol{C}$ are zero matrices, the inversion of this matrix simplifies as follows. This relationship is known as the "Matrix Inversion Lemma."

$$\begin{bmatrix} \boldsymbol{A} & 0 \\ \boldsymbol{C} & \boldsymbol{D} \end{bmatrix}^{-1} = \begin{bmatrix} \boldsymbol{A}^{-1} & 0 \\ -\boldsymbol{D}^{-1}\boldsymbol{C}\boldsymbol{A}^{-1} & \boldsymbol{D}^{-1} \end{bmatrix} \tag{A.25}$$

$$\begin{bmatrix} \boldsymbol{A} & \boldsymbol{B} \\ 0 & \boldsymbol{D} \end{bmatrix}^{-1} = \begin{bmatrix} \boldsymbol{A}^{-1} & -\boldsymbol{A}^{-1}\boldsymbol{B}\boldsymbol{D}^{-1} \\ 0 & \boldsymbol{D}^{-1} \end{bmatrix} \tag{A.26}$$

The null space of matrix $\boldsymbol{A}$, denoted by $\mathcal{N}$, is defined as follows.

$$\mathcal{N}(\boldsymbol{A}) = \{\boldsymbol{x} \in \mathbb{R}^n : \boldsymbol{A}\boldsymbol{x} = \boldsymbol{0}\} \tag{A.27}$$

If the matrix $\boldsymbol{A}$ has dimensions $n \times n$, then

$$\text{rank}(\boldsymbol{A}) + \text{dim}\mathcal{N}(\boldsymbol{A}) = n \tag{A.28}$$

Thus, a matrix is invertible, if and only if, its null space consists solely of the zero vector.

$$\text{dim}\mathcal{N}(\boldsymbol{A}) = 0$$

The induced norm of a square matrix is defined as follows.

$$\|\boldsymbol{A}\| = \sup_{\|\boldsymbol{x}\| \neq 0} \frac{\|\boldsymbol{A}\boldsymbol{x}\|}{\|\boldsymbol{x}\|} \tag{A.29}$$

With this definition, it can be shown that:

$$\|\boldsymbol{A}\boldsymbol{x}\| \leq \|\boldsymbol{A}\| \, \|\boldsymbol{x}\| \tag{A.30}$$

$$\|\boldsymbol{A}\boldsymbol{B}\| \leq \|\boldsymbol{A}\| \, \|\boldsymbol{B}\| \tag{A.31}$$

A.3 Eigenvalues and Singular Values

Eigenvalues of a square matrix $\boldsymbol{A}$ are obtained from the solution of the characteristic equation of the system, which is defined as follows.

$$\det(\lambda \boldsymbol{I} - \boldsymbol{A}) = 0 \tag{A.32}$$

The solutions to this equation, $\lambda_1, \lambda_2, \ldots, \lambda_n$, determine the eigenvalues of the matrix $\boldsymbol{A}$. The eigenvalues of the matrix $\boldsymbol{A}$ and its transpose $\boldsymbol{A}^T$ are equal. The vectors that satisfy the equation

$$(\lambda_i \boldsymbol{I} - \boldsymbol{A})\boldsymbol{v}_i = \boldsymbol{0} \tag{A.33}$$

are called the eigenvectors of the matrix $\boldsymbol{A}$. If the eigenvalues of matrix $\boldsymbol{A}$ are distinct, then there exist n independent eigenvectors that satisfy equation A.33. The matrix $\boldsymbol{T}$, composed of the eigenvectors of matrix $\boldsymbol{A}$ as its columns, is referred to as the homogeneous transformation matrix, which is used to convert matrix $\boldsymbol{A}$ into diagonal form. Using this matrix, the converted matrix $\boldsymbol{\Lambda} = \boldsymbol{T}^{-1}\boldsymbol{A}\boldsymbol{T}$ will be diagonal, and its diagonal elements will be the eigenvalues of matrix $\boldsymbol{A}$.

Eigenvalues and eigenvectors are not defined for non-square matrices. For this type of matrices, the concept of singular values is utilized. Consider a matrix $\boldsymbol{A} \in \mathbb{R}^{m \times n}$ where $m < n$. In this case, a square matrix with dimensions $m \times m$ can be formed using the product $\boldsymbol{A}\boldsymbol{A}^T$. The singular values of matrix $\boldsymbol{A}$ denoted by σ_i are obtained from the square root of the eigenvalues of the square matrix $\boldsymbol{A}\boldsymbol{A}^T$.

$$\sigma_i(\boldsymbol{A}) = \sqrt{\lambda_i(\boldsymbol{A}\boldsymbol{A}^T)} \tag{A.34}$$

In many applications, there is a need to decompose a matrix based on its singular values. In singular value decomposition (SVD), a non-square matrix $\boldsymbol{A}$ is decomposed as follows.

$$\boldsymbol{A} = \boldsymbol{U}\boldsymbol{\Sigma}\boldsymbol{V}^T \tag{A.35}$$

In which the orthogonal matrices $\boldsymbol{U} \in \mathbb{R}^{m \times m}$ and $\boldsymbol{V} \in \mathbb{R}^{n \times n}$ are defined as follows.

$$\boldsymbol{U} = \begin{bmatrix} \boldsymbol{u}_1 & \boldsymbol{u}_2 & \cdots & \boldsymbol{u}_m \end{bmatrix}, \boldsymbol{V} = \begin{bmatrix} \boldsymbol{v}_1 & \boldsymbol{v}_2 & \cdots & \boldsymbol{v}_n \end{bmatrix} \tag{A.36}$$

Furthermore, the matrix $\boldsymbol{\Sigma} \in \mathbb{R}^{m \times n}$ is calculated based on the singular values of the matrix as follows.

$$\boldsymbol{\Sigma} = \left[\begin{array}{cccc|ccc} \sigma_1 & 0 & \cdots & 0 & 0 & \cdots & 0 \\ 0 & \sigma_2 & & 0 & 0 & \cdots & 0 \\ \vdots & & & \vdots & \vdots & & \vdots \\ 0 & 0 & \cdots & \sigma_m & 0 & \cdots & 0 \end{array}\right] \tag{A.37}$$

Thus, to decompose the singular value matrix $\boldsymbol{A}$, first, the singular values are calculated using equation A.34 as follows.

$$\det(\boldsymbol{A}\boldsymbol{A}^T - \lambda_i \boldsymbol{I}) = 0; \;\; \sigma_i = \sqrt{\lambda_i} \tag{A.38}$$

Next, using the obtained singular values, eigenvectors $\boldsymbol{u}_1, \boldsymbol{u}_2, \ldots, \boldsymbol{u}_m$ are calculated, which satisfy the following equation.

$$(\boldsymbol{A}\boldsymbol{A}^T - \sigma_i^2 \boldsymbol{I})\boldsymbol{u}_i = \boldsymbol{0} \tag{A.39}$$

Now, the orthogonal matrix $\boldsymbol{U}$ is formed using the determined eigenvectors as $\boldsymbol{U} = \begin{bmatrix} \boldsymbol{u}_1 & \boldsymbol{u}_2 & \cdots & \boldsymbol{u}_m \end{bmatrix}$, and equation A.39 can be rewritten as follows.

$$\boldsymbol{A}\boldsymbol{A}^T \boldsymbol{U} = \boldsymbol{U}\boldsymbol{\Sigma}_m^2 \tag{A.40}$$

In this equation, $\boldsymbol{\Sigma}_m$ is the diagonal part of the matrix $\boldsymbol{\Sigma}$, which is obtained from:

$$\boldsymbol{\Sigma}_m = \begin{bmatrix} \sigma_1 & 0 & \cdots & 0 \\ 0 & \sigma_2 & & 0 \\ \vdots & & & \vdots \\ 0 & 0 & \cdots & \sigma_m \end{bmatrix} \tag{A.41}$$

With this equation, one can obtain the m columns of matrix $\boldsymbol{V}$, denoted by $\boldsymbol{V}_m$, by:

$$\boldsymbol{V}_m = \boldsymbol{A}^T \boldsymbol{U} \boldsymbol{\Sigma}_m^{-1} \tag{A.42}$$

The remaining $n - m$ columns of matrix $\boldsymbol{V}$, denoted by $\boldsymbol{V}_{n-m}$, can be determined from the orthogonal matrix $\boldsymbol{V} = [\boldsymbol{V}_m \mid \boldsymbol{V}_{n-m}]$, although this part of the matrix directly does not affect the singular value decomposition. The singular value decomposition of matrices is widely utilized in various applications, particularly in solving inverse problems for matrices used in robotics.

A.4 Pseudo-Inverse

The pseudo-inverse of matrices is extensively used in the context of solving linear inverse problems. The solution to the matrix equation $\boldsymbol{A}\boldsymbol{x} = \boldsymbol{y}$ can be easily obtained using the pseudo-inverse of matrix $\boldsymbol{A}$: $\boldsymbol{x} = \boldsymbol{A}^{\dagger}\boldsymbol{y}$. This holds true even when matrix $\boldsymbol{A}$ is not square, as long as it has full rank. The pseudo-inverse of matrix $\boldsymbol{A}$ generalizes this relationship to cases where matrix $\boldsymbol{A}$ is not square or is not full rank. In robotics, the pseudo-inverse of the Jacobian matrix is commonly used in redundancy resolution and robot calibration problems.

Theorem A.1 (Pseudo-Inverse) *Consider a real, full-rank matrix $\boldsymbol{A} \in \mathbb{R}^{m\times n}$. The pseudo-inverse of matrix $\boldsymbol{A}$, denoted by $\boldsymbol{A}^\dagger$, is uniquely defined by a matrix that satisfies the following four conditions.*

$$\boldsymbol{A}\boldsymbol{A}^\dagger\boldsymbol{A} = \boldsymbol{A} \tag{A.43}$$
$$\boldsymbol{A}^\dagger\boldsymbol{A}\boldsymbol{A}^\dagger = \boldsymbol{A}^\dagger \tag{A.44}$$
$$(\boldsymbol{A}\boldsymbol{A}^\dagger)^T = \boldsymbol{A}\boldsymbol{A}^\dagger \tag{A.45}$$
$$(\boldsymbol{A}^\dagger\boldsymbol{A})^T = \boldsymbol{A}^\dagger\boldsymbol{A} \tag{A.46}$$

To explore the proof of this theorem, please refer to [190].

A.4.1 Pseudo-Inverse Properties

The pseudo-inverse possesses the following properties:

$$(\boldsymbol{A}^\dagger)^\dagger = \boldsymbol{A}$$

$$(\boldsymbol{A}^T)^\dagger = \boldsymbol{A}^{\dagger^T},\ \ (\boldsymbol{A}\boldsymbol{A}^T)^\dagger = \boldsymbol{A}^{\dagger^T}\boldsymbol{A}^\dagger$$

$$\boldsymbol{A}^\dagger = (\boldsymbol{A}^T\boldsymbol{A})^\dagger\boldsymbol{A}^T = \boldsymbol{A}^T(\boldsymbol{A}\boldsymbol{A}^T)^\dagger$$

$$\boldsymbol{A}^T\boldsymbol{A}\boldsymbol{A}^\dagger = \boldsymbol{A}^\dagger\boldsymbol{A}\boldsymbol{A}^T = \boldsymbol{A}^T$$

- Suppose matrix $\boldsymbol{A}$ has dimensions $m \times n$ and matrix $\boldsymbol{B}$ has dimensions $n \times p$. Then, in general, $(\boldsymbol{A}\boldsymbol{B})^\dagger$ is not equal to $\boldsymbol{B}^\dagger\boldsymbol{A}^\dagger$. However, if the ranks of the two matrices are equal and given by $\text{rank}(\boldsymbol{A}) = \text{rank}(\boldsymbol{B}) = n$, then the equality $(\boldsymbol{A}\boldsymbol{B})^\dagger = \boldsymbol{B}^\dagger\boldsymbol{A}^\dagger$ holds.

- In a matrix $\boldsymbol{A}$ with dimensions $m \times n$, if $m < n$ and $\text{rank}(\boldsymbol{A}) = m$, then the matrix $\boldsymbol{A}\boldsymbol{A}^T$ is an invertible matrix with dimensions $m \times m$. In this case, the pseudo-inverse is obtained by the following equation

$$\boldsymbol{A}^\dagger = \boldsymbol{A}^T(\boldsymbol{A}\boldsymbol{A}^T)^{-1} \tag{A.47}$$

  This pseudo-inverse is referred to as the "Right Inverse."

- In a matrix $\boldsymbol{A}$ with dimensions $m \times n$, if $m > n$ and $\text{rank}(\boldsymbol{A}) = n$, then the matrix $\boldsymbol{A}^T\boldsymbol{A}$ is an invertible matrix with dimensions $n \times n$. In this case, the pseudo-inverse is obtained by the following relationship:

$$\boldsymbol{A}^\dagger = (\boldsymbol{A}^T\boldsymbol{A})^{-1}\boldsymbol{A}^T \tag{A.48}$$

  This pseudo-inverse is referred to as the "Left Inverse."

- If we express the singular value decomposition of the matrix $\boldsymbol{A}$ as $\boldsymbol{A} = \boldsymbol{U}\boldsymbol{\Sigma}\boldsymbol{V}^T$ (Equation A.37), then the pseudo-inverse $\boldsymbol{A}^\dagger \in \mathbb{R}^{m\times m}$ can be obtained using the following equation.

$$\boldsymbol{A}^\dagger = \boldsymbol{V}\ \left[\boldsymbol{\Sigma}_m^{-1}\ \mid\ \boldsymbol{0}\right]\ \boldsymbol{U}^T \tag{A.49}$$

A.4.2 Linear Inverse Problem

In this section, we use the pseudo-inverse to determine the solution to two inverse linear problems. Suppose we have a set of linear equations represented by the matrix equation $\boldsymbol{A}\boldsymbol{x} = \boldsymbol{y}$, where $\boldsymbol{A}$ is an $m \times n$ matrix. In this case, the solution to this system of linear equations for an unknown vector $\boldsymbol{x}$ with dimensions $n \times 1$ is the vector $\boldsymbol{y}$ with dimensions

$m \times 1$. Now, if the matrix $\boldsymbol{A}$ is square ($m = n$) and non-singular, then with rank($\boldsymbol{A}$) $= n$, the solution to the inverse problem can be determined using the inverse matrix $\boldsymbol{A}^{-1}$, and it is given by $\boldsymbol{x} = \boldsymbol{A}^{-1}\boldsymbol{y}$.

In the case where the matrix $\boldsymbol{A}$ is non-square, with $m > n$, where the rank of matrix $\boldsymbol{A}$ is n, there is no direct inverse for the matrix $\boldsymbol{A}$. In this scenario, the system of linear equations is over-determined, having more equations than unknowns. In general, such a system of equations does not have a solution that satisfies all equations, and the use of the pseudo-inverse matrix $\boldsymbol{A}^\dagger$ provides a solution that minimizes the Euclidean norm of the error vector $\boldsymbol{e} = \boldsymbol{A}\boldsymbol{x} - \boldsymbol{y}$.

$$V = \|\boldsymbol{A}\boldsymbol{x} - \boldsymbol{y}\|^2 = \|\boldsymbol{e}\|^2 = \boldsymbol{e}^T\boldsymbol{e} \tag{A.50}$$

To prove this claim, since in the optimal response, the derivative of the cost function must be equal to zero $\partial V/\partial \boldsymbol{x} = \boldsymbol{0}$, compute:

$$\begin{aligned} \frac{\partial(\boldsymbol{e}^T\boldsymbol{e})}{\partial \boldsymbol{x}} &= 2\boldsymbol{A}^T\boldsymbol{e} = 0 \\ &= 2\boldsymbol{A}^T(\boldsymbol{A}\boldsymbol{x} - \boldsymbol{y}) = 0 \end{aligned} \tag{A.51}$$

As a result, in the optimal response, we will have:

$$\boldsymbol{A}^T\boldsymbol{A}\boldsymbol{x} = \boldsymbol{A}^T\boldsymbol{y} \tag{A.52}$$

Since $\boldsymbol{A}^T\boldsymbol{A}$ is a square matrix with dimensions $n \times n$, it is invertible. Therefore, the solution to the equation A.52 can be obtained as follows.

$$\boldsymbol{x}_{opt} = (\boldsymbol{A}^T\boldsymbol{A})^{-1}\boldsymbol{A}^T\boldsymbol{y} \tag{A.53}$$

As seen, the optimal solution to the inverse problem $\boldsymbol{A}\boldsymbol{x} = \boldsymbol{y}$ can be determined by the left inverse of the matrix $\boldsymbol{A}$, as previously defined in equation A.48.

Now, if the matrix $\boldsymbol{A}$ is non-square with $m < n$, and the rank of the matrix $\boldsymbol{A}$ is equal to m, in this case the direct inverse of the matrix $\boldsymbol{A}$ does not exist, and the number of equations m is less than the number of unknowns n. This system of linear equations is under-determined. In this situation, there are countless solutions to the system of linear equations, and one can choose a solution that optimizes a desired criterion. In general, it can be shown that the solution to this category of algebraic equations is obtained from the following equation.

$$\boldsymbol{x} = \boldsymbol{A}^\dagger\boldsymbol{y} + (\boldsymbol{I} - \boldsymbol{A}^\dagger\boldsymbol{A})\boldsymbol{z} \tag{A.54}$$

In which $\boldsymbol{z}$ is an arbitrary vector with a dimension of n, and $\boldsymbol{A}^\dagger$ is the right inverse of the matrix $\boldsymbol{A}$. According to the following calculations, this solution satisfies the system of linear equations, and therefore, it is considered as a general solution for this system.

$$\begin{aligned} \boldsymbol{A}\boldsymbol{x} &= \boldsymbol{A}\boldsymbol{A}^\dagger\boldsymbol{y} + \boldsymbol{A}(\boldsymbol{I} - \boldsymbol{A}^\dagger\boldsymbol{A})\boldsymbol{z} \\ &= \boldsymbol{A}\boldsymbol{A}^T(\boldsymbol{A}\boldsymbol{A}^T)^{-1}\boldsymbol{y} + (\boldsymbol{A} - \boldsymbol{A}\boldsymbol{A}^T(\boldsymbol{A}\boldsymbol{A}^T)^{-1}\boldsymbol{A})\boldsymbol{z} = \boldsymbol{y} + (\boldsymbol{A} - \boldsymbol{A})\boldsymbol{z} = \boldsymbol{y} \end{aligned}$$

Moreover, it can be shown that the base solution of this equation, determined by the relation $\boldsymbol{x}_o = \boldsymbol{A}^\dagger\boldsymbol{y}$, minimizes the norm $\|\boldsymbol{x}\|$, among countless solutions to this system of linear equations. The proof of this claim can be derived by considering that other solutions to this system of equations add an expression to the optimal base solution, which is orthogonal to the base solution, thereby increasing the length of the resulting vector. To prove the orthogonality of these two vectors, it can be shown that their dot product is equal to zero.

$$\begin{aligned} (\boldsymbol{A}^\dagger\boldsymbol{y})^T(\boldsymbol{I} - \boldsymbol{A}^\dagger\boldsymbol{A}) &= \boldsymbol{y}^T\boldsymbol{A}^{\dagger T}(\boldsymbol{I} - \boldsymbol{A}^\dagger\boldsymbol{A}) \\ &= \boldsymbol{y}^T\left[(\boldsymbol{I} - \boldsymbol{A}^\dagger\boldsymbol{A})\boldsymbol{A}^\dagger\right]^T\boldsymbol{z} \\ &= \boldsymbol{y}^T(\boldsymbol{A}^\dagger - \boldsymbol{A}^\dagger\boldsymbol{A}\boldsymbol{A}^\dagger)^T\boldsymbol{z} = 0 \end{aligned}$$

In these calculations, the symmetry property of the matrix $\boldsymbol{I} - \boldsymbol{A}^{\dagger}\boldsymbol{A}$ has been utilized to simplify the first equation, and equation A.44 has been used to simplify the final equation. In this way, the right inverse is employed to solve under-determined inverse problems.

A.5 Kronecker Product

The Kronecker product, denoted by $\otimes$, defines an operation for multiplying two matrices of arbitrary dimensions, resulting in a block matrix. If matrix $\boldsymbol{A}$ has dimensions $m \times n$ and matrix $\boldsymbol{B}$ has dimensions $p \times q$, then the Kronecker product of these two matrices, denoted as $\boldsymbol{A} \otimes \boldsymbol{B}$, is a matrix with dimensions $mp \times nq$. It is formed by creating a block matrix from the element-wise multiplication of the first matrix by the second matrix.

$$\boldsymbol{A} \otimes \boldsymbol{B} = \begin{bmatrix} a_{11}\boldsymbol{B} & a_{12}\boldsymbol{B} & \cdots & a_{1n}\boldsymbol{B} \\ a_{21}\boldsymbol{B} & a_{22}\boldsymbol{B} & \cdots & a_{2n}\boldsymbol{B} \\ \vdots & \vdots & & \vdots \\ a_{m1}\boldsymbol{B} & a_{m2}\boldsymbol{B} & \cdots & a_{mn}\boldsymbol{B} \end{bmatrix}. \tag{A.55}$$

The Kronecker product possesses properties of bilinearity and associativity:

$$\alpha(\boldsymbol{A} \otimes \boldsymbol{B}) = (\alpha\boldsymbol{A}) \otimes \boldsymbol{B} = \boldsymbol{A} \otimes (\alpha\boldsymbol{B}) \ ; \ \alpha \in R \tag{A.56}$$
$$\boldsymbol{A} \otimes (\boldsymbol{B} + \boldsymbol{C}) = \boldsymbol{A} \otimes \boldsymbol{B} + \boldsymbol{A} \otimes \boldsymbol{C} \tag{A.57}$$
$$(\boldsymbol{A} + \boldsymbol{B}) \otimes \boldsymbol{C} = \boldsymbol{A} \otimes \boldsymbol{C} + \boldsymbol{B} \otimes \boldsymbol{C} \tag{A.58}$$
$$(\boldsymbol{A} \otimes \boldsymbol{B}) \otimes \boldsymbol{C} = \boldsymbol{A} \otimes (\boldsymbol{B} \otimes \boldsymbol{C}) \tag{A.59}$$

Furthermore, for matrices with appropriate dimensions, their mixed product has the following property.

$$(\boldsymbol{A} \otimes \boldsymbol{B})(\boldsymbol{C} \otimes \boldsymbol{D}) = \boldsymbol{AC} \otimes \boldsymbol{BD} \tag{A.60}$$

The matrix $(\boldsymbol{A} \otimes \boldsymbol{B})$ is invertible, if and only if, matrices $\boldsymbol{A}$ and $\boldsymbol{B}$ are invertible. In this case, the inverse of this matrix is calculated as follows.

$$(\boldsymbol{A} \otimes \boldsymbol{B})^{-1} = \boldsymbol{A}^{-1} \otimes \boldsymbol{B}^{-1} \tag{A.61}$$

Also, the transpose in this product has the property of distributivity.

$$(\boldsymbol{A} \otimes \boldsymbol{B})^T = \boldsymbol{A}^T \otimes \boldsymbol{B}^T \tag{A.62}$$

Another property of the Kronecker product used in simplifying dynamic equations is as follows. With the definition of the vector $\boldsymbol{x} \in \mathbb{R}^n$, since the Kronecker product is not commutative in general, i.e., $\boldsymbol{I}_{n\times n} \otimes \boldsymbol{x} \neq \boldsymbol{x} \otimes \boldsymbol{I}_{n\times n}$, in which $\boldsymbol{I}_{n\times n}$ denotes the identity matrix of size $n \times n$. However, it can be shown that the following relationship holds for any arbitrary vector $\boldsymbol{x} \in \mathbb{R}^n$.

$$(\boldsymbol{I}_{n\times n} \otimes \boldsymbol{x})\, \boldsymbol{x} = (\boldsymbol{x} \otimes \boldsymbol{I}_{n\times n})\, \boldsymbol{x} \tag{A.63}$$

Let us define the time derivative and partial derivative of matrices using the Kronecker product and investigate how the Kronecker product is used in the gradients of composite matrices. Consider the matrix $\boldsymbol{A}(\boldsymbol{q})$, according to the convention, the partial derivative of this matrix with respect to the variable $\boldsymbol{q}$ can be written in the form of a column-block matrix as follows.

$$\frac{\partial \boldsymbol{A}}{\partial \boldsymbol{q}} = \begin{bmatrix} \frac{\partial \boldsymbol{A}}{\partial q_1} \\ \vdots \\ \frac{\partial \boldsymbol{A}}{\partial q_n} \end{bmatrix} \tag{A.64}$$

With this definition, the time derivative of a matrix can be obtained using the Kronecker product, as described in [109].

$$\frac{d\boldsymbol{A}}{dt} = \frac{\partial \boldsymbol{A}^T}{\partial \boldsymbol{q}} (\boldsymbol{I}_{n \times n} \otimes \dot{\boldsymbol{q}}) \tag{A.65}$$

If the relative derivative of the product of two matrices is of interest, use the following relationship.

$$\frac{\partial}{\partial \boldsymbol{q}} (\boldsymbol{A}(\boldsymbol{q})\boldsymbol{B}(\boldsymbol{q})) = (\boldsymbol{I}_{n \times n} \otimes \boldsymbol{A}) \frac{\partial \boldsymbol{B}}{\partial \boldsymbol{q}} + \frac{\partial \boldsymbol{A}}{\partial \boldsymbol{q}} \boldsymbol{B} \tag{A.66}$$

B

Review on Nonlinear Control

In this appendix, a brief introduction to the representation of dynamic systems in state space, fundamental definitions, and stability theorems for nonlinear systems are presented. This introduction specifically reviews the stability of time-invariant or autonomous systems, and interested readers are recommended to refer to the following books [134, 191] for further study.

B.1 Dynamical Systems

As detailed in Chapter 8, the dynamics formulation of serial robots are expressed by a set of nonlinear second-order differential equations. It has also been demonstrated in Chapter 9, that the inertia matrix in robots is positive-definite and invertible. With this consideration and by forming the state variable vector of the robot as the augmentation of the motion variables and their derivatives, $\boldsymbol{x} = [\boldsymbol{q} \;\; \dot{\boldsymbol{q}}]^T$, the dynamic equations of the robot can be rewritten in state space.

$$\dot{\boldsymbol{x}} = \boldsymbol{f}(\boldsymbol{x}, \boldsymbol{u}) \tag{B.1}$$

In this equation, $\boldsymbol{x}$ is the state vector of the system, $\boldsymbol{u}$ is the input vector, and $\boldsymbol{f}$ is a continuous and multivariate function $\boldsymbol{f} : \mathbb{R}^n \to \mathbb{R}^n$, called as the vector field.

Equation B.1 represents the dynamic equations of the open-loop robot in state space. In this representation, the dynamic equations of robot motion are shown as a set of first-order differential equations, where the vector of joint torques appears as the input vector in the system. If the joint torques are computed based on control methods discussed in the book and applied to the system, the input vector $\boldsymbol{u}$ becomes a function of the robot state vector $\boldsymbol{x}$, i.e., $\boldsymbol{u} = \boldsymbol{g}(\boldsymbol{x})$. With this description, the closed-loop system can be simplified into the following equation.

$$\dot{\boldsymbol{x}} = \boldsymbol{f}(\boldsymbol{x}, \boldsymbol{g}(\boldsymbol{x})) = \boldsymbol{f}(\boldsymbol{x}) \tag{B.2}$$

It is evident that in the general case, the dynamic equations of the closed-loop system are multivariable and nonlinear. To analyze stability in such a system, it is necessary to utilize the theory of nonlinear systems.

B.2 Stability Definitions

Definition B.1 (Equilibrium Point) *Consider the nonlinear system B.2 represented in the state space $\mathbb{R}^n$. An equilibrium point of the system refers to a point in the $\mathbb{R}^n$ space where the variations in the state variables are equal to zero, or in other words, the system comes to rest at this point, $\dot{\boldsymbol{x}} = \boldsymbol{f}(\boldsymbol{x}) = \boldsymbol{0}$. If at the origin, $\dot{\boldsymbol{x}}$ becomes zero, then $\boldsymbol{f}(\boldsymbol{0}) = \boldsymbol{0}$. In this case, the origin of the state space $\mathbb{R}^n$ is defined as an equilibrium point of the system.*

DOI: 10.1201/9781003491415-B

Note that if the equilibrium point of the system $\boldsymbol{f}$ is not at the origin, it can easily be transferred to the origin by a change of coordinates. Therefore, assuming $\boldsymbol{f}(\mathbf{0}) = \mathbf{0}$ constrain the generality of this definition. If the initial conditions of the state variables are at the origin, $\boldsymbol{x}(0) = \mathbf{0}$, then the time response of the state variables, known as the trajectory of the system, remains at the origin for all times; we call this the trivial trajectory.

$$\boldsymbol{x}(t) \equiv \mathbf{0} \qquad \forall t > 0$$

The stability analysis of a system essentially examines the characteristics of nontrivial trajectories in the system. In general terms, it can be said that an equilibrium point in a system is stable if the trajectory created under initial conditions near the origin remains close to the equilibrium point and does not diverge from it. Stability in the sense of *Lyapunov* can be precisely defined as follows.

Definition B.2 (Lyapunov Stability) *The equilibrium point of the system B.2 at the origin is said to be:*

- ***Stable:*** *If and only if, for every $\epsilon > 0$, there exists a $\delta(\epsilon) > 0$ such that*

$$\|\boldsymbol{x}(0)\| < \delta \Rightarrow \|\boldsymbol{x}(t)\| < \epsilon \qquad \forall t > 0 \tag{B.3}$$

- ***Asymptotic Stable:*** *If and only if, it is stable and, in addition, there exists a $\delta > 0$ such that*

$$\|\boldsymbol{x}(0)\| < \delta \Rightarrow \lim \|\boldsymbol{x}(t)\| \to 0 \quad as \ t \to \infty \tag{B.4}$$

- ***Unstable:*** *If it is not stable.*

Lyapunov stability can be interpreted as every trajectory starting from a neighborhood of the origin remains in the vicinity of the origin for all times. Furthermore, if the trajectory tends toward the origin as time approaches infinity, the equilibrium point is asymptotically stable. With this interpretation, if the system's state is perturbed slightly from the origin, it returns to its equilibrium state. An equilibrium point is only unstable when, with the smallest deviation from the equilibrium point, the trajectory diverges and does not return toward the equilibrium.

It should be noted that these definitions are interpreted locally and are not global. This means that it may satisfy these conditions in a small neighborhood of the origin but not hold farther away. Stability is only *globally* achieved when these conditions hold for all initial conditions.

Another important concept of stability can be defined by bounding the trajectory. This definition of stability is used when the system is nonlinear and time-varying. In such a case, initial conditions are defined at time t_0, and the trajectory's properties are examined from that time onward. In time-varying systems, the concept of uniform stability is defined, where the behavioral properties of the trajectory are not dependent on the time t_0.

Definition B.3 (Uniformly Ultimately Bounded (UUB)) *The trajectory of a system $\boldsymbol{x}(t) : [t_0, \infty) \to \mathbb{R}^n$ with initial conditions $\boldsymbol{x}(t_0) = x_0$ is said to be uniformly ultimately bounded (UUB) with respect to the set $\boldsymbol{S}$ if there exists a time $T(\boldsymbol{x}_0, \boldsymbol{S})$ such that for all $t \geq T(\boldsymbol{x}_0, \boldsymbol{S})$, the trajectory $\boldsymbol{x}(t)$ remains within the set $\boldsymbol{S}$.*

$$\boldsymbol{x}(t) \in \boldsymbol{S} \quad ; \quad \forall t \geq t_0 + T \tag{B.5}$$

This type of trajectory bounds means that, starting from the initial time t_0, the state variables $\boldsymbol{x}(t)$ will eventually enter and remain within the set $\boldsymbol{S}$. Additionally, the uniformity in this definition implies that this boundedness is not a function of time t_0. If the set $\boldsymbol{S}$ is a small neighborhood near the equilibrium point, this type of boundedness is very close to the concept of stability, but it lacks the asymptotic property. This means that the trajectory remains in the vicinity of the equilibrium point with a limited but computable steady-state error.

B.3 Lyapunov Stability

Alexander Mikhailovich Lyapunov (1857-1918) was a Russian mathematician and physicist renowned for advancing the theory of stability of dynamic systems and presenting a mathematical framework for this purpose [192]. The fundamental philosophy of Lyapunov's method is a mathematical generalization of a fundamental physical observation. In a dynamical system, if its total energy is continuously dissipated, then the system, whether linear or nonlinear, must converge to a stationary equilibrium at some point. Since the energy is a scalar function, stability of a system can be inferred solely by examining changes in a energy-type scalar function.

To provide a brief overview of Lyapunov's direct method in the analysis of nonlinear systems, we will first present a physical interpretation on a simple system and then articulate the mathematical generalization of this idea, precisely described by Lyapunov, in the form of some theorems.

To this end, consider a simple mechanical system composed of a mass, a spring, and a damper, as illustrated in Figure 14.1(b). Assume that in this system, both the spring and the damper exhibit nonlinear behavior and are modeled by the relationships $F_s = k_0x + k_1x^3$ and $F_d = b\dot{x}|\dot{x}|$, respectively. In this model, k_0, k_1 are the stiffness coefficients of the spring, and b is the damping coefficient of the damper, all of which are scalar and positive values. Furthermore, assume that no external forces, such as driving forces or disturbances, are applied to the system. In this case, the dynamics formulation of the mass motion can be described using the following equation:

$$m\ddot{x} + b\dot{x}|\dot{x}| + k_0x + k_1x^3 = 0 \tag{B.6}$$

In this equation, m represents the mass of the body, and x is its displacement. This dynamic equation can be rewritten in the state space by considering the state vector $\boldsymbol{x} = [x\ \dot{x}]^T$, which augment the position and velocity of the mass. Stability analysis of the origin equilibrium point in this system can be interpreted by examining the convergence or divergence of the state vector $\boldsymbol{x}(t)$ with respect to its initial conditions $\boldsymbol{x}(0)$ in the vicinity of the equilibrium point. It is evident that directly utilizing the stability definition is very challenging in determining the stability of the equilibrium point. For this reason, here, we analyze the total energy of the system and its behavior over time.

The total energy of the system, which is the sum of kinetic and potential energy and is denoted by the scalar variable $V(\boldsymbol{x})$, can be calculated in terms of the system's state vector as follows:

$$\begin{aligned} V(\boldsymbol{x}) &= \frac{1}{2}m\dot{x}^2 + \int_0^x (k_0y + k_1y^3)dy \\ &= \frac{1}{2}m\dot{x}^2 + \frac{1}{2}k_0x^2 + \frac{1}{4}k_1x^4 \end{aligned}$$

Now, examine the concept of the stability of motion based on the total energy of the system. Using physical observations, it can be stated that:

- The total energy of the system becomes zero only when the system is in its equilibrium state at the origin, $x = \dot{x} = 0$.
- Asymptotic stability implies the convergence of the total energy of the system toward zero over time.
- Instability is directly associated with an increase in the total energy within the system.

With these observations, it can be concluded that the stability of the motion trajectory is related to the rate of change of energy within the system, rather than the energy itself. By calculating the rate of change of energy with respect to time, it can be expressed as:

$$\dot{V}(\boldsymbol{x}) = m\ddot{x}\dot{x} + (k_0x + k_1x^3)\dot{x} = (m\ddot{x} + k_0x + k_1x^3)\dot{x} \tag{B.7}$$

In order to analyze the stability of the motion trajectory of the system $\boldsymbol{x}(t)$, it is essential to examine the solution of the state equations concerning its initial conditions $\boldsymbol{x}(0) = \boldsymbol{x}_0$. Note that the solution of the system's state equations satisfies the relation in B.6. Therefore, substitute equation B.6 into B.7, and obtain the changes in the total energy of the system as follows:

$$\dot{V}(\boldsymbol{x}) = (-b\,\dot{x}\,|\dot{x}|)\dot{x} = -b\,|\dot{x}^3| \leq 0 \tag{B.8}$$

This equation indicates that over time, the total energy of the system is consistently decreasing until $\dot{x} = 0$. This means that, regardless of the initial excitation, the system will lose energy until it reaches a stationary state, representing the concept of asymptotic stability at the origin.

To mathematically generalize these observations, we analyze the characteristics of the total energy function of the system.

- The total energy function of the system is a scalar function and strictly positive, becoming zero only at the origin: $V(\boldsymbol{x}) > 0$; $\forall \boldsymbol{x} \neq \boldsymbol{0}$.
- The total energy function of the system, $V(\boldsymbol{x})$, is zero at the equilibrium origin: $V(\boldsymbol{0}) = 0$.
- The time derivative of the total energy function of the system, $\dot{V}(\boldsymbol{x})$, is uniformly decreasing.

Definition B.4 (Locally Positive-Definite:) *A continuous scalar function $V(\boldsymbol{x})$ is locally positive-definite, if and only if, $V(\boldsymbol{0}) = 0$, and within a neighborhood B_{R_0} of the origin with radius R_0, it is strictly positive: $V(\boldsymbol{x}) > 0$.*

Definition B.5 (Lyapunov Function Candidate:) *A locally positive-definite function $V(\boldsymbol{x})$ is called a Lyapunov function candidate, if it is continuous and has first-order partial derivatives with respect to the state vector $\boldsymbol{x}$ within a close neighborhood of the origin.*

Note that if the above conditions are satisfied throughout the entire state space, we call the scalar function globally positive-definite. Furthermore, we define $V(\boldsymbol{x})$ as globally negative-definite if $-V(\boldsymbol{x})$ is globally positive-definite. The function $V(\boldsymbol{x})$ is termed as positive semi-definite, if it can become zero at points other than the origin but remains non-negative elsewhere: $V(\boldsymbol{x}) \geq 0$.

With this introduction, we can present the Lyapunov direct method for analyzing the stability of nonlinear autonomous systems in the form of the Lyapunov stability theorem.

Theorem B.1 (Lyapunov Local Theorem) *Suppose $\boldsymbol{x} = \boldsymbol{0}$ is the equilibrium point of the nonlinear system B.2, and $D \subset \mathbb{R}^n$ is a domain containing the equilibrium point. Furthermore, assume that the scalar function $V : D \to \mathbb{R}$ is continuous and locally positive-definite, such that*

$$V(\boldsymbol{0}) = 0 \;\; and \;\; V(\boldsymbol{x}) > 0 \;\; in \;\; D - \{\boldsymbol{0}\}$$
$$\dot{V}(\boldsymbol{x}) \leq 0 \;\; in \;\; D$$

Then, the equilibrium point $\boldsymbol{x} = \boldsymbol{0}$ is stable. *If*

$$\dot{V}(\boldsymbol{x}) < 0 \;\; in \;\; D - \{\boldsymbol{0}\}$$

Then, the equilibrium point $\boldsymbol{x} = \boldsymbol{0}$ is asymptotically stable.

Note that, despite using the total energy of the system as the Lyapunov candidate function in the presented example, this theorem imposes no specific constraints on the choice of this function other than its local positive-definite property within the domain D. Furthermore, the proof of the Lyapunov theorem does not have a direct connection to the physical interpretation provided in this section, and it establishes the stability of the motion trajectory through a mathematical description rather than energy considerations [134]. The proof of this theorem is based on Lyapunov level sets $\boldsymbol{S}(c)$, defined as

$$\boldsymbol{S}(c) = \{\boldsymbol{x} \in \mathbb{R}^n \mid V(\boldsymbol{x}) = c\} \tag{B.9}$$

The general principles of proving this theorem are based on the fact that if $V(\boldsymbol{x})$ satisfies the theorem conditions, then definite level sets can be defined in the state space that do not intersect with each other and converge to the origin as the value of c decreases. Moreover, these level sets also indicate the convergence direction of the motion trajectory, and with decreasing level sets, the motion trajectory asymptotically converges to the origin.

This mathematical interpretation of the stability of the motion trajectory and the Lyapunov stability theorem is very powerful; however, it is both straightforward and challenging. It is straightforward in the sense that if a Lyapunov candidate function satisfying these conditions can be found, the stability of the equilibrium point is easily ensured. However, since this theorem only provides sufficient conditions, the failure to satisfy these conditions does not necessarily imply instability. Instead, it means that an appropriate candidate for the Lyapunov function has not been chosen, and the search for a suitable candidate function must continue.

On the other hand, there is currently no comprehensive method for determining a suitable Lyapunov function candidate in nonlinear systems, and the developed methods are all limited to specific types of nonlinear systems [193]. Although the total energy of the system is always an appropriate initial choice for the Lyapunov function, if it does not satisfy the conditions of the theorem, a systematic method for tuning it and reaching a suitable Lyapunov function has not been found.

In the direct Lyapunov method, the stability and asymptotic stability of the equilibrium point are distinguished with a slight difference in the Lyapunov theorem conditions. This is done while finding a Lyapunov function that has a negative-definite derivative and satisfies asymptotic stability conditions is much more challenging than ensuring stability. Fortunately, this limitation has been reasonably addressed in the Krasovskii-Lasalle's invariance theorem, details of which will be reviewed in Section B.4.

Many methods and theorems derived from the Lyapunov stability theorem are applicable to various systems. In the rest of this chapter, only two cases of the extension of the Lyapunov direct method are presented: the Lyapunov's golbal stability theorem and the UUB Stability theorem.

Theorem B.2 (Global Lyapunov Theorem) *Suppose $\boldsymbol{x} = \mathbf{0}$ is the equilibrium point of the nonlinear system B.2, and there exists a continuous, locally definite positive scalar function $V : \mathbb{R}^n \to \mathbb{R}$ such that*

$$\begin{gathered} V(\mathbf{0}) = 0 \ \ and \ \ V(\boldsymbol{x}) > 0 \ , \ \ \forall \boldsymbol{x} \neq \mathbf{0} \\ If \ \ \|\boldsymbol{x}\| \to \infty \Rightarrow V(\boldsymbol{x}) \to \infty \\ \dot{V}(\boldsymbol{x}) \leq 0, \ \ \forall \boldsymbol{x} \neq \mathbf{0} \end{gathered} \tag{B.10}$$

Then, the equilibrium point $\boldsymbol{x} = \mathbf{0}$ is "globally stable." Moreover, if

$$\dot{V}(\boldsymbol{x}) < 0 \ \ in \ \ D - \{\mathbf{0}\}$$

Then, the equilibrium point $\boldsymbol{x} = \mathbf{0}$ is "asymptotically globally stable."

Theorem B.3 (Uniformly Ultimately Bounded (UUB)) *Suppose $\boldsymbol{x} = \mathbf{0}$ is the equilibrium point of the nonlinear system B.2, and $D \subset \mathbb{R}^n$ is a domain containing the equilibrium point. Furthermore, assume that the scalar function $V : D \to \mathbb{R}$ is continuous and locally positive-definite, and $\boldsymbol{S}(c)$ is an arbitrary level set of the Lyapunov function V defined for $c > 0$ as in Equation B.9. In this case, the motion trajectory of the system $\boldsymbol{x}(t)$ will be uniformly ultimately bounded with respect to the set $\boldsymbol{S}$ if:*

$$\dot{V}(\boldsymbol{x}) < 0 \ : \forall \boldsymbol{x} \ \ Out \ of \ Set \ \ \boldsymbol{S}$$

If $\dot{V}$ is strictly negative outside the set $\boldsymbol{S}$, the system's trajectory outside this set will converge toward it. When the motion trajectory is within this set, convergence to the origin cannot be guaranteed. However, leaving the motion trajectory from this set is also not possible, because the direction of motion is from outside the set toward its interior for all trajectories. Therefore, the motion trajectory will remain uniformly within the set $\boldsymbol{S}$ until the ultimate time.

B.4 The Krasovskii-Lasalle Theorem

As mentioned earlier, in many cases, the trajectory of a nonlinear system is asymptotically stable, but it is not always easy to find a Lyapunov candidate function whose derivative is strictly negative. In cases where $\dot{V}$ is negative semi-definite, the concept of *invariant set* in dynamical systems can be used to analyze the asymptotic stability of motion trajectories and estimate the convergence region.

In this regard, this section focuses on the definition of invariant sets and how to use them to analyze the asymptotic stability of motion trajectories in nonlinear systems. The Krasovskii-Lasalle theorem is employed for this purpose. This theorem has broader applications and is used for analyzing the convergence of trajectories toward limit cycles, as well as estimating the convergence region of these trajectories [134].

Definition B.6 (Invarient Set) *The set $\boldsymbol{M}$ in the dynamical system described by the equation B.2 is called an invariant set if any motion trajectory that starts inside this set remains within it for future times.*

$$\boldsymbol{x}(0) \in \boldsymbol{M} \Rightarrow \boldsymbol{x}(t) \in \boldsymbol{M} \ \ , \ \ \forall t > 0$$

An invariant set is a generalization of the concept of an equilibrium point with higher dimensions in a dynamical system. If a set contains only one equilibrium point of the system, it is called an invariant set, because if the initial conditions of the system conform to the equilibrium point, no new trajectory is generated, and the system remains on the equilibrium point in future times.

A set that includes multiple equilibrium points of the system is also an invariant set. Additionally, a set of points that forms a limit cycle of a dynamic system is also considered an invariant set. A set of points that describes a specific motion trajectory of the system from time zero to infinity is also an invariant set. In both cases, if we start the motion from a point in these sets, the points forming the motion trajectory will remain on the limit cycle or predefined trajectory in future times, and thus they will not leave the invariant set.

The Krasovskii-Lasalle theorem introduces conditions under which all trajectories of the system will converge toward an invariant set. A direct consequence of this theorem in the case where the invariant set only contains one equilibrium point is the asymptotic stability of the equilibrium point, even if the derivative of the Lyapunov candidate function is not strictly negative-definite.

Theorem B.4 (Krasovskii-Lasalle) *For the nonlinear system introduced in equation B.2, assume $D \subset \mathbb{R}^n$ is a bounded region in the state space, and $\Omega \subset D$ is a compact set that is invariant with respect to the dynamical system. Additionally, suppose $V : D \to \mathbb{R}$ is a continuously differentiable Lyapunov candidate function such that it is negative semi-definite in the region Ω ($\dot{V} \leq 0$). In this case, let E be a set of points in Ω where the Lyapunov function's derivative is zero ($\dot{V} = 0$), and let M be the largest invariant set in E. Then, any trajectory starting from within Ω will converge to the invariant set M as time goes to infinity, i.e., $t \to \infty$.*

This theorem does not directly address asymptotic stability of the equilibrium point in cases where $\dot{V}(\boldsymbol{x})$ is negatively semi-definite, and it is not limited to such applications. However, when the origin $\boldsymbol{x} = \mathbf{0}$ is the only point within the invariant set M, asymptotic stability of the equilibrium point is guaranteed even when $V(\boldsymbol{x})$ is negatively semi-definite. This implies that if the Lyapunov function V does not become zero along any trajectory, it must converge to the equilibrium point. In this case, the only solution of the dynamical system satisfying $\dot{V} = 0$ is the equilibrium point at the origin. The result of this theorem is used to establish asymptotic stability at the origin in cases where $V(\boldsymbol{x})$ is negative semi-definite.

Bibliography

[1] Hamid D. Taghirad. *Parallel robots: mechanics and control*, second edition, CRC press, 2025.

[2] John J. Craig. *Introduction to Robotics.* Pearson Education Limited, 2021.

[3] Lung-Wen Tsai. *Robot analysis: the mechanics of serial and parallel manipulators.* John Wiley & Sons, 1999.

[4] M. W. Spong, S. Hutchinson, and M. Vidyasagar. *Robot modeling and control*, volume 3. Wiley New York, 2020.

[5] L. A. Ballard, S. Sabanovic, J. Kaur, and S. Milojevic. George Charles Devol Jr. [History]. *IEEE Robotics & Automation Magazine*, 19(3):114–119, 2012.

[6] Johanna Wallén. *The history of the industrial robot.* Linköping University Electronic Press, 2008.

[7] S. Saeedvand, M. Jafari, H. S. Aghdasi, and J. Baltes. A comprehensive survey on humanoid robot development. *The Knowledge Engineering Review*, 34, 2019.

[8] J. F. Engelberger. *Robotics in Practice.* AMACOM, New York, 1980.

[9] International Federation of Robotics. Standardization, Robot definitions at ISO. URL `https://ifr.org/standardisation`.

[10] Karel Čapek. *Rossum's Universal Robots.* Doubleday, New York, 1923. translated by Paul Selver.

[11] B. Dasgupta and T. Mruthyunjaya. The stewart platform manipulator: a review. *Mechanism and machine theory*, 35(1):15–40, 2000.

[12] A. Gasparetto and L. Scalera. From the unimate to the delta robot: the early decades of industrial robotics. In *Explorations in the history and heritage of machines and mechanisms*, pages 284–295. Springer, 2019.

[13] S. Zimmermann, R. Poranne, and S. Coros. Go fetch! Dynamic grasps using boston dynamics spot with external robotic arm. In *2021 IEEE International Conference on Robotics and Automation (ICRA)*, pages 4488–4494. IEEE, 2021.

[14] G. Bai, Y. Ge, D. Scoby, B. Leavitt, V. Stoerger, N. Kirchgessner, S. Irmak, G. Graef, J. Schnable, and T. Awada. Nu-spidercam: A large-scale, cable-driven, integrated sensing and robotic system for advanced phenotyping, remote sensing, and agronomic research. *Computers and Electronics in Agriculture*, 160:71–81, 2019.

[15] B. A. Aikenhead, R. G. Daniell, and F. M. Davis. Canadarm and the space shuttle. *Journal of Vacuum Science & Technology A: Vacuum, Surfaces, and Films*, 1(2): 126–132, 1983.

[16] S. Sachdev, W. Harvey, G. Gibbs, B. Marcotte, N. Buckley, T. Braithwaite, and D. Rey. Canada and the international space station program: Overview and status since IAC 2005. In *57th International Astronautical Congress*, pages B4–1, 2006.

[17] E. Coleshill, L. Oshinowo, R. Rembala, B. Bina, D. Rey, and Sh. Sindelar. Dextre: Improving maintenance operations on the international space station. *Acta Astronautica*, 64(9-10):869–874, 2009.

[18] C. Freschi, V. Ferrari, F. Melfi, M. Ferrari, F. Mosca, and A. Cuschieri. Technical review of the da Vinci surgical telemanipulator. *The International Journal of Medical Robotics and Computer Assisted Surgery*, 9(4):396–406, 2013.

[19] M. D. De Smet, G. J. L. Naus, K. Faridpooya, and M. Mura. Robotic-assisted surgery in ophthalmology. *Current Opinion in Ophthalmology*, 29(3):248–253, 2018.

[20] R. Gopura, D. Bandara, K. Kiguchi, and G. Mann. Developments in hardware systems of active upper-limb exoskeleton robots: A review. *Robotics and Autonomous Systems*, 75:203–220, 2016.

[21] K. A. Farley, K. H. Williford, K. M. Stack, R. Bhartia, A. l. Chen, M. de la Torre, K. Hand, et al. Mars 2020 mission overview. *Space Science Reviews*, 216(8):1–41, 2020.

[22] M. Shamiyeh, J. Bijewitz, and M. Hornung. A review of recent personal air vehicle concepts. In *Aerospace Europe 6th CEAS conference*, volume 913, pages 1–18, 2017.

[23] O. Stasse and T. Flayols. An overview of humanoid robots technologies. *Biomechanics of Anthropomorphic Systems*, pages 281–310, 2019.

[24] Y. Sakagami, R. Watanabe, C. Aoyama, S. Matsunaga, N. Higaki, and K. Fujimura. The intelligent ASIMO: System overview and integration. In *IEEE/RSJ international conference on intelligent robots and systems*, volume 3, pages 2478–2483. IEEE, 2002.

[25] U. Pagallo. Vital, sophia, and Co.? the quest for the legal personhood of robots. *Information*, 9(9):230, 2018.

[26] E. R. Davies. *Machine vision: theory, algorithms, practicalities.* Elsevier, 2004.

[27] L. Rabiner and B. H. Juang. *Fundamentals of speech recognition.* Prentice-Hall, Inc., 1993.

[28] T. Dean, J. Allen, and Y. Aloimonos. *Artificial intelligence: theory and practice.* Benjamin-Cummings Publishing Co., Inc., 1995.

[29] Y. Wang and F. Zhang. *Trends in control and decision-making for human-robot collaboration systems.* Springer, 2017.

[30] J. P. Laumond et al. *Robot motion planning and control*, volume 229. Springer, 1998.

[31] S. G. Tzafestas. *Introduction to mobile robot control.* Elsevier, 2013.

[32] R. Siegwart, I. R. Nourbakhsh, and D. Scaramuzza. *Introduction to autonomous mobile robots.* MIT press, 2011.

[33] S. Thrun, W. Burgard, and D. Fox. *Probalistic robotics.* Emerald Group Publishing Limited, 2006.

[34] C. W. De Silva. *Sensors and actuators: Engineering system instrumentation.* CRC Press, 2015.

[35] S. T. Venkataraman and T. Iberall. *Dextrous robot hands.* Springer Science & Business Media, 2012.

[36] Z. Kappassov, J. A. Corrales, and V. Perdereau. Tactile sensing in dexterous robot hands. *Robotics and Autonomous Systems*, 74:195–220, 2015.

[37] L. Euler. Nova methodus motum corporum rigidorum degerminandi. *Novi commentarii academiae scientiarum Petropolitanae*, pages 208–238, 1776.

[38] M. Nitschke and E. H. Knickmeyer. Rotation parameters: A survey of techniques. *Journal of surveying engineering*, 126(3):83–105, 2000.

[39] M. Ceccarelli. Screw axis defined by giulio mozzi in 1763 and early studies on helicoidal motion. *Mechanism and Machine Theory*, 35(6):761–770, 2000.

[40] J. M. McCarthy and G. S. Soh. *Geometric design of linkages*, volume 11. Springer Science & Business Media, 2010.

[41] Giulio Mozzi. *Discorso matematico sopra il rotamento momentaneo dei corpi [Mathematical treatise on the temporally revolving of bodies].* nella stamperia di Donato Campo, 1763.

[42] M. Chasles. Notes on the general properties of a system of two identical bodies randomly located in space; and on the finite or infinitesimal motion of a free solid body. *Bull. des Sci. Math., Astron., Phys. et Chim*, 14:321–326, 1830.

[43] R. S. Ball. *A Treatise on the Theory of Screws.* Cambridge University Press, Cambridge, 1900.

[44] J. Gallardo-Alvarado. *Kinematic analysis of parallel manipulators by algebraic screw theory.* Springer, 2016.

[45] C. D Crane, M. Griffis, and J. Duffy. *Screw theory and its application to spatial robot manipulators.* Cambridge University Press, 2022.

[46] J. Pardos-Gotor. *Screw theory in robotics: an illustrated and practicable introduction to modern mechanics.* CRC Press, 2021.

[47] H. Cheng and K. C. Gupta. An historical note on finite rotations. *Journal of Applied Mechanics*, 56:139–145, 1989.

[48] Z. Huang, Q. Li, and H. Ding. Basics of screw theory. In *Theory of Parallel Mechanisms*, pages 1–16. Springer, 2013.

[49] T. Y. Lam. Hamilton's quaternions. In *Handbook of algebra*, volume 3, pages 429–454. Elsevier, 2003.

[50] S. L. Altmann. *Rotations, quaternions, and double groups.* Courier Corporation, 2005.

[51] S. Sarabandi and F. Thomas. A survey on the computation of quaternions from rotation matrices. *Journal of mechanisms and robotics*, 11(2), 2019.

[52] S. Zhao. Time derivative of rotation matrices: A tutorial. *arXiv preprint arXiv:1609.06088*, 2016.

[53] John Voight. *Quaternion algebras.* Springer Nature, 2021.

[54] K. W. Spring. Euler parameters and the use of quaternion algebra in the manipulation of finite rotations: a review. *Mechanism and machine theory*, 21(5):365–373, 1986.

[55] J. L. Meriam and L. G. Kraige. *Engineering Mechanics: Dynamics.* Wiley, New York, 5th edition, 2007.

[56] Bernard Roth. Screws, motors, and wrenches that cannot be bought in a hardware store. In *Proc. Int. Symp. Robotics Research*, pages 679–693, 1984.

[57] R. P. Paul. *Robot manipulators: mathematics, programming, and control.* The MIT Press, Cambridge. MA., 1981.

[58] Saeed B. Niku. *Introduction to robotics: analysis, control, applications.* John Wiley & Sons, 2020.

[59] J. Denavit and R. S. Hartenberg. A kinematic notation for lower-pair mechanisms based on matrices. *Journal of Applied Mechanics*, 22(2):215–221, 1955.

[60] Bernard Roth. Performance evaluation of manipulators from a kinematic viewpoint. *NBS Special Publication*, pages 39–61, 1975.

[61] Jorge Angeles. On the numerical solution of the inverse kinematic problem. *The International Journal of Robotics Research*, 4(2):21–37, 1985.

[62] Y. Nakamura and H. Hanafusa. Inverse Kinematic Solutions With Singularity Robustness for Robot Manipulator Control. *Journal of Dynamic Systems, Measurement, and Control*, 108(3):163–171, 09 1986. ISSN 0022-0434.

[63] D. R. Baker and C. W. Wampler. On the inverse kinematics of redundant manipulators. *The International journal of robotics research*, 7(2):3–21, 1988.

[64] J. Angeles. *Fundamentals of Robotic Mechanical Systems: Theory, Methods and Algorithms.* Springer, 2003.

[65] J. Angeles and K. E. Zanganeh. The semigraphical determination of all real inverse kinematic solutions of general six-revolute manipulators. In *RoManSy 9*, pages 23–32. Springer, 1993.

[66] D. Manocha and J. F. Canny. Efficient inverse kinematics for general 6R manipulators. *IEEE transactions on robotics and automation*, 10(5):648–657, 1994.

[67] Donald Lee Pieper. *The kinematics of manipulators under computer control.* Unpublished Ph.D. Thesis, Stanford University, 1969.

[68] K. Levenberg. A method for the solution of certain non-linear problems in least squares. *Quarterly of applied mathematics*, 2(2):164–168, 1944.

[69] D. W. Marquardt. An algorithm for least-squares estimation of nonlinear parameters. *Journal of the society for Industrial and Applied Mathematics*, 11(2):431–441, 1963.

[70] J. J. Moré. The levenberg-marquardt algorithm: implementation and theory. In *Numerical analysis*, pages 105–116. Springer, 1978.

[71] R. Bernard and S. Albright. *Robot calibration.* Springer Science & Business Media, 1993.

[72] H. Zhuang and Z. S. Roth. *Camera-aided robot calibration.* CRC press, 2018.

[73] F. C. Park. Optimal Robot Design and Differential Geometry. *Journal of Mechanical Design*, 117(B):87–92, 06 1995.

[74] C. A. Klein and B. E. Blaho. Dexterity measures for the design and control of kinematically redundant manipulators. *The International Journal of Robotics Research*, 6(2):72–83, 1987.

[75] T. Yoshikawa. Manipulability of robotic mechanisms. *The International Journal of Robotics Research*, 4(2):3–9, 1985.

[76] A. Karger. Singularity Analysis of Serial Robot-Manipulators. *Journal of Mechanical Design*, 118(4):520–525, 12 1996.

[77] J. Lenarcic and C. Galletti. *Advances in robot kinematics.* Kluwer Academic, 2004.

[78] K. L. Doty, C. Melchiorri, E. M. Schwartz, and C. Bonivento. Robot manipulability. *IEEE Transactions on Robotics and Automation*, 11(3):462–468, 1995.

[79] W. Xu, J. Zhang, B. Liang, and B. Li. Singularity analysis and avoidance for robot manipulators with nonspherical wrists. *IEEE Transactions on Industrial Electronics*, 63(1):277–290, 2015.

[80] T. Yoshikawa. Dynamic manipulability of robot manipulators. *Transactions of the Society of Instrument and Control Engineers*, 21(9):970–975, 1985.

[81] L. El Ghaoui. Inversion error, condition number, and approximate inverses of uncertain matrices. *Linear algebra and its applications*, 343:171–193, 2002.

[82] C. Gosselin and J. Angeles. A global performance index for the kinematic optimization of robotic manipulators. *Journal of Mechanical Design*, 113(3):220–226, 09 1991.

[83] J. Angeles and A. A. Rojas. Manipulator inverse kinematics via condition number minimization and constitution. *Inter J Robotics and Automation*, 2(2):61–69, 1987.

[84] C. A. Klein and C. H. Huang. Review of pseudoinverse control for use with kinematically redundant manipulators. *IEEE Transactions on Systems, Man and Cybernetics*, 13(2):245–250, 1983.

[85] K. Glass, R. Colbaugh, D. Lim, and H. Seraji. Real-time collision avoidance for redundant manipulators. *IEEE Transactions on Robotics and Automation*, 11(3): 448–457, 1995.

[86] H. Seraji. Task-based configuration control of redundant manipulators. *Journal of Robotic Systems*, 9(3):411–451, 1992.

[87] D. E. Whitney. Resolved motion rate control of manipulators and human prostheses. *IEEE Transactions on Man-Machine Systems*, 10(2):47–53, 1969.

[88] A. Liégeois. Automatic supervisory control of the configuration and behavior of multibody mechanisms. *IEEE Transactions Syst., Man, Cybern.*, 7(12):868–871, 1977.

[89] J. Hollerbach and K. Suh. Redundancy resolution of manipulators through torque optimization. *IEEE Journal of Robotics and Automation*, 3(4):308–316, 1988.

[90] O. Khatib. Unified approach for motion and force control of robot manipulators: The operational space formulation. *IEEE Transactions on Robotics and Automation*, 3(1): 43–53, 1987.

[91] R. Dubey and J. Y. S. Luh. Redundant robot control using task based performance measures. *Journal of Robotic Systems*, 5(5):409–432, 1988.

[92] S. L. Chiu. Task compatibility of manipulator postures. *Int. J. of Robotics Research*, 7(5):13–21, 1988.

[93] H. Seraji. Configuration control of redundant manipulators: theory and implementation. *IEEE Transactions on Robotics and Automation*, 5(4):472–490, 1989.

[94] S. Boyd and L. Vandenberghe. *Convex optimization.* Cambridge university press, 2004.

[95] Y. Nakamura. *Advanced robotics: redundancy and optimization.* Addison-Wesley Longman Publishing Co., Inc., 1990.

[96] T. Yoshikawa. *Foundations of robotics: analysis and control.* MIT press, 1990.

[97] S. S. Rao. *Engineering optimization: theory and practice.* John Wiley & Sons, 2019.

[98] M. J. Kochenderfer and T. A. Wheeler. *Algorithms for optimization.* Mit Press, 2019.

[99] A. Hassani, S. A. Khalilpour, A. Bataleblu, and Hamid D. Taghirad. Full dynamic model of 3-UPU translational parallel manipulator for model-based control schemes. *Robotica*, page 1"C16, 2022.

[100] A. Hassani, M. R. Dindarloo, R. Khorrambakht, A. Bataleblu, R. Heidari, M. Motaharifar, S. F. Mohammadi, and H. D. Taghirad. On the dynamic calibration and trajectory control of ARASH:ASiST. In *2022 8th International Conference on Control, Instrumentation and Automation (ICCIA)*, pages 1–6. IEEE Xplore, 2022.

[101] A. Hassani, A. Bataleblu, S. A. Khalilpour, and Hamid D. Taghirad. Dynamic modeling and identification of aras-diamond: A vitreoretinal eye surgery robot. *Modares Mechanical Engineering*, 21(11), 2021.

[102] T. R. Kane and D. A. Levinson. *Dynamics, theory and applications.* McGraw Hill, 1985.

[103] J. H. Ginsberg. *Advanced engineering dynamics.* Cambridge University Press, 1998.

[104] E. A. Desloge. The gibbs–appell equations of motion. *American Journal of Physics*, 56(9):841–846, 1988.

[105] E. A. Desloge. Relationship between kane's equations and the gibbs-appell equations. *Journal of Guidance, Control, and Dynamics*, 10(1):120–122, 1987.

[106] T. R. Kane and D. A. Levinson. The use of kane's dynamical equations in robotics. *The International Journal of Robotics Research*, 2(3):3–21, 1983.

[107] J. L. Meriam, L. G. Kraige, and J. N. Bolton. *Engineering mechanics: dynamics.* John Wiley & Sons, 2020.

[108] A. R. Abdulghany. Generalization of parallel axis theorem for rotational inertia. *American Journal of Physics*, 85(10):791–795, 2017.

[109] V. K. Nguyen. Consistent definition of partial derivatives of matrix functions in dynamics of mechanical systems. *Mechanism and Machine Theory*, 45(7):981–988, 2010.

[110] C. G. Atkeson, C. H. An, and J. M. Hollerbach. Estimation of inertial parameters of manipulator loads and links. *Int. J. Robotics Res.*, 5(3):101–119, 1986.

[111] H. Mayeda, K. Yoshida, and K. Osuka. Base parameters of manipulator dynamic models. *IEEE Transactions on Robotics and Automation*, 6(3):312–321, 1990.

[112] W. S. Lu and Q. H. Meng. Regressor formulation of robot dynamics: computation and applications. *IEEE Transactions on Robotics and Automation*, 9(3):323–333, 1993.

[113] G. Calafiore, M. Indri, and B. Bona. Robot dynamic calibration: Optimal excitation trajectories and experimental parameter estimation. *Journal of robotic systems*, 18 (2):55–68, 2001.

[114] Hamid D. Taghirad. *Robust torque control of harmonic drive systems.* PhD thesis, McGill University, Montreal, Canada, 1997.

[115] C. Alexopoulos and P. M. Griffin. Path planning for a mobile robot. *IEEE Transactions on systems, man, and cybernetics*, 22(2):318–322, 1992.

[116] H. Zhang, W. Lin, and A. Chen. Path planning for the mobile robot: A review. *Symmetry*, 10(10):450, 2018.

[117] M. N. Ab Wahab, S. Nefti-Meziani, and A. Atyabi. A comparative review on mobile robot path planning: Classical or meta-heuristic methods? *Annual Reviews in Control*, 50:233–252, 2020.

[118] A. Gasparetto, P. Boscariol, A. Lanzutti, and R. Vidoni. *Path Planning and Trajectory Planning Algorithms: A General Overview*, pages 3–27. Springer International Publishing, Cham, 2015.

[119] L. F. Shampine. *Numerical solution of ordinary differential equations.* Routledge, 2018.

[120] K. Atkinson, W. Han, and D. E. Stewart. *Numerical solution of ordinary differential equations*, volume 108. John Wiley & Sons, 2011.

[121] J. C. Butcher. *Numerical methods for ordinary differential equations.* John Wiley & Sons, 2016.

[122] M. Gautier. Numerical calculation of the base inertial parameters of robots. *Journal of robotic systems*, 8(4):485–506, 1991.

[123] J. Wu, J. Wang, and Z. You. An overview of dynamic parameter identification of robots. *Robotics and computer-integrated manufacturing*, 26(5):414–419, 2010.

[124] Hamid D. Taghirad and P. R. Belanger. Modeling and parameter identification of harmonic drive systems. *Journal of Dynamic Systems, Measurements, and Control, ASME Pub.*, 120(4):439–444, 1998.

[125] B. Siciliano, L. Sciavicco, L. Villani, and G. Oriolo. *Force control.* Springer, 2009.

[126] C. C. Cheah. On duality of inverse jacobian and transpose jacobian in task-space regulation of robots. In *Proceedings 2006 IEEE International Conference on Robotics and Automation, 2006. ICRA 2006.*, pages 2571–2576. IEEE, 2006.

[127] S. A. A. Moosavian and E. Papadopoulos. Modified transpose jacobian control of robotic systems. *Automatica*, 43(7):1226–1233, 2007.

[128] D. Bevly, S. Dubowsky, and C. Mavroidis. A simplified cartesian-computed torque controller for highly geared systems and its application to an experimental climbing robot. *J. Dyn. Sys., Meas., Control*, 122(1):27–32, 2000.

[129] F. Golnaraghi and B. C. Kuo. *Automatic control systems.* McGraw-Hill, 10th edition, 2017.

[130] K. J. Åstrom and T. Hägglund. *Advanced PID Control.* ISA, Instrumentation Systems and Automation Society, 2006.

[131] A. O'dwyer. *Handbook of PI and PID controller tuning rules.* World Scientific, 2009.

[132] J. C. Doyle, B. A. Francis, and A. R. Tannenbaum. *Feedback control theory.* Courier Corporation, 2013.

[133] C. An, C. Atkeson, J. Griffiths, and J. M. Hollerbach. Experimental evaluation of feedforward and computed torque control. In *Proceedings. 1987 IEEE International Conference on Robotics and Automation*, volume 4, pages 165–168. IEEE, 1987.

[134] Hassan K. Khalil. *Nonlinear control.* Pearson New York, 2015.

[135] J. J. E. Slotine and W. Li. *Applied nonlinear control.* Prentice hall Englewood Cliffs, NJ, 1991.

[136] H. Asada, Z. D. Ma, and H. Tokumaru. Inverse dynamics of flexible robot arms: Modeling and computation for trajectory control. *Journal of Dynamics Systems, Measurements, and Control*, 122(2):177–185, 1990.

[137] D. Nguyen-Tuong, J. Peters, M. Seeger, and B. Schölkopf. Learning inverse dynamics: a comparison. In *European symposium on artificial neural networks*, 2008.

[138] D. S. Kwon and W. J. Book. A Time-Domain Inverse Dynamic Tracking Control of a Single-Link Flexible Manipulator. *Journal of Dynamic Systems, Measurement, and Control*, 116(2):193–200, 06 1994.

[139] W. L. Stout and M. E. Sawan. Application of h-infinity theory to robot manipulator control. In *[Proceedings 1992] The First IEEE Conference on Control Applications*, pages 148–153. IEEE, 1992.

[140] P. A. Ioannou and J. Sun. *Robust adaptive control.* Courier Corporation, 2012.

[141] S. Roy and I. N. Kar. *Adaptive-Robust Control with Limited Knowledge on Systems Dynamics.* Springer, 2020.

[142] M. T. Ravichandran and A. D. Mahindrakar. Robust stabilization of a class of underactuated mechanical systems using time scaling and lyapunov redesign. *IEEE Transactions on Industrial Electronics*, 58(9):4299–4313, 2010.

[143] E. R. Alphonsus and M. O. Abdullah. A review on the applications of programmable logic controllers (PLCs). *Renewable and Sustainable Energy Reviews*, 60:1185–1205, 2016.

[144] C. Silva and E. Trélat. Smooth regularization of bang-bang optimal control problems. *IEEE Transactions on Automatic Control*, 55(11):2488–2499, 2010.

[145] R. A. De Carlo, S. H. Zak, and G. P. Matthews. Variable structure control of nonlinear multivariable systems: a tutorial. *Proceedings of the IEEE*, 76(3):212–232, 1988.

[146] C. Edwards and S. Spurgeon. *Sliding mode control: theory and applications.* Crc Press, 1998.

[147] N. K. Sharma and J. Sivaramakrishnan. *Discrete-time higher order sliding mode: The concept and the control.* Springer, 2018.

[148] Y. Shtessel, C. Edwards, L. Fridman, A. Levant, et al. *Sliding mode control and observation*, volume 10. Springer, 2014.

[149] N. B. Cheng, L. W. Guan, L. P. Wang, and J. Han. Chattering reduction of sliding mode control by adopting nonlinear saturation function. In *Advanced Materials Research*, volume 143, pages 53–61. Trans Tech Publ, 2011.

[150] Q. P. Ha, Q. H. Nguyen, D. C. Rye, and Hugh F Durrant-Whyte. Fuzzy sliding-mode controllers with applications. *IEEE Transactions on industrial electronics*, 48 (1):38–46, 2001.

[151] L. Fridman, A. Levant, et al. Higher order sliding modes. *Sliding mode control in engineering*, 11:53–102, 2002.

[152] V. Utkin. Discussion aspects of high-order sliding mode control. *IEEE Transactions on Automatic Control*, 61(3):829–833, 2015.

[153] R. Hýl and R. Wagnerová. Design and implementation of cascade control structure for superheated steam temperature control. In *2016 17th International Carpathian Control Conference (ICCC)*, pages 253–258. IEEE, 2016.

[154] M. Idres, O. Mustapha, and M. Okasha. Quadrotor trajectory tracking using pid cascade control. In *IOP Conference Series: Materials Science and Engineering*, volume 270, page 012010. IOP Publishing, 2017.

[155] F. Kulakov, G. Alferov, B. Sokolov, P. Gorovenko, and A. Sharlay. Dynamic analysis of space robot remote control system. In *AIP Conference Proceedings*, volume 1959, page 080014. AIP Publishing LLC, 2018.

[156] R. Rembala and C. Ower. Robotic assembly and maintenance of future space stations based on the ISS mission operations experience. *Acta Astronautica*, 65(7-8):912–920, Oct.-Nov. 2009.

[157] V. K. Bohns. (mis) understanding our influence over others: A review of the underestimation-of-compliance effect. *Current Directions in Psychological Science*, 25(2):119–123, 2016.

[158] U. Mohrlen, D. Weber, R. Gonzalez, D. M. Schmid, T. Sulser, and R. Gobet. Robot-assisted minimal invasive pediatric urology. *Journal of Pediatric Urology*, 5(1):S45–S46, 2009.

[159] Hamid D. Taghirad and P. R. Belanger. H-infinity based robust torque control of harmonic drive systems. *Journal of Dynamic Systems, Measurements, and Control, ASME Pub.*, 123(3):338–345, 2001.

[160] S. G. Khan, G. Herrmann, M. Al Grafi, T. Pipe, and C. Melhuish. Compliance control and human–robot interaction: Part 1 Survey. *International journal of humanoid robotics*, 11(03):1430001, 2014.

[161] Y. Li and I. Kao. A review of modeling of soft-contact fingers and stiffness control for dextrous manipulation in robotics. In *Proceedings 2001 ICRA. IEEE International Conference on Robotics and Automation (Cat. No. 01CH37164)*, volume 3, pages 3055–3060. IEEE, 2001.

[162] R. C. Dorf and R. H. Bishop. *Modern control systems.* Prentice Hall, 12th edition, 2010.

[163] G. Zeng and A. Hemami. An overview of robot force control. *Robotica*, 15(5):473–482, 1997.

[164] J. Roy and L. L. Whitcomb. Adaptive force control of position/velocity controlled robots: theory and experiment. *IEEE Transactions on Robotics and Automation*, 18 (2):121–137, 2002.

[165] B. Siciliano, L. Sciavicco, L. Villani, and G. Oriolo. *Robotics: Modelling, Planning and Control.* Springer Publishing Company, Incorporated, 2010. ISBN 1849966346.

[166] M. H. Raibert and J. J. Craig. Hybrid position/force control of manipulators. *IEEE Trans. ASME, J. Dyn. Syst. Meas. Control*, 102:126, 1981.

[167] A. De Luca and R. Mattone. Sensorless robot collision detection and hybrid force/motion control. In *Proceedings of the 2005 IEEE international conference on robotics and automation*, pages 999–1004. IEEE, 2005.

[168] N. Simaan, A. Bajo, J. L Netterville, C. G. Garrett, and R. E. Goldman. Systems and methods for safe compliant insertion and hybrid force/motion telemanipulation of continuum robots, January 10 2017. US Patent 9,539,726.

[169] Neville Hogan. Impedance control: An approach to manipulation: Part 1 theory. *Journal of Dynamics Systems, Measurements, and Control*, 107:1–16, 1985.

[170] Neville Hogan. Impedance control: An approach to manipulation: Part 2 implementation. *Journal of Dynamics Systems, Measurements, and Control*, 107:17–24, 1985.

[171] P. Song, Y. Yu, and X. Zhang. A tutorial survey and comparison of impedance control on robotic manipulation. *Robotica*, 37(5):801–836, 2019.

[172] A. Hassani, A. Bataleblu, S. A. Khalilpour, and Hamid D. Taghirad. Dynamic modeling and identification of aras-diamond: A vitreoretinal eye surgery robot. *Modares Mechanical Engineering*, 21(11):783–795, 2021.

[173] S. P. Chan, B. Yao, W. B. Gao, and M. Cheng. Robust impedance control of robot manipulators. *International Journal of Robotics & Automation*, 6(4):220–227, 1991.

[174] Z. Lu and A. A. Goldenberg. Robust impedance control and force regulation: Theory and experiments. *The International journal of robotics research*, 14(3):225–254, 1995.

[175] K. J. Xu, C. Li, and Z. N. Zhu. Dynamic modeling and compensation of robot six-axis wrist force/torque sensor. *IEEE Transactions on Instrumentation and Measurement*, 56(5):2094–2100, 2007.

[176] S. Gholami, A. Arjmandi, and Hamid D. Taghirad. An observer-based adaptive impedance-control for robotic arms: Case study in smos robot. In *2016 4th International Conference on Robotics and Mechatronics (ICROM)*, pages 49–54. IEEE, 2016.

[177] L. G. García-Valdovinos, V. Parra-Vega, and M. A. Arteaga. Observer-based sliding mode impedance control of bilateral teleoperation under constant unknown time delay. *Robotics and Autonomous Systems*, 55(8):609–617, 2007.

[178] L. Peternel, T. Petrič, and J. Babič. Human-in-the-loop approach for teaching robot assembly tasks using impedance control interface. In *2015 IEEE international conference on robotics and automation (ICRA)*, pages 1497–1502. IEEE, 2015.

[179] C. Tzafestas, S. Velanas, and G. Fakiridis. Adaptive impedance control in haptic teleoperation to improve transparency under time-delay. In *2008 IEEE International Conference on Robotics and Automation*, pages 212–219. IEEE, 2008.

[180] L. J. Love and W. J. Book. Force reflecting teleoperation with adaptive impedance control. *IEEE Transactions on Systems, Man, and Cybernetics, Part B (Cybernetics)*, 34(1):159–165, 2004.

[181] F. Ficuciello, L. Villani, and B. Siciliano. Variable impedance control of redundant manipulators for intuitive human–robot physical interaction. *IEEE Transactions on Robotics*, 31(4):850–863, 2015.

[182] S. Y. Lo, C. A. Cheng, and H. P. Huang. Virtual impedance control for safe human-robot interaction. *Journal of Intelligent & Robotic Systems*, 82(1):3–19, 2016.

[183] Mojtaba Sharifi, Hassan Salarieh, Saeed Behzadipour, and Mahdi Tavakoli. Beating-heart robotic surgery using bilateral impedance control: Theory and experiments. *Biomedical signal processing and control*, 45:256–266, 2018.

[184] R. Heidari, M. Motaharifar, and H. D. Taghirad. Robust impedance control for dual user haptic training system. In *2019 7th International Conference on Robotics and Mechatronics (ICRoM)*, pages 181–185. IEEE, 2019.

[185] M. Sharifi, S. Behzadipour, H. Salarieh, and M. Tavakoli. Cooperative modalities in robotic tele-rehabilitation using nonlinear bilateral impedance control. *Control Engineering Practice*, 67:52–63, 2017.

[186] A. Rashvand, M. J. Ahmadi, M. Motaharifar, M. Tavakoli, and Hamid D. Taghirad. Adaptive robust impedance control of haptic systems for skill transfer. In *2021 9th RSI International Conference on Robotics and Mechatronics (ICRoM)*, pages 586–591. IEEE, 2021.

[187] C. Yang, C. Zeng, P. Liang, Z. Li, R. Li, and C. Y. Su. Interface design of a physical human–robot interaction system for human impedance adaptive skill transfer. *IEEE Transactions on Automation Science and Engineering*, 15(1):329–340, 2017.

[188] Gilbert Strang. *Introduction to linear algebra.* Wellesley-Cambridge Press, MA, 3rd edition, 2005.

[189] N. Johnston. *Introduction to Linear and Matrix Algebra.* Springer Nature, 2021.

[190] G. Wang, Y. Wei, S. Qiao, P. Lin, and Y. Chen. *Generalized inverses: theory and computations*, volume 53. Springer, 2018.

[191] A. Isidori. *Nonlinear Control Systems.* Springer Verlag, New York, 1999.

[192] P. S. Shcherbakov. Alexander mikhailovitch lyapunov: On the centenary of his doctoral dissertation on stability of motion. *Automatica*, 28(5):865–871, 1992.

[193] Amir Ali Ahmadi and Pablo A Parrilo. On higher order derivatives of lyapunov functions. In *Proceedings of the 2011 American Control Conference*, pages 1313–1314. IEEE, 2011.

Index